CRC HANDBOOK OF

Phase Equilibria and Thermodynamic Data of Polymer Solutions at Elevated Pressures

CRC HANDBOOK OF

Phase Equilibria and Thermodynamic Data of Polymer Solutions at Elevated Pressures

Christian Wohlfarth

CRC Press is an imprint of the
Taylor & Francis Group, an **informa** business

CRC Press
Taylor & Francis Group
6000 Broken Sound Parkway NW, Suite 300
Boca Raton, FL 33487-2742

First issued in paperback 2021

ISBN 13: 978-1-03-209882-1 (pbk)
ISBN 13: 978-1-4987-0320-8 (hbk)

Publisher's Note
The publisher has gone to great lengths to ensure the quality of this reprint but points out that some imperfections in the original copies may be apparent.

Visit the Taylor & Francis Web site at
http://www.taylorandfrancis.com

and the CRC Press Web site at
http://www.crcpress.com

CONTENTS

4. HIGH-PRESSURE FLUID PHASE EQUILIBRIUM (HPPE) DATA OF POLYMER SOLUTIONS

5. PVT DATA OF POLYMERS AND SOLUTIONS

APPENDICES

PREFACE

Today, there is still a strong and continuing interest in thermodynamic properties of polymer solutions at elevated pressures. Thus, about ten years after the *CRC Handbook of Thermodynamic Data of Polymer Solutions at Elevated Pressures* was published, necessity as well as desire arises for a supplementary book that includes and provides newly published experimental data from the last decade.

There are about **500** newly published references containing about **175** new vapor-liquid equilibrium data sets, **25** new liquid-liquid equilibrium data sets, **540** new high-pressure fluid phase equilibrium data sets, **60** new data sets describing PVT-properties of polymers, and **20** new data sets with densities or excess volumes. So, in comparison to the original handbook, the new supplementary volume contains even a larger amount of data and will be a useful as well as necessary completion of the original handbook.

The *Supplement* will be divided into the five chapters: (1) Introduction, (2) Vapor-Liquid Equilibrium (VLE) Data and Gas Solubilities at Elevated Pressures, (3) Liquid-Liquid Equilibrium (LLE) Data of Polymer Solutions at Elevated Pressures, (4) High-Pressure Fluid Phase Equilibrium (HPPE) Data of Polymer Solutions, (5) PVT Data of Polymers and Solutions. Finally, appendices quickly route the user to the desired data sets.

Additionally, tables of systems are provided where results were published only in graphical form in the original literature to lead the reader to further sources. Data are included only if numerical values were published or authors provided their numerical results by personal communication (and I wish to thank all those who did so). No digitized data have been included in this data collection.

The closing date for the data compilation was June 30, 2014. However, the user who is in need of new additional data sets is kindly invited to ask for new information beyond this book via e-mail at christian.wohlfarth@chemie.uni-halle.de. Additionally, the author will be grateful to all users who call his attention to mistakes and make suggestions for improvements.

The new *CRC Handbook of Phase Equilibria and Thermodynamic Data of Polymer Solutions at Elevated Pressures* will again be useful to researchers, specialists, and engineers working in the fields of polymer science, physical chemistry, chemical engineering, materials science, biological science and technology, and those developing computerized predictive packages. The book should also be of use as a data source to Ph.D. students and faculty in chemistry, physics, chemical engineering, biotechnology, and materials science departments at universities.

Christian Wohlfarth

About the Author

Christian Wohlfarth is associate professor for physical chemistry at Martin-Luther University Halle-Wittenberg, Germany. He earned his degree in chemistry in 1974 and wrote his Ph.D. thesis in 1977 on investigations of the second dielectric virial coefficient and the intermolecular pair potential, both at Carl Schorlemmer Technical University Merseburg. In 1985, he wrote his habilitation thesis, *Phase Equilibria in Systems with Polymers and Copolymers*, at the Technical University Merseburg.

Since then, Dr. Wohlfarth's main research has been related to polymer systems. Currently, his research topics are molecular thermodynamics, continuous thermodynamics, phase equilibria in polymer mixtures and solutions, polymers in supercritical fluids, PVT behavior and equations of state, and sorption properties of polymers, about which he has published approximately 100 original papers. He has written the following books: *Vapor-Liquid Equilibria of Binary Polymer Solutions*, *CRC Handbook of Thermodynamic Data of Copolymer Solutions*, *CRC Handbook of Thermodynamic Data of Aqueous Polymer Solutions*, *CRC Handbook of Thermodynamic Data of Polymer Solutions at Elevated Pressures*, *CRC Handbook of Enthalpy Data of Polymer-Solvent Systems*, *CRC Handbook of Liquid-Liquid Equilibrium Data of Polymer Solutions*, *CRC Handbook of Phase Equilibria and Thermodynamic Data of Copolymer Solutions*, and *CRC Handbook of Phase Equilibria and Thermodynamic Data of Aqueous Polymer Solutions*.

He is working on the evaluation, correlation, and calculation of thermophysical properties of pure compounds and binary mixtures resulting in eleven volumes of the *Landolt-Börnstein New Series*. He is a contributor to the *CRC Handbook of Chemistry and Physics*.

1. INTRODUCTION

1.1. Objectives of the handbook

Knowledge of thermodynamic data of polymer solutions is a necessity for industrial and laboratory processes. Furthermore, such data serve as essential tools for understanding the physical behavior of polymer solutions, for studying intermolecular interactions, and for gaining insights into the molecular nature of mixtures. They also provide the necessary basis for any developments of theoretical thermodynamic models. Scientists and engineers in academic and industrial research need such data and will benefit from a careful collection of existing data. However, the database for polymer solutions at elevated pressures is still modest in comparison with the enormous amount of data for low-molecular mixtures. On the other hand, especially polymer solutions in supercritical fluids are gaining increasing interest (1994MCH, 1997KIR) because of their unique physical properties, and thermodynamic data at elevated pressures are needed for optimizing applications, e.g., separation operations of complex mixtures in the high-pressure synthesis of polymers, recovery of polymer wastes, precipitation, fractionation and purification of polymers, and polymers in green chemistry processes. During the last ten years after the former *CRC Handbook of Thermodynamic Data of Polymer Solutions at Elevated Pressures* was published, a large amount of new experimental data has been published, which is now compiled in this new *CRC Handbook of Phase Equilibria and Thermodynamic Data of Polymer Solutions at Elevated Pressures*.

Basic information on polymers can still be found in the *Polymer Handbook* (1999BRA), and there is also a chapter on properties of polymers and polymer solutions in the *CRC Handbook of Chemistry and Physics* (2011HAY). Older data for polymer solutions at elevated pressures can partly be found in the data books written by the author of this *Handbook* (1994WOH, 2001WOH, 2004WOH, 2005WOH, 2006WOH, 2008WOH, 2009WOH, 2011WOH, and 2013WOH). At least, polymer solution data are available from the Dortmund Data Bank (2006DDB). The new *Handbook* provides scientists and engineers with an up-to-date compilation from the literature of the available thermodynamic data on polymer solutions at elevated pressures. The *Handbook* does not present theories and models for polymer solution thermodynamics. Other publications (1971YAM, 1990FUJ, 1990KAM, 1999KLE, 1999PRA, and 2001KON) can serve as starting points for investigating those issues.

The data within this book are divided into four chapters:

- Vapor-liquid equilibrium (VLE) data and gas solubilities at elevated pressures
- Liquid-liquid equilibrium (LLE) data of polymer solutions at elevated pressures
- High-pressure fluid phase equilibrium (HPPE) data of polymer solutions
- PVT data of polymers and solutions

Data from investigations applying to more than one chapter are divided and appear in the relevant chapters. Data are included only if numerical values were published or authors provided their results by personal communication (and I wish to thank all those who did so). No digitized data have been included in this data collection, but a number of tables include systems based on data published in graphical form as phase diagrams or related figures.

1.2. Experimental methods involved

Besides the common progress in instrumentation and computation, no remarkable new developments have been made with respect to the experimental methods involved here. So, a short summary of this chapter should be sufficient for the *Handbook*. The necessary equations are given together with some short explanations only.

At higher pressures, the classification of experimental data into vapor-liquid, liquid-liquid, or fluid-fluid equilibrium data is not always simple. Also the expression "elevated pressure" is more or less relative. With respect to the latter, data are included here if at least some data points of a system were measured at a pressure above normal pressure. With respect to a classification of experimental data into the chapters 2, 3, or 4, practical reasons as well as theoretical considerations are applied. Theoretical considerations are taken into account for most polymer systems with subcritical solvents and supercritical fluids. Van Konynenburg and Scott (1980VAN) cited six classes of phase diagrams and showed that almost all known types of phase equilibria of binary mixtures can be classified within their scheme. Yelash and Kraska (1999YEL) developed global phase diagrams for mixtures of spherical molecules with non-spherical molecules also including polymers. So, most of the systems in this book could at least be categorized into the corresponding chapters in accordance to their rules. Nevertheless, sometimes problems remain for a number of systems with subcritical or supercritical fluid solvents. Therefore, in Chapter 4, the type of equilibrium within one data set is sometimes stated individually for each data point.

Accordingly, a classical discussion of experimental methods for vapor-liquid or liquid-liquid equilibrium measurements does not really fit here. Information about experimental methods for polymer solutions at ordinary pressures can be found in (1975BON) and (2000WOH) or in former handbooks (e.g., 2001WOH and 2004WOH). Here, the classification also used by Christov and Dohrn (2002CHR) is chosen where experimental methods for the investigation of high-pressure phase equilibria are divided into two main classes, depending on how the composition is determined: analytical (direct sampling methods) and synthetic (indirect sampling methods).

Analytical methods

Analytical methods involve the determination of the composition of the coexisting phases. This can be done by taking samples from each phase and analyzing them outside the equilibrium cell at normal pressure or by using physicochemical methods of analysis inside the equilibrium cell under pressure. If one needs the determination of more information than the total polymer composition, i.e., if one wants to characterize the polymer with respect to molar mass (distribution) or chemical composition (distribution), the sampling technique is unavoidable.

Withdrawing a large sample from an autoclave causes a considerable pressure drop, which disturbs the phase equilibrium significantly. This pressure drop can be avoided by using a variable-volume cell, by using a buffer autoclave in combination with a syringe pump or by blocking off a large sampling volume from the equilibrium cell before pressure reduction. If only a small sample is withdrawn or if a relatively large equilibrium cell is used, the slight pressure drop does not affect the phase composition significantly. Small samples can be withdrawn using capillaries or special sampling valves. Often sampling valves are directly coupled to analytical equipment. Analytical methods can be classified as isothermal methods, isobaric-isothermal methods, and isobaric methods.

Using the *isothermal mode*, an equilibrium cell is charged with the system of interest, the mixture is heated to the desired temperature, and this temperature is then kept constant. The pressure is adjusted *in the heterogeneous region* above or below the desired equilibrium value depending on how the equilibrium will change pressure. After intensive mixing over the time necessary for equilibrating the system, the pressure reaches a plateau value. The equilibration time is the most important point for polymer solutions. Due to their (high) viscosity and slow diffusion, equilibration will often need many hours or even days. The pressure can be readjusted by adding or withdrawing of material or by changing the volume of the cell if necessary. Before analyzing the compositions of the coexisting phases, the mixture is usually given some time for a clear phase separation. Sampling through capillaries can lead to differential vaporization (especially for mixtures containing gases and high-boiling solvents as well) when no precautions have been taken to prevent a pressure drop all along the capillary. This problem can be avoided with an experimental design that ensures that most of the pressure drop occurs at the end of the capillary. Sometimes, one or more phases will be recirculated to reduce sampling problems. The use of physicochemical methods of analysis inside the equilibrium cell, e.g., by a spectrometer, avoids the problems related to sampling. On the other hand, time-consuming calibrations can be necessary. At the end, isothermal methods need relatively simple and inexpensive laboratory equipment. If carried out carefully, they can produce reliable results.

Isobaric-isothermal methods are often also called dynamic methods. One or more fluid streams are pumped continuously into a thermostated equilibrium cell. The pressure is kept constant during the experiment by controlling an effluent stream, usually of the vapor phase. One can distinguish between continuous-flow methods and semi-flow methods. In continuous-flow methods, both phases flow through the equilibrium cell. They can be used only for systems where the time needed to attain phase equilibrium is sufficiently short. Therefore, such equipment is usually not applied to polymer solutions. In semi-flow methods, only one phase is flowing while the other stays in the equilibrium phase. They are sometimes called gas-saturation methods or pure-gas circulation methods and can be used to measure gas solubilities in liquids and melts or solubilities of liquid or solid substances in supercritical fluids.

Isobaric methods provide an alternative to direct measurements of pressure-temperature compositions in liquid phase compositions in vapor phase (P-T-w^L-w^V) data. This is the measurement of (P-T-w^L) data followed by a consistent thermodynamic analysis. Such isobaric analytical methods are usually made in dynamic mode, too.

The main advantages of the analytical methods are that systems with more than two components can be studied, several isotherms or isobars can be studied with one filling, and the coexistence data are determined directly. The main disadvantage is that the method is not suitable near critical states or for systems where the phases do not separate well. Furthermore, dynamic methods can be difficult in their application to highly viscous media like concentrated polymer solutions where foaming may cause further problems.

Synthetic methods

In synthetic methods, a mixture of known composition is prepared and the phase equilibrium is observed subsequently in an equilibrium cell (the problem of analyzing fluid mixtures is replaced by the problem of "synthesizing" them). After known amounts of the components have been placed into an equilibrium cell, pressure and temperature are adjusted so that the mixture is homogeneous. Then temperature or pressure is varied until formation of a new phase is observed. This is the common way to observe cloud points in demixing polymer systems. No sampling is necessary. The experimental equipment is relatively simple and inexpensive. For multicomponent systems, experiments with synthetic methods yield less information than with analytical methods, because the tie lines cannot be determined without additional experiments. This is specially true when fractionation accompanies demixing.

The appearance of a new phase is either detected visually or by monitoring physical properties. Visually, the beginning of turbidity in the system or the meniscus in a view cell can be observed. Otherwise, light scattering is the common method to detect the formation of the new phase. Both visual and non-visual synthetic methods are widely used in investigations on polymer systems at elevated pressures. The problem of isorefractive systems (where the coexisting phases have approximately the same refractive index) does not belong to polymer solutions where the usually strong concentration dependence of their refractive index prevents such a behavior. If the total volume of a variable-volume cell can be measured accurately, the appearance of a new phase can be observed from the abrupt change in the slope of a pressure-volume plot. An example is given in Chapter 4. The changes of other physical properties like viscosity, ultrasonic absorption, thermal expansion, dielectric constant, heat capacity, or UV and IR absorption are also applied as non-visual synthetic methods for polymer solutions.

A common synthetic method for polymer solutions is the (P-T-w^L) experiment. An equilibrium cell is charged with a known amount of polymer, evacuated and thermostated to the measuring temperature. Then the second component (gas, fluid, solvent) is added and the pressure increases. The second component dissolves into the (amorphous or molten) polymer and the pressure in the equilibrium cell decreases. Therefore, this method is sometimes called the pressure-decay method. Pressure and temperature are registered after equilibration. No samples are taken. The composition of the vapor phase is calculated using a phase equilibrium model if two or more gases or solvents are involved. The determination of the volume of gas or solvent vaporized in the unoccupied space of the apparatus is important as it can cause serious errors in the determination of their final concentrations. The composition of the liquid phase is often obtained by weighing and using the material balance. By repeating the addition of the second component into the cell, several points along the vapor/gas-liquid equilibrium line can be measured. This method is usually applied for all gas solubility/vapor sorption/vapor pressure investigations in systems with polymers. The synthetic method is particularly suitable for measurements near critical states. Simultaneous determination of *PVT* data is possible.

Some problems related to systems with polymers

Details of experimental equipment can be found in the original papers compiled for this book and will not be presented here. Only some problems should be summarized that have to be obeyed and solved during the experiment.

The polymer solution is often of an amount of some cm^3 and may contain about 1g of polymer or even more. Therefore, the equilibration of prepared solutions can be difficult and equilibration is usually very time consuming (liquid oligomers do not need so much time, of course). Increasing viscosity makes the preparation of concentrated solutions more and more difficult with further increasing the amount of polymer. Solutions above 50-60 wt% can hardly be prepared (depending on the solvent/polymer pair under investigation).

All impurities in the pure solvents have to be eliminated. Degassing of solvents (and sometimes of polymers too) is absolutely necessary. Polymers and solvents must keep dry. Sometimes, inhibitors and antioxidants are added to polymers. They may probably influence the position of the equilibrium. The thermal stability of polymers must be obeyed, otherwise, depolymerization or formation of networks by chemical processes can change the sample during the experiment.

Certain principles must be obeyed for experiments where liquid-liquid equilibrium is observed in polymer-solvent (or supercritical fluid) systems. To understand the results of LLE experiments in polymer solutions, one has to take into account the strong influence of polymer distribution functions on LLE, because fractionation occurs during demixing. Fractionation takes place with respect to molar mass distribution as well as to chemical distribution if copolymers are involved. Fractionation during demixing leads to some effects by which the LLE phase behavior differs from that of an ordinary, strictly binary mixture, because a common polymer solution is a multicomponent system. Cloud-point curves are measured instead of binodals; and per each individual feed concentration of the mixture, two parts of a coexistence curve occur below (for upper critical solution temperature, UCST, behavior) or above the cloud-point curve (for lower critical solution temperature, LCST, behavior), i.e., produce an infinite number of coexistence data.

Distribution functions of the feed polymer belong only to cloud-point data. On the other hand, each pair of coexistence points is characterized by two new and different distribution functions in each coexisting phase. The critical concentration is the only feed concentration where both parts of the coexistence curve meet each other on the cloud-point curve at the critical point that belongs to the feed polymer distribution function. The threshold point (maximum or minimum corresponding to either UCST or LCST behavior) temperature (or pressure) is not equal to the critical point, since the critical point is to be found at a shoulder of the cloud-point curve. Details were discussed by Koningsveld (1968KON, 1972KON). Thus, LLE data have to be specified in the tables as cloud-point or coexistence data, and coexistence data make sense only if the feed concentration is given. This is not always the case, however.

Special methods are necessary to measure the critical point. Only for solutions of monodisperse polymers, the critical point is the maximum (or minimum) of the binodal. Binodals of polymer solutions can be rather broad and flat. Then, the exact position of the critical point can be obtained by the method of the rectilinear diameter:

$$\frac{(\varphi_B^{\ I} - \varphi_B^{\ II})}{2} - \varphi_{B,crit} \propto (1 - \frac{T}{T_{crit}})^{1-\alpha} \tag{1}$$

where:

$\varphi_B^{\ I}$	volume fraction of the polymer B in coexisting phase I
$\varphi_B^{\ II}$	volume fraction of the polymer B in coexisting phase II
$\varphi_{B,crit}$	volume fraction of the polymer B at the critical point
T	(measuring) temperature
T_{crit}	critical temperature (LLE)
α	critical exponent

For solutions of polydisperse polymers, such a procedure cannot be used because the critical concentration must be known in advance to measure its corresponding coexistence curve. Two different methods were developed to solve this problem: the phase-volume-ratio method (1968KON) where one uses the fact that this ratio is exactly equal to one only at the critical point, and the coexistence concentration plot (1969WOL) where an isoplethal diagram of values of φ_B^{I} and φ_B^{II} vs. the feed concentration, φ_{0B}, gives the critical point as the intersection of cloud-point curve and shadow curve. Details will not be discussed here. Treating polymer solutions with distribution functions by continuous thermodynamics is reviewed in (1989RAE) and (1990RAE).

PVT/density measurement for polymer melt and solution

There are two widely practiced methods for the *PVT* measurement of polymers and polymer solutions:

1. Piston-die technique
2. Confining fluid technique

which were described in detail by Zoller in papers and books (1986ZOL and 1995ZOL). Thus, a short summary is sufficient here.

In the piston-die technique, the material is confined in a rigid die or cylinder, which it has to fill completely. A pressure is applied to the sample as a load on a piston, and the movement of the piston with pressure and temperature changes is used to calculate the specific volume of the sample. Experimental problems concerning solid samples need not be discussed here, since only data for the liquid/molten (equilibrium) state are taken into consideration for this handbook. A typical practical complication is leakage around the piston when low-viscosity melts or solutions are tested. Seals cause an amount of friction leading to uncertainties in the real pressure applied. There are commercial devices as well as laboratory-built machines which have been used in the literature.

In the confining fluid technique, the material is surrounded at all times by a confining (inert) fluid, often mercury, and the combined volume changes of sample and fluid are measured by a suitable technique as a function of temperature and pressure. The volume change of the sample is determined by subtracting the volume change of the confining fluid. A problem with this technique lies in potential interactions between fluid and sample. Precise knowledge of the *PVT* properties of the confining fluid is additionally required. The above-mentioned problems for the piston-die technique can be avoided.

For both techniques, the absolute specific volume of the sample must be known at a single condition of pressure and temperature. Normally, these conditions are chosen to be 298.15 K and normal pressure (101.325 kPa). There are a number of methods to determine specific volumes (or densities) under these conditions. For polymeric samples, hydrostatic weighing or density gradient columns were often used.

The tables in Chapter 5 do not provide specific volumes below the melting transition of semicrystalline materials or below the glass transition of amorphous samples, because *PVT* data of solid polymer samples are non-equilibrium data and depend on sample history and experimental procedure (which will not be discussed here). *PVT* data for many polymers are given in (1995ZOL), for copolymers in (2001WOH) and (2011WOH). Parameters of the Tait equation for a number of polymers are given in (1999BRA) and (2003WOH).

Measurement of densities for polymer solutions at elevated pressures can be made today by U-tube vibrating densimeters. Such instruments are commercially available. Calibration is often made with pure water. Otherwise, densities can also be measured by the above discussed *PVT* equipment or in the equilibrium cell of a synthetic method.

Excess volumes are determined by

$$V^E_{spec} = V_{spec} - (w_A V_{0A,\,spec} + w_B V_{0B,\,spec}) \tag{2}$$

or

$$V^E = (x_A M_A + x_B M_B)/\rho - (x_A M_A/\rho_A + x_B M_B/\rho_B) \tag{3}$$

where:

V^E	molar excess volume at temperature T
V^E_{spec}	specific excess volume at temperature T
$V_{0A,\,spec}$	specific volume of pure solvent A at temperature T
$V_{0B,\,spec}$	specific volume of pure polymer B at temperature T
ρ	density of the mixture at temperature T
ρ_A	density of pure solvent A at temperature T
ρ_B	density of pure polymer B at temperature T

1.3. Guide to the data tables

Characterization of the polymers

Polymers vary by a number of characterization variables (1991BAR, 1999PET). The molar mass and their distribution function are the most important variables. However, tacticity, sequence distribution, branching, and end groups determine their thermodynamic behavior in solution too. Unfortunately, much less information is provided with respect to the polymers that were applied in most of the thermodynamic investigations in the original literature. For copolymers, the chemical distribution and the average chemical composition are also to be given. But, in many cases, the samples are characterized only by one or two molar mass averages and some additional information (e.g., T_g, T_m, ρ_B, or how and where they were synthesized). Sometimes even this information is missed.

The molar mass averages are defined as follows:

number average M_n

$$M_n = \frac{\sum_i n_{B_i} M_{B_i}}{\sum_i n_{B_i}} = \frac{\sum_i w_{B_i}}{\sum_i w_{B_i} / M_{B_i}} \tag{4}$$

mass average M_w

$$M_w = \frac{\sum_i n_{B_i} M_{B_i}^2}{\sum_i n_{B_i} M_{B_i}} = \frac{\sum_i w_{B_i} M_{B_i}}{\sum_i w_{B_i}} \quad (5)$$

z-average M_z

$$M_z = \frac{\sum_i n_{B_i} M_{B_i}^3}{\sum_i n_{B_i} M_{B_i}^2} = \frac{\sum_i w_{B_i} M_{B_i}^2}{\sum_i w_{B_i} M_{B_i}} \quad (6)$$

viscosity average M_η

$$M_\eta = \left(\frac{\sum_i w_{B_i} M_{B_i}^a}{\sum_i w_{B_i}} \right)^{1/a} \quad (7)$$

where:

a	exponent in the viscosity-molar mass relationship
M_{Bi}	relative molar mass of the polymer species B_i
n_{Bi}	amount of substance of polymer species B_i
w_{Bi}	mass fraction of polymer species B_i

Measures for the polymer concentration

The following concentration measures are used in the tables of this handbook (where B always denotes the main polymer, A denotes the solvent, and in ternary systems C denotes the third component):

mass/volume concentration

$$c_A = m_A/V \qquad c_B = m_B/V \quad (8)$$

mass fraction

$$w_A = m_A/\Sigma\, m_i \qquad w_B = m_B/\Sigma\, m_i \quad (9)$$

mole fraction

$$x_A = n_A/\Sigma\, n_i \qquad x_B = n_B/\Sigma\, n_i \qquad \text{with} \qquad n_i = m_i/M_i \quad (10)$$

volume fraction

$$\varphi_A = (m_A/\rho_A)/\Sigma\,(m_i/\rho_i) \qquad \varphi_B = (m_B/\rho_B)/\Sigma\,(m_i/\rho_i) \qquad (11)$$

segment fraction

$$\psi_A = x_A r_A/\Sigma\,x_i r_i \quad \psi_B = x_B r_B/\Sigma\,x_i r_i \quad \text{usually with } r_A = 1 \qquad (12)$$

where:

c_A	(mass/volume) concentration of solvent A
c_B	(mass/volume) concentration of polymer B
m_A	mass of solvent A
m_B	mass of polymer B
M_A	relative molar mass of the solvent A
M_B	relative molar mass of the polymer B
M_n	number-average relative molar mass
M_0	molar mass of a basic unit of the polymer B
n_A	amount of substance of solvent A
n_B	amount of substance of polymer B
r_A	segment number of the solvent A, usually $r_A = 1$
r_B	segment number of the polymer B
V	volume of the liquid solution at temperature T
w_A	mass fraction of solvent A
w_B	mass fraction of polymer B
x_A	mole fraction of solvent A
x_B	mole fraction of polymer B
φ_A	volume fraction of solvent A
φ_B	volume fraction of polymer B
ρ_A	density of solvent A
ρ_B	density of polymer B
ψ_A	segment fraction of solvent A
ψ_B	segment fraction of polymer B

For high-molecular polymers, a mole fraction is not an appropriate unit to characterize composition. However, for oligomeric products with rather low molar masses, mole fractions were sometimes used. In the common case of a distribution function for the molar mass, $M_B = M_n$ is to be chosen. Mass fraction and volume fraction can be considered as special cases of segment fractions depending on the way by which the segment size is actually determined: $r_i/r_A = M_i/M_A$ or $r_i/r_A = V_i/V_A = (M_i/\rho_i)/(M_A/\rho_A)$, respectively. Classical segment fractions are calculated by applying $r_i/r_A = V_i^{vdW}/V_A^{vdW}$ ratios where hard-core van-der-Waals volumes, V_i^{vdW}, are taken into account. Their special values depend on the chosen equation of state or simply some group contribution schemes, e.g., (1968BON, 1990KRE) and have to be specified.

Volume fractions imply a temperature dependence and, as they are defined in the equations above, neglect excess volumes of mixing and, very often, the densities of the polymer in the state of the solution are not known correctly. However, volume fractions can be calculated without the exact knowledge of the polymer molar mass (or its averages).

Tables of experimental data

The data tables in each chapter are provided there in order of the names of the polymers. In this data book, mostly source-based polymer names are applied. These names are more common in use, and they are usually given in the original sources, too. Structure-based names, for which details about their nomenclature can be found in the *Polymer Handbook* (1999BRA), are chosen in some single cases only. CAS index names for polymers are not applied here. Latest *IUPAC Recommendations* for class names of polymers based on chemical structure and molecular architecture are given in (2009BAR). A list of systems and properties in order of the polymers in Appendix 1 utilizes the names as given in the chapters of this book.

Within types of polymers the individual samples are ordered by their increasing average molar mass, and, when necessary, systems are ordered by increasing temperature. In ternary systems, ordering is additionally made subsequently according to the name of the third component in the system. Each data set begins with the lines for the solution components, e.g., in binary systems

Polymer (B):	**1,2-polybutadiene**		**2005STR**
Characterization:	M_n/g.mol^{-1} = 97000, M_w/g.mol^{-1} = 105000, ρ/(kg m^{-3})= 898 − 0.453(θ/°C), Bayer AG, Germany		
Solvent (A):	**n-butane**	C_4H_{10}	**106-97-8**

where the polymer sample is given in the first line together with the reference. The second line then provides the characterization available for the polymer sample. The following line gives the solvent's chemical name, molecular formula, and CAS registry number.

In ternary and quaternary systems, the following lines are either for a second solvent or a second polymer or a salt or another chemical compound, e.g., in ternary systems with two solvents

Polymer (B):	**polyethylene**		**2009HAR**
Characterization:	M_n/g.mol^{-1} = 13200, M_w/g.mol^{-1} = 15400, Scientific Polymer Products, Inc., Ontario, NY		
Solvent (A):	**cyclohexane**	C_6H_{12}	**110-82-7**
Solvent (C):	**n-hexane**	C_6H_{14}	**110-54-3**

or, e.g., in quaternary (or higher) systems like

Polymer (B):	**poly(ethylene-*co*-norbornene)**		**2013SAT**
Characterization:	M_n/g.mol^{-1} = 38800, M_w/g.mol^{-1} = 74600, 52 mol% ethylene, T_g/K = 411.15, Topas Advanced Polymers GmbH		
Solvent (A):	**ethene**	C_2H_4	**74-85-1**
Solvent (C):	**toluene**	C_7H_8	**108-88-3**
Solvent (D):	**bicyclo[2,2,1]-2-heptene (norbornene)**	C_7H_{10}	**498-66-8**

There are some exceptions from this type of presentation within the tables for the *PVT* data of pure polymers. These tables are prepared in the forms as chosen in 2004WOH.

The originally measured data for each single system are sometimes listed together with some comment lines if necessary. The data are usually given as published, but temperatures are always given in K. Pressures are sometimes recalculated into kPa or MPa.

Because many investigations on liquid-liquid or fluid-fluid equilibrium in polymer solutions at elevated pressures are provided in the literature in figures only, chapters 3 and 4 contain additional tables referring to types of systems, included components, and references.

Final day for including data into this *Handbook* was June, 30, 2014.

1.4. List of symbols

a	exponent in the viscosity-molar mass relationship
a_A	activity of solvent A
A, B, C	parameters of the Tait equation
c_A	(mass/volume) concentration of solvent A
c_B	(mass/volume) concentration of polymer B
m_A	mass of solvent A
m_B	mass of polymer B
M	relative molar mass
M_A	molar mass of the solvent A
M_B	molar mass of the polymer B
M_n	number-average relative molar mass
M_w	mass-average relative molar mass
M_η	viscosity-average relative molar mass
M_z	z-average relative molar mass
M_0	molar mass of a basic unit of the polymer B
MI	melting index
n_A	amount of substance of solvent A
n_B	amount of substance of polymer B
P	pressure
P_0	standard pressure (= 0.101325 MPa)
P_A	partial vapor pressure of the solvent A at temperature T
P_{0A}	vapor pressure of the pure liquid solvent A at temperature T
P_{crit}	critical pressure
R	gas constant
r_A	segment number of the solvent A, usually $r_A = 1$
r_B	segment number of the polymer B
T	(measuring) temperature
T_{crit}	critical temperature
T_g	glass transition temperature
T_m	melting transition temperature
V, V_{spec}	volume or specific volume at temperature T
V_0	reference volume
V^E	excess volume at temperature T

V^{vdW}	hard-core van-der-Waals volume
ΔV	volume difference $V - V_0$ (in swelling experiments)
w_A	mass fraction of solvent A
w_B	mass fraction of polymer B
$w_{B,crit}$	mass fraction of the polymer B at the critical point
x_A	mole fraction of solvent A
x_B	mole fraction of polymer B
α	critical exponent
γ_A	activity coefficient of the solvent A in the liquid phase with activity $a_A = x_A\gamma_A$
λ	wavelength
φ_A	volume fraction of solvent A
φ_B	volume fraction of polymer B
$\varphi_{B,crit}$	volume fraction of the polymer B at the critical point
ρ	density (of the mixture) at temperature T
ρ_A	density of solvent A
ρ_B	density of polymer B
ψ_A	segment fraction of solvent A
ψ_B	segment fraction of polymer B

1.5. References

1968BON Bondi, A., *Physical Properties of Molecular Crystals, Liquids and Glasses*, J. Wiley & Sons, New York, 1968.

1968KON Koningsveld, R. and Staverman, A.J., Liquid-liquid phase separation in multicomponent polymer solutions I and II, *J. Polym. Sci.*, Pt. A-2, 6, 305, 325, 1968.

1969WOL Wolf, B.A., Zur Bestimmung der kritischen Konzentration von Polymerlösungen, *Makromol. Chem.*, 128, 284, 1969.

1971YAM Yamakawa, H., *Modern Theory of Polymer Solutions*, Harper & Row, New York, 1971.

1972KON Koningsveld, R., Polymer solutions and fractionation, in *Polymer Science*, Jenkins, E.D., Ed., North-Holland, Amsterdam, 1972, 1047.

1975BON Bonner, D.C., Vapor-liquid equilibria in concentrated polymer solutions, *Macromol. Sci. Rev. Macromol. Chem.*, C13, 263, 1975.

1980VAN Van Konynenburg, P.H. and Scott, R.L., Critical lines and phase equilibria in binary van-der-Waals mixtures, *Philos. Trans. Roy. Soc.*, 298, 495, 1980.

1986ZOL Zoller, P., Dilatometry, in *Encyclopedia of Polymer Science and Engineering*, Vol. 5, 2nd ed., Mark, H. et al., Eds., J. Wiley & Sons, New York, 1986, 69.

1987COO Cooper, A.R., Molecular weight determination, in *Encyclopedia of Polymer Science and Engineering*, Vol. 10, 2nd ed., Mark, H. et al., Eds., J. Wiley & Sons, New York, 1986, 1.

1989RAE Rätzsch, M.T. and Kehlen, H., Continuous thermodynamics of polymer systems, *Prog. Polym. Sci.*, 14, 1, 1989.

1990BAR Barton, A.F.M., *CRC Handbook of Polymer-Liquid Interaction Parameters and Solubility Parameters*, CRC Press, Boca Raton, 1990.

1990FUJ Fujita, H., *Polymer Solutions*, Elsevier, Amsterdam, 1990.

1990KAM Kamide, K., *Thermodynamics of Polymer Solutions*, Elsevier, Amsterdam, 1990.

1990KRE [Van] Krevelen, D.W., *Properties of Polymers*, 3rd ed., Elsevier, Amsterdam, 1990.

1990RAE Rätzsch, M.T. and Wohlfarth, C., Continuous thermodynamics of copolymer systems, *Adv. Polym. Sci.*, 98, 49, 1990.

1991BAR Barth, H.G. and Mays, J.W., Eds., *Modern Methods of Polymer Characterization*, J. Wiley & Sons, New York, 201, 1991.

1993DAN Danner, R.P. and High, M.S., *Handbook of Polymer Solution Thermodynamics*, American Institute of Chemical Engineers, New York, 1993.

1994MCH McHugh, M.A. and Krukonis, V.J., *Supercritical Fluid Extraction: Principles and Practice*, 2nd ed., Butterworth Publishing, Stoneham, 1994.

1994WOH Wohlfarth, C., *Vapour-Liquid Equilibrium Data of Binary Polymer Solutions: Physical Science Data*, 44, Elsevier, Amsterdam, 1994.

1995ZOL Zoller, P. and Walsh, D.J., *Standard Pressure-Volume-Temperature Data for Polymers*, Technomic Publishing, Lancaster, 1995.

1997KIR Kiran, E. and Zhuang, W., *Miscibility and Phase Separation of Polymers in Near- and Supercritical Fluids*, ACS Symposium Series 670, 2, 1997.

1999BRA Brandrup, J., Immergut, E.H., and Grulke, E.A., Eds., *Polymer Handbook*, 4th ed., J. Wiley & Sons, New York, 1999.

1999KIR Kirby, C.F. and McHugh, M.A., Phase behavior of polymers in supercritical fluid solvents, *Chem. Rev.*, 99, 565, 1999.

1999KLE Klenin, V.J., *Thermodynamics of Systems Containing Flexible-Chain Polymers*, Elsevier, Amsterdam, 1999.

1999PET Pethrick, R.A. and Dawkins, J.V., Eds., *Modern Techniques for Polymer Characterization*, J. Wiley & Sons, Chichester, 1999.

1999PRA Prausnitz, J.M., Lichtenthaler, R.N., and de Azevedo, E.G., *Molecular Thermodynamics of Fluid Phase Equilibria*, 3rd ed., Prentice Hall, Upper Saddle River, NJ, 1999.

1999YEL Yelash, L.V. and Kraska, T., The global phase behavior of binary mixtures of chain molecules. Theory and application, *Phys. Chem. Chem. Phys.*, 1, 4315, 1999.

2000WOH Wohlfarth, C., Methods for the measurement of solvent activity of polymer solutions, in *Handbook of Solvents*, Wypych, G., Ed., ChemTec Publishing, Toronto, 146, 2000.

2001KON Koningsveld, R., Stockmayer, W.H., and Nies, E., *Polymer Phase Diagrams*, Oxford University Press, Oxford, 2001.

2001WOH Wohlfarth, C., *CRC Handbook of Thermodynamic Data of Copolymer Solutions*, CRC Press, Boca Raton, 2001.

2002CHR Christov, M. and Dohrn, R., High-pressure fluid phase equilibria. Experimental methods and systems investigated (1994-1999), *Fluid Phase Equil.*, 202, 153, 2002.

2003WOH Wohlfarth, C., Pressure-volume-temperature relationship for polymer melts, in *CRC Handbook of Chemistry and Physics*, Lide, D.R., Ed., 84th ed., pp. 13-16 to 13-20, 2003.

2004WOH Wohlfarth, C., *CRC Handbook of Thermodynamic Data of Aqueous Polymer Solutions*, CRC Press, Boca Raton, 2004.

2005WOH Wohlfarth, C., *CRC Handbook of Thermodynamic Data of Polymer Solutions at Elevated Pressures*, Taylor & Francis, CRC Press, Boca Raton, 2005.

2006DDB Dortmund Data Bank, Polymer Subset, www.ddbst.de, 2006.

2006WOH Wohlfarth, C., *CRC Handbook of Enthalpy Data of Polymer-Solvent Systems*, Taylor & Francis, CRC Press, Boca Raton, 2006.

2008WOH Wohlfarth, C., *CRC Handbook of Liquid-Liquid Equilibrium Data of Polymer Solutions*, Taylor & Francis, CRC Press, Boca Raton, 2008.

2009WOH Wohlfarth, C.: Landolt-Börnstein New Series, Group VIII, Vol. 6, *Polymers*, Subvolume D, *Polymer Solutions*, Part 1, *Physical Properties and Their Relations* I (*Thermodynamic Properties: Phase Equilibria*), Springer Verlag, Berlin, Heidelberg 2009.

2011HAY Haynes, W.M., Ed., *CRC Handbook of Chemistry and Physics, Section 13: Polymer Properties*, 92nd ed., Taylor & Francis, CRC Press, Boca Raton, 2011.

2011WOH Wohlfarth, C., *CRC Handbook of Phase Equilibria and Thermodynamic Data of Copolymer Solutions*, CRC Press, Boca Raton, 2011.

2013WOH Wohlfarth, C., *CRC Handbook of Phase Equilibria and Thermodynamic Data of Aqueous Polymer Solutions*, CRC Press, Boca Raton, 2013.

2. VAPOR-LIQUID EQUILIBRIUM (VLE) DATA AND GAS SOLUBILITIES AT ELEVATED PRESSURES

2.1. Binary polymer solutions

Polymer (B): **polyamide-11** **2005SO2**
Characterization: commercial sample, NKT Flexibles
Solvent (A): **carbon dioxide** CO_2 **124-38-9**

Type of data: gas solubility

T/K = 323.15

P/bar	21.0	21.0	22.1	31.1	31.1	39.8	39.8
w_A	0.0117	0.0132	0.0126	0.0175	0.0185	0.0216	0.0226

T/K = 343.15

P/bar	20.3	26.4	37.5
w_A	0.0117	0.0141	0.0184

T/K = 363.15

P/bar	20.0	21.8	27.9	28.2	30.4	42.4
w_A	0.00688	0.00776	0.00984	0.0146	0.0119	0.0138

Polymer (B): **polyamide-11** **2005SO2**
Characterization: commercial sample, NKT Flexibles
Solvent (A): **methane** CH_4 **74-82-8**

Type of data: gas solubility

T/K = 323.15

P/bar	51.2	107.2	156.6
w_A	0.00317	0.00449	0.00344

T/K = 343.15

P/bar	55.3	104.0	158.1
w_A	0.00260	0.00495	0.00530

T/K = 363.15

P/bar	53.1	106.5	144.4
w_A	0.00237	0.00511	0.00648

Polymer (B):	**1,2-polybutadiene**	**2005STR**
Characterization:	M_n/g.mol^{-1} = 97000, M_w/g.mol^{-1} = 105000, ρ /(kg m^{-3})= 898 − 0.453(θ/°C), Bayer AG, Germany	
Solvent (A):	**n-butane** C_4H_{10}	**106-97-8**

Type of data: vapor-liquid equilibrium

T/K = 298.15

φ_B	0.9107	0.8747	0.7997	0.6589	0.5692	0.5292	0.4593
P_A/kPa	58.92	95.16	139.44	195.48	215.50	221.64	233.00

T/K = 303.15

φ_B	0.9164	0.8872	0.8198	0.6971	0.6718	0.6061	0.5685	0.4807	0.3573
P_A/kPa	61.20	100.56	147.96	215.38	226.50	243.00	250.56	267.50	279.00

T/K = 313.15

φ_B	0.9303	0.9052	0.8553	0.7649	0.7295	0.6865	0.6523	0.5031	0.3921
P_A/kPa	66.84	111.12	164.04	249.84	278.00	297.72	306.00	353.50	371.00

T/K = 323.15

φ_B	0.9400	0.9179	0.8817	0.8163	0.7774	0.7431	0.7265	0.5744	0.4494
P_A/kPa	71.64	118.36	177.72	277.08	321.50	342.12	351.00	441.50	477.00

φ_B	0.4097
P_A/kPa	481.50

T/K = 333.15

φ_B	0.9382	0.9274	0.9008	0.8530	0.8111	0.7930	0.7806	0.6490	0.5144
P_A/kPa	73.40	125.16	189.84	299.52	361.50	374.16	385.80	521.50	603.00

φ_B	0.4741	0.2262
P_A/kPa	602.00	638.30

T/K = 343.15

φ_B	0.9598	0.9366	0.9140	0.8800	0.8434	0.8330	0.8199	0.7151	0.5800
P_A/kPa	82.50	132.24	200.52	319.08	391.50	402.36	414.60	588.00	721.50

φ_B	0.5578	0.2415
P_A/kPa	723.50	806.50

T/K = 348.15

φ_B	0.9582	0.9407	0.9188	0.8904	0.8562	0.8480	0.8339	0.7402	0.6252
P_A/kPa	83.40	135.72	205.44	327.96	405.00	414.72	427.08	615.00	773.50

φ_B	0.5997	0.2569
P_A/kPa	776.00	902.00

Polymer (B): **1,4-*cis*-polybutadiene** **2005STR**
Characterization: M_n/g.mol^{-1} = 224000, M_w/g.mol^{-1} = 650000, 98% *cis*, ρ /(kg m^{-3})= 909 − 0.558(θ/°C), Bayer AG, Germany
Solvent (A): **n-butane** C_4H_{10} **106-97-8**

Type of data: vapor-liquid equilibrium

T/K = 298.15

φ_B	0.9000	0.7472	0.6137	0.5662	0.5394	0.5049
P_A/kPa	95.40	191.64	225.48	234.00	237.00	240.50

T/K = 303.15

φ_B	0.9094	0.7794	0.6585	0.6109	0.5669	0.5295
P_A/kPa	99.50	205.80	250.44	266.16	272.50	277.00

T/K = 313.15

φ_B	0.9254	0.8322	0.7416	0.6898	0.6304	0.5913	0.4713
P_A/kPa	107.04	229.32	291.96	325.56	343.50	355.00	375.50

T/K = 323.15

φ_B	0.9371	0.8694	0.7989	0.6839	0.6452	0.5130	0.3732
P_A/kPa	113.64	247.80	321.24	413.00	433.50	483.50	493.00

T/K = 333.15

φ_B	0.9459	0.8961	0.8397	0.7446	0.7133	0.6083	0.5294	0.2967
P_A/kPa	119.64	263.40	344.64	460.50	489.00	579.00	620.00	637.50

T/K = 343.15

φ_B	0.9525	0.9166	0.8696	0.7890	0.7627	0.6898	0.6384	0.4473	0.3994
P_A/kPa	125.16	277.44	364.80	498.00	531.50	649.50	721.00	804.00	806.00

T/K = 348.15

φ_B	0.9547	0.9246	0.8814	0.8067	0.7824	0.7215	0.6835	0.4998	0.4527
P_A/kPa	127.68	283.92	374.04	514.50	550.00	677.50	759.00	884.00	902.00

Polymer (B): **poly(ε-caprolactone)** **2006LEE**
Characterization: M_n/g.mol^{-1} = 69000, M_w/g.mol^{-1} = 120000
Solvent (A): **carbon dioxide** CO_2 **124-38-9**

Type of data: gas solubility

T/K = 313.15

P/bar	100	125	153	200
w_A	0.4655	0.7079	0.7390	0.7622

T/K = 333.15

P/bar	80	100	125	153	200
w_A	0.0285	0.3328	0.5196	0.5783	0.6209

Polymer (B): **poly(dimethylsiloxane)** **2009MIL**
Characterization: M_n/g.mol^{-1} = 237, Dow Corning
Solvent (A): **carbon dioxide** **CO_2** **124-38-9**

Type of data: vapor-liquid equilibrium

T/K = 298.15

w_B	0.00	0.05	0.10	0.15	0.20	0.25	0.30	0.35	0.40
P/MPa	6.43	6.23	6.05	5.95	5.86	5.55	5.40	5.13	4.91
w_B	0.45	0.50	0.55	0.60	0.65	0.70	0.75	0.80	0.85
P/MPa	4.70	4.38	4.12	3.70	3.32	2.90	2.59	2.24	1.80
w_B	0.90	0.95							
P/MPa	1.15	0.55							

Polymer (B): **polyester (hyperbranched)** **2009KOZ**
Characterization: M_n/g.mol^{-1} = 4300, M_w/g.mol^{-1} = 6500, Boltorn U3000, dendritic polyester with an average of 14 unsaturated fatty ester end groups, Perstorp Speciality Chemicals AB, Sweden
Solvent (A): **carbon dioxide** **CO_2** **124-38-9**

Type of data: vapor-liquid equilibrium

w_A	0.05	0.05	0.05	0.05	0.05	0.05	0.05	0.05	0.05
T/K	328.49	328.56	333.52	333.56	338.47	338.60	343.53	343.62	348.58
P/MPa	2.87	2.86	3.04	3.06	3.21	3.26	3.42	3.43	3.61
w_A	0.05	0.05	0.05	0.05	0.05	0.10	0.10	0.10	0.10
T/K	348.60	353.60	358.68	363.61	368.69	288.50	293.53	298.50	303.49
P/MPa	3.63	3.83	4.00	4.19	4.39	2.85	3.13	3.44	3.74
w_A	0.10	0.10	0.10	0.10	0.10	0.10	0.10	0.10	0.10
T/K	308.51	313.52	318.50	323.53	328.55	333.62	338.54	343.51	348.61
P/MPa	4.11	4.46	4.81	5.13	5.54	5.94	6.29	6.72	7.08
w_A	0.10	0.10	0.10	0.10	0.10	0.10	0.10	0.10	0.10
T/K	353.62	358.65	363.70	368.72	372.27	377.35	377.85	382.26	387.23
P/MPa	7.48	7.91	8.36	8.77	9.01	9.55	9.56	9.92	10.30
w_A	0.10	0.10	0.10	0.10	0.10	0.10	0.15	0.15	0.15
T/K	392.18	397.37	402.17	407.12	412.05	416.99	283.44	288.47	293.49
P/MPa	10.68	11.15	11.60	11.97	12.32	12.75	3.55	3.95	4.36
w_A	0.15	0.15	0.15	0.15	0.15	0.15	0.15	0.15	0.15
T/K	298.40	303.50	308.46	313.51	318.44	318.45	323.46	323.51	328.52
P/MPa	4.82	5.35	5.84	6.40	7.01	6.99	7.58	7.74	8.34
w_A	0.15	0.15	0.15	0.15	0.15	0.15	0.15		
T/K	333.56	338.56	343.47	348.59	353.66	358.63	363.72		
P/MPa	8.99	9.62	10.27	10.90	11.58	12.26	12.97		

Polymer (B): **polyester (hyperbranched)** **2009POR**
Characterization: M_n/g.mol^{-1} = 6560, M_w/g.mol^{-1} = 10500, Boltorn H3200, solid, fatty acid (C20/C22) modified, hyperbranched polyester, Perstorp Speciality Chemicals AB, Sweden
Solvent (A): **propane** **C_3H_8** **74-98-6**

Type of data: vapor-liquid equilibrium

w_B	0.701	0.701	0.701	0.701	0.701	0.701	0.801	0.801	0.801
T/K	314.33	313.23	322.71	332.31	342.02	345.92	316.32	322.69	332.34
P/MPa	1.41	1.37	1.70	2.09	2.55	2.76	1.32	1.52	1.89

w_B	0.801	0.801	0.801	0.801	0.801	0.801	0.801	0.801	0.801
T/K	341.98	351.51	361.07	366.72	370.58	380.60	390.15	400.16	410.28
P/MPa	2.24	2.69	3.21	3.56	3.83	4.54	5.37	6.36	7.44

w_B	0.801	0.801	0.801	0.801	0.801	0.901	0.901	0.901	0.901
T/K	420.27	430.38	440.29	450.24	460.28	322.77	332.23	341.92	351.82
P/MPa	8.46	9.54	10.52	11.45	12.34	1.06	1.27	1.51	1.78

w_B	0.901	0.901	0.901	0.901	0.901	0.901	0.901	0.901	0.901
T/K	361.54	371.30	380.90	390.62	400.63	410.61	420.55	430.53	440.50
P/MPa	2.08	2.40	2.75	3.12	3.52	3.95	4.40	4.85	5.31

w_B	0.901	0.901	0.901
T/K	450.33	460.16	470.07
P/MPa	5.76	6.22	6.70

Comments: VLLE data are given in Chapter 3.

Polymer (B): **polyester resin** **2010SKE**
Characterization: M_n/g.mol^{-1} = 3000, M_w/g.mol^{-1} = 7500, T_g/K = 326.9, CPE55, Helios, Slovenia
mole fractions of the monomeric components:
isophthalic acid 0.09, terephthalic acid 0.37
adipic acid 0.04, ethylen glycol 0.02
neopentyl glycol 0.46, trimethylolpropane 0.01
Solvent (A): **carbon dioxide** **CO_2** **124-38-9**

Type of data: gas solubility

T/K = 306

P/MPa	0.14	0.69	1.27	2.25	4.68	11.05	13.04
c_A/(g CO_2/g polymer)	0.0003	0.0068	0.0188	0.0362	0.0643	0.1399	0.2194

P/MPa	15.44	18.40	19.95	21.24	22.15	25.34	28.60
c_A/(g CO_2/g polymer)	0.3145	0.4086	0.4668	0.5037	0.5345	0.5878	0.6183

P/MPa	31.02
c_A/(g CO_2/g polymer)	0.6366

continued

continued

T/K = 343

P/MPa	0.18	0.69	1.30	2.41	4.76	10.37	13.47
c_A/(g CO_2/g polymer)	0.0191	0.0237	0.0299	0.0385	0.0553	0.0984	0.1556
P/MPa	16.20	19.00	21.13	23.62	25.67	25.98	29.05
c_A/(g CO_2/g polymer)	0.2394	0.3214	0.3645	0.3957	0.4149	0.4189	0.4461
P/MPa	31.02						
c_A/(g CO_2/g polymer)	0.4636						

Polymer (B): **polyester resin** **2010SKE**

Characterization: M_n/g.mol^{-1} = 3700, M_w/g.mol^{-1} = 8000, T_g/K = 333.9,
CPE67, Helios, Slovenia
mole fractions of the monomeric components:
isophthalic acid 0.10, terephthalic acid 0.38
adipic acid 0.03, neopentyl glycol 0.48
trimethylolpropane 0.01

Solvent (A): **carbon dioxide** **CO_2** **124-38-9**

Type of data: gas solubility

T/K = 306.2

P/MPa	0.14	0.50	1.12	1.99	4.09	5.06	10.15
c_A/(g CO_2/g polymer)	0.0023	0.0040	0.0117	0.0272	0.0498	0.0793	0.1509
P/MPa	12.04	14.89	17.88	19.80	22.73	25.26	28.05
c_A/(g CO_2/g polymer)	0.2723	0.3694	0.4535	0.4887	0.5316	0.5467	0.5613
P/MPa	29.88						
c_A/(g CO_2/g polymer)	0.5684						

T/K = 344.0

P/MPa	0.24	0.56	1.06	2.02	3.98	5.09	10.33
c_A/(g CO_2/g polymer)	0.0032	0.0033	0.0092	0.0165	0.0361	0.0475	0.1421
P/MPa	12.07	15.19	18.10	20.11	23.31	24.95	28.13
c_A/(g CO_2/g polymer)	0.1626	0.2184	0.3140	0.3652	0.4154	0.4351	0.4671
P/MPa	29.91						
c_A/(g CO_2/g polymer)	0.4833						

Polymer (B):	**polyester resin**	**2010SKE**
Characterization:	M_n/g.mol^{-1} = 3900, M_w/g.mol^{-1} = 8600, T_g/K = 332.3, CPE60, Helios, Slovenia mole fractions of the monomeric components: isophthalic acid 0.07, terephthalic acid 0.36 adipic acid 0.06, ethylen glycol 0.10 neopentyl glycol 0.37, trimethylolpropane 0.04	
Solvent (A):	**carbon dioxide** CO_2	**124-38-9**

Type of data: gas solubility

T/K = 309.2

P/MPa	0.13	0.60	0.99	2.19	4.10	5.22	10.38
c_A/(g CO_2/g polymer)	0.0049	0.0093	0.0105	0.0303	0.0416	0.0748	0.1281
P/MPa	12.42	15.09	18.01	21.35	23.53	26.05	28.73
c_A/(g CO_2/g polymer)	0.2278	0.3023	0.3590	0.4458	0.4838	0.5436	0.5812
P/MPa	31.00						
c_A/(g CO_2/g polymer)	0.5895						

T/K = 343.3

P/MPa	0.25	0.62	1.30	1.96	4.74	11.62	14.80
c_A/(g CO_2/g polymer)	0.0008	0.0031	0.0093	0.0138	0.0402	0.1296	0.2071
P/MPa	16.50	18.43	20.27	22.21	25.00	28.14	30.43
c_A/(g CO_2/g polymer)	0.2611	0.3163	0.3487	0.3830	0.4151	0.4444	0.4623

Polymer (B):	**polyether-polycarbonate-polyol**	**2012FON**
Characterization:	M_w/g.mol^{-1} = 3046, M_w/M_n = 1.11 PPP, Bayer Material Science AG, Germany	
Solvent (A):	**carbon dioxide** CO_2	**124-38-9**

Type of data: gas solubility

T/K = 363

P/MPa	1.547	3.644	5.828	7.848	9.929
w_A	0.02507	0.05324	0.08402	0.11555	0.15015

T/K = 383

P/MPa	1.410	2.672	4.351	5.659	7.099	9.114
w_A	0.01776	0.03308	0.05372	0.06954	0.08741	0.11547

T/K = 413

P/MPa	1.493	3.474	4.959	6.958	8.487	9.813
w_A	0.01513	0.03408	0.04944	0.06763	0.08191	0.09554

Polymer (B): **polyether-polycarbonate-polyol** **2012FON**
Characterization: M_w/g.mol^{-1} = 3046, M_w/M_n = 1.11
PPP, Bayer Material Science AG, Germany
Solvent (A): **propylene oxide** C_3H_6O **75-56-9**

Type of data: gas solubility

T/K = 363

P/MPa	0.057	0.104	0.155	0.201	0.241	0.263
w_A	0.03371	0.06660	0.09748	0.12599	0.15396	0.17916

T/K = 383

P/MPa	0.106	0.199	0.278	0.334	0.386	0.432	0.455
w_A	0.03486	0.06733	0.09861	0.12283	0.14748	0.17029	0.18352

T/K = 413

P/MPa	0.184	0.338	0.478	0.598	0.700	0.790
w_A	0.03179	0.06351	0.09402	0.12337	0.15067	0.17642

Polymer (B): **polyethylene** **2005WAN**
Characterization: LDPE, T_m/K = 382.7, Asahi Kasei Company, Japan
Solvent (A): **n-butane** C_4H_{10} **106-97-8**

Type of data: gas solubility

T/K = 383.15

P/MPa	0.373	0.521	0.805	1.077	1.193	1.386	1.455
c_A/(g C_4H_{10}/g polymer)	0.0420	0.0632	0.111	0.179	0.213	0.292	0.322

Comments: Concentrations of C_4H_{10} are corrected for polymer swelling.

T/K = 413.15

P/MPa	0.529	0.752	1.192	1.632	1.837	2.203	2.347
c_A/(g C_4H_{10}/g polymer)	0.0389	0.0572	0.103	0.165	0.199	0.273	0.302

Comments: Concentrations of C_4H_{10} are corrected for polymer swelling.

T/K = 443.15

P/MPa	0.698	1.005	1.624	2.260	2.601	3.185	3.447
c_A/(g C_4H_{10}/g polymer)	0.0360	0.0550	0.0957	0.152	0.184	0.251	0.277

Comments: Concentrations of C_4H_{10} are corrected for polymer swelling.

T/K = 473.15

P/MPa	0.871	1.266	2.075	2.923	3.427
c_A/(g C_4H_{10}/g polymer)	0.0334	0.0516	0.0891	0.140	0.170

Comments: Concentrations of C_4H_{10} are corrected for polymer swelling.

Polymer (B): **polyethylene** **2006PAR**
Characterization: M_n/g.mol^{-1} = 8200, M_w/g.mol^{-1} = 111000, HDPE, T_m/K = 407, Japan Polyolefines
Solvent (A): **carbon dioxide** **CO_2** **124-38-9**

Type of data: gas solubility

T/K = 453.15

P/MPa	3.4	7.0	10.5	14.0	17.4	20.9	24.4
c_A/(g CO_2/g polymer)	0.020	0.042	0.064	0.088	0.113	0.141	0.172
P/MPa	27.9	31.3	34.5				
c_A/(g CO_2/g polymer)	0.204	0.249	0.300				

Polymer (B): **polyethylene** **2007BOY**
Characterization: MDPE, T_m/K = 400, 49% crystallinity, Finathene 3802, Institut Francais du Petrole, IFP, France
Solvent (A): **carbon dioxide** **CO_2** **124-38-9**

Type of data: gas solubility

T/K = 333.15

P/MPa	3.51	5.09	7.65	10.02	12.37	15.26	20.25
c_A/[cm^3 (STP)/cm^3 polymer]	5.74	9.07	14.95	19.68	26.19	35.22	40.71
P/MPa	25.06	35.14					
c_A/[cm^3 (STP)/cm^3 polymer]	47.38	55.04					

T/K = 338.15

P/MPa	5.84	9.12	10.99	12.46	13.73	15.14	16.91
c_A/[cm^3 (STP)/cm^3 polymer]	9.61	13.29	12.34	16.99	18.12	19.29	20.89
P/MPa	18.97	21.23					
c_A/[cm^3 (STP)/cm^3 polymer]	22.41	22.59					

Polymer (B): **polyethylene** **2013CUC**
Characterization: M_w/g.mol^{-1} = 244000, M_w/M_n = 10.14, LDPE, ρ = 0.923 g/cm^3, Helios, Domzale, Slovenia
Solvent (A): **carbon dioxide** **CO_2** **124-38-9**

Type of data: gas solubility

T/K = 373.15

P/MPa	0.58	1.11	2.17	4.41	5.13	11.94	14.28
vc_A/(g CO_2/g polymer)	0.00000	0.00000	0.01076	0.03530	0.03514	0.06954	0.06652
P/MPa	16.60	18.16	20.14	22.16	24.80	27.53	30.82
c_A/(g CO_2/g polymer)	0.09492	0.10902	0.15512	0.18675	0.21446	0.27377	0.36377

Comments: Concentrations of CO_2 are corrected for polymer swelling.

Polymer (B): **polyethylene** **2013CUC**
Characterization: M_w/g.mol^{-1} = 179000, M_w/M_n = 6.379,
LDPE, ρ = 0.9225 g/cm^3, Helios, Domzale, Slovenia
Solvent (A): **carbon dioxide** CO_2 **124-38-9**

Type of data: gas solubility

T/K = 373.15

P/MPa	0.26	0.61	1.09	2.09	4.02	5.31	11.56
c_A/(g CO_2/g polymer)	0.00109	0.00291	0.00456	0.00917	0.02410	0.03414	0.04016
P/MPa	13.02	15.36	18.27	20.02	21.81	24.63	28.01
c_A/(g CO_2/g polymer)	0.04726	0.04847	0.04404	0.05251	0.07428	0.10970	0.14199
P/MPa	30.71						
c_A/(g CO_2/g polymer)	0.15975						

Comments: Concentrations of CO_2 are corrected for polymer swelling.

Polymer (B): **polyethylene** **2013CUC**
Characterization: M_w/g.mol^{-1} = 73000, M_w/M_n = 8.72,
HDPE, ρ = 0.964 g/cm^3, Helios, Domzale, Slovenia
Solvent (A): **carbon dioxide** CO_2 **124-38-9**

Type of data: gas solubility

T/K = 373.15

P/MPa	0.60	1.08	2.12	4.26	5.66	11.83	13.17
c_A/(g CO_2/g polymer)	0.00000	0.00000	0.00652	0.00891	0.01004	0.01966	0.02135
P/MPa	15.74	17.99	20.19	22.14	25.20	27.64	30.46
c_A/(g CO_2/g polymer)	0.02028	0.03902	0.05145	0.07851	0.09208	0.12269	0.13768

Comments: Concentrations of CO_2 are corrected for polymer swelling.

Polymer (B): **polyethylene** **2005WAN**
Characterization: LDPE, T_m/K = 382.7, Asahi Kasei Company, Japan
Solvent (A): **1-chloro-1,1-difluoroethane** $C_2H_3ClF_2$ **75-68-3**

Type of data: gas solubility

T/K = 383.15

P/MPa	0.335	0.727	1.089	1.602
c_A/(g $C_2H_3ClF_2$/g polymer)	0.0061	0.0136	0.0201	0.0314

Comments: Concentrations of $C_2H_3ClF_2$ are corrected for polymer swelling.

T/K = 413.15

P/MPa	0.373	0.815	1.226	1.823
c_A/(g $C_2H_3ClF_2$/g polymer)	0.0057	0.0127	0.0192	0.0300

Comments: Concentrations of $C_2H_3ClF_2$ are corrected for polymer swelling.

continued

continued

T/K = 443.15

P/MPa	0.414	0.903	1.364	2.038
c_A/(g $C_2H_3ClF_2$/g polymer)	0.0055	0.0125	0.0187	0.0296

Comments: Concentrations of $C_2H_3ClF_2$ are corrected for polymer swelling.

T/K = 473.15

P/MPa	0.453	0.989	1.500	2.256
c_A/(g $C_2H_3ClF_2$/g polymer)	0.0054	0.0124	0.0185	0.0296

Comments: Concentrations of $C_2H_3ClF_2$ are corrected for polymer swelling.

Polymer (B): **polyethylene** **2009HAR**
Characterization: M_n/g.mol^{-1} = 13200, M_w/g.mol^{-1} = 15400,
Scientific Polymer Products, Inc., Ontario, NY
Solvent (A): **cyclohexane** **C_6H_{12}** **110-82-7**

Type of data: vapor-liquid equilibrium

w_B	0.019	0.019	0.019	0.019	0.019	0.019	0.019	0.019	0.096
T/K	373.08	393.21	413.22	433.10	443.15	453.21	463.16	473.12	373.12
P_A/MPa	0.07	0.18	0.32	0.54	0.67	0.82	1.00	1.18	0.06

w_B	0.096	0.096	0.096	0.096	0.096	0.096	0.096	0.152	0.152
T/K	393.08	413.05	433.13	443.07	453.14	463.16	473.10	373.20	393.16
P_A/MPa	0.21	0.29	0.51	0.63	0.82	0.96	1.18	0.09	0.20

w_B	0.152	0.152	0.152	0.152	0.152	0.152
T/K	412.92	433.02	443.07	453.14	463.07	473.05
P_A/MPa	0.35	0.55	0.68	0.83	1.02	1.20

Polymer (B): **polyethylene** **2011HAR**
Characterization: M_n/g.mol^{-1} = 2720, M_w/g.mol^{-1} = 10900, M_z/g.mol^{-1} = 35300
Solvent (A): **n-hexane** **C_6H_{14}** **110-54-3**

Type of data: vapor-liquid equilibrium

w_B	0.0025	0.0025	0.0025	0.0025	0.0025	0.0051	0.0051	0.0051	0.0051
T/K	373.34	393.14	413.24	433.27	453.23	373.14	393.27	413.17	433.09
P_A/MPa	0.15	0.26	0.47	0.80	1.11	0.09	0.25	0.47	0.75

w_B	0.0051	0.0099	0.0099	0.0099	0.0099	0.0196	0.0196	0.0196	0.0196
T/K	453.07	373.17	393.24	413.20	433.25	373.21	393.09	413.21	433.15
P_A/MPa	1.20	0.11	0.29	0.48	0.79	0.11	0.28	0.50	0.78

w_B	0.0292	0.0292	0.0292	0.0292	0.0512	0.0512	0.0512	0.0512	0.0715
T/K	372.99	393.13	413.16	433.21	373.25	393.38	413.21	433.32	373.19
P_A/MPa	0.12	0.29	0.48	0.76	0.14	0.29	0.55	0.81	0.15

continued

continued

w_B	0.0715	0.0715	0.0715	0.0715	0.0984	0.0984	0.0984	0.0984	0.0984
T/K	393.33	413.26	433.17	453.16	373.16	393.25	413.16	433.12	453.10
P_A/MPa	0.36	0.52	0.76	1.11	0.12	0.25	0.49	0.74	1.10

Polymer (B): **polyethylene** **2011HAR**
Characterization: M_n/g.mol^{-1} = 4460, M_w/g.mol^{-1} = 13100, M_z/g.mol^{-1} = 25100
Solvent (A): **n-hexane** **C_6H_{14}** **110-54-3**

Type of data: vapor-liquid equilibrium

w_B	0.0025	0.0025	0.0025	0.0025	0.0025	0.0047	0.0047	0.0047	0.0047
T/K	373.74	393.34	413.15	433.42	452.87	373.64	393.07	413.32	433.21
P_A/MPa	0.20	0.34	0.54	0.79	1.09	0.24	0.43	0.58	0.79
w_B	0.0047	0.0081	0.0081	0.0081	0.0081	0.0167	0.0167	0.0167	0.0167
T/K	453.28	373.49	393.14	412.94	433.48	373.27	393.14	413.21	433.26
P_A/MPa	1.24	0.17	0.31	0.50	0.74	0.11	0.24	0.48	0.78
w_B	0.0306	0.0306	0.0306	0.0306	0.0390	0.0390	0.0390	0.0390	0.0489
T/K	373.26	393.27	413.28	433.04	373.38	393.14	413.29	433.30	373.65
P_A/MPa	0.12	0.25	0.48	0.73	0.15	0.25	0.51	0.77	0.18
w_B	0.0489	0.0489	0.0489	0.0715	0.0715	0.0715	0.0715	0.0892	0.0892
T/K	393.61	413.05	433.29	373.11	393.14	413.32	433.16	373.20	393.43
P_A/MPa	0.29	0.51	0.74	0.17	0.31	0.55	0.78	0.12	0.26
w_B	0.0892	0.0892	0.1019	0.1019	0.1019	0.1019			
T/K	413.16	433.30	373.10	393.04	413.08	432.94			
P_A/MPa	0.55	0.75	0.13	0.33	0.49	0.82			

Polymer (B): **polyethylene** **2008HAR**
Characterization: M_n/g.mol^{-1} = 14400, M_w/g.mol^{-1} = 15500,
Scientific Polymer Products, Inc., Ontario, NY
Solvent (A): **n-hexane** **C_6H_{14}** **110-54-3**

Type of data: vapor-liquid equilibrium

w_B	0.0193	0.0193	0.0193	0.0193	0.0193	0.0193	0.0496	0.0496	0.0496
T/K	373.13	393.24	413.20	433.14	443.09	453.14	373.18	393.14	413.11
P_A/MPa	0.19	0.32	0.52	0.82	0.98	1.15	0.19	0.32	0.51
w_B	0.0496	0.0496	0.0923	0.0923	0.0923	0.0923	0.0923	0.1287	0.1287
T/K	433.17	443.12	373.18	393.14	413.18	433.14	443.14	373.11	393.12
P_A/MPa	0.79	0.91	0.20	0.34	0.58	0.80	0.99	0.19	0.33
w_B	0.1287	0.1287	0.1287						
T/K	413.17	433.15	443.16						
P_A/MPa	0.53	0.78	0.96						

Polymer (B): **polyethylene** **2006NAG**
Characterization: M_n/g.mol^{-1} = 43700, M_w/g.mol^{-1} = 52000, M_z/g.mol^{-1} = 59000, 2.05 ethyl branches per 100 backbone C-atomes, hydrogenated polybutadiene PBD 50000, was denoted as LLDPE, DSM, The Netherlands
Solvent (A): **n-hexane** **C_6H_{14}** **110-54-3**

Type of data: vapor-liquid equilibrium

w_B	0.0024	was kept constant				
T/K	411.396	416.30	421.22	426.08	430.92	436.09
P_A/MPa	0.618	0.703	0.753	0.823	0.903	1.008

w_B	0.0049	was kept constant			
T/K	410.94	415.91	425.79	430.74	435.63
P_A/MPa	0.652	0.707	0.827	0.907	0.992

w_B	0.0078	was kept constant			
T/K	410.94	415.92	425.79	430.74	435.64
P_A/MPa	0.652	0.707	0.747	0.827	0.992

w_B	0.0223	was kept constant			
T/K	411.58	416.42	421.41	426.12	431.12
P_A/MPa	0.613	0.673	0.743	0.808	0.888

w_B	0.0452	was kept constant		
T/K	411.199	416.162	421.063	425.985
P_A/MPa	0.614	0.679	0.749	0.819

w_B	0.0923	was kept constant		
T/K	411.65	416.54	421.46	431.31
P_A/MPa	0.609	0.674	0.739	0.889

w_B	0.1946	was kept constant				
T/K	411.092	415.975	420.932	425.915	430.949	435.888
P_A/MPa	0.613	0.678	0.753	0.828	0.898	0.988

w_B	0.2435	was kept constant			
T/K	420.924	425.802	430.584	435.601	440.595
P_A/MPa	0.754	0.814	0.899	0.984	1.069

w_B	0.3031	was kept constant						
T/K	411.50	416.36	421.27	426.22	431.21	436.10	441.10	446.00
P_A/MPa	0.667	0.752	0.837	0.907	0.972	1.047	1.147	1.252

Polymer (B): **polyethylene** **2008HAR**
Characterization: M_n/g.mol^{-1} = 82000, M_w/g.mol^{-1} = 108000,
Scientific Polymer Products, Inc., Ontario, NY
Solvent (A): **n-hexane** **C_6H_{14}** **110-54-3**

Type of data: vapor-liquid equilibrium

w_B	0.0100	0.0100	0.0100	0.0194	0.0194	0.0194	0.0487	0.0487	0.0487
T/K	373.17	393.18	413.11	373.12	393.10	413.15	373.17	393.12	413.14
P_A/MPa	0.26	0.41	0.62	0.22	0.34	0.55	0.22	0.35	0.57

w_B	0.0885	0.0885	0.0885	0.1249	0.1249	0.1249
T/K	373.11	393.17	413.13	373.18	393.13	413.13
P_A/MPa	0.21	0.35	0.55	0.19	0.31	0.52

Polymer (B): **polyethylene** **2010HA2**
Characterization: M_n/g.mol^{-1} = 13200, M_w/g.mol^{-1} = 15400,
Scientific Polymer Products, Inc., Ontario, NY
Solvent (A): **1-hexene** **C_6H_{12}** **592-41-6**

Type of data: vapor-liquid equilibrium

w_B	0.020	0.020	0.020	0.020	0.020	0.049	0.049	0.049	0.049
T/K	373.23	393.58	413.42	433.21	443.32	373.29	393.35	413.19	433.40
P_A/MPa	0.09	0.25	0.49	0.78	0.98	0.33	0.49	0.70	1.00

w_B	0.049	0.080	0.080	0.080	0.080	0.080	0.099	0.099	0.099
T/K	443.21	373.22	393.12	413.01	433.36	443.31	373.45	393.65	413.48
P_A/MPa	1.19	0.07	0.22	0.45	0.79	1.01	0.05	0.26	0.46

w_B	0.099	0.120	0.120	0.120	0.120
T/K	433.66	373.35	393.52	413.20	433.80
P_A/MPa	0.78	0.05	0.25	0.44	0.73

Polymer (B): **polyethylene** **2010HA2**
Characterization: M_n/g.mol^{-1} = 82000, M_w/g.mol^{-1} = 108000,
Scientific Polymer Products, Inc., Ontario, NY
Solvent (A): **1-hexene** **C_6H_{12}** **592-41-6**

Type of data: vapor-liquid equilibrium

w_B	0.020	0.020	0.049	0.049	0.049	0.080	0.080	0.099	0.099
T/K	393.19	413.37	373.68	393.35	413.37	373.23	393.44	373.34	393.25
P_A/MPa	0.13	0.34	0.16	0.33	0.56	0.21	0.33	0.22	0.34

w_B	0.120	0.120
T/K	373.14	393.36
P_A/MPa	0.20	0.32

Polymer (B): **polyethylene** **2007NAG**
Characterization: M_n/g.mol^{-1} = 43700, M_w/g.mol^{-1} = 52000, M_z/g.mol^{-1} = 59000, 2.05 ethyl branches per 100 backbone C-atomes, hydrogenated polybutadiene PBD 50000, was denoted as LLDPE, DSM, The Netherlands
Solvent (A): **2-methylpentane** C_6H_{14} **107-83-5**

Type of data: vapor-liquid equilibrium

w_B	0.0020	was kept constant	
T/K	391.09	401.08	411.06
P_A/MPa	0.480	0.595	0.725

w_B	0.0053	was kept constant	
T/K	390.98	400.93	411.03
P_A/MPa	0.489	0.604	0.739

w_B	0.0103	was kept constant			
T/K	391.28	396.13	401.06	406.02	410.89
P_A/MPa	0.520	0.575	0.630	0.700	0.765

w_B	0.0503	was kept constant	
T/K	391.31	396.11	401.19
P_A/MPa	0.524	0.574	0.639

w_B	0.0984	was kept constant	
T/K	391.31	396.11	400.97
P_A/MPa	0.518	0.568	0.623

w_B	0.1507	was kept constant		
T/K	390.75	395.68	400.46	403.50
P_A/MPa	0.512	0.567	0.617	0.687

Polymer (B): **polyethylene** **2005WAN**
Characterization: LDPE, T_m/K = 382.7, Asahi Kasei Company, Japan
Solvent (A): **2-methylpropane** C_4H_{10} **75-28-5**

Type of data: gas solubility

T/K = 383.15

P/MPa	0.380	0.709	0.944	1.119	1.383	1.525
c_A/(g C_4H_{10}/g polymer)	0.0305	0.0626	0.0901	0.113	0.158	0.187

Comments: Concentrations of C_4H_{10} are corrected for polymer swelling.

T/K = 413.15

P/MPa	0.501	0.976	1.326	1.571	2.032	2.240
c_A/(g C_4H_{10}/g polymer)	0.0285	0.0577	0.0832	0.102	0.146	0.172

Comments: Concentrations of C_4H_{10} are corrected for polymer swelling.

continued

continued

T/K = 443.15

P/MPa	0.639	1.258	1.731	2.048	2.765	3.025
c_A/(g C_4H_{10}/g polymer)	0.0266	0.0535	0.0772	0.0933	0.140	0.160

Comments: Concentrations of C_4H_{10} are corrected for polymer swelling.

T/K = 473.15

P/MPa	0.795	1.542	2.142	2.535
c_A/(g C_4H_{10}/g polymer)	0.0246	0.0499	0.0721	0.0860

Comments: Concentrations of C_4H_{10} are corrected for polymer swelling.

Polymer (B): **polyethylene** **2005WAN**
Characterization: LDPE, T_m/K = 382.7, Asahi Kasei Company, Japan
Solvent (A): **1,1,1,2-tetrafluoroethane** $C_2H_2F_4$ **811-97-2**

Type of data: gas solubility

T/K = 383.15

P/MPa	0.270	0.436	0.581	0.857	1.228	1.525
c_A/(g $C_2H_2F_4$/g polymer)	0.0193	0.0313	0.0457	0.0725	0.118	0.163

Comments: Concentrations of $C_2H_2F_4$ are corrected for polymer swelling.

T/K = 413.15

P/MPa	0.324	0.514	0.719	1.065	1.584	2.017
c_A/(g $C_2H_2F_4$/g polymer)	0.0179	0.0295	0.0415	0.0657	0.104	0.144

Comments: Concentrations of $C_2H_2F_4$ are corrected for polymer swelling.

T/K = 443.15

P/MPa	0.388	0.618	0.865	1.292	1.937	2.500
c_A/(g $C_2H_2F_4$/g polymer)	0.0164	0.0269	0.0378	0.0594	0.0943	0.130

Comments: Concentrations of $C_2H_2F_4$ are corrected for polymer swelling.

T/K = 473.15

P/MPa	0.451	0.719	1.006	1.508	2.280	2.979
c_A/(g $C_2H_2F_4$/g polymer)	0.0152	0.0249	0.0350	0.0546	0.0866	0.118

Comments: Concentrations of $C_2H_2F_4$ are corrected for polymer swelling.

Polymer (B): **poly(ethylene-*co*-norbornene)** **2013SAT**
Characterization: M_n/g.mol^{-1} = 38800, M_w/g.mol^{-1} = 74600, 52 mol% ethylene, T_g/K = 411.15, Topas Advanced Polymers GmbH
Solvent (A): **ethene** **C_2H_4** **74-85-1**

Type of data: gas solubility

T/K = 413.15

P/MPa	2.044	4.016	5.994	7.986	9.989
c_A/(g C_2H_4/g polymer)	0.00699	0.0145	0.0217	0.0286	0.0331

Comments: Concentrations of C_2H_4 are corrected for polymer swelling.

T/K = 443.15

P/MPa	2.012	4.013	5.994	8.040	10.00
c_A/(g C_2H_4/g polymer)	0.0063	0.0124	0.0190	0.0250	0.0300

Comments: Concentrations of C_2H_4 are corrected for polymer swelling.

T/K = 473.15

P/MPa	2.011	4.093	6.005	8.021	9.979
c_A/(g C_2H_4/g polymer)	0.0056	0.0112	0.0166	0.0222	0.0279

Comments: Concentrations of C_2H_4 are corrected for polymer swelling.

Polymer (B): **poly(ethylene-*co*-1-octene)** **2005LEE**
Characterization: M_n/g.mol^{-1} = 64810, M_w/g.mol^{-1} = 135500, 15.3 mol% 1-octene, T_m/K = 323.2, T_g/K = 214.2, ρ = 0.86 g/cm^3, DuPont Dow Elastomers Corporation
Solvent (A): **cyclohexane** **C_6H_{12}** **110-82-7**

Type of data: vapor-liquid equilibrium

w_B	0.0501	was kept constant			
T/K	322.85	348.45	373.55	397.95	423.05
P_A/bar	1.601	2.302	3.282	4.752	7.231

Polymer (B): **poly(ethylene-*co*-1-octene)** **2005LEE**
Characterization: M_n/g.mol^{-1} = 64810, M_w/g.mol^{-1} = 135500, 15.3 mol% 1-octene, T_m/K = 323.2, T_g/K = 214.2, ρ = 0.86 g/cm^3, DuPont Dow Elastomers Corporation
Solvent (A): **cyclopentane** **C_5H_{10}** **287-92-3**

Type of data: vapor-liquid equilibrium

w_B	0.0501	was kept constant			
T/K	322.85	348.75	373.15	398.35	423.15
P_A/bar	1.870	3.399	5.531	8.809	13.537

Polymer (B): **poly(ethylene-*co*-1-octene)** **2005LEE**
Characterization: M_n/g.mol^{-1} = 64810, M_w/g.mol^{-1} = 135500,
15.3 mol% 1-octene, T_m/K = 323.2, T_g/K = 214.2,
ρ = 0.86 g/cm^3, DuPont Dow Elastomers Corporation
Solvent (A): **n-heptane** **C_7H_{16}** **142-82-5**

Type of data: vapor-liquid equilibrium

w_B	0.0500	was kept constant			
T/K	323.35	348.25	373.45	398.15	422.85
P_A/bar	1.572	2.101	2.797	3.855	5.354

Polymer (B): **poly(ethylene-*co*-1-octene)** **2005LEE**
Characterization: M_n/g.mol^{-1} = 64810, M_w/g.mol^{-1} = 135500,
15.3 mol% 1-octene, T_m/K = 323.2, T_g/K = 214.2,
ρ = 0.86 g/cm^3, DuPont Dow Elastomers Corporation
Solvent (A): **n-hexane** **C_6H_{14}** **110-54-3**

Type of data: vapor-liquid equilibrium

w_B	0.0500	was kept constant			
T/K	322.45	353.15	383.45	412.55	442.75
P_A/bar	1.699	3.056	5.472	8.623	13.371

Polymer (B): **poly(ethylene-*co*-1-octene)** **2005LEE**
Characterization: M_n/g.mol^{-1} = 64810, M_w/g.mol^{-1} = 135500,
15.3 mol% 1-octene, T_m/K = 323.2, T_g/K = 214.2,
ρ = 0.86 g/cm^3, DuPont Dow Elastomers Corporation
Solvent (A): **n-octane** **C_8H_{18}** **111-65-9**

Type of data: vapor-liquid equilibrium

w_B	0.0499	was kept constant			
T/K	323.05	348.65	373.15	398.05	421.85
P_A/bar	1.121	1.513	2.027	2.630	3.375

Polymer (B): **poly(ethylene-*co*-1-octene)** **2005LEE**
Characterization: M_n/g.mol^{-1} = 64810, M_w/g.mol^{-1} = 135500,
15.3 mol% 1-octene, T_m/K = 323.2, T_g/K = 214.2,
ρ = 0.86 g/cm^3, DuPont Dow Elastomers Corporation
Solvent (A): **n-pentane** **C_5H_{12}** **109-66-0**

Type of data: vapor-liquid equilibrium

w_B	0.0099	was kept constant		
T/K	323.05	352.55	378.15	402.95
P_A/bar	2.385	4.757	8.049	12.626

continued

continued

w_B	0.0202	was kept constant			
T/K	323.05	342.95	368.35	393.15	
P_A/bar	2.125	3.581	6.383	10.421	
w_B	0.0504	was kept constant			
T/K	322.95	343.45	368.25	394.15	
P_A/bar	2.248	3.875	6.692	10.955	
w_B	0.1133	was kept constant			
T/K	322.95	342.85	368.05	393.05	
P_A/bar	2.189	3.644	6.359	10.411	
w_B	0.2000	was kept constant			
T/K	327.95	352.55	377.95	403.85	
P_A/bar	3.081	5.217	8.500	13.341	
w_B	0.3940	was kept constant			
T/K	325.65	347.85	376.55	400.35	424.25
P_A/bar	2.728	4.463	7.976	12.283	18.339

Comments: Some LLE data are given in Chapter 3.

Polymer (B): **poly(ethylene glycol)** **2012LIJ**
Characterization: M_n/g.mol^{-1} = 150, tri(ethylene glycol),
Sinopharm Chemical Reagent Co., Ltd., Shanghai, China
Solvent (A): **carbon dioxide** **CO_2** **124-38-9**

Type of data: gas solubility

T/K = 303.15

P/kPa	137.7	264.2	417.5	558.0	828.0	1104.5
x_A	0.0103	0.0175	0.0300	0.0401	0.0614	0.0850

T/K = 313.15

P/kPa	123.8	167.6	281.2	418.0	566.0	853.4	1134.0
x_A	0.0071	0.0099	0.0169	0.0267	0.0339	0.0551	0.0713

T/K = 323.15

P/kPa	117.9	281.3	346.3	571.6	867.8	1143.5
x_A	0.0055	0.0152	0.0192	0.0304	0.0468	0.0600

T/K = 333.15

P/kPa	164.8	295.2	436.5	579.0	856.5	1130.0
x_A	0.0076	0.0137	0.0202	0.0268	0.0397	0.0523

Polymer (B): **poly(ethylene glycol)** **2012LIJ**
Characterization: M_n/g.mol^{-1} = 200,
Sinopharm Chemical Reagent Co., Ltd., Shanghai, China
Solvent (A): **carbon dioxide** **CO_2** **124-38-9**

Type of data: gas solubility

T/K = 303.15

P/kPa	140.8	268.0	398.3	531.2	832.5	1102.5
x_A	0.0140	0.0303	0.0436	0.0586	0.0900	0.1162

T/K = 313.15

P/kPa	72.2	91.9	154.8	273.3	497.3	772.0	1143.0
x_A	0.0062	0.0077	0.0132	0.0242	0.0422	0.0676	0.0994

T/K = 323.15

P/kPa	163.0	286.1	424.0	551.2	830.0	1116.0
x_A	0.0116	0.0213	0.0323	0.0428	0.0625	0.0833

T/K = 333.15

P/kPa	154.8	292.6	445.5	580.0	849.5	1103.5
x_A	0.0092	0.0206	0.0298	0.0392	0.0600	0.0756

Polymer (B): **poly(ethylene glycol)** **2007HOU**
Characterization: M_n/g.mol^{-1} = 200, Beijing Chemical Reagent Plant, China
Solvent (A): **carbon dioxide** **CO_2** **124-38-9**

Type of data: gas solubility

T/K = 303.15

P/MPa	1.00	2.00	3.00	4.00	5.00	6.00	7.00
w_A	0.0137	0.0341	0.0611	0.0931	0.120	0.152	0.182

T/K = 313.15

P/MPa	1.10	2.04	2.99	4.15	5.11	6.11	7.11	8.10
w_A	0.0130	0.0334	0.0532	0.0710	0.106	0.129	0.147	0.167

T/K = 323.15

P/MPa	1.28	2.15	3.04	4.11	5.05	6.06	6.94	8.07
w_A	0.0149	0.0296	0.0431	0.0627	0.0781	0.0979	0.113	0.130

Polymer (B): **poly(ethylene glycol)** **2012LIJ**
Characterization: M_n/g.mol^{-1} = 300,
Sinopharm Chemical Reagent Co., Ltd., Shanghai, China
Solvent (A): **carbon dioxide** **CO_2** **124-38-9**

Type of data: gas solubility

T/K = 303.15

P/kPa	148.3	278.5	424.2	663.0	804.0	1066.0
x_A	0.0226	0.0483	0.0693	0.1045	0.1352	0.1735

T/K = 313.15

P/kPa	156.5	278.5	419.0	586.0	835.0	1102.0
x_A	0.0190	0.0363	0.0581	0.0764	0.1117	0.1470

T/K = 323.15

P/kPa	141.3	307.9	421.6	576.0	840.5	1107.0
x_A	0.0154	0.0321	0.0504	0.0672	0.0935	0.1286

T/K = 333.15

P/kPa	155.2	311.0	437.0	591.2	880.5	1130.5
x_A	0.0155	0.0312	0.0462	0.0605	0.0940	0.1168

Polymer (B): **poly(ethylene glycol)** **2012LIJ**
Characterization: M_n/g.mol^{-1} = 400,
Aladdin Chemistry Co., Ltd., Shanghai, China
Solvent (A): **carbon dioxide** **CO_2** **124-38-9**

Type of data: gas solubility

T/K = 303.15

P/kPa	145.3	282.2	431.2	561.5	815.5	1070.0
x_A	0.0319	0.0622	0.0897	0.1244	0.1684	0.2220

T/K = 313.15

P/kPa	166.0	300.5	414.4	578.0	834.5	1087.5
x_A	0.0279	0.0567	0.0759	0.1027	0.1467	0.1905

T/K = 323.15

P/kPa	154.1	217.4	443.0	564.0	893.8	1120.5
x_A	0.0237	0.0333	0.0690	0.0877	0.1362	0.1701

T/K = 333.15

P/kPa	147.8	282.9	434.5	557.2	831.0	1104.0
x_A	0.0213	0.0421	0.0625	0.0775	0.1184	0.1545

Polymer (B): **poly(ethylene glycol)** **2004GUA**
Characterization: M_w/g.mol^{-1} = 400, Sigma Chemical Co., Inc., St. Louis, MO
Solvent (A): **carbon dioxide** **CO_2** **124-38-9**

Type of data: gas solubility and swelling

T/K = 313.15

P/bar	53	80	116
w_A	0.111	0.159	0.226
$(\Delta V/V)$/%	16	25	35

Polymer (B): **poly(ethylene glycol)** **2014HRN**
Characterization: M_n/g.mol^{-1} = 1000, T_m/K = 412.56, Merck, Darmstadt, Germany
Solvent (A): **carbon dioxide** **CO_2** **124-38-9**

Type of data: gas solubility and swelling

T/K = 343.15

P/MPa	0.58	0.98	1.87	4.16	5.53	11.82	14.62
c_A/(g CO_2/g polymer)	0.00752	0.01621	0.03303	0.08081	0.09729	0.19409	0.24190
$(\Delta V/V)$/%	4.47	16.52	20.54	28.57	28.57	48.66	52.68

P/MPa	17.17	18.75	21.06	22.77	26.36
c_A/(g CO_2/g polymer)	0.29534	0.33835	0.39571	0.41633	0.47392
$(\Delta V/V)$/%	52.68	56.70	60.72	60.72	64.73

Polymer (B): **poly(ethylene glycol)** **2014HRN**
Characterization: M_n/g.mol^{-1} = 1500, T_m/K = 420.31, Merck, Darmstadt, Germany
Solvent (A): **carbon dioxide** **CO_2** **124-38-9**

Type of data: gas solubility and swelling

T/K = 343.15

P/MPa	0.74	1.08	2.05	46.0	12.69	13.13	14.54
c_A/(g CO_2/g polymer)	0.00998	0.01600	0.03383	0.08064	0.17311	0.17538	0.21695
$(\Delta V/V)$/%	5.26	10.53	15.79	21.05	52.63	52.63	57.89

P/MPa	17.68	20.18	21.93	24.32	29.15	30.55
c_A/(g CO_2/g polymer)	0.28934	0.36044	0.37969	0.43613	0.47131	0.48189
$(\Delta V/V)$/%	57.89	57.89	63.16	68.42	68.42	68.42

Polymer (B): **poly(ethylene glycol)** **2008AI1**
Characterization: M_n/g.mol^{-1} = 1500, Merck, Darmstadt, Germany
Solvent (A): **carbon dioxide** **CO_2** **124-38-9**

Type of data: gas solubility

T/K = 298.2

P/MPa	7.46	10.20	12.31	15.66	20.64	25.42
w_A	0.0120	0.0902	0.1251	0.1580	0.1858	0.2106

T/K = 323.2

P/MPa	5.65	8.22	10.14	11.93	13.66	16.16	18.13	20.24	21.99
w_A	0.1193	0.1565	0.1893	0.2071	0.2320	0.2450	0.2616	0.2906	0.2975

P/MPa	24.87
w_A	0.3130

Polymer (B): **poly(ethylene glycol)** **2014HRN**
Characterization: M_n/g.mol^{-1} = 3000, T_m/K = 431.32, Merck, Darmstadt, Germany
Solvent (A): **carbon dioxide** **CO_2** **124-38-9**

Type of data: gas solubility and swelling

T/K = 343.15

P/MPa	0.59	1.14	2.08	4.77	5.41	11.83	12.68
c_A/(g CO_2/g polymer)	0.00645	0.01642	0.03350	0.07769	0.09403	0.14358	0.14037
$(\Delta V/V)$/%	5.56	11.11	16.67	22.22	27.78	44.44	44.44

P/MPa	14.93	17.79	19.60	21.42	25.21	28.09	29.66
c_A/(g CO_2/g polymer)	0.18570	0.24079	0.29982	0.34354	0.38610	0.43859	0.45061
$(\Delta V/V)$/%	50.00	50.00	55.56	61.11	61.11	66.67	66.67

Polymer (B): **poly(ethylene glycol)** **2014HRN**
Characterization: M_n/g.mol^{-1} = 4000, T_m/K = 432.68, Merck, Darmstadt, Germany
Solvent (A): **carbon dioxide** **CO_2** **124-38-9**

Type of data: gas solubility and swelling

T/K = 343.15

P/MPa	0.56	1.09	2.18	4.68	5.42	12.11	13.23
c_A/(g CO_2/g polymer)	0.00920	0.01367	0.03293	0.07279	0.09622	0.11803	0.14835
$(\Delta V/V)$/%	5.26	10.53	15.79	21.05	26.32	36.84	42.11

P/MPa	14.85	17.86	20.01	21.38	23.86	27.77	30.47
c_A/(g CO_2/g polymer)	0.17074	0.22922	0.28980	0.33567	0.36314	0.42677	0.44472
$(\Delta V/V)$/%	42.11	42.11	47.37	52.63	52.63	57.89	57.89

Polymer (B): **poly(ethylene glycol)** **2008AI1**
Characterization: M_n/g.mol^{-1} = 4000, Merck, Darmstadt, Germany
Solvent (A): **carbon dioxide** CO_2 **124-38-9**

Type of data: gas solubility

T/K = 297.9

P/MPa	7.63	10.50	12.10	15.82	18.46	19.97	22.57	24.59
w_A	0.0157	0.0515	0.0780	0.1144	0.1280	0.1378	0.1470	0.1533

T/K = 323.2

P/MPa	5.69	8.53	10.61	12.26	15.27	18.21	19.62	21.87	24.66
w_A	0.1124	0.1511	0.1798	0.1973	0.2281	0.2494	0.2569	0.2674	0.2750

Polymer (B): **poly(ethylene glycol)** **2014HRN**
Characterization: M_n/g.mol^{-1} = 6000, T_m/K = 434.17, Merck, Darmstadt, Germany
Solvent (A): **carbon dioxide** CO_2 **124-38-9**

Type of data: gas solubility and swelling

T/K = 343.15

P/MPa	0.57	1.09	2.11	4.41	5.21	11.30	12.72
c_A/(g CO_2/g polymer)	0.00755	0.01754	0.03674	0.07947	0.09508	0.11470	0.13599
$(\Delta V/V)$/%	9.09	13.63	13.63	18.18	18.18	18.18	31.82

P/MPa	15.36	18.30	19.74	21.56	24.95	28.35	30.93
c_A/(g CO_2/g polymer)	0.18009	0.25350	0.27327	0.31789	0.37863	0.43249	0.44564
$(\Delta V/V)$/%	36.36	40.91	40.91	45.45	50.00	54.55	54.55

Polymer (B): **poly(ethylene glycol)** **2003KUK**
Characterization: M_n/g.mol^{-1} = 6000
Solvent (A): **carbon dioxide** CO_2 **124-38-9**

Type of data: gas solubility

T/K = 336.15

P/bar	52.7	57.1	108.8	154.1	205.8	253.5	296.9	324.4
w_A	0.095	0.102	0.154	0.197	0.218	0.245	0.255	0.257

T/K = 353.15

P/bar	32.0	62.0	89.0	117.0	139.0	169.0	192.0	222.0	263.0
w_A	0.036	0.053	0.077	0.093	0.119	0.128	0.130	0.137	0.154

P/bar	61	100	128	150	179	210	239	267	324
w_A	0.098	0.144	0.177	0.190	0.206	0.224	0.233	0.250	0.276

P/bar	37.3	59.4	106.9	139.5	176.4	202.7	230.4
w_A	0.050	0.082	0.141	0.183	0.218	0.231	0.244

continued

continued

T/K = 373.15

P/bar	46	87	123	160	187	211	269	298	328
w_A	0.061	0.108	0.142	0.168	0.189	0.200	0.227	0.234	0.256
P/bar	36.2	59.9	104.2	126.9	160.7	200.2	258.4	290.5	337.7
w_A	0.043	0.066	0.114	0.139	0.175	0.197	0.223	0.238	0.243

T/K = 393.15

P/bar	41.0	63.0	84.0	103.0	123.0	141.0	160.0	175.0	210.0
w_A	0.027	0.057	0.070	0.082	0.089	0.097	0.103	0.114	0.125
P/bar	28	56	93	125	157	186	214	274	328
w_A	0.036	0.065	0.099	0.119	0.147	0.166	0.176	0.206	0.229
P/bar	62.9	110.4	164.3	206.1	251.3	292.4			
w_A	0.044	0.100	0.157	0.181	0.201	0.211			

Polymer (B): **poly(ethylene glycol)** **2003KUK**
Characterization: M_n/g.mol^{-1} = 8000
Solvent (A): **carbon dioxide** CO_2 **124-38-9**

Type of data: gas solubility

T/K = 353.15

P/bar	29	55	83	143	170	200	258	319	
w_A	0.010	0.075	0.121	0.192	0.213	0.220	0.243	0.264	
P/bar	33.5	59.4	97.3	126.1	157.3	211.5	241.1	298.4	335.8
w_A	0.049	0.066	0.138	0.175	0.206	0.228	0.237	0.250	0.260

T/K = 373.15

P/bar	25	55	94	126	154	187	273	325
w_A	0.012	0.059	0.112	0.138	0.156	0.186	0.224	0.233
P/bar	36.8	64.5	114.5	168.7	212.9	329.3		
w_A	0.045	0.071	0.136	0.174	0.204	0.223		

T/K = 393.15

P/bar	51	77	112	146	178	212	275	308	
w_A	0.034	0.081	0.116	0.140	0.154	0.194	0.223	0.227	
P/bar	34.0	61.3	93.6	110.8	123.6	134.0	157.5	182.8	215.9
w_A	0.034	0.058	0.087	0.107	0.109	0.128	0.137	0.147	0.167
P/bar	245.0	272.5	303.2	327.7					
w_A	0.179	0.189	0.215	0.233					

Polymer (B): **poly(ethylene glycol)** **2003KUK**
Characterization: M_n/g.mol^{-1} = 9000
Solvent (A): **carbon dioxide** **CO_2** **124-38-9**

Type of data: gas solubility

T/K = 353.15

P/bar	31	69	100	128	169	192	218	236	267
w_A	0.053	0.102	0.140	0.171	0.196	0.211	0.219	0.228	0.241

P/bar	296	326
w_A	0.241	0.259

T/K = 373.15

P/bar	28	57	97	127	160	192	219	243	275
w_A	0.036	0.076	0.120	0.153	0.167	0.177	0.200	0.209	0.220

P/bar	295	326
w_A	0.228	0.235

T/K = 393.15

P/bar	32	61	93	127	158	188	216	238	271
w_A	0.037	0.072	0.097	0.128	0.144	0.161	0.178	0.184	0.203

P/bar	303	325
w_A	0.225	0.230

Polymer (B): **poly(ethylene glycol)** **2014HRN**
Characterization: M_n/g.mol^{-1} = 10000, T_m/K = 435.87,
Merck, Darmstadt, Germany
Solvent (A): **carbon dioxide** **CO_2** **124-38-9**

Type of data: gas solubility and swelling

T/K = 343.15

P/MPa	0.63	1.05	2.04	4.16	4.95	10.50	13.43
c_A/(g CO_2/g polymer)	0.00970	0.01757	0.03504	0.07515	0.09140	0.12955	0.13727
($\Delta V/V$)/%	0.00	4.35	8.69	13.0	13.0	13.0	26.01

P/MPa	15.10	18.00	20.14	22.42	25.13	28.16	30.71
c_A/(g CO_2/g polymer)	0.16203	0.23217	0.25869	0.31365	0.36416	0.41552	0.45908
($\Delta V/V$)/%	26.01	30.43	30.43	34.78	39.13	43.48	47.83

Polymer (B): **poly(ethylene glycol)** **2003KUK**
Characterization: M_n/g.mol^{-1} = 12000
Solvent (A): **carbon dioxide** CO_2 **124-38-9**

Type of data: gas solubility

T/K = 335.15

P/bar	55.7	102.9	157.1	195.6	239.8	292.4
w_A	0.101	0.193	0.288	0.286	0.277	0.278

T/K = 353.15

P/bar	24	57	84	114	143	172	227	280	307
w_A	0.017	0.078	0.110	0.153	0.173	0.214	0.243	0.263	0.287
P/bar	58.3	107.4	143.3	175.8	210.7	235.2	260.4	283.9	311.7
w_A	0.081	0.149	0.151	0.170	0.216	0.226	0.250	0.253	0.263

T/K = 373.15

P/bar	33	62	91	119	123	154	180	209	236
w_A	0.030	0.062	0.088	0.126	0.134	0.171	0.175	0.195	0.204
P/bar	266	297	322						
w_A	0.221	0.237	0.270						
P/bar	39.5	61.4	93.3	93.5	121.1	153.3	185.8	218.1	246.0
w_A	0.043	0.066	0.104	0.106	0.136	0.153	0.187	0.200	0.202
P/bar	268.8	311.0	334.1						
w_A	0.218	0.219	0.232						

T/K = 393.15

P/bar	19	30	49	76	106	139	161	169	187
w_A	0.015	0.025	0.035	0.071	0.089	0.119	0.136	0.148	0.159
P/bar	199	231	255	265	282	298	306	315	330
w_A	0.171	0.190	0.199	0.203	0.212	0.216	0.219	0.219	0.225
P/bar	34.9	61.1	93.7	122.2	152.5	195.2	227.0	258.9	319.8
w_A	0.042	0.081	0.149	0.151	0.170	0.216	0.226	0.250	0.263

Polymer (B): **poly(ethylene glycol)** **2014HRN**
Characterization: M_n/g.mol^{-1} = 15000, T_m/K = 429.75,
Merck, Darmstadt, Germany
Solvent (A): **carbon dioxide** CO_2 **124-38-9**

Type of data: gas solubility and swelling

T/K = 343.15

P/MPa	0.53	1.07	2.13	4.16	5.10	10.72	13.11
c_A/(g CO_2/g polymer)	0.00667	0.01585	0.03338	0.06987	0.09022	0.10983	0.09272
$(\Delta V/V)$/%	0.00	5.26	10.53	15.79	21.05	21.05	26.32

continued

continued

P/MPa	14.85	18.01	19.50	22.17	24.69	28.83	30.79
c_A/(g CO_2/g polymer)	0.11186	0.16680	0.21982	0.24602	0.28599	0.31342	0.32342
$(\Delta V/V)$/%	31.58	31.58	36.84	36.84	42.11	42.11	42.11

Polymer (B): **poly(ethylene glycol)** **2014HRN**
Characterization: M_n/g.mol^{-1} = 20000, T_m/K = 435.81,
Merck, Darmstadt, Germany
Solvent (A): **carbon dioxide** **CO_2** **124-38-9**

Type of data: gas solubility and swelling

T/K = 343.15

P/MPa	0.49	1.09	2.11	4.23	5.10	11.70	12.80
c_A/(g CO_2/g polymer)	0.00656	0.01608	0.03387	0.06660	0.08551	0.07250	0.06965
$(\Delta V/V)$/%	0.00	5.00	10.0	10.0	15.0	15.0	20.0

P/MPa	15.00	17.70	19.90	21.70	24.90	27.90	30.15
c_A/(g CO_2/g polymer)	0.08678	0.14444	0.16755	0.20865	0.25964	0.27458	0.28272
$(\Delta V/V)$/%	25.0	30.0	30.0	35.0	40.0	40.0	40.0

Polymer (B): **poly(ethylene glycol)** **2003KUK**
Characterization: M_n/g.mol^{-1} = 20000
Solvent (A): **carbon dioxide** **CO_2** **124-38-9**

Type of data: gas solubility

T/K = 340.15

P/bar	58.1	114.1	161.3	204.7	251.4
w_A	0.105	0.159	0.189	0.227	0.249

T/K = 353.15

P/bar	55.5	87.0	122.0	151.3	176.0	202.0	247.0	288.0	319.0
w_A	0.067	0.118	0.152	0.193	0.209	0.215	0.234	0.255	0.262

P/bar	43.8	60.7	96.2	125.4	189.4	220.4	250.1	303.0	342.8
w_A	0.076	0.106	0.145	0.191	0.257	0.259	0.260	0.262	0.264

T/K = 373.15

P/bar	24.0	54.0	87.0	119.0	148.0	176.0	213.0	247.0	325.0
w_A	0.016	0.052	0.082	0.116	0.133	0.165	0.178	0.214	0.240

T/K = 393.15

P/bar	29.0	61.0	91.0	124.0	151.0	180.0	209.0	238.0	267.0
w_A	0.021	0.038	0.057	0.078	0.101	0.135	0.156	0.175	0.190

P/bar	298.0	326.0
w_A	0.210	0.221

continued

continued

P/bar	34.6	60.0	101.1	115.0	148.5	186.2	229.2	275.7	312.4
w_A	0.037	0.064	0.101	0.129	0.155	0.181	0.204	0.226	0.248

Polymer (B): **poly(ethylene glycol)** **2014HRN**
Characterization: M_n/g.mol^{-1} = 35000, T_m/K = 438.19,
Merck, Darmstadt, Germany
Solvent (A): **carbon dioxide** CO_2 **124-38-9**

Type of data: gas solubility and swelling

T/K = 343.15

P/MPa	0.59	1.01	2.11	4.07	5.00	11.39	13.39
c_A/(g CO_2/g polymer)	0.00825	0.01620	0.03724	0.07039	0.09176	0.10991	0.08949
($\Delta V/V$)/%	0.00	4.35	8.70	8.70	13.04	13.04	13.04
P/MPa	15.24	17.57	19.86	22.24	25.60	28.03	30.31
c_A/(g CO_2/g polymer)	0.11061	0.17628	0.20233	0.25226	0.27414	0.31704	0.32862
($\Delta V/V$)/%	17.39	21.74	21.74	26.09	26.09	30.43	30.43

Polymer (B): **poly(ethylene glycol)** **2003KUK**
Characterization: M_n/g.mol^{-1} = 35000
Solvent (A): **carbon dioxide** CO_2 **124-38-9**

Type of data: gas solubility

T/K = 342.15

P/bar	54.4	105.1	156.2	203.0	245.2	293.8	322.0
w_A	0.097	0.165	0.198	0.232	0.252	0.279	0.294

T/K = 353.15

P/bar	34.0	71.0	91.0	121.0	147.0	173.0	203.0	236.0	251.0
w_A	0.032	0.039	0.065	0.081	0.098	0.128	0.148	0.161	0.181
P/bar	277.0	291.0	330.0						
w_A	0.195	0.207	0.230						
P/bar	36.8	56.5	92.7	106.2	133.1	165.5	197.1	226.4	252.1
w_A	0.055	0.089	0.139	0.158	0.194	0.224	0.234	0.240	0.239
P/bar	279.8	311.8	334.3						
w_A	0.243	0.247	0.255						

T/K = 373.15

P/bar	59.5	87.0	121.0	151.0	179.0	210.5	268.0	297.0	324.0
w_A	0.013	0.051	0.145	0.171	0.180	0.200	0.218	0.227	0.243
P/bar	37.6	62.5	85.9	116.1	149.0	183.9	214.7	243.9	274.6
w_A	0.044	0.080	0.106	0.142	0.166	0.187	0.211	0.232	0.240

continued

continued

P/bar	300.5	336.2
w_A	0.245	0.256

T/K = 393.15

P/bar	33.0	76.0	121.0	151.0	212.0	243.0	270.0	299.0	330.0
w_A	0.009	0.068	0.102	0.120	0.154	0.176	0.195	0.208	0.225
P/bar	43.2	82.9	112.8	143.9	176.5	219.5	245.8	281.9	304.6
w_A	0.043	0.085	0.119	0.144	0.168	0.188	0.212	0.228	0.233
P/bar	339.5								
w_A	0.243								

Polymer (B): **poly(ethylene glycol)** **2014HRN**
Characterization: M_n/g.mol^{-1} = 100000, T_m/K = 438.98, Fluka
Solvent (A): **carbon dioxide** CO_2 **124-38-9**

Type of data: gas solubility and swelling

T/K = 343.15

P/MPa	0.52	1.09	2.08	3.98	5.15	11.40	13.38
c_A/(g CO_2/g polymer)	0.00595	0.01436	0.02854	0.05729	0.07295	0.03827	0.04730
$(\Delta V/V)$/%	0.00	5.22	5.22	6.69	6.69	2.69	17.11
P/MPa	15.10	20.50	22.10	25.30	27.90	30.17	
c_A/(g CO_2/g polymer)	0.06284	0.14355	0.15183	0.19483	0.23544	0.24172	
$(\Delta V/V)$/%	21.61	26.12	26.12	30.62	35.12	35.12	

Polymer (B): **poly(ethylene glycol) diacetate** **2007CHA**
Characterization: M_n/g.mol^{-1} = 300
Solvent (A): **methanol** CH_4O **67-56-1**

Type of data: vapor-liquid equilibrium

T/K = 333.15

x_A	0.000	0.229	0.398	0.609	0.700	0.800	0.900
P/kPa	0.86	6.80	10.78	15.23	16.95	17.40	18.13

T/K = 343.15

x_A	0.000	0.229	0.398	0.609	0.700	0.800	0.900
P/kPa	1.25	10.63	17.98	23.58	23.60	25.96	29.65

T/K = 353.15

x_A	0.000	0.229	0.398	0.609	0.700	0.800	0.900
P/kPa	1.82	15.92	25.25	36.00	35.43	39.21	41.60

continued

continued

T/K = 363.15

x_A	0.000	0.229	0.398	0.609	0.700	0.800	0.900
P/kPa	2.45	22.62	36.72	46.43	51.07	54.79	60.74

T/K = 373.15

x_A	0.000	0.229	0.398	0.609	0.700	0.800	0.900
P/kPa	3.19	27.78	44.79	59.51	67.82	80.00	88.57

T/K = 383.15

x_A	0.000	0.229	0.398	0.609	0.700	0.800	0.900
P/kPa	4.02	32.12	54.21	81.32	90.97	112.79	125.81

T/K = 393.15

x_A	0.000	0.229	0.398	0.609	0.700	0.800	0.900
P/kPa	4.96	42.75	66.22	102.76	124.76	154.91	175.60

Comments: PEGDAE possesses a non-negligible vapor pressure, so that the pressure given in the table is the total vapor pressure of the system.

Polymer (B): **poly(ethylene glycol) diacetate** **2007CHA**
Characterization: M_n/g.mol^{-1} = 300
Solvent (A): **water** H_2O **7732-18-5**

Type of data: vapor-liquid equilibrium

T/K = 333.15

x_A	0.000	0.205	0.399	0.502	0.602	0.698	0.800
P/kPa	0.86	18.13	35.24	45.09	55.05	66.09	75.01

T/K = 343.15

x_A	0.000	0.205	0.399	0.502	0.602	0.698	0.800
P/kPa	1.25	24.46	50.68	64.62	76.94	94.84	100.59

T/K = 353.15

x_A	0.000	0.205	0.399	0.502	0.602	0.698	0.800
P/kPa	1.82	34.09	70.67	90.03	107.79	129.31	143.37

T/K = 363.15

x_A	0.000	0.205	0.399	0.502	0.602	0.698	0.800
P/kPa	2.45	46.05	95.78	120.42	149.97	179.59	201.84

T/K = 373.15

x_A	0.000	0.205	0.399	0.502	0.602	0.698	0.800
P/kPa	3.19	61.09	123.84	162.44	202.99	252.79	280.30

continued

continued

T/K = 383.15

x_A	0.000	0.205	0.399	0.502	0.602	0.698	0.800
P/kPa	4.02	79.34	167.91	209.95	275.83	317.18	377.45

T/K = 393.15

x_A	0.000	0.205	0.399	0.502	0.602
P/kPa	4.96	101.04	216.55	284.03	500.09

Comments: PEGDAE possesses a non-negligible vapor pressure, so that the pressure given in the table is the total vapor pressure of the system.

Polymer (B): **poly(ethylene glycol) dimethyl ether** **2009MIL**
Characterization: M_n/g.mol^{-1} = 250, Sigma-Aldrich, Inc., St. Louis, MO
Solvent (A): **carbon dioxide** CO_2 **124-38-9**

Type of data: vapor-liquid equilibrium

T/K = 298.15

w_B	0.00	0.05	0.10	0.15	0.20	0.30	0.35	0.40	0.45
P/MPa	6.43	6.22	6.17	6.17	6.15	6.06	5.94	5.68	5.39
w_B	0.50	0.55	0.60	0.65	0.70	0.75	0.80	0.85	0.90
P/MPa	5.09	4.73	4.29	3.72	3.18	2.63	2.10	1.54	0.908
w_B	0.95								
P/MPa	0.533								

Polymer (B): **poly(ethylene glycol) dimethyl ether** **2010REV**
Characterization: M_n/g.mol^{-1} = 250, Sigma-Aldrich, Inc., St. Louis, MO
Solvent (A): **carbon dioxide** CO_2 **124-38-9**

Type of data: gas solubility (bubble points)

x_A	0.210	0.210	0.210	0.210	0.210	0.210	0.210	0.210	0.210
T/K	293.35	303.25	311.95	323.55	333.25	343.15	353.45	363.05	373.35
P/bar	11.5	13.4	15.0	17.6	21.0	23.5	25.6	28.1	31.7
x_A	0.286	0.286	0.286	0.286	0.286	0.286	0.286	0.286	0.286
T/K	293.55	303.25	312.95	322.95	333.35	342.95	352.95	363.15	373.35
P/bar	14.2	16.8	19.8	24.4	26.5	29.1	31.5	33.6	37.0
x_A	0.384	0.384	0.384	0.384	0.384	0.384	0.384	0.384	
T/K	294.15	304.35	313.25	322.95	333.15	342.45	362.45	373.35	
P/bar	18.5	23.1	27.1	31.0	35.2	39.0	47.5	52.0	
x_A	0.445	0.445	0.445	0.445	0.445	0.445	0.445	0.445	0.445
T/K	293.95	303.05	312.55	322.95	333.25	343.45	353.15	362.05	373.15
P/bar	21.5	27.6	32.7	38.1	43.8	50.1	56.2	60.1	64.4

continued

continued

x_A	0.550	0.550	0.550	0.550	0.550	0.550	0.550	0.550	0.550
T/K	292.85	303.15	312.25	323.05	333.05	343.25	353.35	363.65	373.45
P/bar	28.4	35.5	42.0	48.8	56.8	64.8	73.0	81.5	88.5

x_A	0.630	0.630	0.630	0.630	0.630	0.630	0.630	0.630	0.630
T/K	292.95	303.25	312.95	322.95	333.25	344.25	353.45	363.55	373.55
P/bar	33.0	41.0	49.8	58.9	69.0	80.6	90.6	101.0	112.0

Polymer (B): **poly(ethylene glycol) dimethyl ether** **2012RA2**
Characterization: M_n/g.mol^{-1} = 340, Genosorb 1753, Clariant International Ltd., Frankfurt am Main, Germany
Solvent (A): **carbon dioxide** **CO_2** **124-38-9**

Type of data: gas solubility

T/K = 298.15

P_A/kPa	219	584	997	1222	1842	2125	2811	3624	4403
x_A	0.064	0.184	0.267	0.330	0.455	0.565	0.622	0.691	0.719
P_A/kPa	5349	6410							
x_A	0.738	0.747							

T/K = 313.15

P_A/kPa	273	731	1055	1365	2020	2491	3100	3987	4947
x_A	0.058	0.144	0.222	0.266	0.340	0.395	0.475	0.542	0.612
P_A/kPa	5755	6166	6326	6494					
x_A	0.675	0.684	0.695	0.699					

T/K = 323.15

P_A/kPa	279	550	1424	2213	2795	3149	3977
x_A	0.060	0.110	0.230	0.333	0.385	0.424	0.474

T/K = 343.15

P_A/kPa	329	536	906	1552	2135	3622	3921	4319	5018
x_A	0.052	0.100	0.145	0.202	0.288	0.414	0.443	0.485	0.523
P_A/kPa	5743	6049	7598	7937					
x_A	0.564	0.582	0.625	0.636					

Polymer (B): **poly(ethylene glycol) dimethyl ether** **2012RA1**
Characterization: M_n/g.mol^{-1} = 340, Genosorb 1753, Clariant International Ltd., Frankfurt am Main, Germany
Solvent (A): **ethane** C_2H_6 **74-84-0**

Type of data: gas solubility

T/K = 298.15

P_A/kPa	6358.9	4310.8	4292.3	4187.3	3653.3	3439.4	2833.8	2322.9	2127.3
x_A	0.470	0.460	0.450	0.450	0.420	0.400	0.340	0.300	0.280
P_A/kPa	1815.1	1343.1	939.4	771.7	565.4	411.5	283.8	220.7	170.0
x_A	0.250	0.200	0.160	0.150	0.130	0.120	0.090	0.080	0.070

T/K = 313.15

P_A/kPa	6628.0	5863.5	5539.6	4847.0	4319.7	3516.7	3110.1	2716.7	1801.8
x_A	0.451	0.450	0.455	0.434	0.396	0.353	0.312	0.284	0.209
P_A/kPa	1209.9	738.9	376.9	258.9	194.0				
x_A	0.161	0.125	0.085	0.063	0.056				

T/K = 333.15

P_A/kPa	7655.12	7154.34	7096.65	6851.73	6585.44	6247.38	5921.38	5465.86	4973.00
x_A	0.481	0.454	0.444	0.433	0.432	0.419	0.399	0.399	0.374
P_A/kPa	4553.03	4177.79	3843.81	3308.95	2907.10	2395.01	1969.92	1510.90	1131.31
x_A	0.353	0.340	0.318	0.304	0.254	0.233	0.195	0.159	0.136
P_A/kPa	791.01	386.95	173.09						
x_A	0.105	0.070	0.051						

Polymer (B): **poly(ethylene glycol) dimethyl ether** **2012RA1**
Characterization: M_n/g.mol^{-1} = 340, Genosorb 1753, Clariant International Ltd., Frankfurt am Main, Germany
Solvent (A): **methane** CH_4 **74-82-8**

Type of data: gas solubility

T/K = 298.15

P_A/kPa	7210.7	6898.8	6804.5	6393.0	5345.8	4895.5	4422.2	4237.2	3929.6
x_A	0.134	0.131	0.131	0.125	0.113	0.106	0.099	0.100	0.094
P_A/kPa	3380.8	2773.9	1625.3	1098.8	697.4	638.6	509.7	383.4	314.9
x_A	0.085	0.074	0.052	0.039	0.028	0.027	0.020	0.018	0.016
P_A/kPa	185.5								
x_A	0.013								

continued

continued

T/K = 313.15

P_A/kPa	7406.3	6375.2	5326.1	4328.4	4147.5	3996.9	3464.7	3396.7	3061.2
x_A	0.140	0.124	0.119	0.102	0.090	0.086	0.082	0.080	0.073
P_A/kPa	2266.3	2135.7	1674.7	1615.9	1112.6	1069.1	644.8	641.9	403.3
x_A	0.065	0.057	0.047	0.041	0.038	0.039	0.024	0.032	0.014
P_A/kPa	378.6	205.8	169.4	141.8					
x_A	0.022	0.011	0.013	0.011					

T/K = 333.15

P_A/kPa	6730.6	6230.7	5981.4	5713.1	5453.6	5096.6	4712.1	4369.7	3416.3
x_A	0.135	0.129	0.123	0.120	0.115	0.112	0.106	0.104	0.090
P_A/kPa	2674.8	2458.6	1697.9	1265.2	1129.9	705.8	378.9	238.8	186.8
x_A	0.076	0.075	0.056	0.047	0.046	0.031	0.020	0.013	0.014

Polymer (B): **poly(ethylene glycol) monododecanoate** **2013KHO1**
Characterization: M_n/g.mol^{-1} = 400, Tokyo Chemical Industry Co., Tokyo, Japan
Solvent (A): **2-butanol** $C_4H_{10}O$ **78-92-2**

Type of data: vapor-liquid equilibrium

x_B = 0.100 was kept constant

T/K	334.6	343.4	353.5	362.7	372.4	374.7	381.0	383.0	393.2
P_A/kPa	18.2	27.9	42.2	59.7	85.5	92.7	120.2	126.6	174.9
T/K	402.7	413.0	423.4						
P_A/kPa	233.0	308.4	404.2						

x_B = 0.200 was kept constant

T/K	333.4	345.7	353.7	359.9	363.1	371.2	372.0	383.0	393.2
P_A/kPa	16.9	29.4	38.7	48.8	54.2	70.1	72.7	105.9	140.9
T/K	403.5	404.1	414.0	422.9					
P_A/kPa	185.0	187.9	243.5	303.4					

x_B = 0.299 was kept constant

T/K	333.4	344.7	351.6	352.5	363.6	362.3	370.5	370.9	380.7
P_A/kPa	15.7	24.1	31.3	32.1	47.9	45.8	57.9	58.5	76.7
T/K	392.5	392.9	404.0	413.0	414.1	424.5			
P_A/kPa	102.4	103.4	140.0	169.8	172.4	213.9			

x_B = 0.399 was kept constant

T/K	334.6	343.7	352.4	354.4	364.2	371.9	373.5	381.6	382.4
P_A/kPa	14.8	19.6	26.3	27.9	40.3	49.7	50.7	62.1	62.7
T/K	393.9	394.7	403.3	404.0	412.8	424.2			
P_A/kPa	79.4	80.9	99.0	100.6	119.2	147.9			

Polymer (B): **poly(ethylene glycol) monododecanoate** **2012KHO**
Characterization: M_n/g.mol^{-1} = 400, Tokyo Chemical Industry Co., Tokyo, Japan
Solvent (A): **ethanol** C_2H_6O **64-17-5**

Type of data: vapor-liquid equilibrium

x_B = 0.100 was kept constant

T/K	343.2	345.2	352.0	353.2	354.2	363.2	365.1	373.2	375.2
P_A/kPa	61.0	66.1	81.5	85.7	88.0	119.8	127.5	167.8	178.9
T/K	383.2	384.9	393.2	403.2	3.2	422.3	423.2		
P_A/kPa	227.5	239.0	307.9	399.6	537.8	676.2	693.0		

x_B = 0.200 was kept constant

T/K	343.2	344.2	353.2	354.2	355.2	363.2	364.1	365.0	365.1
P_A/kPa	54.0	56.2	79.5	81.6	84.0	106.7	110.1	112.4	113.1
T/K	372.2	373.2	374.2	382.2	383.2	384.2	393.2	394.1	403.2
P_A/kPa	141.7	145.1	149.5	192.7	198.0	203.7	273.8	281.0	357.9
T/K	404.1	405.0	413.2	422.3	423.2				
P_A/kPa	368.8	376.8	473.7	586.3	595.8				

x_B = 0.300 was kept constant

T/K	343.2	344.2	351.2	353.2	362.7	363.2	373.2	374.2	376.5
P_A/kPa	53.1	54.6	69.7	73.4	96.0	97.4	118.7	122.5	127.0
T/K	381.3	382.2	383.2	393.2	394.1	403.6	411.2	413.2	423.2
P_A/kPa	146.9	149.2	151.8	187.6	192.2	232.8	265.0	275.5	329.0

x_B = 0.400 was kept constant

T/K	343.2	353.2	354.2	363.2	373.2	383.2	393.2	394.5	403.2
P_A/kPa	42.0	57.2	58.2	72.6	87.5	101.1	129.6	130.7	152.1
T/K	404.5	409.6	413.2	423.2					
P_A/kPa	153.8	167.6	175.6	201.8					

Polymer (B): **poly(ethylene glycol) monododecanoate** **2012KHO**
Characterization: M_n/g.mol^{-1} = 400, Tokyo Chemical Industry Co., Tokyo, Japan
Solvent (A): **methanol** CH_4O **67-56-1**

Type of data: vapor-liquid equilibrium

x_B = 0.100 was kept constant

T/K	343.2	353.2	354.1	362.2	363.2	372.2	373.2	374.3	383.2
P_A/kPa	102.0	145.5	149.4	192.7	200.1	259.5	265.5	274.8	345.3
T/K	393.2	395.1	403.2	403.2	412.3	412.9	413.2	422.6	423.2
P_A/kPa	462.3	487.8	606.7	607.8	770.9	793.9	801.1	1012.3	1041.5

continued

continued

x_B = 0.150 was kept constant

T/K	343.2	353.2	363.2	372.4	373.2	374.2	375.2	383.2	384.1
P_A/kPa	90.4	124.5	167.2	205.9	209.3	214.8	220.7	259.5	268.0
T/K	393.2	403.2	413.2	423.2					
P_A/kPa	335.4	452.3	575.5	739.3					

x_B = 0.200 was kept constant

T/K	343.2	343.6	352.5	353.2	355.6	361.8	363.2	365.2	373.2
P_A/kPa	84.2	85.0	110.5	111.5	117.9	142.2	143.0	154.1	175.3
T/K	374.2	382.7	383.2	392.8	393.2	394.5	403.2	413.2	423.0
P_A/kPa	183.8	222.9	227.5	292.3	293.5	303.5	365.1	403.5	510.6

x_B = 0.250 was kept constant

T/K	343.2	344.2	350.6	353.2	354.2	363.2	372.4	372.4
P_A/kPa	70.8	72.7	88.4	92.6	95.3	118.6	146.9	146.9
T/K	373.2	383.2	393.2	403.2	413.2	423.2		
P_A/kPa	147.5	175.4	204.3	234.2	267.6	297.5		

Polymer (B): **poly(ethylene glycol) monododecanoate** **2013KHO1**
Characterization: M_n/g.mol^{-1} = 400, Tokyo Chemical Industry Co., Tokyo, Japan
Solvent (A): **2-methyl-2-propanol** **$C_4H_{10}O$** **75-65-0**

Type of data: vapor-liquid equilibrium

x_B = 0.100 was kept constant

T/K	323.7	331.5	343.2	351.1	361.2	371.6	381.3	391.1	400.8
P_A/kPa	22.4	31.6	52.3	72.0	105.1	150.4	199.7	262.9	345.4

x_B = 0.200 was kept constant

T/K	322.5	333.5	341.9	352.1	360.5	371.2	380.8	391.0	401.5
P_A/kPa	19.8	29.8	42.9	64.1	85.5	120.8	159.7	210.1	276.0

x_B = 0.300 was kept constant

T/K	321.1	334.2	343.8	352.7	361.6	371.2	381.8	392.1	401.5
P_A/kPa	18.6	28.0	40.9	56.9	75.4	98.2	129.4	165.7	203.3

x_B = 0.432 was kept constant

T/K	321.9	334.5	343.0	353.0	362.8	372.5	382.9	388.4	400.6
P_A/kPa	16.9	25.2	33.3	48.2	64.2	80.9	103.7	116.8	149.3

Polymer (B): **poly(ethylene glycol) monododecanoate** **2013KHO1**
Characterization: M_n/g.mol^{-1} = 400, Tokyo Chemical Industry Co., Tokyo, Japan
Solvent (A): **1-pentanol** **$C_5H_{12}O$** **71-41-0**

Type of data: vapor-liquid equilibrium

x_B = 0.099 was kept constant

T/K	340.2	352.3	361.4	370.1	380.4	390.6	399.6	407.1	417.2
P_A/kPa	4.4	8.2	12.0	18.7	28.6	42.0	56.1	68.9	94.9

x_B = 0.199 was kept constant

T/K	340.8	352.3	361.9	369.8	380.3	390.9	400.4	409.0	417.9
P_A/kPa	4.1	7.4	11.6	17.3	26.5	37.5	54.5	70.7	82.9

x_B = 0.300 was kept constant

T/K	342.1	351.6	361.0	370.5	380.4	391.6	399.8	409.4	418.1
P_A/kPa	4.2	6.9	10.4	16.7	24.1	33.3	42.1	55.0	69.8

x_B = 0.397 was kept constant

T/K	341.3	353.4	362.1	370.6	380.3	390.4	399.9	410.1	419.4
P_A/kPa	3.6	6.5	9.1	13.3	20.0	28.2	38.3	50.7	60.1

Polymer (B): **poly(ethylene glycol) monododecanoate** **2012KHO**
Characterization: M_n/g.mol^{-1} = 400, Tokyo Chemical Industry Co., Tokyo, Japan
Solvent (A): **2-propanol** **C_3H_8O** **67-63-0**

Type of data: vapor-liquid equilibrium

x_B = 0.100 was kept constant

T/K	343.2	353.2	354.2	363.2	363.7	371.8	373.2	383.2	384.2
P_A/kPa	51.5	78.8	81.5	115.8	118.0	160.2	168.6	238.4	245.0
T/K	393.2	394.1	403.2	413.2	423.2				
P_A/kPa	318.0	327.2	428.0	576.2	752.8				

x_B = 0.200 was kept constant

T/K	343.2	343.6	353.2	354.2	360.8	363.2	370.4	373.2	383.2
P_A/kPa	47.0	47.4	67.0	70.9	88.8	94.9	120.7	132.3	174.8
T/K	384.9	393.2	403.2	413.2	423.2				
P_A/kPa	182.2	232.6	300.6	415.8	526.8				

x_B = 0.300 was kept constant

T/K	343.2	344.2	353.2	354.2	363.2	364.1	371.4	373.2	381.6
P_A/kPa	41.2	42.8	59.2	60.8	80.0	81.9	98.3	102.6	124.1
T/K	383.2	393.2	394.1	403.2	413.2	422.3	423.2		
P_A/kPa	128.5	167.8	171.0	213.2	266.7	325.8	332.2		

continued

continued

x_B = 0.400 was kept constant

T/K	343.2	344.2	353.2	354.2	363.2	373.2	383.2	393.2	403.2
P_A/kPa	38.4	39.5	53.4	54.5	70.4	92.2	115.4	133.3	159.3

T/K	413.2	423.2
P_A/kPa	186.3	213.0

Polymer (B): **poly(ethylene glycol) mono-4-nonylphenyl ether 2011KHO2**
Characterization: M_n/g.mol^{-1} = 303, Tokyo Chemical Industry Co., Tokyo, Japan
Solvent (A): **ethanol** **C_2H_6O** **64-17-5**

Type of data: vapor-liquid equilibrium

x_B = 0.100 was kept constant

T/K	342.8	352.2	361.5	363.2	372.9	382.1	383.2	393.2	413.5
P_A/kPa	62.9	94.7	134.8	147.4	199.8	259.4	268.7	352.5	615.9

T/K	414.0
P_A/kPa	619.7

x_B = 0.200 was kept constant

T/K	343.2	353.2	354.2	360.6	362.3	363.2	371.2	372.0	373.1
P_A/kPa	61.9	90.2	92.3	112.1	119.1	126.6	167.1	171.3	177.1

T/K	383.5	393.8	394.2	403.2	404.2	413.1
P_A/kPa	232.8	300.7	305.6	388.9	399.0	497.3

x_B = 0.300 was kept constant

T/K	341.9	344.7	353.2	353.5	354.5	363.1	364.6	373.1	374.1
P_A/kPa	61.5	65.5	85.3	84.5	87.3	108.8	115.0	150.5	154.3

T/K	383.2	383.3	384.2	394.8	404.3	404.5	413.2
P_A/kPa	189.9	190.4	194.7	245.6	305.8	307.5	381.9

x_B = 0.400 was kept constant

T/K	343.6	353.4	364.1	365.1	374.1	375.1	383.8	384.2	394.6
P_A/kPa	57.5	74.3	93.3	94.8	117.1	119.3	146.2	147.6	177.6

T/K	395.2	406.9	413.5	414.1
P_A/kPa	180.9	225.6	246.6	248.1

Polymer (B): **poly(ethylene glycol) mono-4-nonylphenyl ether 2011KHO2**
Characterization: M_n/g.mol^{-1} = 303, Tokyo Chemical Industry Co., Tokyo, Japan
Solvent (A): **methanol** **CH_4O** **67-56-1**

Type of data: vapor-liquid equilibrium

x_B = 0.100 was kept constant

T/K	340.6	342.1	342.6	343.2	353.2	353.3	353.4	362.1	363.1
P_A/kPa	105.3	111.5	114.4	117.3	166.9	167.5	168.1	218.4	224.7
T/K	373.1	373.3	382.3	383.1	393.2	403.1	413.2	413.5	
P_A/kPa	302.2	303.5	391.8	400.2	513.5	675.5	870.9	877.9	

x_B = 0.200 was kept constant

T/K	343.2	343.6	344.6	353.2	353.5	362.3	362.7	371.6	373.1
P_A/kPa	107.3	109.1	113.7	154.7	156.1	202.2	204.8	261.6	271.6
T/K	373.3	382.2	383.2	394.2	403.2	413.2	414.7		
P_A/kPa	272.9	342.3	351.4	453.5	556.4	712.3	736.3		

x_B = 0.300 was kept constant

T/K	343.2	343.6	344.6	361.4	363.2	372.4	373.1	383.2	393.2
P_A/kPa	94.7	95.6	97.7	166.9	173.2	211.1	213.9	262.3	296.3
T/K	402.6	403.2	413.1	414.0					
P_A/kPa	362.7	368.8	435.0	442.1					

x_B = 0.400 was kept constant

T/K	343.1	354.1	363.2	375.5	384.5	393.8	403.2	413.8
P_A/kPa	86.7	103.1	136.3	175.0	208.9	231.0	261.6	287.6

Polymer (B): **poly(ethylene glycol) mono-4-nonylphenyl ether 2011KHO2**
Characterization: M_n/g.mol^{-1} = 303, Tokyo Chemical Industry Co., Tokyo, Japan
Solvent (A): **2-propanol** **C_3H_8O** **67-63-0**

Type of data: vapor-liquid equilibrium

x_B = 0.100 was kept constant

T/K	351.9	353.8	354.4	363.2	364.2	365.0	383.3	384.3	393.0
P_A/kPa	82.3	86.5	88.1	123.2	127.7	132.7	251.3	259.5	340.7
T/K	403.2	413.4	423.3	423.9	432.8	433.2	441.6	441.8	451.2
P_A/kPa	457.9	588.5	762.4	772.9	961.8	969.0	1171.4	1174.1	1429.3

x_B = 0.200 was kept constant

T/K	353.2	354.2	355.2	363.2	364.2	365.2	373.2	374.2	374.9
P_A/kPa	76.3	79.1	81.4	105.3	109.1	115.0	157.6	160.9	164.5
T/K	384.2	385.2	386.1	393.8	394.1	395.2	403.7	413.2	414.3
P_A/kPa	215.6	221.5	227.4	290.2	291.4	302.1	378.1	483.3	493.7

continued

continued

T/K	423.2	423.5	424.0	433.2	433.3	442.8	452.5		
P_A/kPa	616.6	619.1	628.2	795.4	797.1	991.7	1206.8		

$x_B = 0.300$ was kept constant

T/K	353.0	354.2	355.5	360.7	363.4	375.0	375.6	381.8	382.4
P_A/kPa	67.1	69.8	73.1	86.9	91.0	139.0	142.0	172.6	175.0
T/K	384.8	404.7	413.8	414.0	423.2	424.3	424.4	433.1	442.7
P_A/kPa	184.8	300.1	367.2	369.4	441.0	453.4	454.8	553.4	710.6
T/K	452.1	452.3	452.6						
P_A/kPa	853.7	860.2	862.1						

$x_B = 0.400$ was kept constant

T/K	351.5	361.4	363.4	365.2	371.5	372.3	374.1	374.8	382.3
P_A/kPa	54.9	75.8	78.7	82.1	95.5	97.7	101.2	103.3	129.9
T/K	383.6	395.5	396.2	406.9	414.8	415.3	426.6	426.9	435.3
P_A/kPa	134.4	173.9	176.7	219.3	260.8	263.2	320.3	323.6	427.2
T/K	445.0	445.3	454.4	454.5					
P_A/kPa	513.3	525.4	610.5	613.9					

Polymer (B): **poly(ethylene glycol) mono-4-octylphenyl ether** **2013KHO2**
Characterization: M_n/g.mol^{-1} = 647, Tokyo Chemical Industry Co., Tokyo, Japan
Solvent (A): **2-butanol** **$C_4H_{10}O$** **78-92-2**

Type of data: vapor-liquid equilibrium

$x_B = 0.100$ was kept constant

T/K	333.3	342.8	353.2	363.6	374.1	374.4	382.3	385.3	394.0
P_A/kPa	16.5	26.0	42.3	62.9	92.3	89.2	121.3	133.8	176.8
T/K	403.1	413.7	422.4						
P_A/kPa	232.7	309.5	388.9						

$x_B = 0.200$ was kept constant

T/K	333.6	343.1	354.2	364.5	372.2	381.7	382.5	395.6	403.8
P_A/kPa	16.4	25.3	37.9	53.9	68.7	89.2	90.7	127.2	155.7
T/K	404.6	413.7	422.8						
P_A/kPa	157.6	196.8	238.7						

$x_B = 0.307$ was kept constant

T/K	331.0	341.4	350.6	360.8	369.7	378.7	388.7	397.4	406.0
P_A/kPa	12.5	19.7	27.1	37.9	49.9	62.5	77.4	92.1	106.1
T/K	415.1								
P_A/kPa	125.2								

continued

continued

x_B = 0.400 was kept constant

T/K	335.5	343.0	356.4	363.1	364.4	373.5	381.7	393.2	394.0
P_A/kPa	14.7	19.5	29.1	35.2	36.1	46.1	55.0	70.3	72.0
T/K	404.1	404.7	405.1	413.7	424.5				
P_A/kPa	83.0	83.0	84.0	96.8	109.3				

Polymer (B): **poly(ethylene glycol) mono-4-octylphenyl ether** **2011KHO1**
Characterization: M_n/g.mol^{-1} = 647, Tokyo Chemical Industry Co., Tokyo, Japan
Solvent (A): **ethanol** **C_2H_6O** **64-17-5**

Type of data: vapor-liquid equilibrium

x_B = 0.100 was kept constant

T/K	345.1	353.0	353.1	353.5	363.1	373.1	373.2	373.3	373.7
P_A/kPa	74.5	92.3	92.4	93.5	132.0	179.4	179.9	180.4	181.9
T/K	381.4	383.1	383.2	393.0	402.4	404.6	413.0	413.1	414.5
P_A/kPa	232.4	244.4	245.5	317.1	407.8	431.8	516.4	517.4	537.4
T/K	423.2	423.9	434.3	444.3	454.1				
P_A/kPa	670.9	681.5	864.7	1083.5	1322.6				

x_B = 0.200 was kept constant

T/K	342.6	343.1	353.1	353.8	361.2	363.1	363.4	364.8	370.9
P_A/kPa	61.1	61.9	82.9	84.5	98.1	100.8	101.7	105.3	126.8
T/K	373.1	373.2	373.3	383.2	384.3	393.2	403.2	404.1	405.1
P_A/kPa	133.0	133.4	134.3	172.2	175.8	215.2	261.2	265.9	269.0
T/K	413.3	414.0	423.0	433.2	443.3	452.9			
P_A/kPa	326.5	329.5	400.2	482.6	585.4	709.4			

x_B = 0.300 was kept constant

T/K	343.2	343.6	344.3	353.0	354.1	363.1	364.1	365.4	372.9
P_A/kPa	54.1	54.6	55.9	70.0	71.9	87.9	89.4	91.2	108.7
T/K	373.8	374.1	383.2	384.0	385.5	402.5	403.2	413.6	423.2
P_A/kPa	109.4	110.5	131.5	134.3	136.3	178.6	181.7	218.1	250.1
T/K	423.3	432.9	433.4	443.0					
P_A/kPa	251.9	288.0	286.7	346.3					

Polymer (B): **poly(ethylene glycol) mono-4-octylphenyl ether** **2011KHO1**
Characterization: M_n/g.mol^{-1} = 647, Tokyo Chemical Industry Co., Tokyo, Japan
Solvent (A): **methanol** **CH_4O** **67-56-1**

Type of data: vapor-liquid equilibrium

x_B = 0.100 was kept constant

T/K	341.0	341.6	342.6	352.5	353.5	354.5	361.4	362.4	363.4
P_A/kPa	93.3	94.5	98.2	145.7	151.2	156.9	202.0	208.0	214.6

T/K	372.4	382.3	392.2	394.2
P_A/kPa	276.3	368.4	479.6	500.2

x_B = 0.143 was kept constant

T/K	343.6	344.6	352.5	353.1	353.5	362.6	363.1	363.4	373.1
P_A/kPa	99.4	103.5	137.0	139.2	141.0	184.5	186.0	187.2	235.3

T/K	373.3	383.2	394.4	403.1
P_A/kPa	236.2	298.8	383.4	462.0

x_B = 0.200 was kept constant

T/K	343.2	343.8	352.5	353.2	362.8	363.2	383.2	393.2
P_A/kPa	75.9	76.6	98.0	99.5	124.0	125.6	177.2	220.7

x_B = 0.300 was kept constant

T/K	343.1	345.3	353.2	353.5	354.5	363.2	363.4	364.8	371.4
P_A/kPa	72.2	74.8	85.9	86.1	86.7	96.2	6.8	99.4	116.2

T/K	373.1	374.3	383.3	384.0
P_A/kPa	119.3	122.1	142.2	143.1

Polymer (B): **poly(ethylene glycol) mono-4-octylphenyl ether** **2013KHO2**
Characterization: M_n/g.mol^{-1} = 647, Tokyo Chemical Industry Co., Tokyo, Japan
Solvent (A): **2-methyl-2-propanol** **$C_4H_{10}O$** **75-65-0**

Type of data: vapor-liquid equilibrium

x_B = 0.100 was kept constant

T/K	322.6	333.4	343.3	352.9	362.7	72.9	382.6	390.4	401.3
P_A/kPa	21.7	34.1	52.3	76.1	107.9	151.9	205.1	258.9	341.3

x_B = 0.200 was kept constant

T/K	322.8	331.6	341.4	350.0	361.2	370.3	379.4	389.1	398.9
P_A/kPa	19.1	28.5	42.7	59.8	86.8	113.0	147.4	190.3	241.4

x_B = 0.300 was kept constant

T/K	322.9	332.2	342.4	351.1	362.3	370.8	381.4	390.8	400.3
P_A/kPa	17.4	24.6	36.1	48.8	65.7	82.1	103.1	128.3	162.9

continued

continued

x_B = 0.432 was kept constant

T/K	321.0	331.2	340.9	351.2	360.1	370.8	379.3	389.8	399.2
P_A/kPa	14.1	20.8	27.5	39.1	49.1	64.0	74.4	89.6	105.3

Polymer (B): **poly(ethylene glycol) mono-4-octylphenyl ether** **2013KHO2**
Characterization: M_n/g.mol^{-1} = 647, Tokyo Chemical Industry Co., Tokyo, Japan
Solvent (A): **1-pentanol** **$C_5H_{12}O$** **71-41-0**

Type of data: vapor-liquid equilibrium

x_B = 0.100 was kept constant

T/K	332.4	343.3	352.7	363.5	373.6	381.5	393.1	402.4	412.0
P_A/kPa	2.7	4.7	7.7	14.1	21.4	28.8	43.1	62.6	86.0
T/K	422.0								
P_A/kPa	118.1								

x_B = 0.200 was kept constant

T/K	331.9	339.8	350.0	360.6	370.2	379.7	389.1	398.8	408.9
P_A/kPa	2.4	3.9	6.5	10.7	16.0	23.4	33.5	44.2	59.8
T/K	419.6								
P_A/kPa	80.5								

x_B = 0.300 was kept constant

T/K	332.5	342.0	350.5	360.8	372.4	381.4	389.8	400.3	410.8
P_A/kPa	2.0	3.3	5.2	8.9	13.7	18.5	26.8	34.9	46.5
T/K	420.6								
P_A/kPa	59.2								

x_B = 0.400 was kept constant

T/K	334.8	344.7	354.1	363.2	371.2	382.1	391.6	400.7	411.5
P_A/kPa	2.3	3.9	6.1	8.4	11.5	17.1	22.6	26.6	32.2
T/K	420.2								
P_A/kPa	38.0								

Polymer (B): **poly(ethylene glycol) mono-4-octylphenyl ether** **2011KHO1**
Characterization: M_n/g.mol^{-1} = 647, Tokyo Chemical Industry Co., Tokyo, Japan
Solvent (A): **2-propanol** **C_3H_8O** **67-63-0**

Type of data: vapor-liquid equilibrium

x_B = 0.100 was kept constant

T/K	351.6	360.9	363.0	363.8	371.6	373.3	384.0	393.4	402.7
P_A/kPa	77.1	101.9	112.0	113.8	156.5	165.4	229.6	308.3	406.0

continued

continued

T/K	403.6	412.9	414.2	422.9	423.5	433.2	433.6	442.8	451.6
P_A/kPa	414.4	522.7	532.9	680.0	693.2	873.8	880.3	1086.8	1306.3

x_B = 0.145 was kept constant

T/K	351.8	363.3	373.7	383.8	394.8	404.5	414.1	423.9	434.1
P_A/kPa	69.5	98.1	150.2	204.8	279.0	362.1	462.6	572.4	734.3
T/K	444.2	453.2							
P_A/kPa	905.6	1082.7							

x_B = 0.200 was kept constant

T/K	350.1	354.7	362.7	371.4	373.7	382.6	384.0	393.1	395.3
P_A/kPa	60.1	70.3	88.3	113.8	122.7	161.0	166.1	208.8	219.3
T/K	405.1	406.1	415.0	415.4	424.1	425.7	434.3	434.7	443.3
P_A/kPa	274.9	277.9	348.4	349.7	433.2	443.7	522.2	526.6	637.6
T/K	443.8	452.8	453.3						
P_A/kPa	641.9	777.7	776.8						

x_B = 0.300 was kept constant

T/K	351.7	352.7	364.5	374.9	385.6	396.2	405.4	417.6	427.7
P_A/kPa	49.8	51.4	71.4	88.0	110.1	146.6	174.6	212.6	251.4
T/K	436.8	442.0	451.4						
P_A/kPa	279.2	310.9	380.2						

Polymer (B): **poly(ethylene terephthalate)** **2008LIZ**
Characterization: 35.39% crystallinity, commercial film sample, Taicang Changfa Mapping Materials, China
Solvent (A): **carbon dioxide** **CO_2** **124-38-9**

Type of data: gas solubility

T/K = 383.15

P/MPa	7.1	13.3	18.0	22.8	24.8	26.5	29.0
c_A/(g CO_2/g polymer)	0.0135	0.0229	0.0291	0.0350	0.0357	0.0376	0.0394

T/K = 393.15

P/MPa	5.0	7.0	13.0	18.0	22.4	25.4	27.0
c_A/(g CO_2/g polymer)	0.0082	0.0111	0.0204	0.0256	0.0315	0.0337	0.0348
P/MPa	30.0						
c_A/(g CO_2/g polymer)	0.0374						

continued

continued

T/K = 403.15

P/MPa	6.0	9.0	14.0	18.0	22.0	25.0	27.5
c_A/(g CO_2/g polymer)	0.0085	0.0135	0.0205	0.0241	0.0276	0.0307	0.0329
P/MPa	30.0						
c_A/(g CO_2/g polymer)	0.0344						

Polymer (B): **poly(hexafluoropropylene oxide) carboxylic acid end-capped** **2006AGU**
Characterization: M_w/g.mol^{-1} = 2500, M_w/M_n = 1.108, Krytox 157FSL, Dupont
Solvent (A): **carbon dioxide** **CO_2** **124-38-9**

Type of data: critical points of vapor-liquid equilibrium

x_A	1.0000	0.9983	0.9844	0.9714	0.9636
T/K	304.18	304.35	304.55	304.60	304.65
P/MPa	7.40	7.43	7.46	7.49	7.50

Polymer (B): **poly(hexafluoropropylene oxide) carboxylic acid end-capped** **2004CAS, 2005CAS**
Characterization: M_w/g.mol^{-1} = 2500, M_w/M_n = 1.108, Krytox 157FSL, Dupont
Solvent (A): **carbon dioxide** **CO_2** **124-38-9**

Type of data: vapor-liquid equilibrium

T/K = 313.15

x_A (liquid phase)	0.0362	0.0582	0.0702	0.0745	0.0962	0.1136	0.1265	0.1662
P/MPa	2.408	3.046	3.884	4.158	5.182	5.923	6.019	7.064
x_A (liquid phase)	0.2045	0.2351	0.3269	0.3946	0.4171	0.5506	0.5990	0.6932
P/MPa	7.700	8.083	8.742	9.406	9.577	9.860	10.189	10.310
x_A (liquid phase)	0.7532							
P/MPa	10.409							
x_A (gas phase)	0.9812	0.9540	0.9189	0.9262	0.9022	0.9149	0.9284	0.9299
P/MPa	2.4031	3.041	3.876	4.159	5.184	5.882	6.020	7.060
x_A (gas phase)	0.9585	0.9713	0.9596	0.9226	0.9270	0.8735	0.8114	0.7949
P/MPa	7.694	8.043	8.737	9.401	9.521	9.810	10.179	10.321
x_A (gas phase)	0.7521							
P/MPa	10.408							

continued

continued

T/K = 323.15

x_A (liquid phase)	0.0780	0.1228	0.1623	0.2106	0.2482	0.2767	0.4177	0.4211
P/MPa	5.483	7.090	8.065	9.345	9.969	10.446	11.658	11.781
x_A (liquid phase)	0.5687	0.6251	0.6353	0.7271				
P/MPa	12.325	12.536	12.550	12.586				
x_A (gas phase)	0.9927	0.9574	0.9634	0.9502	0.9663	0.9597	0.9201	0.9163
P/MPa	5.483	7.080	8.023	9.342	9.948	10.426	11.738	11.780
x_A (gas phase)	0.8870	0.8275	0.8021	0.7369				
P/MPa	12.315	12.506	12.500	12.585				

T/K = 333.15

x_A (liquid phase)	0.0621	0.0980	0.1152	0.1435	0.2034	0.2341	0.2854	0.2873
P/MPa	4.919	6.356	7.397	8.026	9.346	10.328	11.771	12.102
x_A (liquid phase)	0.4076	0.4753	0.5341	0.6326				
P/MPa	13.398	14.076	14.443	14.649				
x_A (gas phase)	0.9644	0.9670	0.9490	0.9385	0.9604	0.9675	0.9694	0.9678
P/MPa	4.911	6.362	7.395	8.044	9.346	10.319	11.767	12.099
x_A (gas phase)	0.9211	0.8848	0.7858	0.6587				
P/MPa	3.467	14.071	14.434	14.649				

T/K = 343.15

x_A (liquid phase)	0.0930	0.1487	0.1760	0.2242	0.2876	0.3021	0.4324	0.4963
P/MPa	7.297	9.074	9.643	11.355	13.454	14.235	15.573	16.589
x_A (liquid phase)	0.5459	0.5979						
P/MPa	16.892	17.027						
x_A (gas phase)	0.9571	0.9650	0.9760	0.9678	0.9650	0.9540	0.9235	0.8059
P/MPa	7.299	9.075	9.647	11.367	13.542	14.233	15.567	16.578
x_A (gas phase)	0.7035	–						
P/MPa	16.882	17.019						

Polymer (B): **poly(DL-lactic acid)** **2008PIN**

Characterization: M_n/g.mol^{-1} = 6400, M_w/g.mol^{-1} = 13000, T_g/K = 314.95, ρ = 1.283 g/cm^3

Solvent (A): **carbon dioxide** **CO_2** **124-38-9**

Type of data: gas solubility

T/K = 308.15

P/bar	23	52	78	99	121	152	204
c_A/(g CO_2/g polymer)	0.074	0.209	0.389	0.445	0.475	0.506	0.549

Polymer (B): **poly(DL-lactic acid)** **2008PIN**
Characterization: M_n/g.mol^{-1} = 27800, M_w/g.mol^{-1} = 52000, T_g/K = 320.05, ρ = 1.275 g/cm^3
Solvent (A): **carbon dioxide** **CO_2** **124-38-9**

Type of data: gas solubility

T/K = 308.15

P/bar	20	21	50	52	79	100	103
c_A/(g CO_2/g polymer)	0.063	0.066	0.188	0.199	0.363	0.426	0.422
P/bar	119	150	151	206	207		
c_A/(g CO_2/g polymer)	0.446	0.492	0.480	0.538	0.525		

Polymer (B): **poly(DL-lactic acid)** **2008KAS**
Characterization: M_w/g.mol^{-1} = 75000-120000, Aldrich Chem. Co., Inc., Milwaukee, WI
Solvent (A): **carbon dioxide** **CO_2** **124-38-9**

Type of data: gas solubility

T/K = 313.15

P/MPa	0.862	2.213	4.103	5.269	7.931
w_A	0.0575	0.1109	0.1855	0.2316	0.3367

T/K = 333.15

P/MPa	1.241	1.813	2.393	4.172	5.096	5.917	6.386	7.800
w_A	0.0292	0.0463	0.0609	0.1145	0.1421	0.1650	0.1830	0.2150

T/K = 344.15

P/MPa	1.100	2.213	3.400	4.103	5.269	6.896	7.800
w_A	0.0163	0.0464	0.0786	0.0977	0.1292	0.1734	0.1978

Polymer (B): **poly(DL-lactic acid)** **2004OLI**
Characterization: M_w/g.mol^{-1} = 102800, T_g/K = 329.9, L:D ratio = 80:20, Cargill–Dow Polymers
Solvent (A): **carbon dioxide** **CO_2** **124-38-9**

Type of data: gas solubility

T/K = 293.2

P/bar	0.142	0.230	0.361	0.462	0.578	0.656	0.818
c_A/[cm^3 (STP)/cm^3 polymer]	0.78	1.19	1.63	2.05	2.23	2.58	2.93
P/bar	0.893	0.980					
c_A/[cm^3 (STP)/cm^3 polymer]	3.20	3.49					

Comments: The measuring temperature is below T_g.

continued

continued

T/K = 303.2

P/bar	0.226	0.512	0.677	0.805	0.857	0.960	0.961
c_A/[cm^3 (STP)/cm^3 polymer]	0.98	1.65	1.94	2.28	2.39	2.63	2.66

Comments: The measuring temperature is below T_g.

T/K = 313.2

P/bar	0.197	0.271	0.334	0.359	0.487	0.524	0.634
c_A/[cm^3 (STP)/cm^3 polymer]	0.46	0.74	0.85	0.93	1.13	1.23	1.50

P/bar	0.719	0.973
c_A/[cm^3 (STP)/cm^3 polymer]	1.59	1.84

Comments: The measuring temperature is below T_g.

Polymer (B): **poly(DL-lactic acid)** **2014MAH**
Characterization: M_n/g.mol^{-1} = 72000, M_w/g.mol^{-1} = 136000, 1.4 % D-content, Nature Works
Solvent (A): **carbon dioxide** **CO_2** **124-38-9**

Type of data: gas solubility

T/K = 453.15

P/MPa	6.89	10.34	13.79	17.24	20.68
c_A/(g CO_2/g polymer)	0.0444	0.0667	0.0883	0.1098	0.1314

Comments: Concentrations of CO_2 are corrected for polymer swelling.

T/K = 463.15

P/MPa	6.89	10.34	13.79	17.24	20.68
c_A/(g CO_2/g polymer)	0.0426	0.0635	0.0842	0.1041	0.1248

Comments: Concentrations of CO_2 are corrected for polymer swelling.

T/K = 473.15

P/MPa	6.89	10.34	13.79	17.24	20.68
c_A/(g CO_2/g polymer)	0.0387	0.0587	0.0767	0.0936	0.1103

Comments: Concentrations of CO_2 are corrected for polymer swelling.

Polymer (B): **poly(DL-lactic acid)** **2014MAH**
Characterization: M_n/g.mol^{-1} = 85000, M_w/g.mol^{-1} = 161000, 4.2 % D-content, NatureWorks™ LLC
Solvent (A): **carbon dioxide** **CO_2** **124-38-9**

Type of data: gas solubility

T/K = 453.15

P/MPa	6.89	10.34	13.79	17.24	20.68
c_A/(g CO_2/g polymer)	0.0442	0.0666	0.0890	0.1107	0.1306

Comments: Concentrations of CO_2 are corrected for polymer swelling.

T/K = 463.15

P/MPa	6.89	10.34	13.79	17.24	20.68
c_A/(g CO_2/g polymer)	0.0407	0.0623	0.0824	0.1019	0.1215

Comments: Concentrations of CO_2 are corrected for polymer swelling.

T/K = 473.15

P/MPa	6.89	10.34	13.79	17.24	20.68
c_A/(g CO_2/g polymer)	0.0384	0.0579	0.0769	0.0953	0.1123

Comments: Concentrations of CO_2 are corrected for polymer swelling.

Polymer (B): **poly(DL-lactic acid)** **2014MAH**
Characterization: M_n/g.mol^{-1} = 100000, M_w/g.mol^{-1} = 190000, 1.2 % D-content, NatureWorks™ LLC
Solvent (A): **carbon dioxide** **CO_2** **124-38-9**

Type of data: gas solubility

T/K = 453.15

P/MPa	6.89	10.34	13.79	17.24	20.68
c_A/(g CO_2/g polymer)	0.0449	0.0670	0.0885	0.1094	0.1292

Comments: Concentrations of CO_2 are corrected for polymer swelling.

T/K = 463.15

P/MPa	6.89	10.34	13.79	17.24	20.68
c_A/(g CO_2/g polymer)	0.0419	0.0626	0.0831	0.1033	0.1232

Comments: Concentrations of CO_2 are corrected for polymer swelling.

T/K = 473.15

P/MPa	6.89	10.34	13.79	17.24	20.68
c_A/(g CO_2/g polymer)	0.0375	0.0574	0.0768	0.0953	0.1124

Comments: Concentrations of CO_2 are corrected for polymer swelling.

Polymer (B): **poly(DL-lactic acid)** **2006LIG1**
Characterization: M_n/g.mol^{-1} = 660000, PLA 3001D, NatureWorks™ LLC
Solvent (A): **carbon dioxide** **CO_2** **124-38-9**

Type of data: gas solubility

T/K = 453.15

P/MPa	3.47	6.94	10.44	13.93	17.42	20.92	24.42
c_A/(g CO_2/g polymer)	0.022	0.045	0.070	0.095	0.120	0.15	0.17

P/MPa	27.89
c_A/(g CO_2/g polymer)	0.20

Comments: Solubilities were averaged from the data in the original source.

T/K = 473.15

P/MPa	3.50	6.98	10.44	13.92	17.41	20.89	24.38
c_A/(g CO_2/g polymer)	0.020	0.039	0.059	0.080	0.10	0.12	0.145

P/MPa	27.86
c_A/(g CO_2/g polymer)	0.17

Comments: Solubilities were averaged from the data in the original source.

Polymer (B): **poly(L-lactic acid)** **2008AI2**
Characterization: M_w/g.mol^{-1} = 42000, Boehringer Ingelheim, Germany
Solvent (A): **carbon dioxide** **CO_2** **124-38-9**

Type of data: gas solubility

T/K = 308.15

P/MPa	9.62	12.12	15.14	18.53	20.92	22.92	25.55	28.67	30.29
w_A	0.2851	0.3575	0.3874	0.4062	0.4139	0.4199	0.4244	0.4264	0.4301

T/K = 313.15

P/MPa	10.44	12.29	15.69	18.46	21.06	22.98	25.75	28.28	31.46
w_A	0.2450	0.3258	0.3698	0.3864	0.3985	0.4051	0.4125	0.4174	0.4228

T/K = 323.15

P/MPa	10.73	12.66	15.65	17.84	20.18	22.46	25.31	27.50	29.90
w_A	0.1652	0.2534	0.3294	0.3538	0.3718	0.3840	0.3971	0.4044	0.4116

Polymer (B): **poly(L-lactic acid)** **2004OLI**
Characterization: T_g/K = 326.0, L:D ratio = 98:02, 10% crystallinity, Cargill–Dow Polymers
Solvent (A): **carbon dioxide** CO_2 **124-38-9**

Type of data: gas solubility

T/K = 283.1

P/bar	0.123	0.278	0.402	0.472	0.628	0.735	0.921
c_A/[cm^3 (STP)/cm^3 polymer]	0.98	1.95	2.59	2.98	3.73	4.10	4.95
c_A/[cm^3 (STP)/cm^3 polymer]*)	1.08	2.15	2.85	3.28	4.10	4.51	5.45

P/bar	1.007	1.047
c_A/[cm^3 (STP)/cm^3 polymer]	5.33	5.34
c_A/[cm^3 (STP)/cm^3 polymer]*)	5.86	5.88

Comments: *) Values are calculated per cm^3 amorphous polymer considering a 10% crystallinity. The measuring temperature is below T_g.

T/K = 293.2

P/bar	0.130	0.280	0.487	0.681	0.922	1.065
c_A/[cm^3 (STP)/cm^3 polymer]	0.71	1.39	2.23	2.92	3.73	4.20
c_A/[cm^3 (STP)/cm^3 polymer]*)	0.78	1.53	2.45	3.21	4.11	4.62

Comments: *) Values are calculated per cm^3 amorphous polymer considering a 10% crystallinity. The measuring temperature is below T_g.

T/K = 303.2

P/bar	0.136	0.283	0.468	0.638	0.814	1.098
c_A/[cm^3 (STP)/cm^3 polymer]	0.49	0.98	1.54	2.03	2.51	3.23
c_A/[cm^3 (STP)/cm^3 polymer]*)	0.54	1.08	1.70	2.23	2.76	3.56

Comments: *) Values are calculated per cm^3 amorphous polymer considering a 10% crystallinity. The measuring temperature is below T_g.

T/K = 313.4

P/bar	0.145	0.297	0.459	0.631	0.798	1.082
c_A/[cm^3 (STP)/cm^3 polymer]	0.36	0.72	1.09	1.47	1.84	2.43
c_A/[cm^3 (STP)/cm^3 polymer]*)	0.40	0.79	1.20	1.62	2.02	2.67

Comments: *) Values are calculated per cm^3 amorphous polymer considering a 10% crystallinity. The measuring temperature is below T_g.

Polymer (B): **poly(L-lactic acid)** **2006OL1**
Characterization: T_g/K = 326.0, L:D ratio = 98:02, 10% crystallinity, Cargill–Dow Polymers
Solvent (A): **ethene** C_2H_4 **74-85-1**

Type of data: gas solubility

T/K = 283.0

P/bar	0.158	0.327	0.472	0.636	0.816	1.026	1.092
c_A/[cm^3 (STP)/cm^3 polymer]	0.42	0.77	1.03	1.33	1.64	2.01	2.08
c_A/[cm^3 (STP)/cm^3 polymer]*$^)$	0.46	0.85	1.14	1.47	1.81	2.21	2.29

Comments: *$^)$ Values are calculated per cm^3 amorphous polymer considering a 10% crystallinity. The measuring temperature is below T_g.

T/K = 294.3

P/bar	0.129	0.275	0.435	0.579	0.763	1.044
c_A/[cm^3 (STP)/cm^3 polymer]	0.23	0.47	0.74	1.02	1.30	1.69
c_A/[cm^3 (STP)/cm^3 polymer]*$^)$	0.25	0.52	0.82	1.12	1.43	1.85

Comments: *$^)$ Values are calculated per cm^3 amorphous polymer considering a 10% crystallinity. The measuring temperature is below T_g.

T/K = 303.3

P/bar	0.122	0.317	0.511	0.722	1.025
c_A/[cm^3 (STP)/cm^3 polymer]	0.17	0.47	0.73	1.00	1.32
c_A/[cm^3 (STP)/cm^3 polymer]*$^)$	0.19	0.52	0.80	1.10	1.46

Comments: *$^)$ Values are calculated per cm^3 amorphous polymer considering a 10% crystallinity. The measuring temperature is below T_g.

T/K = 313.1

P/bar	0.315	0.469	0.640	0.840	1.070
c_A/[cm^3 (STP)/cm^3 polymer]	0.40	0.59	0.80	1.03	1.30
c_A/[cm^3 (STP)/cm^3 polymer]*$^)$	0.44	0.65	0.88	1.14	1.43

Comments: *$^)$ Values are calculated per cm^3 amorphous polymer considering a 10% crystallinity. The measuring temperature is below T_g.

Polymer (B): **poly(DL-lactic acid)** **2004OLI**
Characterization: M_w/g.mol^{-1} = 102800, T_g/K = 329.9, L:D ratio = 80:20, Cargill–Dow Polymers
Solvent (A): **nitrogen** N_2 **7727-37-9**

Type of data: gas solubility

T/K = 283.2

P/bar	0.107	0.138	0.322	0.591	0.685	0.863	1.022
c_A/[cm^3 (STP)/cm^3 polymer]	0.15	0.24	0.39	0.86	0.99	1.18	1.32

continued

continued

P/bar	1.258
c_A/[cm^3 (STP)/cm^3 polymer]	1.54

Comments: The measuring temperature is below T_g.

T/K = 293.2

P/bar	0.116	0.302	0.478	0.530	0.755	0.912	1.046
c_A/[cm^3 (STP)/cm^3 polymer]	0.08	0.21	0.32	0.32	0.46	0.55	0.68

Comments: The measuring temperature is below T_g.

T/K = 303.2

P/bar	0.077	0.173	0.302	0.398	0.501	0.696	0.904
c_A/[cm^3 (STP)/cm^3 polymer]	0.06	0.11	0.14	0.16	0.20	0.245	0.32

P/bar	1.026
c_A/[cm^3 (STP)/cm^3 polymer]	0.36

Comments: The measuring temperature is below T_g.

T/K = 313.2

P/bar	0.190	0.273	0.413	0.605	0.717	0.810	1.022
c_A/[cm^3 (STP)/cm^3 polymer]	0.05	0.06	0.10	0.14	0.17	0.19	0.23

Comments: The measuring temperature is below T_g.

Polymer (B): **poly(DL-lactic acid)** **2006LIG1**
Characterization: M_n/g.mol^{-1} = 660000, PLA 3001D, NatureWorks™ LLC
Solvent (A): **nitrogen** N_2 **7727-37-9**

Type of data: gas solubility

T/K = 453.15

P/MPa	3.48	6.97	10.44	13.92	17.41	20.89	24.38
c_A/(g N_2/g polymer)	0.0021	0.0043	0.0065	0.0086	0.0107	0.0128	0.0148

P/MPa	27.87
c_A/(g N_2/g polymer)	0.0170

Comments: Solubilities were averaged from the data in the original source.

T/K = 473.15

P/MPa	3.45	6.94	10.45	13.93	17.41	20.90	24.39
c_A/(g N_2/g polymer)	0.0022	0.0045	0.0066	0.0087	0.0108	0.0129	0.0150

P/MPa	27.88
c_A/(g N_2/g polymer)	0.0171

Comments: Solubilities were averaged from the data in the original source.

Polymer (B): **poly(DL-lactic acid)** **2004OLI**
Characterization: M_w/g.mol^{-1} = 102800, T_g/K = 329.9, L:D ratio = 80:20, Cargill–Dow Polymers
Solvent (A): **oxygen** **O_2** **7782-44-7**

Type of data: gas solubility

T/K = 293.2

P/bar	0.110	0.188	0.188	0.312	0.454	0.468	0.610
c_A/[cm^3 (STP)/cm^3 polymer]	0.50	0.62	0.58	0.72	0.88	0.89	1.04

P/bar	0.765	0.925	0.995
c_A/[cm^3 (STP)/cm^3 polymer]	1.25	1.48	1.57

Comments: The measuring temperature is below T_g.

T/K = 303.2

P/bar	0.103	0.106	0.214	0.351	0.352	0.396	0.473
c_A/[cm^3 (STP)/cm^3 polymer]	0.19	0.20	0.27	0.38	0.32	0.44	0.49

P/bar	0.663	0.889	0.987
c_A/[cm^3 (STP)/cm^3 polymer]	0.76	1.04	1.05

Comments: The measuring temperature is below T_g.

T/K = 313.2

P/bar	0.030	0.063	0.098	0.184	0.270	0.337	0.404
c_A/[cm^3 (STP)/cm^3 polymer]	0.06	0.10	0.12	0.21	0.23	0.30	0.26

P/bar	0.488	0.547	0.643	0.711	0.793	0.799	0.944
c_A/[cm^3 (STP)/cm^3 polymer]	0.43	0.39	0.51	0.50	0.55	0.54	0.58

P/bar	0.955
c_A/[cm^3 (STP)/cm^3 polymer]	0.66

Comments: The measuring temperature is below T_g.

Polymer (B): **poly(DL-lactic acid-*co*-glycolic acid)** **2008PIN**
Characterization: M_n/g.mol^{-1} = 45300, M_w/g.mol^{-1} = 77000, T_g/K = 321.75, 15 mol% glycolide, ρ = 1.293 g/cm^3
Solvent (A): **carbon dioxide** **CO_2** **124-38-9**

Type of data: gas solubility

T/K = 308.15

P/bar	22	52	78	100	121	151	200
c_A/(g CO_2/g polymer)	0.068	0.190	0.343	0.407	0.431	0.459	0.499

Polymer (B): **poly(DL-lactic acid-*co*-glycolic acid)** **2008KAS**
Characterization: M_w/g.mol^{-1} = 50000-75000, 15 mol% glycolide, Aldrich Chem. Co., Inc., Milwaukee, WI
Solvent (A): **carbon dioxide** **CO_2** **124-38-9**

Type of data: gas solubility

T/K = 313.15

P/MPa	1.448	2.427	3.786	4.482	5.055	5.662	6.600	7.700
w_A	0.0236	0.0614	0.1138	0.1407	0.1628	0.1862	0.2224	0.2649

T/K = 333.15

P/MPa	2.248	3.136	3.931	4.669	5.343	6.137	7.800
w_A	0.0348	0.0586	0.0799	0.0997	0.1178	0.1391	0.1836

T/K = 344.15

P/MPa	2.317	3.448	4.669	5.593	6.496	8.103
w_A	0.0344	0.0627	0.0932	0.1163	0.1389	0.1790

Polymer (B): **poly(DL-lactic acid-*co*-glycolic acid)** **2008PIN**
Characterization: M_n/g.mol^{-1} = 41000, M_w/g.mol^{-1} = 72000, T_g/K = 323.55, 25 mol% glycolide, ρ = 1.353 g/cm^3
Solvent (A): **carbon dioxide** **CO_2** **124-38-9**

Type of data: gas solubility

T/K = 308.15

P/bar	24	52	77	100	120	151	200
c_A/(g CO_2/g polymer)	0.070	0.176	0.314	0.393	0.420	0.455	0.518

Polymer (B): **poly(DL-lactic acid-*co*-glycolic acid)** **2008PIN**
Characterization: M_n/g.mol^{-1} = 30800, M_w/g.mol^{-1} = 52000, T_g/K = 322.25, 35 mol% glycolide, ρ = 1.346 g/cm^3
Solvent (A): **carbon dioxide** **CO_2** **124-38-9**

Type of data: gas solubility

T/K = 308.15

P/bar	21	50	74	101	120	152	201
c_A/(g CO_2/g polymer)	0.056	0.160	0.268	0.336	0.357	0.377	0.397

Polymer (B): **poly(DL-lactic acid-*co*-glycolic acid)** **2008KAS**
Characterization: M_w/g.mol^{-1} = 40000-75000, 35 mol% glycolide, Sigma Chemical Co., Inc., St. Louis, MO
Solvent (A): **carbon dioxide** **CO_2** **124-38-9**

Type of data: gas solubility

T/K = 313.15

P/MPa	1.380	2.240	3.130	3.896	4.848	5.696	6.351	7.900
w_A	0.0178	0.0487	0.0804	0.1074	0.1413	0.1715	0.1948	0.2499

T/K = 333.15

P/MPa	1.213	1.848	2.069	2.503	3.172	4.241	4.813	5.627	6.500
w_A	0.0054	0.0175	0.0217	0.0299	0.0426	0.0629	0.0738	0.0893	0.1059

P/MPa	7.900
w_A	0.1325

T/K = 344.15

P/MPa	2.393	3.206	3.793	4.703	5.593	6.206	6.862	7.689
w_A	0.0235	0.0348	0.0431	0.0558	0.0683	0.0768	0.0860	0.0976

Polymer (B): **poly(DL-lactic acid-*co*-glycolic acid)** **2005ELV**
Characterization: M_n/g.mol^{-1} = 15300, M_w/g.mol^{-1} = 26000, T_g/K = 317.61, 46 mol% glycolide, Alkermes Inc., Cincinnati, OH
Solvent (A): **carbon dioxide** **CO_2** **124-38-9**

Type of data: gas solubility

T/K = 313.15

P/MPa	1.0	2.0	2.4	3.0	3.9
w_A	0.0169	0.0368	0.0442	0.0626	0.0776

Polymer (B): **poly(DL-lactic acid-*co*-glycolic acid)** **2005ELV**
Characterization: M_n/g.mol^{-1} = 15000, M_w/g.mol^{-1} = 27000, T_g/K = 314.14, 47 mol% glycolide, Alkermes Inc., Cincinnati, OH
Solvent (A): **carbon dioxide** **CO_2** **124-38-9**

Type of data: gas solubility

T/K = 313.15

P/MPa	0.6	2.0	3.0
w_A	0.0109	0.0334	0.0508

Polymer (B):	**poly(DL-lactic acid-*co*-glycolic acid)**	**2005ELV**
Characterization:	M_n/g.mol^{-1} = 26100, M_w/g.mol^{-1} = 47000, T_g/K = 316.75, 47 mol% glycolide, Alkermes Inc., Cincinnati, OH	
Solvent (A):	**carbon dioxide** CO_2	**124-38-9**

Type of data: gas solubility

T/K = 313.15

P/MPa	1.0	2.0	3.0
w_A	0.0143	0.0306	0.0502

Polymer (B):	**poly(DL-lactic acid-*co*-glycolic acid)**	**2008PIN**
Characterization:	M_n/g.mol^{-1} = 33300, M_w/g.mol^{-1} = 53000, T_g/K = 320.15, 50 mol% glycolide, ρ = 1.407 g/cm^3	
Solvent (A):	**carbon dioxide** CO_2	**124-38-9**

Type of data: gas solubility

T/K = 308.15

P/bar	22	53	80	103	123	156	200
c_A/(g CO_2/g polymer)	0.054	0.153	0.249	0.287	0.309	0.330	0.348

Polymer (B):	**poly(DL-lactic acid-*co*-glycolic acid)**	**2008AI2**
Characterization:	M_w/g.mol^{-1} = 70000, 50 mol% glycolide, Boehringer Ingelheim, Germany	
Solvent (A):	**carbon dioxide** CO_2	**124-38-9**

Type of data: gas solubility

T/K = 308.15

P/MPa	10.14	11.92	15.31	18.40	21.44	23.11	26.21	29.49	31.47
w_A	0.1464	0.1968	0.2380	0.2577	0.2754	0.2794	0.2880	0.2929	0.2963

T/K = 313.15

P/MPa	10.22	12.09	15.30	18.19	20.61	22.99	25.59	28.46	30.70
w_A	0.0945	0.1666	0.2188	0.2407	0.2555	0.2636	0.2730	0.2823	0.2876

T/K = 323.15

P/MPa	12.43	15.06	19.30	21.02	22.97	24.07	26.44	28.10	30.29
w_A	0.0903	0.1793	0.2236	0.1339	0.2445	0.2492	0.2588	0.2638	0.2702

Polymer (B): **poly(DL-lactic acid-*co*-glycolic acid)** **2005ELV**
Characterization: M_w/g.mol^{-1} = 14000, T_g/K = 309.45,
52 mol% glycolide, Alkermes Inc., Cincinnati, OH
Solvent (A): **carbon dioxide** **CO_2** **124-38-9**

Type of data: gas solubility

T/K = 313.15

P/MPa	1.0	2.0	3.0
w_A	0.0169	0.0288	0.0462

Polymer (B): **poly{[2-(methacryloyloxy)ethyl]trimethylammonium tetrafluoroborate}** **2009TAN**
Characterization: P[MATMA][BF4], density =1.18 g/cm^3, T_g/K = 489,
synthesized in the laboratory
Solvent (A): **carbon dioxide** **CO_2** **124-38-9**

Type of data: gas solubility

T/K = 295.15

P/bar	0.10	0.20	0.40	0.60	0.80
c_A/(mg CO_2/g polymer)	2.677	4.482	7.330	9.592	11.501

Polymer (B): **poly(methyl methacrylate)** **2011RUI**
Characterization: M_w/g.mol^{-1} = 50000, MSC Industrial Supply Co.
Solvent (A): **carbon dioxide** **CO_2** **124-38-9**

Type of data: gas solubility

T/K	323.45	323.65	323.55	323.55	322.95	323.55	323.65
P/bar	22.7	54.0	95.5	115.4	68.1	147.7	175.6
c_A/(g CO_2/g polymer)	0.02183	0.10817	0.20812	0.27264	0.15237	0.36261	0.42297
T/K	323.65	323.35	323.25	323.45	323.75	323.75	323.45
P/bar	175.9	189.1	51.1	95.1	115.6	147.6	173.1
c_A/(g CO_2/g polymer)	0.42123	0.45332	0.12001	0.21459	0.29112	0.40809	0.48747
T/K	323.35						
P/bar	188.9						
c_A/(g CO_2/g polymer)	0.53306						

Polymer (B): **poly(methyl methacrylate)** **2006GAL**
Characterization: M_w/g.mol^{-1} = 350000, T_g/K = 395
Solvent (A): **carbon dioxide** **CO_2** **124-38-9**

Type of data: gas solubility

T/K = 323.15

P/MPa	6.6	9.5	11.6	14.1	15.7	17.1
c_A/(g CO_2/g polymer)	0.106	0.166	0.156	0.230	0.239	0.244

Polymer (B):	**poly(methyl methacrylate)**	**2005RAJ**
Characterization:	M_n/g.mol^{-1} = 600000, T_g/K = 378, ρ = 1.17 g/cm^3	
Solvent (A):	**carbon dioxide** CO_2	**124-38-9**

Type of data: gas solubility

T/K = 323.15

P/bar	54	78	95	116	128	135	150
c_A/(g CO_2/g polymer)	0.0999	0.1377	0.1650	0.1886	0.1981	0.2065	0.2155
P/bar	176	189					
c_A/(g CO_2/g polymer)	0.2256	0.2307					

T/K = 338.15

P/bar	64	90	116	136	153	175	193
c_A/(g CO_2/g polymer)	0.0799	0.1161	0.1501	0.1685	0.1819	0.1969	0.2059
P/bar	207	237					
c_A/(g CO_2/g polymer)	0.2138	0.2277					

T/K = 353.15

P/bar	50	103	108	162	183	207	238
c_A/(g CO_2/g polymer)	0.0479	0.1082	0.1146	0.1663	0.1786	0.1916	0.2049

Polymer (B):	**poly(perfluoropropylene oxide) ether**	**2009MIL**
Characterization:	M_n/g.mol^{-1} = 960, Krytox GPL 100	
Solvent (A):	**carbon dioxide** CO_2	**124-38-9**

Type of data: vapor-liquid equilibrium

T/K = 298.15

w_B	0.00	0.05	0.10	0.15	0.20	0.25	0.30	0.35	0.40
P/MPa	6.43	6.26	6.15	6.11	6.03	6.10	6.00	5.92	5.90
w_B	0.45	0.50	0.55	0.60	0.65	0.70	0.75	0.80	0.85
P/MPa	5.87	5.76	5.64	5.51	5.41	5.21	4.93	4.45	3.80
w_B	0.90	0.95							
P/MPa	2.80	1.60							

Polymer (B):	**polypropylene (isotactic)**	**2004SAT**
Characterization:	M_n/g.mol^{-1} = 79500, M_w/g.mol^{-1} = 350000, Idemitsu Petrochemical Co., Ltd., Ichihara, Japan	
Solvent (A):	**n-butane** C_4H_{10}	**106-97-8**

Type of data: gas solubility

T/K = 438.15

P/MPa	0.3879	0.795	1.085	1.291	1.617	1.978	2.074
c_A/(g C_4H_{10}/g polymer)	0.0227	0.0501	0.0698	0.0880	0.116	0.149	0.159
w_A	0.0222	0.0477	0.0652	0.0809	0.1039	0.1297	0.1372

continued

continued

T/K = 453.15

P/MPa	0.3588	0.4412	0.912	0.933	1.248	1.490	1.846
c_A/(g C_4H_{10}/g polymer)	0.0175	0.0209	0.0484	0.0493	0.0668	0.0849	0.110
w_A	0.0172	0.0205	0.0462	0.0470	0.0626	0.0783	0.0991
P/MPa	2.274	2.411					
c_A/(g C_4H_{10}/g polymer)	0.142	0.153					
w_A	0.1243	0.1327					

T/K = 468.15

P/MPa	0.3972	0.4955	1.030	1.038	1.414	1.690	2.074
c_A/(g C_4H_{10}/g polymer)	0.0165	0.0208	0.0469	0.0469	0.0648	0.0822	0.104
w_A	0.0162	0.0204	0.0448	0.0448	0.0609	0.0760	0.0942
P/MPa	2.574	2.753					
c_A/(g C_4H_{10}/g polymer)	0.132	0.148					
w_A	0.1166	0.1289					

T/K = 483.15

P/MPa	0.4372	0.5494	1.146	1.150	1.582	1.893	2.302
c_A/(g C_4H_{10}/g polymer)	0.0166	0.0211	0.0454	0.0455	0.0641	0.0781	0.0971
w_A	0.0163	0.0207	0.0434	0.0435	0.0602	0.0724	0.0885
P/MPa	2.873	3.103					
c_A/(g C_4H_{10}/g polymer)	0.125	0.143					
w_A	0.1111	0.1251					

Polymer (B): **polypropylene (isotactic)** **2009LID**

Characterization: M_n/g.mol^{-1} = 42000, M_w/g.mol^{-1} = 197000,
T_m/K = 441.75, crystallinity = 47.1%,
Shanghai Petrochemical Company, China

Solvent (A): **carbon dioxide** CO_2 **124-38-9**

Type of data: gas solubility

T/K = 373.15

P/MPa	2.93	3.81	4.89	5.52	6.31	7.44	8.48
c_A/(g CO_2/kg polymer)	26.18	35.85	46.82	56.74	3.31	74.40	84.89
P/MPa	9.08	9.57	10.06	10.73	11.76	12.55	
c_A/(g CO_2/kg polymer)	92.61	98.42	106.55	113.47	125.03	136.99	

Comments: Gas solubility with swelling correction in the amorphous regions of the solid-state iPP.

continued

continued

T/K = 398.15

P/MPa	2.95	4.12	5.66	5.98	6.92	8.10	10.07
c_A/(g CO_2/kg polymer)	23.73	35.30	47.99	51.88	59.94	70.53	90.97
P/MPa	10.60	12.19	12.23	13.41	13.90		
c_A/(g CO_2/kg polymer)	94.43	112.84	113.81	126.28	133.06		

Comments: Gas solubility with swelling correction in the amorphous regions of the solid-state iPP.

T/K = 423.15

P/MPa	4.39	5.30	6.35	7.46	8.12	8.50	9.27
c_A/(g CO_2/kg polymer)	31.16	37.68	47.47	55.23	62.08	63.67	70.58
P/MPa	10.75	12.52	13.62	14.96			
c_A/(g CO_2/kg polymer)	83.70	98.94	116.48	129.81			

Comments: Gas solubility with swelling correction in the amorphous regions of the solid-state iPP.

Polymer (B): **polypropylene (isotactic)** **2012CHE**
Characterization: M_w/g.mol^{-1} = 197000, M_w/M_n = 5.1, T_m/K = 441, crystallinity = 47.8%, Shanghai Petrochemical Co., China
Solvent (A): **carbon dioxide** **CO_2** **124-38-9**

Type of data: gas solubility

T/K = 473.15

P/MPa	5	7	10	13	15	17	20
exp. swelling/(cm^3/cm^3 melt)	0.0548	0.0697	0.0900	0.1077	0.1180	0.1271	0.1384
c_A/(g CO_2/g polymer, corr.)	0.0333	0.0528	0.0736	0.1008	0.1168	0.1363	0.1562
P/MPa	22	25	27	30			
exp. swelling/(cm^3/cm^3 melt)	0.1444	0.1509	0.1536	0.1551			
c_A/(g CO_2/g polymer, corr.)	0.1691	0.1865	0.1961	0.2096			

Comments: Gas solubility with swelling correction.

T/K = 493.15

P/MPa	5	7	10	13	15	17	20
exp. swelling/(cm^3/cm^3 melt)	0.0399	0.0548	0.0761	0.0946	0.1050	0.1138	0.1237
c_A/(g CO_2/g polymer, corr.)	0.0352	0.0526	0.0722	0.0915	0.1078	0.1210	0.1362
P/MPa	22	25	27	30			
exp. swelling/(cm^3/cm^3 melt)	0.1285	0.1337	0.1366	0.1417			
c_A/(g CO_2/g polymer, corr.)	0.1456	0.1601	0.1712	0.1876			

Comments: Gas solubility with swelling correction.

Polymer (B): **polypropylene (isotactic)** **2012CHE**
Characterization: M_w/g.mol^{-1} = 231800, M_w/M_n = 14.8, T_m/K = 434, crystallinity = 45.5%, RS1684, Basell Co.
Solvent (A): **carbon dioxide** **CO_2** **124-38-9**

Type of data: gas solubility

T/K = 473.15

P/MPa	5	7	10	13	15	17	20
exp. swelling/(cm^3/cm^3 melt)	0.0349	0.0486	0.0702	0.0911	0.1036	0.1145	0.1273
c_A/(g CO_2/g polymer, corr.)	0.0311	0.0445	0.0645	0.0884	0.1027	0.1160	0.1341

P/MPa	22	25	27	30
exp. swelling/(cm^3/cm^3 melt)	0.1332	0.1382	0.1392	0.1379
c_A/(g CO_2/g polymer, corr.)	0.1450	0.1629	0.1741	0.1868

Comments: Gas solubility with swelling correction.

T/K = 493.15

P/MPa	5	7	10	13	15	17	20
exp. swelling/(cm^3/cm^3 melt)	0.0237	0.0362	0.0575	0.0786	0.0913	0.1023	0.1147
c_A/(g CO_2/g polymer, corr.)	0.0253	0.0402	0.0602	0.0753	0.0893	0.1018	0.1145

P/MPa	22	25	27	30
exp. swelling/(cm^3/cm^3 melt)	0.1203	0.1251	0.1266	0.1279
c_A/(g CO_2/g polymer, corr.)	0.1258	0.1417	0.1525	0.1650

Comments: Gas solubility with swelling correction.

Polymer (B): **polypropylene (isotactic)** **2007LEI**
Characterization: M_w/g.mol^{-1} = 985000, T_g/K = 275.6, T_m/K = 432.2, crystallinity = 62 wt%
Solvent (A): **carbon dioxide** **CO_2** **124-38-9**

Type of data: gas solubility

T/K = 313.2

P/MPa	5.02	10.02
c_A/(g CO_2/g polymer)	0.0299	0.0542
w_A	0.0290	0.0514

Comments: Gas solubility with swelling correction, polymer in rubbery state below T_m.

T/K = 333.2

P/MPa	5.03	10.05
c_A/(g CO_2/g polymer)	0.0254	0.0475
w_A	0.0248	0.0453

Comments: Gas solubility with swelling correction, polymer in rubbery state below T_m.

continued

continued

T/K = 353.2

P/MPa	5.04	10.03
c_A/(g CO_2/g polymer)	0.0222	0.0430
w_A	0.0217	0.0412

Comments: Gas solubility with swelling correction, polymer in rubbery state below T_m.

T/K = 373.2

P/MPa	5.03	10.05
c_A/(g CO_2/g polymer)	0.0205	0.0395
w_A	0.0201	0.0380

Comments: Gas solubility with swelling correction, polymer in rubbery state below T_m.

T/K = 393.2

P/MPa	5.07	10.05
c_A/(g CO_2/g polymer)	0.0226	0.0438
w_A	0.0221	0.0420

Comments: Gas solubility with swelling correction, polymer in rubbery state below T_m.

T/K = 443.7

P/MPa	4.96	9.94	14.94	19.92	24.91
c_A/(g CO_2/g polymer)	0.0422	0.0851	0.1363	0.1938	0.2617
w_A	0.0405	0.0784	0.1200	0.1623	0.2074

Comments: Gas solubility with swelling correction, polymer in molten state.

T/K = 464.2

P/MPa	4.97	9.96	14.94	19.93	24.91
c_A/(g CO_2/g polymer)	0.0401	0.0807	0.1301	0.1850	0.2552
w_A	0.0386	0.0747	0.1151	0.1561	0.2033

Comments: Gas solubility with swelling correction, polymer in molten state.

T/K = 483.7

P/MPa	4.97	9.96	14.93	19.93	24.91
c_A/(g CO_2/g polymer)	0.0380	0.0761	0.1228	0.1752	0.2429
w_A	0.0366	0.0707	0.1094	0.1491	0.1954

Comments: Gas solubility with swelling correction, polymer in molten state.

Polymer (B): **polypropylene (isotactic)** **2004SAT**
Characterization: M_n/g.mol^{-1} = 79500, M_w/g.mol^{-1} = 350000,
Idemitsu Petrochemical Co., Ltd., Ichihara, Japan
Solvent (A): **2-methylpropane** C_4H_{10} **75-28-5**

Type of data: gas solubility

T/K = 438.15

P/MPa	0.3314	0.6017	0.993	1.442	1.743	2.054
c_A/(g C_4H_{10}/g polymer)	0.0141	0.0288	0.0483	0.0746	0.0952	0.115
w_A	0.0139	0.0280	0.0461	0.0694	0.0869	0.1031

T/K = 453.15

P/MPa	0.3692	0.4035	0.672	1.115	1.628	1.983	2.347
c_A/(g C_4H_{10}/g polymer)	0.0134	0.0156	0.0277	0.0468	0.0718	0.0924	0.111
w_A	0.0132	0.0154	0.0270	0.0447	0.0670	0.0846	0.0999

T/K = 468.15

P/MPa	0.4076	0.4489	0.743	1.242	1.814	2.224	2.643
c_A/(g C_4H_{10}/g polymer)	0.0129	0.0151	0.0266	0.0452	0.0692	0.0900	0.108
w_A	0.0127	0.0149	0.0259	0.0432	0.0647	0.0826	0.0975

T/K = 483.15

P/MPa	0.4434	0.4940	0.812	1.368	2.001	2.468	2.947
c_A/(g C_4H_{10}/g polymer)	0.0125	0.0147	0.0257	0.0439	0.0668	0.0878	0.105
w_A	0.0123	0.0145	0.0251	0.0421	0.0626	0.0807	0.0950

Polymer (B): **polypropylene (isotactic)** **2011YAO**
Characterization: M_n/g.mol^{-1} = 46000, M_w/g.mol^{-1} = 229200, 50,2% crystallinity,
Sinopec Yangzi Petrochemcial Company Co., Nanjin, China
Solvent (A): **propene** C_3H_6 **115-07-1**

Type of data: gas solubility

T/K = 348.15

P/MPa	1.54	1.68	1.82	1.97	2.09	2.21	2.41
c_A/(g/g amorph. polymer)	0.0598	0.0651	0.0690	0.0763	0.0862	0.0910	0.1070
P/MPa	2.53	2.71	2.98				
c_A/(g/g amorph. polymer)	0.1123	0.1251	0.1381				

T/K = 358.15

P/MPa	1.13	1.72	1.85	1.99	2.18	2.38	2.75
c_A/(g/g amorph. polymer)	0.0336	0.0534	0.0604	0.0695	0.0700	0.0854	0.0984
P/MPa	2.94	3.09	3.29	3.64	3.77		
c_A/(g/g amorph. polymer)	0.1090	0.1113	0.1264	0.1456	0.1590		

continued

continued

T/K = 368.15

P/MPa	1.19	1.41	1.73	1.91	2.42	2.78	3.11
c_A/(g/g amorph. polymer)	0.0275	0.0387	0.0497	0.0484	0.0750	0.0866	0.0932
P/MPa	3.48	3.83	4.43	4.68	4.87	5.05	5.27
c_A/(g/g amorph. polymer)	0.1193	0.1287	0.1462	0.1645	0.1753	0.1881	0.1911
P/MPa	5.44	5.58	5.72	6.01	6.22	6.43	6.62
c_A/(g/g amorph. polymer)	0.1941	0.1965	0.1982	0.2056	0.2087	0.2103	0.2128
P/MPa	6.81	7.16					
c_A/(g/g amorph. polymer)	0.2157	0.2183					

T/K = 383.15

P/MPa	0.94	1.12	1.90	2.38	2.75	3.06	3.56
c_A/(g/g amorph. polymer)	0.0184	0.0260	0.0401	0.0609	0.0677	0.0782	0.0981
P/MPa	3.58	4.17	4.42	5.35	5.50	5.79	6.51
c_A/(g/g amorph. polymer)	0.1020	0.1197	0.1283	0.1505	0.1641	0.1723	0.1841
P/MPa	6.53	6.95	7.41	8.18			
c_A/(g/g amorph. polymer)	0.1879	0.1967	0.2034	0.2061			

Polymer (B): **poly(propylene glycol)** **2004GUA**
Characterization: M_w/g.mol^{-1} = 2700, Aldrich Chem. Co., Inc., Milwaukee, WI
Solvent (A): **carbon dioxide** CO_2 **124-38-9**

Type of data: gas solubility and swelling

T/K = 298.15

P/bar	20	40	60
w_A	0.050	0.110	0.184
($\Delta V/V$)/%	11.2	21	41

T/K = 308.15

P/bar	20	40	60
w_A	0.035	0.086	0.118
($\Delta V/V$)/%	6.95	17	24.5

Polymer (B): **poly(propylene glycol) dimethyl ether** **2009MIL**
Characterization: M_n/g.mol^{-1} = 230, Polymer Source, Inc.
Solvent (A): **carbon dioxide** CO_2 **124-38-9**

Type of data: vapor-liquid equilibrium

T/K = 298.15

w_B	0.00	0.05	0.10	0.15	0.20	0.25	0.30	0.35	0.40
P/MPa	6.43	6.11	6.01	5.95	5.98	5.88	5.64	5.48	5.20

continued

continued

w_B	0.45	0.50	0.55	0.60	0.65	0.70	0.75	0.80	0.85
P/MPa	4.92	4.52	4.25	3.88	3.50	2.95	2.53	1.82	1.32

w_B	0.90	0.935
P/MPa	0.79	0.59

Polymer (B): **polystyrene** **2004SAT**
Characterization: M_n/g.mol^{-1} = 70000, M_w/g.mol^{-1} = 187000, T_g/K = 373.6, Mitsui Toatsu Chemicals, Inc., Yokohama, Japan
Solvent (A): **n-butane** C_4H_{10} **106-97-8**

Type of data: gas solubility

T/K = 348.15

P/MPa	0.1615	0.3308	0.342	0.425	0.561	0.673	0.703
c_A/(g C_4H_{10}/g polymer)	0.0139	0.0315	0.0328	0.0427	0.0673	0.0886	0.0960
w_A	0.0137	0.0305	0.0318	0.0410	0.0631	0.0814	0.0876

P/MPa	0.769
c_A/(g C_4H_{10}/g polymer)	0.113
w_A	0.1015

T/K = 373.15

P/MPa	0.2210	0.4665	0.480	0.592	0.815	1.004	1.051
c_A/(g C_4H_{10}/g polymer)	0.0125	0.0286	0.0296	0.0384	0.0606	0.0796	0.0859
w_A	0.0123	0.0278	0.0287	0.0370	0.0571	0.0737	0.0791

P/MPa	1.155
c_A/(g C_4H_{10}/g polymer)	0.103
w_A	0.0934

T/K = 423.15

P/MPa	0.3456	0.7591	0.774	0.944	1.374	1.744	1.827
c_A/(g C_4H_{10}/g polymer)	0.0101	0.0236	0.0241	0.0311	0.0485	0.0626	0.0668
w_A	0.0100	0.0231	0.0235	0.0302	0.0463	0.0589	0.0626

P/MPa	2.083
c_A/(g C_4H_{10}/g polymer)	0.0814
w_A	0.0753

T/K = 473.15

P/MPa	0.4650	1.0511	1.062	1.283	1.917	2.447	2.578
c_A/(g C_4H_{10}/g polymer)	0.00846	0.0200	0.0202	0.0261	0.0403	0.0520	0.0545
w_A	0.0084	0.0196	0.0198	0.0254	0.0387	0.0494	0.0517

P/MPa	3.009
c_A/(g C_4H_{10}/g polymer)	0.0671
w_A	0.0629

Polymer (B): **polystyrene** **2004OLI**
Characterization: M_w/g.mol^{-1} = 197000, T_g/K = 376.2, Polysciences, Inc., Warrington, PA
Solvent (A): **carbon dioxide** CO_2 **124-38-9**

Type of data: gas solubility

T/K = 303.2

P/bar	0.214	0.510	0.767	0.824	0.983
c_A/[cm^3 (STP)/cm^3 polymer]	0.46	1.04	1.53	1.63	1.95

T/K = 313.2

P/bar	0.361	0.711	0.769	0.968
c_A/[cm^3 (STP)/cm^3 polymer]	0.72	1.30	1.42	1.73

Polymer (B): **polystyrene** **2008LIG**
Characterization: M_w/g.mol^{-1} = 196000, T_g/K = 381, ρ = 1.048 g/cm^3, Styron PS865D, Dow Chemical Company
Solvent (A): **1,1-difluoroethane** $C_2H_4F_2$ **75-37-6**

Type of data: gas solubility

T/K = 403.15

P/MPa	1.37	2.81	4.20
c_A/(g $C_2H_4F_2$/g polymer)	0.0275	0.0621	0.1042

T/K = 423.15

P/MPa	1.46	2.79	4.18	5.60
c_A/(g $C_2H_4F_2$/g polymer)	0.0238	0.0513	0.0815	0.1061

T/K = 463.15

P/MPa	1.41	2.81	4.20	5.59
c_A/(g $C_2H_4F_2$/g polymer)	0.0165	0.0334	0.0507	0.0680

Polymer (B): **polystyrene** **2004SAT**
Characterization: M_n/g.mol^{-1} = 70000, M_w/g.mol^{-1} = 187000, T_g/K = 373.6, Mitsui Toatsu Chemicals, Inc., Yokohama, Japan
Solvent (A): **2-methylpropane** C_4H_{10} **75-28-5**

Type of data: gas solubility

T/K = 348.15

P/MPa	0.1380	0.2055	0.315	0.477	0.514	0.675	0.803
c_A/(g C_4H_{10}/g polymer)	0.00692	0.0113	0.0184	0.0305	0.0337	0.0469	0.0619
w_A	0.0069	0.0112	0.0181	0.0296	0.0326	0.0448	0.0583

continued

continued

P/MPa	0.875	0.895					
c_A/(g C_4H_{10}/g polymer)	0.0699	0.0737					
w_A	0.0653	0.0686					

T/K = 373.15

P/MPa	0.1750	0.2663	0.416	0.631	0.707	0.906	1.088
c_A/(g C_4H_{10}/g polymer)	0.00651	0.00952	0.0171	0.0247	0.0309	0.0424	0.0544
w_A	0.0065	0.0094	0.0168	0.0241	0.0300	0.0407	0.0516
P/MPa	1.254	1.263					
c_A/(g C_4H_{10}/g polymer)	0.0641	0.0651					
w_A	0.0602	0.0611					

T/K = 423.15

P/MPa	0.2680	0.3870	0.643	0.935	1.114	1.359	1.675
c_A/(g C_4H_{10}/g polymer)	0.00559	0.00802	0.0149	0.0216	0.0262	0.0322	0.0424
w_A	0.0056	0.0080	0.0147	0.0211	0.0255	0.0312	0.0407
P/MPa	2.000	2.082					
c_A/(g C_4H_{10}/g polymer)	0.0521	0.0541					
w_A	0.0495	0.0513					

T/K = 473.15

P/MPa	0.3615	0.5006	0.868	1.217	1.521	1.781	2.229
c_A/(g C_4H_{10}/g polymer)	0.00491	0.00668	0.0128	0.0169	0.0229	0.0268	0.0352
w_A	0.0049	0.0066	0.0126	0.0166	0.0224	0.0261	0.0340
P/MPa	2.705	2.928					
c_A/(g C_4H_{10}/g polymer)	0.0445	0.0476					
w_A	0.0426	0.0454					

Polymer (B): **polystyrene** **2012GRO**

Characterization: M_w/g.mol^{-1} = 190000, M_w/M_n = 2.7, T_g/K = 378, Lacqrene 1450N, ATOFINA, France

Solvent (A): **1,1,1,2,2-pentafluoroethane** **C_2HF_5** **354-33-6**

Type of data: gas solubility

T/K = 413.15

P/MPa	3.68	5.98	7.24	8.31
c_A/(g C_2HF_5/g polymer)	0.05972	0.17181	0.37763	0.37460

Polymer (B): **polystyrene** **2012GRO**
Characterization: M_w/g.mol^{-1} = 190000, M_w/M_n = 2.7, T_g/K = 378, Lacqrene 1450N, ATOFINA, France
Solvent (A): **1,1,1,2-tetrafluoroethane** **$C_2H_2F_4$** **811-97-2**

Type of data: gas solubility

T/K = 385.34

P/MPa	2.99	4.39	7.18	9.04	11.89	15.48
c_A/(g $C_2H_2F_4$/g polymer)	0.05955	0.10420	0.12698	0.14411	0.14469	0.15454

T/K = 402.94

P/MPa	3.40	5.02	8.87	15.22	19.88
c_A/(g $C_2H_2F_4$/g polymer)	0.06155	0.08153	0.11543	0.14080	0.15847

Polymer (B): **polystyrene** **2008LIG**
Characterization: M_w/g.mol^{-1} = 196000, T_g/K = 381, ρ = 1.048 g/cm^3, Styron PS865D, Dow Chemical Company
Solvent (A): **1,1,1,2-tetrafluoroethane** **$C_2H_2F_4$** **811-97-2**

Type of data: gas solubility

T/K = 403.15

P/MPa	1.44	2.79	4.19
c_A/(g $C_2H_2F_4$/g polymer)	0.0249	0.0458	0.0656

T/K = 423.15

P/MPa	1.39	2.73	4.13	5.53
c_A/(g $C_2H_2F_4$/g polymer)	0.0193	0.0385	0.0571	0.0675

T/K = 463.15

P/MPa	1.36	2.81	4.19	5.59
c_A/(g $C_2H_2F_4$/g polymer)	0.0130	0.0261	0.0386	0.0502

Polymer (B): **polysulfone** **2004TA1**
Characterization: M_w/g.mol^{-1} = 26000, Aldrich Chem. Co., Inc., Milwaukee, WI
Solvent (A): **carbon dioxide** **CO_2** **124-38-9**

Type of data: gas solubility

T/K	313.15	313.15	313.15	323.15	323.15	323.15	333.15	333.15	333.15
P/MPa	20	30	40	20	30	40	20	30	40
w_A	0.106	0.118	0.120	0.097	0.106	0.115	0.091	0.097	0.106

Polymer (B): **poly[1-(4-vinylbenzyl)-4-methylimidazolium tetrafluoroborate]** **2009TAN**
Characterization: P[VBMI][BF4], density =1.38 g/cm^3, T_g/K = 383, synthesized in the laboratory
Solvent (A): **carbon dioxide** **CO_2** **124-38-9**

Type of data: gas solubility

T/K = 295.15

P/bar	0.10	0.20	0.33	0.40	0.60	0.79
c_A/(mg CO_2/g polymer)	0.811	1.527	2.335	2.747	3.807	4.689

Polymer (B): **poly[1-(4-vinylbenzyl)pyridinium tetrafluoroborate]** **2009TAN**
Characterization: P[VBP][BF4], density =1.34 g/cm^3, T_g/K = 428, synthesized in the laboratory
Solvent (A): **carbon dioxide** **CO_2** **124-38-9**

Type of data: gas solubility

T/K = 295.15

P/bar	0.10	0.21	0.30	0.42	0.59	0.78
c_A/(mg CO_2/g polymer)	1.214	2.359	3.224	4.165	5.408	6.603

Polymer (B): **poly[(4-vinylbenzyl)triethylammonium tetrafluoroborate]** **2009TAN**
Characterization: P[VBTEA][BF4], density =1.26 g/cm^3, T_g/K = 458, synthesized in the laboratory
Solvent (A): **carbon dioxide** **CO_2** **124-38-9**

Type of data: gas solubility

T/K = 295.15

P/bar	0.10	0.20	0.30	0.40	0.59	0.78
c_A/(mg CO_2/g polymer)	1.305	2.522	3.655	4.620	6.259	7.696

Polymer (B): **poly[(4-vinylbenzyl)triethylphosphonium tetrafluoroborate]** **2009TAN**
Characterization: P[VBTEP][BF4], density =1.27 g/cm^3, T_g/K = 460, synthesized in the laboratory
Solvent (A): **carbon dioxide** **CO_2** **124-38-9**

Type of data: gas solubility

T/K = 295.15

P/bar	0.10	0.20	0.30	0.40	0.60	0.79
c_A/(mg CO_2/g polymer)	0.983	1.904	2.733	3.509	4.877	6.051

Polymer (B): **poly[(4-vinylbenzyl)trimethylammonium hexafluorophosphate]** **2009TAN**
Characterization: P[VBTMA][PF6], density =1.03 g/cm^3, T_g/K = 528, synthesized in the laboratory
Solvent (A): **carbon dioxide** **CO_2** **124-38-9**

Type of data: gas solubility

T/K = 295.15

P/bar	0.10	0.20	0.30	0.40	0.60	0.79
c_A/(mg CO_2/g polymer)	2.543	4.797	6.586	7.987	10.572	12.634

Polymer (B): **poly[(4-vinylbenzyl)trimethylammonium tetrafluoroborate]** **2009TAN**
Characterization: P[VBTMA][BF4], density =1.04 g/cm^3, T_g/K = 508, synthesized in the laboratory
Solvent (A): **carbon dioxide** **CO_2** **124-38-9**

Type of data: gas solubility

T/K = 295.15

P/bar	0.19	0.40	0.61	0.79
c_A/(mg CO_2/g polymer)	5.814	9.698	12.875	15.200

Polymer (B): **poly[(4-vinylbenzyl)trimethylammonium trifluoromethanesulfonamide]** **2009TAN**
Characterization: P[VBTMA][Tf2N], density =1.29 g/cm^3, T_g/K = 347, synthesized in the laboratory
Solvent (A): **carbon dioxide** **CO_2** **124-38-9**

Type of data: gas solubility

T/K = 295.15

P/bar	0.12	0.20	0.30	0.40	0.60	0.79
c_A/(mg CO_2/g polymer)	0.478	0.794	1.235	1.642	2.390	3.106

Polymer (B): **poly(vinylidene fluoride)** **2006GAL**
Characterization: M_n/g.mol^{-1} = 120600, M_w/g.mol^{-1} = 298300, T_m/K = 446.4, crystallinity 55%, Solef 1010, Solvay Solexis
Solvent (A): **carbon dioxide** **CO_2** **124-38-9**

Type of data: gas solubility

T/K = 313.15

P/MPa	8.6	9.5	10.5	12.9	15.5	21.3	21.3
c_A/(g CO_2/g polymer)	0.074	0.066	0.126	0.112	0.239	0.270	0.286

continued

continued

P/MPa	23.8
c_A/(g CO_2/g polymer)	0.344

Comments: Concentrations of CO_2 are in g/g of amorphous PVDF considering a 55% crystallinity of the polymer.

T/K = 343.15

P/MPa	10.7	11.3	14.6	14.8	17.9	18.5	25.3
c_A/(g CO_2/g polymer)	0.038	0.051	0.078	0.108	0.207	0.206	0.294
P/MPa	26.4	30.2					
c_A/(g CO_2/g polymer)	0.318	0.354					

Comments: Concentrations of CO_2 are in g/g of amorphous PVDF considering a 55% crystallinity of the polymer.

Polymer (B): **starch** **2011MUL**
Characterization: native potato starch, AVEBE, The Netherlands
Solvent (A): **carbon dioxide** **CO_2** **124-38-9**

Type of data: gas solubility

T/K = 323.15

P/MPa	2.0	4.1	5.1	6.1	8.3	9.3	11.0
c_A/(mg CO_2/g polymer)	6.4	13.9	16.3	26.0	29.4	25.4	15.3
P/MPa	12.1						
c_A/(mg CO_2/g polymer)	7.1						

Comments: Concentrations of CO_2 are corrected for polymer swelling.

T/K = 393.15

P/MPa	2.0	3.1	4.1	5.1	6.3	7.7	8.3
c_A/(mg CO_2/g polymer)	9.4	11.7	13.7	15.8	17.7	20.1	21.2
P/MPa	9.0	10.6	11.1	12.0	13.1	14.3	
c_A/(mg CO_2/g polymer)	25.0	24.7	25.5	27.1	28.9	31.0	

Comments: Concentrations of CO_2 are corrected for polymer swelling.

Polymer (B): **starch acetate** **2011MUL**
Characterization: D.S. = 0.5, acetylated potato starch
Solvent (A): **carbon dioxide** **CO_2** **124-38-9**

Type of data: gas solubility

T/K = 323.15

P/MPa	2.1	3.1	4.0	5.0	6.3	7.5	8.2
c_A/(mg CO_2/g polymer)	7.0	11.8	16.7	21.5	38.1	43.0	45.4

continued

continued

P/MPa	9.1	10.1	13.2	15.1
c_A/(mg CO_2/g polymer)	45.9	49.1	46.0	43.2

Comments: Concentrations of CO_2 are corrected for polymer swelling.

T/K = 393.15

P/MPa	2.3	3.1	4.1	5.1	6.5	7.0	8.5
c_A/(mg CO_2/g polymer)	14.2	18.2	23.8	26.7	30.4	33.5	36.7
P/MPa	9.1	10.6	11.0	15.1	17.1	21.0	25.1
c_A/(mg CO_2/g polymer)	37.8	45.6	47.1	58.2	65.1	72.2	79.4

Comments: Concentrations of CO_2 are corrected for polymer swelling.

2.2. Table of binary systems where data were published only in graphical form as phase diagrams or related figures

Polymer (B)	Solvent (A)	Ref.
Cellulose acetate		
	ammonia	2006VOR
	ammonia	2009VOR
	nitrogen	2006VOR
Chitosan		
	carbon dioxide	2006WEI
Methylcellulose		
	carbon dioxide	2011PER
Natural rubber		
	carbon dioxide	2005NUN
	ethene	2005NUN
	methane	2005NUN
	oxygen	2005NUN
	propene	2005NUN
Nitrocellulose		
	carbon dioxide	2012SUN
Poly(acrylonitrile-*co*-butadiene-*co*-styrene)		
	carbon dioxide	2007NAW
Poly(alkylene glycol)		
	carbon dioxide	2008GAR
Polyamide		
	carbon dioxide	2005TA1

Polymer (B)	**Solvent (A)**	**Ref.**
Poly(benzoxazole-*co*-imide)		
	carbon dioixde	2014KIM
	hydrogen	2014KIM
	methane	2014KIM
	nitrogen	2014KIM
	oxygen	2014KIM
Poly(bisphenol A carbonate-*co*-4,40-(3,3,5-trimethylcyclohexylidene) diphenol carbonate)		
	nitrogen	2005LOP
	oxygen	2005LOP
Poly(*tert*-butylacetylene)		
	carbon dioxide	2005NAG
Poly(butylene adipate-*co*-terephthalate)		
	carbon dioxide	2010LIN
Poly(butylene succinate)		
	carbon dioxide	2006LIA
Poly[2-(1-butylimidazolium-3-yl)ethyl methacrylate tetrafluoroborate])		
	carbon dioxide	2005TA4
Poly(ε-caprolactone)		
	carbon dioxide	2005COT
	carbon dioxide	2006LIA
	carbon dioxide	2012CAR
	carbon dioxide	2012TAK
	carbon dioxide	2013KNA
	carbon dioxide	2013MAR
Polycarbonate		
	carbon dioxide	2004TA2
	carbon dioxide	2008ZHA
Poly(diallyldimethylammonium chloride)		
	carbon dioxide	2012BHA

Polymer (B)	Solvent (A)	Ref.
Poly(1H,1H-dihydroperfluorooctyl methacrylate-*ran*-2-tetrahydropyranyl methacrylate)		
	carbon dioxide	2007HUS
Poly(2,6-dimethyl-1,4-phenylene oxide)		
	carbon dioxide	2012HOR
	methanol	2013GAL
	propane	2013GAL
Poly(dimethylsiloxane)		
	1-bromo-1-chloro-2,2,2-trifluoroethane	2005PR1
	n-butane	2007RA1
	carbon dioxide	2005THU
	carbon dioxide	2013LIN
	hexafluoroethane	2010VEG
	methane	2007RA1
	methane	2013LIN
	nitrogen	2013LIN
	octafluoropropane	2005PR2
	ozone	2008DIN
	propane	2005PR1
	propane	2005PR2
	tetrafluoromethane	2010VEG
Polyester (hyperbranched)		
	carbon dioxide	2007ROL
	nitrogen	2007ROL
Polyester resin		
	carbon dioxide	2006NAL
Polyether (hyperbranched)		
	carbon dioxide	2007ROL
	nitrogen	2007ROL
Poly(ether block amide)		
	1-butene	2006LI1
	ethane	2006LI1
	ethene	2006LI1
	propane	2006LI1
	propene	2006LI1

Polymer (B)	Solvent (A)	Ref.
Poly(ether carbonate)		
	carbon dioxide	2005TA3
Poly(ether ester)		
	carbon dioxide	2005TA3
Poly(ether ether ketone)		
	carbon dioxide	2006WAN
Poly(etherimide)		
	carbon dioxide	2007MOO
	methane	2007MOO
	nitrogen	2007MOO
	oxygen	2007MOO
Poly(ethoxylated bisphenol)		
	carbon dioxide	2006NAL
Polyethylene		
	carbon dioxide	1998CHA
	carbon dioxide	2004ARE
	carbon dioxide	2005BOY
	carbon dioxide	2006ABB
	carbon dioxide	2006BOY
	carbon dioxide	2007BOY
	carbon dioxide	2007COM
	carbon dioxide	2007LOP
	carbon dioxide	2009HEI
	difluoromethane	2006ABB
	ethene	2007COM
	ethene	2007LOP
	ethene	2007YA1
	n-hexane	2007YA1
	methane	2009HEI
	2-methylbutane	2007YA1
	2-methylpropane	1998CHA
	nitrogen	1998CHA
	nitrogen	2009HEI
	nitrogen	2010WEI

Polymer (B)	**Solvent (A)**	**Ref.**
	ozone	2008DIN
	propane	2007COM
	propene	2007LOP
Poly(ethylene-*co*-ethyl acrylate)		
	carbon dioxide	2004ARE
Poly(ethylene-*co*-1-hexene)		
	ethene	2006NOV
	1-hexene	2006NOV
Poly(ethylene-*co*-1-octene)		
	carbon dioxide	2007LI1
	nitrogen	2007LI1
Poly(ethylene-*co*-propylene-*co*-norbornene)		
	argon	2005RUT
	carbon dioxide	2005RUT
	methane	2005RUT
	oxygen	2005RUT
Poly(ethylene glycol)		
	carbon dioxide	2004KUK
	carbon dioxide	2005TA2
	carbon dioxide	2008PA1
	carbon dioxide	2008PA2
Poly(ethylene glycol diacrylate) cross-linked		
	carbon dioxide	2005LIN
	ethane	2005LIN
	ethene	2005LIN
	methane	2005LIN
	propane	2005LIN
	propene	2005LIN

Polymer (B)	Solvent (A)	Ref.
Poly(ethylene oxide)		
	carbon dioxide	2004LIN
	carbon dioxide	2006MAD
	carbon dioxide	2008RIB
	carbon dioxide	2014CHA
	ethane	2004LIN
	ethane	2008RIB
	ethene	2004LIN
	helium	2004LIN
	hydrogen	2004LIN
	methane	2004LIN
	nitrogen	2004LIN
	oxygen	2004LIN
	propane	2004LIN
	propene	2004LIN
Poly(ethylene oxide-*b*-1,1'-dihydroperfluorooctyl methacrylate)		
	carbon dioxide	2007LIY
Poly(ethylene terephthalate)		
	argon	2007FAR
	n-butane	2007FAR
	carbon dioxide	2005HIR
	carbon dioxide	2007FAR
	carbon dioxide	2010SOR
	carbon dioxide	2014CHA
	cyclopentane	2007FAR
	1,1-difluoroethane	2007FAR
	n-heptane	2007FAR
	n-hexane	2007FAR
	2-methylbutane	2007FAR
	nitrogen	2007FAR
	nitrogen	2010SOR
	n-pentane	2007FAR
Poly(3-hydroxybutyrate)		
	carbon dioxide	2012TAK
Poly(3-hydroxybutyrate-*co*-3-hydroxyvalerate)		
	carbon dioxide	2007CRA
	carbon dioxide	2012TAK

Polymer (B)	Solvent (A)	Ref.
Polyimide		
	argon	2007VIS
	carbon dioxide	2005MAD
	carbon dioxide	2007MOO
	carbon dioxide	2007VIS
	carbon dioxide	2012HOR
	carbon dioxide	2014LIP
	dichloromethane	2013MIN
	ethane	2014LIP
	helium	2014LIP
	hydrogen	2014LIP
	hydrogen	2013SMI
	krypton	2007VIS
	methane	2014LIP
	methane	2007MOO
	methanol	2013MIN
	methyl acetate	2013MIN
	nitrogen	2014LIP
	nitrogen	2007MOO
	oxygen	2014LIP
	oxygen	2007MOO
	propane	2007VIS
	propane	2014LIP
	2-propanone	2013MIN
	propene	2007VIS
	propene	2014LIP
	water	2013MIN
	xenon	2007VIS
Polyisoprene		
	carbon dioxide	2005KOJ
Poly(DL-lactic acid)		
	carbon dioxide	2005FUJ
	carbon dioxide	2006OL2
	carbon dioxide	2007OLI
	carbon dioxide	2007PIN
	carbon dioxide	2011KAS
	carbon dioxide	2011MAR
	carbon dioxide	2012LIA
	carbon dioxide	2013NOF
	carbon dioxide	2014CHA

Polymer (B)	Solvent (A)	Ref.
	ethane	2008GON
	ethene	2008GON
	helium	2013NOF
	nitrogen	2013NOF
Poly(L-lactic acid)		
	carbon dioxide	2012LIA
	carbon dioxide	2013YUJ
Poly(DL-lactic acid-*co*-glycolic acid)		
	carbon dioxide	2006LI2
	carbon dioxide	2007PIN
	carbon dioxide	2011MAR
Poly{1-[2-(methacryloyloxy)ethyl]-3-butylimidazolium tetrafluoroborate}		
	carbon dioxide	2005TA5
Poly{[2-(methacryloyloxy)ethyl]trimethylammonium tetrafluoroborate}		
	carbon dioxide	2005TA6
	carbon dioxide	2007BL1
Poly(methyl methacrylate)		
	carbon dioxide	2005CAR
	carbon dioxide	2005LIU
	carbon dioxide	2005MAN
	carbon dioxide	2006PAN
	carbon dioxide	2006UEZ
	carbon dioxide	2007NAW
	carbon dioxide	2007PAN
	carbon dioxide	2007TAN
	carbon dioxide	2009CAR
	carbon dioxide	2009CHE
	carbon dioxide	2011KAS
	carbon dioxide	2011TSI
	carbon dioxide	2013INC
	carbon dioxide	2013LIX
	carbon dioxide	2014PIN

Polymer (B)	**Solvent (A)**	**Ref.**
Poly(methyl methacrylate-*co*-ethylhexyl acrylate)		
	carbon dioxide	2006DU1
Poly(methyl methacrylate-*co*-ethylhexyl acrylate-*co*-ethylene glycol dimethacrylate)		
	carbon dioxide	2005DUA
	carbon dioxide	2006DU1
	carbon dioxide	2006DU2
Poly(4-methyl-2-pentyne)		
	carbon dioxide	2005NAG
Poly[oligo(ethylene glycol) methyl ether acrylate-*co*-oligo(ethylene glycol) diacrylate]		
	carbon dioxide	2010RI1
	ethane	2010RI1
Poly(pentyl methacrylate)		
	argon	2013KIM
	helium	2013KIM
	nitrogen	2013KIM
Poly(1-phenyl-1-propyne)		
	carbon dioxide	2005NAG
Poly(propoxylated bisphenol)		
	carbon dioxide	2006NAL
Polypropylene		
	carbon dioxide	2004ARE
	carbon dioxide	2006LIG2
	carbon dioxide	2007LI1
	carbon dioxide	2007LI2
	carbon dioxide	2008LIY
	carbon dioxide	2009HEI
	carbon dioxide	2009LIY
	carbon dioxide	2010HAS
	carbon dioxide	2011KUN
	carbon dioxide	2013DIN
	carbon dioxide	2014CHA

Polymer (B)	**Solvent (A)**	**Ref.**
	ethene	2005PAL
	ethene	2007BAR
	ethene	2013KRO
	nitrogen	2007LI1
	nitrogen	2007LI2
	nitrogen	2009HEI
	propene	2005PAL
	propene	2006PAT
	propene	2007BAR
	propene	2013KRO
Polystyrene		
	argon	2013KIM
	carbon dioxide	2004ARE
	carbon dioxide	2005TA2
	carbon dioxide	2006ABB
	carbon dioxide	2006OL2
	carbon dioxide	2006PAN
	carbon dioxide	2007PAN
	carbon dioxide	2008JON
	carbon dioxide	2010PER
	carbon dioxide	2012GRO
	carbon dioxide	2012HOR
	1-chloro-1,1-difluoroethane	2005CHO
	difluoromethane	2006ABB
	ethene	2008JON
	helium	2013KIM
	2-methylbutane	2011CHM
	2-methylpropane	2010PER
	nitrogen	2012GRO
	nitrogen	2013KIM
	n-pentane	2013HAJ
	n-pentane	2011CHM
	1,1,1,2-tetrafluoroethane	2005CHO
	1,1,1,2-tetrafluoroethane	2010PER
Poly(styrene-*co*-acrylonitrile)		
	carbon dioxide	2013INC
	1,1,1,2-tetrafluoroethane	2010PER

Polymer (B)	Solvent (A)	Ref.
Poly(styrene-*co*-methyl methacrylate)		
	carbon dioxide	2013AZI
Poly(styrene-*b*-pentyl methacrylate)		
	argon	2013KIM
	helium	2013KIM
	nitrogen	2013KIM
Polysulfone		
	carbon dioxide	2004TA1
	carbon dioxide	2006HOE
	carbon dioxide	2007BL1
	carbon dioxide	2012HOR
Poly(tetrafluoroethylene)		
	ozone	2008DIN
Poly[tetrafluoroethylene-*co*-perfluoro(vinyl methyl ether)]		
	carbon dioxide	2005PR3
	carbon dioxide	2006BON
	carbon dioxide	2007FOS
	ethane	2005PR3
	hexfluoroethane	2005PR3
	methane	2005PR3
	nitrogen	2005PR3
	octafluoropropane	2005PR3
	propane	2005PR3
	tetrafluoromethane	2005PR3
Poly(tetrafluoroethylene-*co*-2,2,4-trifluoro-5-trifluoromethoxy-1,3-dioxole)		
	carbon dioxide	2004PRA
	ethane	2004PRA
	methane	2004PRA
	nitrogen	2004PRA
	octafluoropropane	2004PRA
	ozone	2008DIN
	propane	2004PRA

Polymer (B)	**Solvent (A)**	**Ref.**
Poly(tetrahydropyranyl methacrylate-*co*-1H,1H-perfluorooctyl methacrylate)		
	carbon dioxide	2004PHA
Poly(1-trimethylsilyl-1-propyne)		
	n-butane	2014VOP
	n-butane	2007RA2
	carbon dioxide	2014VOP
	carbon dioxide	2005NAG
	methane	2014VOP
	methane	2007RA2
	octafluoropropane	2005PR2
	propane	2005PR2
Poly(vinyl acetate)		
	carbon dioxide	2008JON
	ethene	2008JON
	nitrogen	2008POU
Poly[1-(4-vinylbenzyl)-3-butylimidazolium o-benzoic sulfimide]		
	carbon dioxide	2005TA5
Poly[1-(4-vinylbenzyl)-3-butylimidazolium hexafluorophosphate]		
	carbon dioxide	2005TA4
	carbon dioxide	2005TA5
Poly[1-(4-vinylbenzyl)-3-butylimidazolium tetrafluoroborate]		
	carbon dioxide	2005TA4
	carbon dioxide	2005TA5
	carbon dioxide	2007BL2
Poly[1-(4-vinylbenzyl)-3-butylimidazolium trifluoromethane sulfonamide]		
	carbon dioxide	2005TA5

Polymer (B)	Solvent (A)	Ref.
Poly[1-(4-vinylbenzyl)-3-methylimidazolium tetrafluoroborate]	carbon dioxide	2005TA5
Poly[(4-vinylbenzyl)trimethylammonium) o-benzoic sulphimide]	carbon dioxide	2005TA6
Poly[(4-vinylbenzyl)trimethylammonium chloride]	carbon dioxide	2012BHA
Poly[(4-vinylbenzyl)trimethylammonium hexafluorophosphate]	carbon dioxide	2005TA6
Poly[(4-vinylbenzyl)trimethylammonium tetrafluoroborate]	carbon dioxide	2005TA6
	carbon dioxide	2007BL1
	carbon dioxide	2007BL2
	nitrogen	2005TA6
	oxygen	2005TA6
Poly[(4-vinylbenzyl)trimethylammonium trifluoromethanesulfonamide]	carbon dioxide	2005TA6
Poly(vinylidene fluoride)	carbon dioxide	2005BOY
	carbon dioxide	2006BOY
	carbon dioxide	2007BOY
	carbon dioxide	2008GAL
	vinylidene fluoride	2008GAL
Sodium carboxymethylcellulose	carbon dioxide	2011PER

2.3. Ternary and quaternary polymer solutions

Polymer (B):	**polyether-polycarbonate-polyol**		**2012FON**
Characterization:	M_w/g.mol^{-1} = 3046, M_w/M_n = 1.11 PPP, Bayer Material Science AG, Germany		
Solvent (A):	**carbon dioxide**	CO_2	**124-38-9**
Solvent (C):	**propylene carbonate**	$C_4H_6O_3$	**108-32-7**

Type of data: gas solubility

T/K = 383

P/MPa	1.528	3.782	5.994	8.190	10.07
w_A	0.01519	0.04042	0.07260	0.10511	0.12798
w_C/w_B	1/9	1/9	1/9	1/9	1/9

P/MPa	1.794	3.592	5.626	7.864	10.03
w_A	0.02249	0.04447	0.07363	0.10770	0.14084
w_C/w_B	1/4	1/4	1/4	1/4	1/4

Polymer (B):	**polyether-polycarbonate-polyol**		**2012FON**
Characterization:	M_w/g.mol^{-1} = 3046, M_w/M_n = 1.11 PPP, Bayer Material Science AG, Germany		
Solvent (A):	**propylene oxide**	C_3H_6O	**75-56-9**
Solvent (C):	**propylene carbonate**	$C_4H_6O_3$	**108-32-7**

Type of data: gas solubility

T/K = 383

P/MPa	1.06	1.98	2.77	3.44	4.02	4.50
w_A	0.03730	0.07315	0.10659	0.13792	0.16781	0.19546
w_C/w_B	1/9	1/9	1/9	1/9	1/9	1/9

P/MPa	0.89	1.63	2.35	3.06	3.66	4.12
w_A	0.03449	0.07355	0.10940	0.14436	0.17694	0.20502
w_C/w_B	1/4	1/4	1/4	1/4	1/4	1/4

Polymer (B):	**polyethylene**		**2009HAR**
Characterization:	M_n/g.mol^{-1} = 13200, M_w/g.mol^{-1} = 15400, Scientific Polymer Products, Inc., Ontario, NY		
Solvent (A):	**cyclohexane**	C_6H_{12}	**110-82-7**
Solvent (C):	**n-hexane**	C_6H_{14}	**110-54-3**

Type of data: vapor-liquid equilibrium

continued

continued

w_A	0.098	0.098	0.098	0.098	0.098	0.098	0.096	0.096	0.096
w_B	0.019	0.019	0.019	0.019	0.019	0.019	0.047	0.047	0.047
w_C	0.883	0.883	0.883	0.883	0.883	0.883	0.857	0.857	0.857
T/K	373.68	393.46	413.47	433.51	443.75	453.46	373.54	393.40	413.66
P/MPa	0.21	0.29	0.56	0.76	0.97	1.16	0.18	0.30	0.55
w_A	0.096	0.096	0.096	0.091	0.091	0.091	0.091	0.091	0.091
w_B	0.047	0.047	0.047	0.090	0.090	0.090	0.090	0.090	0.090
w_C	0.857	0.857	0.857	0.819	0.819	0.819	0.819	0.819	0.819
T/K	433.25	443.41	453.60	373.61	393.53	413.56	433.51	443.61	453.38
P/MPa	0.78	0.97	1.16	0.18	0.30	0.55	0.78	0.97	1.17
w_A	0.089	0.089	0.089	0.089	0.089	0.089	0.196	0.196	0.196
w_B	0.112	0.112	0.112	0.112	0.112	0.112	0.020	0.020	0.020
w_C	0.799	0.799	0.799	0.799	0.799	0.799	0.784	0.784	0.784
T/K	373.67	393.36	413.32	433.75	443.36	453.57	373.48	393.68	413.44
P/MPa	0.18	0.30	0.54	0.77	0.95	1.16	0.23	0.31	0.52
w_A	0.196	0.196	0.196	0.196	0.190	0.190	0.190	0.190	0.190
w_B	0.020	0.020	0.020	0.020	0.050	0.050	0.050	0.050	0.050
w_C	0.784	0.784	0.784	0.784	0.760	0.760	0.760	0.760	0.760
T/K	433.54	443.45	453.57	463.34	373.11	393.65	413.62	433.27	443.78
P/MPa	0.77	0.97	1.17	1.38	0.23	0.30	0.51	0.76	0.96
w_A	0.190	0.190	0.182	0.182	0.182	0.182	0.182	0.182	0.182
w_B	0.050	0.050	0.089	0.089	0.089	0.089	0.089	0.089	0.089
w_C	0.760	0.760	0.729	0.729	0.729	0.729	0.729	0.729	0.729
T/K	453.31	463.93	373.09	393.43	413.17	433.68	443.55	453.25	463.90
P/MPa	1.12	1.32	0.23	0.30	0.50	0.77	0.97	1.12	1.32
w_A	0.177	0.177	0.177	0.177	0.177	0.177			
w_B	0.114	0.114	0.114	0.114	0.114	0.114			
w_C	0.709	0.709	0.709	0.709	0.709	0.709			
T/K	373.54	393.46	413.33	433.18	443.73	453.26			
P/MPa	0.23	0.30	0.50	0.77	0.97	1.12			

Polymer (B): **polyethylene** **2009HAR**
Characterization: M_n/g.mol^{-1} = 13200, M_w/g.mol^{-1} = 15400,
Scientific Polymer Products, Inc., Ontario, NY
Solvent (A): **ethene** C_2H_4 **74-85-1**
Solvent (C): **cyclohexane** C_6H_{12} **110-82-7**

Type of data: vapor-liquid equilibrium

w_A	0.100	0.100	0.100	0.100	0.100	0.100	0.100	0.100	0.100
w_B	0.017	0.017	0.017	0.017	0.017	0.017	0.017	0.043	0.043
w_C	0.883	0.883	0.883	0.883	0.883	0.883	0.883	0.857	0.857
T/K	373.72	393.51	413.72	433.71	453.46	473.23	483.27	373.68	393.74
P/MPa	3.36	3.88	4.48	4.93	5.50	6.04	6.29	3.55	4.05

continued

continued

w_A	0.100	0.100	0.100	0.100	0.099	0.099	0.099	0.099	0.099
w_B	0.043	0.043	0.043	0.043	0.081	0.081	0.081	0.081	0.081
w_C	0.857	0.857	0.857	0.857	0.820	0.820	0.820	0.820	0.820
T/K	413.65	433.69	453.23	473.20	373.48	393.64	413.73	433.71	453.61
P/MPa	4.66	5.16	5.72	6.32	3.64	4.26	4.81	5.37	6.01

w_A	0.099	0.099	0.099	0.099	0.099	0.099	0.099
w_B	0.081	0.101	0.101	0.101	0.101	0.101	0.101
w_C	0.820	0.800	0.800	0.800	0.800	0.800	0.800
T/K	473.64	373.72	393.62	413.48	433.69	453.61	473.52
P/MPa	6.62	3.64	4.32	4.82	5.42	6.04	6.70

Polymer (B): **polyethylene** **2008HAR**
Characterization: M_n/g.mol^{-1} = 14400, M_w/g.mol^{-1} = 15500,
Scientific Polymer Products, Inc., Ontario, NY
Solvent (A): **ethene** **C_2H_4** **74-85-1**
Solvent (C): **n-hexane** **C_6H_{14}** **110-54-3**

Type of data: vapor-liquid equilibrium data

w_A	0.0200	0.0200	0.0200	0.0200	0.0183	0.0183	0.0183	0.0183	0.0197
w_B	0.0177	0.0177	0.0177	0.0177	0.0462	0.0462	0.0462	0.0462	0.0928
w_C	0.9623	0.9623	0.9623	0.9623	0.9355	0.9355	0.9355	0.9355	0.8875
T/K	373.22	393.21	413.19	433.09	373.10	393.00	413.12	433.00	373.06
P/MPa	0.93	1.16	1.40	1.79	0.68	0.92	1.17	1.76	0.68

w_A	0.0197	0.0197	0.0201	0.0201	0.0201	0.0481	0.0481	0.0481	0.0499
w_B	0.0928	0.0928	0.1247	0.1247	0.1247	0.0187	0.0187	0.0187	0.0452
w_C	0.8875	0.8875	0.8552	0.8552	0.8552	0.9332	0.9332	0.9332	0.8049
T/K	393.00	413.02	373.00	393.00	412.96	373.19	393.08	413.15	373.03
P/MPa	0.85	1.12	0.97	1.21	1.50	1.53	1.77	2.14	1.67

w_A	0.0499	0.0499	0.0515	0.0515	0.0512	0.0512	0.0946	0.0946	0.0932
w_B	0.0452	0.0452	0.0890	0.0890	0.1111	0.1111	0.0175	0.0175	0.0450
w_C	0.8049	0.8049	0.8595	0.8595	0.8377	0.8377	0.8879	0.8879	0.8618
T/K	393.02	413.03	373.07	393.04	373.17	393.07	373.10	393.17	373.10
P/MPa	2.20	2.40	1.75	2.06	1.52	1.77	2.71	3.14	2.70

w_A	0.0959	0.0949
w_B	0.0834	0.1165
w_C	0.8207	0.7886
T/K	373.17	373.20
P/MPa	2.94	2.93

Polymer (B): **polyethylene** **2006NAG**
Characterization: M_n/g.mol^{-1} = 43700, M_w/g.mol^{-1} = 52000, M_z/g.mol^{-1} = 59000, 2.05 ethyl branches per 100 backbone C-atomes, hydrogenated polybutadiene PBD 50000, was denoted as LLDPE, DSM, The Netherlands
Solvent (A): **ethene** C_2H_4 **74-85-1**
Solvent (C): **n-hexane** C_6H_{14} **110-54-3**

Type of data: vapor-liquid equilibrium data

w_A	0.0118	0.0118	0.0118	0.0205	0.0205	0.0298	0.0298	0.0298	0.0099
w_B	0.0502	0.0502	0.0502	0.0501	0.0501	0.0499	0.0499	0.0499	0.0998
w_C	0.9380	0.9380	0.9380	0.9294	0.9294	0.9203	0.9203	0.9203	0.8903
T/K	411.46	416.37	421.20	406.00	410.93	396.49	401.10	406.30	406.11
P/MPa	1.052	1.116	1.181	1.354	1.434	1.529	1.604	1.699	0.969
w_A	0.0099	0.0099	0.0099	0.0196	0.0196	0.0196	0.0196	0.0296	0.0296
w_B	0.0998	0.0998	0.0998	0.1007	0.1007	0.1007	0.1007	0.1002	0.1002
w_C	0.8903	0.8903	0.8903	0.8797	0.8797	0.8797	0.8797	0.8702	0.8702
T/K	411.11	416.06	421.13	401.65	406.43	411.45	416.30	396.26	401.23
P/MPa	1.039	1.109	1.189	1.309	1.374	1.479	1.549	1.580	1.660
w_A	0.0296	0.0098	0.0098	0.0098	0.0210	0.0210	0.0210	0.0210	0.0293
w_B	0.1002	0.1508	0.1508	0.1508	0.1503	0.1503	0.1503	0.1503	0.1506
w_C	0.8702	0.8394	0.8394	0.8394	0.8287	0.8287	0.8287	0.8287	0.8201
T/K	406.21	416.15	421.11	426.07	400.89	405.73	410.70	415.66	396.32
P/MPa	1.750	1.139	1.219	1.299	1.354	1.429	1.504	1.589	1.664
w_A	0.0293	0.0293							
w_B	0.1506	0.1506							
w_C	0.8201	0.8201							
T/K	401.21	406.10							
P/MPa	1.744	1.834							

Polymer (B): **polyethylene** **2010HA2**
Characterization: M_n/g.mol^{-1} = 13200, M_w/g.mol^{-1} = 15400, Scientific Polymer Products, Inc., Ontario, NY
Solvent (A): **1-hexene** C_6H_{12} **592-41-6**
Solvent (C): **n-hexane** C_6H_{14} **110-54-3**

Type of data: vapor-liquid equilibrium data

w_A	0.493	0.493	0.493	0.493	0.493	0.478	0.478	0.478	0.478
w_B	0.020	0.020	0.020	0.020	0.020	0.050	0.050	0.050	0.050
w_C	0.487	0.487	0.487	0.487	0.487	0.472	0.472	0.472	0.472
T/K	373.08	393.04	413.38	433.54	443.56	393.25	413.41	433.15	443.04
P/MPa	0.03	0.17	0.36	0.66	0.83	0.11	0.33	0.61	0.77

continued

continued

w_A	0.462	0.462	0.462	0.462	0.453	0.453	0.453	0.453	0.453
w_B	0.082	0.082	0.082	0.082	0.100	0.100	0.100	0.100	0.100
w_C	0.456	0.456	0.456	0.456	0.447	0.447	0.447	0.447	0.447
T/K	393.22	413.31	433.27	443.23	373.68	393.34	413.51	433.40	443.34
P/MPa	0.11	0.31	0.56	0.80	0.02	0.14	0.34	0.62	0.81

w_A	0.443	0.443	0.443	0.443
w_B	0.120	0.120	0.120	0.120
w_C	0.437	0.437	0.437	0.437
T/K	373.46	393.14	413.33	433.38
P/MPa	0.03	0.17	0.37	0.65

Polymer (B): **polyethylene** **2010HA2**
Characterization: M_n/g.mol^{-1} = 82000, M_w/g.mol^{-1} = 108000, Scientific Polymer Products, Inc., Ontario, NY
Solvent (A): **1-hexene** C_6H_{12} **592-41-6**
Solvent (C): **n-hexane** C_6H_{14} **110-54-3**

Type of data: vapor-liquid equilibrium data

w_A	0.490	0.490	0.490	0.475	0.475	0.475	0.460	0.460	0.460
w_B	0.020	0.020	0.020	0.050	0.050	0.050	0.080	0.080	0.080
w_C	0.490	0.490	0.490	0.475	0.475	0.475	0.460	0.460	0.460
T/K	373.34	393.59	413.37	373.62	393.43	413.62	373.29	393.12	413.73
P/MPa	0.16	0.31	0.53	0.15	0.31	0.53	0.17	0.30	0.54

w_A	0.450	0.450	0.450	0.440	0.440	0.440
w_B	0.100	0.100	0.100	0.121	0.121	0.121
w_C	0.450	0.450	0.450	0.439	0.439	0.439
T/K	373.04	393.10	413.22	373.32	393.09	413.26
P/MPa	0.25	0.35	0.54	0.19	0.34	0.55

Polymer (B): **poly(ethylene-*co*-norbornene)** **2013SAT**
Characterization: M_n/g.mol^{-1} = 38800, M_w/g.mol^{-1} = 74600, 52 mol% ethylene, T_g/K = 411.15, Topas Advanced Polymers GmbH
Solvent (A): **ethene** C_2H_4 **74-85-1**
Solvent (C): **toluene** C_7H_8 **108-88-3**

Type of data: bubble point pressure

T/K = 323.12

x_A	0.1101	0.1425	0.1830	0.2135	0.2527
x_C	0.8893	0.8570	0.8165	0.7860	0.7468
P/MPa	1.13	1.46	1.78	2.16	2.59

continued

continued

T/K = 343.16

x_A	0.1101	0.1425	0.1830	0.2135	0.2527
x_C	0.8893	0.8570	0.8165	0.7860	0.7468
P/MPa	1.34	1.74	2.17	2.56	3.15

T/K = 363.11

x_A	0.1101	0.1425	0.1830	0.2135	0.2527
x_C	0.8893	0.8570	0.8165	0.7860	0.7468
P/MPa	1.48	2.00	2.62	3.13	3.78

Polymer (B): **poly(ethylene-*co*-norbornene)** **2013SAT**
Characterization: M_n/g.mol^{-1} = 38800, M_w/g.mol^{-1} = 74600, 52 mol% ethylene, T_g/K = 411.15, Topas Advanced Polymers GmbH
Solvent (A): **ethene** C_2H_4 **74-85-1**
Solvent (C): **toluene** C_7H_8 **108-88-3**
Solvent (D): **bicyclo[2,2,1]-2-heptene (norbornene)** C_7H_{10} **498-66-8**

Type of data: bubble point pressure

T/K = 323.12

x_A	0.1309	0.1640	0.1973	0.2425	0.2679	0.1266	0.1713	0.2141	0.2561
x_C	0.7800	0.7503	0.7204	0.6799	0.6571	0.4314	0.4093	0.3882	0.3674
x_D	0.0886	0.0852	0.0818	0.0772	0.0746	0.4415	0.4189	0.3972	0.3760
P/MPa	1.40	1.72	2.13	2.56	2.87	1.32	1.77	2.29	2.71

x_A	0.3362	0.1208	0.1454	0.1839	0.2047	0.2414
x_C	0.3279	0.0861	0.0837	0.0800	0.0779	0.0743
x_D	0.3356	0.7926	0.7703	0.7356	0.7169	0.6839
P/MPa	3.54	1.23	1.43	1.95	2.19	2.61

T/K = 343.16

x_A	0.1309	0.1640	0.1973	0.2425	0.2679	0.1266	0.1713	0.2141	0.2561
x_C	0.7800	0.7503	0.7204	0.6799	0.6571	0.4314	0.4093	0.3882	0.3674
x_D	0.0886	0.0852	0.0818	0.0772	0.0746	0.4415	0.4189	0.3972	0.3760
P/MPa	1.64	2.09	2.57	3.17	3.48	1.59	2.16	2.75	3.28

x_A	0.3362	0.1208	0.1454	0.1839	0.2047	0.2414
x_C	0.3279	0.0861	0.0837	0.0800	0.0779	0.0743
x_D	0.3356	0.7926	0.7703	0.7356	0.7169	0.6839
P/MPa	4.37	1.52	1.74	2.36	2.64	3.15

T/K = 363.11

x_A	0.1309	0.1640	0.1973	0.2425	0.2679	0.1266	0.1713	0.2141	0.2561
x_C	0.7800	0.7503	0.7204	0.6799	0.6571	0.4314	0.4093	0.3882	0.3674
x_D	0.0886	0.0852	0.0818	0.0772	0.0746	0.4415	0.4189	0.3972	0.3760
P/MPa	1.98	2.46	3.07	3.72	4.15	1.87	2.51	3.23	3.86

continued

continued

x_A	0.3362	0.1208	0.1454	0.1839	0.2047	0.2414
x_C	0.3279	0.0861	0.0837	0.0800	0.0779	0.0743
x_D	0.3356	0.7926	0.7703	0.7356	0.7169	0.6839
P/MPa	5.10	1.76	2.09	2.78	3.08	3.69

Polymer (B): **poly(ethylene glycol)** **2007HOU**
Characterization: M_n/g.mol^{-1} = 200, Beijing Chemical Reagent Plant, China
Solvent (A): **carbon dioxide** CO_2 **124-38-9**
Solvent (C): **1-octanol** $C_8H_{18}O$ **111-87-5**

Type of data: gas solubility

T/K = 303.15

P/MPa	1.00	2.00	3.00	4.00	5.00	6.00	7.00	2.11	3.09
w_A	0.0137	0.0341	0.0611	0.0931	0.120	0.152	0.182	0.0587	0.0979
w_B/w_C	100/0	100/0	100/0	100/0	100/0	100/0	100/0	75/25	75/25
P/MPa	4.05	5.14	6.09	1.15	2.07	3.05	4.08	5.01	6.08
w_A	0.131	0.169	0.218	0.0339	0.0664	0.105	0.147	0.180	0.246
w_B/w_C	75/25	75/25	75/25	50/50	50/50	50/50	50/50	50/50	50/50
P/MPa	7.10	2.02	3.07	4.04	4.98	6.11	7.05	1.07	2.11
w_A	0.311	0.0584	0.100	0.141	0.180	0.249	0.341	0.0260	0.0562
w_B/w_C	50/50	25/75	25/75	25/75	25/75	25/75	25/75	0/100	0/100
P/MPa	3.12	4.10	5.10	6.13					
w_A	0.0884	0.134	0.181	0.256					
w_B/w_C	0/100	0/100	0/100	0/100					

T/K = 313.15

P/MPa	1.10	2.04	2.99	4.15	5.11	6.11	7.11	8.10	1.13
w_A	0.0130	0.0334	0.0532	0.0710	0.106	0.129	0.147	0.167	0.0248
w_B/w_C	100/0	100/0	100/0	100/0	100/0	100/0	100/0	100/0	75/25
P/MPa	2.09	3.08	4.17	5.13	6.13	7.16	8.16	1.07	2.12
w_A	0.0515	0.0791	0.1012	0.147	0.171	0.204	0.236	0.0249	0.0574
w_B/w_C	75/25	75/25	75/25	75/25	75/25	75/25	75/25	50/50	50/50
P/MPa	2.95	4.04	5.14	6.17	7.16	8.01	1.58	3.02	3.98
w_A	0.0834	0.106	0.155	0.197	0.235	0.284	0.0347	0.0829	0.113
w_B/w_C	50/50	50/50	50/50	50/50	50/50	50/50	25/75	25/75	25/75
P/MPa	5.01	5.97	7.04	8.05	1.13	2.15	3.15	4.18	5.14
w_A	0.148	0.188	0.247	0.299	0.0264	0.0545	0.0894	0.107	0.166
w_B/w_C	25/75	25/75	25/75	25/75	0/100	0/100	0/100	0/100	0/100
P/MPa	6.15	7.01	7.87						
w_A	0.212	0.261	0.295						
w_B/w_C	0/100	0/100	0/100						

continued

continued

T/K = 323.15

P/MPa	1.28	2.15	3.04	4.11	5.05	6.06	6.94	8.07	0.96
w_A	0.0149	0.0296	0.0431	0.0627	0.0781	0.0979	0.113	0.130	0.0167
w_B/w_C	100/0	100/0	100/0	100/0	100/0	100/0	100/0	100/0	75/25
P/MPa	2.03	3.10	4.03	5.11	6.07	7.11	8.21	0.92	1.90
w_A	0.0422	0.0669	0.0759	0.114	0.141	0.163	0.185	0.0190	0.0414
w_B/w_C	75/25	75/25	75/25	75/25	75/25	75/25	75/25	50/50	50/50
P/MPa	2.98	4.08	4.93	5.92	7.02	7.94	1.00	2.11	3.16
w_A	0.0705	0.0896	0.126	0.151	0.183	0.208	0.0191	0.0522	0.0809
w_B/w_C	50/50	50/50	50/50	50/50	50/50	50/50	25/75	25/75	25/75
P/MPa	4.07	5.09	6.08	7.09	8.13	1.08	2.03	2.98	4.18
w_A	0.106	0.141	0.169	0.207	0.244	0.0188	0.0448	0.0696	0.094
w_B/w_C	25/75	25/75	25/75	25/75	25/75	0/100	0/100	0/100	0/100
P/MPa	5.19	6.62	8.03						
w_A	0.138	0.186	0.254						
w_B/w_C	0/100	0/100	0/100						

Polymer (B): **poly(ethylene glycol)** **2008LIX**
Characterization: M_w/g.mol^{-1} = 200, Beijing Chemical Reagent Plant, China
Solvent (A): **carbon dioxide** CO_2 **124-38-9**
Solvent (C): **1-octene** C_8H_{16} **111-66-0**

Type of data: vapor-liquid equilibrium data

T/K = 308.15

w_A	0.0300	0.0666	0.0957	0.122	0.169	0.194	0.225
w_B	0.956	0.916	0.884	0.857	0.813	0.793	0.767
w_C	0.0140	0.0174	0.0203	0.021	0.018	0.013	0.008
P/MPa	1.24	2.70	4.14	5.13	7.09	8.07	9.55

T/K = 318.15

w_A	0.0222	0.0528	0.0840	0.113	0.145	0.174	0.210
w_B	0.962	0.927	0.892	0.862	0.832	0.807	0.776
w_C	0.0158	0.0202	0.0240	0.025	0.023	0.019	0.014
P/MPa	1.01	2.49	4.02	5.47	7.00	8.08	9.37

T/K = 328.15

w_A	0.0184	0.0441	0.0676	0.0995	0.125	0.149	0.188
w_B	0.961	0.931	0.904	0.871	0.847	0.827	0.794
w_C	0.0206	0.0249	0.0284	0.0295	0.028	0.024	0.018
P/MPa	1.05	2.56	3.97	5.53	7.08	8.01	9.61

Polymer (B): **poly(ethylene glycol)** **2008LIX**
Characterization: M_w/g.mol^{-1} = 400, Beijing Chemical Reagent Plant, China
Solvent (A): **carbon dioxide** **CO_2** **124-38-9**
Solvent (C): **1-octene** **C_8H_{16}** **111-66-0**

Type of data: vapor-liquid equilibrium data

T/K = 308.15

w_A	0.0317	0.0468	0.0878	0.119	0.140	0.182	0.230
w_B	0.951	0.932	0.883	0.849	0.828	0.793	0.753
w_C	0.0173	0.0212	0.0291	0.032	0.032	0.025	0.017
P/MPa	1.27	1.87	3.52	5.15	5.95	7.67	9.17

T/K = 318.15

w_A	0.0261	0.0564	0.0881	0.119	0.149	0.183	0.220
w_B	0.953	0.914	0.877	0.845	0.819	0.791	0.762
w_C	0.0209	0.0296	0.0349	0.036	0.032	0.026	0.018
P/MPa	1.06	2.55	4.07	5.66	6.85	8.30	9.80

T/K = 328.15

w_A	0.0242	0.0522	0.0803	0.113	0.139	0.175	0.196
w_B	0.949	0.914	0.881	0.847	0.825	0.796	0.779
w_C	0.0268	0.0338	0.0387	0.040	0.036	0.029	0.025
P/MPa	1.22	2.71	4.22	5.87	7.21	8.70	9.45

Polymer (B): **poly(ethylene glycol)** **2008LIX**
Characterization: M_w/g.mol^{-1} = 600, Beijing Chemical Reagent Plant, China
Solvent (A): **carbon dioxide** **CO_2** **124-38-9**
Solvent (C): **1-octene** **C_8H_{16}** **111-66-0**

Type of data: vapor-liquid equilibrium data

T/K = 308.15

w_A	0.0357	0.0588	0.0887	0.110	0.132	0.178	0.208	0.249
w_B	0.943	0.915	0.878	0.850	0.826	0.786	0.762	0.734
w_C	0.0213	0.0262	0.0343	0.040	0.042	0.036	0.030	0.017
P/MPa	1.21	2.22	3.44	4.60	5.48	6.98	7.96	9.15

T/K = 318.15

w_A	0.0321	0.0654	0.0965	0.128	0.168	0.199	0.223
w_B	0.940	0.900	0.861	0.826	0.788	0.765	0.746
w_C	0.0279	0.0346	0.0425	0.046	0.044	0.036	0.031
P/MPa	1.17	2.67	4.15	5.59	7.11	8.17	9.12

T/K = 328.15

w_A	0.0282	0.0604	0.0910	0.127	0.139	0.156	0.182	0.208
w_B	0.936	0.899	0.862	0.823	0.810	0.795	0.773	0.753
w_C	0.0358	0.0406	0.0470	0.050	0.051	0.049	0.045	0.039
P/MPa	1.18	2.64	4.15	5.91	6.38	7.14	8.10	9.13

Polymer (B):	**poly(ethylene glycol)**		**2007HOU**
Characterization:	M_n/g.mol^{-1} = 200, Beijing Chemical Reagent Plant, China		
Solvent (A):	**carbon dioxide**	CO_2	**124-38-9**
Solvent (C):	**1-pentanol**	$C_5H_{12}O$	**71-41-0**

Type of data: gas solubility

T/K = 303.15

P/MPa	1.00	2.00	3.00	4.00	5.00	6.00	7.00	1.00	1.97
w_A	0.0137	0.0341	0.0611	0.0931	0.120	0.152	0.182	0.0273	0.0601
w_B/w_C	100/0	100/0	100/0	100/0	100/0	100/0	100/0	75/25	75/25
P/MPa	2.80	3.92	5.10	6.08	7.15	1.09	2.05	3.08	4.17
w_A	0.0956	0.136	0.182	0.229	0.267	0.0319	0.0699	0.119	0.171
w_B/w_C	75/25	75/25	75/25	75/25	75/25	50/50	50/50	50/50	50/50
P/MPa	5.00	6.09	6.95	0.99	1.96	2.98	4.01	4.98	6.11
w_A	0.217	0.284	0.363	0.0355	0.0686	0.117	0.164	0.228	0.313
w_B/w_C	50/50	50/50	50/50	25/75	25/75	25/75	25/75	25/75	25/75
P/MPa	6.95	1.71	3.03	3.96	5.07	5.96	6.65	6.95	
w_A	0.486	0.0591	0.129	0.166	0.236	0.317	0.451	0.720	
w_B/w_C	25/75	0/100	0/100	0/100	0/100	0/100	0/100	0/100	

T/K = 313.15

P/MPa	1.10	2.04	2.99	4.15	5.11	6.11	7.11	8.10	1.00
w_A	0.0130	0.0334	0.0532	0.0710	0.106	0.129	0.147	0.167	0.0252
w_B/w_C	100/0	100/0	100/0	100/0	100/0	100/0	100/0	100/0	75/25
P/MPa	2.00	3.00	4.00	5.00	6.00	7.00	8.00	1.00	2.00
w_A	0.0531	0.0804	0.101	0.143	0.164	0.211	0.234	0.0303	0.0548
w_B/w_C	75/25	75/25	75/25	75/25	75/25	75/25	75/25	50/50	50/50
P/MPa	3.00	4.00	5.00	6.00	7.00	8.26	1.00	2.00	3.00
w_A	0.0925	0.126	0.151	0.211	0.262	0.323	0.0440	0.0677	0.104
w_B/w_C	50/50	50/50	50/50	50/50	50/50	50/50	25/75	25/75	25/75
P/MPa	4.00	5.00	6.00	7.00	8.00	1.00	2.00	3.00	4.00
w_A	0.146	0.174	0.251	0.306	0.353	0.0287	0.0610	0.108	0.149
w_B/w_C	25/75	25/75	25/75	25/75	25/75	0/100	0/100	0/100	0/100
P/MPa	5.00	6.00	7.00	8.00					
w_A	0.184	0.236	0.322	0.531					
w_B/w_C	0/100	0/100	0/100	0/100					

T/K = 323.15

P/MPa	1.28	2.15	3.04	4.11	5.05	6.06	6.94	8.07	1.04
w_A	0.0149	0.0296	0.0431	0.0627	0.0781	0.0979	0.113	0.130	0.0178
w_B/w_C	100/0	100/0	100/0	100/0	100/0	100/0	100/0	100/0	75/25

continued

continued

P/MPa	2.11	2.99	4.02	5.03	6.08	7.10	8.10	1.06	1.98
w_A	0.0391	0.0615	0.0861	0.112	0.1412	0.170	0.194	0.0229	0.0459
w_B/w_C	75/25	75/25	75/25	75/25	75/25	75/25	75/25	50/50	50/50
P/MPa	2.99	4.01	4.91	6.03	7.11	8.04	1.17	2.01	3.03
w_A	0.0743	0.104	0.130	0.172	0.210	0.242	0.0246	0.0497	0.0860
w_B/w_C	50/50	50/50	50/50	50/50	50/50	50/50	25/75	25/75	25/75
P/MPa	4.06	4.96	6.01	7.04	8.04	1.03	1.98	3.09	3.97
w_A	0.118	0.147	0.193	0.240	0.284	0.0223	0.0513	0.0869	0.122
w_B/w_C	25/75	25/75	25/75	25/75	25/75	0/100	0/100	0/100	0/100
P/MPa	5.11	6.01	7.09	8.10					
w_A	0.160	0.208	0.260	0.332					
w_B/w_C	0/100	0/100	0/100	0/100					

Polymer (B): **poly(ethylene glycol)** **2009MAR**
Characterization: M_w/g.mol^{-1} = 6000, Clariant, Burghausen, Germany
Solvent (A): **carbon dioxide** **CO_2** **124-38-9**
Solvent (C): **water** **H_2O** **7732-18-5**

Type of data: coexistence data (VLE)

T/K = 353.15

P/MPa	10	10	10	10	10	10	10
w_A(liquid phase)	0.144	0.116	0.080	0.059	0.044	0.041	0.041
w_B(liquid phase)	0.8570	0.7990	0.6510	0.4750	0.2853	0.0969	0.0000
w_C(liquid phase)	0.0000	0.0852	0.2697	0.4660	0.6710	0.8617	0.9589
w_A(gas phase)	0.9995	0.9983	0.9982	0.9980	0.9967	0.9964	0.9968
w_B(gas phase)	0.00048	0.00029	0.00026	0.00025	0.00010	0.00005	0.00000
w_C(gas phase)	0.00000	0.00140	0.00160	0.00210	0.00320	0.00360	0.00320
P/MPa	20	20	20	20	20	20	20
w_A(liquid phase)	0.220	0.180	0.102	0.069	0.055	0.053	0.051
w_B(liquid phase)	0.7800	0.7460	0.6416	0.4700	0.2840	0.0950	0.0000
w_C(liquid phase)	0.0000	0.0710	0.2564	0.4610	0.6610	0.8520	0.9486
w_A(gas phase)	0.9996	0.9981	0.9979	0.9970	0.9965	0.9962	0.9964
w_B(gas phase)	0.00040	0.00033	0.00035	0.00034	0.00025	0.00011	0.00000
w_C(gas phase)	0.00000	0.00160	0.00180	0.00270	0.00320	0.00370	0.00360
P/MPa	30	30	30	30	30	30	30
w_A(liquid phase)	0.240	0.215	0.125	0.083	0.069	0.067	0.055
w_B(liquid phase)	0.7600	0.7160	0.6290	0.4649	0.2810	0.0980	0.0000
w_C(liquid phase)	0.0000	0.0692	0.2460	0.4520	0.6500	0.8350	0.9450
w_A(gas phase)	0.9997	0.9977	0.9973	0.9955	0.9955	0.9956	0.9960
w_B(gas phase)	0.00032	0.00053	0.00054	0.00041	0.00028	0.00013	0.00000
w_C(gas phase)	0.00000	0.00170	0.00220	0.00410	0.00420	0.00430	0.00400

continued

continued

T/K = 393.15

P/MPa	10	10	10	10	10	10
w_A(liquid phase)	0.119	0.060	0.051	0.043	0.034	0.032
w_B(liquid phase)	0.881	0.673	0.487	0.289	0.107	0.000
w_C(liquid phase)	0.000	0.267	0.463	0.668	0.859	0.968
w_A(gas phase)	0.9995	0.9977	0.9975	0.9965	0.9963	0.9870
w_B(gas phase)	0.00050	0.00028	0.00022	0.00009	0.00003	0.00000
w_C(gas phase)	0.00000	0.00206	0.00226	0.00342	0.00366	0.01300

P/MPa	30	30	30	30	30	30	30
w_A(liquid phase)	0.220	0.190	0.130	0.097	0.079	0.080	0.058
w_B(liquid phase)	0.790	0.744	0.632	0.453	0.368	0.341	0.000
w_C(liquid phase)	0.000	0.068	0.235	0.444	0.553	0.580	0.942
w_A(gas phase)	0.9996	0.9919	0.9894	0.9890	0.9870	0.9868	0.9873
w_B(gas phase)	0.00036	0.00028	0.00026	0.00022	0.00020	0.00007	0.00000
w_C(gas phase)	0.00000	0.00778	0.01034	0.01090	0.01280	0.01312	0.01264

Polymer (B): **poly(ethylene glycol)** **2013NI2**
Characterization: M_n/g.mol^{-1} = 200, analytical grade, Beijing Reagent Company, China
Solvent (A): **15-crown-5 ether** $C_{10}H_{20}O_5$ **33100-27-5**
Solvent (C): **sulfur dioxide** SO_2 **7446-09-5**
Solvent (D): **nitrogen** N_2 **7727-37-9**

Type of data: gas solubility

T/K = 308.15
P/kPa = 122.66 (total gas pressure)

φ_A = 0.0000 was kept constant.

$10^6\varphi_C$	170	295	431	557	696	824
c_C/(mg/L)	151.9	259.2	343.4	457.3	584.4	736.3
P_C/Pa	20.8	36.2	52.9	68.4	85.4	101

φ_A = 1.0000 was kept constant.

$10^6\varphi_C$	176	312	475	555	611	751	804
c_C/(mg/L)	234.4	388.0	604.2	723.1	850.2	992.2	1159
P_C/Pa	21.6	38.2	58.2	68.1	74.9	92.1	98.6

Comments: φ_A denotes the volume fraction of 15-crown-5 ether in the liquid phase.
φ_C denotes the volume fraction of SO_2 in the gas phase (i.e., the remaining part is nitrogen).
c_C/(mol/m^3) denotes the concentration of SO_2 in the liquid phase.
P_C denotes the partial pressure of SO_2 in the gas phase.

Polymer (B):	**poly(ethylene glycol)**		**2013NI2**
Characterization:	M_n/g.mol^{-1} = 300, analytical grade, Beijing Reagent Company, China		
Solvent (A):	**15-crown-5 ether**	$C_{10}H_{20}O_5$	**33100-27-5**
Solvent (C):	**sulfur dioxide**	SO_2	**7446-09-5**
Solvent (D):	**nitrogen**	N_2	**7727-37-9**

Type of data: gas solubility

T/K = 308.15
P/kPa = 122.66 (total gas pressure)

φ_A = 0.0000	was kept constant.						
$10^6\varphi_C$	174	266	417	711	800	938	
c_C/(mg/L)	435.8	505.2	615.8	767.7	828.7	944.3	
P_C/Pa	21.4	32.6	51.2	87.2	98.2	115	
φ_A = 0.2000	was kept constant.						
$10^6\varphi_C$	60.0	176	326	430	618	774	916
c_C/(mg/L)	323.6	521.7	629.0	761.1	904.7	1060	1245
P_C/Pa	7.35	21.6	40.0	52.7	75.8	94.9	112
φ_A = 0.3000	was kept constant.						
$10^6\varphi_C$	111	216	303	491	622	742	
c_C/(mg/L)	168.4	264.1	376.4	566.2	695.0	850.2	
P_C/Pa	13.6	26.5	37.2	60.3	76.3	91.0	
φ_A = 0.5000	was kept constant.						
$10^6\varphi_C$	111	270	454	570	680	824	
c_C/(mg/L)	161.8	330.2	511.8	642.2	772.6	995.5	
P_C/Pa	13.6	33.2	55.7	69.9	83.4	101	
φ_A = 1.0000	was kept constant.						
$10^6\varphi_C$	176	312	475	555	611	751	804
c_C/(mg/L)	234.4	388.0	604.2	723.1	850.2	992.2	1159
P_C/Pa	21.6	38.2	58.2	68.1	74.9	92.1	98.6

Comments: φ_A denotes the volume fraction of 15-crown-5 ether in the liquid phase.
φ_C denotes the volume fraction of SO_2 in the gas phase (i.e., the remaining part is nitrogen).
c_C/(mol/m^3) denotes the concentration of SO_2 in the liquid phase.
P_C denotes the partial pressure of SO_2 in the gas phase.

Polymer (B):	**poly(ethylene glycol)**	**2013NI1**
Characterization:	M_n/g.mol^{-1} = 400, analytical grade, Beijing Reagent Company, China	
Solvent (A):	***N,N*-dimethylformamide** C_3H_7NO	**68-12-2**
Solvent (C):	**sulfur dioxide** SO_2	**7446-09-5**
Solvent (D):	**nitrogen** N_2	**7727-37-9**

Type of data: gas solubility

T/K = 308.15
P/kPa = 122.66 (total gas pressure)

w_A = 0.0000	was kept constant.							
$10^6 \varphi_C$	81.5	148	250	362	504	606	711	796
c_C/(mol/m^3)	7.22	8.89	10.8	12.0	13.4	15.2	16.8	18.1
P_C/Pa	10.02	18.2	30.7	44.4	61.8	74.4	87.2	97.7
$10^6 \varphi_C$	886							
c_C/(mol/m^3)	19.3							
P_C/Pa	108							
w_A = 0.1001	was kept constant.							
$10^6 \varphi_C$	234	468	598	722	781	826		
c_C/(mol/m^3)	11.5	15.9	17.6	21.5	23.5	26.0		
P_C/Pa	28.7	57.4	73.3	88.6	95.8	101		
w_A = 0.1950	was kept constant.							
$10^6 \varphi_C$	177	378	566	727	771	825		
c_C/(mol/m^3)	10.9	15.0	19.7	25.0	25.8	27.6		
P_C/Pa	21.7	46.4	69.4	89.1	94.6	101		
w_A = 0.2998	was kept constant.							
$10^6 \varphi_C$	137	407	619	781	820	871		
c_C/(mol/m^3)	9.91	17.8	23.7	29.5	31.5	34.0		
P_C/Pa	16.8	49.9	76.0	95.8	100	106		
w_A = 0.4000	was kept constant.							
$10^6 \varphi_C$	159	379	603	645	701	821		
c_C/(mol/m^3)	10.6	19.5	28.9	30.5	33.0	38.2		
P_C/Pa	19.5	46.5	74.0	79.1	86.0	100		
w_A = 0.4973	was kept constant.							
$10^6 \varphi_C$	232	452	586	706	781			
c_C/(mol/m^3)	13.0	23.1	30.3	37.6	39.2			
P_C/Pa	28.4	55.5	71.9	86.6	95.8			
w_A = 0.6013	was kept constant.							
$10^6 \varphi_C$	233	433	549	653	722	818		
c_C/(mol/m^3)	13.2	25.4	30.2	38.9	43.5	50.8		
P_C/Pa	28.6	53.1	67.3	80.1	88.6	100		

continued

continued

w_A = 0.7000 was kept constant.

$10^6 \varphi_C$	184	372	512	648	748	824
c_C/(mol/m^3)	5.45	10.4	14.8	19.1	22.5	26.3
P_C/Pa	22.6	45.6	62.8	79.5	91.7	101

w_A = 0.7986 was kept constant.

$10^6 \varphi_C$	172	323	447	583	706	819
c_C/(mol/m^3)	4.89	10.3	14.8	20.7	25.9	32.5
P_C/Pa	21.1	39.6	54.8	71.5	86.6	100

w_A = 0.9000 was kept constant.

$10^6 \varphi_C$	156	334	526	658	772	863
c_C/(mol/m^3)	6.06	11.9	22.6	29.8	37.4	42.5
P_C/Pa	19.1	40.9	64.6	80.7	94.7	105

w_A = 1.0000 was kept constant.

$10^6 \varphi_C$	223	302	337	412	578	657	765	941
c_C/(mol/m^3)	20.3	26.9	31.6	38.4	52.7	61.3	72.3	90.8
P_C/Pa	27.3	37.1	41.4	50.5	70.9	80.5	93.8	115

Comments: w_A denotes the mass fraction of DMF in the liquid phase.
φ_C denotes the volume fraction of SO_2 in the gas phase (i.e., the remaining part is nitrogen).
c_C/(mol/m^3) denotes the concentration of SO_2 in the liquid phase.
P_C denotes the partial pressure of SO_2 in the gas phase.

Polymer (B): **poly(ethylene glycol)** **2013NI2**
Characterization: M_n/g.mol^{-1} = 200, analytical grade, Beijing Reagent Company, China
Solvent (A): **1,4-dioxane** **$C_4H_8O_2$** **123-91-1**
Solvent (C): **sulfur dioxide** **SO_2** **7446-09-5**
Solvent (D): **nitrogen** **N_2** **7727-37-9**

Type of data: gas solubility

T/K = 308.15
P/kPa = 122.66 (total gas pressure)

φ_A = 0.0000 was kept constant.

$10^6 \varphi_C$	170	295	431	557	696	824
c_C/(mg/L)	151.9	259.2	343.4	457.3	584.4	736.3
P_C/Pa	20.8	36.2	52.9	68.4	85.4	101

φ_A = 0.2000 was kept constant.

$10^6 \varphi_C$	147	272	382	470	607	713	815
c_C/(mg/L)	163.4	298.8	401.2	529.9	655.4	817.2	927.8
P_C/Pa	18.1	33.4	46.8	57.7	74.4	87.5	99.9

continued

continued

$\varphi_A = 0.3000$	was kept constant.							
$10^6\varphi_C$	151	275	441	620	711	803	906	
c_C/(mg/L)	227.8	361.5	559.6	789.1	936.0	1137	1315	
P_C/Pa	18.5	33.7	54.1	76.1	87.2	98.5	111	
$\varphi_A = 0.5000$	was kept constant.							
$10^6\varphi_C$	159	263	396	480	616	788	872	
c_C/(mg/L)	264.1	421.0	658.7	804.0	1088	1403	1586	
P_C/Pa	19.5	32.2	48.5	58.8	75.6	97.0	107	
$\varphi_A = 1.0000$	was kept constant.							
$10^6\varphi_C$	153	277	380	98	503	574	648	799
c_C/(mg/L)	314.4	510.8	700.4	710.4	905.0	1070	1278	1707
P_C/Pa	18.8	34.0	46.6	48.8	61.7	70.4	79.5	98.0
$10^6\varphi_C$	964							
c_C/(mg/L)	2010							
P_C/Pa	118							

Comments: φ_A denotes the volume fraction of dioxane in the liquid phase.
φ_C denotes the volume fraction of SO_2 in the gas phase (i.e., the remaining part is nitrogen).
$c_C/(mol/m^3)$ denotes the concentration of SO_2 in the liquid phase.
P_C denotes the partial pressure of SO_2 in the gas phase.

Polymer (B): **poly(ethylene glycol)** **2013NI2**
Characterization: M_n/g.mol^{-1} = 300, analytical grade,
Beijing Reagent Company, China
Solvent (A): **1,4-dioxane** **$C_4H_8O_2$** **123-91-1**
Solvent (C): **sulfur dioxide** **SO_2** **7446-09-5**
Solvent (D): **nitrogen** **N_2** **7727-37-9**

Type of data: gas solubility

T/K = 308.15
P/kPa = 122.66 (total gas pressure)

$\varphi_A = 0.0000$	was kept constant.						
$10^6\varphi_C$	174	266	417	711	800	938	
c_C/(mg/L)	435.8	505.2	615.8	767.7	828.7	944.3	
P_C/Pa	21.4	32.6	51.2	87.2	98.2	115	
$\varphi_A = 0.1000$	was kept constant.						
$10^6\varphi_C$	73.9	211	451	553	675	802	892
c_C/(mg/L)	345.0	374.8	468.8	582.8	690.1	808.9	914.6
P_C/Pa	9.06	25.8	55.3	67.8	82.8	98.4	109.4

continued

continued

φ_A = 0.2000 was kept constant.

$10^6\varphi_C$	73.5	166	274	379	573	710	788
c_C/(mg/L)	254.2	317.0	432.5	551.4	749.5	901.4	1012
P_C/Pa	9.01	20.3	33.6	46.4	70.3	87.0	96.7

φ_A = 0.3000 was kept constant.

$10^6\varphi_C$	131	282	418	587	739	809	875
c_C/(mg/L)	523.3	610.8	721.4	939.4	1179	1271	1395
P_C/Pa	16.1	34.6	51.2	72.0	90.7	99.3	107

φ_A = 0.5000 was kept constant.

$10^6\varphi_C$	67.8	193	342	513	580	714	863
c_C/(mg/L)	376.4	523.3	737.9	983.9	1015	1332	1659
P_C/Pa	8.31	23.7	42.0	62.97	71.2	87.5	106

φ_A = 1.0000 was kept constant.

$10^6\varphi_C$	153	277	380	98	503	574	648	799
c_C/(mg/L)	314.4	510.8	700.4	710.4	905.0	1070	1278	1707
P_C/Pa	18.8	34.0	46.6	48.8	61.7	70.4	79.5	98.0

$10^6\varphi_C$	964
c_C/(mg/L)	2010
P_C/Pa	118

Comments: φ_A denotes the volume fraction of dioxane in the liquid phase.
φ_C denotes the volume fraction of SO_2 in the gas phase (i.e., the remaining part is nitrogen).
c_C/(mol/m^3) denotes the concentration of SO_2 in the liquid phase.
P_C denotes the partial pressure of SO_2 in the gas phase.

Polymer (B): **poly(ethylene glycol)** **2013ZHA**
Characterization: M_n/g.mol^{-1} = 300, analytical grade, Beijing Reagent Company, China
Solvent (A): **nitrogen** N_2 **7727-37-9**
Solvent (C): **sulfur dioxide** SO_2 **7446-09-5**

Type of data: gas solubility

T/K = 298.15
P/kPa = 110.34 (total gas pressure)

$10^6\varphi_C$	24	31	58	102	180	200	286	344
c_C/(mol/m^3)	3.47	3.54	3.76	4.53	5.64	6.69	7.27	8.49
P_C/Pa	2.65	3.42	6.40	11.3	19.9	22.1	31.6	38.0

$10^6\varphi_C$	368	405	471	502	570	635	699	774
c_C/(mol/m^3)	9.01	11.4	10.7	11.0	12.4	13.5	14.7	16.3
P_C/Pa	40.6	44.7	52.0	55.4	62.9	70.1	77.1	85.4

continued

continued

$10^6\varphi_C$	836	900	971
$c_C/(mol/m^3)$	16.6	17.5	18.0
P_C/Pa	92.2	99.3	107

$T/K = 298.15$
$P/kPa = 115.83$ (total gas pressure)

$10^6\varphi_C$	35	87	157	206	271	364	409	477
$c_C/(mol/m^3)$	4.04	4.64	5.01	5.78	6.37	6.79	7.53	8.60
P_C/Pa	4.05	10.1	18.2	23.9	31.4	42.2	47.4	55.3

$10^6\varphi_C$	580	614	719	802	874	958
$c_C/(mol/m^3)$	9.57	10.2	10.7	12.0	13.1	13.8
P_C/Pa	67.2	71.1	83.3	92.9	101	111

$T/K = 298.15$
$P/kPa = 132.87$ (total gas pressure)

$10^6\varphi_C$	11	73	146	209	316	385	490	575
$c_C/(mol/m^3)$	3.40	4.35	5.10	5.96	6.96	8.66	8.98	9.40
P_C/Pa	1.46	9.70	19.4	27.8	42.0	51.2	65.1	76.4

$10^6\varphi_C$	710	804	842	918	1106
$c_C/(mol/m^3)$	11.6	13.0	13.5	13.8	16.5
P_C/Pa	94.3	107	112	122	147

$T/K = 298.15$
$P/kPa = 136.47$ (total gas pressure)

$10^6\varphi_C$	90	127	198	231	297	372	459	535
$c_C/(mol/m^3)$	4.24	4.84	5.66	6.07	7.34	7.95	9.08	9.25
P_C/Pa	12.3	17.3	27.0	31.5	40.5	50.8	62.6	73.0

$10^6\varphi_C$	667	725	806	867	925	959
$c_C/(mol/m^3)$	11.0	11.9	12.9	13.7	13.9	15.0
P_C/Pa	91.0	98.9	110	118	126	131

$T/K = 298.15$
$P/kPa = 142.03$ (total gas pressure)

$10^6\varphi_C$	84	157	219	286	319	448	501	552
$c_C/(mol/m^3)$	3.78	4.99	5.55	6.60	7.40	8.81	9.58	9.74
P_C/Pa	11.9	22.3	31.1	40.6	45.3	63.6	71.2	78.4

$10^6\varphi_C$	619	716	806	907	1000
$c_C/(mol/m^3)$	10.9	11.0	12.8	13.7	15.0
P_C/Pa	87.9	102	114	129	142

continued

continued

T/K = 303.15
P/kPa = 110.34 (total gas pressure)

$10^6 \varphi_C$	3	84	117	159	236	282	355	414
c_C/(mol/m^3)	1.18	3.61	4.00	4.07	4.70	6.21	7.10	7.53
P_C/Pa	0.33	9.27	12.9	17.5	26.0	31.1	39.2	45.7
$10^6 \varphi_C$	479	551	587	614	662	757	851	883
c_C/(mol/m^3)	7.77	8.48	9.20	9.82	10.9	11.3	13.6	14.2
P_C/Pa	52.9	60.8	64.8	67.7	73.0	83.5	93.9	97.4
$10^6 \varphi_C$	942	1017						
c_C/(mol/m^3)	14.6	15.0						
P_C/Pa	104	112						

T/K = 308.15
P/kPa = 110.34 (total gas pressure)

$10^6 \varphi_C$	15	42	108	173	266	356	471	531
c_C/(mol/m^3)	3.40	3.75	4.68	5.36	6.41	7.71	8.34	9.06
P_C/Pa	1.66	4.63	11.9	19.1	29.4	39.3	52.0	58.6
$10^6 \varphi_C$	597	637	674	778	817	895	914	949
c_C/(mol/m^3)	9.55	9.64	10.2	10.1	10.9	11.1	11.2	11.4
P_C/Pa	65.9	70.3	74.4	85.8	90.1	98.8	101	105

T/K = 313.15
P/kPa = 110.34 (total gas pressure)

$10^6 \varphi_C$	8	62	131	234	387	474	507	531
c_C/(mol/m^3)	2.74	3.58	4.06	4.89	6.07	6.61	6.55	6.41
P_C/Pa	0.88	6.84	14.5	25.8	42.7	52.3	55.9	58.6
$10^6 \varphi_C$	591	656	730	796	846	902	988	
c_C/(mol/m^3)	6.75	6.68	8.03	8.35	8.11	9.32	9.45	
P_C/Pa	65.2	72.4	80.5	87.8	93.3	99.5	109	

T/K = 318.15
P/kPa = 110.34 (total gas pressure)

$10^6 \varphi_C$	20	72	116	158	193	287	399	496
c_C/(mol/m^3)	0.25	0.75	1.05	1.63	2.23	2.43	3.32	3.78
P_C/Pa	2.21	9.49	17.4	21.3	31.7	44.0	54.7	61.0
$10^6 \varphi_C$	553	640	798	840	909	959		
c_C/(mol/m^3)	4.07	4.52	4.91	5.11	5.90	6.15		
P_C/Pa	65.5	70.6	88.1	92.7	100	106		

Comments: φ_C denotes the volume fraction of SO_2 in the gas phase (i.e., the remaining part is nitrogen).
c_C/(mol/m^3) denotes the concentration of SO_2 in the liquid phase.
P_C denotes the partial pressure of SO_2 in the gas phase.

Polymer (B):	**poly(ethylene glycol)**		**2012ZHA**
Characterization:	M_n/g.mol^{-1} = 200, analytical grade, Beijing Reagent Company, China		
Solvent (A):	**water**	H_2O	**7732-18-5**
Solvent (C):	**sulfur dioxide**	SO_2	**7446-09-5**
Solvent (D):	**nitrogen**	N_2	**7727-37-9**

Type of data: gas solubility

T/K = 298.15
P/kPa = 122.66 (total gas pressure)

w_B/w_A = 0.0/100.0 was kept constant.

$10^6\varphi_C$	6.40	25.6	53.8	79.1	137	234	381	502
c_C/(mg/L)	61.8	70.3	77.6	84.9	94.	115	153	176
P_C/Pa	0.78	3.14	6.59	9.70	16.8	28.6	46.7	61.5

$10^6\varphi_C$	580	627
c_C/(mg/L)	200	219
P_C/Pa	71.2	76.9

w_B/w_A = 20.05/79.95 was kept constant.

$10^6\varphi_C$	108	208	276	318	439	489	554	625
c_C/(mg/L)	69.8	98.7	132	162	176	189	227	252
P_C/Pa	13.2	25.5	33.8	39.0	53.8	60.0	67.9	76.6

$10^6\varphi_C$	782	837	898
c_C/(mg/L)	298	332	357
P_C/Pa	95.9	103	110

w_B/w_A = 40.02/59.98 was kept constant.

$10^6\varphi_C$	135	190	228	288	382	528	658	705
c_C/(mg/L)	55.3	57.5	63.2	74.0	88.0	108	128	132
P_C/Pa	16.6	23.3	28.0	35.3	46.8	64.7	80.7	86.4

$10^6\varphi_C$	835	879	904
c_C/(mg/L)	163	170	175
P_C/Pa	102	108	111

w_B/w_A = 60.90/39.10 was kept constant.

$10^6\varphi_C$	103	214	359	426	540	620	682	702
c_C/(mg/L)	42.9	64.8	75.3	92.7	106	133	144	149
P_C/Pa	12.6	26.2	44.0	52.2	66.2	76.0	83.6	86.1

$10^6\varphi_C$	761	902
c_C/(mg/L)	171	199
P_C/Pa	93.3	110.6

continued

continued

w_B/w_A = 70.00/30.00 was kept constant.

$10^6\varphi_C$	71.9	141	296	405	502	574	672	733
c_C/(mg/L)	29.2	50.5	83.2	110	120	139	159	178
P_C/Pa	8.81	17.3	36.3	49.7	61.5	70.4	82.3	89.9
$10^6\varphi_C$	774							
c_C/(mg/L)	196							
P_C/Pa	94.9							

w_B/w_A = 80.10/19.90 was kept constant.

$10^6\varphi_C$	105	173	221	328	419	519	601	632
c_C/(mg/L)	42.4	59.4	81.9	107	132	159	171	179
P_C/Pa	12.9	21.2	27.1	40.2	51.4	63.6	73.7	77.5
$10^6\varphi_C$	746	812	859					
c_C/(mg/L)	208	230	254					
P_C/Pa	91.5	99.5	105					

w_B/w_A = 90.20/09.80 was kept constant.

$10^6\varphi_C$	98.2	175	309	456	513	574	699	818
c_C/(mg/L)	81.0	116	165	218	243	268	320	383
P_C/Pa	12.0	21.5	37.9	55.9	62.9	70.4	85.7	100
$10^6\varphi_C$	834	968						
c_C/(mg/L)	417	420						
P_C/Pa	102	119						

w_B/w_A = 100.0/0.0 was kept constant.

$10^6\varphi_C$	36.5	75.0	81.8	158	306	378	478	558
c_C/(mg/L)	49.2	65.7	73.7	130	255	275	358	368
P_C/Pa	4.47	9.20	10.0	19.4	37.5	46.3	58.6	68.4
$10^6\varphi_C$	644	722	884	1037				
c_C/(mg/L)	492	539	749	800				
P_C/Pa	79.0	88.5	108	127				

Comments: w_B/w_A denotes the mass fraction ratio of PEG/water in the liquid phase.
φ_C denotes the volume fraction of SO_2 in the gas phase (i.e., the remaining part is nitrogen).
c_C/(mg/L) denotes the concentration of SO_2 in the liquid phase.
P_C denotes the partial pressure of SO_2 in the gas phase.

Polymer (B):	**poly(ethylene glycol)**		**2013ZHA**
Characterization:	M_n/g.mol^{-1} = 300, analytical grade, Beijing Reagent Company, China		
Solvent (A):	**water**	H_2O	**7732-18-5**
Solvent (C):	**sulfur dioxide**	SO_2	**7446-09-5**
Solvent (D):	**nitrogen**	N_2	**7727-37-9**

Type of data: gas solubility

T/K = 298.15
P/kPa = 110.34 (total gas pressure)

w_B/w_A = 0.0/100.0 was kept constant.

$10^6\varphi_C$	50	113	165	200	280	365	410	523
c_C/(mol/m^3)	0.76	1.18	1.45	1.81	2.09	2.46	2.65	2.93
P_C/Pa	5.52	12.5	18.2	22.1	30.9	40.3	45.2	57.7
$10^6\varphi_C$	589	664	728	864	949	1017		
c_C/(mol/m^3)	3.24	3.47	3.70	4.21	4.42	4.73		
P_C/Pa	65.0	73.3	80.3	95.3	105	112		

w_B/w_A = 5.0/95.0 was kept constant.

$10^6\varphi_C$	8	98	167	294	346	429	532	672
c_C/(mol/m^3)	1.27	2.05	2.36	3.19	3.65	4.18	4.50	5.09
P_C/Pa	0.88	10.8	18.4	32.4	38.2	47.3	58.7	74.1
$10^6\varphi_C$	696	723	828	960	1001			
c_C/(mol/m^3)	5.50	5.75	5.97	6.91	7.08			
P_C/Pa	76.8	79.8	91.4	106	110			

w_B/w_A = 10.0/90.0 was kept constant.

$10^6\varphi_C$	55	85	156	195	320	412	488	560
c_C/(mol/m^3)	0.80	1.30	1.93	2.30	2.76	3.25	3.81	4.01
P_C/Pa	6.07	9.38	17.2	21.5	35.3	45.5	53.8	61.8
$10^6\varphi_C$	619	704	752	886	944	1006		
c_C/(mol/m^3)	4.41	4.76	4.81	5.31	5.61	6.24		
P_C/Pa	68.3	77.7	83.0	97.8	104	111		

w_B/w_A = 20.0/80.0 was kept constant.

$10^6\varphi_C$	19	33	68	126	239	279	341	362
c_C/(mol/m^3)	0.37	0.72	1.03	1.50	1.98	2.20	2.50	2.72
P_C/Pa	2.10	3.64	7.50	13.9	26.4	30.8	37.6	39.9
$10^6\varphi_C$	468	612	707	797	935	1092		
c_C/(mol/m^3)	3.34	3.76	4.01	4.26	5.37	5.93		
P_C/Pa	51.6	67.5	78.0	87.9	103	120		

continued

continued

w_B/w_A = 40.0/60.0 was kept constant.

$10^6\varphi_C$	19	58	113	166	235	341	387	499
c_C/(mol/m^3)	0.01	0.29	0.54	0.92	1.00	1.53	1.95	2.40
P_C/Pa	2.10	6.40	12.5	18.3	25.9	37.6	42.7	55.1

$10^6\varphi_C$	609	759	875	908	988
c_C/(mol/m^3)	2.97	3.37	3.61	3.85	4.29
P_C/Pa	67.2	83.7	96.5	100	109

w_B/w_A = 50.0/50.0 was kept constant.

$10^6\varphi_C$	16	69	93	132	202	273	366	449
c_C/(mol/m^3)	0.09	0.57	0.67	0.88	1.34	1.30	1.96	2.24
P_C/Pa	1.76	7.61	10.3	14.6	22.3	30.1	40.4	49.5

$10^6\varphi_C$	509	627	671	705	843	887	923	1098
c_C/(mol/m^3)	2.41	2.21	2.50	2.89	3.21	3.69	3.80	3.93
P_C/Pa	56.2	69.2	74.0	77.8	93.0	97.9	102	121

w_B/w_A = 60.0/40.0 was kept constant.

$10^6\varphi_C$	88	102	127	211	376	409	436	546
c_C/(mol/m^3)	0.65	0.97	0.90	1.35	1.87	2.05	2.30	2.52
P_C/Pa	9.71	11.3	14.0	23.3	41.5	45.1	48.1	60.2

$10^6\varphi_C$	691	725	806	904	1008
c_C/(mol/m^3)	2.83	3.19	3.56	3.73	4.11
P_C/Pa	76.2	80.0	88.9	99.7	111

w_B/w_A = 80.0/20.0 was kept constant.

$10^6\varphi_C$	8	57	89	182	204	265	300	414
c_C/(mol/m^3)	1.21	1.78	2.24	4.30	3.51	2.38	2.38	3.19
P_C/Pa	0.88	6.29	9.82	20.1	22.5	29.2	33.1	45.7

$10^6\varphi_C$	496	611	793	941
c_C/(mol/m^3)	4.24	5.49	7.53	8.48
P_C/Pa	54.7	67.4	87.5	104

w_B/w_A = 100.0/0.0 was kept constant.

$10^6\varphi_C$	24	31	58	102	180	200	286	344
c_C/(mol/m^3)	3.47	3.54	3.76	4.53	5.64	6.69	7.27	8.49
P_C/Pa	2.65	3.42	6.40	11.3	19.9	22.1	31.6	38.0

$10^6\varphi_C$	368	405	471	502	570	635	699	774
c_C/(mol/m^3)	9.01	11.4	10.7	11.0	12.4	13.5	14.7	16.3
P_C/Pa	40.6	44.7	52.0	55.4	62.9	70.1	77.1	85.4

$10^6\varphi_C$	836	900	971
c_C/(mol/m^3)	16.6	17.5	18.0
P_C/Pa	92.2	99.3	107

Comments: w_B/w_A denotes the mass fraction ratio of PEG/water in the liquid phase.
φ_C denotes the volume fraction of SO_2 in the gas phase (i.e., the remaining part is nitrogen).
$c_C/(mol/m^3)$ denotes the concentration of SO_2 in the liquid phase.
P_C denotes the partial pressure of SO_2 in the gas phase.

Polymer (B): **polyglycerol (hyperbranched)** **2009KOZ**
Characterization: $M_n/g.mol^{-1} = 2700$, $M_w/g.mol^{-1} = 4050$, viscous liquid possessing an inert polyether scaffold, HyperPolymers GmbH
Solvent (A): **carbon dioxide** **CO_2** **124-38-9**
Solvent (C): **methanol** **CH_4O** **67-56-1**

Type of data: vapor-liquid equilibrium data

w_A	0.020	0.020	0.020	0.020	0.020	0.020	0.020	0.020	0.020
w_B	0.245	0.245	0.245	0.245	0.245	0.245	0.245	0.245	0.245
w_C	0.735	0.735	0.735	0.735	0.735	0.735	0.735	0.735	0.735
T/K	313.77	322.97	332.98	343.00	353.03	363.03	373.05	383.03	393.03
P/MPa	0.427	0.504	0.599	0.704	0.827	0.966	1.13	1.31	1.53

w_A	0.020	0.020	0.020	0.020	0.020	0.050	0.050	0.050	0.050
w_B	0.245	0.245	0.245	0.245	0.245	0.2375	0.2375	0.2375	0.2375
w_C	0.735	0.735	0.735	0.735	0.735	0.7125	0.7125	0.7125	0.7125
T/K	403.03	413.00	422.97	432.97	442.93	313.82	323.00	333.00	343.00
P/MPa	1.79	2.08	2.43	2.83	3.28	1.03	1.20	1.39	1.61

w_A	0.050	0.050	0.050	0.050	0.050	0.050	0.050	0.050	0.050
w_B	0.2375	0.2375	0.2375	0.2375	0.2375	0.2375	0.2375	0.2375	0.2375
w_C	0.7125	0.7125	0.7125	0.7125	0.7125	0.7125	0.7125	0.7125	0.7125
T/K	353.03	358.03	363.03	372.97	383.04	393.07	403.10	413.09	423.04
P/MPa	1.83	1.95	2.07	2.33	2.61	2.91	3.27	3.64	4.05

w_A	0.050	0.050	0.050	0.100	0.100	0.100	0.100	0.100	0.100
w_B	0.2375	0.2375	0.2375	0.225	0.225	0.225	0.225	0.225	0.225
w_C	0.7125	0.7125	0.7125	0.675	0.675	0.675	0.675	0.675	0.675
T/K	433.01	442.96	452.95	313.02	323.00	333.00	343.02	353.05	362.99
P/MPa	4.51	5.02	5.59	1.97	2.31	2.67	3.05	3.45	3.86

w_A	0.100	0.100	0.100	0.100	0.100	0.100	0.100	0.100	0.100
w_B	0.225	0.225	0.225	0.225	0.225	0.225	0.225	0.225	0.225
w_C	0.675	0.675	0.675	0.675	0.675	0.675	0.675	0.675	0.675
T/K	363.05	373.05	383.04	393.05	403.03	403.06	413.05	423.05	433.02
P/MPa	3.86	4.28	4.72	5.17	5.63	5.63	6.11	6.62	7.15

w_A	0.100	0.100	0.100	0.100	0.100	0.150	0.150	0.150	0.150
w_B	0.225	0.225	0.225	0.225	0.225	0.2125	0.2125	0.2125	0.2125
w_C	0.675	0.675	0.675	0.675	0.675	0.6375	0.6375	0.6375	0.6375
T/K	442.95	445.43	447.91	447.94	450.43	316.48	323.01	333.03	343.03
P/MPa	7.74	7.85	8.00	8.39	9.10	3.13	3.46	4.00	4.56

continued

continued

w_A	0.150	0.150	0.150	0.150	0.150	0.150	0.150	0.150	0.150
w_B	0.2125	0.2125	0.2125	0.2125	0.2125	0.2125	0.2125	0.2125	0.2125
w_C	0.6375	0.6375	0.6375	0.6375	0.6375	0.6375	0.6375	0.6375	0.6375
T/K	353.08	363.09	373.10	382.98	393.09	403.07	413.04	418.04	423.04
P/MPa	5.14	5.71	6.30	6.88	7.45	8.03	8.61	8.75	8.69

w_A	0.150	0.020	0.020	0.020	0.020	0.020	0.020	0.020	0.020
w_B	0.2125	0.490	0.490	0.490	0.490	0.490	0.490	0.490	0.490
w_C	0.6375	0.490	0.490	0.490	0.490	0.490	0.490	0.490	0.490
T/K	428.04	313.56	322.97	332.97	342.99	353.03	363.04	373.08	383.05
P/MPa	8.60	0.639	0.749	0.881	1.02	1.18	1.36	1.56	1.78

w_A	0.020	0.020	0.020	0.020	0.020	0.020	0.020	0.050	0.050
w_B	0.490	0.490	0.490	0.490	0.490	0.490	0.490	0.475	0.475
w_C	0.490	0.490	0.490	0.490	0.490	0.490	0.490	0.475	0.475
T/K	393.07	403.08	413.08	423.07	433.03	442.99	452.98	313.09	322.96
P/MPa	2.03	2.32	2.65	3.02	3.45	3.94	4.50	1.55	1.82

w_A	0.050	0.050	0.050	0.050	0.050	0.050	0.050	0.050	0.050
w_B	0.475	0.475	0.475	0.475	0.475	0.475	0.475	0.475	0.475
w_C	0.475	0.475	0.475	0.475	0.475	0.475	0.475	0.475	0.475
T/K	332.99	343.01	353.16	363.06	373.11	383.10	393.10	403.11	403.14
P/MPa	2.11	2.42	2.76	3.11	3.47	3.86	4.27	4.77	4.71

w_A	0.050	0.050	0.050	0.050	0.050	0.050	0.100	0.100	0.100
w_B	0.475	0.475	0.475	0.475	0.475	0.475	0.450	0.450	0.450
w_C	0.475	0.475	0.475	0.475	0.475	0.475	0.450	0.450	0.450
T/K	413.12	413.13	423.11	433.06	443.03	453.02	312.97	323.01	333.02
P/MPa	5.19	5.24	5.75	6.30	6.90	7.58	3.07	3.61	4.19

w_A	0.100	0.100	0.100	0.100	0.100	0.100	0.100	0.100	0.100
w_B	0.450	0.450	0.450	0.450	0.450	0.450	0.450	0.450	0.450
w_C	0.450	0.450	0.450	0.450	0.450	0.450	0.450	0.450	0.450
T/K	343.04	353.08	363.13	373.08	383.06	393.08	403.12	413.08	413.10
P/MPa	4.80	5.43	6.06	6.75	7.41	8.10	8.80	9.54	9.51

w_A	0.100	0.100	0.100	0.100	0.100	0.100	0.100	0.100	0.150
w_B	0.450	0.450	0.450	0.450	0.450	0.450	0.450	0.450	0.425
w_C	0.450	0.450	0.450	0.450	0.450	0.450	0.450	0.450	0.425
T/K	423.07	425.97	428.05	433.03	433.05	438.02	442.98	452.93	313.14
P/MPa	10.27	10.40	10.57	10.84	10.82	11.06	11.31	11.78	4.59

w_A	0.150	0.150	0.150	0.150	0.150	0.150	0.150	0.150	0.150
w_B	0.425	0.425	0.425	0.425	0.425	0.425	0.425	0.425	0.425
w_C	0.425	0.425	0.425	0.425	0.425	0.425	0.425	0.425	0.425
T/K	322.98	332.99	353.04	363.06	373.05	383.04	393.01	403.05	413.08
P/MPa	5.40	6.28	7.93	8.59	9.19	9.70	10.09	10.38	10.57

continued

continued

w_A	0.150	0.150
w_B	0.425	0.425
w_C	0.425	0.425
T/K	423.05	433.05
P/MPa	10.67	10.68

Comments: Cloud points and VLLE-data are given in Chapter 3.

Polymer (B): **polyglycerol (hyperbranched)** **2010SCH**
Characterization: M_n/g.mol^{-1} = 5700, M_w/M_n = 1.7, synthesized in the laboratory
Solvent (A): **carbon dioxide** **CO_2** **124-38-9**
Solvent (C): **methanol** **CH_4O** **67-56-1**

Type of data: vapor-liquid equilibrium data

w_A	0.020	0.020	0.020	0.020	0.020	0.020	0.020	0.020	0.020
w_B	0.489	0.489	0.489	0.489	0.489	0.489	0.489	0.489	0.489
w_C	0.491	0.491	0.491	0.491	0.491	0.491	0.491	0.491	0.491
T/K	332.70	342.47	352.20	361.89	371.57	381.29	390.99	400.73	410.48
P/MPa	0.941	1.081	1.246	1.407	1.594	1.812	2.042	2.317	2.622
w_A	0.020	0.050	0.050	0.050	0.050	0.050	0.050	0.050	0.050
w_B	0.489	0.474	0.474	0.474	0.474	0.474	0.474	0.474	0.474
w_C	0.491	0.476	0.476	0.476	0.476	0.476	0.476	0.476	0.476
T/K	420.26	332.66	342.41	352.12	352.17	361.82	361.89	371.57	371.57
P/MPa	2.978	2.218	2.523	2.867	2.848	3.202	3.183	3.552	3.539
w_A	0.050	0.050	0.050	0.050	0.050	0.050	0.050	0.050	0.100
w_B	0.474	0.474	0.474	0.474	0.474	0.474	0.474	0.474	0.449
w_C	0.476	0.476	0.476	0.476	0.476	0.476	0.476	0.476	0.451
T/K	381.29	381.31	390.88	400.76	410.51	420.27	439.79	449.55	332.99
P/MPa	3.914	3.933	4.398	4.818	5.268	5.768	6.879	7.494	4.330
w_A	0.100	0.100	0.100	0.100	0.100	0.100	0.100	0.100	0.020
w_B	0.449	0.449	0.449	0.449	0.449	0.449	0.449	0.449	0.245
w_C	0.451	0.451	0.451	0.451	0.451	0.451	0.451	0.451	0.735
T/K	343.00	353.03	363.03	373.04	383.03	393.04	403.05	408.02	332.86
P/MPa	4.691	5.611	6.282	6.952	7.633	9.454	12.141	13.461	0.641
w_A	0.020	0.020	0.020	0.020	0.020	0.020	0.020	0.020	0.020
w_B	0.245	0.245	0.245	0.245	0.245	0.245	0.245	0.245	0.245
w_C	0.735	0.735	0.735	0.735	0.735	0.735	0.735	0.735	0.735
T/K	342.54	352.22	361.90	371.56	381.24	390.96	400.68	405.65	410.45
P/MPa	0.742	0.858	0.993	1.148	1.328	1.528	1.768	1.888	2.043
w_A	0.020	0.020	0.020	0.020	0.050	0.050	0.050	0.050	0.050
w_B	0.245	0.245	0.245	0.245	0.2375	0.2375	0.2375	0.2375	0.2375
w_C	0.735	0.735	0.735	0.735	0.7125	0.7125	0.7125	0.7125	0.7125
T/K	420.19	429.91	439.66	449.37	333.11	343.09	353.12	363.14	373.16
P/MPa	2.366	2.734	3.161	3.889	1.457	1.672	1.902	2.149	2.413

continued

continued

w_A	0.050	0.050	0.050	0.050	0.050	0.050	0.100	0.100	0.100
w_B	0.2375	0.2375	0.2375	0.2375	0.2375	0.2375	0.225	0.225	0.225
w_C	0.7125	0.7125	0.7125	0.7125	0.7125	0.7125	0.675	0.675	0.675
T/K	383.15	393.15	403.14	413.14	423.09	433.07	313.10	323.09	333.12
P/MPa	2.745	3.052	3.381	3.705	4.116	4.573	1.961	2.301	2.639

w_A	0.100	0.100	0.100	0.100	0.100	0.100	0.100
w_B	0.225	0.225	0.225	0.225	0.225	0.225	0.225
w_C	0.675	0.675	0.675	0.675	0.675	0.675	0.675
T/K	353.15	373.24	393.18	398.13	403.15	413.10	423.06
P/MPa	3.452	4.288	5.177	5.396	5.637	6.113	6.613

Comments: Cloud points and VLLE-data are given in Chapter 3.

Polymer (B): **polyglycerol (hyperbranched)** **2010SCH**
Characterization: M_n/g.mol^{-1} = 10000, M_w/M_n = 1.4, synthesized in the laboratory
Solvent (A): **carbon dioxide** CO_2 **124-38-9**
Solvent (C): **methanol** CH_4O **67-56-1**

Type of data: vapor-liquid equilibrium data

w_A	0.023	0.023	0.023	0.023	0.023	0.023	0.023	0.023	0.023
w_B	0.4895	0.4895	0.4895	0.4895	0.4895	0.4895	0.4895	0.4895	0.4895
w_C	0.4875	0.4875	0.4875	0.4875	0.4875	0.4875	0.4875	0.4875	0.4875
T/K	312.25	317.49	322.18	326.92	332.06	337.17	341.98	346.77	351.95
P/MPa	0.993	1.078	1.168	1.258	1.354	1.460	1.559	1.674	1.779

w_A	0.023	0.023	0.023	0.023	0.023	0.023	0.023	0.023	0.023
w_B	0.4895	0.4895	0.4895	0.4895	0.4895	0.4895	0.4895	0.4895	0.4895
w_C	0.4875	0.4875	0.4875	0.4875	0.4875	0.4875	0.4875	0.4875	0.4875
T/K	356.69	361.78	366.60	371.20	381.09	396.24	401.35	411.19	421.16
P/MPa	1.904	2.014	2.146	2.270	2.555	3.035	3.205	3.591	4.011

w_A	0.023	0.023	0.047	0.047	0.047	0.047	0.047	0.047	0.047
w_B	0.4895	0.4895	0.4775	0.4775	0.4775	0.4775	0.4775	0.4775	0.4775
w_C	0.4875	0.4875	0.4755	0.4755	0.4755	0.4755	0.4755	0.4755	0.4755
T/K	431.00	440.91	312.25	322.16	332.12	342.01	351.95	361.83	371.83
P/MPa	4.486	5.032	1.804	2.089	2.409	2.764	3.130	3.511	3.911

w_A	0.047	0.047	0.047	0.047	0.047	0.020	0.020	0.020	0.020
w_B	0.4775	0.4775	0.4775	0.4775	0.4775	0.243	0.243	0.243	0.243
w_C	0.4755	0.4755	0.4755	0.4755	0.4755	0.737	0.737	0.737	0.737
T/K	381.81	391.74	401.65	411.61	421.43	313.22	322.81	332.48	337.26
P/MPa	4.327	4.777	5.242	5.742	6.284	0.525	0.626	0.736	0.790

w_A	0.020	0.020	0.020	0.020	0.020	0.020	0.020	0.020	0.020
w_B	0.243	0.243	0.243	0.243	0.243	0.243	0.243	0.243	0.243
w_C	0.737	0.737	0.737	0.737	0.737	0.737	0.737	0.737	0.737
T/K	342.14	351.83	361.46	370.40	380.92	390.65	400.46	410.25	419.99
P/MPa	0.856	1.006	1.146	1.316	1.512	1.737	1.992	2.287	2.623

continued

continued

w_A	0.020	0.020	0.020	0.047	0.047	0.047	0.047	0.047	0.047
w_B	0.243	0.243	0.243	0.2363	0.2363	0.2363	0.2363	0.2363	0.2363
w_C	0.737	0.737	0.737	0.7167	0.7167	0.7167	0.7167	0.7167	0.7167
T/K	430.45	439.53	449.38	313.37	323.16	332.06	337.99	342.71	352.61
P/MPa	3.008	3.454	3.969	1.167	1.377	1.597	1.714	1.833	2.088
w_A	0.047	0.047	0.047	0.047	0.047	0.047	0.047	0.047	0.047
w_B	0.2363	0.2363	0.2363	0.2363	0.2363	0.2363	0.2363	0.2363	0.2363
w_C	0.7167	0.7167	0.7167	0.7167	0.7167	0.7167	0.7167	0.7167	0.7167
T/K	362.39	366.42	372.20	382.04	391.90	401.72	411.84	420.55	430.36
P/MPa	2.348	2.449	2.639	2.949	3.269	3.625	4.015	4.380	4.826
w_A	0.047	0.098	0.098	0.098	0.098	0.098	0.098	0.098	0.098
w_B	0.2363	0.2237	0.2237	0.2237	0.2237	0.2237	0.2237	0.2237	0.2237
w_C	0.7167	0.6783	0.6783	0.6783	0.6783	0.6783	0.6783	0.6783	0.6783
T/K	435.22	313.17	323.03	332.92	337.87	342.83	347.79	352.76	357.71
P/MPa	5.066	2.221	2.601	3.001	3.216	3.422	3.642	3.857	4.082
w_A	0.098	0.098	0.098						
w_B	0.2237	0.2237	0.2237						
w_C	0.6783	0.6783	0.6783						
T/K	362.66	372.51	382.41						
P/MPa	4.302	4.748	5.213						

Comments: Cloud points and VLLE-data are given in Chapter 3.

Polymer (B): **polyglycerol (hyperbranched)** **2010SCH**
Characterization: M_n/g.mol^{-1} = 18000, M_w/M_n = 1.4, synthesized in the laboratory
Solvent (A): **carbon dioxide** **CO_2** **124-38-9**
Solvent (C): **methanol** **CH_4O** **67-56-1**

Type of data: vapor-liquid equilibrium data

w_A	0.021	0.021	0.021	0.021	0.021	0.021	0.021	0.021	0.021
w_B	0.4885	0.4885	0.4885	0.4885	0.4885	0.4885	0.4885	0.4885	0.4885
w_C	0.4905	0.4905	0.4905	0.4905	0.4905	0.4905	0.4905	0.4905	0.4905
T/K	313.13	327.95	335.35	342.78	357.68	372.49	387.37	402.21	417.12
P/MPa	0.832	1.067	1.177	1.323	1.618	1.954	2.344	2.804	3.355
w_A	0.021	0.021	0.021	0.055	0.055	0.055	0.055	0.055	0.055
w_B	0.4885	0.4885	0.4885	0.4716	0.4716	0.4716	0.4716	0.4716	0.4716
w_C	0.4905	0.4905	0.4905	0.4734	0.4734	0.4734	0.4734	0.4734	0.4734
T/K	432.02	446.93	451.89	313.15	327.93	342.79	347.77	357.68	372.46
P/MPa	4.006	4.801	5.097	1.167	1.377	1.597	1.714	1.833	2.088
w_A	0.055	0.055	0.055	0.021	0.021	0.021	0.021	0.021	0.021
w_B	0.4716	0.4716	0.4716	0.2448	0.2448	0.2448	0.2448	0.2448	0.2448
w_C	0.4734	0.4734	0.4734	0.7342	0.7342	0.7342	0.7342	0.7342	0.7342
T/K	387.35	402.23	417.08	327.88	342.75	350.23	357.63	372.43	387.33
P/MPa	2.348	2.449	2.639	0.651	0.826	0.927	1.037	1.287	1.592

continued

continued

w_A	0.021	0.021	0.021	0.021	0.021	0.049	0.049	0.049	0.049
w_B	0.2448	0.2448	0.2448	0.2448	0.2448	0.2378	0.2378	0.2378	0.2378
w_C	0.7342	0.7342	0.7342	0.7342	0.7342	0.7132	0.7132	0.7132	0.7132
T/K	402.17	417.08	431.97	439.46	446.90	312.11	326.86	341.65	348.75
P/MPa	1.968	2.439	3.014	3.355	3.735	1.167	1.377	1.597	1.714
w_A	0.049	0.049	0.049	0.049	0.049	0.049	0.102	0.102	0.102
w_B	0.2378	0.2378	0.2378	0.2378	0.2378	0.2378	0.2245	0.2245	0.2245
w_C	0.7132	0.7132	0.7132	0.7132	0.7132	0.7132	0.6735	0.6735	0.6735
T/K	356.38	371.12	385.89	400.53	415.15	422.34	312.08	326.77	341.43
P/MPa	1.833	2.088	2.348	2.449	2.639	2.949	2.297	2.887	3.523
w_A	0.102								
w_B	0.2245								
w_C	0.6735								
T/K	348.57								
P/MPa	3.837								

Comments: Cloud points and VLLE-data are given in Chapter 3.

Polymer (B): **polyglycerol (linear)** **2011SCH**
Characterization: M_n/g.mol^{-1} = 3900, M_w/M_n = 1.4, synthesized in the laboratory
Solvent (A): **carbon dioxide** CO_2 **124-38-9**
Solvent (C): **methanol** CH_4O **67-56-1**

Type of data: vapor-liquid equilibrium data

w_A	0.053	0.053	0.053	0.053	0.053	0.053	0.053	0.053	0.105
w_B	0.473	0.473	0.473	0.473	0.473	0.473	0.473	0.473	0.447
w_C	0.474	0.474	0.474	0.474	0.474	0.474	0.474	0.474	0.448
T/K	333.16	343.12	353.00	363.04	373.11	383.08	393.10	403.09	331.53
P/MPa	2.412	2.755	3.122	3.512	3.913	4.348	4.798	5.286	4.382
w_A	0.105	0.105	0.105	0.105	0.105	0.105	0.105	0.050	0.050
w_B	0.447	0.447	0.447	0.447	0.447	0.447	0.447	0.2375	0.2375
w_C	0.448	0.448	0.448	0.448	0.448	0.448	0.448	0.7125	0.7125
T/K	341.46	351.47	361.43	371.43	381.35	391.29	401.22	333.03	343.06
P/MPa	5.042	5.722	6.418	7.133	7.866	8.602	9.353	1.579	1.797
w_A	0.050	0.050	0.050	0.050	0.050	0.050	0.050	0.050	0.100
w_B	0.2375	0.2375	0.2375	0.2375	0.2375	0.2375	0.2375	0.2375	0.225
w_C	0.7125	0.7125	0.7125	0.7125	0.7125	0.7125	0.7125	0.7125	0.675
T/K	353.09	363.15	373.22	383.23	393.21	403.09	413.09	423.02	331.59
P/MPa	2.030	2.280	2.555	2.838	3.156	3.499	3.864	4.289	2.948
w_A	0.100	0.100	0.100	0.100	0.100	0.100	0.100	0.100	0.100
w_B	0.225	0.225	0.225	0.225	0.225	0.225	0.225	0.225	0.225
w_C	0.675	0.675	0.675	0.675	0.675	0.675	0.675	0.675	0.675
T/K	341.56	351.55	361.54	371.58	381.52	391.51	401.48	411.46	421.17
P/MPa	3.340	3.748	4.166	4.599	5.042	5.500	5.968	6.458	6.959

Polymer (B): **polyglycerol (linear)** **2011SCH**
Characterization: M_n/g.mol^{-1} = 4800, M_w/M_n = 1.2, synthesized in the laboratory
Solvent (A): **carbon dioxide** CO_2 **124-38-9**
Solvent (C): **methanol** CH_4O **67-56-1**

Type of data: vapor-liquid equilibrium data

w_A	0.052	0.052	0.052	0.052	0.052	0.052	0.052	0.052	0.052
w_B	0.475	0.475	0.475	0.475	0.475	0.475	0.475	0.475	0.475
w_C	0.473	0.473	0.473	0.473	0.473	0.473	0.473	0.473	0.473
T/K	333.12	343.10	353.04	363.03	373.04	383.01	393.00	403.01	412.98
P/MPa	2.430	2.790	3.181	3.581	3.999	4.442	4.907	5.408	5.943
w_A	0.052	0.100	0.100	0.100	0.100	0.100	0.100	0.100	0.100
w_B	0.475	0.451	0.451	0.451	0.451	0.451	0.451	0.451	0.451
w_C	0.473	0.449	0.449	0.449	0.449	0.449	0.449	0.449	0.449
T/K	422.99	331.74	341.70	351.68	361.64	371.63	381.56	391.19	401.13
P/MPa	6.517	4.537	5.258	5.944	6.682	7.432	8.200	8.976	9.776
w_A	0.050	0.050	0.050	0.050	0.050	0.050	0.050	0.050	0.050
w_B	0.2375	0.2375	0.2375	0.2375	0.2375	0.2375	0.2375	0.2375	0.2375
w_C	0.7125	0.7125	0.7125	0.7125	0.7125	0.7125	0.7125	0.7125	0.7125
T/K	333.04	343.03	353.07	363.12	373.15	383.13	393.11	403.10	413.05
P/MPa	1.403	1.614	1.839	2.096	2.364	2.654	2.965	3.305	3.681
w_A	0.050	0.100	0.100	0.100	0.100	0.100	0.100	0.100	0.100
w_B	0.2375	0.225	0.225	0.225	0.225	0.225	0.225	0.225	0.225
w_C	0.7125	0.675	0.675	0.675	0.675	0.675	0.675	0.675	0.675
T/K	423.02	331.68	341.66	351.65	361.76	371.62	381.67	391.31	401.27
P/MPa	4.101	2.748	3.131	3.541	3.957	4.382	4.825	5.283	5.751
w_A	0.100	0.100							
w_B	0.225	0.225							
w_C	0.675	0.675							
T/K	411.25	421.23							
P/MPa	6.234	6.744							

Comments: Cloud points and VLLE-data are given in Chapter 3.

Polymer (B): **polyglycerol (linear)** **2011SCH**
Characterization: M_n/g.mol^{-1} = 7800, M_w/M_n = 1.3, synthesized in the laboratory
Solvent (A): **carbon dioxide** CO_2 **124-38-9**
Solvent (C): **methanol** CH_4O **67-56-1**

Type of data: vapor-liquid equilibrium data

w_A	0.050	0.050	0.050	0.050	0.050	0.050	0.050	0.050	0.050
w_B	0.476	0.476	0.476	0.476	0.476	0.476	0.476	0.476	0.476
w_C	0.474	0.474	0.474	0.474	0.474	0.474	0.474	0.474	0.474
T/K	333.12	353.07	363.07	373.05	383.06	393.04	403.04	413.04	423.01
P/MPa	2.391	3.082	3.467	3.858	4.257	4.708	5.182	5.693	6.244

continued

continued

w_A	0.100	0.100	0.100	0.100	0.100	0.100	0.050	0.050	0.050
w_B	0.451	0.451	0.451	0.451	0.451	0.451	0.2375	0.2375	0.2375
w_C	0.449	0.449	0.449	0.449	0.449	0.449	0.7125	0.7125	0.7125
T/K	331.75	341.40	351.71	361.63	371.61	381.56	333.07	343.04	353.02
P/MPa	4.243	4.884	5.559	6.245	6.945	7.660	1.480	1.697	1.930

w_A	0.050	0.050	0.050	0.050	0.050	0.050	0.050	0.100	0.100
w_B	0.2375	0.2375	0.2375	0.2375	0.2375	0.2375	0.2375	0.225	0.225
w_C	0.7125	0.7125	0.7125	0.7125	0.7125	0.7125	0.7125	0.675	0.675
T/K	362.97	372.97	382.96	392.97	402.99	412.97	422.96	331.83	341.68
P/MPa	2.173	2.441	2.731	3.041	3.382	3.757	4.202	2.694	3.077

w_A	0.100	0.100	0.100	0.100	0.100	0.100	0.100
w_B	0.225	0.225	0.225	0.225	0.225	0.225	0.225
w_C	0.675	0.675	0.675	0.675	0.675	0.675	0.675
T/K	351.62	361.59	371.56	381.51	391.48	401.44	411.45
P/MPa	3.485	3.903	4.336	4.778	5.236	5.712	6.197

Comments: Cloud points and VLLE-data are given in Chapter 3.

Polymer (B): **poly(methyl methacrylate)** **2008MAT**
Characterization: M_w/g.mol^{-1} = 15000, Aldrich Chem. Co., Inc., Milwaukee, WI
Solvent (A): **carbon dioxide** **CO_2** **124-38-9**
Solvent (C): **ethanol** **C_2H_6O** **64-17-5**

Type of data: vapor-liquid equilibrium data

w_A	0.114	0.114	0.114	0.114	0.114	0.114	0.114	0.114	0.114
w_B	0.045	0.045	0.045	0.045	0.045	0.045	0.045	0.045	0.045
w_C	0.841	0.841	0.841	0.841	0.841	0.841	0.841	0.841	0.841
T/K	293.3	298.4	303.2	313.2	323.4	333.2	343.4	353.5	359.7
P/MPa	2.0	2.2	2.3	2.7	2.9	3.4	3.8	4.1	4.8

w_A	0.414	0.414	0.414	0.414	0.414	0.414	0.414	0.414	0.414
w_B	0.029	0.029	0.029	0.029	0.029	0.029	0.029	0.029	0.029
w_C	0.557	0.557	0.557	0.557	0.557	0.557	0.557	0.557	0.557
T/K	308.7	310.4	311.4	313.3	313.8	318.4	323.0	328.5	333.0
P/MPa	5.6	5.6	5.7	5.8	5.9	6.1	6.5	6.9	7.2

w_A	0.414	0.414	0.414	0.512	0.512	0.512	0.512	0.512	0.512
w_B	0.029	0.029	0.029	0.025	0.025	0.025	0.025	0.025	0.025
w_C	0.557	0.557	0.557	0.463	0.463	0.463	0.463	0.463	0.463
T/K	343.0	348.3	353.2	313.2	318.7	323.4	333.4	343.5	348.5
P/MPa	8.1	8.6	9.0	6.7	7.2	7.7	8.7	9.9	10.3

w_A	0.512
w_B	0.025
w_C	0.463
T/K	353.0
P/MPa	10.7

Polymer (B): **polypropylene (isotactic)** **2013CHE1**
Characterization: M_w/g.mol^{-1} = 197000, M_w/M_n = 5.1, T_m/K = 441, crystallinity = 47.8%, Shanghai Petrochemical Co., China
Solvent (A): **carbon dioxide** **CO_2** **124-38-9**
Polymer (C): **poly(propylene-*g*-maleic anhydride)**
Characterization: synthesized in the laboratory

Type of data: gas solubility (with swelling correction)

T/K = 473.15

P/MPa	5	6	7	8	9	10	13
c_A/(g CO_2/g polymer B+C)	0.0356	0.0459	0.0536	0.0617	0.0701	0.0794	0.0985

P/MPa	15	17	20	22
c_A/(g CO_2/g polymer B+C)	0.1127	0.1300	0.1475	0.1571

T/K = 493.15

P/MPa	5	6	7	8	9	10	13
c_A/(g CO_2/g polymer B+C)	0.0307	0.0388	0.0460	0.0534	0.0593	0.0667	0.0835

P/MPa	15	17	20	22
c_A/(g CO_2/g polymer B+C)	0.0942	0.1064	0.1227	0.1312

Comments: CO_2-solubilities in various PP/$CaCO_3$ composites are additionally given in the original source.

Polymer (B): **polystyrene** **2013GUT**
Characterization: M_w/g.mol^{-1} = 298000
Solvent (A): **carbon dioxide** **CO_2** **124-38-9**
Solvent (C): **R-(+)-limonene** **$C_{10}H_{16}$** **5989-27-5**

Type of data: vapor-liquid equilibrium data

T/K = 298.15

P/bar	48.01	79.15	96.56	140.94	54.33	70.86	101.40	163.64
w_A(gas phase)	0.9869	0.9755	0.9372	0.9267	0.9887	0.9610	0.9449	0.9407
w_B(gas phase)	0.000	0.000	0.000	0.000	0.000	0.000	0.000	0.000
w_C(gas phase)	0.0131	0.0245	0.0628	0.0733	0.0113	0.0390	0.0551	0.0593
w_A(liquid phase)	0.0519	–	0.0320	0.0381	0.0385	–	0.0353	–
w_B(liquid phase)	0.0480	–	0.2105	0.1925	0.1335	–	0.5069	–
w_C(liquid phase)	0.9002	–	0.7576	0.7694	0.8279	–	0.4579	–

P/bar	52.06	85.30	91.21	144.00
w_A(gas phase)	0.9900	0.9641	0.9575	0.9451
w_B(gas phase)	0.000	0.000	0.000	0.000
w_C(gas phase)	0.0100	0.0359	0.0425	0.0549
w_A(liquid phase)	0.0258	–	0.0409	0.0714
w_B(liquid phase)	0.2597	–	0.6390	0.5409
w_C(liquid phase)	0.7145	–	0.3202	0.3877

continued

continued

T/K = 303.15

P/bar	54.86	82.11	90.90	156.69	50.78	79.03	105.28	145.41
w_A(gas phase)	0.9873	0.9724	0.9335	0.9235	0.9897	0.9741	0.9716	0.9648
w_B(gas phase)	0.000	0.000	0.000	0.000	0.000	0.000	0.000	0.000
w_C(gas phase)	0.0127	0.0276	0.0665	0.0765	0.0103	0.0259	0.0284	0.0352
w_A(liquid phase)	0.0235	0.0563	0.0480	–	0.1130	0.0217	0.0247	–
w_B(liquid phase)	0.0796	0.1233	0.1353	–	0.1377	0.1729	0.2139	–
w_C(liquid phase)	0.8969	0.8203	0.8167	–	0.7493	0.8054	0.7614	–

P/bar	54.76	74.42	105.49	155.46
w_A(gas phase)	0.9938	0.9802	0.9708	0.9686
w_B(gas phase)	0.000	0.000	0.000	0.000
w_C(gas phase)	0.0062	0.0198	0.0292	0.0314
w_A(liquid phase)	0.0450	–	0.0635	–
w_B(liquid phase)	0.2586	–	0.6118	–
w_C(liquid phase)	0.6964	–	0.3248	–

T/K = 313.15

P/bar	46.50	76.61	101.00	137.79	53.47	80.44	102.75	50.47
w_A(gas phase)	0.9792	0.9610	0.9212	0.9005	0.9860	0.9794	0.9623	0.9871
w_B(gas phase)	0.000	0.000	0.000	0.000	0.000	0.000	0.000	0.000
w_C(gas phase)	0.0208	0.0390	0.0788	0.0995	0.0140	0.0206	0.0377	0.0129
w_A(liquid phase)	0.0210	–	0.0463	0.0154	0.0549	0.0311	0.1081	0.0667
w_B(liquid phase)	0.0618	–	0.2842	0.1174	0.1380	0.1053	0.2300	0.2229
w_C(liquid phase)	0.9172	–	0.6695	0.8672	0.8072	0.8635	0.6619	0.7104

P/bar	74.92	97.89	124.68	57.40	100.12	155.42	56.35	102.31
w_A(gas phase)	0.9840	0.9743	0.9110	0.9968	0.9892	0.9829	0.9970	0.9955
w_B(gas phase)	0.000	0.000	0.000	0.000	0.000	0.000	0.000	0.000
w_C(gas phase)	0.0160	0.0257	0.0890	0.0032	0.0108	0.0171	0.0030	0.0045
w_A(liquid phase)	0.0498	0.0302	0.0391	0.0818	0.2215	0.3036	0.0989	0.1033
w_B(liquid phase)	0.1838	0.4586	0.4986	0.3837	0.6616	0.5769	0.4187	0.7519
w_C(liquid phase)	0.7664	0.5112	0.4623	0.5345	0.1169	0.1195	0.4824	0.1448

P/bar	149.90	54.73	95.28	149.22	46.76	103.26	148.34	54.00
w_A(gas phase)	0.9843	0.9981	0.9973	0.9910	0.9984	0.9977	0.9903	0.9987
w_B(gas phase)	0.000	0.000	0.000	0.000	0.000	0.000	0.000	0.000
w_C(gas phase)	0.0157	0.0019	0.0027	0.0090	0.0016	0.0023	0.0097	0.0013
w_A(liquid phase)	0.0554	0.1420	0.0752	0.0786	0.1404	0.0708	0.0552	0.1026
w_B(liquid phase)	0.8063	0.4329	0.7911	0.7700	0.4893	0.8311	0.8634	0.5479
w_C(liquid phase)	0.1383	0.4251	0.1337	0.1514	0.3703	0.0981	0.0814	0.3495

P/bar	103.58	150.77
w_A(gas phase)	0.9981	0.9923
w_B(gas phase)	0.000	0.000
w_C(gas phase)	0.0019	0.0077
w_A(liquid phase)	0.0491	0.1573
w_B(liquid phase)	0.8965	0.7537
w_C(liquid phase)	0.0544	0.0890

Polymer (B): **polystyrene** **2008LIG**
Characterization: M_w/g.mol^{-1} = 196000, T_g/K = 381, ρ = 1.048 g/cm^3, Styron PS865D, Dow Chemical Company
Solvent (A): **1,1-difluoroethane** $C_2H_4F_2$ **75-37-6**
Solvent (C): **1,1,1,2-tetrafluoroethane** $C_2H_2F_4$ **811-97-2**

Type of data: gas solubility

T/K = 403.15

P/MPa	1.47	2.79	1.24	2.75	1.53	2.82
c_{A+C}/(g (A+C)/g polymer)	0.0289	0.0593	0.0227	0.0550	0.0276	0.0523
w_A/w_C (in the gas phase)	75/25	75/25	50/50	50/50	25/75	25/75

T/K = 423.15

P/MPa	1.50	2.72	4.12	5.50	1.32	2.79	4.19
c_{A+C}/(g (A+C)/g polymer)	0.0263	0.0490	0.0790	0.0965	0.0208	0.0459	0.0705
w_A/w_C (in the gas phase)	75/25	75/25	75/25	75/25	50/50	50/50	50/50

P/MPa	5.58	1.35	2.79	4.18	5.58
c_{A+C}/(g (A+C)/g polymer)	0.0847	0.0204	0.0432	0.0629	0.0733
w_A/w_C (in the gas phase)	50/50	25/75	25/75	25/75	25/75

T/K = 463.15

P/MPa	1.37	2.76	4.19	5.59	1.41	2.81	4.20
c_{A+C}/(g (A+C)/g polymer)	0.0156	0.0329	0.0499	0.0659	0.0157	0.0320	0.0475
w_A/w_C (in the gas phase)	75/25	75/25	75/25	75/25	50/50	50/50	50/50

P/MPa	5.60	1.39	2.78	4.19	5.59
c_{A+C}/(g (A+C)/g polymer)	0.0608	0.0137	0.0283	0.0426	0.0538
w_A/w_C (in the gas phase)	50/50	25/75	25/75	25/75	25/75

Comments: Calculated values of theoretical gas solubilities of the single gases A or C are given in the original source. The used thermodynamic model is based on the Simha-Somcynsky equation of state.

Polymer (B): **poly(styrene-*co*-butadiene)** **2005WUH**
Characterization: M_w/g.mol^{-1} = 160000, 18 wt% styrene, SBR, Petrofina, Inc., Brussels, Belgium
Solvent (A): **ethene** C_2H_4 **74-85-1**
Solvent (C): **toluene** C_7H_8 **108-88-3**

Type of data: gas solubility

Comments: w_B/w_C = 10/90 was kept constant.

T/K = 293.15

P/bar	3.45	5.52	6.89	10.34	12.41
w_A	0.0142	0.0259	0.0313	0.0452	0.0596

continued

continued

T/K = 303.15

P/bar	3.45	5.52	6.89	10.34	12.41
w_A	0.0118	0.0213	0.0255	0.0411	0.0545

T/K = 313.15

P/bar	3.45	5.52	6.89	10.34	12.41
w_A	0.0119	0.0143	0.0231	0.0322	0.0434

T/K = 323.15

P/bar	3.45	5.52	6.89	10.34	12.41
w_A	0.0093	0.0124	0.0202	0.0247	0.0389

T/K = 333.15

P/bar	3.45	5.52	6.89	10.34	12.41
w_A	0.0063	0.0123	0.0182	0.0253	0.0319

T/K = 343.15

P/bar	3.45	5.52	6.89	10.34	12.41
w_A	0.0054	0.0128	0.0180	0.0259	0.0293

Polymer (B): **poly(vinylidene fluoride-*co*-chlorotrifluoro-ethylene)** **2005SO1**
Characterization: see comments
Solvent (A): **carbon dioxide** **CO_2** **124-38-9**
Polymer (C): **poly(vinylidene fluoride)**
Characterization: see comments

Comments: The polymer mixture is a commercial polymer SOLEF VF2-CTFE copolymer grade 60512 consisting of 67% (B) and 33% (C) plus a small amount of high-density polyethylene. The density of this alloy is 1.77 g/cm^3, its crystallinity is 45% (by DSC).

Type of data: gas solubility

T/K = 353.15

P/bar	19.1	19.4	31.2	31.0	38.9	38.78
w_A	0.00633	0.00675	0.00848	0.00883	0.01031	0.01388

T/K = 373.15

P/bar	20.1	20.1	28.1	28.1	37.6	37.6
w_A	0.00712	0.00614	0.00890	0.00867	0.01098	0.01106

T/K = 393.15

P/bar	18.6	18.6	29.0	28.5	28.5	29.0	38.4	38.4
w_A	0.00570	0.00411	0.00759	0.00743	0.00599	0.00538	0.00932	0.00854

Polymer (B): **poly(vinylidene fluoride-*co*-chlorotrifluoro-ethylene)** **2005SO1**

Characterization: see comments

Solvent (A): **methane** CH_4 **74-82-8**

Polymer (C): **poly(vinylidene fluoride)**

Characterization: see comments

Comments: The polymer mixture is a commercial polymer SOLEF VF2-CTFE copolymer grade 60512 consisting of 67% (B) and 33% (C) plus a small amount of high-density polyethylene. The density of this alloy is 1.77 g/cm^3, its crystallinity is 45% (by DSC).

Type of data: gas solubility

T/K = 353.15

P/bar	53.7	53.5	104.8	104.8	158.0	158.3
w_A	0.00148	0.00147	0.00387	0.00220	0.00474	0.00336

T/K = 373.15

P/bar	51.4	51.4	50.0	50.0	100.1	100.1	158.2	158.2
w_A	0.00131	0.00051	0.00157	0.00069	0.00255	0.00120	0.00341	0.00276

T/K = 393.15

P/bar	42.4	54.2	54.2	71.9	97.9	97.9	158.5	158.3
w_A	0.00116	0.00111	0.00061	0.00140	0.00217	0.00139	0.00331	0.00250

2.4. Table of ternary or quaternary systems where data were published only in graphical form as phase diagrams or related figures

Polymer (B)	Second and third component	Ref.
Cellulose		
	carbon dioxide and natural rubber	2005NUN
	ethene and natural rubber	2005NUN
	methane and natural rubber	2005NUN
	oxygen and natural rubber	2005NUN
	propene and natural rubber	2005NUN
Natural rubber		
	carbon dioxide and cellulose	2005NUN
	ethene and cellulose	2005NUN
	methane and cellulose	2005NUN
	oxygen and cellulose	2005NUN
	propene and cellulose	2005NUN
Nitrocellulose		
	carbon dioxide and ethanol	2012SUN
	carbon dioxide and ethyl acetate	2012SUN
Poly(acrylamide)		
	carbon dioxide and monoethanolamine	2007PAR
Poly(*tert*-butylacetylene)		
	carbon dioxide and poly(1-trimethylsilyl-1-propyne)	2005NAG
Polycarbonate-bisphenol A		
	carbon dioxide and ethanol/2-propanone	2009HE2
	carbon dioxide and 2-propanone	2009HE2

Polymer (B)	Second and third component	Ref.
Poly(dimethylsiloxane)		
	n-butane and methane	2007RA1
	carbon dioxide and toluene	2002SHI
Poly(etherimide)		
	carbon dioxide and dichloromethane	2009LAW
Polyethylene		
	carbon dioxide and methane	2012MEM
	ethene and n-hexane	2007YA2
	ethene and isopentane	2007YA2
	hydrogen and methane	2012MEM
Poly(ethylene-*co*-1-hexene)		
	ethene and 1-hexene	2006NOV
Poly(ethylene glycol)		
	carbon dioxide and polystyrene	2005TA2
	carbon dioxide and poly(1-vinyl-2-pyrrolidinone)	2006FLE
	sulfur dioxide/nitrogen and water	2013HEZ
Poly(ethylene oxide-*b*-propylene oxide-*b*-ethylene oxide)		
	ethene and 1,4-dimethylbenzene	2005ZHA
	ethene and water/1,4-dimethylbenzene	2005ZHA
Poly(ethylene terephthalate)		
	carbon dioxide and 1-butanol	2005HIR
	carbon dioxide and ethanol	2005HIR
	carbon dioxide and methanol	2005HIR
	carbon dioxide and nitrogen	2007FAR
	carbon dioxide and n-pentane	2007FAR
	carbon dioxide and 1-propanol	2005HIR
Poly(DL-lactic acid)		
	carbon dioxide and poly(L-lactic acid)	2012LIA
	carbon dioxide and tetrahydrofuran	2004XUQ
	carbon dioxide and trichloromethane	2004XUQ
	carbon dioxide and starch	2014MAR

Polymer (B)	Second and third component	Ref.
Poly(L-lactic acid)		
	carbon dioxide and poly(DL-lactic acid)	2012LIA
	carbon dioxide and poly(methyl methacrylate)	2007YA3
Poly(methyl methacrylate)		
	carbon dioxide and methyl methacrylate	2009HE2
	carbon dioxide and methyl methacrylate/poly(vinylidene fluoride)	2009HE2
	carbon dioxide and naphthalene	2006UEZ
	carbon dioxide and poly(L-lactic acid)	2007YA3
	carbon dioxide and poly(methyl methacrylate-*b*-butyl acrylate-*b*-methyl methacrylate)	2014PIN
	carbon dioxide and polystyrene	2013INC
	carbon dioxide and 2-propanone	2009HE1
Poly(methyl methacrylate-*b*-butyl acrylate-*b*-methyl methacrylate)		
	carbon dioxide and poly(methyl methacrylate)	2014PIN
Poly(4-methyl-2-pentyne)		
	carbon dioxide and poly(1-trimethylsilyl-1-propyne)	2005NAG
Poly[oligo(ethylene glycol) methyl ether acrylate-*co*-oligo(ethylene glycol) diacrylate]		
	carbon dioxide and ethane	2010RI2
Poly(1-phenyl-1-propyne)		
	carbon dioxide and poly(1-trimethylsilyl-1-propyne)	2005NAG
Polypropylene		
	carbon dioxide and nitrogen	2009HEI
Polystyrene		
	benzene and carbon dioxide	2008JON
	benzene and ethene	2008JON
	carbon dioxide and decahydronapthalene	2005XUD
	carbon dioxide and 1,4-dimethylbenzene	2008JON
	carbon dioxide and ethylbenzene	2008JON
	carbon dioxide and methanol	2008JON

Polymer (B)	**Second and third component**	**Ref.**
	carbon dioxide and poly(ethylene glycol)	2005TA2
	carbon dioxide and poly(methyl methacrylate)	2013INC
	carbon dioxide and 2-propanol	2007DON
	carbon dioxide and 2-propanone	2009HE1
	carbon dioxide and toluene	2008JON
	carbon dioxide and n-pentane	2011CHM
	ethene and 1,4-dimethylbenzene	2008JON
	ethene and ethylbenzene	2008JON
	ethene and methanol	2008JON
	ethene and toluene	2008JON
	isopentane and n-pentane	2011CHM
	nitrogen and n-pentane	2011CHM
Poly(1-trimethylsilyl-1-propyne)		
	n-butane and methane	2014VOP
	n-butane and methane	2007RA2
	carbon dioxide and methane	2014VOP
	carbon dioxide and poly(*tert*-butylacetylene)	2005NAG
	carbon dioxide and poly(4-methyl-2-pentyne)	2005NAG
	carbon dioxide and poly(1-phenyl-1-propyne)	2005NAG
Poly(vinyl acetate)		
	carbon dioxide and methyl acetate	2008JON
	carbon dioxide and vinyl acetate	2008JON
	ethene and methyl acetate	2008JON
	ethene and vinyl acetate	2008JON
	nitrogen and toluene	2008POU
Poly(vinylidene fluoride)		
	carbon dioxide and methyl methacrylate	2009HE2
	carbon dioxide and methyl methacrylate/ poly(methyl methacrylate)	2009HE2
	carbon dioxide and vinylidene fluoride	2008GAL
Poly(1-vinyl-2-pyrrolidinone)		
	carbon dioxide and poly(ethylene glycol)	2006FLE
Starch		
	carbon dioxide and poly(DL-lactic acid)	2014MAR
	carbon dioxide and water	2006CHE

2.5. References

1998CHA Chaudhary, B.I. and Johns, A.I., Solubilities of nitrogen, isobutane and carbon dioxide in polyethylene, *J. Cell. Plast.*, 34, 312, 1998.

2002SHI Shim, J.J., Distribution of solutes between polymer and supercritical fluid by inverse supercritical fluid chromatography, *Korean J. Chem. Eng.*, 19, 146, 2002.

2003KUK Kukova, E., Phasenverhalten und Transporteigenschaften binärer Systeme aus hochviskosen Polyethylenglykolen und Kohlendioxid, *Dissertation*, Ruhr-Universität Bochum, 2003.

2004ARE Areerat, S., Funami, E., Hayata, Y., Nakagawa, D., and Ohshima, M., Measurement and prediction of diffusion coefficients of supercritical CO_2 in molten polymers, *Polym. Eng. Sci.*, 44, 1915, 2004.

2004CAS Casimiro, T., Shariati, A., Peters, C.J., Nunes da Ponte, M., and Aguiar-Ricardo, A., Phase behavior studies of a perfluoropolyether in high-pressure carbon dioxide, *Fluid Phase Equil.*, 224, 257, 2004.

2004GUA Guadagno, T. and Kazarian, S.G., High-pressure CO_2-expanded solvents: simultaneous measurement of CO_2 sorption and swelling of liquid polymers with *in-situ* near-IR spectroscopy, *J. Phys. Chem. B*, 108, 13995, 2004.

2004KUK Kukova, E., Petermann, M., and Weidner, E., Phasenverhalten (S-L-G) und Transporteigenschaften binaerer Systeme aus hochviskosen Polyethylenglycolen und komprimiertem Kohlendioxid, *Chem. Ing. Techn.*, 76, 280, 2004.

2004LIN Lin, H. and Freeman, B.D., Gas solubility, diffusivity and permeability in poly(ethylene oxide), *J. Membrane Sci.*, 239, 105, 2004.

2004OLI Oliveira, N.S., Oliveira, J., Gomes, T., Ferreira, A., Dorgan, J., and Marrucho, I.M., Gas sorption in poly(lactic acid) and packaging materials, *Fluid Phase Equil.*, 222-223, 317, 2004.

2004PHA Pham, V.Q., Rao, N., and Ober, C.K., Swelling and dissolution rate measurements of polymer thin films in supercritical carbon dioxide, *J. Supercrit. Fluids*, 31, 323, 2004.

2004PRA Prabhakar, R.S., Freeman, B.D., and Roman, I., Gas and vapor sorption and permeation in poly(2,2,4-trifluoro-5-trifluoromethoxy-1,3-dioxole-*co*-tetrafluoroethylene), *Macromolecules*, 37, 7688, 2004.

2004SAT Sato, Y., Wang, M., Takishima, S., Masuoka, H., Watanabe, T., and Fukasawa, Y., Solubility of butane and isobutane in molten polypropylene and polystyrene, *Polym. Eng. Sci.*, 44, 2083, 2004.

2004TA1 Tang, M., Huang,Y.-C., and Chen, Y.-P., Sorption and diffusion of supercritical carbon dioxide into polysulfone, *J. Appl. Polym. Sci.*, 94, 474, 2004.

2004TA2 Tang, M., Du, T.-B., and Chen, Y.-P., Sorption and diffusion of supercritical carbon dioxide in polycarbonate, *J. Supercrit. Fluids*, 28, 207, 2004.

2004XUQ Xu, Q., Pang, P., Peng, Q., Li, J., and Jiang, Y., Application of supercritical carbon dioxide in the preparation of biodegradable polylactide membranes, *J. Appl. Polym. Sci.*, 94, 2158, 2004.

2005BOY Boyer, S.A.E. and Grolier, J.-P.E., Simultaneous measurement of the concentration of a supercritical gas absorbed in a polymer and of the concomitant change in volume of the polymer. The coupled VW-*pVT* technique revisited, *Polymer*, 46, 3737, 2005.

2005CAR Carla, V., Wang, K., Hussain, Y., Efimenko, K., Genzer, J., Grant, C., Sarti, G.C., Carbonell, R.G., and Doghieri, F., Nonequilibrium model for sorption and swelling of bulk glassy polymer films with supercritical carbon dioxide, *Macromolecules*, 38, 10299, 2005.

2005CAS Casimiro, T., Shariati, A., Peters, C.J., Nunes da Ponte, M., and Aguiar-Ricardo, A., Phase behavior studies of a perfluoropolyether in high-pressure carbon dioxide, *Fluid Phase Equil.*, 228-229, 367, 2005.

2005CHO Choudhary, M., Delaviz, Y., Loh, R., Polaksy, M., Wan, C., Todd, D.B. Hyun, K.S., Dey, S., and Wu, F., Measurement of shear viscosity and solubility of polystyrene melts containing various blowing agents, *J. Cell. Plast.*, 41, 589, 2005.

2005COT Cotugno, S., DiMaio, E., Mensitieri, G., Iannace, S., Roberts, G.W., Carbonell, R.G., and Hopfenberg, H.B., Characterization of microcellular biodegradable polymeric foams produced from supercritical carbon dioxide solutions, *Ind. Eng. Chem. Res.*, 44, 1795, 2005.

2005DUA Duarte, A.R.C., Anderson, L.E., Duarte, C.M.M., and Kazarian, S.G., A comparison between gravimetric and *in-situ* spectroscopic methods to measure the sorption of CO_2 in a biocompatible polymer, *J. Supercrit. Fluids*, 36, 160, 2005.

2005ELV Elvassore, N., Vezzu, K., and Bertucco, A., Measurement and modeling of CO_2 absorption in poly(lactic-*co*-glycolic acid), *J. Supercrit. Fluids*, 33, 1, 2005.

2005FUJ Fujiwara, T., Yamaoka, T., Kimura, Y, and Wynne, K.J., Poly(lactide) swelling and melting behavior in supercritical carbon dioxide and post-venting porous material, *Biomacromolecules*, 6, 2370, 2005.

2005HIR Hirogaki, K., Tabata, I., Hisada, K., and Hori, T., An investigation of the interaction of supercritical carbon dioxide with poly(ethylene terephthalate) and the effects of some additive modifiers on the interaction, *J. Supercrit. Fluids*, 36, 166, 2005.

2005KOJ Kojima, M., Tosaka, M., Funami, E., Nitta, K., Ohshima, M., and Kohjiya, S., Phase behavior of crosslinked polyisoprene rubber and supercritical carbon dioxide, *J. Supercrit. Fluids*, 35, 175, 2005.

2005LEE Lee, S.-H., Phase behavior of binary and ternary mixtures of poly(ethylene-*co*-octene)–hydrocarbons (experimental data by S.-H. Lee), *J. Appl. Polym. Sci.*, 95, 161, 2005.

2005LIN Lin, H. and Freeman, B.D., Gas and vapor solubility in cross-linked poly(ethylene glycol diacrylate), *Macromolecules*, 38, 8394, 2008.

2005LIU Liu, D., Li, H., Noon, M.S., and Tomasko, D.L., CO_2-induced PMMA swelling and multiple thermodynamic property analysis using Sanchez-Lacombe EOS, *Macromolecules*, 38, 4416, 2005.

2005LOP Lopez-Gonzalez, M.M., Saiz, E., and Riande, E., Experimental and simulation studies of gas sorption processes in polycarbonate films, *Polymer*, 46, 4322, 2005.

2005MAD Madden, W.C., Punsalan, D., and Koros, W.J., Age-dependent CO_2 sorption in Matrimid asymmetric hollow fiber membranes, *Polymer*, 46, 5433, 2005.

2005MAN Manninen, A.R., Naguib, H.E., A. Nawaby, V., and Day, M., CO_2 sorption and diffusion in polymethyl methacrylate–clay nanocomposites, *Polym. Eng. Sci.*, 45, 904, 2005.

2005NAG Nagai, K., Kanehashi, S., Tabei, S., and Nakagawa, T., Nitrogen permeability and carbon dioxide solubility in poly(1-trimethylsilyl-1-propyne)-based binary substituted polyacetylene blends, *J. Membrane Sci.*, 251, 101, 2005.

2005NUN Nunes, R.C.R., Lopez-Gonzalez, M., and Riande, E., Basic studies on gas solubility in natural rubber–cellulose composites, *J. Polym. Sci.: Part B: Polym. Phys.*, 43, 2131, 2005.

2005PAL Palamara, J.E., Mulcahy, K.A., Jones, A.T., Danner, R.P., Duda, J.L., Solubility and diffusivity of propylene and ethylene in atactic polypropylene by the static sorption technique, *Ind. Eng. Chem. Res.*, 44, 9943, 2005.

2005PR1 Prabhakar, R.S., Raharjo, R., Toy, L.G., Lin, H., and Freeman, B.D., Self-consistent model of concentration and temperature dependence of permeability in rubbery polymers, *Ind. Eng. Chem. Res.*, 44, 1547, 2005.

2005PR2 Prabhakar, R.S., Merkel, T.C., Freeman, B.D., Imizu, T., and Higuchi, A., Sorption and transport properties of propane and perfluoropropane in poly(dimethylsiloxane) and poly(1-trimethylsilyl-1-propyne), *Macromolecules*, 38, 1899, 2005.

2005PR3 Prabhakar, R.S., DeAngelis, M.G., Sarti, G.C., Freeman, B.D., and Coughlin, M.C., Gas and vapor sorption, permeation, and diffusion in poly(tetrafluoroethylene-*co*-perfluoromethyl vinyl ether), *Macromolecules*, 38, 7043, 2005.

2005RAJ Rajendran, A., Bonavoglia, B., Forrer, N., Storti, G., Mazzotti, M., and Morbidelli, M., Simultaneous measurement of swelling and sorption in a supercritical CO_2-poly(methyl methacrylate) system, *Ind. Eng. Chem. Res.*, 44, 2549, 2005.

2005RUT Rutherford, S.W., Kurtz, R.E., Smith, M.G., Honnell, K.G., and Coons, J.E., Measurement and correlation of sorption and transport properties of ethylene-propylene-diene monomer (EPDM) elastomers, *J. Membrane Sci.*, 263, 57, 2005.

2005SO1 Solms, N. von, Zecchin, N., Rubin, A., Andersen, S.I., and Stenby, E.H., Direct measurement of gas solubility and diffusivity in poly(vinylidene fluoride) with a high-pressure microbalance, *Eur. Polym. J.*, 41, 341, 2005.

2005SO2 Solms, N. von, Rubin, A., Andersen, S.I., and Stenby, E.H., Directs measurement of high temperature/high pressure solubility of methane and carbon dioxide in polyamide (PA-11) using a high-pressure microbalance, *Int. J. Thermophys.*, 26, 115, 2005.

2005STR Stryuk, S. and Wolf, B.A., Liquid/gas and liquid/liquid phase behavior of n-butane/ 1,4-polybutadiene versus n-butane/1,2-polybutadiene (experimental data by S. Stryuk and B.A. Wolf), *Macromolecules*, 38, 812, 2005.

2005TA1 Takashima, H., Okamoto, S., Yoshimizu, H., and Tsujita, Y., Structure and gas sorption properties of an aromatic polyamide with long alkyl side chains, *J. Appl. Polym. Sci.*, 97, 1771, 2005.

2005TA2 Taki, K., Nitta, K., Kihara, S.-I., and Ohshima, M., CO_2 foaming of poly(ethylene glycol)/polystyrene blends: relationship of the blend morphology, CO_2 mass transfer, and cellular structure, *J. Appl. Polym. Sci.*, 97, 1899, 2005.

2005TA3 Tan, B., Woods, H.M., Licence, P., Howdle, S.M., and Cooper, A.I., Synthesis and CO_2 solubility studies of poly(ether carbonate)s and poly(ether ester)s produced by step growth polymerization, *Macromolecules*, 38, 1691, 2005.

2005TA4 Tang, J., Sun, W., Tang, H., Radosz, M., and Shen, Y., Enhanced CO_2 absorption of poly(ionic liquid)s, *Macromolecules*, 38, 2037, 2005.

2005TA5 Tang, J., Tang, H., Sun, W., Radosz, M., and Shen, Y., Poly(ionic liquid)s as new materials for CO_2 absorption, *J. Polym. Sci.: Part A: Polym. Chem.*, 43, 5477, 2005.

2005TA6 Tang, J., Tang, H., Sun, W., Radosz, M., Shen, Y., Low-pressure CO_2 sorption in ammonium-based poly(ionic liquid)s, *Polymer*, 46, 12460, 2005.

2005THU Thurecht, K.J., Hill, D.J.T., and Whittaker, A.K., Equilibrium swelling measurements of network and semicrystalline polymers in supercritical carbon dioxide using high-pressure NMR, *Macromolecules*, 38, 3731, 2005.

2005WAN Wang, M., Sato, Y., Iketani, T., Takishima, S., Masuoka, H., Watanabe, T., and Fukasawa, Y., Solubility of HFC-134a, HCFC-142b, butane, and isobutane in low-density polyethylene at temperatures from 383.15 to 473.15K and at pressures up to 3.4MPa, *Fluid Phase Equil.*, 232, 1, 2005.

2005WUH Wu, J., Pan, Q., and Rempel, G.L., Solubility of ethylene in toluene and toluene/styrene–butadiene rubber solutions, *J. Appl. Polym. Sci.*, 96, 645, 2005.

2005XUD Xu, D., Carbonell, R.G., Roberts, G.W., and Kiserow, D.J., Phase equilibrium for the hydrogenation of polystyrene in CO_2-swollen solvents, *J. Supercrit. Fluids*, 34, 1, 2005.

2005ZHA Zhang, R., Liu, J., Han, B., Wang, B., Sun, D., and He, J., Effect of PEO–PPO–PEO structure on the compressed ethylene-induced reverse micelle formation and water solubilization, *Polymer*, 46, 3936, 2005.

2006ABB Abbott, A.P., Brooks, N., Eltringham, W., Hillman, A.R., and Hope, E.G., Polymer modification using difluoromethane (HFC 32) and carbon dioxide, *J. Polym. Sci.: Part B: Polym. Phys.*, 44, 1072, 2006.

2006AGU Aguiar-Ricardo, A., Casimiro, T., Costa, T., Leandro, J., and Ribeiro, N., Visual and acoustic investigation of the critical behavior of mixtures of CO_2 with a perfluorinated polyether, *Fluid Phase Equil.*, 239, 26, 2006.

2006BON Bonavoglia, B., Storti, G., Morbidelli, M., Rajendran, A., and Mazotti, M., Sorption and swelling of semicrystalline polymers in supercritical CO_2, *J. Polym. Sci.: Part B: Polym. Phys.*, 44, 1531, 2006.

2006BOY Boyer, S.A.E., Randzio, S.L., and Grolier, J.-P.E., Thermal expansion of polymers submitted to supercritical CO_2 as a function of pressure, *J. Polym. Sci.: Part B: Polym. Phys.*, 44, 185, 2006.

2006CHE Chen, K.-H.J. and Rizvi, S.S.H., Measurement and prediction of solubilities and diffusion coefficients of carbon dioxide in starch-water mixtures at elevated pressures, *J. Polym. Sci.: Part B: Polym. Phys.*, 44, 607, 2006.

2006DU1 Duarte, R.C., Sampaio de Sousa, A.R., de Sousa, H.C., Gil, M.H.M., Jespersen, H.T., and Duarte, C.M.M., Solubility of dense CO_2 in two biocompatible acrylate copolymers, *Brazil. J. Chem. Eng.*, 23, 191, 2006.

2006DU2 Duarte, A.R.C., Martins, C., Coimbra, P., Gil, M.H.M., de Sousa, H.C., and Duarte, C.M.M., Sorption and diffusion of dense carbon dioxide in a biocompatible polymer, *J. Supercrit. Fluids*, 38, 392, 2006.

2006FLE Fleming, O.S., Chan, K.L.A., and Kazarian, S.G., High-pressure CO_2-enhanced polymer interdiffusion and dissolution studied with *in-situ* ATR-FTIR spectroscopic imaging, *Polymer*, 47, 4649, 2006.

2006GAL Galia, A., Abduljawad, M., Scialdone, O., and Filardo, G., A novel gas chromatographic method to measure sorption of dense gases into polymers, *AIChE-J.*, 52, 2243, 2006.

2006HOE Hoelck, O., Siegert, M.R., Heuchel, M., and Böhning, M., CO_2 sorption induced dilation in polysulfone: comparative analysis of experimental and molecular modeling results, *Macromolecules*, 39, 9590, 2006.

2006LEE Leeke, G.A., Cai, J., and Jenkins, M., Solubility of supercritical carbon dioxide in polycaprolactone (CAPA 6800) at 313 and 333 K, *J. Chem. Eng. Data*, 51, 1877, 2006.

2006LIA Lian, Z., Epstein, S.A., Blenk, C.W., and Shine, A.D., Carbon dioxide-induced melting point depression of biodegradable semicrystalline polymers, *J. Supercrit. Fluids*, 39, 107, 2006.

2006LIG1 Li, G., Li, H., Turng, L.S., Gong, S., and Zhang, C., Measurement of gas solubility and diffusivity in polylactide, *Fluid Phase Equil.*, 246, 158, 2006.

2006LIG2 Li, G., Li, H., Wang, J., and Park, C.B., Investigating the solubility of CO_2 in polypropylene using various eos models, *Cell. Polym.*, 25, 237, 2006.

2006LI1 Liu, L., Chakma, A., and Feng, X., Sorption, diffusion, and permeation of light olefins in poly(ether block amide) membranes, *Chem. Eng. Sci.*, 61, 6142, 2006.

2006LI2 Liu, D. and Tomasko, D.L., Carbon dioxide sorption and dilation of poly(lactide-*co*-glycolide), *J. Supercrit. Fluids*, 39, 416, 2006.

2006MAD Madsen, L.A., Plasticization of poly(ethylene oxide) in fluid CO_2 measured by *in-situ* NMR, *Macromolecules*, 39, 1483, 2006.

2006NAG Nagy, I., Loos, Th.W.de, Krenz, R.A., and Heidemann, R.A., High pressure phase equilibria in the systems linear low density polyethylene + n-hexane and linear low density polyethylene + n-hexane + ethylene: Experimental results and modelling with the Sanchez-Lacombe equation of state, *J. Supercrit. Fluids*, 37, 115, 2006.

2006NAL Nalawade, S.P., Picchioni, F., Janssen, L.P.B.M., Patil, V.E., Keurentjes, J.T.F., and Staudt, R., Solubilities of sub- and supercritical carbon dioxide in polyester resins, *Polym. Eng. Sci.*, 46, 643, 2006.

2006NOV Novak, A., Bobak, M., Kosek, J., Banaszak, B.J., Lo, D., Widya, T., Ray, W.H., Pablo, J.J.de, Ethylene and 1-hexene sorption in LLDPE under typical gas-phase reactor conditions: Experiments, *J. Appl. Polym. Sci.*, 100, 1124, 2006.

2006OL1 Oliveira, N.S., Goncalves, C.M., Coutinho, J.A.P., Ferreira, A., Dorgan, J., and Marrucho, I.M., Carbon dioxide, ethylene and water vapor sorption in poly(lactic acid), *Fluid Phase Equil.*, 250, 116, 2006.

2006OL2 Oliveira, N.S., Dorgan, J., Coutinho, J.A.P., Ferreira, A., Daridon, J.L., and Marrucho, I.M., Gas solubility of carbon dioxide in poly(lactic acid) at high pressures, *J. Polym. Sci.: Part B: Polym. Phys.*, 44, 1010, 2006.

2006PAN Pantoula, M. and Panayiotou, C., Sorption and swelling in glassy polymer/carbon dioxide systems Part I. Sorption, *J. Supercrit. Fluids*, 37, 254, 2006.

2006PAR Park, H.E. and Dealy, J.M., Effects of pressure and supercritical fluids on the viscosity of polyethylene, *Macromolecules*, 39, 5438, 2006.

2006PAT Patzlaff, M., Wittebrock, A., and Reichert, K.-H., Sorption studies of propylene in polypropylene. Diffusivity in polymer particles formed by different polymerization processes, *J. Appl. Polym. Sci.*, 100, 2642, 2006.

2006UEZ Üzer, S., Akman, U., and Hortacsu, Ö., Polymer swelling and impregnation using supercritical CO_2: A model-component study towards producing controlled-release drugs, *J. Supercrit. Fluids*, 38, 119, 2006.

2006VOR Vorotyntsev, I.V., Drozdov, P.N., Mochalov, G.M., Smirnova, N.N., and Suvorov, S.S., Sorption of ammonia and nitrogen on cellulose acetate, *Russ. J. Phys. Chem.*, 80, 2020, 2006.

2006WAN Wang, D., Jiang, W., Gao, H., and Jiang, Z., Diffusion and swelling of carbon dioxide in amorphous poly(ether ether ketone)s, *J. Membrane Sci.*, 281, 203, 2006.

2006WEI Weinstein, R.D. and Papatolis, J., Diffusion of liquid and supercritical carbon dioxide into a chitosan sphere, *Ind. Eng. Chem. Res.*, 45, 8651, 2006.

2007BAR Bartke, M., Kröner, S., Wittebrock, A., Reichert, K.-H., Illiopoulus, I., and Dittrich, C.J., Sorption and diffusion of propylene and ethylene in heterophasic polypropylene copolymers, *Macromol. Symp.*, 259, 327, 2007.

2007BL1 Blasig, A., Tang, J., Hu, X., Shen, Y., and Radosz, M., Magnetic suspension balance study of carbon dioxide solubility in ammonium-based polymerized ionic liquids: Poly(p-vinylbenzyltrimethyl ammonium tetrafluoroborate) and poly([2-(methacryl-oyloxy)ethyl] trimethyl ammonium tetrafluoroborate), *Fluid Phase Equil.*, 256, 75, 2007.

2007BL2 Blasig, A., Tang, J., Hu, X., Tan, S.P., Shen, Y., and Radosz, M., Carbon dioxide solubility in polymerized ionic liquids containing ammonium and imidazolium cations from magnetic suspension balance: P[VBTMA][BF_4] and P[VBMI][BF_4], *Ind. Eng. Chem. Res.*, 46, 5542, 2007.

2007BOY Boyer, S.A.E., Klopffer, M.-H., Martin, J., and Grolier, J.-P.E., Supercritical gas–polymer interactions with applications in the petroleum industry. Determination of thermophysical properties, *J. Appl. Polym. Sci.*, 103, 1706, 2007.

2007CHA Cha, D.-H., Lee, J., Im, J., and Kim, H., (Vapor + liquid) equilibria for the {water + poly(ethylene glycol diacetyl ether) (PEGDAE) and methanol + PEGDAE} systems, *J. Chem. Thermodyn.*, 39, 483, 2007.

2007COM Compan, V., Del Castillo, L.F., Hernandez, S.I., Lopez-Gonzalez, M.M., and Riande, E., On the crystallinity effect on the gas sorption in semicrystalline linear low density polyethylene (LLDPE), *J. Polym. Sci.: Part B: Polym. Phys.*, 45, 1798, 2007.

2007CRA Cravo, C., Duarte, A.R.C., and Duarte, C.M.M., Solubility of carbon dioxide in a natural biodegradable polymer: Determination of diffusion coefficients, *J. Supercrit. Fluids*, 40, 194, 2007.

2007DON Dondero, M., Carella, J., and Borrajo, J., (CO_2 + 2-propanol) mixture as a foaming agent for polystyrene: a simple thermodynamic model for the high pressure VLE-phase diagrams taking into account the foam vitrification, *J. Appl. Polym. Sci.*, 104, 2663, 2007.

2007FAR Faridi, N. and Todd, D., Solubility measurements of blowing agents in polyethylene terephthalate, *J. Cell. Plast.*, 43, 345, 2007.

2007FOS Fossati, P., Sanguineti, A, DeAngelis, M.G., Baschetti, M.G., Doghieri, F., and Sarti, G.C., Gas solubility and permeability in MFA, *J. Polym. Sci.: Part B: Polym. Phys.*, 45, 1637, 2007.

2007HOU Hou, M., Liang, S., Zhang, Z., Song, J., Jiang, T., and Han, B., Determination and modeling of solubility of CO_2 in PEG200 + 1-pentanol and PEG200 + 1-octanol mixtures, *Fluid Phase Equil.*, 258, 108, 2007.

2007HUS Hussain, Y., Wu, Y.-T., Ampaw, P.-J., and Grant, C.S., Dissolution of polymer films in supercritical carbon dioxide using a quartz crystal microbalance, *J. Supercrit. Fluids*, 42, 255, 2007.

2007LEI Lei, Z., Ohyabu, H., Sato, Y., Inomata, H., and Smith Jr., R.L., Solubility, swelling degree and crystallinity of carbon dioxide-polypropylene system, *J. Supercrit. Fluids*, 40, 452, 2007.

2007LI1 Li, G., Gunkel, F., Wang, J., Park, C.B., and Altstaedt, V., Solubility measurements of N_2 and CO_2 in polypropylene and ethene/octene copolymer, *J. Appl. Polym. Sci.*, 103, 2945, 2007.

2007LI2 Li, G., Wang, J., Park, C.B., and Simha, R., Measurement of gas solubility in linear/branched PP melts, *J. Polym. Sci.: Part B: Polym. Phys.*, 45, 2497, 2007.

2007LIY Li, Y., Wang, X., Sanchez, I.C., Johnston, K.P., and Green, P.F, Ordering in asymmetric block copolymer films by a compressible fluid, *J. Phys. Chem. B*, 111, 16, 2007.

2007LOP Lopez-Gonzalez, M., Compan, V., and Riande, E., Gas sorption in semicrystalline rubbery polymers revisited, *J. Appl. Polym. Sci.*, 105, 903, 2007.

2007MOO Moore, T.T. and Koros, W.J., Gas sorption in polymers, molecular sieves, and mixed matrix membranes, *J. Appl. Polym. Sci.*, 104, 4053, 2007.

2007NAG Nagy, I., Krenz, R.A., Heidemann, R.A., and de Loos, Th.W., High-pressure phase equilibria in the system linear low density polyethylene + isohexane: Experimental results and modelling, *J. Supercrit. Fluids*, 40, 125, 2007.

2007NAW Nawaby, A.V., Handa, Y.P., Liao, X., Yoshitaka, Y., and Tomohiro, M., Polymer-CO_2 systems exhibiting retrograde behavior and formation of nanofoams, *Polym. Int.*, 56, 67, 2007.

2007OLI Oliveira, N.S., Dorgan, J., Coutinho, J.A.P., Ferreira, A., Daridon, J.L., and Marrucho, I.M., Gas solubility of carbon dioxide in poly(lactic acid) at high pressures: thermal treatment effect, *J. Polym. Sci.: Part B: Polym. Phys.*, 45, 616, 2007.

2007PAN Pantoula, M., Schnitzler, J. von, Eggers, R., and Panayiotou, C., Sorption and swelling in glassy polymer/carbon dioxide systems. Part II—Swelling, *J. Supercrit. Fluids*, 39, 426, 2007.

2007PAR Park, S.-W., Choi, B.-S., and Lee, J.-W., Effect of polyacrylamide on absorption rate of carbon dioxide in aqueous polyacrylamide solution containing monoethanolamine, *J. Ind. Eng. Chem.*, 13, 7, 2007.

2007PIN Pini, R., Storti, G., Mazzotti, M., Tai, H., Shakesheff, K.M., and Howdle, S.M., Sorption and swelling of poly(DL-lactic acid) and poly(lactic-*co*-glycolic acid) in supercritical CO_2, *Macromol. Symp.*, 259, 197, 2007.

2007RA1 Raharjo, R.D., Freeman, B.D., and Sanders, E.S., Pure and mixed gas CH4 and n-C_4H_{10} sorption and dilation in poly(dimethylsiloxane), *J. Membrane Sci.*, 292, 45, 2007.

2007RA2 Raharjo, R.D., Freeman, B.D., and Sanders, E.S., Pure and mixed gas CH_4 and n-C_4H_{10} sorption and dilation in poly(1-trimethylsilyl-1-propyne), *Polymer*, 48, 6097, 2007.

2007ROL Rolker, J., Seiler, M., Mokrushina, L., and Arlt, W., Potential of branched polymers in the field of gas absorption: experimental gas solubilities and modeling, *Ind. Eng. Chem. Res.*, 46, 6572, 2007.

2007TAN Tang, Q., Yang, B, Zhao, Y., and Zhao, L., Sorption and diffusion of sub/supercritical carbon dioxide in poly(methyl methacrylate), *J. Macromol. Sci., Part B: Polym. Phys.*, 46, 275, 2007.

2007VIS Visser, T. and Wessling, M., When do sorption-induced relaxations in glassy polymers set in?, *Macromolecules*, 40, 4992, 2007.

2007YA1 Yao, W., Hu, X., and Yang, Y., Modeling solubility of gases in semicrystalline polyethylene, *J. Appl. Polym. Sci.*, 103, 1737, 2007.

2007YA2 Yao, W., Hu, X., and Yang, Y., Modeling the solubility of ternary mixtures of ethylene, iso-pentane, n-hexane in semicrystalline polyethylene, *J. Appl. Polym. Sci.*, 104, 3654, 2007.

2007YA3 Yao, B., Nawaby, A.V., Liao, X., and Burk, R., Physical characteristics of PLLA/PMMA blends and their CO_2 blowing foams, *J. Cell. Plast.*, 43, 385, 2007.

2008AI1 Aionicesei, E., Skerget, M., and Knez, Z., Measurement and modeling of the CO_2 solubility in poly(ethylene glycol) of different molecular weights, *J. Chem. Eng. Data*, 53, 185, 2008.

2008AI2 Aionicesei, E., Skerget, M., and Knez, Z., Measurement of CO_2 solubility and diffusivity in poly(L-lactide) and poly(DL-lactide-*co*-glycolide) by magnetic suspension balance, *J. Supercrit. Fluids*, 47, 296, 2008.

2008DIN Dingemans, M., Dewulf, J., Van Hecke, W., and Van Langenhove, H., Determination of ozone solubility in polymeric materials, *Chem. Eng. J.*, 138, 172, 2008.

2008GAL Galia, A., Cipollina, A., Scialdone, O., and Filardo, G., Investigation of multicomponent sorption in polymers from fluid mixtures at supercritical conditions: The case of the carbon dioxide/vinylidene fluoride/poly(vinylidene fluoride) system, *Macromolecules*, 41, 1521, 2008.

2008GAR Garcia, J., Youbi-Idrissi, M., Bonjour, J., and Fernandez, J., Experimental and PC-SAFT volumetric and phase behavior of carbon dioxide + PAG or POE lubricant systems, *J. Supercrit. Fluids*, 47, 8, 2008.

2008GON Goncalves, C.M.B., Coutinho, J.A.P., and Marrucho, I.M., Light olefins/paraffins sorption in poly(lactic acid) films, *J. Polym. Sci.: Part B: Polym. Phys.*, 46, 1312, 2008.

2008HAR Haruki, M., Takakura, Y., Sugiura, H., Kihara, S., and Takishima, S., Phase behavior for the supercritical ethylene + hexane + polyethylene systems, *J. Supercrit. Fluids*, 44, 284, 2008.

2008JON Jones, A.T., Danner, R.P., and Duda, J.L., Influence of high pressure gases on polymer-solvent thermodynamic and transport behavior, *J. Appl. Polym. Sci.*, 110, 1632, 2008.

2008KAS Kasturirangan, A., Grant, C., and Teja, A.S., Compressible lattice model for phase equilibria in CO_2 + polymer systems, *Ind. Eng. Chem. Res.*, 47, 645, 2008.

2008LIG Li, G., Leung, S.N., Hasan, M.M., Wang, J., Park, C.B., and Simha, R., A thermodynamic model for ternary mixture systems − gas blends in a polymer melt, *Fluid Phase Equil.*, 266, 129, 2008.

2008LIX Li, X., Hou, M., Han, B., Wang, X., Yang, G., and Zou, L., Vapor-liquid equilibria of CO_2 + 1-octene + polyethylene glycol systems, *J. Chem. Eng. Data*, 53, 1216, 2008.

2008LIY Li, Y.G., Park., C.B., Li, H.B., and Wang, J., Measurement of the *PVT* property of PP/CO_2 solution, *Fluid Phase Equil.*, 270, 15, 2008.

2008LIZ Li, Z.-Y., Meng, T.-Y., Liu, X.-W., Xia, Y.-J., and Hu, D.-P., Phase equilibrium characteristics of supercritical CO_2/poly(ethylene terephthalate) binary system, *J. Appl. Polym. Sci.*, 109, 2836, 2008.

2008MAT Matsuyama, K. and Mishima, K., Phase behavior of the mixtures of CO_2 + poly(methyl methacrylate) + ethanol at high pressure, *J. Chem. Eng. Data*, 53, 1151, 2008.

2008PA1 Pasquali, I., Andanson, J.M., Kazarian, S.G., and Bettini, R., Measurement of CO_2 sorption and PEG 1500 swelling by ATR-IR spectroscopy, *J. Supercrit. Fluids*, 45, 384, 2008.

2008PA2 Pasquali, I., Comi, L., Pucciarelli, F., and Bettini, R., Swelling, melting point reduction and solubility of PEG 1500 in supercritical CO_2, *Int. J. Pharmaceutics*, 356, 76, 2008.

2008PIN Pini, R., Storti, G., Mazzotti, M., Tai, H., Shakesheff, K.M., and Howdle, S.M., Sorption and swelling of poly(DL-lactic acid) and poly(lactic-*co*-glycolic acid) in supercritical CO_2: An experimental and modeling study, *J. Polym. Sci.: Part B: Polym. Phys.*, 46, 483, 2008.

2008POU Pourdarvish, R., Danner, R.P., and Duda, J.L., Solubility and diffusivity measurements in nitrogen–poly(vinyl acetate) and nitrogen–toluene–poly(vinyl acetate) systems with the differential pressure decay technique, *J. Appl. Polym. Sci.*, 108, 1407, 2008.

2008RIB Ribeiro, C.P. and Freeman, B.D., Sorption, dilation, and partial molar volumes of carbon dioxide and ethane in cross-linked poly(ethylene oxide), *Macromolecules*, 41, 9458, 2008.

2008ZHA Zhao, J.J., Zhao, Y.P., and Yang, B., Investigation of sorption and diffusion of supercritical carbon dioxide in polycarbonate, *J. Appl. Polym. Sci.*, 109, 1661, 2008.

2009CAR Carla, V., Hussain, Y., Grant, C., Sarti, G.C., Carbonell, R.G., and Doghieri, F., Modeling sorption kinetics of carbon dioxide in glassy polymeric films using the nonequilibrium thermodynamics approach, *Ind. Eng. Chem. Res.*, 48, 3844, 2009.

2009CHE Chen, L., Cao, G., Zhang, R., He, F., Liu, T., Zhao, L., Yuan, W., and Roberts, G.W., *In-situ* visual measurement of poly(methyl methacrylate) swelling in supercritical carbon dioxide and interrelated thermodynamic modeling, *Huagong Xuebao*, 60, 2351, 2009.

2009HAR Haruki, M., Sato, K., Kihara, S.-I., and Takishima, S., High pressure phase behavior for the supercritical ethylene + cyclohexane + hexane + polyethylene systems, *J. Supercrit. Fluids*, 49, 125, 2009.

2009HE1 He, J. and Wang, B., Acetone influence on glass transition of poly(methyl methacrylate) and polystyrene in compressed carbon dioxide, *Ind. Eng. Chem. Res.*, 48, 5093, 2009.

2009HE2 He, J. and Wang, B., The temperature, cosolvent, and blending effects on the partitions between polymer and compressed carbon dioxide, *Ind. Eng. Chem. Res.*, 48, 7359, 2009.

2009HEI Heinrich, H., Jaeger, P., and Eggers, R., Diffusion von Gasen und Gasgemischen in Polymere unter hohen Drücken, *Chem.-Ing. Techn.*, 81, 1607, 2009.

2009KOZ Kozlowska, M.K., Jürgens, B.F., Schacht, C.S., Gross, J., and Loos, Th.W. de, Phase behavior of hyperbranched polymer systems: Experiments and application of the perturbed-chain polar SAFT equation of state, *J. Phys. Chem. B*, 113, 1022, 2009.

2009LAW Law, Y.Y., Balashova, I.M., and Danner, R.P., Effect of high pressure carbon dioxide on the solubility and diffusivity of dichloromethane in polyetherimide, *J. Appl. Polym. Sci.*, 114, 2497, 2009.

2009LID Li, D., Liu, T., Zhao, L., and Yuan, W., Solubility and diffusivity of carbon dioxide in solid-state isotactic polypropylene by the pressure-decay method, *Ind. Eng. Chem. Res.*, 48, 7117, 2009.

2009LIY Li, Y.G. and Park, C.B., Effects of branching on the pressure-volume-temperature behaviors of PP/CO_2 solutions, *Ind. Eng. Chem. Res.*, 48, 6633, 2009.

2009MAR Martin, A., Pham, H.M., Kilzer, A., Kareth, S., and Weidner, E., Phase equilibria of carbon dioxide + poly(ethylene glycol) + water mixtures at high pressure: Measurements and modelling, *Fluid Phase Equil.*, 286, 162, 2009.

2009MIL Miller, M.B.,Chen, D.-L., Xie, H.-B., Luebke, D.R., Johnson, J.K., and Enick, R.M., Solubility of CO_2 in CO_2-philic oligomers; COSMOtherm predictions and experimental results, *Fluid Phase Equil.*, 287, 26, 2009.

2009POR Portela, V.M., Straver, E.J.M., and de Loos, Th.W., High-pressure phase behavior of the system propane-Boltorn H3200, *J. Chem. Eng. Data*, 54, 2593, 2009.

2009TAN Tang, J., Shen, Y., Radosz, M., and Sun, W., Isothermal carbon dioxide sorption in poly(ionic liquid)s, *Ind. Eng. Chem. Res.*, 48, 9113, 2009.

2009VOR Vorotyntsev, I.V. and Gamayunova, T.V., The thermal equation for ammonia sorption by cellulose acetate, *Russ. J. Phys. Chem. A*, 83, 818, 2009.

2010HA2 Haruki, M., Mano, S., Koga, Y., Kihara, S., and Takishima, S., Phase behaviors for the supercritical ethylene + 1-hexene + hexane + polyethylene systems at elevated temperatures and pressures, *Fluid Phase Equil.*, 295, 137, 2010.

2010HAS Hasan, M.M., Li, Y.G., Li, G., Park, C.B., and Chen, P., Determination of solubilities of CO_2 in linear and branched polypropylene using a magnetic suspension balance and a PVT apparatus, *J. Chem. Eng. Data*, 55, 4885, 2010.

2010LIN Lin, S., Yang, J., Yan, J., Zhao, Y., and Yang, B., Sorption and diffusion of supercritical carbon dioxide in a biodegradable polymer, *J. Macromol. Sci.: Part B: Phys.*, 49, 286, 2010.

2010PER Perez-Blanco, M., Hammons, J.R., and Danner, R.P., Measurement of the solubility and diffusivity of blowing agents in polystyrene, *J. Appl. Polym. Sci.*, 116, 2359, 2010.

2010REV Revelli, A.-L., Mutelet, F., and Jaubert, J.-N., High carbon dioxide solubilities in imidazolium-based ionic liquids and in poly(ethylene glycol) dimethyl ether, *J. Phys. Chem. B*, 114, 12908, 2010.

2010RI1 Ribeiro, C.P. and Freeman, B.D., Solubility and partial molar volume of carbon dioxide and ethane in crosslinked poly(ethylene oxide) copolymer, *J. Polym. Sci.: Part B: Polym. Phys.*, 48, 456, 2010.

2010RI2 Ribeiro, C.P. and Freeman, B.D., Carbon dioxide/ethane mixed-gas sorption and dilation in a cross-linked poly(ethylene oxide) copolymer, *Polymer*, 51, 1156, 2010.

2010SCH Schacht, C.S., Bahramali, S., Wilms, D., Frey, H., Gross, J., and de Loos, Th.W., Phase behavior of the system hyperbranched polyglycerol + methanol + carbon dioxide, *Fluid Phase Equil.*, 299, 252, 2010.

2010SKE Skerget, M., Mandzuka, Z., Aionicesei, E., Knez, Z., Jese, R., Znoj, B., and Venturini, P., Solubility and diffusivity of CO_2 in carboxylated polyesters, *J. Supercrit. Fluids*, 51, 306, 2010.

2010SOR Sorrentino, L., DiMaio, E., and Iannace, S., Poly(ethylene terephthalate) foams: Correlation between the polymer properties and the foaming process, *J. Appl. Polym. Sci.*, 116, 27, 2010.

2010VEG Vegt, N. F. A. van der, Kusuma, V.A., and Freeman, B.D., Basis of solubility versus T_C correlations in polymeric gas separation membranes, *Macromolecules*, 43, 1473, 2010.

2010WEI Wei, H., Thompson, R.B., Park, C.B., and Chen, P., Surface tension of high density polyethylene (HDPE) in supercritical nitrogen: Effect of polymer crystallization, *Colloids Surfaces A*, 354, 347, 2010.

2011CHM Chmelar, J., Gregor, T., Hajova, H., Nistor, A., and Kosek, J., Experimental study and PC-SAFT simulations of sorption equilibria in polystyrene, *Polymer*, 52, 3082, 2011.

2011HAR Haruki, M., Nakanishi, K., Mano, S., Kihara, S.-I., and Takishima, S., Effect of molecular weight distribution on the liquid-liquid phase separation behavior of polydispersed polyethylene solutions at high temperatures, *Fluid Phase Equil.*, 305, 152, 2011.

2011KAS Kasturirangan, A., Koh, C.A., and Teja, A.S., Glass-transition temperatures in CO_2 + polymer systems: Modeling and experiment, *Ind. Eng. Chem. Res.*, 50, 158, 2011.

2011KHO1 Khoiroh, I. and Lee, M.-J., Isothermal vapor-liquid equilibrium for binary mixtures of polyoxyethylene 4-octylphenyl ether with methanol, ethanol, or propan-2-ol, *J. Chem. Eng. Data*, 56, 1178, 2011.

2011KHO2 Khoiroh, I. and Lee, M.-J., Isothermal (vapour + liquid) equilibrium for binary mixtures of polyethylene glycol mono-4-nonylphenyl ether (PEGNPE) with methanol, ethanol, or 2-propanol, *J. Chem. Thermodyn.*, 43, 1417, 2011.

2011KUN Kundra, P., Upreti, S.R., Lohi, A., and Wu, J., Experimental determination of composition-dependent diffusivity of carbon dioxide in polypropylene, *J. Chem. Eng. Data*, 53, 21, 2011.

2011MAR Markocic, E., Skerget, M., and Knez, Z., Solubility and diffusivity of CO_2 in poly(L-lactide)–hydroxyapatite and poly(DL-lactide-*co*-glycolide)–hydroxyapatite composite biomaterials, *J. Supercrit. Fluids*, 55, 1046, 2011.

2011MUL Muljana, H., Picchioni, F., Heeres, H.J., and Janssen, L.P.B.M., Experimental and modeling studies on the solubility of sub- and supercritical carbon dioxide ($scCO_2$) in potato starch and derivatives, *Polym. Eng. Sci.*, 51, 28, 2011.

2011PER Perko, T., Markocic, E., Knez, Z., and Skerget, M., Solubility and diffusivity of CO_2 in natural methyl cellulose and sodium carboxymethyl cellulose, *J. Chem. Eng. Data*, 56, 4040, 2011.

2011RUI Ruiz-Alsop, R.N., Mueller, P.A., Richards, J.R., Kao, C.-P.C., and Brown, G.G., Simultaneous *in-situ* measurement of sorption and swelling of polymers in gases and supercritical fluids, *J. Polym. Sci.: Part B: Polym. Phys.*, 49, 574, 2011.

2011SCH Schacht, C.S., Schuell, C., Frey, H., Loos, Th.W. de, and Gross, J., Phase behavior of the system linear polyglycerol + methanol + carbon dioxide, *J. Chem. Eng. Data*, 56, 2927, 2011.

2011TSI Tsioptsias, C. and Panayiotou, C., Simultaneous determination of sorption, heat of sorption, diffusion coefficient and glass transition depression in polymer-CO_2 systems, *Thermochim. Acta*, 521, 98, 2011.

2011YAO Yao, Z., Zhu, F.-J., Chen, Z.-H., Zeng, C., and Cao, K., Solubility of subcritical and supercritical propylene in semicrystalline polypropylene, *J. Chem. Eng. Data*, 56, 1174, 2011.

2012BHA Bhavsar, R.S., Kumbharkar, S.C., and Kharul, U.K., Polymeric ionic liquids (PILs): Effect of anion variation on their CO_2 sorption, *J. Membrane Sci.*, 389, 305, 2012.

2012CHE Chen, J., Liu, T., Zhao, L., and Yuan, W.-K., Determination of CO_2 solubility in isotactic polypropylene melts with different polydispersities using magnetic suspension balance combined with swelling correction, *Thermochim. Acta*, 530, 79, 2012.

2012CAR Carbone, M.G.P., Di Maio, E., Scherillo, G., Mensitieri, G., and Iannace, S., Solubility, mutual diffusivity, specific volume and interfacial tension of molten PCL/CO_2 solutions by a fully experimental procedure: Effect of pressure and temperature, *J. Supercrit. Fluids*, 67, 131, 2012.

2012FON Fonseca, J.M.S., Dohrn, R., Wolf, A., and Bachmann, R., The solubility of carbon dioxide and propylene oxide in polymers derived from carbon dioxide, *Fluid Phase Equil.*, 318, 83, 2012.

2012GRO Grolier, J.-P.E. and Randzio, S.L., Simple gases to replace non-environmentally friendly polymer foaming agents. A thermodynamic investigation, *J. Chem. Thermodyn.*, 46, 42, 2012.

2012HOR Horn, N.R. and Paul, D.R., Carbon dioxide sorption and plasticization of thin glassy polymer films tracked by optical methods, *Macromolecules*, 45, 2820, 2012.

2012KHO Khoiroh, I. and Lee, M.-J., Isothermal vapor-liquid equilibrium for binary mixtures of polyoxyethylene dodecanoate with methanol, ethanol, or propan-2-ol, *J. Chem. Eng. Data*, 57, 545, 2012.

2012LIA Liao, X. and Nawaby, A.V., Solvent free generation of open and skinless foam in poly(L-lactic acid)/poly(DL-lactic acid) blends using carbon dioxide, *Ind. Eng. Chem. Research*, 51, 6722, 2012.

2012LIJ Li, J., Ye, Y., Chen, L., and Qi, Z., Solubilities of CO_2 in poly(ethylene glycols) from (303.15 to 333.15) K, *J. Chem. Eng. Data*, 57, 610, 2012.

2012MEM Memari, P., Lachet, V., Klopffer, M.-H., Flaconneche, B., and Rousseau, B., Gas mixture solubilities in polyethylene below its melting temperature: Experimental and molecular simulation studies, *J. Membrane Sci.*, 390-391, 194, 2012.

2012RA1 Rayer, A.V., Henni, A., and Tontiwachwuthikul, P., High-pressure solubility of methane (CH_4) and ethane (C_2H_6) in mixed polyethylene glycol dimethyl ethers (Genosorb 1753) and its selectivity in natural gas sweetening operations, *J. Chem. Eng. Data*, 57, 764, 2012.

2012RA2 Rayer, A.V., Henni, A., and Tontiwachwuthikul, P., High pressure physical solubility of carbon dioxide (CO_2) in mixed polyethylene glycol dimethyl ethers (Genosorb 1753), *Can. J. Chem. Eng.*, 90, 576, 2012.

2012SUN Sun, M. and Ying, S., Sorption and diffusion of supercritical carbon dioxide into nitrocellulose with or without cosolvents, *Polym.-Plast. Technol. Eng.*, 51, 1346, 2012.

2012TAK Takahashi, S., Hassler, J.C., and Kiran, E., Melting behavior of biodegradable polyesters in carbon dioxide at high pressures, *J. Supercrit. Fluids*, 72, 278, 2012.

2012ZHA Zhang, J., Zheng, Y., Wu, S., Hu, Y., and Yan, L., Absorption of dilute sulfur dioxide in aqueous poly(ethylene glycol)200 solutions at 298.15 K and 122.60 kPa, *Asian J. Chem.*, 24, 4729, 2012.

2013AZI Azimi, H.R. and Rezaei, M., Solubility and diffusivity of carbon dioxide in St-MMA copolymers, *J. Chem. Thermodyn.*, 58, 279, 2013.

2013CHE1 Chen, J., Liu, T., Zhao, L., and Yuan, W.-K., Experimental measurements and modeling of solubility and diffusivity of CO_2 in polypropylene/micro- and nanocalcium carbonate composites, *Ind. Eng. Chem. Res.*, 52, 5100, 2013.

2013CHE2 Chen, J., Liu, T., Yuan, W.-K., Zhao, L., Solubility and diffusivity of CO_2 in polypropylene/micro-calcium carbonate composites, *J. Supercrit. Fluids*, 77, 33, 2013.

2013CUC Cucek, D., Perko, T., Ilic, L., Znoj, B., Venturini, P., Knez, Z., and Skerget, M., Phase equilibria and diffusivity of dense gases in various polyethylenes, *J. Supercrit. Fluids*, 78, 54, 2013.

2013DIN Ding, J., Ma, W., Song, F., and Zhong, Q., Foaming of polypropylene with supercritical carbon dioxide: An experimental and theoretical study on a new process, *J. Appl. Polym. Sci.*, 130, 2877, 2013.

2013GAL Galizia, M., Daniel, C., Guerra, G., Mensitieri, G., Solubility and diffusivity of low molecular weight compounds in semi-crystalline poly(2,6-dimethyl-1,4-phenylene oxide): The role of the crystalline phase, *J. Membrane Sci.*, 443, 100, 2013.

2013GUT Gutiérrez, C., Rodríguez, J.F., Gracia, I., Lucas, A. de, and García, M.T., High-pressure phase equilibria of polystyrene dissolutions in limonene in presence of CO_2, *J. Supercrit. Fluids*, 84, 211, 2013.

2013HAJ Hajova, H., Chmelar, J., Nistor, A., Gregor, T., and Kosek, J., Experimental study of sorption and diffusion of n-pentane in polystyrene, *J. Chem. Eng. Data*, 58, 851, 2013.

2013HEZ He, Z., Liu, J., and Zhang, N., Solubility of dilute sulfur dioxide in aqueous poly(ethylene glycol) 1000 solutions at 298.15 K and 123.15 kPa, *Asian J. Chem.*, 25, 8614, 2013.

2013INC Inceoglu, S., Young, N.P., Jackson, A.J., Kline, S.R., Costeux, S., and Balsara, N.P., Effect of supercritical carbon dioxide on the thermodynamics of model blends of styrene-acrylonitrile copolymer and poly(methyl methacrylate) studied by small-angle neutron scattering, *Macromolecules*, 46, 6345, 2013.

2013KHO1 Khoiroh, I. and Lee, M.-J., Isothermal vapour–liquid equilibrium of binary systems containing polyoxyethylene dodecanoate and alcohols, *J. Chem. Thermodyn.*, 56, 99, 2013.

2013KHO2 Khoiroh, I. and Lee, M.-J., Vapor–liquid equilibria of binary systems composed of polyoxyethylene 4-octylphenyl ether and alcohols: Experimental measurements and correlation, *Fluid Phase Equil.*, 360, 111, 2013.

2013KIM Kim, H.J., Moon, H.C., Kim, H., Kim, K., Kim, J.K., and Cho, J., Pressure effect of various inert gases on the phase behavior of polystyrene-*block*-poly(n-pentyl methacrylate) copolymer, *Macromolecules*, 46, 493, 2013.

2013KNA Knauer, O.S., Carbone, M.G.P., Braeuer, A., Di Maio, E., and Leipertz, A., Investigation of CO_2 sorption in molten polymers at high pressures using Raman line imaging, *Polymer*, 54, 812, 2013.

2013KRO Kröner, T. and Bartke, M., Sorption of olefins in high impact polypropylene: Experimental determination and mass transport modeling, *Macromol. React. Eng.*, 7, 453, 2013.

2013LIN Lin, D., Ding, Z., Liu, L., and Ma, R., A method to obtain gas-PDMS membrane interaction parameters for UNIQUAC model, *Chin. J. Chem. Eng.*, 21, 485, 2013.

2013LIX Li, X.-K., Cao, G.-P., Chen, L.-H., Zhang, R.-H., Liu, H.-L., and Shi, Y.-H., Study of the anomalous sorption behavior of CO_2 into poly(methyl methacrylate) films in the vicinity of the critical pressure and temperature using a quartz crystal microbalance (QCM), *Langmuir*, 29, 14089, 2013.

2013MAR Markocic, E., Skerget, M., and Knez, Z., Effect of temperature and pressure on the behavior of poly(ε-caprolactone) in the presence of supercritical carbon dioxide, *Ind. Eng. Chem. Res.*, 52, 15594, 2013.

2013MIN Minelli, M., Cocchi, G., Ansaloni, L., Baschetti, M.G., De Angelis, M.G., Doghieri, F., Vapor and liquid sorption in matrimid polyimide: Experimental characterization and modeling, *Ind. Eng. Chem. Res.*, 52, 8936, 2013.

2013NI1 Niu, Y., Gao, F., Zhu, R., Sun, S., and Wei, X., Solubility of dilute SO_2 in mixtures of *N,N*-dimethylformamide + polyethylene glycol 400 and the density and viscosity of the mixtures, *J. Chem. Eng. Data*, 58, 639, 2013.

2013NI2 Niu, Y., Gao, F., Sun, S., Xiao, J., and Wei, X., Solubility of dilute SO_2 in 1,4-dioxane, 15-crown-5 ether, polyethylene glycol 200, polyethylene glycol 300, and their binary mixtures at 308.15 K and 122.66 kPa, *Fluid Phase Equil.*, 344, 65, 2013.

2013NOF Nofar, M., Tabatabaei, A., Ameli, A., and Park, C.B., Comparison of melting and crystallization behaviors of polylactide under high-pressure CO_2, N_2, and He, *Polymer*, 54, 6471, 2013.

2013SAT Sato, Y., Hosaka, N., Inomata, H., and Kanaka, K., Solubility of ethylene in norbornene + toluene + cyclic olefin copolymer systems, *Fluid Phase Equil.*, 352, 80, 2013.

2013SMI Smith, Z.P., Tiwari, R.R., Murphy, T.M., Sanders, D.F., Gleason, K.L., Paul, D.R., Freeman, B.D., Hydrogen sorption in polymers for membrane applications, *Polymer*, 54, 3026, 2013.

2013YUJ Yu, J., Tang, C., Guan, Y., Yao, S., and Zhu, Z., Sorption and diffusion behavior of carbon dioxide into poly(L-lactic acid) films at elevated pressures, *Chin. J. Chem. Eng.*, 21, 1296, 2013.

2013ZHA Zhang, N., Zhang, J., Zhang, Y., Bai, J., and Wei, X., Solubility and Henry's law constant of sulfur dioxide in aqueous poly(ethylene glycol) 300 solution at different temperatures and pressures, *Fluid Phase Equil.*, 348, 9, 2013.

2014CHA Champeau, M., Thomassin, J.-M., Jérome, C., and Tassaing, T., In situ FTIR micro-spectroscopy to investigate polymeric fibers under supercritical carbon dioxide: CO_2 sorption and swelling measurements, *J. Supercrit. Fluids*, 90, 44, 2014.

2014HRN Hrncic, M.K., Markocic, E., Trupej, N., Skerget, M., and Knez, Z., Investigation of thermodynamic properties of the binary system poly(ethylene glycol)/CO_2 using new methods, *J. Supercrit. Fluids*, 87, 50, 2014.

2014KIM Kim, S., Woo, K.T., Lee, J.M., Quay, J.R., Murphy, M.K., and Lee, Y.M., Gas sorption, diffusion, and permeation in thermally rearranged poly(benzoxazole-*co*-imide) membranes, *J. Membrane Sci.*, 453, 556, 2014.

2014LIP Li, P., Chung, T.S., and Paul, D.R., Temperature dependence of gas sorption and permeation in PIM-1, *J. Membrane Sci.*, 450, 380, 2014.

2014MAH Mahmood, S.H., Keshtkar, M., and Park, C.B., Determination of carbon dioxide solubility in polylactide acid with accurate PVT properties, *J. Chem. Thermodyn.*, 70, 13, 2014.

2014MAR Martins, M., Craveiro, R., Paiva, A., Duarte, A.R.C., and Reis, R.L., Supercritical fluid processing of natural based polymers doped with ionic liquids, *Chem. Eng. J.*, 241, 122, 2014.

2014PIN Pinto, J., Dumon, M., Pedros, M., Reglero, J., and Rodriguez-Perez, M.A., Nanocellular CO_2 foaming of PMMA assisted by block copolymer nanostructuration, *Chem. Eng. J.*, 243, 428, 2014.

2014VOP Vopièka, O., DeAngelis, M.G., and Sarti, G.C., Mixed gas sorption in glassy polymeric membranes: I. CO_2/CH_4 and n-C_4/CH_4 mixtures sorption in poly(1-trimethylsilyl-1-propyne) (PTMSP), *J. Membrane Sci.*, 449, 97, 2014.

3. LIQUID-LIQUID EQUILIBRIUM (LLE) DATA OF POLYMER SOLUTIONS AT ELEVATED PRESSURES

3.1. Cloud-point and/or coexistence curves of quasibinary solutions

Polymer (B): **polybutadiene** **2010MIL**
Characterization: M_n/g.mol^{-1} = 15800, M_w/g.mol^{-1} = 16600,
15% 1,2-vinyl content, synthesized in the laboratory
Solvent (A): **diethyl ether** **$C_4H_{10}O$** **60-29-7**

Type of data: cloud points

w_B 0.042 was kept constant

T/K	402.15	412.15	423.15	433.15	443.15	453.15	463.15
P/bar	13	33	54	73	90	106	123

Type of data: liquid-liquid-vapor three phase equilibrium

T/K	393.15	462.15
P/bar	10	39

Polymer (B): **polybutadiene** **2010MIL**
Characterization: M_n/g.mol^{-1} = 40000, M_w/g.mol^{-1} = 41200,
15% 1,2-vinyl content, synthesized in the laboratory
Solvent (A): **diethyl ether** **$C_4H_{10}O$** **60-29-7**

Type of data: cloud points

w_B	0.043	0.043	0.043	0.043	0.043	0.043	0.043	0.043	0.043
T/K	383.15	393.15	403.15	413.15	423.15	432.15	444.15	453.15	463.15
P/bar	15	31	43	61	78	95	113	129	145

w_B	0.059	0.059	0.059	0.059	0.059	0.059	0.059	0.059	0.059
T/K	384.15	387.15	392.15	407.15	417.15	426.15	439.25	446.15	457.15
P/bar	40	41	66	84	102	119	137	152	170

w_B	0.075	0.075	0.075	0.075	0.075	0.075	0.075	0.075
T/K	377.15	389.15	397.15	407.15	419.15	428.15	437.15	451.15
P/bar	21	37	57	75	91	108	125	146

Polymer (B): **polyethylene** **2010MIL**
Characterization: M_n/g.mol^{-1} = 22000, M_w/g.mol^{-1} = 53000,
HDPE, Du Pont de Nemours, USA
Solvent (A): **diethyl ether** $C_4H_{10}O$ **60-29-7**

Type of data: cloud points

w_B	0.027	0.027	0.027	0.027	0.027	0.027	0.027	0.027	0.027
T/K	397.15	400.15	406.15	416.15	426.15	431.15	438.15	443.15	446.15
P/bar	232	228	225	219	218	218	230	230	232

w_B	0.027	0.027	0.027	0.040	0.040	0.040	0.040	0.040	0.040
T/K	450.15	456.15	460.65	392.15	400.15	405.15	409.15	415.15	420.15
P/bar	235	237	240	224	215	210	205	204	204

w_B	0.040	0.040	0.040	0.040	0.040	0.040	0.040	0.040
T/K	424.15	429.15	436.15	440.15	445.15	450.15	455.15	462.15
P/bar	207	209	213	217	222	224	229	233

Polymer (B): **polyethylene** **2011HAR**
Characterization: M_n/g.mol^{-1} = 2720, M_w/g.mol^{-1} = 10900, M_z/g.mol^{-1} = 35300
Solvent (A): **n-hexane** C_6H_{14} **110-54-3**

Type of data: cloud points

w_B	0.0025	0.0025	0.0025	0.0025	0.0051	0.0051	0.0051	0.0051	0.0051
T/K	463.10	473.33	483.04	493.13	453.12	463.06	473.22	483.09	493.12
P/MPa	2.22	3.70	4.80	5.87	1.62	2.91	4.36	5.64	6.84

w_B	0.0099	0.0099	0.0099	0.0099	0.0099	0.0196	0.0196	0.0196	0.0196
T/K	453.11	463.11	473.11	483.21	493.27	453.04	463.17	473.15	483.18
P/MPa	1.74	3.14	4.46	5.73	6.92	1.76	3.20	4.62	5.79

w_B	0.0196	0.0292	0.0292	0.0292	0.0292	0.0292	0.0512	0.0512	0.0512
T/K	493.12	453.02	463.03	473.10	483.16	493.18	452.94	463.33	473.12
P/MPa	6.97	1.63	3.07	4.27	5.55	6.80	1.57	3.05	4.20

w_B	0.0512	0.0512	0.0715	0.0715	0.0715	0.0715	0.0984	0.0984	0.0984
T/K	483.07	493.20	463.17	473.22	483.17	493.17	463.16	473.15	483.12
P/MPa	5.42	6.60	2.40	3.68	4.87	6.05	1.82	3.16	4.41

w_B	0.0984
T/K	493.20
P/MPa	5.60

Type of data: liquid-liquid-vapor three phase equilibrium

w_B	0.0025	0.0025	0.0025	0.0025	0.0051	0.0051	0.0051	0.0051	0.0099
T/K	463.10	473.34	483.12	493.18	463.04	473.21	483.06	493.12	453.14
P/MPa	1.35	1.58	1.85	2.20	1.40	1.58	1.87	2.17	1.15

continued

continued

w_B	0.0099	0.0099	0.0099	0.0099	0.0196	0.0196	0.0196	0.0196	0.0196
T/K	463.09	473.09	483.23	493.22	453.01	463.17	473.09	483.14	493.19
P/MPa	1.37	1.62	1.88	2.21	1.12	1.39	1.62	1.90	2.22
w_B	0.0292	0.0292	0.0292	0.0292	0.0292	0.0512	0.0512	0.0512	0.0512
T/K	453.02	463.06	473.11	483.05	493.13	452.93	463.32	473.09	483.06
P/MPa	1.12	1.34	1.61	1.92	2.25	1.13	1.37	1.57	1.85
w_B	0.0512	0.0715	0.0715	0.0715	0.0715	0.0984	0.0984	0.0984	0.0984
T/K	493.13	463.16	473.23	483.19	493.20	463.18	473.13	483.12	493.20
P/MPa	2.18	1.32	1.54	1.85	2.19	1.33	1.58	1.88	2.19

Polymer (B): **polyethylene** **2011HAR**
Characterization: M_n/g.mol^{-1} = 4460, M_w/g.mol^{-1} = 13100, M_z/g.mol^{-1} = 25100
Solvent (A): **n-hexane** C_6H_{14} **110-54-3**

Type of data: cloud points

w_B	0.0025	0.0025	0.0025	0.0025	0.0047	0.0047	0.0047	0.0047	0.0081
T/K	463.69	473.44	483.25	493.20	463.56	473.40	482.85	493.13	453.14
P/MPa	2.13	3.54	4.56	5.82	2.35	3.61	4.67	5.96	1.44
w_B	0.0081	0.0081	0.0081	0.0081	0.0167	0.0167	0.0167	0.0167	0.0167
T/K	463.50	473.00	482.99	493.22	453.12	463.14	473.23	483.29	493.22
P/MPa	2.69	3.75	4.70	5.88	2.02	3.39	4.63	5.82	6.87
w_B	0.0306	0.0306	0.0306	0.0306	0.0306	0.0390	.0390	0.0390	0.0390
T/K	452.99	462.99	473.26	483.23	493.21	453.19	463.13	473.13	483.17
P/MPa	2.03	3.43	4.66	5.89	7.06	1.93	3.30	4.61	5.84
w_B	0.0390	0.0489	0.0489	0.0489	0.0715	0.0715	0.0715	0.0715	0.0715
T/K	493.14	453.44	463.42	473.25	452.92	462.86	472.99	482.92	493.22
P/MPa	7.00	1.76	3.15	4.41	1.51	2.78	4.17	5.35	6.67
w_B	0.0892	0.0892	0.0892	0.0892	0.0892	.1019	0.1019	0.1019	0.1019
T/K	453.09	463.29	473.15	483.15	493.14	453.12	463.37	473.31	483.13
P/MPa	1.38	2.73	4.02	5.19	6.43	1.34	2.57	3.75	4.95
w_B	0.1019								
T/K	492.90								
P/MPa	6.08								

Type of data: liquid-liquid-vapor three phase equilibrium

w_B	0.0025	0.0025	0.0025	0.0025	0.0047	0.0047	0.0047	0.0047	0.0081
T/K	463.66	473.41	483.22	493.17	463.42	473.25	482.85	493.03	453.12
P/MPa	1.34	1.57	1.84	2.13	1.43	1.63	1.87	2.18	1.09
w_B	0.0081	0.0081	0.0081	0.0081	0.0167	0.0167	0.0167	0.0167	0.0167
T/K	463.33	472.95	483.01	493.18	453.12	463.13	473.15	483.22	493.10
P/MPa	1.34	1.57	1.84	2.14	1.10	1.34	1.58	1.88	2.21

continued

continued

w_B	0.0306	0.0306	0.0306	0.0306	0.0306	0.0390	0.0390	0.0390	0.0390
T/K	453.00	462.97	473.28	483.18	493.26	453.18	463.14	473.10	483.07
P/MPa	1.10	1.31	1.57	1.85	2.18	1.12	1.35	1.60	1.87
w_B	0.0390	0.0489	0.0489	0.0489	0.0715	0.0715	0.0715	0.0715	0.0715
T/K	493.09	453.19	463.36	473.38	452.91	462.84	473.02	482.89	493.30
P/MPa	2.20	1.12	1.35	1.57	1.08	1.33	1.53	1.81	2.16
w_B	0.0892	0.0892	0.0892	0.0892	0.0892	0.1019	0.1019	0.1019	0.1019
T/K	453.09	463.31	473.13	483.09	493.07	453.10	463.47	473.42	483.27
P/MPa	1.10	1.33	1.52	1.85	2.12	1.08	1.33	1.54	1.86
w_B	0.1019								
T/K	492.84								
P/MPa	2.14								

Polymer (B): **polyethylene** **2008HAR**
Characterization: M_n/g.mol^{-1} = 14400, M_w/g.mol^{-1} = 15500,
Scientific Polymer Products, Inc., Ontario, NY
Solvent (A): **n-hexane** C_6H_{14} **110-54-3**

Type of data: cloud points

w_B	0.0193	0.0193	0.0496	0.0496	0.0496	0.0923	0.0923	0.0923	0.1287
T/K	463.13	473.16	453.17	463.14	473.17	453.15	463.18	473.19	453.18
P/MPa	2.3	3.6	1.6	2.9	4.1	2.0	3.4	4.7	2.0
w_B	0.1287	0.1287							
T/K	463.12	473.13							
P/MPa	3.2	4.5							

Type of data: liquid-liquid-vapor three phase equilibrium

w_B	0.0193	0.0193	0.0496	0.0496	0.0496	0.0923	0.0923	0.0923	0.1287
T/K	463.12	473.10	453.14	463.12	473.15	453.14	463.12	473.17	453.18
P/MPa	1.38	1.61	1.13	1.32	1.57	1.16	1.40	1.63	1.16
w_B	0.1287	0.1287							
T/K	463.10	473.17							
P/MPa	1.37	1.52							

Polymer (B): **polyethylene** **2006NAG**
Characterization: M_n/g.mol^{-1} = 43700, M_w/g.mol^{-1} = 52000, M_z/g.mol^{-1} = 59000,
2.05 ethyl branches per 100 backbone C-atomes,
hydrogenated polybutadiene PBD 50000, was denoted
as LLDPE, DSM, The Netherlands
Solvent (A): **n-hexane** C_6H_{14} **110-54-3**

continued

continued

Type of data: cloud points

w_B	0.0005	0.0005	0.0005	0.0005	0.0005	0.0005	0.0005	0.0005	0.0005
T/K	450.65	455.60	460.52	465.47	470.26	475.12	480.14	485.02	490.13
P/MPa	1.706	2.386	3.051	3.746	4.336	4.911	5.611	6.136	6.806

w_B	0.0005	0.0024	0.0024	0.0024	0.0024	0.0024	0.0024	0.0024	0.0024
T/K	495.03	440.92	445.88	450.92	455.86	460.76	466.02	470.62	475.56
P/MPa	7.281	1.198	1.898	2.668	3.408	4.048	4.703	5.333	5.948

w_B	0.0024	0.0024	0.0024	0.0024	0.0049	0.0049	0.0049	0.0049	0.0049
T/K	480.46	485.21	490.07	494.94	440.48	445.50	450.36	455.27	460.43
P/MPa	6.498	7.068	7.643	8.223	1.572	2.322	3.017	3.662	4.402

w_B	0.0049	0.0049	0.0049	0.0049	0.0049	0.0049	0.0049	0.0078	0.0078
T/K	465.40	470.32	475.12	480.08	485.17	490.14	494.91	435.55	440.60
P/MPa	5.052	5.652	6.242	6.867	7.437	8.017	8.547	1.099	1.874

w_B	0.0078	0.0078	0.0078	0.0078	0.0078	0.0078	0.0078	0.0078	0.0078
T/K	445.56	450.47	455.40	460.33	465.27	470.19	475.15	480.05	484.91
P/MPa	2.599	3.274	3.964	4.719	5.269	5.899	6.494	7.099	7.699

w_B	0.0078	0.0078	0.0100	0.0100	0.0100	0.0100	0.0100	0.0100	0.0100
T/K	490.19	495.08	435.95	440.86	445.93	450.76	455.68	460.64	465.65
P/MPa	8.274	8.809	1.223	1.973	2.738	3.453	4.103	4.798	5.428

w_B	0.0100	0.0100	0.0100	0.0100	0.0100	0.0100	0.0223	0.0223	0.0223
T/K	470.57	475.52	480.43	485.28	490.18	495.20	435.86	440.94	445.88
P/MPa	6.023	6.648	7.228	7.788	8.353	8.913	1.563	2.348	3.048

w_B	0.0223	0.0223	0.0223	0.0223	0.0223	0.0223	0.0223	0.0223	0.0223
T/K	450.77	455.64	460.68	465.64	470.62	475.53	480.52	485.32	490.24
P/MPa	3.748	4.408	5.098	5.763	6.388	6.988	7.588	8.168	8.738

w_B	0.0223	0.0452	0.0452	0.0452	0.0452	0.0452	0.0452	0.0452	0.0452
T/K	495.24	435.79	440.73	445.66	450.62	455.59	460.51	465.56	470.53
P/MPa	9.298	1.750	2.490	3.200	3.900	4.590	5.260	5.920	6.560

w_B	0.0452	0.0452	0.0452	0.0452	0.0452	0.0585	0.0585	0.0585	0.0585
T/K	475.45	480.39	485.32	490.13	494.97	435.70	440.62	445.52	450.36
P/MPa	7.165	7.765	8.345	8.905	9.445	2.024	2.719	3.394	4.119

w_B	0.0585	0.0585	0.0585	0.0585	0.0585	0.0585	0.0585	0.0585	0.0606
T/K	460.25	465.20	470.10	475.03	480.03	485.07	490.13	495.02	435.90
P/MPa	5.449	6.084	6.674	7.334	7.969	8.569	9.124	9.699	1.964

w_B	0.0606	0.0606	0.0606	0.0606	0.0606	0.0606	0.0606	0.0606	0.0606
T/K	440.70	445.68	450.67	455.37	460.64	465.45	470.40	475.34	480.24
P/MPa	2.614	3.369	4.099	4.724	5.459	6.094	6.724	7.344	7.944

continued

continued

w_B	0.0606	0.0606	0.0606	0.06242	0.06242	0.06242	0.06242	0.06242	0.06242
T/K	485.23	490.26	495.15	435.72	440.65	445.61	450.46	455.52	460.47
P/MPa	8.519	9.084	9.619	1.894	2.624	3.384	4.119	4.789	5.494

w_B	0.06242	0.06242	0.06242	0.06242	0.06242	0.06242	0.06242	0.065445	0.065445
T/K	465.71	470.67	475.56	480.50	485.43	490.46	495.72	435.89	440.81
P/MPa	6.144	6.734	7.374	7.994	8.559	9.134	9.709	1.924	2.749

w_B	0.065445	0.065445	0.065445	0.065445	0.065445	0.065445	0.065445	0.065445	0.065445
T/K	445.79	450.61	455.61	460.50	465.30	470.25	475.23	480.14	484.88
P/MPa	3.399	4.069	4.774	5.474	6.044	6.724	7.319	7.899	8.469

w_B	0.065445	0.065445	0.06690	0.06690	0.06690	0.06690	0.06690	0.06690	0.06690
T/K	489.85	494.89	435.93	440.80	445.73	450.66	455.55	460.52	465.40
P/MPa	9.069	9.624	1.969	2.594	3.349	4.034	4.684	5.394	5.994

w_B	0.06690	0.06690	0.06690	0.06690	0.06690	0.06690	0.0706	0.0706	0.0706
T/K	470.31	475.21	480.28	485.25	490.00	495.00	436.14	441.05	445.86
P/MPa	6.644	7.244	7.869	8.454	9.014	9.589	1.909	2.684	3.374

w_B	0.0706	0.0706	0.0706	0.0706	0.0706	0.0706	0.0706	0.0706	0.0706
T/K	450.805	455.70	460.81	465.77	470.61	475.57	480.38	485.36	490.26
P/MPa	4.084	4.794	5.434	6.029	6.709	7.324	7.899	8.469	9.029

w_B	0.0706	0.0734	0.0734	0.0734	0.0734	0.0734	0.0734	0.0734	0.0734
T/K	495.323	430.71	435.74	440.62	445.60	450.52	455.46	460.35	465.27
P/MPa	9.629	0.973	1.748	2.503	3.223	3.918	4.603	5.263	5.913

w_B	0.0734	0.0734	0.0734	0.0734	0.0734	0.0734	0.0826	0.0826	0.0826
T/K	470.15	475.21	480.11	485.29	490.16	495.18	436.21	441.16	446.06
P/MPa	6.538	7.188	7.768	8.443	9.018	9.543	1.814	2.564	3.279

w_B	0.0826	0.0826	0.0826	0.0826	0.0826	0.0826	0.0826	0.0826	0.0826
T/K	450.95	455.85	460.80	465.70	470.83	475.71	480.57	485.49	490.59
P/MPa	3.989	4.664	5.329	5.969	6.624	7.229	7.834	8.414	8.989

w_B	0.0826	0.0923	0.0923	0.0923	0.0923	0.0923	0.0923	0.0923	0.0923
T/K	495.31	431.30	436.40	441.11	446.17	451.05	456.00	458.55	460.95
P/MPa	9.529	0.985	1.790	2.475	3.210	3.905	4.590	4.940	5.255

w_B	0.0923	0.0923	0.0923	0.0923	0.0923	0.0923	0.0923	0.0923	0.1534
T/K	463.34	465.91	473.36	475.73	482.45	487.53	491.14	495.73	435.805
P/MPa	5.575	5.905	6.900	7.175	7.990	8.585	8.980	9.485	1.363

w_B	0.1534	0.1534	0.1534	0.1534	0.1534	0.1534	0.1534	0.1534	0.1534
T/K	440.700	445.574	450.492	455.459	460.303	465.374	470.264	475.266	480.228
P/MPa	2.103	2.798	3.513	4.233	4.898	5.563	6.213	6.843	7.438

w_B	0.1534	0.1534	0.1534	0.19463	0.19463	0.19463	0.19463	0.19463	0.19463
T/K	485.118	489.966	495.032	435.888	440.916	446.062	451.113	455.543	460.414
P/MPa	8.028	8.598	9.178	0.989	1.624	2.449	3.149	3.774	4.424

continued

continued

w_B	0.19463	0.19463	0.19463	0.19463	0.19463	0.19463	0.19463	0.2435	0.2435
T/K	465.513	470.451	475.400	480.519	485.453	490.548	495.514	445.479	450.554
P/MPa	5.144	5.799	6.419	7.074	7.649	8.244	8.849	1.644	2.374
w_B	0.2435	0.2435	0.2435	0.2435	0.2435	0.2435	0.2435	0.2435	0.2435
T/K	455.445	460.460	465.535	470.421	475.462	480.029	484.864	489.887	494.885
P/MPa	3.094	3.849	4.499	5.104	5.774	6.199	6.894	7.499	8.099
w_B	0.3031	0.3031	0.3031	0.3031	0.3031	0.3031	0.3031	0.3031	0.3031
T/K	451.011	455.705	460.899	465.863	470.829	475.762	480.675	485.650	490.540
P/MPa	1.597	2.297	3.067	3.847	4.547	4.897	5.642	6.247	6.847
w_B	0.3031								
T/K	495.435								
P/MPa	7.497								

Type of data: liquid-liquid-vapor three phase equilibrium

w_B	0.0005	0.0005	0.0005	0.0005	0.0005	0.0024	0.0024	0.0024	0.0024
T/K	450.58	460.52	470.32	480.30	490.07	440.94	450.82	460.74	470.75
P/MPa	1.251	1.481	1.736	2.026	2.341	1.068	1.278	1.603	1.758
w_B	0.0024	0.0024	0.0049	0.0049	0.0049	0.0049	0.0049	0.0049	0.0078
T/K	480.45	490.02	440.62	450.62	460.54	470.46	480.31	490.20	440.61
P/MPa	2.053	2.363	1.087	1.292	1.517	1.762	2.052	2.367	1.081
w_B	0.0078	0.0078	0.0078	0.0078	0.0078	0.0223	0.0223	0.0223	0.0223
T/K	450.62	460.54	470.46	480.32	490.20	440.970	450.959	460.669	470.571
P/MPa	1.320	1.519	1.782	2.057	2.371	1.068	1.273	1.538	1.753
w_B	0.0223	0.0223	0.0452	0.0452	0.0452	0.0452	0.0452	0.0452	0.0452
T/K	480.380	490.182	431.016	440.959	450.757	460.601	470.530	480.378	490.215
P/MPa	2.128	2.353	0.899	1.079	1.274	1.499	1.759	2.049	2.364
w_B	0.0923	0.0923	0.0923	0.0923	0.0923	0.0923	0.0923	0.1946	0.1946
T/K	431.31	441.38	451.33	460.96	470.67	480.46	490.32	441.080	450.976
P/MPa	0.890	1.075	1.275	1.490	1.740	2.030	2.355	1.098	1.288
w_B	0.1946	0.1946	0.1946	0.1946	0.2435	0.2435	0.2435	0.3031	0.3031
T/K	460.856	470.696	480.547	491.112	445.488	450.430	460.276	460.793	470.728
P/MPa	1.513	1.773	2.058	2.418	1.164	1.269	1.494	1.521	1.748
w_B	0.3031								
T/K	480.391								
P/MPa	2.087								

Polymer (B): **polyethylene** **2008HAR**
Characterization: M_n/g.mol^{-1} = 82000, M_w/g.mol^{-1} = 108000,
Scientific Polymer Products, Inc., Ontario, NY
Solvent (A): **n-hexane** C_6H_{14} **110-54-3**

continued

continued

Type of data: cloud points

w_B	0.0100	0.0100	0.0100	0.0100	0.0100	0.0194	0.0194	0.0194	0.0194
T/K	433.14	443.11	453.14	463.12	473.15	433.15	443.14	453.16	463.10
P/MPa	2.5	3.9	5.3	6.5	7.8	2.6	4.0	5.3	6.6

w_B	0.0194	0.0487	0.0487	0.0487	0.0487	0.0487	0.0885	0.0885	0.0885
T/K	473.18	433.17	443.10	453.11	463.11	473.13	433.20	443.12	453.19
P/MPa	7.9	3.2	4.6	6.0	7.3	8.5	2.6	4.1	5.5

w_B	0.0885	0.0885	0.1249	0.1249	0.1249	0.1249	0.1249
T/K	463.18	473.16	433.19	443.10	453.14	463.11	473.16
P/MPa	6.8	8.3	2.6	3.9	5.3	6.7	7.9

Type of data: liquid-liquid-vapor three phase equilibrium

w_B	0.0100	0.0100	0.0100	0.0100	0.0100	0.0194	0.0194	0.0194	0.0194
T/K	433.13	443.09	453.12	463.10	473.13	433.12	443.12	453.13	463.12
P/MPa	0.83	1.02	1.22	1.44	1.73	0.82	1.00	1.20	1.43

w_B	0.0194	0.0487	0.0487	0.0487	0.0487	0.0487	0.0885	0.0885	0.0885
T/K	473.17	433.15	443.12	453.12	463.15	473.11	433.21	443.11	453.13
P/MPa	1.74	0.82	0.98	1.19	1.42	1.70	0.82	0.98	1.17

w_B	0.0885	0.0885	0.1249	0.1249	0.1249	0.1249	0.1249
T/K	463.14	473.14	433.19	443.10	453.15	463.13	473.15
P/MPa	1.42	1.72	0.80	0.98	1.17	1.45	1.69

Polymer (B): **polyethylene** **2010HA2**
Characterization: M_n/g.mol^{-1} = 13200, M_w/g.mol^{-1} = 15400,
Scientific Polymer Products, Inc., Ontario, NY
Solvent (A): **1-hexene** C_6H_{12} **592-41-6**

Type of data: cloud points

w_B	0.020	0.020	0.020	0.049	0.049	0.049	0.080	0.080	0.080
T/K	453.36	462.91	472.97	453.10	463.46	473.40	453.13	463.16	472.99
P/MPa	1.87	3.18	4.49	2.24	3.62	5.00	2.41	3.78	5.08

w_B	0.080	0.099	0.099	0.099	0.120	0.120	0.120	0.120
T/K	473.33	443.35	453.18	463.55	43.58	453.50	463.21	473.29
P/MPa	5.22	1.44	2.63	4.00	1.31	2.64	3.92	5.30

Type of data: liquid-liquid-vapor three phase equilibrium

w_B	0.020	0.020	0.020	0.049	0.049	0.049	0.080	0.080	0.080
T/K	453.14	462.85	472.93	453.24	463.07	473.18	453.13	463.14	472.92
P/MPa	1.20	1.46	1.71	1.41	1.63	1.91	1.22	1.44	1.75

w_B	0.099	0.099	0.099	0.099	0.120	0.120	0.120	0.120
T/K	443.35	453.25	463.45	473.25	443.52	453.41	463.23	473.56
P/MPa	0.95	1.16	1.42	1.69	0.94	1.15	1.40	1.69

Polymer (B): **polyethylene** **2010HA2**
Characterization: M_n/g.mol^{-1} = 82000, M_w/g.mol^{-1} = 108000,
Scientific Polymer Products, Inc., Ontario, NY
Solvent (A): **1-hexene** C_6H_{12} **592-41-6**

Type of data: cloud points

w_B	0.020	0.020	0.020	0.020	0.020	0.020	0.049	0.049	0.049
T/K	433.45	443.45	453.22	463.08	473.41	473.19	433.49	443.35	453.34
P/MPa	2.55	3.93	5.25	6.60	7.90	8.63	3.25	4.63	6.07

w_B	0.049	0.080	0.080	0.080	0.080	0.080	0.080	0.099	0.099
T/K	463.55	413.50	433.62	443.18	453.11	463.05	472.96	413.51	433.49
P/MPa	7.37	1.04	3.63	4.72	6.21	7.41	8.67	1.57	3.77

w_B	0.099	0.099	0.099	0.099	0.120	0.120	0.120	0.120	0.120
T/K	443.33	453.06	463.36	473.43	413.54	433.02	443.38	453.21	463.59
P/MPa	4.82	6.34	7.47	8.65	1.53	3.66	4.87	6.37	7.41

w_B	0.120
T/K	473.44
P/MPa	8.60

Type of data: liquid-liquid-vapor three phase equilibrium

w_B	0.020	0.020	0.020	0.020	0.020	0.049	0.049	0.049	0.049
T/K	433.36	443.33	453.33	463.26	473.51	433.54	443.39	453.10	463.49
P/MPa	0.67	0.86	1.04	1.30	1.62	0.87	1.06	1.26	1.51

w_B	0.049	0.080	0.080	0.080	0.080	0.080	0.080	0.099	0.099
T/K	473.19	413.61	433.67	443.33	453.22	463.06	473.00	413.44	433.37
P/MPa	1.80	0.57	0.87	1.07	1.31	1.53	1.81	0.56	0.86

w_B	0.099	0.099	0.099	0.099	0.120	0.120	0.120	0.120	0.120
T/K	443.36	452.98	463.32	473.51	413.56	433.08	443.09	452.99	463.49
P/MPa	1.05	1.27	1.53	1.79	0.58	0.87	1.10	1.28	1.54

w_B	0.120
T/K	473.22
P/MPa	1.81

Polymer (B): **polyethylene** **2007NAG**
Characterization: M_n/g.mol^{-1} = 43700, M_w/g.mol^{-1} = 52000, M_z/g.mol^{-1} = 59000,
2.05 ethyl branches per 100 backbone C-atomes,
hydrogenated polybutadiene PBD 50000, was denoted
as LLDPE, DSM, The Netherlands
Solvent (A): **2-methylpentane** C_6H_{14} **107-83-5**

continued

continued

Type of data: cloud points

w_B	0.0020	0.0020	0.0020	0.0020	0.0020	0.0020	0.0020	0.0020	0.0020
T/K	420.92	425.84	430.77	435.89	440.81	445.84	450.72	455.58	460.45
P/MPa	1.060	1.765	2.350	3.460	4.185	4.795	5.410	6.040	6.645
w_B	0.0020	0.0020	0.0020	0.0020	0.0020	0.0020	0.0020	0.0053	0.0053
T/K	465.35	470.40	475.24	480.25	485.13	490.17	495.10	421.03	425.96
P/MPa	7.220	7.820	8.345	8.890	9.410	9.935	10.450	1.975	2.729
w_B	0.0053	0.0053	0.0053	0.0053	0.0053	0.0053	0.0053	0.0053	0.0053
T/K	430.96	435.86	440.86	445.74	450.66	455.60	460.58	465.43	470.34
P/MPa	3.434	4.119	4.784	5.444	6.059	6.664	7.274	7.894	8.419
w_B	0.0053	0.0053	0.0053	0.0053	0.0053	0.0103	0.0103	0.0103	0.0103
T/K	475.28	480.17	484.99	489.91	494.85	415.69	420.61	425.57	430.83
P/MPa	8.969	9.499	10.019	10.519	11.039	1.570	2.385	3.060	3.940
w_B	0.0103	0.0103	0.0103	0.0103	0.0103	0.0103	0.0103	0.0103	0.0103
T/K	435.61	440.61	445.54	450.46	455.16	460.09	465.04	470.01	474.83
P/MPa	4.640	5.275	5.900	6.535	7.130	7.750	8.285	8.645	9.450
w_B	0.0103	0.0103	0.0103	0.0103	0.0288	0.0288	0.0288	0.0288	0.0288
T/K	479.72	484.62	489.48	494.39	410.72	416.06	420.97	425.91	430.79
P/MPa	9.935	10.465	10.965	11.465	1.519	2.284	3.009	3.694	4.374
w_B	0.0288	0.0288	0.0288	0.0288	0.0288	0.0288	0.0288	0.0288	0.0288
T/K	435.70	440.68	445.60	450.57	455.50	460.40	465.21	470.09	475.03
P/MPa	5.049	5.699	6.334	6.969	7.584	8.174	8.724	9.284	9.844
w_B	0.0288	0.0288	0.0288	0.0288	0.0503	0.0503	0.0503	0.0503	0.0503
T/K	479.98	484.97	489.97	494.89	411.02	415.89	420.79	425.72	430.65
P/MPa	10.364	10.894	11.419	11.894	1.874	2.569	3.344	4.049	4.774
w_B	0.0503	0.0503	0.0503	0.0503	0.0503	0.0503	0.0503	0.0503	0.0503
T/K	435.61	440.59	445.52	450.34	455.49	460.44	465.38	470.35	475.32
P/MPa	5.394	6.069	6.844	7.449	8.099	8.669	9.224	9.799	10.294
w_B	0.0503	0.0503	0.0503	0.0503	0.0604	0.0604	0.0604	0.0604	0.0604
T/K	480.11	485.09	490.01	494.96	410.57	415.12	420.77	425.77	430.71
P/MPa	10.809	11.339	11.769	12.309	1.497	2.347	3.072	3.767	4.467
w_B	0.0604	0.0604	0.0604	0.0604	0.0604	0.0604	0.0604	0.0604	0.0604
T/K	435.64	440.68	445.55	450.57	455.68	460.69	465.38	470.22	475.23
P/MPa	5.132	5.787	6.417	7.067	7.667	8.377	8.867	9.397	9.937
w_B	0.0604	0.0604	0.0604	0.0604	0.0638	0.0638	0.0638	0.0638	0.0638
T/K	480.11	485.07	489.96	494.86	411.71	416.61	421.60	426.50	431.40
P/MPa	10.462	11.057	11.492	11.992	1.893	2.598	3.368	4.033	4.743

continued

continued

w_B	0.0638	0.0638	0.0638	0.0638	0.0638	0.0638	0.0638	0.0638	0.0638
T/K	436.35	441.27	446.24	451.19	456.18	461.06	465.84	470.83	475.70
P/MPa	5.443	6.048	6.718	7.323	7.968	8.548	9.043	9.693	10.173

w_B	0.0638	0.0638	0.0638	0.0638	0.0647	0.0647	0.0647	0.0647	0.0647
T/K	480.63	485.60	490.54	495.43	410.78	415.70	420.63	425.71	430.72
P/MPa	10.798	11.273	11.818	12.273	1.718	2.523	3.208	3.983	4.718

w_B	0.0647	0.0647	0.0647	0.0647	0.0647	0.0647	0.0647	0.0647	0.0647
T/K	435.77	440.66	445.56	450.62	455.58	460.40	465.39	470.27	475.19
P/MPa	5.448	6.118	6.673	7.323	7.933	8.393	8.973	9.543	10.068

w_B	0.0647	0.0647	0.0647	0.0647	0.0984	0.0984	0.0984	0.0984	0.0984
T/K	480.25	485.08	489.96	494.84	405.92	410.92	415.77	420.68	425.67
P/MPa	10.623	11.143	11.633	12.143	0.873	1.698	2.498	3.143	3.968

w_B	0.0984	0.0984	0.0984	0.0984	0.0984	0.0984	0.0984	0.0984	0.0984
T/K	435.76	440.72	445.69	450.51	455.36	460.31	465.21	470.15	475.06
P/MPa	5.298	5.933	6.643	7.243	7.828	8.408	9.068	9.718	10.263

w_B	0.0984	0.0984	0.0984	0.0984	0.1507	0.1507	0.1507	0.1507	0.1507
T/K	479.97	484.85	489.77	494.64	411.08	416.02	420.98	425.94	430.88
P/MPa	10.758	11.198	11.743	12.218	1.197	1.957	2.682	3.402	4.107

w_B	0.1507	0.1507	0.1507	0.1507	0.1507	0.1507	0.1507	0.1507	0.1507
T/K	435.94	440.72	445.60	450.41	455.48	460.49	465.40	470.20	474.94
P/MPa	4.807	5.452	6.097	6.717	7.342	7.937	8.542	9.087	9.717

w_B	0.1507	0.1507	0.1507	0.1507	0.2494	0.2494	0.2494	0.2494	0.2494
T/K	480.04	485.01	489.92	494.85	425.78	430.67	435.64	440.57	445.51
P/MPa	10.207	10.797	11.322	11.772	2.047	2.797	3.417	4.212	4.867

w_B	0.2494	0.2494	0.2494	0.2494	0.2494	0.2494	0.2494	0.2494	0.2494
T/K	450.39	455.39	460.34	465.24	470.14	475.12	480.10	484.96	489.82
P/MPa	5.542	6.132	6.707	7.272	7.942	8.442	9.017	9.542	10.022

w_B	0.2494
T/K	494.77
P/MPa	10.642

Type of data: liquid-liquid-vapor three phase equilibrium

w_B	0.0020	0.0020	0.0020	0.0020	0.0020	0.0020	0.0020	0.0020	0.0053
T/K	420.92	430.65	440.88	450.74	460.55	470.39	480.26	490.13	420.75
P/MPa	0.875	1.045	1.250	1.480	1.730	2.020	2.340	2.710	0.884

w_B	0.0053	0.0053	0.0053	0.0053	0.0053	0.0053	0.0053	0.0103	0.0103
T/K	430.70	440.52	450.50	460.48	470.21	480.12	489.90	420.61	430.71
P/MPa	1.064	1.264	1.489	1.749	2.029	2.354	2.719	0.920	1.100

continued

continued

w_B	0.0103	0.0103	0.0103	0.0103	0.0103	0.0103	0.0288	0.0288	0.0288
T/K	440.50	450.27	460.19	469.97	479.83	489.62	411.02	421.01	430.77
P/MPa	1.295	1.520	1.775	2.060	2.385	2.740	0.729	0.884	1.065
w_B	0.0288	0.0288	0.0288	0.0288	0.0288	0.0288	0.0503	0.0503	0.0503
T/K	440.72	450.46	460.45	470.31	480.24	490.00	411.00	420.86	430.94
P/MPa	1.269	1.479	1.739	2.029	2.349	2.714	0.774	0.924	1.104
w_B	0.0503	0.0503	0.0503	0.0503	0.0503	0.0503	0.0984	0.0984	0.0984
T/K	440.81	450.71	460.47	470.34	480.11	489.94	410.89	420.75	440.55
P/MPa	1.304	1.534	1.784	2.074	2.394	2.774	0.753	0.908	1.283
w_B	0.0984	0.0984	0.0984	0.0984	0.0984	0.1507	0.1507	0.1507	0.1507
T/K	450.39	460.24	470.15	479.86	489.69	410.42	420.34	430.45	440.21
P/MPa	1.508	1.763	2.048	2.373	2.728	0.752	0.902	1.077	1.277
w_B	0.1507	0.1507	0.1507	0.1507	0.1507				
T/K	450.09	459.98	469.86	479.81	489.73				
P/MPa	1.502	1.757	2.047	2.367	2.667				

Polymer (B): **poly(ethylene-*co*-1-butene), deuterated** **2008KOS**

Characterization: M_n/g.mol^{-1} = 222700, M_w/g.mol^{-1} = 245000, 20.2 mol% 1-butene, 10 ethyl branches per 100 backbone C-atomes, completely deuterated polybutadiene, synthesized in the laboratory

Solvent (A): **2,2-dimethylpropane** C_5H_{12} **463-82-1**

Type of data: cloud points

w_B 0.003 was kept constant

T/K	285.95	289.15	292.75	296.65	299.45	304.15	306.55	310.35	314.75
P/MPa	35.3	32.4	30.3	27.1	26.3	25.0	23.8	22.6	21.7
T/K	317.85	322.65	327.35	329.65	333.15	337.05	339.55	341.35	345.45
P/MPa	21.3	20.8	20.4	20.2	20.0	19.6	19.5	19.4	19.4
T/K	349.45	352.75	358.55	361.65	364.65	366.95	369.35	372.75	375.45
P/MPa	19.4	19.5	19.8	19.9	20.0	20.1	20.2	20.4	20.6
T/K	379.15	381.75	385.35	389.55	391.35	394.85	398.15	401.85	406.45
P/MPa	20.8	20.9	21.1	21.4	21.5	21.7	22.0	22.3	22.6
T/K	409.05	412.95	415.55	418.75	423.05	425.75	428.75	431.25	434.85
P/MPa	22.9	23.1	23.3	23.5	23.9	24.1	24.3	24.5	24.7
T/K	435.85	440.05	441.25	445.65	450.15	454.35	456.05	460.65	461.35
P/MPa	24.7	25.0	25.1	25.3	25.7	26.2	26.4	27.1	27.2

Comments: Solid-liquid equilibrium data are additionally given in the original source.

Polymer (B): **poly(ethylene-*co*-1-butene), deuterated** **2008KOS**
Characterization: M_n/g.mol^{-1} = 222700, M_w/g.mol^{-1} = 245000,
20.2 mol% 1-butene, 10 ethyl branches per 100 backbone
C-atomes, completely deuterated polybutadiene,
synthesized in the laboratory
Solvent (A): **2-methylbutane** C_5H_{12} **78-78-4**

Type of data: cloud points

w_B 0.005 was kept constant

T/K	371.15	391.75	414.45	437.05
P/MPa	1.5	4.6	8.6	12.0

w_B 0.019 was kept constant

T/K	367.95	378.25	386.25	410.85	426.95	448.15
P/MPa	1.6	3.4	5.1	8.8	11.1	14.3

Comments: Solid-liquid equilibrium data are additionally given in the original source.

Polymer (B): **poly(ethylene-*co*-1-butene), deuterated** **2008KOS**
Characterization: M_n/g.mol^{-1} = 222700, M_w/g.mol^{-1} = 245000,
20.2 mol% 1-butene, 10 ethyl branches per 100 backbone
C-atomes, completely deuterated polybutadiene,
synthesized in the laboratory
Solvent (A): **n-pentane** C_5H_{12} **109-66-0**

Type of data: cloud points

w_B	0.005	0.005	0.005	0.005	0.018	0.018
T/K	395.65	408.45	420.95	439.55	389.85	408.45
P/MPa	2.6	4.6	6.8	9.5	1.8	5.0

Comments: Solid-liquid equilibrium data are additionally given in the original source.

Polymer (B): **poly(ethylene-*co*-1-octene)** **2005LEE**
Characterization: M_n/g.mol^{-1} = 64810, M_w/g.mol^{-1} = 135500,
15.3 mol% 1-octene, T_m/K = 323.2, T_g/K = 214.2,
ρ = 0.86 g/cm^3, DuPont Dow Elastomers Corporation
Solvent (A): **n-pentane** C_5H_{12} **109-66-0**

Type of data: cloud points

w_B	0.0099	0.0202	0.0202	0.0504	0.0504	0.1133	0.1133	0.2000
T/K	413.55	404.35	413.45	403.35	416.05	403.15	412.95	414.95
P/bar	23.553	25.268	40.458	26.542	46.730	21.348	37.763	26.297

Polymer (B): **poly(methyl methacrylate)** **2008MAT**
Characterization: M_w/g.mol^{-1} = 15000, Aldrich Chem. Co., Inc., Milwaukee, WI
Solvent (A): **ethanol** C_2H_6O **64-17-5**

Type of data: cloud points

P/MPa = vapor pressure

w_B	0.0257	0.0492	0.0903	0.100	0.129	0.179
T/K	344.3	345.8	346.6	346.5	343.7	341.6

P/MPa = 5

w_B	0.0257	0.0492	0.0903	0.0995	0.129	0.179
T/K	342.2	343.7	345.6	345.4	342.1	340.3

P/MPa = 10

w_B	0.0257	0.0492	0.0903	0.0995	0.129	0.179
T/K	340.9	342.1	344.4	343.3	340.7	338.9

P/MPa = 15

w_B	0.0257	0.0492	0.0903	0.0995	0.129	0.179
T/K	337.3	340.4	342.5	341.8	337.8	339.4

Polymer (B): **poly(vinyl methyl ether)** **2006LOO**
Characterization: M_n/g.mol^{-1} = 8000, M_w/g.mol^{-1} = 20000,
Aldrich Chem. Co., Inc., Milwaukee, WI
Solvent (A): **water** H_2O **7732-18-5**

Type of data: cloud points

w_B	0.02	0.02	0.02	0.02	0.02	0.02	0.02	0.02	0.02
T/K	310.75	310.65	309.55	305.45	301.75	301.15	300.15	292.85	291.65
P/MPa	0.1	49.2	56.2	87.2	108	124	130	142	143
	(a)	(b)	(b)	(b)	(b)	(b)	(b)	(b)	(b)
w_B	0.1	0.1	0.1	0.1	0.1	0.1	0.1	0.1	0.1
T/K	307.35	306.95	303.85	300.15	298.05	294.85	295.35	293.25	290.15
P/MPa	0.1	32.4	61.6	116	140	144	152	163	190
	(a)	(b)	(b)	(b)	(b)	(b)	(b)	(b)	(b)
w_B	0.1	0.1	0.1	0.2	0.2	0.2	0.2	0.2	0.2
T/K	289.25	288.25	284.75	308.65	308.25	308.15	307.75	307.65	306.05
P/MPa	196	199	202	0.1	15.5	77.2	94.4	111	141
	(b)	(b)	(b)	(a)	(b)	(b)	(b)	(b)	(b)
w_B	0.2	0.2	0.2	0.2	0.2	0.2	0.3	0.3	0.3
T/K	303.95	303.15	302.45	299.15	294.15	292.95	309.65	307.75	307.85
P/MPa	174	181	188	198	217	218	0.1	83.8	84.2
	(b)	(b)	(b)	(b)	(b)	(b)	(a)	(b)	(b)

continued

continued

w_B	0.3	0.3	0.3	0.3	0.3	0.3	0.3	0.3	0.3
T/K	308.45	305.85	303.55	302.15	301.65	297.95	298.25	295.15	289.95
P/MPa	95.1	157	177	202	225	254	265	289	323
	(b)	(b)	(b)	(b)	(b)	(b)	(b)	(b)	(b)
w_B	0.35	0.35	0.35	0.35	0.35	0.35	0.4	0.4	0.4
T/K	309.45	310.85	306.25	297.25	292.35	290.35	307.15	310.75	310.25
P/MPa	0.1	62.3	264	379	454	473	0.1	90.9	229
	(a)	(b)	(b)	(b)	(b)	(b)	(a)	(b)	(b)
w_B	0.4	0.4	0.4	0.4	0.4	0.5	0.5	0.5	0.5
T/K	306.95	305.75	300.65	296.35	291.15	302.65	303.35	310.35	316.15
P/MPa	294	334	396	513	587	0.1	14.6	75.4	160
	(b)	(b)	(b)	(b)	(b)	(a)	(b)	(b)	(b)
w_B	0.5	0.5	0.5	0.5	0.5	0.5	0.5	0.5	0.5
T/K	318.15	318.85	318.95	317.65	316.45	312.35	309.85	308.65	306.25
P/MPa	213	271	274	355	478	541	612	664	731.5
	(b)	(b)	(b)	(b)	(b)	(b)	(b)	(b)	(b)
w_B	0.5	0.5	0.5	0.5	0.5	0.5	0.5	0.5	0.6
T/K	300.85	296.15	315.65	318.15	316.15	318.15	322.15	322.15	301.55
P/MPa	816	892	150	150	150	400	400	400	0.1
	(b)	(b)	(b)	(c)	(c)	(b)	(c)	(c)	(a)
w_B	0.6	0.6	0.6	0.6	0.6	0.6	0.6	0.6	0.6
T/K	307.75	308.35	312.15	313.15	317.65	321.45	323.95	323.95	325.85
P/MPa	59.5	74.3	103	127	220	315	342	363	628
	(b)	(b)	(b)	(b)	(b)	(b)	(b)	(b)	(b)
w_B	0.6	0.6	0.6	0.6	0.6	0.6	0.6	0.6	0.7
T/K	326.15	326.15	311.55	316.85	320.95	324.75	320.15	323.35	302.15
P/MPa	856	1030	137.72	144.51	301.4	572.45	311.02	486.74	0.1
	(b)	(b)	(b)	(b)	(b)	(b)	(b)	(b)	(a)
w_B	0.7	0.7	0.7						
T/K	308.65	318.15	328.15						
P/MPa	103	176	308						
	(b)	(b)	(b)						

Comments: Experimental cloud point temperatures of PVME/water solutions for different compositions at different pressures determined by (a) small angle laser light scattering, (b) optical microscopy and (c) IR spectroscopy.

3.2. Table of binary systems where data were published only in graphical form as phase diagrams or related figures

Polymer (B)	Solvent (A)	Ref.
Poly(butadiene-*b*-styrene)		
	dichlorobenzene	2004ABB
Poly[2-(2-ethoxy)ethoxyethyl vinyl ether-*b*-(2-methoxyethyl vinyl ether)]		
	deuterium oxide	2006OSA
	deuterium oxide	2010SHI
	water	2006OSA
Polyethylene		
	n-butane	2009KOJ
	2-chloropropane	2009KOJ
	dichloromethane	2009KOJ
	2,2-dimethylbutane	2009KOJ
	2,3-dimethylbutane	2009KOJ
	n-heptane	2009KOJ
	n-hexane	2004GHO
	n-hexane	2005HEI
	n-hexane	2009KOJ
	2-methylbutane	2009KOJ
	2-methylpentane	2009KOJ
	3-methylpentane	2009KOJ
	n-pentane	2006UPP
	n-pentane	2009KOJ
Poly(ethylene-*co*-1-octene)		
	n-heptane	2005LEE
	n-hexane	2005LEE
	n-octane	2005LEE
	n-pentane	2005LEE

Polymer (B)	**Solvent (A)**	**Ref.**
Poly(ethylene oxide-*b*-propylene oxide-*b*-ethylene oxide)		
	deuterium oxide	2010KLO
	water	2009KOS
Poly(*N*-(1-hydroxymethyl)propylmethacrylamide)		
	water	2005SET
Poly(*N*-isopropylacrylamide)		
	deuterium oxide	2010SHI
	water	2010SHI
	water	2014EBE
Poly(*N*-isopropylmethacrylamide)		
	water	2006KIR
	water	2007SHI
Poly(4-methyl-1-pentene)		
	n-pentane	2006FA1
Polystyrene		
	cyclohexane	2005XUD
	cyclohexylbenzene	2005XUD
	decahydroquinoline	2005XUD
	decahydronaphthalene	2005XUD
	dicyclohexyl	2005XUD
	ethylcyclohexane	2005XUD
	n-hexane	2005XUD
	hexylcyclohexane	2005XUD
	isopropyl benzoate	2005XUD
	methylcyclohexane	2005XUD
	perhydrofluorene	2005XUD
	2-propanone	2006FA2
	n-tetradecane	2005XUD
	tetrahydrofuran	2005XUD
	1,2,3,4-tetrahydronapthalene	2005XUD

3.3. Cloud-point and/or coexistence curves of quasiternary and/or quasiquaternary solutions

Polymer (B): **polyethylene** **2009HAR**
Characterization: M_n/g.mol^{-1} = 13200, M_w/g.mol^{-1} = 15400,
Scientific Polymer Products, Inc., Ontario, NY
Solvent (A): **cyclohexane** C_6H_{12} **110-82-7**
Solvent (C): **n-hexane** C_6H_{14} **110-54-3**

Type of data: cloud points

w_A	0.098	0.098	0.098	0.098	0.098	0.096	0.096	0.096	0.096
w_B	0.019	0.019	0.019	0.019	0.019	0.047	0.047	0.047	0.047
w_C	0.883	0.883	0.883	0.883	0.883	0.857	0.857	0.857	0.857
T/K	463.65	468.61	473.11	478.63	483.55	463.67	468.71	473.31	478.42
P/MPa	1.82	2.48	3.05	3.79	4.32	2.13	2.79	3.42	4.03
w_A	0.096	0.091	0.091	0.091	0.091	0.091	0.089	0.089	0.089
w_B	0.047	0.090	0.090	0.090	0.090	0.090	0.112	0.112	0.112
w_C	0.857	0.819	0.819	0.819	0.819	0.819	0.799	0.799	0.799
T/K	483.78	463.70	468.61	473.67	478.42	483.77	463.76	468.79	473.67
P/MPa	4.73	2.25	2.89	3.58	4.21	4.86	2.14	2.82	3.53
w_A	0.089	0.089	0.196	0.196	0.196	0.196	0.190	0.190	0.190
w_B	0.112	0.112	0.020	0.020	0.020	0.020	0.050	0.050	0.050
w_C	0.799	0.799	0.784	0.784	0.784	0.784	0.760	0.760	0.760
T/K	478.71	483.66	469.75	473.97	478.12	483.18	468.15	473.29	478.52
P/MPa	4.14	4.73	1.60	2.17	2.68	3.33	1.70	2.35	3.01
w_A	0.190	0.182	0.182	0.182	0.182	0.177	0.177	0.177	0.177
w_B	0.050	0.089	0.089	0.089	0.089	0.114	0.114	0.114	0.114
w_C	0.760	0.729	0.729	0.729	0.729	0.709	0.709	0.709	0.709
T/K	483.41	468.56	473.52	478.61	483.28	468.37	473.19	478.61	483.42
P/MPa	3.63	1.70	2.33	3.01	3.60	1.70	2.31	2.99	3.60

Type of data: liquid-liquid-vapor three phase equilibrium

w_A	0.098	0.098	0.098	0.098	0.098	0.096	0.096	0.096	0.096
w_B	0.019	0.019	0.019	0.019	0.019	0.047	0.047	0.047	0.047
w_C	0.883	0.883	0.883	0.883	0.883	0.857	0.857	0.857	0.857
T/K	463.66	468.55	473.17	478.63	483.55	463.65	468.59	473.29	478.63
P/MPa	1.35	1.50	1.62	1.69	1.83	1.35	1.47	1.60	1.70
w_A	0.096	0.091	0.091	0.091	0.091	0.091	0.089	0.089	0.089
w_B	0.047	0.090	0.090	0.090	0.090	0.090	0.112	0.112	0.112
w_C	0.857	0.819	0.819	0.819	0.819	0.819	0.799	0.799	0.799
T/K	483.78	463.69	468.59	473.64	478.52	483.75	463.69	468.68	473.64
P/MPa	1.87	1.36	1.47	1.60	1.72	1.87	1.33	1.44	1.57

continued

continued

w_A	0.089	0.089	0.196	0.196	0.196	0.196	0.190	0.190	0.190
w_B	0.112	0.112	0.020	0.020	0.020	0.020	0.050	0.050	0.050
w_C	0.799	0.799	0.784	0.784	0.784	0.784	0.760	0.760	0.760
T/K	478.71	483.66	469.87	474.09	478.21	483.36	468.22	473.74	478.38
P/MPa	1.69	1.85	1.53	1.60	1.69	1.80	1.40	1.52	1.66
w_A	0.190	0.182	0.182	0.182	0.182	0.177	0.177	0.177	0.177
w_B	0.050	0.089	0.089	0.089	0.089	0.114	0.114	0.114	0.114
w_C	0.760	0.729	0.729	0.729	0.729	0.709	0.709	0.709	0.709
T/K	483.36	468.57	473.51	478.34	483.45	464.12	468.31	473.32	478.34
P/MPa	1.77	1.42	1.52	1.66	1.78	1.32	1.42	1.52	1.66
w_A	0.177								
w_B	0.114								
w_C	0.709								
T/K	483.40								
P/MPa	1.76								

Polymer (B): **polyethylene** **2010HA2**
Characterization: M_n/g.mol^{-1} = 13200, M_w/g.mol^{-1} = 15400,
Scientific Polymer Products, Inc., Ontario, NY
Solvent (A): **1-hexene** C_6H_{12} **592-41-6**
Solvent (C): **n-hexane** C_6H_{14} **110-54-3**

Type of data: cloud points

w_A	0.493	0.493	0.493	0.478	0.478	0.478	0.462	0.462	0.462
w_B	0.020	0.020	0.020	0.050	0.050	0.050	0.082	0.082	0.082
w_C	0.487	0.487	0.487	0.472	0.472	0.472	0.456	0.456	0.456
T/K	453.55	463.45	473.43	453.14	463.44	473.44	453.62	463.23	473.20
P/MPa	1.66	3.02	4.30	1.97	3.38	4.71	2.31	3.51	4.87
w_A	0.462	0.453	0.453	0.443	0.443	0.443	0.443		
w_B	0.082	0.100	0.100	0.120	0.120	0.120	0.120		
w_C	0.456	0.447	0.447	0.437	0.437	0.437	0.437		
T/K	473.10	453.19	463.16	443.53	453.32	463.25	473.10		
P/MPa	4.80	2.23	3.53	0.94	2.30	3.65	4.91		

Type of data: liquid-liquid-vapor three phase equilibrium

w_A	0.493	0.493	0.493	0.478	0.478	0.478	0.462	0.462	0.462
w_B	0.020	0.020	0.020	0.050	0.050	0.050	0.082	0.082	0.082
w_C	0.487	0.487	0.487	0.472	0.472	0.472	0.456	0.456	0.456
T/K	453.39	463.19	473.35	453.12	463.22	473.30	453.37	463.63	473.06
P/MPa	1.02	1.28	1.56	0.99	1.28	1.56	1.00	1.22	1.51
w_A	0.453	0.453	0.453	0.443	0.443	0.443	0.443		
w_B	0.100	0.100	0.100	0.120	0.120	0.120	0.120		
w_C	0.447	0.447	0.447	0.437	0.437	0.437	0.437		
T/K	453.15	463.12	472.83	443.40	453.39	463.16	473.39		
P/MPa	1.02	1.29	1.55	0.83	1.07	1.29	1.57		

Polymer (B):	**polyethylene**		**2010HA2**
Characterization:	M_n/g.mol^{-1} = 82000, M_w/g.mol^{-1} = 108000, Scientific Polymer Products, Inc., Ontario, NY		
Solvent (A):	**1-hexene**	C_6H_{12}	**592-41-6**
Solvent (C):	**n-hexane**	C_6H_{14}	**110-54-3**

Type of data: cloud points

w_A	0.490	0.490	0.490	0.490	0.490	0.475	0.475	0.475	0.475
w_B	0.020	0.020	0.020	0.020	0.020	0.050	0.050	0.050	0.050
w_C	0.490	0.490	0.490	0.490	0.490	0.475	0.475	0.475	0.475
T/K	433.42	443.18	453.27	463.11	473.08	433.34	443.17	453.26	463.29
P/MPa	2.53	3.89	5.30	6.64	7.92	3.28	4.67	6.10	7.37

w_A	0.475	0.460	0.460	0.460	0.460	0.460	0.450	0.450	0.450
w_B	0.050	0.080	0.080	0.080	0.080	0.080	0.100	0.100	0.100
w_C	0.475	0.460	0.460	0.460	0.460	0.460	0.450	0.450	0.450
T/K	473.17	433.42	443.06	453.27	463.39	473.52	433.12	443.27	453.04
P/MPa	8.62	2.87	4.40	5.70	7.02	8.37	2.87	4.40	5.77

w_A	0.450	0.450	0.440	0.440	0.440	0.440	0.440
w_B	0.100	0.100	0.121	0.121	0.121	0.121	0.121
w_C	0.450	0.450	0.439	0.439	0.439	0.439	0.439
T/K	463.09	473.22	433.22	442.99	453.20	463.10	473.24
P/MPa	7.02	8.59	2.90	4.43	5.80	7.05	8.57

Type of data: liquid-liquid-vapor three phase equilibrium

w_A	0.490	0.490	0.490	0.490	0.490	0.475	0.475	0.475	0.475
w_B	0.020	0.020	0.020	0.020	0.020	0.050	0.050	0.050	0.050
w_C	0.490	0.490	0.490	0.490	0.490	0.475	0.475	0.475	0.475
T/K	433.51	443.20	453.37	463.19	473.01	433.45	443.06	453.37	463.50
P/MPa	0.80	0.98	1.21	1.44	1.69	0.84	1.03	1.24	1.47

w_A	0.475	0.460	0.460	0.460	0.460	0.460	0.450	0.450	0.450
w_B	0.050	0.080	0.080	0.080	0.080	0.080	0.100	0.100	0.100
w_C	0.475	0.460	0.460	0.460	0.460	0.460	0.450	0.450	0.450
T/K	473.37	433.53	443.10	453.34	463.01	473.52	433.05	443.33	453.23
P/MPa	1.73	0.81	0.99	1.22	1.44	1.70	0.80	1.00	1.19

w_A	0.450	0.450	0.440	0.440	0.440	0.440	0.440
w_B	0.100	0.100	0.121	0.121	0.121	0.121	0.121
w_C	0.450	0.450	0.439	0.439	0.439	0.439	0.439
T/K	463.30	473.15	433.32	443.14	453.23	463.18	472.98
P/MPa	1.44	1.68	0.86	1.04	1.21	1.47	1.75

Polymer (B):	**poly[ethylene-*co*-propylene-*co*-(ethylene norbornene)]**		**2013KIR**
Characterization:	M_w/g.mol^{-1} = 150000, 70% ethylene, 0.5% diene, T_m/K = 314.2, T_g/K = 229.2, EPDM 3745, Dow Chemical Co.		
Solvent (A):	**n-octane**	C_8H_{18}	**111-65-9**
Solvent (C):	**propane**	C_3H_8	**74-98-6**

Type of data: cloud points

w_A	0.619	was kept constant
w_B	0.117	was kept constant
w_C	0.264	was kept constant

T/K	397.95	398.65	423.15	446.35	457.85	459.85	460.15
P/bar	40	45	78	120	130	135	135

Type of data: vapor-liquid-liquid equilibrium data

w_A	0.619	was kept constant
w_B	0.117	was kept constant
w_C	0.264	was kept constant

T/K	397.95	398.65	423.15	446.35	460.15
P/bar	24	25	53	72	90

3.4. Table of ternary or quaternary systems where data were published only in graphical form as phase diagrams or related figures

Polymer (B)	Second and third component	Ref.
Polyethylene		
	1-bromo-1-chloro-2,2,2-trifluoroethane and n-pentane	2009KOJ
	bromoethane and n-pentane	2009KOJ
	1-bromopropane and n-pentane	2009KOJ
	2-bromopropane and n-pentane	2009KOJ
	1-butanol and n-pentane	2009KOJ
	2-butanone and n-pentane	2009KOJ
	chloroethane and n-pentane	2009KOJ
	1-chloropropane and n-pentane	2009KOJ
	cyclohexane and 1,2-dichloro-1,2-difluoroethane	2009KOJ
	cyclohexane and ethene/1-butene	2007BUC
	cyclohexane and n-pentane	2009KOJ
	n-decane and n-pentane	2009KOJ
	dibromodifluoromethane and n-pentane	2009KOJ
	dibutyl ether and n-pentane	2009KOJ
	1,2-dichloro-1,2-difluoroethane and n-pentane	2009KOJ
	1,1-dichloroethane and n-pentane	2009KOJ
	1,2-dichloroethane and n-pentane	2009KOJ
	1,2-dichloroethene and n-pentane	2009KOJ
	cis-1,2-dichloroethene and n-pentane	2009KOJ
	dichloromethane and n-pentane	2009KOJ
	1,2-dichloropropane and n-pentane	2009KOJ
	1,4-dioxane and n-pentane	2009KOJ
	n-heptane and n-pentane	2009KOJ
	n-hexane and n-pentane	2009KOJ
	n-hexane and ethene/1-butene	2007BUC
	iodoethane and n-pentane	2009KOJ
	n-octane and n-pentane	2009KOJ
	n-pentane and 2-pentanone	2009KOJ
	n-pentane and 1-propanol	2009KOJ

Polymer (B)	Second and third component	Ref.
	n-pentane and 2-propanol	2009KOJ
	n-pentane and 2-propanone	2009KOJ
	n-pentane and 1,1,2,2-tetrachloro-1,2-difluoroethane	2009KOJ
	n-pentane and tetrahydrofurane	2009KOJ
	n-pentane and tetrahydropyrane	2009KOJ
	n-pentane and 1,1,2-trichloroethane	2009KOJ
	n-pentane and trichloroethene	2009KOJ
	n-pentane and trichlorofluoromethane	2009KOJ
	n-pentane and trichloromethane	2009KOJ
Poly(ethylene glycol)		
	water and carbon dioxide	2014CHO
	water and oxygen	2014CHO
Poly(*N*-isopropylacrylamide)		
	1-butanol and water	2014EBE
	dimethylsulfoxide and water	2012OSA
	ethanol and water	2014EBE
	methanol and water	2014EBE
	1-pentanol and water	2014EBE
	1-propanol and water	2014EBE
	sodium chloride and water	2014EBE

3.5. References

2004ABB Abbas, B., Schwahn, D., and Willner, L., Phase behavior of the polybutadiene–polystyrene diblock copolymer with the addition of the nonselective solvent dichlorobenzene in temperature and pressure fields, *J. Polym. Sci.: Part B: Polym. Phys.*, 42, 3179, 2004.

2004GHO Ghosh, A., Ting, P.D., and Chapman, W.G., Thermodynamic stability analysis and pressure-temperature flash for polydisperse polymer solutions, *Ind. Eng. Chem. Res.*, 43, 6222, 2004.

2005HEI Heidemann, R.A. , Krenz, R.A., and Laursen, T., Spinodal curves and critical points in mixtures containing polydisperse polymers with many components, *Fluid Phase Equil.*, 228-229, 239, 2005.

2005LEE Lee, S.-H., Phase behavior of binary and ternary mixtures of poly(ethylene-*co*-octene)–hydrocarbons (experimental data by S.-H. Lee), *J. Appl. Polym. Sci.*, 95, 161, 2005.

2005SET Seto, Y., Aoki, T., and Kunugi, S., Temperature- and pressure-responsive properties of L- and DL-forms of poly(*N*-(1-hydroxymethyl)propylmethacrylamide) in aqueous solutions, *Colloid Polym. Sci.*, 283, 1137, 2005.

2005XUD Xu, D., Carbonell, R.G., Roberts, G.W., and Kiserow, D.J., Phase equilibrium for the hydrogenation of polystyrene in CO_2-swollen solvents, *J. Supercrit. Fluids*, 34, 1, 2005.

2006FA1 Fang, J. and Kiran, E., Crystallization and gelation of isotactic poly(4-methyl-1-pentene) in n-pentane and in n-pentane + CO_2 at high pressures, *J. Supercrit. Fluids*, 38, 132, 2006.

2006FA2 Fang, J. and Kiran, E., Kinetics of pressure-induced phase separation in polystyrene + acetone solutions at high pressures, *Polymer*, 47, 7943, 2006.

2006KIR Kirpach, A. and Adolf, D., High pressure induced coil-globule transitions of smart polymers, *Macromol. Symp.*, 237, 7, 2006.

2006LOO Loozen, E., Nies, E., Heremans, K., and Berghmans, H., The influence of pressure on the lower critical solution temperature miscibility behavior of aqueous solutions of poly(vinyl methyl ether) and the relation to the compositional curvature of the volume of mixing, *J. Phys. Chem. B*, 110, 7793, 2006.

2006NAG Nagy, I., Loos, Th.W.de, Krenz, R.A., and Heidemann, R.A., High pressure phase equilibria in the systems linear low density polyethylene + n-hexane and linear low density polyethylene + n-hexane + ethylene: Experimental results and modelling with the Sanchez-Lacombe equation of state, *J. Supercrit. Fluids*, 37, 115, 2006.

2006OSA Osaka, N., Okabe, S., Karino, T., Hirabaru, Y., Aoshima, S., and Shibayama, M., Micro- and macrophase separations of hydrophobically solvated block copolymer aqueous solutions induced by pressure and temperature, *Macromolecules*, 39, 5875, 2006.

2006UPP Upper, G., Zhang, W., Beckel, D., Sohn, S., Liu, K., and Kiran, E., Phase boundaries and crystallization of polyethylene in n-pentane and n-pentane + carbon dioxide fluid mixtures, *Ind. Eng. Chem. Res.*, 45, 1478, 2006.

2007BUC Buchelli, A. and Todd, W.G., On-line liquid-liquid phase separation predictor in the high-density polyethylene solution polymerization process, *Ind. Eng. Chem. Res.*, 46, 4307, 2007.

2007NAG Nagy, I., Krenz, R.A., Heidemann, R.A., and de Loos, Th.W., High-pressure phase equilibria in the system linear low density polyethylene + isohexane: Experimental results and modelling, *J. Supercrit. Fluids*, 40, 125, 2007.

2007SHI Shibayama, M., Studies on pressure induced phase separation and hydrophobic interaction of polymer solutions and gels by neutron scattering and light scattering (Jap.), *Koatsuryoko Kagaku Gijutsu*, 17, 131, 2007.

2008HAR Haruki, M., Takakura, Y., Sugiura, H., Kihara, S., and Takishima, S., Phase behavior for the supercritical ethylene + hexane + polyethylene systems, *J. Supercrit. Fluids*, 44, 284, 2008.

2008KOS Kostko, A.F., Lee, S.H., Liu, J., DiNoia, T.P., Kim, Y., and McHugh, M.A., Cloud-point behavior of poly(ethylene-*co*-20.2 mol% 1-butene) (PEB10) in ethane and deuterated ethane and of deuterated PEB10 in pentane isomers, *J. Chem. Eng. Data*, 53, 1626, 2008.

2008MAT Matsuyama, K. and Mishima, K., Phase behavior of the mixtures of CO_2 + poly(methyl methacrylate) + ethanol at high pressure, *J. Chem. Eng. Data*, 53, 1151, 2008.

2009HAR Haruki, M., Sato, K., Kihara, S.-I., and Takishima, S., High pressure phase behavior for the supercritical ethylene + cyclohexane + hexane + polyethylene systems, *J. Supercrit. Fluids*, 49, 125, 2009.

2009KOJ Kojima, J., Takenaka, M., Nakayama, Y., and Saeki, S., Measurements of phase behavior for polyethylene in hydrocarbons, halogenated hydrocarbons, and oxygen-containing hydrocarbons, at high pressure and high temperature, *J. Chem. Eng. Data*, 54, 1585, 2009.

2009KOS Kostko, A.F., Harden, J.L., and McHugh, M.A., Dynamic light scattering study of concentrated triblock copolymer micellar solutions under pressure, *Macromolecules*, 42, 5328, 2009.

2010HA2 Haruki, M., Mano, S., Koga, Y., Kihara, S., and Takishima, S., Phase behaviors for the supercritical ethylene + 1-hexene + hexane + polyethylene systems at elevated temperatures and pressures, *Fluid Phase Equil.*, 295, 137, 2010.

2010KLO Kloxin, C.J. and Zanten, J.H. van, High pressure phase diagram of an aqueous PEO-PPO-PEO triblock copolymer system via probe diffusion measurements, *Macromolecules*, 43, 2084, 2010.

2010MIL Milanesio, J.M., Mabe, G.D.B., Ciolino, A.E., Quinzani, L.M., and Zabaloy, M.S., Experimental cloud points for polybutadiene + light solvent and polyethylene + light solvent systems at high pressure, *J. Supercrit. Fluids*, 55, 363, 2010.

2010SHI Shibayama, M. and Osaka, N., Pressure- and temperature-induced phase separation transition in homopolymer, block copolymer, and protein in water, *Macromol. Symp.*, 291-292, 115, 2010.

2011HAR Haruki, M., Nakanishi, K., Mano, S., Kihara, S.-I., and Takishima, S., Effect of molecular weight distribution on the liquid-liquid phase separation behavior of polydispersed polyethylene solutions at high temperatures, *Fluid Phase Equil.*, 305, 152, 2011.

2012OSA Osaka, N. and Shibayama, M., Pressure effects on cononsolvency behavior of poly(*N*-isopropylacrylamide) in water/DMSO mixed solvents, *Macromolecules*, 45, 2171, 2012.

2013KIR Kiran, E., Hassler, J.C., and Srivastava, R., Miscibility, phase separation, and phase settlement dynamics in solutions of ethylene-propylene-diene monomer elastomer in propane + n-octane binary fluid mixtures at high pressures, *Ind. Eng. Chem. Res.*, 52, 1806, 2013.

2014CHO Choi, J.Y., Kim, J.Y., Moon, H.J., Park, M.H., and Jeong, B., CO_2- and O_2-sensitive fluorophenyl end-capped poly(ethylene glycol), *Macromol. Rapid Commun.*, 35, 66, 2014.

2014EBE Ebeling, B., Eggers, S., Hendrich, M., Nitschke, A., and Vana, P., Flipping the pressure- and temperature-dependent cloud-point behavior in the cononsolvency system of poly(*N*-isopropylacrylamide) in water and ethanol, *Macromolecules*, 47, 1462, 2014.

4. HIGH-PRESSURE FLUID PHASE EQUILIBRIUM (HPPE) DATA OF POLYMER SOLUTIONS

4.1. Cloud-point and/or coexistence curves of quasibinary solutions

Polymer (B): **poly(acetoacetoxyethyl methacrylate-*b*-1,1,2,2-tetrahydroperfluorodecyl acrylate)** **2011RIB**

Characterization: M_n/g.mol^{-1} = 53200, 12.6 mol% acetoacetoxyethyl methacrylate, synthesized in the laboratory

Solvent (A): **carbon dioxide** **CO_2** **124-38-9**

Type of data: cloud points

w_B 0.040 was kept constant

T/K	293.45	299.25	303.85	308.75	313.95	318.55	23.55	328.25	333.65
P/bar	98.3	120.0	136.9	157.4	172.2	187.9	203.3	219.1	234.1
ρ/(g/cm^3)	0.8520	0.8378	0.8273	0.8203	0.8056	0.7967	0.7860	0.7783	0.7669

T/K	337.65
P/bar	248.4
ρ/(g/cm^3)	0.7627

Polymer (B): **poly(acetoacetoxyethyl methacrylate-*b*-1,1,2,2-tetrahydroperfluorodecyl acrylate)** **2011RIB**

Characterization: M_n/g.mol^{-1} = 86500, 26.2 mol% acetoacetoxyethyl methacrylate, synthesized in the laboratory

Solvent (A): **carbon dioxide** **CO_2** **124-38-9**

Type of data: cloud points

w_B 0.040 was kept constant

T/K	293.65	298.85	303.85	309.45	314.85	318.75	323.75	328.85	333.95
P/bar	103.8	132.0	148.0	181.8	207.0	227.3	246.7	263.4	277.0
ρ/(g/cm^3)	0.8573	0.8545	0.8405	0.8425	0.8374	0.8359	0.8286	0.8192	0.8079

Polymer (B): **poly(acetoacetoxyethyl methacrylate-*co*-1,1,2,2-tetrahydroperfluorodecyl acrylate)** **2011RIB**

Characterization: M_n/g.mol^{-1} = 63200, 10.8 mol% acetoacetoxyethyl methacrylate, gradient copolymer, synthesized in the laboratory

Solvent (A): **carbon dioxide** **CO_2** **124-38-9**

continued

continued

Type of data: cloud points

w_B 0.040 was kept constant

T/K	292.55	298.45	303.25	308.05	312.75	317.55	322.65	327.15	331.95
P/bar	71.4	92.9	111.5	127.5	143.8	160.4	176.6	188.8	203.3
ρ/(g/cm^3)	0.8177	0.8025	0.7940	0.7822	0.7731	0.7650	0.7552	0.7444	0.7362

T/K	336.85
P/bar	216.7
ρ/(g/cm^3)	0.7267

Polymer (B): **poly(acetoacetoxyethyl methacrylate-*co*-1,1,2,2-tetrahydroperfluorodecyl acrylate)** **2011RIB**
Characterization: M_n/g.mol^{-1} = 89200, 26.8 mol% acetoacetoxyethyl methacrylate, gradient copolymer, synthesized in the laboratory
Solvent (A): **carbon dioxide** **CO_2** **124-38-9**

Type of data: cloud points

w_B 0.040 was kept constant

T/K	293.65	298.05	303.35	307.95	312.95	317.75	322.45	327.25	332.35
P/bar	105.5	123.6	143.6	164.6	179.5	197.2	213.5	230.3	246.0
ρ/(g/cm^3)	0.8593	0.8504	0.8387	0.8335	0.8201	0.8119	0.8035	0.7959	0.7864

T/K	337.55
P/bar	262.7
ρ/(g/cm^3)	0.7782

Polymer (B): **poly[4,4'-bis(trifluorovinyloxy)biphenyl]** **2005SHE**
Characterization: M_n/g.mol^{-1} = 37500, M_w/g.mol^{-1} = 78800, T_g/K = 417, synthesized in the laboratory
Solvent (A): **dimethyl ether** **C_2H_6O** **115-10-6**

Type of data: cloud points

w_B 0.030 was kept constant

T/K	295.35	311.85	314.55	317.15	326.65	343.15	357.65	382.55	404.15
P/bar	5.5	9.0	9.7	15.2	58.8	119.3	173.7	252.4	311.0
ρ/(g/cm^3)	0.673	0.654	0.656	0.658	0.655	0.641	0.638	0.625	0.613
	VLE	VLE	VLE	LLE	LLE	LLE	LLE	LLE	LLE

T/K	315.85	326.75	360.45	384.15	401.95
P/bar	9.7	12.8	25.9	39.7	53.8
ρ/(g/cm^3)	0.546	0.556	0.505	0.489	0.355
	VLLE	VLLE	VLLE	VLLE	VLLE

Polymer (B): **poly{1,1-bis[4-(trifluorovinyloxy)phenyl]-hexafluoroisopropylidene}** **2005SHE**
Characterization: M_n/g.mol^{-1} = 31300, M_w/g.mol^{-1} = 62600, T_g/K = 386, T_m/K = 392, synthesized in the laboratory
Solvent (A): **carbon dioxide** **CO_2** **124-38-9**

Type of data: cloud points

w_B 0.028 was kept constant

T/K	307.95	320.85	327.95	336.05	344.55	364.55	384.05
P/bar	870.0	839.7	828.3	824.1	826.9	840.0	859.3
ρ/(g/cm^3)	1.110	1.088	1.065	1.055	1.032	0.998	0.967

Polymer (B): **poly{1,1-bis[4-(trifluorovinyloxy)phenyl]-hexafluoroisopropylidene}** **2005SHE**
Characterization: M_n/g.mol^{-1} = 31300, M_w/g.mol^{-1} = 62600, T_g/K = 386, T_m/K = 392, synthesized in the laboratory
Solvent (A): **dimethyl ether** **C_2H_6O** **115-10-6**

Type of data: cloud points

w_B 0.051 was kept constant

T/K	327.25	340.15	356.55	363.65	366.25	369.05	372.95	375.25	385.85
P/bar	13.6	18.1	25.2	28.6	30.0	35.5	49.9	53.1	83.4
ρ/(g/cm^3)	0.649	0.629	0.591	0.574	0.568	0.571	0.568	0.568	0.560
	VLE	VLE	VLE	VLE	LLE	LLE	LLE	LLE	LLE

T/K	299.65	406.95	369.85	376.85	383.85	393.35	401.75
P/bar	113.6	129.3	32.8	36.2	41.7	48.6	55.2
ρ/(g/cm^3)	0.555	0.553	0.519	0.503	0.501	0.428	0.354
	LLE	LLE	VLLE	VLLE	VLLE	VLLE	VLLE

Polymer (B): **poly{1,1-bis[4-(trifluorovinyloxy)phenyl]-hexafluoroisopropylidene}** **2005SHE**
Characterization: M_n/g.mol^{-1} = 31300, M_w/g.mol^{-1} = 62600, T_g/K = 386, T_m/K = 392, synthesized in the laboratory
Solvent (A): **propane** **C_3H_8** **74-98-6**

Type of data: cloud points

w_B 0.036 was kept constant

T/K	377.75	380.35	386.25	399.55	409.25
P/bar	1291.8	1220.3	1096.3	849.2	759.9
ρ/(g/cm^3)	0.624	0.617	0.606	0.577	0.562

Polymer (B): **polybutadiene** **2010MIL**
Characterization: M_n/g.mol^{-1} = 15800, M_w/g.mol^{-1} = 16600, 15% 1,2-vinyl content, synthesized in the laboratory
Solvent (A): **dimethyl ether** C_2H_6O **115-10-6**

Type of data: cloud points

w_B	0.019	0.019	0.019	0.019	0.019	0.019	0.019	0.019	0.019
T/K	308.15	323.15	333.15	342.15	353.15	361.15	371.15	380.15	394.15
P/bar	34	62	87	106	128	147	167	186	202
w_B	0.019	0.019	0.028	0.028	0.028	0.028	0.028	0.028	0.028
T/K	401.15	411.15	322.15	332.15	343.15	353.15	363.15	371.15	381.15
P/bar	220	237	92	111	134	154	171	190	208
w_B	0.028	0.028	0.028	0.028	0.046	0.046	0.046	0.046	0.046
T/K	390.15	401.15	411.15	419.15	313.15	323.15	334.15	343.15	353.15
P/bar	226	245	263	276	79	102	122	135	163
w_B	0.046	0.046	0.046	0.046	0.046	0.046	0.046	0.046	0.046
T/K	363.15	370.15	380.15	394.15	401.15	321.15	333.15	341.15	351.15
P/bar	184	196	208	223	232	99	122	138	156
w_B	0.046	0.046	0.046	0.046	0.046	0.046	0.046		
T/K	361.15	371.15	383.15	391.15	401.15	410.15	420.15		
P/bar	175	194	213	228	245	260	277		

Polymer (B): **polybutadiene** **2012MIL**
Characterization: M_n/g.mol^{-1} = 15800, M_w/g.mol^{-1} = 16600, 15% 1,2-vinyl content, synthesized in the laboratory
Solvent (A): **n-pentane** C_5H_{12} **109-66-0**

Type of data: cloud points

w_B	0.011	0.011	0.011	0.011	0.011	0.011	0.011	0.011	0.011
T/K	396.15	400.15	406.15	409.15	414.15	418.15	422.15	427.15	431.15
P/bar	19	28	36	43	48	53	62	69	72
w_B	0.011	0.011	0.011	0.011	0.033	0.033	0.033	0.033	0.033
T/K	438.15	441.15	446.15	451.15	384.15	389.15	395.15	398.15	401.15
P/bar	77	83	89	96	26	35	44	50	55
w_B	0.033	0.033	0.033	0.033	0.033	0.033	0.033	0.033	0.033
T/K	405.15	411.15	417.15	420.15	425.15	431.15	436.15	443.15	447.15
P/bar	63	72	82	87	93	101	109	118	122
w_B	0.033	0.033	0.036	0.036	0.036	0.036	0.036	0.036	0.036
T/K	451.15	457.15	391.15	399.15	411.15	419.15	428.15	438.15	447.15
P/bar	126	134	25	39	55	68	79	94	108
w_B	0.036								
T/K	456.15								
P/bar	117								

Polymer (B): **1,4-*cis*-polybutadiene** **2007TAN**
Characterization: M_n/g.mol^{-1} = 1750, M_w/M_n = 1.1,
Polymer Source, Inc., Quebec, Canada
Solvent (A): **propane** C_3H_8 **74-98-6**

Type of data: cloud points

w_B	0.005	was kept constant			
T/K	353.0	373.0	393.2	413.1	433.0
P/bar	34.9	67.1	94.9	116.8	135.7

Polymer (B): **1,4-*cis*-polybutadiene** **2007TAN**
Characterization: M_n/g.mol^{-1} = 5300, M_w/M_n = 1.04,
Polymer Source, Inc., Quebec, Canada
Solvent (A): **propane** C_3H_8 **74-98-6**

Type of data: cloud points

w_B	0.005	was kept constant							
T/K	293.1	303.2	313.2	333.1	353.2	373.2	393.0	413.1	433.2
P/bar	98.5	112.7	126.3	153.6	180.4	205.7	229.4	250.0	267.2

Polymer (B): **1,4-*cis*-polybutadiene** **2007TAN**
Characterization: M_n/g.mol^{-1} = 8800, M_w/M_n = 1.04,
Polymer Source, Inc., Quebec, Canada
Solvent (A): **propane** C_3H_8 **74-98-6**

Type of data: cloud points

w_B	0.005	was kept constant							
T/K	293.1	303.2	313.2	333.3	353.2	373.2	393.1	413.1	433.2
P/bar	267.9	265.2	266.3	275.7	289.3	304.3	320.5	335.3	348.8

Polymer (B): **polybutadiene** **2010MIL**
Characterization: M_n/g.mol^{-1} = 14200, M_w/g.mol^{-1} = 14800,
22% 1,2-vinyl content, synthesized in the laboratory
Solvent (A): **propane** C_3H_8 **74-98-6**

Type of data: cloud points

w_B	0.022	0.022	0.022	0.022	0.022	0.022	0.022	0.034	0.034
T/K	332.15	343.15	356.15	368.15	386.15	394.15	401.15	316.15	326.15
P/bar	518	515	514	516	520	520	517	558	542
w_B	0.034	0.034	0.034	0.034	0.034	0.034	0.034	0.034	0.043
T/K	337.15	347.15	357.15	370.15	381.15	392.15	403.15	415.15	329.15
P/bar	540	540	538	537	534	533	537	537	536
w_B	0.043	0.043	0.043	0.043	0.043	0.043	0.043		
T/K	341.15	353.15	356.15	371.15	374.15	390.15	400.15		
P/bar	522	515	507	500	502	500	500		

Polymer (B): **1,4-*cis*-polybutadiene** **2007TAN**
Characterization: M_n/g.mol^{-1} = 18000, M_w/M_n = 1.04,
Polymer Source, Inc., Quebec, Canada
Solvent (A): **propane** **C_3H_8** **74-98-6**

Type of data: cloud points

w_B 0.005 was kept constant

T/K	293.1	303.2	313.2	333.2	353.1	373.2	393.3	413.2	433.2
P/bar	495.1	470.2	453.1	435.3	427.9	428.8	434.1	440.9	447.6

Polymer (B): **1,4-*cis*-polybutadiene** **2007TAN**
Characterization: M_n/g.mol^{-1} = 58000, M_w/M_n = 1.05,
Polymer Source, Inc., Quebec, Canada
Solvent (A): **propane** **C_3H_8** **74-98-6**

Type of data: cloud points

w_B 0.005 was kept constant

T/K	293.4	313.3	353.2	393.2	433.0	452.9
P/bar	872.1	753.6	641.3	598.1	585.3	579.7

Polymer (B): **polybutadiene** **2009WI2**
Characterization: M_n/g.mol^{-1} = 38200, M_w/M_n = 1.05,
synthesized in the laboratory
Solvent (A): **propane** **C_3H_8** **74-98-6**

Type of data: cloud points

w_B 0.005 was kept constant

T/K	293.15	293.15	313.15	313.25	353.05	352.85	393.25	393.25	433.25
P/bar	675	674	600	601	537	537	522	522	522
T/K	433.15	453.25	453.15						
P/bar	522	555	551						

Polymer (B): **polybutadiene (deuterated)** **2009WI1**
Characterization: M_n/g.mol^{-1} = 38300, M_w/g.mol^{-1} = 39800,
synthesized in the laboratory
Solvent (A): **propane** **C_3H_8** **74-98-6**

Type of data: cloud points

w_B 0.005 was kept constant

T/K	293.25	313.25	323.05	393.15	433.15	453.25
P/bar	496	460	440	448	463	485

Polymer (B): **polybutadiene (hydrogenated)** **2009WI2**
Characterization: M_n/g.mol^{-1} = 39600, M_w/M_n = 1.05, synthesized in the laboratory
Solvent (A): **propane** C_3H_8 **74-98-6**

Type of data: cloud points

w_B 0.005 was kept constant

T/K	353.25	353.15	373.15	373.15	393.15	393.15	413.15	413.15	453.15
P/bar	462	462	460	460	461	462	466	467	457

T/K	453.15
P/bar	455

Polymer (B): **polybutadiene** **2009WI2**
Characterization: M_n/g.mol^{-1} = 38200, M_w/M_n = 1.05, synthesized in the laboratory
Solvent (A): **propene** C_3H_6 **115-07-1**

Type of data: cloud points

w_B 0.005 was kept constant

T/K	293.35	293.35	314.85	314.65	333.35	333.45	352.85	352.95	373.05
P/bar	343	343	365	365	385	385	405	406	428

T/K	372.95	392.55	393.25	413.35	413.55	433.25	433.35	453.15	453.25
P/bar	428	447	448	466	466	480	480	493	493

Polymer (B): **polybutadiene (hydrogenated)** **2009WI2**
Characterization: M_n/g.mol^{-1} = 39600, M_w/M_n = 1.05, synthesized in the laboratory
Solvent (A): **propene** C_3H_6 **115-07-1**

Type of data: cloud points

w_B 0.005 was kept constant

T/K	372.85	373.15	393.25	393.25	413.15	413.15	433.15	433.15	453.15
P/bar	489	489	488	488	492	492	497	497	502

T/K	453.15
P/bar	503

Polymer (B): **poly(2-butoxyethyl acrylate)** **2014JAN**
Characterization: M_w/g.mol^{-1} = 80000, M_w/M_n = 1.6, Scientific Polymer Products, Inc., Ontario, NY
Solvent (A): **n-butane** C_4H_{10} **106-97-8**

Type of data: cloud points

continued

continued

w_B	0.050	was kept constant					
T/K	333.6	353.5	373.2	393.6	413.3	433.3	453.2
P/MPa	23.62	15.35	13.97	14.31	15.69	17.41	19.14

Polymer (B): **poly(2-butoxyethyl acrylate)** **2014JAN**
Characterization: M_w/g.mol^{-1} = 80000, M_w/M_n = 1.6
Scientific Polymer Products, Inc., Ontario, NY
Solvent (A): **1-butene** C_4H_8 **106-98-9**

Type of data: cloud points

w_B	0.049	was kept constant			
T/K	373.3	393.4	413.8	433.2	453.3
P/MPa	4.14	7.41	10.35	12.93	15.69

Polymer (B): **poly(2-butoxyethyl acrylate)** **2014JAN**
Characterization: M_w/g.mol^{-1} = 80000, M_w/M_n = 1.6
Scientific Polymer Products, Inc., Ontario, NY
Solvent (A): **carbon dioxide** CO_2 **124-38-9**

Type of data: cloud points

w_B	0.050	was kept constant					
T/K	363.9	368.7	373.3	393.4	414.3	433.9	454.2
P/MPa	224.31	183.10	163.10	134.83	124.31	117.41	115.17

Polymer (B): **poly(2-butoxyethyl acrylate)** **2014JAN**
Characterization: M_w/g.mol^{-1} = 80000, M_w/M_n = 1.6
Scientific Polymer Products, Inc., Ontario, NY
Solvent (A): **dimethyl ether** C_2H_6O **115-10-6**

Type of data: cloud points

w_B	0.050	was kept constant				
T/K	353.8	373.5	393.7	414.5	433.7	453.7
P/MPa	2.93	8.28	11.90	16.38	18.79	20.86

Polymer (B): **poly(2-butoxyethyl acrylate)** **2014JAN**
Characterization: M_w/g.mol^{-1} = 80000, M_w/M_n = 1.6
Scientific Polymer Products, Inc., Ontario, NY
Solvent (A): **propane** C_3H_8 **74-98-6**

Type of data: cloud points

w_B	0.051	was kept constant					
T/K	333.4	353.8	373.7	393.9	413.8	433.8	454.2
P/MPa	56.03	41.88	36.38	34.31	33.97	34.31	34.66

Polymer (B): **poly(2-butoxyethyl acrylate)** **2014JAN**
Characterization: M_w/g.mol^{-1} = 80000, M_w/M_n = 1.6
Scientific Polymer Products, Inc., Ontario, NY
Solvent (A): **propene** C_3H_6 **115-07-1**

Type of data: cloud points

w_B	0.048	was kept constant					
T/K	333.6	353.4	373.4	393.5	413.6	433.3	453.6
P/MPa	16.38	19.48	22.93	25.69	28.10	30.17	31.90

Polymer (B): **poly(*tert*-butyl acrylate)** **2006BY1**
Characterization: M_n/g.mol^{-1} = 72000, M_w/g.mol^{-1} = 250000,
Polysciences, Inc., Warrington, PA
Solvent (A): **carbon dioxide** CO_2 **124-38-9**

Type of data: cloud points

w_B	0.048	was kept constant				
T/K	395.55	401.85	411.35	427.45	442.55	457.65
P/bar	2063.8	1711.4	1427.2	1294.5	1240.7	1215.2

Polymer (B): **poly(butyl methacrylate)** **2004BEC, 2004LAT**
Characterization: M_n/g.mol^{-1} = 38100, M_w/g.mol^{-1} = 65500,
synthesized in the laboratory
Solvent (A): **ethene** C_2H_4 **74-85-1**

Type of data: cloud points

w_B	0.05	was kept constant						
T/K	373.15	393.15	413.15	433.15	453.15	473.15	488.15	508.15
P/bar	1380	1320	1215	1161	1147	1111	1077	1031

Polymer (B): **poly(*tert*-butyl methacrylate)** **2006BY1**
Characterization: M_n/g.mol^{-1} = 69800, M_w/g.mol^{-1} = 180000,
Polysciences, Inc., Warrington, PA
Solvent (A): **carbon dioxide** CO_2 **124-38-9**

Type of data: cloud points

w_B	0.046	was kept constant			
T/K	434.95	437.75	444.75	456.15	463.95
P/bar	2575.5	2377.6	2186.6	1933.8	1913.1

Polymer (B): **poly(ε-caprolactone)** **2006BY2**
Characterization: M_w/g.mol^{-1} = 170000, Honam Petrochemical Co., South Korea
Solvent (A): **1-butene** C_4H_8 **106-98-9**

Type of data: cloud points

w_B 0.051 was kept constant

T/K	425.25	430.85	435.15	444.95	455.65	464.15	473.35	484.85
P/bar	1610.3	1734.5	1825.9	1979.3	2298.3	2650.0	2746.6	2836.2

Polymer (B): **poly(ε-caprolactone)** **2006BY2**
Characterization: M_w/g.mol^{-1} = 170000, Honam Petrochemical Co., South Korea
Solvent (A): **carbon dioxide** CO_2 **124-38-9**

Type of data: cloud points

w_B 0.040 was kept constant

T/K	458.25	463.05	467.75	476.15	485.65	494.95	502.15
P/bar	2832.8	2781.0	2750.0	2677.6	2622.4	2560.4	2522.4

Polymer (B): **poly(ε-caprolactone)** **2006PAR**
Characterization: M_w/g.mol^{-1} = 14000, Aldrich Chem. Co., Inc., Milwaukee, WI
Solvent (A): **chlorodifluoromethane** $CHClF_2$ **75-45-6**

Type of data: cloud points

w_B	0.0299	0.0299	0.0299	0.0299	0.0299	0.0299	0.0299	0.0299	0.0299
T/K	322.4	333.6	343.3	354.6	363.2	373.3	383.1	393.8	402.8
P/MPa	1.9	5.9	9.1	12.8	15.4	18.4	21.3	24.4	26.7
w_B	0.0299	0.0501	0.0501	0.0501	0.0501	0.0501	0.0501	0.0501	0.0501
T/K	410.3	323.8	336.1	342.9	353.2	365.2	373.3	383.5	394.1
P/MPa	28.8	2.5	6.4	9.0	12.4	16.2	18.6	21.6	24.6
w_B	0.0501	0.0501	0.0695	0.0695	0.0695	0.0695	0.0695	0.0695	0.0695
T/K	403.1	411.7	322.4	332.3	344.1	352.3	364.4	373.1	383.7
P/MPa	27.0	29.3	1.5	5.0	8.9	11.6	15.3	17.9	20.9
w_B	0.0695	0.0695	0.0695						
T/K	393.5	403.2	412.5						
P/MPa	23.2	25.9	28.1						

Polymer (B): **poly(ε-caprolactone)** **2006BY2**
Characterization: M_w/g.mol^{-1} = 170000, Honam Petrochemical Co., South Korea
Solvent (A): **chlorodifluoromethane** $CHClF_2$ **75-45-6**

Type of data: cloud points

w_B 0.047 was kept constant

continued

continued

T/K	312.75	320.55	333.85	344.25	351.05	360.85	371.05	382.75	393.65
P/bar	43.1	74.1	126.9	168.3	194.1	229.0	262.4	297.6	328.3

T/K	403.35	415.25	423.35	444.15
P/bar	359.7	393.5	422.1	454.8

Polymer (B): **poly(ε-caprolactone)** **2006PAR**
Characterization: M_w/g.mol^{-1} = 14000, Aldrich Chem. Co., Inc., Milwaukee, WI
Solvent (A): **dimethyl ether** C_2H_6O **115-10-6**

Type of data: cloud points

w_B	0.0299	0.0299	0.0299	0.0299	0.0299	0.0299	0.0299	0.0299	0.0299
T/K	313.7	324.0	333.2	343.7	353.0	363.5	373.9	383.6	394.3
P/MPa	14.0	16.8	18.9	21.6	23.9	26.2	28.3	30.5	32.7

w_B	0.0299	0.0299	0.0499	0.0499	0.0499	0.0499	0.0499	0.0499	0.0499
T/K	402.9	411.9	314.0	325.3	334.7	345.0	354.4	363.7	374.2
P/MPa	34.0	35.5	13.3	16.3	18.7	21.2	23.4	25.6	27.8

w_B	0.0499	0.0499	0.0499	0.0499	0.0696	0.0696	0.0696	0.0696	0.0696
T/K	383.3	393.6	404.7	413.6	312.5	322.7	332.6	343.2	353.4
P/MPa	29.7	31.7	33.7	35.2	13.1	15.8	18.3	20.8	23.1

w_B	0.0696	0.0696	0.0696	0.0696	0.0696	0.0696
T/K	362.8	373.2	384.2	393.8	403.2	413.6
P/MPa	25.3	27.5	29.8	31.7	33.4	35.2

Polymer (B): **poly(ε-caprolactone)** **2006BY2**
Characterization: M_w/g.mol^{-1} = 170000, Honam Petrochemical Co., South Korea
Solvent (A): **dimethyl ether** C_2H_6O **115-10-6**

Type of data: cloud points

w_B 0.050 was kept constant

T/K	328.95	350.15	368.25	389.55	411.05	427.05	444.95	459.85	471.55
P/bar	261.7	311.4	355.2	388.6	434.1	456.9	481.0	494.8	496.6

Polymer (B): **poly(ε-caprolactone)** **2006BY2**
Characterization: M_w/g.mol^{-1} = 170000, Honam Petrochemical Co., South Korea
Solvent (A): **propene** C_3H_6 **115-07-1**

Type of data: cloud points

w_B 0.053 was kept constant

T/K	419.25	424.05	427.45	435.95	446.25	456.15	465.45	474.25
P/bar	2767.2	2632.8	2556.9	2360.3	2163.8	2015.5	1887.9	1825.9

Polymer (B): **poly(4-chlorostyrene)** **2013YA2**
Characterization: M_n/g.mol^{-1} = 57000, M_w/g.mol^{-1} = 87000,
Scientific Polymer Products, Inc., Ontario, NY
Solvent (A): **1-butene** C_4H_8 **106-98-9**

Type of data: cloud points

w_B 0.046 was kept constant

T/K	446.7	452.2	459.1	467.5	483.5
P/MPa	243.62	227.76	211.55	193.97	168.10

Polymer (B): **poly(4-chlorostyrene)** **2013YA2**
Characterization: M_n/g.mol^{-1} = 57000, M_w/g.mol^{-1} = 87000,
Scientific Polymer Products, Inc., Ontario, NY
Solvent (A): **dimethyl ether** C_2H_6O **115-10-6**

Type of data: cloud points

w_B 0.049 was kept constant

T/K	335.1	355.3	375.2	394.4	414.8	435.3	451.8
P/MPa	40.35	47.24	52.76	57.59	61.55	65.00	67.76

Polymer (B): **poly(4-chlorostyrene)** **2013YA2**
Characterization: M_n/g.mol^{-1} = 57000, M_w/g.mol^{-1} = 87000,
Scientific Polymer Products, Inc., Ontario, NY
Solvent (A): **propene** C_3H_6 **115-07-1**

Type of data: cloud points

w_B 0.046 was kept constant

T/K	477.1	481.5	488.1	494.2
P/MPa	239.14	230.69	220.00	209.83

Polymer (B): **poly(cyclohexene oxide-*co*-carbon dioxide)** **2005SCH**
Characterization: M_w/g.mol^{-1} = 12000, M_w/M_n < 1.2, synthesized in the laboratory
Solvent (A): **carbon dioxide** CO_2 **124-38-9**

Type of data: cloud points

w_B 0.0051 was kept constant

T/K	455.06	459.66	465.20	475.95	479.33	479.93
P/MPa	335	309	292	257	254	252

Polymer (B): **poly(decyl acrylate)** **2006BY3**
Characterization: M_w/g.mol^{-1} = 130000,
Scientific Polymer Products, Inc., Ontario, NY
Solvent (A): **n-butane** C_4H_{10} **106-97-8**

Type of data: cloud points

w_B 0.051 was kept constant

T/K	354.55	375.55	394.05	414.05	435.25	454.75
P/bar	19.0	50.0	68.3	102.4	126.6	143.1

Polymer (B): **poly(decyl acrylate)** **2006BY3**
Characterization: M_w/g.mol^{-1} = 130000,
Scientific Polymer Products, Inc., Ontario, NY
Solvent (A): **1-butene** C_4H_8 **106-98-9**

Type of data: cloud points

w_B 0.065 was kept constant

T/K	395.05	414.85	435.45	454.85
P/bar	53.5	87.9	119.0	139.7

Polymer (B): **poly(decyl acrylate)** **2006BY3**
Characterization: M_w/g.mol^{-1} = 130000,
Scientific Polymer Products, Inc., Ontario, NY
Solvent (A): **carbon dioxide** CO_2 **124-38-9**

Type of data: cloud points

w_B 0.051 was kept constant

T/K	451.45	459.95	468.95	469.05	479.65
P/bar	1936.2	1770.7	1500.3	1485.9	1449.3

Polymer (B): **poly(decyl acrylate)** **2006BY3**
Characterization: M_w/g.mol^{-1} = 130000,
Scientific Polymer Products, Inc., Ontario, NY
Solvent (A): **dimethyl ether** C_2H_6O **115-10-6**

Type of data: cloud points

w_B 0.044 was kept constant

T/K	374.65	392.45	414.85	432.85	454.95
P/bar	44.5	81.0	129.3	156.9	177.6

Polymer (B): **poly(decyl acrylate)** **2006BY3**
Characterization: M_w/g.mol^{-1} = 130000,
Scientific Polymer Products, Inc., Ontario, NY
Solvent (A): **propane** **C_3H_8** **74-98-6**

Type of data: cloud points

w_B 0.046 was kept constant

T/K	332.05	353.75	375.25	392.55	414.65	435.35	453.55
P/bar	169.0	181.0	205.9	228.6	258.6	273.5	285.9

Polymer (B): **poly(decyl acrylate)** **2006BY3**
Characterization: M_w/g.mol^{-1} = 130000,
Scientific Polymer Products, Inc., Ontario, NY
Solvent (A): **propene** **C_3H_6** **115-07-1**

Type of data: cloud points

w_B 0.056 was kept constant

T/K	334.05	356.15	372.75	393.15	415.35	431.75	453.65
P/bar	96.6	142.4	174.1	208.6	245.2	256.9	277.6

Polymer (B): **poly[4-(decyloxymethyl)styrene]** **2008SH3**
Characterization: M_n/g.mol^{-1} = 12100, M_w/g.mol^{-1} = 33700,
synthesized in the laboratory
Solvent (A): **dimethyl ether** **C_2H_6O** **115-10-6**

Type of data: cloud points

w_B 0.039 was kept constant

T/K	372.05	392.35	412.55	432.65	452.35
P/bar	50.0	101.7	150.0	181.0	191.0

Polymer (B): **poly[4-(decylsulfonylmethyl)styrene]** **2008SH3**
Characterization: M_n/g.mol^{-1} = 18400, M_w/g.mol^{-1} = 49300,
synthesized in the laboratory
Solvent (A): **dimethyl ether** **C_2H_6O** **115-10-6**

Type of data: cloud points

w_B 0.045 was kept constant

T/K	325.35	333.55	353.15	373.65	394.05	414.85	435.75	452.15
P/bar	110.3	125.9	162.1	198.3	236.2	270.7	301.7	315.5

Polymer (B): **poly[4-(decylthiomethyl)styrene]** **2008SH3**
Characterization: M_n/g.mol^{-1} = 10700, M_w/g.mol^{-1} = 43700, synthesized in the laboratory
Solvent (A): **dimethyl ether** C_2H_6O **115-10-6**

Type of data: cloud points

w_B 0.037 was kept constant

T/K	330.45	354.05	374.95	394.75	412.95	433.95	453.15
P/bar	17.2	63.8	115.5	170.7	196.6	234.5	258.6

Polymer (B): **poly(dimethylsiloxane)** **2013BYU**
Characterization: M_w/g.mol^{-1} = 38900, M_w/M_n = 2.84, Scientific Polymer Products, Inc., Ontario, NY
Solvent (A): **n-butane** C_4H_{10} **106-97-8**

Type of data: cloud points

w_B 0.067 was kept constant

T/K	413.5	432.6	452.9
P/MPa	3.10	4.83	6.90

Polymer (B): **poly(dimethylsiloxane)** **2013BYU**
Characterization: M_w/g.mol^{-1} = 90200, M_w/M_n = 1.96, Scientific Polymer Products, Inc., Ontario, NY
Solvent (A): **n-butane** C_4H_{10} **106-97-8**

Type of data: cloud points and bubble points

w_B 0.073 was kept constant

T/K	341.6	354.5	373.7	393.9	411.9	433.6	454.0
P/MPa	0.52	0.86	1.55	2.24	3.28	5.35	7.50
	BP	BP	BP	BP	CP	CP	CP

Polymer (B): **poly(dimethylsiloxane)** **2013BYU**
Characterization: M_w/g.mol^{-1} = 170300, M_w/M_n = 1.75, Scientific Polymer Products, Inc., Ontario, NY
Solvent (A): **n-butane** C_4H_{10} **106-97-8**

Type of data: cloud points

w_B 0.074 was kept constant

T/K	411.5	435.8	452.6
P/MPa	3.33	5.86	7.61

Polymer (B): **poly(dimethylsiloxane)** **2013BYU**
Characterization: M_w/g.mol^{-1} = 38900, M_w/M_n = 2.84,
Scientific Polymer Products, Inc., Ontario, NY
Solvent (A): **1-butene** C_4H_8 **106-98-9**

Type of data: cloud points and bubble points

w_B 0.071 was kept constant

T/K	338.2	354.1	372.9	393.1	412.1	35.7	453.6
P/MPa	0.52	0.86	1.55	2.14	2.93	5.86	8.28
	BP	BP	BP	BP	CP	CP	CP

Polymer (B): **poly(dimethylsiloxane)** **2013BYU**
Characterization: M_w/g.mol^{-1} = 90200, M_w/M_n = 1.96,
Scientific Polymer Products, Inc., Ontario, NY
Solvent (A): **1-butene** C_4H_8 **106-98-9**

Type of data: cloud points and bubble points

w_B 0.075 was kept constant

T/K	341.4	354.7	374.6	391.4	413.9	433.1	454.9
P/MPa	0.52	0.86	1.48	2.07	3.28	5.52	7.79
	BP	BP	BP	BP	CP	CP	CP

Polymer (B): **poly(dimethylsiloxane)** **2013BYU**
Characterization: M_w/g.mol^{-1} = 170300, M_w/M_n = 1.75,
Scientific Polymer Products, Inc., Ontario, NY
Solvent (A): **1-butene** C_4H_8 **106-98-9**

Type of data: cloud points and bubble points

w_B 0.073 was kept constant

T/K	340.4	354.4	373.9	393.2	413.9	432.2	452.2
P/MPa	0.52	0.86	1.45	2.11	3.62	5.52	7.36
	BP	BP	BP	BP	CP	CP	CP

Polymer (B): **poly(dimethylsiloxane)** **2013BYU**
Characterization: M_w/g.mol^{-1} = 38900, M_w/M_n = 2.84,
Scientific Polymer Products, Inc., Ontario, NY
Solvent (A): **carbon dioxide** CO_2 **124-38-9**

Type of data: cloud points

w_B 0.050 was kept constant

T/K	315.2	335.0	353.3	374.0	393.7	414.2	433.2	452.6
P/MPa	33.13	34.14	37.07	40.86	43.97	47.41	49.14	51.38

Polymer (B): **poly(dimethylsiloxane)** **2013BYU**
Characterization: M_w/g.mol^{-1} = 90200, M_w/M_n = 1.96,
Scientific Polymer Products, Inc., Ontario, NY
Solvent (A): **carbon dioxide** CO_2 **124-38-9**

Type of data: cloud points

w_B 0.049 was kept constant

T/K	333.7	353.3	374.1	393.6	413.3	432.4	453.8
P/MPa	39.14	40.86	44.48	47.59	50.52	52.41	54.83

Polymer (B): **poly(dimethylsiloxane)** **2013BYU**
Characterization: M_w/g.mol^{-1} = 170300, M_w/M_n = 1.75,
Scientific Polymer Products, Inc., Ontario, NY
Solvent (A): **carbon dioxide** CO_2 **124-38-9**

Type of data: cloud points

w_B 0.055 was kept constant

T/K	334.5	354.2	373.0	391.9	413.9	434.7	454.3
P/MPa	40.52	42.24	45.00	47.76	50.86	53.28	55.52

Polymer (B): **poly(dimethylsiloxane)** **2013BYU**
Characterization: M_w/g.mol^{-1} = 38900, M_w/M_n = 2.84,
Scientific Polymer Products, Inc., Ontario, NY
Solvent (A): **dimethyl ether** C_2H_6O **115-10-6**

Type of data: cloud points and bubble points

w_B 0.052 was kept constant

T/K	334.4	353.2	372.4	393.3	413.9	433.4	452.5
P/MPa	1.21	2.24	2.76	4.14	7.41	10.00	12.41
	BP	BP	BP	CP	CP	CP	CP

Polymer (B): **poly(dimethylsiloxane)** **2013BYU**
Characterization: M_w/g.mol^{-1} = 90200, M_w/M_n = 1.96,
Scientific Polymer Products, Inc., Ontario, NY
Solvent (A): **dimethyl ether** C_2H_6O **115-10-6**

Type of data: cloud points and bubble points

w_B 0.056 was kept constant

T/K	334.1	354.9	372.8	394.2	413.2	432.9	453.0
P/MPa	1.21	2.07	3.10	4.66	7.76	10.52	12.59
	BP	BP	BP	CP	CP	CP	CP

Polymer (B): **poly(dimethylsiloxane)** **2013BYU**
Characterization: M_w/g.mol^{-1} = 170300, M_w/M_n = 1.75,
Scientific Polymer Products, Inc., Ontario, NY
Solvent (A): **dimethyl ether** **C_2H_6O** **115-10-6**

Type of data: cloud points and bubble points

w_B 0.058 was kept constant

T/K	333.8	352.4	372.4	393.1	413.2	434.6	454.5
P/MPa	1.21	1.90	3.28	4.83	8.28	11.21	13.79
	BP	BP	BP	CP	CP	CP	CP

Polymer (B): **poly(dimethylsiloxane)** **2013BYU**
Characterization: M_w/g.mol^{-1} = 38900, M_w/M_n = 2.84,
Scientific Polymer Products, Inc., Ontario, NY
Solvent (A): **propane** **C_3H_8** **74-98-6**

Type of data: cloud points and bubble points

w_B 0.052 was kept constant

T/K	333.5	353.6	373.2	394.5	414.0	33.4	452.7
P/MPa	1.55	2.41	5.35	8.79	11.03	13.28	15.00
	BP	CP	CP	CP	CP	CP	CP

Polymer (B): **poly(dimethylsiloxane)** **2013BYU**
Characterization: M_w/g.mol^{-1} = 90200, M_w/M_n = 1.96,
Scientific Polymer Products, Inc., Ontario, NY
Solvent (A): **propane** **C_3H_8** **74-98-6**

Type of data: cloud points and bubble points

w_B 0.051 was kept constant

T/K	334.0	354.1	373.5	393.5	412.9	434.2	452.7
P/MPa	1.89	3.28	6.21	9.14	11.55	4.31	15.69
	BP	CP	CP	CP	CP	CP	CP

Polymer (B): **poly(dimethylsiloxane)** **2013BYU**
Characterization: M_w/g.mol^{-1} = 170300, M_w/M_n = 1.75,
Scientific Polymer Products, Inc., Ontario, NY
Solvent (A): **propane** **C_3H_8** **74-98-6**

Type of data: cloud points and bubble points

w_B 0.049 was kept constant

T/K	333.5	355.2	373.2	393.2	413.0	434.1	452.7
P/MPa	1.90	3.84	6.72	9.54	12.07	14.61	16.03
	BP	CP	CP	CP	CP	CP	CP

Polymer (B): **poly(dimethylsiloxane)** **2013BYU**
Characterization: M_w/g.mol^{-1} = 38900, M_w/M_n = 2.84,
Scientific Polymer Products, Inc., Ontario, NY
Solvent (A): **propene** C_3H_6 **115-07-1**

Type of data: cloud points

w_B 0.064 was kept constant

T/K	355.7	375.5	394.7	14.6	432.1
P/MPa	3.97	7.24	9.66	12.41	14.48

Polymer (B): **poly(dimethylsiloxane)** **2013BYU**
Characterization: M_w/g.mol^{-1} = 90200, M_w/M_n = 1.96,
Scientific Polymer Products, Inc., Ontario, NY
Solvent (A): **propene** C_3H_6 **115-07-1**

Type of data: cloud points

w_B 0.078 was kept constant

T/K	354.0	372.8	395.0	412.1	435.2
P/MPa	4.14	7.39	10.66	12.76	15.69

Polymer (B): **poly(dimethylsiloxane)** **2013BYU**
Characterization: M_w/g.mol^{-1} = 170300, M_w/M_n = 1.75,
Scientific Polymer Products, Inc., Ontario, NY
Solvent (A): **propene** C_3H_6 **115-07-1**

Type of data: cloud points

w_B 0.073 was kept constant

T/K	355.1	374.2	393.1	412.1	434.6	442.4
P/MPa	4.31	7.76	10.75	13.23	15.86	16.58

Polymer (B): **poly(dimethylsiloxane) monomethacrylate** **2005TAI**
Characterization: M_n/g.mol^{-1} = 10000, Aldrich Chem. Co., Inc., Milwaukee, WI
Solvent (A): **carbon dioxide** CO_2 **124-38-9**

Type of data: cloud points

w_B 0.012 was kept constant

T/K	319.15	328.15	338.15	347.15
P/MPa	19.5	22.1	24.6	26.8

Polymer (B): **poly(dimethylsiloxane) monomethacrylate** **2010LE1**
Characterization: M_n/g.mol^{-1} = 10000
Solvent (A): **carbon dioxide** CO_2 **124-38-9**

Type of data: cloud points

w_B 0.002 was kept constant

T/K	298.15	308.15	318.15	328.15	338.15
P/bar	112	141	166	198	225

Polymer (B): **poly(1,4-dioxan-2-one)** **2012KIR**
Characterization: M_n/g.mol^{-1} = 35700, M_w/g.mol^{-1} = 89400, T_g/K = 259.2, T_m/K = 382.2, Johnson & Johnson Corporate Biomaterials Center, Somerville, NJ
Solvent (A): **2-propanone** C_3H_6O **67-64-1**

Type of data: cloud points

w_B	0.010	0.010	0.010	0.010	0.010	0.010	0.010	0.010	0.010
T/K	441.55	429.75	418.55	403.75	393.85	386.75	377.55	367.35	357.55
P/MPa	16.47	15.15	13.52	13.48	13.55	13.56	13.03	13.42	14.07
w_B	0.010	0.025	0.025	0.025	0.025	0.025	0.025	0.025	0.025
T/K	350.55	440.75	422.35	412.75	401.25	390.15	380.15	372.05	365.15
P/MPa	14.24	22.18	21.22	21.17	21.49	22.25	24.66	27.42	29.98
w_B	0.025	0.025	0.050	0.050	0.050	0.050	0.050	0.050	0.050
T/K	356.05	346.65	439.45	432.75	423.25	406.65	393.85	384.55	375.55
P/MPa	32.73	38.36	25.05	25.20	24.52	25.41	27.39	28.86	31.97
w_B	0.050	0.050	0.050	0.075	0.075	0.075	0.075	0.075	0.075
T/K	368.25	360.25	353.05	440.65	430.25	418.35	405.75	396.05	386.35
P/MPa	34.59	36.89	41.68	25.36	25.09	25.03	25.63	26.83	29.17
w_B	0.075	0.075	0.075	0.075	0.100	0.100	0.100	0.100	0.100
T/K	377.35	369.35	361.05	352.55	439.05	425.45	417.45	409.25	400.95
P/MPa	31.32	34.07	38.74	44.69	22.76	22.31	22.39	22.45	23.08
w_B	0.100	0.100	0.100	0.100	0.100	0.100	0.150	0.150	0.150
T/K	394.05	387.45	376.55	368.45	361.65	352.85	443.15	429.55	420.85
P/MPa	23.74	24.60	26.84	29.84	32.44	36.07	19.69	18.84	18.94
w_B	0.150	0.150	0.150	0.150	0.150	0.150	0.150	0.150	
T/K	411.65	401.55	395.15	388.15	374.85	364.85	357.45	350.35	
P/MPa	18.36	18.32	18.82	19.59	22.70	24.88	28.03	32.86	

Polymer (B): **poly(dodecyl acrylate)** **2010LE2**
Characterization: M_w/g.mol^{-1} = 44000, M_w/M_n = 1.6, T_g/K = 243.2, Scientific Polymer Products, Inc., Ontario, NY
Solvent (A): **n-butane** **C_4H_{10}** **106-97-8**

Type of data: cloud points

w_B 0.050 was kept constant

T/K	392.1	413.5	433.1	454.6
P/MPa	4.31	7.76	10.66	13.28

Polymer (B): **poly(dodecyl acrylate)** **2010LE2**
Characterization: M_w/g.mol^{-1} = 44000, M_w/M_n = 1.6, T_g/K = 243.2, Scientific Polymer Products, Inc., Ontario, NY
Solvent (A): **1-butene** **C_4H_8** **106-98-9**

Type of data: cloud points

w_B 0.054 was kept constant

T/K	394.6	413.5	434.0	453.7
P/MPa	3.97	6.72	9.83	12.72

Polymer (B): **poly(dodecyl acrylate)** **2010LE2**
Characterization: M_w/g.mol^{-1} = 44000, M_w/M_n = 1.6, T_g/K = 243.2, Scientific Polymer Products, Inc., Ontario, NY
Solvent (A): **carbon dioxide** **CO_2** **124-38-9**

Type of data: cloud points

w_B 0.052 was kept constant

T/K	456.9	459.0	460.5	463.4
P/MPa	205.28	169.55	157.07	150.14

Polymer (B): **poly(dodecyl acrylate)** **2010LE2**
Characterization: M_w/g.mol^{-1} = 44000, M_w/M_n = 1.6, T_g/K = 243.2, Scientific Polymer Products, Inc., Ontario, NY
Solvent (A): **dimethyl ether** **C_2H_6O** **115-10-6**

Type of data: cloud points

w_B 0.047 was kept constant

T/K	374.3	392.3	413.2	432.5	454.5
P/MPa	3.62	7.76	12.03	15.35	18.10

Polymer (B): **poly(dodecyl acrylate)** **2010LE2**
Characterization: M_w/g.mol^{-1} = 44000, M_w/M_n = 1.6, T_g/K = 243.2,
Scientific Polymer Products, Inc., Ontario, NY
Solvent (A): **propane** C_3H_8 **74-98-6**

Type of data: cloud points

w_B 0.056 was kept constant

T/K	333.9	354.2	375.2	395.0	413.8	435.1	455.6
P/MPa	8.79	12.93	16.72	19.66	22.24	24.66	26.38

Polymer (B): **poly(dodecyl acrylate)** **2010LE2**
Characterization: M_w/g.mol^{-1} = 44000, M_w/M_n = 1.6, T_g/K = 243.2,
Scientific Polymer Products, Inc., Ontario, NY
Solvent (A): **propene** C_3H_6 **115-07-1**

Type of data: cloud points

w_B 0.051 was kept constant

T/K	333.4	354.1	375.4	395.4	413.3	435.2	454.7
P/MPa	7.76	12.24	16.38	20.17	22.59	25.35	27.76

Polymer (B): **poly(dodecyl methacrylate)** **2009LIU**
Characterization: M_w/g.mol^{-1} = 250000, T_g/K = 208,
Scientific Polymer Products, Inc., Ontario, NY
Solvent (A): **n-butane** C_4H_{10} **106-97-8**

Type of data: cloud points

w_B 0.056 was kept constant

T/K	374.9	393.4	412.3	433.1	454.1
P/MPa	4.66	7.76	10.86	13.62	16.03

Polymer (B): **poly(dodecyl methacrylate)** **2009LIU**
Characterization: M_w/g.mol^{-1} = 250000, T_g/K = 208,
Scientific Polymer Products, Inc., Ontario, NY
Solvent (A): **1-butene** C_4H_8 **106-98-9**

Type of data: cloud points

w_B 0.049 was kept constant

T/K	394.0	413.1	435.0	453.7
P/MPa	5.45	8.79	12.59	14.66

Polymer (B): **poly(dodecyl methacrylate)** **2009LIU**
Characterization: M_w/g.mol^{-1} = 250000, T_g/K = 208,
Scientific Polymer Products, Inc., Ontario, NY
Solvent (A): **dimethyl ether** C_2H_6O **115-10-6**

Type of data: cloud points

w_B 0.042 was kept constant

T/K	374.2	394.2	414.0	434.8
P/MPa	7.07	11.55	15.35	18.79

Polymer (B): **poly(dodecyl methacrylate)** **2009LIU**
Characterization: M_w/g.mol^{-1} = 250000, T_g/K = 208,
Scientific Polymer Products, Inc., Ontario, NY
Solvent (A): **propane** C_3H_8 **74-98-6**

Type of data: cloud points

w_B 0.052 was kept constant

T/K	334.4	355.6	374.4	394.9	414.7	435.5	453.5
P/MPa	14.66	18.45	21.21	24.31	27.07	29.14	30.52

Polymer (B): **poly(dodecyl methacrylate)** **2009LIU**
Characterization: M_w/g.mol^{-1} = 250000, T_g/K = 208,
Scientific Polymer Products, Inc., Ontario, NY
Solvent (A): **propene** C_3H_6 **115-07-1**

Type of data: cloud points

w_B 0.052 was kept constant

T/K	332.9	354.4	375.9	394.2	416.0	434.0	455.1
P/MPa	12.24	17.41	21.21	23.97	27.07	29.48	31.55

Polymer (B): **polyester (hyperbranched)** **2010GRE**
Characterization: M_n/g.mol^{-1} = 1200, Boltorn-type,
OH-endgroups are esterified to -O(CO)CH_3
Solvent (A): **carbon dioxide** CO_2 **124-38-9**

Type of data: cloud points

w_B	0.1052	0.1052	0.1052	0.1052	0.1052	0.1052	0.1052	0.1052	0.1052
T/K	422.85	422.84	422.86	402.43	402.42	402.44	381.94	381.93	381.94
P/MPa	183.59	183.01	183.34	182.66	182.28	182.44	180.86	181.12	181.25

w_B	0.1052	0.1052	0.1052	0.1052	0.1052	0.1052	0.1052	0.1052	0.1052
T/K	360.99	360.99	360.98	340.45	340.45	340.44	320.06	320.07	320.07
P/MPa	176.15	175.87	175.54	168.90	169.14	169.27	160.88	161.12	160.66

continued

continued

w_B	0.1052	0.1052	0.1052	0.1326	0.1326	0.1326	0.1326	0.1326	0.1326
T/K	303.65	303.65	303.64	423.05	423.04	423.03	412.81	412.81	412.83
P/MPa	152.32	152.55	151.91	181.55	181.14	180.90	180.67	180.72	180.57
w_B	0.1326	0.1326	0.1326	0.1326	0.1326	0.1326	0.1326	0.1326	0.1326
T/K	402.61	402.62	402.62	392.38	392.38	392.37	382.09	382.10	382.12
P/MPa	180.94	180.84	181.59	180.49	179.82	180.22	179.08	179.22	179.59
w_B	0.1326	0.1326	0.1326	0.1326	0.1326	0.1326	0.1326	0.1326	0.1326
T/K	371.86	371.86	371.87	361.08	361.09	361.12	350.88	350.85	350.85
P/MPa	176.25	176.10	175.86	173.21	173.88	174.22	171.19	170.64	170.32
w_B	0.1326	0.1326	0.1326	0.1326	0.1326	0.1326	0.1326	0.1326	0.1326
T/K	340.49	340.51	340.52	330.25	330.26	330.27	320.86	320.87	320.87
P/MPa	167.48	167.10	166.87	161.89	162.62	162.57	158.07	158.40	158.75
w_B	0.1326	0.1326	0.1326	0.1326	0.1326	0.1326			
T/K	310.45	310.45	310.50	302.32	302.31	302.31			
P/MPa	152.40	152.66	152.94	147.99	148.57	148.86			

Polymer (B): **polyester (hyperbranched)** **2010GRE**
Characterization: M_n/g.mol^{-1} = 2100, Boltorn-type,
OH-endgroups are esterified to -O(CO)CH_3
Solvent (A): **carbon dioxide** CO_2 **124-38-9**

Type of data: cloud points

w_B 0.0989 was kept constant

T/K	422.84	422.85	422.86	412.65	412.66	412.67	402.56	402.57	402.57
P/MPa	104.68	104.37	104.14	103.33	103.55	103.15	100.87	101.06	101.27
T/K	392.08	392.10	392.11	381.79	381.79	381.80	371.22	371.23	371.23
P/MPa	98.75	98.44	98.61	94.67	94.87	95.14	92.49	92.66	92.88
T/K	361.13	361.13	361.14	350.67	350.69	350.70	340.55	340.54	340.54
P/MPa	89.03	88.77	88.59	85.44	85.24	85.11	80.37	80.12	79.86
T/K	330.15	330.16	330.16	320.94	320.93	320.94	310.58	310.57	310.56
P/MPa	76.79	76.44	76.17	72.58	72.18	72.34	67.47	67.67	67.82

Polymer (B): **polyester (hyperbranched)** **2010GRE**
Characterization: M_n/g.mol^{-1} = 4250, Boltorn-type,
OH-endgroups are esterified to -O(CO)$(CH_2)_5CH_3$
Solvent (A): **carbon dioxide** CO_2 **124-38-9**

Type of data: cloud points

w_B	0.1005	0.1005	0.1005	0.1005	0.1005	0.1005	0.1005	0.1005	0.1005
T/K	422.81	422.80	422.81	402.33	402.32	402.32	381.81	381.81	381.81
P/MPa	83.84	83.62	83.76	81.36	80.41	80.73	78.72	79.02	79.14

continued

continued

w_B	0.1005	0.1005	0.1005	0.1005	0.1005	0.1005	0.1005	0.1005	0.1005
T/K	361.66	361.66	361.67	340.34	340.35	340.35	319.79	319.78	319.79
P/MPa	78.16	78.06	78.28	76.65	76.77	76.88	78.80	78.44	78.62

w_B	0.1005	0.1005	0.1005	0.3515	0.3515	0.3515	0.3515	0.3515	0.3515
T/K	303.63	303.60	303.60	22.92	422.91	422.91	402.47	402.48	402.48
P/MPa	92.09	91.83	92.11	65.09	65.24	64.90	62.60	62.24	62.18

w_B	0.3515	0.3515	0.3515	0.3515	0.3515	0.3515	0.3515	0.3515	0.3515
T/K	381.66	381.67	381.68	361.09	361.09	361.08	340.53	340.53	340.54
P/MPa	59.54	59.33	59.20	56.78	56.52	56.32	53.72	53.98	54.05

w_B	0.3515	0.3515	0.3515	0.3515	0.3515	0.3515
T/K	319.98	319.99	319.99	304.65	304.64	304.63
P/MPa	49.19	48.97	48.72	47.40	47.54	47.66

Polymer (B): **polyester (hyperbranched)** **2010GRE**
Characterization: M_n/g.mol^{-1} = 2150, Boltorn-type,
OH-endgroups are esterified to -O(CO)$(CH_2)_{10}CH_3$
Solvent (A): **carbon dioxide** CO_2 **124-38-9**

Type of data: cloud points

w_B	0.0944	0.0944	0.0944	0.0944	0.0944	0.0944	0.0944	0.0944	0.0944
T/K	422.63	422.64	422.64	402.24	402.23	402.24	381.77	381.76	381.76
P/MPa	87.56	87.28	87.32	89.38	89.22	89.18	95.22	95.15	94.74

w_B	0.1047	0.1047	0.1047	0.1047	0.1047	0.1047	0.1047	0.1047	0.1047
T/K	422.84	422.84	422.83	402.27	402.28	402.28	381.75	381.74	381.74
P/MPa	87.62	87.18	87.24	91.43	91.28	91.08	101.93	102.06	101.86

w_B	0.1047	0.1047	0.1047	0.1047	0.1047	0.1047	0.1047	0.1047	0.1047
T/K	371.08	371.09	371.08	365.99	365.98	365.98	363.89	363.90	363.91
P/MPa	117.10	116.83	117.16	132.00	131.87	131.67	145.23	144.78	144.94

w_B	0.1047	0.1047	0.1047	0.1047	0.3038	0.3038	0.3038	0.3038	0.3038
T/K	362.87	362.88	362.89	362.63	422.57	422.60	422.64	402.23	402.24
P/MPa	160.38	159.74	160.77	185.21	75.18	75.55	75.33	77.62	77.48

w_B	0.3038	0.3038	0.3038	0.3038	0.3038	0.3038	0.3038	0.3038	0.3038
T/K	402.24	381.73	381.75	381.77	360.94	360.94	360.93	350.60	350.63
P/MPa	77.30	80.23	80.36	80.43	90.83	91.04	91.23	110.52	109.52

w_B	0.3038	0.3038	0.3038	0.3038	0.3038	0.3038	0.3038	0.3038	0.3038
T/K	350.65	348.49	348.53	348.55	346.47	346.47	346.48	345.86	345.86
P/MPa	109.93	116.43	115.57	116.23	126.28	127.18	127.34	158.58	158.95

w_B	0.3038	0.3038
T/K	345.86	346.50
P/MPa	159.72	185.66

Polymer (B): **polyester (hyperbranched)** **2010GRE**
Characterization: M_n/g.mol^{-1} = 5600, Boltorn-type,
OH-endgroups are esterified to -O(CO)$(CH_2)_{10}CH_3$
Solvent (A): **carbon dioxide** CO_2 **124-38-9**

Type of data: cloud points

w_B	0.0069	0.0069	0.0069	0.0069	0.0069	0.0069	0.0069	0.0069	0.0069
T/K	422.98	422.97	422.97	412.59	412.59	412.58	402.39	402.40	402.40
P/MPa	94.32	94.55	94.19	96.90	97.13	96.56	98.41	98.67	98.83
w_B	0.0069	0.0069	0.0069	0.0069	0.0069	0.0069	0.0069	0.0069	0.0069
T/K	392.17	392.17	392.16	381.94	381.95	381.95	371.25	371.26	371.27
P/MPa	104.30	104.84	105.12	110.75	110.47	110.22	136.54	136.82	137.03
w_B	0.0069	0.0116	0.0116	0.0116	0.0116	0.0116	0.0116	0.0116	0.0116
T/K	369.35	423.34	423.34	423.34	413.10	413.11	413.12	402.87	402.88
P/MPa	182.56	98.44	98.67	98.87	100.84	101.06	101.34	103.53	103.90
w_B	0.0116	0.0116	0.0116	0.0116	0.0116	0.0116	0.0116	0.0116	0.0116
T/K	402.89	392.63	392.63	392.64	382.16	382.17	382.18	375.76	375.76
P/MPa	104.35	113.86	114.28	114.44	128.74	129.21	129.45	162.49	163.29
w_B	0.0116	0.0116	0.0229	0.0229	0.0229	0.0229	0.0229	0.0229	0.0229
T/K	375.77	374.97	423.10	423.11	423.12	413.68	413.68	413.69	402.73
P/MPa	163.84	187.56	102.95	102.57	102.34	106.18	105.88	105.68	112.77
w_B	0.0229	0.0229	0.0229	0.0229	0.0229	0.0229	0.0229	0.0229	0.0229
T/K	402.72	402.72	392.35	392.35	392.36	382.23	382.24	379.14	379.14
P/MPa	112.24	112.59	122.36	121.89	122.12	144.85	145.38	183.80	184.37
w_B	0.0229	0.0229	0.0567	0.0567	0.0567	0.0567	0.0567	0.0567	0.0567
T/K	379.14	378.45	423.32	423.31	423.32	413.23	413.22	413.22	402.86
P/MPa	184.71	194.56	101.54	101.34	101.19	105.67	105.42	105.31	111.83
w_B	0.0567	0.0567	0.0567	0.0567	0.0567	0.0567	0.0567	0.0567	0.0567
T/K	402.87	402.87	392.46	392.45	392.45	382.35	382.34	382.35	379.25
P/MPa	111.57	111.47	121.14	120.84	120.53	151.49	151.22	150.80	189.77
w_B	0.0567	0.0567	0.0952	0.0952	0.0952	0.0952	0.0952	0.0952	0.0952
T/K	379.25	379.26	423.61	423.60	423.59	413.35	413.36	413.36	405.57
P/MPa	190.64	191.29	96.36	95.88	95.94	99.07	98.64	99.11	102.28
w_B	0.0952	0.0952	0.0952	0.0952	0.0952	0.0952	0.0952	0.0952	0.0952
T/K	405.57	405.58	398.76	398.76	398.75	392.75	392.76	392.76	381.59
P/MPa	101.87	101.71	106.58	105.97	106.04	110.69	110.53	110.85	126.08
w_B	0.0952	0.0952	0.0952	0.0952	0.0952	0.0952	0.0952	0.0952	0.2212
T/K	381.58	381.58	374.03	374.03	374.04	371.92	371.92	371.92	422.96
P/MPa	125.77	125.81	160.31	159.44	159.66	196.68	197.15	197.60	84.85

continued

continued

w_B	0.2212	0.2212	0.2212	0.2212	0.2212	0.2212	0.2212	0.2212	0.2212
T/K	422.96	422.97	412.70	412.71	412.72	402.39	402.38	402.38	392.08
P/MPa	84.47	84.66	86.53	85.97	89.19	88.58	88.35	88.19	91.89
w_B	0.2212	0.2212	0.2212	0.2212	0.2212	0.2212	0.2212	0.2212	0.2212
T/K	392.08	392.09	381.93	381.93	382.93	371.24	371.25	371.26	360.93
P/MPa	91.58	91.34	98.50	97.57	97.33	107.09	106.68	106.88	134.31
w_B	0.2212	0.2212	0.2212	0.3050	0.3050	0.3050	0.3050	0.3050	0.3050
T/K	360.94	360.95	359.20	423.16	423.19	423.21	413.11	413.11	413.12
P/MPa	134.89	135.28	195.59	78.35	78.01	77.94	78.04	78.11	78.09
w_B	0.3050	0.3050	0.3050	0.3050	0.3050	0.3050	0.3050	0.3050	0.3050
T/K	402.75	402.76	402.76	392.54	392.55	392.55	382.23	382.23	382.24
P/MPa	79.53	79.49	79.42	80.15	80.28	80.20	83.37	83.53	83.61
w_B	0.3050	0.3050	0.3050	0.3050	0.3050	0.3050	0.3050	0.3050	0.3050
T/K	372.33	372.33	372.34	361.26	361.26	361.25	350.97	350.96	350.96
P/MPa	87.67	87.75	87.83	97.88	97.95	98.02	136.79	136.90	137.03
w_B	0.3050	0.3050	0.3050	0.3050	0.6036	0.6036	0.6036	0.6036	0.6036
T/K	349.88	349.88	349.87	349.87	423.17	423.19	423.19	402.73	402.72
P/MPa	193.78	194.05	193.59	193.81	49.49	48.91	49.24	47.09	46.55
w_B	0.6036	0.6036	0.6036	0.6036	0.6036	0.6036	0.6036	0.6036	0.6036
T/K	402.73	423.17	382.22	382.23	382.24	361.32	361.32	361.33	382.22
P/MPa	46.77	49.49	44.34	44.12	44.02	41.42	41.39	41.11	44.34

Polymer (B): **polyester (hyperbranched)** **2010GRE**
Characterization: M_n/g.mol^{-1} = 7200, Boltorn-type,
OH-endgroups are esterified to -O(CO)$(CH_2)_{16}CH_3$
Solvent (A): **carbon dioxide** **CO_2** **124-38-9**

Type of data: cloud points

w_B 0.1755 was kept constant

T/K	423.31	423.51	423.51	413.19	413.18	413.18	410.10	410.10	410.11
P/MPa	124.82	124.95	125.02	142.86	143.04	143.34	149.02	149.66	150.13
T/K	407.05	407.06	407.07	404.99	405.00	404.99	403.17	402.18	403.18
P/MPa	158.87	159.58	159.89	167.59	168.25	168.85	180.65	180.92	181.21
T/K	401.89	401.88	401.90						
P/MPa	195.57	196.30	197.24						

Polymer (B): **polyester (hyperbranched)** **2010GRE**
Characterization: M_n/g.mol^{-1} = 1200, Boltorn-type,
OH-endgroups are esterified to -O(CO)CF_3
Solvent (A): **carbon dioxide** **CO_2** **124-38-9**

Type of data: cloud points

w_B	0.0206	0.0206	0.0206	0.0206	0.0206	0.0206	0.0206	0.0206	0.0206
T/K	270.82	270.95	270.99	291.47	291.49	291.46	302.19	302.15	302.18
P/MPa	12.17	12.30	12.22	21.11	21.18	21.17	24.43	24.44	24.46
w_B	0.0206	0.0206	0.0206	0.0206	0.0206	0.0206	0.0206	0.0206	0.0206
T/K	312.25	312.41	312.41	332.78	332.89	332.90	354.36	355.55	354.95
P/MPa	28.38	28.49	28.57	34.71	34.72	34.66	40.23	39.08	39.20
w_B	0.0206	0.0206	0.0206	0.0495	0.0495	0.0495	0.0495	0.0495	0.0495
T/K	375.94	376.03	376.07	282.30	282.17	282.09	282.04	281.99	281.96
P/MPa	45.31	45.35	45.18	13.37	16.96	16.44	15.81	15.67	14.03
w_B	0.0495	0.0495	0.0495	0.0495	0.0495	0.0495	0.0495	0.0495	0.0495
T/K	287.70	287.76	287.83	287.89	287.89	295.10	295.14	295.16	295.18
P/MPa	15.86	18.59	15.75	15.92	18.16	18.30	22.69	21.29	20.69
w_B	0.0495	0.0495	0.0495	0.0495	0.0495	0.0495	0.0495	0.0495	0.0495
T/K	295.18	295.18	301.16	301.14	301.13	301.14	300.74	301.10	301.09
P/MPa	19.91	19.74	22.82	22.67	22.64	22.65	21.35	21.12	21.14
w_B	0.0495	0.0495	0.0495	0.0495	0.0495	0.0495	0.0495	0.0495	0.0495
T/K	311.24	311.30	311.31	311.35	321.28	321.59	321.61	321.63	331.92
P/MPa	25.40	25.80	25.62	25.65	29.35	29.39	29.42	29.38	30.93
w_B	0.0495	0.0495	0.0495	0.0495	0.0495	0.0495	0.0495	0.0495	0.0495
T/K	331.91	331.86	331.82	342.07	342.01	341.97	341.98	352.32	352.41
P/MPa	31.37	31.20	31.13	34.14	34.57	34.52	34.56	37.64	36.78
w_B	0.0495	0.0495	0.0495	0.0495	0.0495	0.0495	0.0495	0.0495	0.0495
T/K	352.39	352.37	362.52	362.58	362.57	362.52	374.98	374.78	374.80
P/MPa	37.78	37.00	40.33	40.57	40.80	40.62	43.19	43.33	43.46
w_B	0.0495	0.1007	0.1007	0.1007	0.1007	0.1007	0.1007	0.1007	0.1007
T/K	374.83	282.01	282.02	282.02	301.99	302.01	301.99	319.33	320.97
P/MPa	43.41	17.47	17.38	17.62	26.05	26.03	26.06	32.52	32.58
w_B	0.1007	0.1007	0.1007	0.1007	0.1007	0.1007	0.1007	0.1007	0.1007
T/K	319.28	339.60	339.57	339.53	354.94	354.89	354.82	375.98	376.00
P/MPa	32.58	39.14	39.10	39.11	43.15	43.13	43.15	47.67	47.89
w_B	0.1007	0.2028	0.2028	0.2028	0.2028	0.2028	0.2028	0.2028	0.2028
T/K	375.99	281.97	282.06	282.07	282.07	282.05	300.10	300.12	300.09
P/MPa	47.95	14.42	14.43	15.23	15.98	15.47	22.71	22.66	22.62
w_B	0.2028	0.2028	0.2028	0.2028	0.2028	0.2028	0.2028	0.2028	0.2028
T/K	320.29	320.51	320.55	340.96	341.02	355.47	355.48	355.48	375.95
P/MPa	30.51	30.49	30.60	37.38	37.39	41.32	41.38	41.38	45.92

continued

continued

w_B	0.2028	0.2028	0.2028
T/K	375.95	375.96	341.06
P/MPa	46.01	46.08	37.39

Polymer (B): **polyester (hyperbranched)** **2010GRE**
Characterization: M_n/g.mol^{-1} = 3950, Boltorn-type,
OH-endgroups are esterified to -O(CO)CF_3
Solvent (A): **carbon dioxide** **CO_2** **124-38-9**

Type of data: cloud points

w_B	0.0218	0.0218	0.0218	0.0218	0.0218	0.0218	0.0218	0.0218	0.0218
T/K	281.15	281.14	281.19	300.61	300.62	300.62	320.24	320.28	320.29
P/MPa	18.32	18.46	18.43	27.80	27.53	27.43	33.38	33.14	33.09
w_B	0.0218	0.0218	0.0218	0.0218	0.0218	0.0218	0.0218	0.0218	0.0218
T/K	340.77	340.74	340.82	361.42	361.42	361.39	376.20	376.36	376.32
P/MPa	38.31	38.52	38.52	44.09	44.22	44.28	47.57	47.76	47.84
w_B	0.0639	0.0639	0.0639	0.0639	0.0639	0.0639	0.0639	0.0639	0.0639
T/K	281.71	281.71	281.71	281.71	291.26	291.23	291.23	291.23	299.89
P/MPa	18.49	18.68	18.70	18.71	22.30	22.80	22.82	22.75	27.01
w_B	0.0639	0.0639	0.0639	0.0639	0.0639	0.0639	0.0639	0.0639	0.0639
T/K	300.12	300.12	310.12	310.40	310.39	320.22	320.67	320.65	330.58
P/MPa	26.98	26.90	30.23	30.25	30.16	33.01	33.57	33.34	36.75
w_B	0.0639	0.0639	0.0639	0.0639	0.0639	0.0639	0.0639	0.0639	0.0639
T/K	331.03	330.95	341.24	341.27	341.29	341.27	351.52	351.56	351.56
P/MPa	36.83	36.84	39.08	39.35	39.84	40.06	42.25	42.42	42.39
w_B	0.0639	0.0639	0.0639	0.0639	0.0639	0.0639	0.1022	0.1022	0.1022
T/K	361.72	361.75	361.70	375.44	375.63	375.42	281.06	281.01	280.99
P/MPa	44.63	44.72	44.76	47.59	47.62	47.72	19.73	19.92	19.94
w_B	0.1022	0.1022	0.1022	0.1022	0.1022	0.1022	0.1022	0.1022	0.1022
T/K	301.74	301.85	301.86	320.47	320.43	320.43	340.93	340.93	340.94
P/MPa	29.41	29.00	28.95	35.06	34.87	34.88	41.11	40.99	40.87
w_B	0.1022	0.1022	0.1022	0.1022	0.1022	0.1022	0.2001	0.2001	0.2001
T/K	361.43	361.45	361.41	375.96	376.06	376.15	281.15	281.08	281.18
P/MPa	46.17	46.08	46.11	49.15	49.37	49.30	15.74	15.97	16.01
w_B	0.2001	0.2001	0.2001	0.2001	0.2001	0.2001	0.2001	0.2001	0.2001
T/K	301.03	301.18	301.29	320.47	320.54	320.55	341.04	341.01	341.00
P/MPa	25.08	25.04	25.06	31.93	31.85	31.80	38.47	38.38	38.35
w_B	0.2001	0.2001	0.2001	0.2001	0.2001	0.2001			
T/K	361.59	361.58	361.53	376.22	376.23	376.18			
P/MPa	43.95	43.93	43.81	46.94	46.91	46.93			

Polymer (B): **polyester (hyperbranched)** **2010GRE**
Characterization: M_n/g.mol^{-1} = 10500, Boltorn-type,
OH-endgroups are esterified to -O(CO)CF_3
Solvent (A): **carbon dioxide** **CO_2** **124-38-9**

Type of data: cloud points

w_B	0.0229	0.0229	0.0229	0.0229	0.0229	0.0229	0.0229	0.0229	0.0229
T/K	280.52	280.52	280.54	301.31	301.74	302.08	320.24	320.47	320.37
P/MPa	21.92	22.04	21.92	30.33	30.56	30.66	37.11	37.15	37.11
w_B	0.0229	0.0229	0.0229	0.0229	0.0229	0.0229	0.0229	0.0229	0.0229
T/K	339.24	339.74	339.81	339.83	361.61	361.68	361.55	376.22	376.19
P/MPa	42.79	44.32	44.21	44.05	48.55	49.01	49.14	51.43	51.57
w_B	0.0229	0.0229	0.0229	0.0229	0.0502	0.0502	0.0502	0.0502	0.0502
T/K	376.17	376.02	376.09	375.92	281.19	281.18	281.18	291.68	291.58
P/MPa	51.65	51.66	51.84	51.79	23.66	23.77	23.75	27.47	27.83
w_B	0.0502	0.0502	0.0502	0.0502	0.0502	0.0502	0.0502	0.0502	0.0502
T/K	291.56	299.10	299.11	299.11	310.12	310.36	310.38	320.43	320.70
P/MPa	27.69	30.06	29.90	29.79	33.92	34.33	34.25	37.62	37.81
w_B	0.0502	0.0502	0.0502	0.0502	0.0502	0.0502	0.0502	0.0502	0.0502
T/K	320.71	330.97	330.96	330.96	341.07	341.10	341.05	351.33	351.34
P/MPa	37.79	41.05	40.71	40.99	43.76	44.08	44.11	47.01	47.19
w_B	0.0502	0.0502	0.0502	0.0502	0.0502	0.0502	0.0502	0.1002	0.1002
T/K	351.33	361.46	361.53	361.39	374.89	374.63	374.56	280.12	280.03
P/MPa	47.32	49.50	49.83	49.69	52.87	52.95	52.94	22.87	23.25
w_B	0.1002	0.1002	0.1002	0.1002	0.1002	0.1002	0.1002	0.1002	0.1002
T/K	280.03	280.02	300.08	300.05	300.04	320.46	320.44	320.46	341.17
P/MPa	23.25	23.05	31.56	31.53	31.40	39.18	39.08	39.07	45.62
w_B	0.1002	0.1002	0.1002	0.1002	0.1002	0.1002	0.1002	0.1002	0.2032
T/K	341.33	341.35	362.04	362.14	362.11	376.81	376.78	376.75	281.16
P/MPa	45.63	45.51	50.52	50.38	50.25	53.56	53.62	53.73	19.33
w_B	0.2032	0.2032	0.2032	0.2032	0.2032	0.2032	0.2032	0.2032	0.2032
T/K	281.17	280.98	301.06	301.19	301.28	319.46	319.36	319.30	340.79
P/MPa	19.17	19.00	27.89	27.84	27.86	34.66	34.56	34.55	41.54
w_B	0.2032	0.2032	0.2032	0.2032	0.2032	0.2032	0.2032	0.2032	
T/K	341.03	341.11	361.58	361.57	361.52	376.24	376.00	376.02	
P/MPa	41.40	41.41	46.85	46.66	46.62	49.61	49.44	49.46	

Polymer (B): **polyester (hyperbranched)** **2010GRE**
Characterization: M_n/g.mol^{-1} = 1400, Boltorn-type,
OH-endgroups are modified to -OSi(CH_3)$_3$
Solvent (A): **carbon dioxide** CO_2 **124-38-9**

Type of data: cloud points

w_B	0.0189	0.0189	0.0189	0.0189	0.0189	0.0189	0.0189	0.0189	0.0189
T/K	280.96	281.06	281.05	281.05	290.49	290.49	290.50	290.49	300.98
P/MPa	5.44	5.45	5.38	5.27	8.88	9.16	9.20	9.26	15.33

w_B	0.0189	0.0189	0.0189	0.0189	0.0189	0.0189	0.0189	0.0189	0.0189
T/K	301.11	301.12	301.13	301.17	301.15	311.26	311.26	311.26	311.25
P/MPa	15.56	15.50	15.42	13.79	13.93	17.36	17.63	17.72	17.64

w_B	0.0189	0.0189	0.0189	0.0189	0.0189	0.0189	0.0189	0.0189	0.0189
T/K	311.48	321.20	321.61	321.66	321.65	331.74	331.81	331.89	331.92
P/MPa	17.05	20.38	20.62	20.71	20.91	24.28	24.55	24.60	24.61

w_B	0.0189	0.0189	0.0189	0.0189	0.0189	0.0189	0.0189	0.0189	0.0189
T/K	342.16	342.14	342.07	342.05	352.22	352.33	352.32	352.33	362.62
P/MPa	27.91	28.17	28.18	28.27	31.20	31.42	31.48	31.46	33.80

w_B	0.0189	0.0189	0.0189	0.0189	0.0189	0.0189	0.0189	0.0575	0.0575
T/K	362.64	362.62	362.59	374.92	375.01	375.06	375.01	279.52	279.26
P/MPa	34.02	34.03	34.01	36.52	36.60	36.64	36.63	6.81	6.71

w_B	0.0575	0.0575	0.0575	0.0575	0.0575	0.0575	0.0575	0.0575	0.0575
T/K	279.24	279.30	291.01	291.25	291.42	291.49	301.10	301.09	301.07
P/MPa	6.71	6.73	11.36	11.41	11.44	11.47	16.20	16.25	16.25

w_B	0.0575	0.0575	0.0575	0.0575	0.0575	0.0575	0.0575	0.0575	0.0575
T/K	301.05	311.35	311.34	311.33	311.32	321.25	321.28	321.23	321.28
P/MPa	16.25	20.83	20.73	20.64	20.60	23.50	23.58	23.66	23.58

w_B	0.0575	0.0575	0.0575	0.0575	0.0575	0.0575	0.0575	0.0575	0.0575
T/K	331.56	331.60	331.53	331.48	341.56	341.58	341.58	341.61	352.07
P/MPa	25.99	26.19	26.18	26.17	29.01	29.17	29.20	29.19	31.19

w_B	0.0575	0.0575	0.0575	0.0575	0.0575	0.0575	0.0575	0.0575	0.0575
T/K	352.04	352.01	352.00	362.05	361.94	361.86	361.92	375.22	375.28
P/MPa	31.18	31.13	31.08	33.32	33.28	33.27	33.26	35.65	35.89

w_B	0.0575	0.0575	0.1012	0.1012	0.1012	0.1012	0.1012	0.1012	0.1012
T/K	375.27	375.32	280.45	280.52	292.11	292.06	292.05	292.07	302.42
P/MPa	35.95	35.98	7.93	7.89	12.99	12.92	12.89	12.87	17.15

w_B	0.1012	0.1012	0.1012	0.1012	0.1012	0.1012	0.1012	0.1012	0.1012
T/K	302.17	302.01	301.92	311.28	311.28	311.28	311.29	321.29	321.53
P/MPa	17.01	16.89	16.84	19.59	19.66	19.63	19.66	23.09	23.28

continued

continued

w_B	0.1012	0.1012	0.1012	0.1012	0.1012	0.1012	0.1012	0.1012	0.1012
T/K	321.51	321.50	321.49	331.64	331.69	331.73	331.71	331.56	331.60
P/MPa	23.29	23.30	22.99	26.59	26.66	26.71	26.72	25.99	26.19
w_B	0.1012	0.1012	0.1012	0.1012	0.1012	0.1012	0.1012	0.1012	0.1012
T/K	341.93	341.92	341.91	341.92	352.04	352.05	352.05	352.06	362.20
P/MPa	29.57	29.80	29.82	29.86	32.65	32.90	32.95	32.97	35.46
w_B	0.1012	0.1012	0.1012	0.1012	0.1012	0.1012	0.1012	0.1889	0.1889
T/K	362.28	362.31	362.24	375.57	375.56	375.59	375.58	280.94	280.90
P/MPa	35.60	35.67	35.67	38.48	38.59	38.61	38.62	4.64	4.61
w_B	0.1889	0.1889	0.1889	0.1889	0.1889	0.1889	0.1889	0.1889	0.1889
T/K	280.87	280.87	290.11	290.18	290.19	290.17	302.09	302.15	302.16
P/MPa	4.51	4.49	8.84	8.91	9.23	8.72	15.39	15.37	15.33
w_B	0.1889	0.1889	0.1889	0.1889	0.1889	0.1889	0.1889	0.1889	0.1889
T/K	302.13	312.01	312.24	312.25	312.27	322.14	322.48	322.51	322.50
P/MPa	15.30	17.43	17.46	17.52	17.55	21.01	21.37	21.42	21.44
w_B	0.1889	0.1889	0.1889	0.1889	0.1889	0.1889	0.1889	0.1889	0.1889
T/K	332.58	332.52	332.54	332.57	342.86	342.89	342.89	342.89	353.09
P/MPa	24.44	24.31	24.24	24.27	27.84	27.87	27.89	27.85	30.81
w_B	0.1889	0.1889	0.1889	0.1889	0.1889	0.1889	0.1889	0.1889	0.1889
T/K	353.10	353.11	353.11	363.18	363.07	362.95	362.92	375.26	375.47
P/MPa	30.98	30.96	30.98	33.37	33.50	33.48	33.47	36.07	36.24
w_B	0.1889	0.1889							
T/K	375.55	375.54							
P/MPa	36.27	36.23							

Polymer (B): **polyester (hyperbranched)** **2010GRE**
Characterization: M_n/g.mol^{-1} = 3450, Boltorn-type,
OH-endgroups are modified to -OSi(CH$_3$)$_3$
Solvent (A): **carbon dioxide** **CO_2** **124-38-9**

Type of data: cloud points

w_B	0.0246	0.0246	0.0246	0.0246	0.0246	0.0246	0.0246	0.0246	0.0246
T/K	280.64	280.70	280.77	299.85	300.27	300.27	320.38	320.45	320.49
P/MPa	21.71	22.09	22.00	25.67	25.71	25.68	31.09	31.13	31.16
w_B	0.0246	0.0246	0.0246	0.0246	0.0246	0.0246	0.0246	0.0246	0.0246
T/K	341.10	341.15	341.15	361.52	361.38	361.31	375.62	375.39	375.29
P/MPa	36.86	36.98	37.00	41.97	42.02	42.10	44.97	45.12	44.99
w_B	0.0510	0.0510	0.0510	0.0510	0.0510	0.0510	0.0510	0.0510	0.0510
T/K	279.92	279.97	279.92	299.97	299.98	299.98	320.06	320.29	320.30
P/MPa	23.06	23.18	23.14	26.66	26.72	26.73	31.85	31.93	31.98

continued

continued

w_B	0.0510	0.0510	0.0510	0.0510	0.0510	0.0510	0.0510	0.0510	0.0510
T/K	340.65	340.81	340.75	361.23	361.23	361.18	375.92	376.02	375.92
P/MPa	37.32	37.36	37.38	42.30	42.34	42.31	45.40	45.41	45.42
w_B	0.1089	0.1089	0.1089	0.1089	0.1089	0.1089	0.1089	0.1089	0.1089
T/K	280.44	280.42	280.42	299.00	298.99	298.98	320.32	320.34	320.32
P/MPa	21.57	21.43	21.31	24.80	24.80	24.88	31.22	31.22	31.22
w_B	0.1089	0.1089	0.1089	0.1089	0.1089	0.1089	0.1089	0.1089	0.1089
T/K	340.71	340.72	340.71	361.16	361.15	361.13	375.62	375.53	375.53
P/MPa	36.93	36.88	36.87	42.01	42.05	42.01	45.07	45.07	45.04
w_B	0.2091	0.2091	0.2091	0.2091	0.2091	0.2091	0.2091	0.2091	0.2091
T/K	280.07	280.00	280.00	299.67	300.00	300.02	320.45	320.38	320.36
P/MPa	17.77	17.78	17.64	22.80	23.09	23.19	28.87	28.67	28.72
w_B	0.2091	0.2091	0.2091	0.2091	0.2091	0.2091	0.2091	0.2091	0.2091
T/K	340.83	340.74	340.75	361.82	361.52	361.38	375.86	375.76	375.72
P/MPa	34.66	34.62	34.62	39.99	39.96	39.83	42.93	42.89	42.97

Polymer (B): **polyester (hyperbranched)** **2010GRE**
Characterization: M_n/g.mol^{-1} = 5200, Boltorn-type,
OH-endgroups are modified to -OSi(CH_3)$_3$
Solvent (A): **carbon dioxide** CO_2 **124-38-9**

Type of data: cloud points

w_B	0.0206	0.0206	0.0206	0.0206	0.0206	0.0206	0.0206	0.0206	0.0206
T/K	278.56	278.24	278.07	277.96	280.24	280.16	280.16	290.28	290.31
P/MPa	38.00	39.34	39.86	39.60	39.93	40.17	40.23	38.14	38.46
w_B	0.0206	0.0206	0.0206	0.0206	0.0206	0.0206	0.0206	0.0206	0.0206
T/K	290.32	300.91	300.97	301.04	320.48	320.53	320.51	340.90	341.02
P/MPa	38.47	39.04	38.97	39.21	42.39	42.48	42.50	46.39	46.31
w_B	0.0206	0.0206	0.0206	0.0206	0.0206	0.0206	0.0206	0.0515	0.0515
T/K	341.02	355.26	355.40	355.41	376.28	376.22	376.32	277.35	277.30
P/MPa	46.23	49.21	49.12	49.04	52.74	52.45	52.71	54.75	54.35
w_B	0.0515	0.0515	0.0515	0.0515	0.0515	0.0515	0.0515	0.0515	0.0515
T/K	277.38	277.40	283.01	282.99	282.99	283.00	289.80	290.31	290.50
P/MPa	54.42	54.21	44.60	45.16	45.54	45.97	40.30	40.21	40.27
w_B	0.0515	0.0515	0.0515	0.0515	0.0515	0.0515	0.0515	0.0515	0.0515
T/K	292.54	292.54	292.54	299.81	300.28	300.59	300.63	300.12	300.14
P/MPa	40.54	40.89	40.74	42.03	41.57	41.57	41.34	39.02	40.41
w_B	0.0515	0.0515	0.0515	0.0515	0.0515	0.0515	0.0515	0.0515	0.0515
T/K	300.14	300.14	310.38	310.32	310.31	310.29	320.19	320.43	320.45
P/MPa	40.20	40.19	40.62	40.73	40.66	40.61	44.46	43.83	43.67

continued

continued

w_B	0.0515	0.0515	0.0515	0.0515	0.0515	0.0515	0.0515	0.0515	0.0515
T/K	320.48	330.78	330.78	330.80	330.79	341.08	341.11	341.12	341.11
P/MPa	43.64	45.62	45.24	45.12	44.84	46.51	46.56	46.75	46.78
w_B	0.0515	0.0515	0.0515	0.0515	0.0515	0.0515	0.0515	0.0515	0.0515
T/K	351.35	351.38	351.35	351.34	361.45	361.60	361.62	375.07	375.07
P/MPa	48.97	48.88	48.89	48.96	51.06	51.06	51.07	53.64	53.65
w_B	0.0515	0.0515	0.1334	0.1334	0.1334	0.1334	0.1334	0.1334	0.1334
T/K	375.07	375.09	284.90	284.49	284.41	284.31	290.58	290.09	289.78
P/MPa	53.61	53.60	58.90	60.77	61.08	61.41	50.73	52.05	52.57
w_B	0.1334	0.1334	0.1334	0.1334	0.1334	0.1334	0.1334	0.1334	0.1334
T/K	289.67	300.26	300.67	300.82	300.92	320.62	320.60	320.58	341.14
P/MPa	52.80	42.25	43.48	43.36	43.30	44.32	44.75	44.78	48.17
w_B	0.1334	0.1334	0.1334	0.1334	0.1334	0.1334	0.1334	0.1334	
T/K	341.20	341.18	361.78	361.80	361.78	376.38	376.41	376.48	
P/MPa	48.23	48.29	52.16	52.21	52.15	54.64	54.57	54.85	

Polymer (B): **polyester (hyperbranched)** **2010GRE**
Characterization: M_n/g.mol^{-1} = 1550, Boltorn-type,
OH-endgroups are esterified to -O[(CO)(CH$_2$)$_5$O]$_5$Si(CH$_3$)$_3$
Solvent (A): **carbon dioxide** CO_2 **124-38-9**

Type of data: cloud points

w_B 0.0511 was kept constant

T/K	280.11	280.15	300.13	300.13	300.13	300.12	320.70	320.72	320.73
P/MPa	58.47	58.55	63.50	63.81	62.98	62.79	69.00	68.86	69.11
T/K	340.51	341.08	341.35	361.97	361.99	362.01	376.38	376.21	376.21
P/MPa	72.54	72.36	72.46	77.57	77.48	77.27	80.82	80.66	80.92

Polymer (B): **polyester (hyperbranched)** **2010GRE**
Characterization: M_n/g.mol^{-1} = 4600, Boltorn-type,
OH-endgroups are esterified to -O[(CO)O(CH$_2$)$_5$O]$_5$(CO)CH$_3$
Solvent (A): **carbon dioxide** CO_2 **124-38-9**

Type of data: cloud points

w_B	0.1052	0.1052	0.1052	0.1052	0.1052	0.1052	0.1052	0.1052	0.1052
T/K	422.85	422.84	422.86	402.43	402.42	402.44	381.94	381.93	381.94
P/MPa	183.59	183.01	183.34	182.66	182.28	182.44	180.86	181.12	181.25
w_B	0.1052	0.1052	0.1052	0.1052	0.1052	0.1052	0.1052	0.1052	0.1052
T/K	360.99	360.99	360.98	340.45	340.45	340.44	320.06	320.07	320.07
P/MPa	176.15	175.87	175.54	168.90	169.14	169.27	160.88	161.12	160.66

continued

continued

w_B	0.1052	0.1052	0.1052	0.1326	0.1326	0.1326	0.1326	0.1326	0.1326
T/K	303.65	303.65	303.64	423.05	423.04	423.03	412.81	412.81	412.83
P/MPa	152.32	152.55	151.91	181.55	181.14	180.90	180.67	180.72	180.57
w_B	0.1326	0.1326	0.1326	0.1326	0.1326	0.1326	0.1326	0.1326	0.1326
T/K	402.61	402.62	402.62	392.38	392.38	392.37	382.09	382.10	382.12
P/MPa	180.94	180.84	181.59	180.49	179.82	180.22	179.08	179.22	179.59
w_B	0.1326	0.1326	0.1326	0.1326	0.1326	0.1326	0.1326	0.1326	0.1326
T/K	371.86	371.86	371.87	361.08	361.09	361.12	350.88	350.85	350.85
P/MPa	176.25	176.10	175.86	173.21	173.88	174.22	171.19	170.64	170.32
w_B	0.1326	0.1326	0.1326	0.1326	0.1326	0.1326	0.1326	0.1326	0.1326
T/K	340.49	340.51	340.52	330.25	330.26	330.27	320.86	320.87	320.87
P/MPa	167.48	167.10	166.87	161.89	162.62	162.57	158.07	158.40	158.75
w_B	0.1326	0.1326	0.1326	0.1326	0.1326	0.1326			
T/K	310.45	310.45	310.50	302.32	302.31	302.31			
P/MPa	152.40	152.66	152.94	147.99	148.57	148.86			

Polymer (B): **polyester (hyperbranched)** **2010GRE**
Characterization: M_n/g.mol^{-1} = 4450, Boltorn-type,
OH-endgroups are esterified to -O[(CO)$(CH_2)_5$O]$_5$(CO)CH_3
Solvent (A): **carbon dioxide** **CO_2** **124-38-9**

Type of data: cloud points

w_B	0.0215	0.0215	0.0215	0.0215	0.0215	0.0215	0.0215	0.0215	0.0215
T/K	423.36	423.37	423.35	12.95	412.95	412.96	402.88	402.89	402.89
P/MPa	159.10	158.59	158.74	159.48	159.62	159.52	159.88	160.24	160.44
w_B	0.0215	0.0215	0.0215	0.0215	0.0215	0.0215	0.0215	0.0215	0.0215
T/K	392.56	392.57	392.57	382.40	382.41	382.40	371.89	371.90	371.91
P/MPa	162.17	161.84	161.91	163.40	163.52	163.86	165.18	165.24	165.32
w_B	0.0215	0.0215	0.0215	0.0215	0.0215	0.0215	0.0215	0.0215	0.0215
T/K	361.44	361.43	361.44	351.17	351.17	351.16	340.88	340.89	330.60
P/MPa	168.34	168.26	168.20	172.55	172.38	172.15	176.73	177.03	184.98
w_B	0.0215	0.0215	0.0859	0.0859	0.0859	0.0859	0.0859	0.0859	0.0859
T/K	330.60	330.60	423.52	423.52	423.51	413.28	413.28	413.28	402.98
P/MPa	185.22	185.57	172.56	171.56	171.87	170.69	170.72	170.66	169.92
w_B	0.0859	0.0859	0.0859	0.0859	0.0859	0.0859	0.0859	0.0859	0.0859
T/K	402.31	402.31	392.47	392.49	392.50	382.52	382.53	382.52	372.33
P/MPa	169.84	170.43	171.55	171.90	172.14	175.52	175.89	176.02	177.27
w_B	0.0859	0.0859	0.0859	0.0859	0.0859	0.0859	0.0859	0.0859	0.0859
T/K	372.33	372.35	361.55	361.55	361.54	351.22	351.22	351.21	340.84
P/MPa	177.08	177.18	178.30	178.47	179.02	181.60	181.86	182.27	185.17

continued

continued

w_B	0.0859	0.0859	0.0859	0.0859	0.0859	0.2882	0.2882	0.2882	0.2882
T/K	340.92	340.93	330.58	330.58	330.57	423.33	423.35	423.34	423.34
P/MPa	185.24	185.53	195.38	195.20	194.87	145.29	144.98	144.57	144.94
w_B	0.2882	0.2882	0.2882	0.2882	0.2882	0.2882	0.2882	0.2882	0.2882
T/K	423.32	413.26	413.25	413.25	402.87	402.88	402.87	392.57	392.57
P/MPa	144.63	144.78	144.83	144.89	145.90	145.07	145.24	145.16	145.07
w_B	0.2882	0.2882	0.2882	0.2882	0.2882	0.2882	0.2882	0.2882	0.2882
T/K	392.56	382.32	382.36	382.33	371.89	371.89	371.88	361.39	361.39
P/MPa	145.14	145.06	145.24	145.44	145.98	146.06	146.11	146.52	146.44
w_B	0.2882	0.2882	0.2882	0.2882	0.2882	0.2882	0.2882	0.2882	0.2882
T/K	361.41	351.15	351.14	351.15	340.87	340.86	340.88	330.37	330.36
P/MPa	146.63	147.08	147.22	147.43	148.35	147.77	147.87	148.56	148.63

Polymer (B): **polyester (hyperbranched)** **2010GRE**
Characterization: M_n/g.mol^{-1} = 5600, Boltorn-type,
OH-endgroups are esterified to -O(CO)$(CH_2)_{10}CH_3$
Solvent (A): **ethane** C_2H_6 **74-84-0**

Type of data: cloud points

w_B 0.0768 was kept constant

T/K	280.47	280.41	280.40	290.13	290.22	290.22	300.03	300.06	300.06
P/MPa	42.34	42.28	42.20	42.41	42.37	42.35	42.92	42.84	42.91
T/K	310.22	310.20	310.21	320.34	320.34	320.35	330.53	330.51	330.52
P/MPa	42.97	43.27	43.29	43.87	43.94	43.91	44.75	44.78	44.82
T/K	340.35	340.32	340.42	350.76	350.78	350.78	360.93	361.00	361.04
P/MPa	45.61	45.67	45.68	46.49	46.53	46.59	47.28	47.31	47.37
T/K	371.30	371.22	371.19						
P/MPa	48.05	48.15	48.14						

Polymer (B): **polyester (hyperbranched)** **2010GRE**
Characterization: M_n/g.mol^{-1} = 7200, Boltorn-type,
OH-endgroups are esterified to -O(CO)$(CH_2)_{16}CH_3$
Solvent (A): **ethane** C_2H_6 **74-84-0**

Type of data: cloud points

w_B 0.0769 was kept constant

T/K	300.00	300.02	300.01	310.09	310.17	310.19	319.97	320.20	320.28
P/MPa	46.47	46.46	46.49	47.59	47.61	47.59	48.31	48.36	48.40
T/K	330.55	330.60	330.61	340.79	340.76	340.78	351.00	351.15	351.17
P/MPa	49.53	49.50	49.46	50.88	50.60	50.48	51.26	51.36	51.28

continued

continued

T/K	361.46	361.45	361.42	371.80	371.75	371.75
P/MPa	52.19	52.16	52.16	52.92	52.81	52.90

Polymer (B): **polyester (hyperbranched)** **2010GRE**
Characterization: M_n/g.mol^{-1} = 3450, Boltorn-type,
OH-endgroups are modified to -OSi(CH_3)$_3$
Solvent (A): **ethane** C_2H_6 **74-84-0**

Type of data: cloud points

w_B 0.0755 was kept constant

T/K	280.48	280.24	280.13	290.70	290.68	290.69	300.08	300.08	300.08
P/MPa	31.28	31.47	31.37	31.84	31.60	31.44	31.24	31.18	31.08
T/K	310.21	310.28	310.29	320.67	320.66	320.65	330.94	330.92	330.94
P/MPa	31.40	31.84	31.67	32.34	32.15	31.81	34.07	33.02	33.71
T/K	330.92	341.23	341.24	341.24	341.23	351.51	351.49	351.49	361.74
P/MPa	33.48	35.22	34.93	34.65	34.52	35.41	35.89	35.87	36.38
T/K	361.76	361.75	372.11	372.11	372.10				
P/MPa	36.66	36.68	37.92	37.71	37.56				

Polymer (B): **polyester (hyperbranched)** **2010GRE**
Characterization: M_n/g.mol^{-1} = 7200, Boltorn-type,
OH-endgroups are esterified to -O(CO)(CH_2)$_{16}$$CH_3$
Solvent (A): **propane** C_3H_8 **74-98-6**

Type of data: cloud points

w_B 0.0780 was kept constant

T/K	310.27	310.30	310.30	320.61	320.63	320.52	330.73	330.77	330.72
P/MPa	3.09	3.06	3.04	5.62	5.56	5.51	7.79	7.84	7.82
T/K	340.97	340.94	340.95	351.21	351.22	351.21	361.45	361.49	361.48
P/MPa	9.95	10.03	10.02	12.05	12.05	12.04	13.99	13.99	14.07
T/K	372.02	372.03	371.96						
P/MPa	15.93	15.92	15.92						

Polymer (B): **polyester (hyperbranched)** **2010GRE**
Characterization: M_n/g.mol^{-1} = 3450, Boltorn-type,
OH-endgroups are modified to -OSi(CH_3)$_3$
Solvent (A): **propane** C_3H_8 **74-98-6**

Type of data: cloud points

w_B 0.1051 was kept constant

continued

continued

T/K	341.09	341.08	341.09	351.23	351.31	351.32	361.56	361.52	361.52
P/MPa	3.31	3.37	3.38	5.36	5.46	5.51	7.55	7.37	7.44
T/K	371.56	371.68	371.70	382.25	382.24	382.24	395.10	395.22	395.16
P/MPa	8.92	8.94	9.00	10.47	10.47	10.46	12.15	12.28	12.20

Polymer (B): **polyester (hyperbranched)** **2009POR**
Characterization: M_n/g.mol^{-1} = 6560, M_w/g.mol^{-1} = 10500,
Boltorn H3200, solid, fatty acid (C_{20}/C_{22}) modified,
Perstorp Speciality Chemicals AB, Sweden
Solvent (A): **propane** **C_3H_8** **74-98-6**

Type of data: cloud points

w_B	0.025	0.025	0.025	0.025	0.025	0.025	0.025	0.025	0.025
T/K	313.43	315.36	318.28	320.22	323.17	325.12	328.10	333.03	337.98
P/MPa	4.06	4.45	5.31	5.71	6.50	6.85	7.27	8.28	9.03
w_B	0.025	0.025	0.025	0.025	0.050	0.050	0.050	0.050	0.050
T/K	342.95	347.88	352.86	357.77	312.96	315.00	316.50	317.91	320.96
P/MPa	9.94	10.78	11.43	11.75	5.15	5.30	5.75	5.76	6.75
w_B	0.050	0.050	0.050	0.050	0.050	0.050	0.050	0.099	0.099
T/K	322.70	327.50	332.11	336.96	341.90	346.85	351.71	308.51	313.21
P/MPa	6.83	7.86	8.64	9.59	10.60	11.65	12.57	4.50	5.72
w_B	0.099	0.099	0.099	0.099	0.099	0.099	0.099	0.200	0.200
T/K	317.91	322.67	327.41	332.21	337.13	342.04	346.93	313.57	318.37
P/MPa	6.82	7.89	8.93	9.97	11.00	11.99	13.00	5.83	6.93
w_B	0.200	0.200	0.200	0.200	0.200	0.200	0.200	0.301	0.301
T/K	323.30	328.19	333.09	337.97	342.86	347.77	352.67	308.54	313.38
P/MPa	8.02	9.08	10.10	11.09	12.06	13.01	13.92	4.45	5.57
w_B	0.301	0.301	0.301	0.301	0.301	0.301	0.301	0.401	0.401
T/K	318.29	323.22	328.18	333.08	338.11	343.07	348.06	313.17	317.64
P/MPa	6.69	7.77	8.83	9.87	10.87	11.84	12.79	5.27	6.30
w_B	0.401	0.401	0.401	0.401	0.401	0.401	0.401	0.501	0.501
T/K	322.10	326.34	330.85	336.12	340.94	345.77	350.74	308.61	310.57
P/MPa	7.35	8.32	9.34	10.32	11.27	12.19	13.08	3.15	3.55
w_B	0.501	0.501	0.501	0.501	0.501	0.501	0.501	0.501	0.501
T/K	312.62	318.73	323.28	328.17	333.12	338.04	342.94	347.88	352.79
P/MPa	4.05	5.39	6.46	7.52	8.55	9.55	10.52	11.46	12.38
w_B	0.501	0.603	0.603	0.603	0.603	0.603	0.603	0.603	0.603
T/K	357.62	315.94	318.38	323.33	328.33	333.24	338.15	343.02	347.90
P/MPa	13.24	1.75	2.32	3.43	4.51	5.55	6.55	7.53	8.48

continued

continued

w_B	0.603	0.603	0.603	0.603	0.701	0.701	0.701	0.701	0.701
T/K	352.80	357.74	362.67	367.63	345.27	346.76	347.69	348.76	350.10
P/MPa	9.41	10.32	11.20	12.06	2.78	3.00	3.21	3.39	3.65
w_B	0.701	0.701	0.701	0.701	0.701	0.701	0.701	0.701	0.701
T/K	351.88	353.94	356.79	358.82	361.64	363.60	366.20	370.92	380.62
P/MPa	4.02	4.47	4.96	5.29	5.90	6.23	6.65	7.47	9.08
w_B	0.701	0.701	0.701						
T/K	390.73	400.82	410.66						
P/MPa	10.66	12.13	13.48						

Type of data: vapor-liquid-liquid equilibrium data

w_B	0.025	0.025	0.025	0.025	0.025	0.025	0.025	0.025	0.050
T/K	312.06	313.42	323.22	333.05	343.04	352.96	362.90	370.04	307.34
P/MPa	1.34	1.38	1.72	2.12	2.59	3.13	3.75	4.25 (*)	1.22
w_B	0.050	0.050	0.050	0.050	0.050	0.050	0.050	0.050	0.050
T/K	308.48	310.84	313.30	314.16	322.99	331.59	341.84	351.62	361.35
P/MPa	1.25	1.32	1.40	1.42	1.73	2.04	2.54	3.06	3.66
w_B	0.050	0.099	0.099	0.099	0.099	0.099	0.099	0.099	0.099
T/K	369.98	303.70	306.25	308.67	313.21	322.77	332.21	341.97	351.72
P/MPa	4.30 (*)	1.14	1.21	1.27	1.41	1.75	2.12	2.58	3.09
w_B	0.099	0.099	0.099	0.200	0.200	0.200	0.200	0.200	0.200
T/K	361.51	369.99	369.82	308.65	309.43	313.52	318.36	323.26	333.08
P/MPa	3.70	4.27	4.27 (*)	1.25	1.27	1.40	1.55	1.73	2.13
w_B	0.200	0.200	0.200	0.200	0.301	0.301	0.301	0.301	0.301
T/K	343.03	352.79	362.58	369.99	308.52	308.66	313.42	323.20	333.12
P/MPa	2.60	3.12	3.73	4.25 (*)	1.24	1.24	1.38	1.72	2.12
w_B	0.301	0.301	0.301	0.301	0.301	0.401	0.401	0.401	0.401
T/K	343.19	353.05	362.92	370.00	369.99	308.48	308.89	312.93	322.19
P/MPa	2.59	3.13	3.75	4.23	4.25 (*)	1.25	1.26	1.38	1.71
w_B	0.401	0.401	0.401	0.401	0.401	0.401	0.501	0.501	0.501
T/K	331.20	340.63	350.04	359.81	369.15	369.34	311.42	313.41	323.28
P/MPa	2.08	2.53	3.02	3.62	4.25	4.25 (*)	1.37	1.43	1.77
w_B	0.501	0.501	0.501	0.501	0.501	0.603	0.603	0.603	0.603
T/K	333.11	342.99	352.78	362.63	369.85	308.58	313.17	312.52	323.31
P/MPa	2.17	2.63	3.16	3.78	4.27 (*)	1.25	1.39	1.40	1.74
w_B	0.603	0.603	0.603	0.603	0.603	0.701	0.701	0.701	
T/K	333.20	343.07	352.90	362.70	369.91	351.99	361.62	369.93	
P/MPa	2.13	2.60	3.13	3.74	4.26 (*)	3.10	3.69	4.29 (*)	

Comments: (*) Critical end point. VLE data are given in Chapter 2. Equilibrium data with a solid phase are not included here and have to be found in the original source.

Polymer (B): **poly[2-(2-ethoxyethoxy)ethyl acrylate]** **2014YOO**
Characterization: M_w/g.mol^{-1} = 100000, T_g/K = 203.2,
Scientific Polymer Products, Inc., Ontario, NY
Solvent (A): **n-butane** C_4H_{10} **106-97-8**

Type of data: cloud points

w_B 0.052 was kept constant

T/K	440.4	441.5	444.3	454.9
P/MPa	206.6	124.1	99.83	63.91

Polymer (B): **poly[2-(2-ethoxyethoxy)ethyl acrylate]** **2014YOO**
Characterization: M_w/g.mol^{-1} = 100000, T_g/K = 203.2,
Scientific Polymer Products, Inc., Ontario, NY
Solvent (A): **1-butene** C_4H_8 **106-98-9**

Type of data: cloud points

w_B 0.047 was kept constant

T/K	354.0	373.2	393.8	14.6	433.6	452.7
P/MPa	68.28	43.97	33.97	30.17	28.79	28.79

Polymer (B): **poly[2-(2-ethoxyethoxy)ethyl acrylate]** **2014YOO**
Characterization: M_w/g.mol^{-1} = 100000, T_g/K = 203.2,
Scientific Polymer Products, Inc., Ontario, NY
Solvent (A): **carbon dioxide** CO_2 **124-38-9**

Type of data: cloud points

w_B 0.047 was kept constant

T/K	333.3	353.5	374.0	393.6	414.1	433.9	453.2
P/MPa	201.6	168.5	151.9	144.3	138.8	135.3	132.6

Polymer (B): **poly[2-(2-ethoxyethoxy)ethyl acrylate]** **2014YOO**
Characterization: M_w/g.mol^{-1} = 100000, T_g/K = 203.2,
Scientific Polymer Products, Inc., Ontario, NY
Solvent (A): **dimethyl ether** C_2H_6O **115-10-6**

Type of data: cloud points

w_B 0.049 was kept constant

T/K	374.2	386.4	392.8	398.2	405.5	413.5	424.3	434.1	453.9
P/MPa	8.45	11.27	12.59	13.97	15.35	16.73	18.79	20.86	24.31

T/K	474.0
P/MPa	27.07

Polymer (B): **poly[2-(2-ethoxyethoxy)ethyl acrylate]** **2014YOO**
Characterization: M_w/g.mol^{-1} = 100000, T_g/K = 203.2,
Scientific Polymer Products, Inc., Ontario, NY
Solvent (A): **propane** C_3H_8 **74-98-6**

Type of data: cloud points

w_B 0.050 was kept constant

T/K	425.4	432.2	435.2	442.3	453.2	465.6
P/MPa	199.66	174.48	157.93	137.59	112.24	97.93

Polymer (B): **poly[2-(2-ethoxyethoxy)ethyl acrylate]** **2014YOO**
Characterization: M_w/g.mol^{-1} = 100000, T_g/K = 203.2,
Scientific Polymer Products, Inc., Ontario, NY
Solvent (A): **propene** C_3H_6 **115-07-1**

Type of data: cloud points

w_B 0.048 was kept constant

T/K	334.5	354.6	374.3	394.3	413.1	433.3	454.1
p/MPa	113.62	76.21	62.59	56.38	52.93	51.03	50.35

Polymer (B): **poly(ethyl acrylate)** **2004BY3**
Characterization: M_w/g.mol^{-1} = 70000, Polysciences, Inc., Warrington, PA
Solvent (A): **1-butene** C_4H_8 **106-98-9**

Type of data: cloud points

w_B 0.054 was kept constant

T/K	333.95	336.45	338.95	341.35	344.55	348.85	353.65	362.35	372.45
P/bar	1892.8	1619.7	1479.7	1338.3	1200.3	1068.6	934.8	759.0	631.0

T/K	383.55	403.05	422.15
P/bar	538.3	431.4	372.8

Polymer (B): **poly(ethyl acrylate)** **2004BY3**
Characterization: M_w/g.mol^{-1} = 70000, Polysciences, Inc., Warrington, PA
Solvent (A): **carbon dioxide** CO_2 **124-38-9**

Type of data: cloud points

w_B 0.054 was kept constant

T/K	324.35	325.25	326.85	329.05	331.65	335.65	347.15	363.95	383.75
P/bar	2070.7	1915.5	1846.6	1762.4	1706.6	1630.0	1482.8	1403.1	1354.5

T/K	403.05	423.75
P/bar	1327.9	1317.2

Polymer (B): **poly(ethyl acrylate)** **2004BEC, 2004LAT**
Characterization: M_n/g.mol^{-1} = 85400, M_w/g.mol^{-1} = 154000, synthesized in the laboratory
Solvent (A): **ethene** C_2H_4 **74-85-1**

Type of data: cloud points

w_B 0.05 was kept constant

T/K	354.15	364.15	373.15	393.15	414.35	434.65	453.85	473.45	492.35
P/bar	1613	1541	1485	1376	1288	1217	1162	1117	1077

T/K	513.55	533.15
P/bar	1038	1010

Polymer (B): **poly(ethyl acrylate)** **2004BY3**
Characterization: M_w/g.mol^{-1} = 70000, Polysciences, Inc., Warrington, PA
Solvent (A): **propene** C_3H_6 **115-07-1**

Type of data: cloud points

w_B 0.054 was kept constant

T/K	311.75	316.55	320.75	324.35	330.65	344.55	364.55	384.45	405.25
P/bar	1647.2	1498.3	1354.8	1233.4	1119.7	903.1	736.9	667.2	621.7

T/K	423.75
P/bar	599.0

Polymer (B): **polyethylene** **2004LAT**
Characterization: M_n/g.mol^{-1} = 6300, M_w/g.mol^{-1} = 19300, LDPE, synthesized in the laboratory
Solvent (A): **ethene** C_2H_4 **74-85-1**

Type of data: cloud points

w_B 0.05 was kept constant

T/K	412.15	431.15	450.15	471.15	492.15	510.15	530.15
P/bar	1590	1487	1403	1334	1276	1236	1194

Polymer (B): **polyethylene** **2004LAT**
Characterization: M_n/g.mol^{-1} = 9500, M_w/g.mol^{-1} = 30000, LDPE, synthesized in the laboratory
Solvent (A): **ethene** C_2H_4 **74-85-1**

Type of data: cloud points

w_B 0.05 was kept constant

T/K	403.15	413.15	422.15	433.15	442.15	453.55	463.65	473.35	483.15
P/bar	1748	1666	1606	1542	1500	1458	1418	1383	1351

continued

continued

T/K	493.15	503.15	513.15	523.15	532.15
P/bar	1324	1297	1274	1255	1239

Polymer (B): **polyethylene** **2004BEC, 2004LAT**
Characterization: M_n/g.mol^{-1} = 12000, M_w/g.mol^{-1} = 45300, LDPE, synthesized in the laboratory
Solvent (A): **ethene** **C_2H_4** **74-85-1**

Type of data: cloud points

w_B 0.05 was kept constant

T/K	403.15	414.15	423.15	434.15	442.15	454.15	463.15	473.15	482.15
P/bar	1637	1585	1522	1468	1428	1389	1356	1323	1299

T/K	492.15	502.15	512.15	521.15
P/bar	1275	1249	1226	1204

Polymer (B): **polyethylene** **2004LAT**
Characterization: M_n/g.mol^{-1} = 19900, M_w/g.mol^{-1} = 58300, LDPE, synthesized in the laboratory
Solvent (A): **ethene** **C_2H_4** **74-85-1**

Type of data: cloud points

w_B 0.05 was kept constant

T/K	398.15	403.15	413.15	422.15	432.15	452.15	471.15	492.15
P/bar	1540	1500	1450	1420	1380	1310	1260	1210

Polymer (B): **polyethylene** **2004LAT**
Characterization: M_n/g.mol^{-1} = 33000, M_w/g.mol^{-1} = 129000, LDPE, synthesized in the laboratory
Solvent (A): **ethene** **C_2H_4** **74-85-1**

Type of data: cloud points

w_B 0.05 was kept constant

T/K	394.15	396.15	401.15	402.15	405.15	412.65	413.15	433.15	445.15
P/bar	1495	1485	1441	1436	1413	1366	1362	1275	1236

T/K	452.15	467.15	472.15	474.15	490.65
P/bar	1211	1175	1161	1153	1122

Polymer (B): **poly(ethylene-*co*-benzyl methacrylate)** **2004LAT**
Characterization: 3.3 mol% benzyl methacrylate, synthesized in the laboratory
Solvent (A): **ethene** **C_2H_4** **74-85-1**

Type of data: cloud points

continued

continued

w_B	0.05	was kept constant				
T/K	389.15	411.15	430.15	450.15	470.15	489.15	510.15
P/bar	1697	1581	1501	1426	1366	1311	1268

Polymer (B): **poly(ethylene-*co*-benzyl methacrylate)** **2004LAT**
Characterization: M_n/g.mol^{-1} = 47100, M_w/g.mol^{-1} = 73500,
12.7 mol% benzyl methacrylate, synthesized in the laboratory
Solvent (A): **ethene** **C_2H_4** **74-85-1**

Type of data: cloud points

w_B 0.05 was kept constant

T/K	373.15	393.15	413.15	433.15	453.15	468.15	488.15	508.15	528.15
P/bar	1903	1748	1656	1579	1512	1459	1410	1361	1302

Polymer (B): **poly(ethylene-*co*-benzyl methacrylate)** **2004LAT**
Characterization: M_n/g.mol^{-1} = 36900, M_w/g.mol^{-1} = 60700,
20.0 mol% benzyl methacrylate, synthesized in the laboratory
Solvent (A): **ethene** **C_2H_4** **74-85-1**

Type of data: cloud points

w_B 0.05 was kept constant

T/K	458.15	473.15	483.15	488.15
P/bar	2006	1830	1747	1722

Polymer (B): **poly(ethylene-*co*-benzyl methacrylate)** **2004LAT**
Characterization: M_n/g.mol^{-1} = 36900, M_w/g.mol^{-1} = 61100,
26.1 mol% benzyl methacrylate, synthesized in the laboratory
Solvent (A): **ethene** **C_2H_4** **74-85-1**

Type of data: cloud points

w_B 0.05 was kept constant

T/K	458.15	473.15	483.15	493.15	503.15	513.15	523.15	528.15
P/bar	2380	2299	2230	2123	2055	2014	1949	1881

Polymer (B): **poly(ethylene-*co*-1-butene)** **2004KER**
Characterization: M_n/g.mol^{-1} = 230000, M_w/g.mol^{-1} = 232500,
20.2 mol% 1-butene, 10 ethyl branches per 100 backbone
C-atomes, completely hydrogenated polybutadiene,
synthesized in the laboratory
Solvent (A): **n-butane** **C_4H_{10}** **106-97-8**

Type of data: cloud points

w_B 0.001-0.006 T/K 403.15 P/bar 175

Polymer (B): **poly(ethylene-*co*-1-butene)** **2005LID**
Characterization: M_n/g.mol^{-1} = 230000, M_w/g.mol^{-1} = 232500, 20.2 mol% 1-butene, 10 ethyl branches per 100 backbone C-atomes, completely hydrogenated polybutadiene, synthesized in the laboratory
Solvent (A): **dimethyl ether** **C_2H_6O** **115-10-6**

Type of data: cloud points

w_B 0.005 was kept constant

T/K	383.15	403.15	423.15	443.15
P/bar	800	586	513	495

Comments: The cloud-point pressures are independent on w_B between w_B = 0.001 and w_B = 0.006. More cloud points at w_B = 0.05 are given in Fig. 2 of the original source.

Polymer (B): **poly(ethylene-*co*-1-butene), deuterated** **2008KOS**
Characterization: M_n/g.mol^{-1} = 222700, M_w/g.mol^{-1} = 245000, 20.2 mol% 1-butene, 10 ethyl branches per 100 backbone C-atomes, completely deuterated polybutadiene, synthesized in the laboratory
Solvent (A): **2,2-dimethylpropane** **C_5H_{12}** **463-82-1**

Type of data: cloud points

w_B 0.003 was kept constant

T/K	285.95	289.15	292.75	296.65	299.45	304.15	306.55	310.35	314.75
P/MPa	35.3	32.4	30.3	27.1	26.3	25.0	23.8	22.6	21.7
T/K	317.85	322.65	327.35	329.65	333.15	337.05	339.55	341.35	345.45
P/MPa	21.3	20.8	20.4	20.2	20.0	19.6	19.5	19.4	19.4
T/K	349.45	352.75	358.55	361.65	364.65	366.95	369.35	372.75	375.45
P/MPa	19.4	19.5	19.8	19.9	20.0	20.1	20.2	20.4	20.6
T/K	379.15	381.75	385.35	389.55	391.35	394.85	398.15	401.85	406.45
P/MPa	20.8	20.9	21.1	21.4	21.5	21.7	22.0	22.3	22.6
T/K	409.05	412.95	415.55	418.75	423.05	425.75	428.75	431.25	434.85
P/MPa	22.9	23.1	23.3	23.5	23.9	24.1	24.3	24.5	24.7
T/K	435.85	440.05	441.25	445.65	450.15	454.35	456.05	460.65	461.35
P/MPa	24.7	25.0	25.1	25.3	25.7	26.2	26.4	27.1	27.2

Comments: Solid-liquid equilibrium data are additionally given in the original source.

Polymer (B): **poly(ethylene-*co*-1-butene)** **2008KOS**
Characterization: M_n/g.mol^{-1} = 230000, M_w/g.mol^{-1} = 232500,
20.2 mol% 1-butene, 10 ethyl branches per 100 backbone
C-atomes, completely hydrogenated polybutadiene,
synthesized in the laboratory
Solvent (A): **ethane-d6** **C_2D_6** **1632-99-1**

Type of data: cloud points

w_B 0.048 was kept constant

T/K	368.25	373.65	375.05	384.35	393.85	395.15	404.65	405.05	424.25
P/MPa	117.0	114.3	113.9	112.0	109.5	108.8	107.8	107.2	103.9

T/K	425.05	427.65
P/MPa	104.3	103.3

Polymer (B): **poly(ethylene-*co*-1-butene)** **2008KOS**
Characterization: M_n/g.mol^{-1} = 230000, M_w/g.mol^{-1} = 232500,
20.2 mol% 1-butene, 10 ethyl branches per 100 backbone
C-atomes, completely hydrogenated polybutadiene,
synthesized in the laboratory
Solvent (A): **ethane** **C_2H_6** **74-84-0**

Type of data: cloud points

w_B 0.043 was kept constant

T/K	367.65	369.85	376.65	383.35	393.15	403.45	423.65
P/MPa	128.4	126.0	123.2	121.2	118.8	115.9	111.6

Polymer (B): **poly(ethylene-*co*-1-butene)** **2004KER**
Characterization: M_n/g.mol^{-1} = 230000, M_w/g.mol^{-1} = 232500,
20.2 mol% 1-butene, 10 ethyl branches per 100 backbone
C-atomes, completely hydrogenated polybutadiene,
synthesized in the laboratory
Solvent (A): **ethane** **C_2H_6** **74-84-0**

Type of data: cloud points

w_B 0.002-0.007 T/K 403.15 P/bar 1040

Polymer (B): **poly(ethylene-*co*-1-butene), deuterated** **2008KOS**
Characterization: M_n/g.mol^{-1} = 222700, M_w/g.mol^{-1} = 245000,
20.2 mol% 1-butene, 10 ethyl branches per 100 backbone
C-atomes, completely deuterated polybutadiene,
synthesized in the laboratory
Solvent (A): **2-methylbutane** **C_5H_{12}** **78-78-4**

continued

continued

Type of data: cloud points

w_B	0.005	was kept constant				
T/K	371.15	391.75	414.45	437.05		
P/MPa	1.5	4.6	8.6	12.0		
w_B	0.019	was kept constant				
T/K	367.95	378.25	386.25	410.85	426.95	448.15
P/MPa	1.6	3.4	5.1	8.8	11.1	14.3

Comments: Solid-liquid equilibrium data are additionally given in the original source.

Polymer (B): **poly(ethylene-*co*-1-butene)** **2004KER**
Characterization: M_n/g.mol^{-1} = 230000, M_w/g.mol^{-1} = 232500, 20.2 mol% 1-butene, 10 ethyl branches per 100 backbone C-atomes, completely hydrogenated polybutadiene, synthesized in the laboratory
Solvent (A): **n-pentane** **C_5H_{12}** **109-66-0**

Type of data: cloud points

w_B 0.002-0.006 T/K 403.15 P/bar 40

Polymer (B): **poly(ethylene-*co*-1-butene), deuterated** **2008KOS**
Characterization: M_n/g.mol^{-1} = 222700, M_w/g.mol^{-1} = 245000, 20.2 mol% 1-butene, 10 ethyl branches per 100 backbone C-atomes, completely deuterated polybutadiene, synthesized in the laboratory
Solvent (A): **n-pentane** **C_5H_{12}** **109-66-0**

Type of data: cloud points

w_B	0.005	0.005	0.005	0.005	0.018	0.018
T/K	395.65	408.45	420.95	439.55	389.85	408.45
P/MPa	2.6	4.6	6.8	9.5	1.8	5.0

Comments: Solid-liquid equilibrium data are additionally given in the original source.

Polymer (B): **poly(ethylene-*co*-1-butene)** **2004KER**
Characterization: M_n/g.mol^{-1} = 230000, M_w/g.mol^{-1} = 232500, 20.2 mol% 1-butene, 10 ethyl branches per 100 backbone C-atomes, completely hydrogenated polybutadiene, synthesized in the laboratory
Solvent (A): **propane** **C_3H_8** **74-98-6**

Type of data: cloud points

w_B 0.002-0.006 T/K 403.15 P/bar 460

Polymer (B): **poly(ethylene-*co*-butyl methacrylate)** **2004LAT**
Characterization: 6.7 mol% butyl methacrylate, synthesized in the laboratory
Solvent (A): **ethene** C_2H_4 **74-85-1**

Type of data: cloud points

w_B 0.05 was kept constant

T/K	374.15	393.15	414.15	434.15	453.15	472.15	491.15	513.15	531.15
P/bar	1414	1330	1264	1208	1169	1134	1106	1079	1052

Polymer (B): **poly(ethylene-*co*-butyl methacrylate)** **2004LAT**
Characterization: M_n/g.mol^{-1} = 23400, M_w/g.mol^{-1} = 39800,
11.7 mol% butyl methacrylate, synthesized in the laboratory
Solvent (A): **ethene** C_2H_4 **74-85-1**

Type of data: cloud points

w_B 0.05 was kept constant

T/K	372.15	393.15	411.15	432.15	451.15	470.15	489.15	510.15	531.15
P/bar	1315	1239	1194	1161	1122	1095	1063	1039	1017

Polymer (B): **poly(ethylene-*co*-butyl methacrylate)** **2004LAT**
Characterization: M_n/g.mol^{-1} = 24800, M_w/g.mol^{-1} = 41000,
18.5 mol% butyl methacrylate, synthesized in the laboratory
Solvent (A): **ethene** C_2H_4 **74-85-1**

Type of data: cloud points

w_B 0.05 was kept constant

T/K	353.15	373.15	393.15	413.15	433.15	453.15	473.15	488.15	508.15
P/bar	1082	1045	1016	993	973	958	946	930	915

T/K	528.15
P/bar	901

Polymer (B): **poly(ethylene-*co*-butyl methacrylate)** **2004LAT**
Characterization: M_n/g.mol^{-1} = 24500, M_w/g.mol^{-1} = 44400,
22.7 mol% butyl methacrylate, synthesized in the laboratory
Solvent (A): **ethene** C_2H_4 **74-85-1**

Type of data: cloud points

w_B 0.05 was kept constant

T/K	353.15	374.15	394.15	413.15	432.15	452.15	472.15	491.15	511.15
P/bar	1121	1085	1054	1028	1008	987	970	951	939

T/K	529.15
P/bar	922

Polymer (B): **poly(ethylene-*co*-butyl methacrylate)** **2004LAT**
Characterization: M_n/g.mol^{-1} = 55000, M_w/g.mol^{-1} = 96700,
33.8 mol% butyl methacrylate, synthesized in the laboratory
Solvent (A): **ethene** **C_2H_4** **74-85-1**

Type of data: cloud points

w_B 0.05 was kept constant

T/K	350.15	372.15	393.15	413.15	433.15	453.15	471.15	491.15	510.15
P/bar	1002	974	954	940	925	912	901	891	879

T/K	530.15
P/bar	876

Polymer (B): **poly(ethylene-*co*-butyl methacrylate)** **2004LAT**
Characterization: M_n/g.mol^{-1} = 58900, M_w/g.mol^{-1} = 101700,
44.0 mol% butyl methacrylate, synthesized in the laboratory
Solvent (A): **ethene** **C_2H_4** **74-85-1**

Type of data: cloud points

w_B 0.05 was kept constant

T/K	362.15	374.15	393.15	413.15	432.15	452.15	471.15	490.15	511.15
P/bar	1064	1038	1023	1005	995	969	957	952	938

T/K	531.15
P/bar	924

Polymer (B): **poly(ethylene-*co*-butyl methacrylate-*co*-methacrylic acid)** **2004LAT, 2007TUM**
Characterization: M_n/g.mol^{-1} = 11500, M_w/g.mol^{-1} = 33200,
0.4 mol% butyl methacrylate, 6.6 mol% methacrylic acid,
synthesized in the laboratory by partial esterification
Solvent (A): **ethene** **C_2H_4** **74-85-1**

Type of data: cloud points

w_B 0.05 was kept constant

T/K	474	483	490	490	492	493	498	501	503
P/bar	2750	2505	2328	2306	2283	2240	2147	2071	2035

T/K	508	512	518	521
P/bar	1946	1869	1795	1742

Polymer (B): **poly(ethylene-*co*-butyl methacrylate-*co*-methacrylic acid)** **2004LAT**
Characterization: 0.6 mol% butyl methacrylate, 6.4 mol% methacrylic acid, synthesized in the laboratory by partial esterification
Solvent (A): **ethene** **C_2H_4** **74-85-1**

Type of data: cloud points

w_B 0.05 was kept constant

T/K	475.85	476.15	477.85	481.65	482.55	489.65	489.95	493.65	493.85
P/bar	2647	2643	2584	2469	2474	2269	2258	2160	2158

T/K	502.15	512.55
P/bar	1991	1810

Polymer (B): **poly(ethylene-*co*-butyl methacrylate-*co*-methacrylic acid)** **2004LAT**
Characterization: 0.8 mol% butyl methacrylate, 6.3 mol% methacrylic acid, synthesized in the laboratory by partial esterification
Solvent (A): **ethene** **C_2H_4** **74-85-1**

Type of data: cloud points

w_B 0.05 was kept constant

T/K	467.15	472.15	478.15	483.15	490.15	492.15	498.15	504.15	514.15
P/bar	2707	2565	2429	2327	2222	2175	2062	1972	1845

T/K	523.15	530.15
P/bar	1762	1681

Polymer (B): **poly(ethylene-*co*-butyl methacrylate-*co*-methacrylic acid)** **2004LAT, 2007TUM**
Characterization: M_n/g.mol^{-1} = 11500, M_w/g.mol^{-1} = 33200, 0.9 mol% butyl methacrylate, 6.1 mol% methacrylic acid, synthesized in the laboratory by partial esterification
Solvent (A): **ethene** **C_2H_4** **74-85-1**

Type of data: cloud points

w_B 0.05 was kept constant

T/K	471	472	478	479	482	487	488	492	497
P/bar	2608	2566	2455	2364	2292	2168	2148	2076	1952

T/K	502	506	508	513	517	518	522	527	528
P/bar	1888	1813	1786	1728	1678	1672	1626	1573	1556

Polymer (B): **poly(ethylene-*co*-butyl methacrylate-*co*-methacrylic acid)** **2004LAT, 2007TUM**

Characterization: M_n/g.mol^{-1} = 11500, M_w/g.mol^{-1} = 33200, 1.7 mol% butyl methacrylate, 5.6 mol% methacrylic acid, synthesized in the laboratory by partial esterification

Solvent (A): **ethene** C_2H_4 **74-85-1**

Type of data: cloud points

w_B 0.05 was kept constant

T/K	479	480	493	500	511	520
P/bar	2070	2076	1875	1735	1680	1605

Polymer (B): **poly(ethylene-*co*-butyl methacrylate-*co*-methacrylic acid)** **2004LAT, 2007TUM**

Characterization: M_n/g.mol^{-1} = 11500, M_w/g.mol^{-1} = 33200, 3.5 mol% butyl methacrylate, 3.8 mol% methacrylic acid, synthesized in the laboratory by partial esterification

Solvent (A): **ethene** C_2H_4 **74-85-1**

Type of data: cloud points

w_B 0.05 was kept constant

T/K	435	451	462	471	481	490	501	521	529
P/bar	2650	2264	2047	1915	1789	1682	1468	1377	1321

Polymer (B): **poly(ethylene-*co*-butyl methacrylate-*co*-methacrylic acid)** **2007TUM**

Characterization: M_n/g.mol^{-1} = 11500, M_w/g.mol^{-1} = 33200, 4.2 mol% butyl methacrylate, 2.9 mol% methacrylic acid, synthesized in the laboratory by partial esterification

Solvent (A): **ethene** C_2H_4 **74-85-1**

Type of data: cloud points

w_B 0.05 was kept constant

T/K	433	446	452	453	465	471	472	482	483
P/bar	2195	2012	1904	1888	1678	1590	1580	1481	1477

T/K	493	501	502	514	522	531	532
P/bar	1387	1323	1327	1258	1225	1223	1180

Polymer (B): **poly(ethylene-*co*-butyl methacrylate-*co*-methacrylic acid)** **2004LAT, 2007TUM**
Characterization: M_n/g.mol^{-1} = 11500, M_w/g.mol^{-1} = 33200, 5.5 mol% butyl methacrylate, 1.8 mol% methacrylic acid, synthesized in the laboratory by partial esterification
Solvent (A): **ethene** C_2H_4 **74-85-1**

Type of data: cloud points

w_B 0.05 was kept constant

T/K	373	393	413	433	444	445	452	456	465
P/bar	2130	1860	1645	1460	1366	1374	1315	1301	1247

T/K	466	472	487	491	507	511
P/bar	1246	1212	1136	1138	1078	1066

Polymer (B): **poly(ethylene-*co*-butyl methacrylate-*co*-methacrylic acid)** **2004LAT**
Characterization: 5.5 mol% butyl methacrylate, 7.3 mol% methacrylic acid, synthesized in the laboratory by partial esterification
Solvent (A): **ethene** C_2H_4 **74-85-1**

Type of data: cloud points

w_B 0.05 was kept constant

T/K	486.15	494.15	498.15	504.15	508.15	513.15	519.15	528.15	532.15
P/bar	2530	2324	2231	2123	2041	1944	1873	1779	1750

Polymer (B): **poly(ethylene-*co*-butyl methacrylate-*co*-methacrylic acid)** **2004LAT**
Characterization: 6.2 mol% butyl methacrylate, 0.9 mol% methacrylic acid, synthesized in the laboratory by partial esterification
Solvent (A): **ethene** C_2H_4 **74-85-1**

Type of data: cloud points

w_B 0.05 was kept constant

T/K	373.15	381.85	392.15	403.95	412.15	429.05	432.05	445.15	453.15
P/bar	1717	1625	1529	1448	1405	1340	1319	1272	1236

T/K	461.15	471.15	480.45	490.75	500.65	518.15	520.15
P/bar	1218	1196	1171	1149	1127	1097	1096

Polymer (B): **poly(ethylene-*co*-butyl methacrylate-*co*-methacrylic acid)** **2004LAT**
Characterization: 6.8 mol% butyl methacrylate, 0.3 mol% methacrylic acid, synthesized in the laboratory by partial esterification
Solvent (A): **ethene** C_2H_4 **74-85-1**

Type of data: cloud points

w_B 0.05 was kept constant

T/K	354.15	363.65	372.35	383.05	391.15	394.45	411.65	428.75	430.45
P/bar	1626	1546	1486	1420	1379	1361	1288	1229	1223

T/K	443.15	451.95	464.65	473.15	485.35	492.35	511.25
P/bar	1190	1170	1144	1126	1104	1092	1064

Polymer (B): **poly(ethylene-*co*-ethyl acrylate)** **2004BEC, 2004LAT**
Characterization: 3.6 mol% ethyl acrylate, synthesized in the laboratory
Solvent (A): **ethene** C_2H_4 **74-85-1**

Type of data: cloud points

w_B 0.05 was kept constant

T/K	384.05	395.15	414.85	435.15	454.15	473.15	492.15	514.15	531.15
P/bar	1631	1568	1471	1394	1336	1288	1247	1207	1175

Polymer (B): **poly(ethylene-*co*-ethyl acrylate)** **2004BEC, 2004LAT**
Characterization: 6.3 mol% ethyl acrylate, synthesized in the laboratory
Solvent (A): **ethene** C_2H_4 **74-85-1**

Type of data: cloud points

w_B 0.05 was kept constant

T/K	372.15	393.15	413.15	433.15	452.15	472.55	492.15	511.15	531.35
P/bar	1510	1420	1351	1295	1252	1212	1181	1147	1122

Polymer (B): **poly(ethylene-*co*-ethyl acrylate)** **2004BEC, 2004LAT**
Characterization: M_n/g.mol^{-1} = 78400, M_w/g.mol^{-1} = 156600, 23.4 mol% ethyl acrylate, synthesized in the laboratory
Solvent (A): **ethene** C_2H_4 **74-85-1**

Type of data: cloud points

w_B 0.05 was kept constant

T/K	353.15	364.15	373.75	393.15	413.15	434.15	454.15	473.15	493.65
P/bar	1280	1256	1236	1190	1155	1123	1095	1072	1047

T/K	513.55	533.15
P/bar	1025	1000

Polymer (B): **poly(ethylene-*co*-ethyl acrylate)** **2004BEC, 2004LAT**
Characterization: M_n/g.mol^{-1} = 64400, M_w/g.mol^{-1} = 116700,
28.4 mol% ethyl acrylate, synthesized in the laboratory
Solvent (A): **ethene** C_2H_4 **74-85-1**

Type of data: cloud points

w_B 0.05 was kept constant

T/K	363.15	373.15	383.15	393.15	413.15	433.15	453.15	473.15	493.15
P/bar	1236	1206	1183	1165	1131	1100	1072	1048	1022

T/K	513.15	533.15
P/bar	1000	982

Polymer (B): **poly(ethylene-*co*-ethyl acrylate)** **2004BEC, 2004LAT**
Characterization: M_n/g.mol^{-1} = 75900, M_w/g.mol^{-1} = 159500,
34.8 mol% ethyl acrylate, synthesized in the laboratory
Solvent (A): **ethene** C_2H_4 **74-85-1**

Type of data: cloud points

w_B 0.05 was kept constant

T/K	366.65	374.65	397.15	412.15	434.15	454.65	474.65	491.15	513.15
P/bar	1296	1272	1219	1191	1151	1120	1090	1066	1039

T/K	532.15
P/bar	1018

Polymer (B): **poly(ethylene-*co*-ethyl acrylate)** **2004LAT**
Characterization: M_n/g.mol^{-1} = 67600, M_w/g.mol^{-1} = 142900,
45.7 mol% ethyl acrylate, synthesized in the laboratory
Solvent (A): **ethene** C_2H_4 **74-85-1**

Type of data: cloud points

w_B 0.05 was kept constant

T/K	372.75	391.95	412.15	433.15	453.15	472.15	491.15	511.15	529.15
P/bar	1324	1265	1217	1173	1135	1104	1076	1048	1025

Polymer (B): **poly(ethylene-*co*-1-hexene)** **2001DOE2**
Characterization: M_n/g.mol^{-1} = 60000, M_w/g.mol^{-1} = 129000, 16.1 wt% 1-hexene
Solvent (A): **ethene** C_2H_4 **74-85-1**

Type of data: coexistence data

T/K = 433.15

w_B(total)	0.15	was kept constant			
P/bar	974	1149	1245	1363	1373
w_B(sol phase)	0.009	0.020	0.029	0.039	–
w_B(gel phase)	0.485	0.390	0.320	–	0.150

Polymer (B): **poly(ethylene-*co*-methacrylic acid)** **2004LAT**
Characterization: 7.0 mol% methacrylic acid, synthesized in the laboratory
Solvent (A): **ethene** C_2H_4 **74-85-1**

Type of data: cloud points

w_B 0.05 was kept constant

T/K	488.15	493.65	494.65	501.15	502.15	505.15	506.15	512.15	518.15
P/bar	2620	2408	2393	2203	2198	2109	2089	1980	1872

T/K	522.15	527.15
P/bar	1807	1707

Polymer (B): **poly(ethylene-*co*-methacrylic acid)** **2004LAT**
Characterization: 7.2 mol% methacrylic acid, synthesized in the laboratory
Solvent (A): **ethene** C_2H_4 **74-85-1**

Type of data: cloud points

w_B 0.05 was kept constant

T/K	495.55	496.35	497.35	502.25	502.65	503.35	507.15	507.35	507.85
P/bar	2380	2485	2366	2252	2257	2224	2160	2151	2137

T/K	513.95	514.35	515.45	518.45	519.45	520.15	524.15	529.15	533.15
P/bar	2032	2014	2001	1951	1943	1981	1912	1837	1780

Polymer (B): **poly(ethylene-*co*-methyl acrylate-*co*-vinyl acetate)** **2007TUM**
Characterization: M_w/g.mol^{-1} = 110000, 23 mol% methyl acrylate, 3.5 mol% vinyl acetate, synthesized in the laboratory
Solvent (A): **ethene** C_2H_4 **74-85-1**

Type of data: cloud points

w_B 0.03 was kept constant

T/K	383	393	403	413	423	433	443	453	463
P/bar	1880	1810	1710	1650	1600	1560	1490	1460	1410

T/K	473	483	493
P/bar	1360	1340	1320

Polymer (B): **poly(ethylene-*co*-methyl acrylate-*co*-vinyl acetate)** **2007TUM**
Characterization: M_w/g.mol^{-1} = 110000, 35 mol% methyl acrylate, 3.5 mol% vinyl acetate, synthesized in the laboratory
Solvent (A): **ethene** C_2H_4 **74-85-1**

Type of data: cloud points

continued

continued

w_B 0.03 was kept constant

T/K	373	383	393	403	413	423	433	443	453
P/bar	2040	1960	1870	1780	1720	1660	1620	1560	1500
T/K	463	473	483	493	503				
P/bar	1450	1410	1380	1340	1310				

Polymer (B): **poly(ethylene-*co*-methyl acrylate-*co*-vinyl acetate)** **2007TUM**
Characterization: M_w/g.mol^{-1} = 110000, 40 mol% methyl acrylate, 3.5 mol% vinyl acetate, synthesized in the laboratory
Solvent (A): **ethene** C_2H_4 **74-85-1**

Type of data: cloud points

w_B 0.03 was kept constant

T/K	383	393	403	413	423	433	443	453	463
P/bar	2120	2020	1920	1870	1800	1740	1680	1640	1590
T/K	473	483	493						
P/bar	1550	1510	1490						

Polymer (B): **poly(ethylene-*co*-methyl methacrylate)** **2004LAT**
Characterization: 9.6 mol% methyl methacrylate, synthesized in the laboratory
Solvent (A): **ethene** C_2H_4 **74-85-1**

Type of data: cloud points

w_B 0.05 was kept constant

T/K	372.45	393.25	402.55	412.05	422.45	432.05	451.65	472.35	491.85
P/bar	1510	1369	1312	1290	1255	1227	1185	1145	1109
T/K	511.55	527.05							
P/bar	1076	1063							

Polymer (B): **poly(ethylene-*co*-methyl methacrylate)** **2004BEC, 2004LAT**
Characterization: M_n/g.mol^{-1} = 41400, M_w/g.mol^{-1} = 67100, 16.9 mol% methyl methacrylate, synthesized in the laboratory
Solvent (A): **ethene** C_2H_4 **74-85-1**

Type of data: cloud points

w_B 0.05 was kept constant

T/K	383.15	393.15	413.15	433.15	453.15	473.15
P/bar	1465	1435	1354	1292	1237	1208

Polymer (B): **poly(ethylene-*co*-methyl methacrylate) 2004BEC, 2004LAT**
Characterization: M_n/g.mol^{-1} = 10900, M_w/g.mol^{-1} = 20000,
18.5 mol% methyl methacrylate, synthesized in the laboratory
Solvent (A): **ethene** **C_2H_4** **74-85-1**

Type of data: cloud points

w_B 0.05 was kept constant

T/K	353.15	373.15	393.15	413.15	433.15	453.15	473.15	488.15	508.15
P/bar	1430	1309	1252	1206	1176	1137	1108	1086	1057

T/K	528.15
P/bar	1039

Polymer (B): **poly(ethylene-*co*-methyl methacrylate) 2004LAT**
Characterization: 29.8 mol% methyl methacrylate, synthesized in the laboratory
Solvent (A): **ethene** **C_2H_4** **74-85-1**

Type of data: cloud points

w_B 0.05 was kept constant

T/K	370.25	390.35	411.55	430.85	450.75	470.25	490.45	509.45	529.45
P/bar	1371	1301	1246	1200	1161	1130	1108	1076	1049

Polymer (B): **poly(ethylene-*co*-methyl methacrylate) 2004BEC, 2004LAT**
Characterization: M_n/g.mol^{-1} = 49800, M_w/g.mol^{-1} = 83500,
35.1 mol% methyl methacrylate, synthesized in the laboratory
Solvent (A): **ethene** **C_2H_4** **74-85-1**

Type of data: cloud points

w_B 0.05 was kept constant

T/K	393.15	413.15	433.15	453.15	473.15	488.15	508.15
P/bar	1500	1419	1341	1286	1238	1202	1152

Polymer (B): **poly(ethylene-*co*-methyl methacrylate) 2004BEC, 2004LAT**
Characterization: M_n/g.mol^{-1} = 28600, M_w/g.mol^{-1} = 52600,
41.5 mol% methyl methacrylate, synthesized in the laboratory
Solvent (A): **ethene** **C_2H_4** **74-85-1**

Type of data: cloud points

w_B 0.05 was kept constant

T/K	393.15	403.15	413.15	423.15	443.15	463.15	473.15	483.15	493.15
P/bar	1673	1585	1533	1478	1378	1306	1271	1242	1210

Polymer (B): **poly(ethylene-*co*-propyl methacrylate)** **2004BEC, 2004LAT**
Characterization: 6.9 mol% propyl acrylate, synthesized in the laboratory
Solvent (A): **ethene** C_2H_4 **74-85-1**

Type of data: cloud points

w_B 0.05 was kept constant

T/K	366.45	376.15	386.65	393.15	413.15	423.15	442.15	452.15	472.15
P/bar	1527	1476	1433	1408	1338	1310	1262	1244	1204

T/K	491.15	511.15	531.15
P/bar	1173	1141	1115

Polymer (B): **poly(ethylene-*co*-propyl methacrylate)** **2004BEC, 2004LAT**
Characterization: M_n/g.mol^{-1} = 83000, M_w/g.mol^{-1} = 147300,
14.1 mol% propyl acrylate, synthesized in the laboratory
Solvent (A): **ethene** C_2H_4 **74-85-1**

Type of data: cloud points

w_B 0.05 was kept constant

T/K	353.15	363.15	373.65	397.85	415.45	434.15	452.15	473.15	492.15
P/bar	1380	1330	1297	1234	1192	1150	1127	1098	1074

T/K	509.15	532.15
P/bar	1056	1035

Polymer (B): **poly(ethylene-*co*-propyl methacrylate)** **2004BEC, 2004LAT**
Characterization: M_n/g.mol^{-1} = 54200, M_w/g.mol^{-1} = 112400,
18.9 mol% propyl acrylate, synthesized in the laboratory
Solvent (A): **ethene** C_2H_4 **74-85-1**

Type of data: cloud points

w_B 0.05 was kept constant

T/K	367.15	373.15	393.15	412.15	431.15	451.15	468.15	491.55	512.15
P/bar	1250	1226	1183	1153	1124	1097	1077	1055	1031

T/K	529.85
P/bar	1018

Polymer (B): **poly(ethylene-*co*-propyl methacrylate)** **2004BEC, 2004LAT**
Characterization: M_n/g.mol^{-1} = 58100, M_w/g.mol^{-1} = 127000,
26.2 mol% propyl acrylate, synthesized in the laboratory
Solvent (A): **ethene** C_2H_4 **74-85-1**

Type of data: cloud points

continued

continued

w_B	0.05	was kept constant						
T/K	354.15	374.15	392.15	413.15	433.15	451.15	471.15	491.15	512.15
P/bar	1208	1182	1147	1112	1083	1062	1038	1017	995

T/K	530.15
P/bar	978

Polymer (B): **poly(ethylene-*co*-propyl methacrylate)** **2004LAT**
Characterization: M_n/g.mol^{-1} = 78400, M_w/g.mol^{-1} = 128600, 31.2 mol% propyl acrylate, synthesized in the laboratory
Solvent (A): **ethene** C_2H_4 **74-85-1**

Type of data: cloud points

w_B 0.05 was kept constant

T/K	351.55	362.45	373.15	393.15	412.35	432.65	452.15	472.65	492.15
P/bar	1215	1186	1163	1129	1100	1069	1046	1024	1003

T/K	510.15	530.15
P/bar	985	964

Polymer (B): **poly(ethylene-*co*-propyl methacrylate)** **2004LAT**
Characterization: M_n/g.mol^{-1} = 49000, M_w/g.mol^{-1} = 126200, 37.4 mol% propyl acrylate, synthesized in the laboratory
Solvent (A): **ethene** C_2H_4 **74-85-1**

Type of data: cloud points

w_B 0.05 was kept constant

T/K	351.45	365.15	375.15	391.15	412.15	434.15	453.15	471.65	492.15
P/bar	1218	1193	1172	1139	1106	1072	1050	1029	1008

T/K	512.15	528.15
P/bar	987	974

Polymer (B): **poly(ethylene-*co*-vinyl acetate)** **2001DOE2**
Characterization: M_n/g.mol^{-1} = 61900, M_w/g.mol^{-1} = 167000, 27.5 wt% vinyl acetate
Solvent (A): **ethene** C_2H_4 **74-85-1**

Type of data: coexistence data

T/K = 433.15

w_B(total)	0.28	was kept constant						
P/bar	622	633	723	737	760	786	854	872
w_B(gel phase)	–	0.61	–	0.57	–	0.52	0.50	–
w_B(sol phase)	0.06	–	0.06	–	0.07	–	–	0.08

continued

continued

P/bar	920	955	1066					
w_B(gel phase)	–	0.40	0.28					
w_B(sol phase)	0.10	–	–					
w_B(total)	0.20	was kept constant						
P/bar	868	966	1064	1098	1136			
w_B(gel phase)	0.508	0.437	0.350	0.297	0.200			
w_B(sol phase)	0.014	0.028	0.044	0.049	–			
w_B(total)	0.15	was kept constant						
P/bar	820	884	1040	1161	1174			
w_B(gel phase)	0.523	0.477	0.360	0.240	0.150			
w_B(sol phase)	0.009	0.012	0.029	0.052	–			
w_B(total)	0.10	was kept constant						
P/bar	915	940	977	1009	1038	1056	1095	1121
w_B(gel phase)	–	0.48	–	0.42	–	0.37	–	0.30
w_B(sol phase)	0.01	–	0.015	–	0.02	–	0.03	–
P/bar	1130	1170	1192					
w_B(gel phase)	–	0.23	–					
w_B(sol phase)	0.045	–	0.098					

Polymer (B): **poly(ethylene glycol)** **2006BY4**
Characterization: M_w/g.mol^{-1} = 400, SFC Co., Yeosu, South Korea
Solvent (A): **1-butene** C_4H_8 **106-98-9**

Type of data: cloud points

w_B 0.05 was kept constant

T/K	396.05	396.55	397.15	398.35	401.75	408.45	423.85	443.65
P/bar	1549.3	1150.7	781.0	528.6	330.0	216.9	138.3	113.4

Polymer (B): **poly(ethylene glycol)** **2006BY4**
Characterization: M_w/g.mol^{-1} = 600, SFC Co., Yeosu, South Korea
Solvent (A): **1-butene** C_4H_8 **106-98-9**

Type of data: cloud points

w_B 0.05 was kept constant

T/K	397.35	398.05	399.35	402.25	408.55	423.85	443.75
P/bar	1477.6	1044.5	746.5	492.8	305.2	229.3	171.4

Polymer (B): **poly(ethylene glycol)** **2006BY4**
Characterization: M_w/g.mol^{-1} = 1000, SFC Co., Yeosu, South Korea
Solvent (A): **1-butene** C_4H_8 **106-98-9**

Type of data: cloud points

continued

continued

w_B 0.05 was kept constant

T/K	396.15	396.55	397.35	398.65	401.25	405.65	423.25	444.55
P/bar	2216.9	1789.3	1501.0	1163.1	885.9	677.6	432.8	320.4

Polymer (B): **poly(ethylene glycol)** **2006BY4**
Characterization: M_w/g.mol^{-1} = 2000, SFC Co., Yeosu, South Korea
Solvent (A): **1-butene** C_4H_8 **106-98-9**

Type of data: cloud points

w_B 0.05 was kept constant

T/K	397.85	398.85	400.95	403.15	410.15	424.55	443.95
P/bar	2172.8	1827.9	1483.1	1116.2	901.0	699.7	534.8

Polymer (B): **poly(ethylene glycol)** **2006BY4**
Characterization: M_w/g.mol^{-1} = 4000, SFC Co., Yeosu, South Korea
Solvent (A): **1-butene** C_4H_8 **106-98-9**

Type of data: cloud points

w_B 0.05 was kept constant

T/K	407.25	410.25	411.85	414.35	418.95	424.35	443.95
P/bar	2141.4	1756.9	1603.1	1427.9	1146.8	1001.7	714.5

Polymer (B): **poly(ethylene glycol)** **2006BY4**
Characterization: M_w/g.mol^{-1} = 200, SFC Co., Yeosu, South Korea
Solvent (A): **carbon dioxide** CO_2 **124-38-9**

Type of data: cloud points

w_B 0.05 was kept constant

T/K	310.45	313.55	324.35	333.35	354.25	374.75	393.55
P/bar	1244.5	1079.0	781.0	651.4	512.2	480.4	459.7

Polymer (B): **poly(ethylene glycol)** **2006BY4**
Characterization: M_w/g.mol^{-1} = 300, SFC Co., Yeosu, South Korea
Solvent (A): **carbon dioxide** CO_2 **124-38-9**

Type of data: cloud points

w_B 0.05 was kept constant

T/K	315.05	323.75	338.45	353.95	373.25	393.95
P/bar	936.9	816.9	712.8	642.4	599.7	556.2

Polymer (B): **poly(ethylene glycol)** **2006BY4**
Characterization: M_w/g.mol^{-1} = 400, SFC Co., Yeosu, South Korea
Solvent (A): **carbon dioxide** CO_2 **124-38-9**

Type of data: cloud points

w_B 0.05 was kept constant

T/K	306.55	310.05	313.75	324.35	339.05	354.65	374.85	393.95
P/bar	1163.1	1079.0	1026.6	903.8	813.5	737.6	678.2	666.5

Polymer (B): **poly(ethylene glycol)** **2006BY4**
Characterization: M_w/g.mol^{-1} = 600, SFC Co., Yeosu, South Korea
Solvent (A): **carbon dioxide** CO_2 **124-38-9**

Type of data: cloud points

w_B 0.05 was kept constant

T/K	308.45	310.15	314.75	323.65	338.95	354.25	373.75
P/bar	1779.7	1698.3	1525.9	1281.7	1055.5	950.0	847.9

Polymer (B): **poly(ethylene glycol)** **2006BY4**
Characterization: M_w/g.mol^{-1} = 960,
Scientific Polymer Products, Inc., Ontario, NY
Solvent (A): **carbon dioxide** CO_2 **124-38-9**

Type of data: cloud points

w_B 0.05 was kept constant

T/K	353.45	363.45	373.15	382.85	394.75	404.85	414.25
P/bar	1158.3	1108.6	1065.5	1036.2	997.9	960.3	928.3

Polymer (B): **poly(ethylene glycol)** **2006BY4**
Characterization: M_w/g.mol^{-1} = 1000, SFC Co., Yeosu, South Korea
Solvent (A): **carbon dioxide** CO_2 **124-38-9**

Type of data: cloud points

w_B 0.05 was kept constant

T/K	310.75	316.15	328.95	344.35	363.95	383.95	404.25	423.35
P/bar	1764.5	1663.8	1459.0	1305.2	1200.3	1132.8	1077.6	1032.8

Polymer (B): **poly(ethylene glycol)** **2006BY4**
Characterization: M_w/g.mol^{-1} = 2000, SFC Co., Yeosu, South Korea
Solvent (A): **carbon dioxide** CO_2 **124-38-9**

Type of data: cloud points

w_B 0.05 was kept constant

T/K	336.75	338.75	342.65	349.95	364.35	384.75	404.25	426.65
P/bar	2363.1	2310.7	2232.1	2121.7	1912.1	1703.8	1557.6	1498.3

Polymer (B): **poly(ethylene glycol)** **2006BY4**
Characterization: M_w/g.mol^{-1} = 4000, SFC Co., Yeosu, South Korea
Solvent (A): **carbon dioxide** **CO_2** **124-38-9**

Type of data: cloud points

w_B 0.05 was kept constant

T/K	407.45	410.15	415.25	422.15	429.05	448.45	468.45
P/bar	2254.1	2214.1	2147.9	2072.1	2003.1	1858.3	1787.2

Polymer (B): **poly(ethylene glycol)** **2006BY4**
Characterization: M_w/g.mol^{-1} = 600, SFC Co., Yeosu, South Korea
Solvent (A): **propene** **C_3H_6** **115-07-1**

Type of data: cloud points

w_B 0.05 was kept constant

T/K	386.35	387.05	388.55	390.65	393.75	397.45	405.05	410.35	423.55
P/bar	2050.0	1605.9	1284.5	1011.4	801.7	667.9	532.8	456.9	377.6

T/K	432.65	443.85
P/bar	332.1	311.4

Polymer (B): **poly(ethylene glycol)** **2006BY4**
Characterization: M_w/g.mol^{-1} = 1000, SFC Co., Yeosu, South Korea
Solvent (A): **propene** **C_3H_6** **115-07-1**

Type of data: cloud points

w_B 0.05 was kept constant

T/K	406.85	407.65	409.75	413.25	418.05	423.05	427.05	443.35	464.45
P/bar	2207.2	1746.6	1386.6	1153.4	987.9	845.9	745.9	671.4	665.9

Polymer (B): **poly(ethylene oxide-*b*-dimethylsiloxane-*b*-ethylene oxide)** **2012STO**
Characterization: M_n/g.mol^{-1} = 1900, 41 wt% PEO, $(EO)_8$-$(PDMS)_{15}$-$(EO)_8$, purified in the laboratory, 2-8692 Fluid, Dow Corning
Solvent (A): **carbon dioxide** **CO_2** **124-38-9**

Type of data: cloud points

w_B	0.0026	0.0026	0.0026	0.0026	0.0026	0.0026	0.0026	0.0026	0.0026
T/K	294.1	298.5	303.9	308.7	313.3	318.3	322.8	327.9	332.1
P/MPa	8.57	10.41	12.15	14.04	15.79	17.31	18.58	19.89	21.25

w_B	0.0408	0.0408	0.0408	0.0408	0.0408	0.0408	0.0408	0.0408	0.0408
T/K	293.3	296.9	299.8	303.3	306.9	307.9	311.1	312.8	317.1
P/MPa	23.5	25.4	25.7	26.7	27.3	27.6	28.2	28.9	29.9

continued

continued

w_B	0.0408	0.0408	0.0408	0.0408	0.0408	0.0408	0.0408	0.0408	0.0408
T/K	317.6	321.4	322.1	326.1	326.6	330.9	331.8	335.5	336.5
P/MPa	30.0	31.1	31.0	32.0	32.3	33.3	32.9	35.1	34.4

Polymer (B): **poly(ethylene oxide-*b*-dimethylsiloxane-*b*-ethylene oxide)** **2012STO**

Characterization: M_n/g.mol^{-1} = 4000, 60 wt% PEO, $(EO)_{25}$-$(PDMS)_{21}$-$(EO)_{25}$, contains < 5 wt% octamethylcyclotetrasiloxane, ABCR® DBE-C25

Solvent (A): **carbon dioxide** **CO_2** **124-38-9**

Type of data: cloud points

w_B 0.0021 was kept constant

T/K	335.6	331.7	326.2	321.6	315.9	312.0	306.7	301.6	297.2
P/MPa	33.52	32.99	31.59	30.39	28.92	28.39	27.85	26.08	26.18

Polymer (B): **poly(ethylene oxide-*b*-propylene oxide-*b*-ethylene oxide)** **2009STO**

Characterization: M_n/g.mol^{-1} = 2000, 10 wt% PEO, L61, $(EO)_2$-$(PO)_{31}$-$(EO)_2$, BASF SE, Germany

Solvent (A): **carbon dioxide** **CO_2** **124-38-9**

Type of data: cloud points

w_B 0.08 was kept constant

T/K	293.1	299.2	303.7	308.4	313.1	318.2	322.3	328.2	332.5
P/MPa	11.5	13.7	15.5	16.9	18.4	19.8	21.1	22.4	23.9

T/K	337.3
P/MPa	25.1

Polymer (B): **poly(ethylene oxide-*b*-propylene oxide-*b*-ethylene oxide)** **2009STO**

Characterization: M_n/g.mol^{-1} = 2750, 10 wt% PEO, L81, $(EO)_3$-$(PO)_{42}$-$(EO)_3$, BASF SE, Germany

Solvent (A): **carbon dioxide** **CO_2** **124-38-9**

Type of data: cloud points

w_B 0.08 was kept constant

T/K	294.1	298.9	303.3	308.1	312.5	318.1	323.6	328.2	332.8
P/MPa	18.4	19.6	20.7	22.2	23.2	24.4	25.8	26.9	28.2

T/K	337.1
P/MPa	29.0

Polymer (B): **poly(ethylene oxide-*b*-propylene oxide-*b*-ethylene oxide)** **2009STO**
Characterization: M_n/g.mol^{-1} = 2500, 20 wt% PEO, L62, $(EO)_5$-$(PO)_{34}$-$(EO)_5$, BASF SE, Germany
Solvent (A): **carbon dioxide** **CO_2** **124-38-9**

Type of data: cloud points

w_B 0.10 was kept constant

T/K	293.4	298.6	303.4	308.3	313.1	319.2	323.3	328.6	335.5
P/MPa	22.7	23.4	24.6	25.6	26.6	27.7	28.6	29.3	31.3

Polymer (B): **poly(ethylene oxide-*b*-1,1,2,2-tetrahydroperfluorodecyl acrylate)** **2004MA2**
Characterization: M_n/g.mol^{-1} = 22800, 10.3 wt% PEO, molar ratio of PFDA/PEO = 39.5/1 by NMR, synthesized in the laboratory
Solvent (A): **carbon dioxide** **CO_2** **124-38-9**

Type of data: cloud points

w_B 0.01 was kept constant

T/K	338.75	333.25	328.35	323.25	318.15	313.25	308.35	303.65	298.55
P/bar	248.2	232.1	216.8	199.5	182.8	165.5	149.0	130.4	108.0

T/K	293.45
P/bar	87.1

Polymer (B): **poly(ethylene oxide-*b*-1,1,2,2-tetrahydroperfluorodecyl acrylate)** **2004MA1**
Characterization: M_n/g.mol^{-1} = 23000, 10.3 wt% PEO, molar ratio of PFDA/PEO = 39.8/1 by NMR, synthesized in the laboratory
Solvent (A): **carbon dioxide** **CO_2** **124-38-9**

Type of data: cloud points

w_B 0.04 was kept constant

T/K	338.15	333.65	328.05	323.15	318.35	313.25	308.35	303.35	298.35
P/bar	241.2	228.8	212.5	196.3	181.3	164.6	148.2	129.9	112.2

T/K	293.15
P/bar	93.6

Polymer (B): **poly(ethylene oxide-*b*-1,1,2,2-tetrahydroperfluorodecyl acrylate)** **2004MA2**
Characterization: M_n/g.mol^{-1} = 23000, 10.3 wt% PEO, molar ratio of PFDA/PEO = 39.8/1 by NMR, synthesized in the laboratory
Solvent (A): **carbon dioxide** **CO_2** **124-38-9**

Type of data: cloud points

w_B 0.01 was kept constant

T/K	338.85	333.95	327.75	323.25	318.45	313.35	308.05	303.65	298.45
P/bar	249.8	234.0	213.0	197.4	181.1	164.9	150.3	131.3	111.5

T/K	293.05
P/bar	93.4

Polymer (B): **poly(2-ethylhexyl acrylate)** **2007LI1**
Characterization: M_w/g.mol^{-1} = 90000, T_g/K = 223, Scientific Polymer Products, Inc., Ontario, NY
Solvent (A): **n-butane** **C_4H_{10}** **106-97-8**

Type of data: cloud points

w_B 0.052 was kept constant

T/K	373.7	393.3	413.5	433.4	453.4
P/MPa	2.35	5.69	8.93	11.90	14.10

Polymer (B): **poly(2-ethylhexyl acrylate)** **2007LI1**
Characterization: M_w/g.mol^{-1} = 90000, T_g/K = 223, Scientific Polymer Products, Inc., Ontario, NY
Solvent (A): **1-butene** **C_4H_8** **106-98-9**

Type of data: cloud points

w_B 0.050 was kept constant

T/K	394.4	412.8	432.9	453.1
P/MPa	4.31	7.41	10.52	12.93

Polymer (B): **poly(2-ethylhexyl acrylate)** **2007LI1**
Characterization: M_w/g.mol^{-1} = 90000, T_g/K = 223, Scientific Polymer Products, Inc., Ontario, NY
Solvent (A): **carbon dioxide** **CO_2** **124-38-9**

Type of data: cloud points

w_B 0.048 was kept constant

T/K	420.9	423.3	434.0	443.8	454.8
P/MPa	248.45	173.41	141.31	129.69	123.10

Polymer (B): **poly(2-ethylhexyl acrylate)** **2007LI1**
Characterization: M_w/g.mol^{-1} = 90000, T_g/K = 223,
Scientific Polymer Products, Inc., Ontario, NY
Solvent (A): **dimethyl ether** C_2H_6O **115-10-6**

Type of data: cloud points

w_B 0.068 was kept constant

T/K	374.3	394.9	414.8	432.4	452.5
P/MPa	4.31	9.19	12.93	16.03	17.59

Polymer (B): **poly(2-ethylhexyl acrylate)** **2007LI1**
Characterization: M_w/g.mol^{-1} = 90000, T_g/K = 223,
Scientific Polymer Products, Inc., Ontario, NY
Solvent (A): **propane** C_3H_8 **74-98-6**

Type of data: cloud points

w_B 0.053 was kept constant

T/K	354.1	372.3	394.2	414.2	433.8	452.5
P/MPa	13.62	16.72	20.52	23.28	25.48	27.41

Polymer (B): **poly(2-ethylhexyl acrylate)** **2007LI1**
Characterization: M_w/g.mol^{-1} = 90000, T_g/K = 223,
Scientific Polymer Products, Inc., Ontario, NY
Solvent (A): **propene** C_3H_6 **115-07-1**

Type of data: cloud points

w_B 0.065 was kept constant

T/K	373.9	394.0	413.5	434.4	453.9
P/MPa	17.07	20.86	23.97	26.72	28.79

Polymer (B): **poly(2-ethylhexyl methacrylate)** **2007LI1**
Characterization: M_w/g.mol^{-1} = 100000, T_g/K = 263,
Scientific Polymer Products, Inc., Ontario, NY
Solvent (A): **n-butane** C_4H_{10} **106-97-8**

Type of data: cloud points

w_B 0.049 was kept constant

T/K	377.1	395.6	415.9	435.7	455.1
P/MPa	4.66	7.41	10.52	13.28	15.35

Polymer (B): **poly(2-ethylhexyl methacrylate)** **2007LI1**
Characterization: M_w/g.mol^{-1} = 100000, T_g/K = 263,
Scientific Polymer Products, Inc., Ontario, NY
Solvent (A): **1-butene** C_4H_8 **106-98-9**

Type of data: cloud points

w_B 0.054 was kept constant

T/K	374.7	393.6	415.4	435.4	455.4
P/MPa	2.93	6.03	9.83	12.59	15.00

Polymer (B): **poly(2-ethylhexyl methacrylate)** **2007LI1**
Characterization: M_w/g.mol^{-1} = 100000, T_g/K = 263,
Scientific Polymer Products, Inc., Ontario, NY
Solvent (A): **carbon dioxide** CO_2 **124-38-9**

Type of data: cloud points

w_B 0.059 was kept constant

T/K	419.9	434.8	445.3	456.7	466.2	477.7
P/MPa	247.07	161.38	141.21	132.24	127.90	126.48

Polymer (B): **poly(2-ethylhexyl methacrylate)** **2007LI1**
Characterization: M_w/g.mol^{-1} = 100000, T_g/K = 263,
Scientific Polymer Products, Inc., Ontario, NY
Solvent (A): **dimethyl ether** C_2H_6O **115-10-6**

Type of data: cloud points

w_B 0.043 was kept constant

T/K	375.5	395.9	415.8	434.5	455.9
P/MPa	6.38	10.35	14.31	18.10	20.69

Polymer (B): **poly(2-ethylhexyl methacrylate)** **2007LI1**
Characterization: M_w/g.mol^{-1} = 100000, T_g/K = 263,
Scientific Polymer Products, Inc., Ontario, NY
Solvent (A): **propane** C_3H_8 **74-98-6**

Type of data: cloud points

w_B 0.052 was kept constant

T/K	335.6	354.2	372.9	396.8	415.5	434.2	454.3
P/MPa	14.31	18.10	20.86	23.97	26.72	28.79	30.17

Polymer (B): **poly(2-ethylhexyl methacrylate)** **2007LI1**
Characterization: M_w/g.mol^{-1} = 100000, T_g/K = 263,
Scientific Polymer Products, Inc., Ontario, NY
Solvent (A): **propene** C_3H_6 **115-07-1**

Type of data: cloud points

w_B 0.051 was kept constant

T/K	334.5	356.0	375.9	393.2	415.6	436.8	453.4
P/MPa	11.48	16.17	20.17	23.55	26.69	29.31	31.21

Polymer (B): **polyglycerol (hyperbranched)** **2010GRE**
Characterization: M_n/g.mol^{-1} = 4400, OH-endgroups are esterified
to -O(CO)CH_3
Solvent (A): **carbon dioxide** CO_2 **124-38-9**

Type of data: cloud points

w_B	0.0952	0.0952	0.0952	0.0952	0.0952	0.0952	0.0952	0.0952	0.0952
T/K	422.74	422.74	422.75	402.32	402.32	402.33	381.65	381.66	381.67
P/MPa	109.86	110.52	110.73	105.98	106.23	106.57	99.18	99.51	99.39
w_B	0.0952	0.0952	0.0952	0.0952	0.0952	0.0952	0.0952	0.0952	0.0952
T/K	360.99	360.99	361.00	340.43	340.44	340.44	320.12	320.13	320.13
P/MPa	92.24	92.57	92.84	85.39	85.12	85.50	78.20	78.35	78.58
w_B	0.0952	0.0952	0.0952	0.1369	0.1369	0.1369	0.1369	0.1369	0.1369
T/K	303.42	303.42	303.43	422.65	422.65	422.66	402.20	402.21	402.19
P/MPa	73.27	72.84	73.46	108.49	107.69	106.67	103.50	102.65	101.82
w_B	0.1369	0.1369	0.1369	0.1369	0.1369	0.1369	0.1369	0.1369	0.1369
T/K	381.74	381.75	381.75	360.77	360.79	360.81	340.20	340.22	340.25
P/MPa	96.49	95.52	95.02	89.54	88.55	88.21	82.32	81.73	81.31
w_B	0.1369	0.1369	0.1369	0.1369	0.1369	0.1369	0.1615	0.1615	0.1615
T/K	319.65	319.62	319.60	302.34	302.37	302.35	423.06	423.07	423.08
P/MPa	75.42	75.09	74.85	68.80	69.48	69.73	103.03	102.89	102.56
w_B	0.1615	0.1615	0.1615	0.1615	0.1615	0.1615	0.1615	0.1615	0.1615
T/K	402.28	402.27	402.28	381.66	381.65	381.65	360.45	360.46	360.46
P/MPa	96.56	96.88	97.17	90.24	90.54	90.75	83.14	82.85	83.10
w_B	0.1615	0.1615	0.1615	0.1615	0.1615	0.1615	0.1615	0.1615	0.1615
T/K	340.24	340.24	340.25	320.08	320.09	320.10	303.13	303.14	303.14
P/MPa	77.26	76.58	76.80	69.17	69.43	69.78	62.98	63.26	63.43

Polymer (B): **polyglycerol (hyperbranched)** **2010GRE**
Characterization: M_n/g.mol^{-1} = 9800, OH-endgroups are esterified to -O(CO)$(CH_2)_5CH_3$
Solvent (A): **carbon dioxide** CO_2 **124-38-9**

Type of data: cloud points

w_B 0.0960 was kept constant

T/K	422.93	422.94	422.95	412.76	412.77	412.77	402.50	402.51	402.52
P/MPa	91.37	90.85	91.03	88.90	88.59	88.87	85.94	85.57	85.71
T/K	392.30	392.30	392.31	382.05	382.06	382.07	371.37	371.37	371.38
P/MPa	82.84	83.18	82.99	80.30	79.68	79.44	80.56	80.22	79.84
T/K	361.09	361.11	361.10	350.77	350.77	350.78	340.47	340.48	340.49
P/MPa	80.27	80.03	79.95	80.85	81.08	81.25	81.87	82.04	81.77
T/K	330.11	330.13	330.15	319.94	319.93	319.95	309.47	309.47	309.48
P/MPa	82.87	83.03	83.44	90.28	90.61	98.92	104.58	104.25	103.84

Polymer (B): **polyglycerol (hyperbranched)** **2010GRE**
Characterization: M_n/g.mol^{-1} = 13500, OH-endgroups are esterified to -O(CO)$(CH_2)_{10}CH_3$
Solvent (A): **carbon dioxide** CO_2 **124-38-9**

Type of data: cloud points

w_B	0.0229	0.0229	0.0229	0.0229	0.0229	0.0229	0.0229	0.0229	0.0229
T/K	423.08	423.07	412.86	412.86	412.87	402.64	402.64	402.65	392.40
P/MPa	105.44	106.14	108.67	109.11	108.97	116.23	116.67	115.95	134.78
w_B	0.0229	0.0229	0.0229	0.0229	0.0229	0.0229	0.0972	0.0972	0.0972
T/K	392.39	392.40	385.26	385.27	385.27	386.63	422.97	423.00	422.95
P/MPa	134.31	135.15	168.57	167.25	167.93	197.17	115.29	114.37	114.83
w_B	0.0972	0.0972	0.0972	0.0972	0.0972	0.0972	0.0972	0.0972	0.0972
T/K	412.76	412.75	412.75	402.52	402.54	402.52	392.28	392.28	392.29
P/MPa	117.31	116.93	116.69	124.05	123.72	124.40	138.79	139.03	139.39
w_B	0.0972	0.0972	0.0972	0.0972	0.0972	0.0972	0.3171	0.3171	0.3171
T/K	385.15	385.16	385.17	383.17	383.18	383.17	402.75	402.79	402.79
P/MPa	162.79	163.87	163.58	193.86	194.77	196.78	80.14	80.54	80.77
w_B	0.3171	0.3171	0.3171	0.3171	0.3171	0.3171	0.3171	0.3171	0.3171
T/K	382.22	382.24	382.26	361.26	361.26	361.25	350.99	350.98	350.97
P/MPa	84.16	83.67	83.95	95.40	95.77	96.27	132.78	133.17	133.52

Polymer (B): **polyglycerol (hyperbranched)** **2010GRE**
Characterization: M_n/g.mol^{-1} = 17950, OH-endgroups are esterified to -O(CO)$(CH_2)_{16}CH_3$
Solvent (A): **carbon dioxide** **CO_2** **124-38-9**

Type of data: cloud points

w_B	0.0931	0.0931	0.0931	0.0931	0.0931	0.0931	0.0931	0.0931	0.0931
T/K	422.65	422.65	422.64	417.59	417.59	417.60	415.44	415.44	415.45
P/MPa	163.46	164.32	163.99	178.86	179.90	178.36	196.13	197.55	196.77
w_B	0.3363	0.3363	0.3363	0.3363	0.3363	0.3363	0.3363	0.3363	0.3363
T/K	420.50	420.51	420.52	400.07	400.08	400.10	379.67	379.63	379.68
P/MPa	79.57	79.20	78.74	87.42	87.05	87.22	112.63	112.11	112.35
w_B	0.3363	0.3363	0.3363	0.3363					
T/K	374.17	374.16	374.17	375.52					
P/MPa	146.12	146.79	177.17	190.64					

Polymer (B): **polyglycerol (hyperbranched)** **2010GRE**
Characterization: M_n/g.mol^{-1} = 17950, OH-endgroups are esterified to -O(CO)$(CH_2)_{16}CH_3$
Solvent (A): **ethane** **C_2H_6** **74-84-0**

Type of data: cloud points

w_B 0.0762 was kept constant

T/K	310.33	310.32	310.32	320.37	320.42	320.56	330.53	330.83	330.83
P/MPa	35.20	35.07	34.88	36.77	36.72	36.70	38.61	38.78	38.74
T/K	341.24	341.18	341.15	351.46	351.44	351.42	361.77	361.77	361.75
P/MPa	40.38	40.42	40.30	41.73	41.68	41.78	42.83	42.94	42.95
T/K	376.22	376.15	376.12						
P/MPa	44.40	44.78	44.56						

Polymer (B): **polyglycerol (hyperbranched)** **2010GRE**
Characterization: M_n/g.mol^{-1} = 7650, OH-endgroups are modified to -OSi$(CH_3)_3$
Solvent (A): **ethane** **C_2H_6** **74-84-0**

Type of data: cloud points

w_B 0.0765 was kept constant

T/K	281.10	281.05	281.01	300.04	300.06	300.06	320.61	320.59	320.59
P/MPa	24.57	24.55	24.50	29.02	29.24	29.07	32.57	32.51	32.54
T/K	340.96	341.03	341.06	361.50	361.55	361.60	375.92	375.80	375.74
P/MPa	36.16	36.31	36.38	38.91	38.87	39.01	39.93	40.28	40.32

Polymer (B): **polyglycerol (hyperbranched)** **2010GRE**
Characterization: M_n/g.mol^{-1} = 7650, OH-endgroups are modified to -OSi(CH$_3$)$_3$
Solvent (A): **propane** **C_3H_8** **74-98-6**

Type of data: cloud points

w_B 0.0756 was kept constant

T/K	331.08	331.16	331.18	341.44	341.52	341.59	351.72	351.75	351.74
P/MPa	2.37	2.46	2.39	4.59	4.71	4.67	6.65	6.69	6.73
T/K	362.16	362.14	362.15	372.48	372.47	372.48	382.92	382.85	382.89
P/MPa	8.56	8.57	8.58	10.33	10.38	10.40	12.17	12.40	12.28
T/K	396.02								
P/MPa	14.47								

Polymer (B): **poly(heptadecafluorodecyl acrylate)** **2006SHI**
Characterization: synthesized in the laboratory
Solvent (A): **carbon dioxide** **CO_2** **124-38-9**

Type of data: cloud points

w_B	0.0094	0.0290	0.0499	0.0737	0.0927	0.0094	0.0290	0.0499	0.0737
T/K	303.67	303.67	303.67	303.67	303.67	308.45	308.45	308.45	308.45
P/MPa	8.18	9.05	9.34	10.24	9.74	8.96	10.41	10.92	11.75
w_B	0.0927	0.0094	0.0290	0.0499	0.0737	0.0927	0.0094	0.0290	0.0499
T/K	308.45	313.65	313.65	313.65	313.65	313.65	318.53	318.53	318.53
P/MPa	11.30	10.34	12.23	12.79	13.61	13.10	11.85	13.47	14.16
w_B	0.0737	0.0927	0.0094	0.0290	0.0499	0.0737	0.0927	0.0094	0.0290
T/K	318.53	318.53	323.47	323.47	323.47	323.47	323.47	328.48	328.48
P/MPa	15.16	14.78	13.39	15.40	15.65	16.68	16.20	14.63	16.58
w_B	0.0499	0.0737	0.0927	0.0094	0.0290	0.0499	0.0737	0.0927	0.0094
T/K	328.48	328.48	328.48	333.54	333.54	333.54	333.54	333.54	338.18
P/MPa	16.99	17.92	17.63	16.16	17.89	17.95	19.40	18.92	17.54
w_B	0.0290	0.0499	0.0737	0.0927	0.0094	0.0290	0.0499	0.0737	0.0927
T/K	338.18	338.18	338.18	338.18	343.54	343.54	343.54	343.54	343.54
P/MPa	19.09	19.40	20.44	20.23	18.64	20.09	20.76	21.99	21.61
w_B	0.0094	0.0290	0.0499	0.0737	0.0927	0.0094	0.0290	0.0499	0.0737
T/K	348.23	348.23	348.23	348.23	348.23	353.41	353.41	353.41	353.41
P/MPa	19.75	21.92	22.05	22.88	22.71	20.99	22.81	22.99	24.16
w_B	0.0927	0.0094	0.0290	0.0499	0.0737	0.0927	0.0094	0.0290	0.0499
T/K	353.41	358.07	358.07	358.07	358.07	358.07	363.30	363.30	363.30
P/MPa	23.85	22.47	23.85	24.05	25.29	24.94	23.50	25.16	25.43
w_B	0.0737	0.0927							
T/K	363.30	363.30							
P/MPa	26.41	25.98							

Polymer (B): **poly(heptadecafluorodecyl methacrylate)** **2008SH1**
Characterization: inherent viscosity in hexafluoroisopropanol at 307.2 K and 0.227 g/dL = 1.217, synthesized in the laboratory
Solvent (A): **carbon dioxide** **CO_2** **124-38-9**

Type of data: cloud points

w_B	0.0095	0.0285	0.0488	0.0695	0.0904
T/K = 303.7					
P/MPa	9.13	9.34	9.68	10.15	10.36
T/K = 308.6					
P/MPa	10.68	11.17	11.65	11.68	12.04
T/K = 313.7					
P/MPa	12.41	12.79	13.20	13.47	13.66
T/K = 318.6					
P/MPa	13.78	14.61	14.89	15.01	15.09
T/K = 323.5					
P/MPa	15.06	15.85	16.51	16.61	16.80
T/K = 328.5					
P/MPa	16.27	17.08	18.06	18.13	18.14
T/K = 333.5					
P/MPa	18.16	18.45	19.37	19.37	19.30
T/K = 338.5					
P/MPa	19.16	19.68	20.44	20.78	20.88
T/K = 343.4					
P/MPa	20.40	21.02	21.78	22.00	22.05
T/K = 348.2					
P/MPa	21.99	22.26	23.02	23.26	23.17
T/K = 353.5					
P/MPa	23.19	23.12	24.29	24.57	24.62
T/K = 358.2					
P/MPa	24.29	24.64	25.36	25.50	25.56
T/K = 363.2					
P/MPa	25.14	26319	26.29	26.57	26.41

Polymer (B): **poly(heptadecafluorodecyl methacrylate)** **2008SH1**
Characterization: inherent viscosity in hexafluoroisopropanol at 307.2 K and 0.5 g/dL = 0.130, synthesized in the laboratory
Solvent (A): **carbon dioxide** **CO_2** **124-38-9**

Type of data: cloud points

w_B 0.05 was kept constant

T/K	303.6	308.8	313.9	318.8	323.7	328.5	333.3	338.3	343.3
P/MPa	12.00	13.86	15.68	17.12	18.82	20.30	21.75	23.06	24.37

T/K	348.2	353.4	358.6	364.0
P/MPa	25.61	26.82	28.06	29.23

Polymer (B): **poly(heptadecafluorodecyl methacrylate)** **2008SH1**
Characterization: inherent viscosity in hexafluoroisopropanol at 307.2 K and 0.5 g/dL = 0.111, synthesized in the laboratory
Solvent (A): **carbon dioxide** **CO_2** **124-38-9**

Type of data: cloud points

w_B 0.05 was kept constant

T/K	303.7	308.3	313.8	318.5	323.4	328.6	333.5	338.8	343.3
P/MPa	11.75	13.10	15.20	16.82	18.18	19.85	21.37	22.74	23.81

T/K	348.4	353.2	358.8	363.4
P/MPa	25.36	26.22	27.57	28.12

Polymer (B): **poly(heptadecafluorodecyl methacrylate)** **2008SH1**
Characterization: inherent viscosity in hexafluoroisopropanol at 307.2 K and 0.5 g/dL = 0.070, synthesized in the laboratory
Solvent (A): **carbon dioxide** **CO_2** **124-38-9**

Type of data: cloud points

w_B 0.05 was kept constant

T/K	303.5	308.5	313.5	318.2	323.8	328.6	333.5	338.5	343.7
P/MPa	10.99	12.99	14.68	16.23	18.16	19.30	20.71	22.05	23.43

T/K	348.6	353.5	358.8	363.5
P/MPa	24.50	25.74	26.84	27.88

Polymer (B): **poly(heptadecafluorodecyl methacrylate)** **2008SH1**
Characterization: inherent viscosity in hexafluoroisopropanol at 307.2 K and 0.5 g/dL = 0.049, synthesized in the laboratory
Solvent (A): **carbon dioxide** **CO_2** **124-38-9**

Type of data: cloud points

continued

continued

w_B 0.05 was kept constant

T/K	303.4	308.5	313.4	318.5	323.1	328.4	333.3	338.3	343.4
P/MPa	9.68	11.65	13.20	14.89	16.51	18.06	19.37	20.44	21.78

T/K	348.3	353.3	358.4	363.0
P/MPa	23.02	24.29	25.36	26.29

Polymer (B): **poly(2-hydroxypropyl acrylate)** **2012YAN**
Characterization: M_w/g.mol^{-1} = 26000, M_w/M_n = 2.47, T_g/K = 266,
Scientific Polymer Products, Inc., Ontario, NY
Solvent (A): **chlorodifluoromethane** **$CHClF_2$** **75-45-6**

Type of data: cloud points

w_B 0.049 was kept constant

T/K	461.3	466.5	473.4	482.9
P/MPa	236.7	215.3	202.2	189.8

Polymer (B): **poly(2-hydroxypropyl acrylate)** **2012YAN**
Characterization: M_w/g.mol^{-1} = 26000, M_w/M_n = 2.47, T_g/K = 266,
Scientific Polymer Products, Inc., Ontario, NY
Solvent (A): **dimethyl ether** **C_2H_6O** **115-10-6**

Type of data: cloud points

w_B 0.048 was kept constant

T/K	323.3	332.7	342.9	348.0	355.0	363.3	375.0	379.3	392.4
P/MPa	175.0	161.7	149.1	141.4	137.9	127.1	121.6	116.4	113.7

T/K	394.4	408.7	413.2	422.4	431.6	437.3	442.7	454.0
P/MPa	110.5	105.2	106.0	101.0	99.3	98.1	96.4	94.5

Polymer (B): **poly(2-hydroxypropyl methacrylate)** **2012YAN**
Characterization: M_w/g.mol^{-1} = 42000, M_w/M_n = 2.48, T_g/K = 346,
Scientific Polymer Products, Inc., Ontario, NY
Solvent (A): **dimethyl ether** **C_2H_6O** **115-10-6**

Type of data: cloud points

w_B 0.048 was kept constant

T/K	337.0	350.6	356.1	364.3	374.8	379.4	395.1	408.8	413.9
P/MPa	222.2	192.9	176.4	172.6	156.9	155.3	142.9	132.8	125.2

T/K	424.1	434.0	439.9	452.0	454.1
P/MPa	123.0	119.6	17.2	113.9	112.6

Polymer (B): **poly(isobornyl acrylate)** **2013YA1**
Characterization: M_w/g.mol^{-1} = 1000000,
Scientific Polymer Products, Inc., Ontario, NY
Solvent (A): **n-butane** C_4H_{10} **106-97-8**

Type of data: cloud points

w_B 0.053 was kept constant

T/K	325.0	333.6	354.8	371.2	392.6	412.8	432.8	452.5	474.1
P/MPa	41.21	40.52	39.55	40.52	41.21	41.90	43.28	44.66	46.03

Polymer (B): **poly(isobornyl acrylate)** **2013YA1**
Characterization: M_w/g.mol^{-1} = 1000000,
Scientific Polymer Products, Inc., Ontario, NY
Solvent (A): **1-butene** C_4H_8 **106-98-9**

Type of data: cloud points

w_B 0.043 was kept constant

T/K	323.8	333.8	351.0	370.2	391.6	411.2	433.5	455.0	474.8
P/MPa	15.00	17.07	20.52	23.97	27.07	29.83	32.93	35.69	38.45

Polymer (B): **poly(isobornyl acrylate)** **2013YA1**
Characterization: M_w/g.mol^{-1} = 100000,
Scientific Polymer Products, Inc., Ontario, NY
Solvent (A): **dimethyl ether** C_2H_6O **115-10-6**

Type of data: cloud points

w_B 0.043 was kept constant

T/K	334.6	354.2	374.7	393.1	415.6	435.2	454.2
P/MPa	31.90	35.35	38.79	41.55	45.69	47.93	45.00

Polymer (B): **poly(isobornyl acrylate)** **2013YA1**
Characterization: M_w/g.mol^{-1} = 1000000,
Scientific Polymer Products, Inc., Ontario, NY
Solvent (A): **dimethyl ether** C_2H_6O **115-10-6**

Type of data: cloud points

w_B 0.049 was kept constant

T/K	330.6	350.4	376.4	395.7	412.1	433.2	453.1	473.8
P/MPa	30.52	33.10	36.72	40.00	42.93	46.03	48.28	49.14

Polymer (B): **poly(isobornyl acrylate)** **2013YA1**
Characterization: M_w/g.mol^{-1} = 1000000,
Scientific Polymer Products, Inc., Ontario, NY
Solvent (A): **propane** C_3H_8 **74-98-6**

Type of data: cloud points

w_B 0.051 was kept constant

T/K	324.0	334.8	353.7	374.5	390.3	413.2	435.6	455.6	474.6
P/MPa	84.31	81.55	78.45	76.03	75.00	74.52	74.31	74.31	73.97

Polymer (B): **poly(isobornyl acrylate)** **2013YA1**
Characterization: M_w/g.mol^{-1} = 1000000,
Scientific Polymer Products, Inc., Ontario, NY
Solvent (A): **propene** C_3H_6 **115-07-1**

Type of data: cloud points

w_B 0.047 was kept constant

T/K	324.9	343.4	363.7	384.6	401.1	423.7	444.2	459.3	475.2
P/MPa	56.03	58.10	60.17	62.59	63.97	66.03	67.76	68.79	69.83

Polymer (B): **poly(isobornyl methacrylate)** **2014JEO**
Characterization: M_w/g.mol^{-1} = 100000, T_g/K = 383.2,
Scientific Polymer Products, Inc., Ontario, NY
Solvent (A): **n-butane** C_4H_{10} **106-97-8**

Type of data: cloud points

w_B 0.049 was kept constant

T/K	334.7	354.3	372.4	394.2	414.9	434.8	455.0
P/MPa	93.28	79.69	71.90	65.52	60.17	56.72	54.83

Polymer (B): **poly(isobornyl methacrylate)** **2014JEO**
Characterization: M_w/g.mol^{-1} = 100000, T_g/K = 383.2,
Scientific Polymer Products, Inc., Ontario, NY
Solvent (A): **1-butene** C_4H_8 **106-98-9**

Type of data: cloud points

w_B 0.048 was kept constant

T/K	332.8	353.8	373.9	395.3	414.7	434.8	454.1
P/MPa	23.97	26.90	30.17	32.93	35.35	37.76	39.48

Polymer (B): **poly(isobornyl methacrylate)** **2014JEO**
Characterization: M_w/g.mol^{-1} = 100000, T_g/K = 383.2,
Scientific Polymer Products, Inc., Ontario, NY
Solvent (A): **dimethyl ether** C_2H_6O **115-10-6**

Type of data: cloud points

w_B 0.051 was kept constant

T/K	332.7	353.8	375.2	394.2	416.4	433.8	452.9
P/MPa	41.03	51.55	43.62	45.69	48.45	50.52	52.59

Polymer (B): **poly(isobornyl methacrylate)** **2014JEO**
Characterization: M_w/g.mol^{-1} = 100000, T_g/K = 383.2,
Scientific Polymer Products, Inc., Ontario, NY
Solvent (A): **propane** C_3H_8 **74-98-6**

Type of data: cloud points

w_B 0.048 was kept constant

T/K	334.4	353.5	373.2	393.3	413.0	432.0	452.0
P/MPa	104.83	93.79	86.38	82.24	78.79	76.72	76.03

Polymer (B): **poly(isobornyl methacrylate)** **2014JEO**
Characterization: M_w/g.mol^{-1} = 100000, T_g/K = 383.2,
Scientific Polymer Products, Inc., Ontario, NY
Solvent (A): **propene** C_3H_6 **115-07-1**

Type of data: cloud points

w_B 0.052 was kept constant

T/K	333.5	354.2	374.4	394.2	412.9	434.3	453.2
P/MPa	66.90	66.90	68.07	69.14	69.48	70.17	71.21

Polymer (B): **poly(isobutyl acrylate)** **2006BY1**
Characterization: M_n/g.mol^{-1} = 22000, M_w/g.mol^{-1} = 120000,
Polysciences, Inc., Warrington, PA
Solvent (A): **carbon dioxide** CO_2 **124-38-9**

Type of data: cloud points

w_B 0.055 was kept constant

T/K	323.95	326.05	338.95	346.55	363.25	378.95	392.95	408.05	423.15
P/bar	1822.4	1384.5	1044.8	970.7	925.9	915.5	915.5	919.0	925.5

Polymer (B): **poly(isobutyl methacrylate)** **2006BY1**
Characterization: M_n/g.mol^{-1} = 114000, M_w/g.mol^{-1} = 200000,
Polysciences, Inc., Warrington, PA
Solvent (A): **carbon dioxide** **CO_2** **124-38-9**

Type of data: cloud points

w_B 0.053 was kept constant

T/K	355.95	359.55	368.45	380.75	393.05	407.65	422.35	435.65
P/bar	2046.6	1784.5	1493.1	1374.5	1302.1	1246.6	1226.9	1221.4

Polymer (B): **poly(isodecyl acrylate)** **2008BY2**
Characterization: M_w/g.mol^{-1} = 60000, T_g/K = 213,
Scientific Polymer Products, Inc., Ontario, NY
Solvent (A): **n-butane** **C_4H_{10}** **106-97-8**

Type of data: cloud points

w_B 0.053 was kept constant

T/K	394.25	412.75	434.45	454.95
P/bar	48.6	77.6	108.6	131.4

Polymer (B): **poly(isodecyl acrylate)** **2008BY2**
Characterization: M_w/g.mol^{-1} = 60000, T_g/K = 213,
Scientific Polymer Products, Inc., Ontario, NY
Solvent (A): **1-butene** **C_4H_8** **106-98-9**

Type of data: cloud points

w_B 0.053 was kept constant

T/K	393.85	411.25	433.35	452.85
P/bar	32.8	63.8	94.8	115.5

Polymer (B): **poly(isodecyl acrylate)** **2008BY2**
Characterization: M_w/g.mol^{-1} = 60000, T_g/K = 213,
Scientific Polymer Products, Inc., Ontario, NY
Solvent (A): **carbon dioxide** **CO_2** **124-38-9**

Type of data: cloud points

w_B 0.056 was kept constant

T/K	418.55	435.45	449.25	463.35	474.35
P/bar	1922.9	1360.3	1232.8	1208.6	1175.9

Polymer (B): **poly(isodecyl acrylate)** **2008BY2**
Characterization: M_w/g.mol^{-1} = 60000, T_g/K = 213,
Scientific Polymer Products, Inc., Ontario, NY
Solvent (A): **dimethyl ether** C_2H_6O **115-10-6**

Type of data: cloud points

w_B 0.057 was kept constant

T/K	375.55	393.55	413.95	434.25	453.25
P/bar	36.2	77.6	101.7	139.7	177.6

Polymer (B): **poly(isodecyl acrylate)** **2008BY2**
Characterization: M_w/g.mol^{-1} = 60000, T_g/K = 213,
Scientific Polymer Products, Inc., Ontario, NY
Solvent (A): **propane** C_3H_8 **74-98-6**

Type of data: cloud points

w_B 0.059 was kept constant

T/K	331.95	354.55	375.05	393.85	414.85	434.95	453.95
P/bar	85.5	125.7	167.2	196.6	232.8	253.5	271.7

Polymer (B): **poly(isodecyl acrylate)** **2008BY2**
Characterization: M_w/g.mol^{-1} = 60000, T_g/K = 213,
Scientific Polymer Products, Inc., Ontario, NY
Solvent (A): **propene** C_3H_6 **115-07-1**

Type of data: cloud points

w_B 0.058 was kept constant

T/K	333.15	352.25	374.85	395.55	414.75	432.85	455.75
P/bar	67.2	115.5	167.2	201.7	225.9	241.7	277.6

Polymer (B): **poly(isodecyl methacrylate)** **2009KIM**
Characterization: M_w/g.mol^{-1} = 165000, T_g/K = 232.2,
Scientific Polymer Products, Inc., Ontario, NY
Solvent (A): **n-butane** C_4H_{10} **106-97-8**

Type of data: cloud points

w_B 0.053 was kept constant

T/K	374.1	394.4	414.1	433.1	453.5
P/MPa	3.62	7.41	10.17	12.93	14.66

Polymer (B): **poly(isodecyl methacrylate)** **2009KIM**
Characterization: M_w/g.mol^{-1} = 165000, T_g/K = 232.2,
Scientific Polymer Products, Inc., Ontario, NY
Solvent (A): **1-butene** C_4H_8 **106-98-9**

Type of data: cloud points

w_B	0.056	was kept constant		
T/K	394.6	411.2	433.4	452.8
P/MPa	5.35	8.45	11.90	14.31

Polymer (B): **poly(isodecyl methacrylate)** **2009KIM**
Characterization: M_w/g.mol^{-1} = 165000, T_g/K = 232.2,
Scientific Polymer Products, Inc., Ontario, NY
Solvent (A): **carbon dioxide** CO_2 **124-38-9**

Type of data: cloud points

w_B	0.054	was kept constant			
T/K	438.2	444.3	453.3	465.0	475.7
P/MPa	255.15	195.52	163.62	145.69	142.76

Polymer (B): **poly(isodecyl methacrylate)** **2009KIM**
Characterization: M_w/g.mol^{-1} = 165000, T_g/K = 232.2,
Scientific Polymer Products, Inc., Ontario, NY
Solvent (A): **dimethyl ether** C_2H_6O **115-10-6**

Type of data: cloud points

w_B	0.046	was kept constant			
T/K	375.6	392.8	412.4	432.7	453.7
P/MPa	5.00	8.79	13.28	17.00	19.31

Polymer (B): **poly(isodecyl methacrylate)** **2009KIM**
Characterization: M_w/g.mol^{-1} = 165000, T_g/K = 232.2,
Scientific Polymer Products, Inc., Ontario, NY
Solvent (A): **propane** C_3H_8 **74-98-6**

Type of data: cloud points

w_B	0.070	was kept constant					
T/K	332.6	353.4	374.6	394.7	412.5	433.5	453.3
P/MPa	12.59	16.72	19.83	22.93	25.52	27.93	29.66

Polymer (B): **poly(isodecyl methacrylate)** **2009KIM**
Characterization: M_w/g.mol^{-1} = 165000, T_g/K = 232.2,
Scientific Polymer Products, Inc., Ontario, NY
Solvent (A): **propene** C_3H_6 **115-07-1**

Type of data: cloud points

w_B 0.046 was kept constant

T/K	333.7	353.9	374.6	393.9	413.6	432.3	455.0
P/MPa	10.04	14.66	19.48	23.14	26.72	29.14	30.17

Polymer (B): **poly(isooctyl acrylate)** **2008BY1**
Characterization: M_w/g.mol^{-1} = 60000,
Scientific Polymer Products, Inc., Ontario, NY
Solvent (A): **n-butane** C_4H_{10} **106-97-8**

Type of data: cloud points

w_B 0.062 was kept constant

T/K	374.35	394.55	413.35	433.95	453.85
P/bar	25.9	57.6	89.0	119.0	139.7

Polymer (B): **poly(isooctyl acrylate)** **2008BY1**
Characterization: M_w/g.mol^{-1} = 60000,
Scientific Polymer Products, Inc., Ontario, NY
Solvent (A): **1-butene** C_4H_8 **106-98-9**

Type of data: cloud points

w_B 0.062 was kept constant

T/K	393.25	413.25	434.95	453.35
P/bar	36.2	70.7	101.7	125.9

Polymer (B): **poly(isooctyl acrylate)** **2008BY1**
Characterization: M_w/g.mol^{-1} = 60000,
Scientific Polymer Products, Inc., Ontario, NY
Solvent (A): **carbon dioxide** CO_2 **124-38-9**

Type of data: cloud points

w_B 0.049 was kept constant

T/K	393.35	399.15	413.45	429.35	444.25	454.95
P/bar	2029.3	1670.7	1315.5	1174.1	1137.9	1132.8

Polymer (B): **poly(isooctyl acrylate)** **2008BY1**
Characterization: M_w/g.mol^{-1} = 60000,
Scientific Polymer Products, Inc., Ontario, NY
Solvent (A): **dimethyl ether** C_2H_6O **115-10-6**

Type of data: cloud points

w_B 0.054 was kept constant

T/K	373.15	394.85	413.85	433.35	454.15
P/bar	36.2	81.0	122.4	160.3	162.1

Polymer (B): **poly(isooctyl acrylate)** **2008BY1**
Characterization: M_w/g.mol^{-1} = 60000,
Scientific Polymer Products, Inc., Ontario, NY
Solvent (A): **propane** C_3H_8 **74-98-6**

Type of data: cloud points

w_B 0.055 was kept constant

T/K	333.25	355.35	373.45	393.55	413.35	434.25	453.45
P/bar	108.6	150.0	174.1	201.7	232.8	256.9	279.3

Polymer (B): **poly(isooctyl acrylate)** **2008BY1**
Characterization: M_w/g.mol^{-1} = 60000,
Scientific Polymer Products, Inc., Ontario, NY
Solvent (A): **propene** C_3H_6 **115-07-1**

Type of data: cloud points

w_B 0.054 was kept constant

T/K	332.95	353.55	373.35	394.25	412.85	432.05	455.85
P/bar	74.1	119.0	153.4	198.3	229.3	250.0	279.7

Polymer (B): **1,4-polyisoprene** **2007TAN**
Characterization: M_n/g.mol^{-1} = 3000, M_w/M_n = 1.06,
Polymer Source, Inc., Quebec, Canada
Solvent (A): **propane** C_3H_8 **74-98-6**

Type of data: cloud points

w_B 0.005 was kept constant

T/K	333.3	353.9	373.3	393.4	413.3	433.2	453.1
P/bar	56.0	89.3	121.2	146.8	169.2	181.4	192.5

Polymer (B): **1,4-polyisoprene** **2007TAN**
Characterization: M_n/g.mol^{-1} = 10100, M_w/M_n = 1.04,
Polymer Source, Inc., Quebec, Canada
Solvent (A): **propane** C_3H_8 **74-98-6**

Type of data: cloud points

w_B 0.005 was kept constant

T/K	293.2	313.1	353.3	373.5	393.4	433.5	453.0
P/bar	95.6	130.0	192.8	218.6	246.1	286.7	301.7

Polymer (B): **1,4-polyisoprene** **2007TAN**
Characterization: M_n/g.mol^{-1} = 30000, M_w/M_n = 1.04,
Polymer Source, Inc., Quebec, Canada
Solvent (A): **propane** C_3H_8 **74-98-6**

Type of data: cloud points

w_B 0.005 was kept constant

T/K	293.3	313.2	354.5	393.9	433.6	453.4
P/bar	252.2	266.4	304.3	339.6	372.3	382.1

Polymer (B): **1,4-polyisoprene** **2007TAN**
Characterization: M_n/g.mol^{-1} = 76500, M_w/M_n = 1.07,
Polymer Source, Inc., Quebec, Canada
Solvent (A): **propane** C_3H_8 **74-98-6**

Type of data: cloud points

w_B 0.005 was kept constant

T/K	293.5	313.4	353.5	393.7	433.3
P/bar	366.4	365.1	384.0	412.4	436.2

Polymer (B): **poly(isopropyl acrylate)** **2007BY2**
Characterization: M_w/g.mol^{-1} = 120000,
Scientific Polymer Products, Inc., Ontario, NY
Solvent (A): **n-butane** C_4H_{10} **106-97-8**

Type of data: cloud points

w_B 0.049 was kept constant

T/K	446.45	464.35	476.05	484.35
P/bar	2333	1488	850	738

Polymer (B): **poly(isopropyl acrylate)** **2007BY2**
Characterization: M_w/g.mol^{-1} = 120000,
Scientific Polymer Products, Inc., Ontario, NY
Solvent (A): **1-butene** C_4H_8 **106-98-9**

Type of data: cloud points

w_B 0.049 was kept constant

T/K	353.35	367.85	384.25	397.95	413.95	430.95	444.25	460.45	474.75
P/bar	2347	1559	909	690	543	429	385	357	340

Polymer (B): **poly(isopropyl acrylate)** **2007BY2**
Characterization: M_w/g.mol^{-1} = 120000,
Scientific Polymer Products, Inc., Ontario, NY
Solvent (A): **carbon dioxide** CO_2 **124-38-9**

Type of data: cloud points

w_B 0.042 was kept constant

T/K	406.65	420.85	432.25	443.05	452.85	460.75	471.85	485.85
P/bar	2685	2219	1814	1621	1519	1455	1405	1352

Polymer (B): **poly(isopropyl acrylate)** **2007BY2**
Characterization: M_w/g.mol^{-1} = 120000,
Scientific Polymer Products, Inc., Ontario, NY
Solvent (A): **dimethyl ether** C_2H_6O **115-10-6**

Type of data: cloud points

w_B 0.051 was kept constant

T/K	373.25	392.85	413.55	434.45	452.45
P/bar	41	85	129	164	181

Polymer (B): **poly(isopropyl acrylate)** **2007BY2**
Characterization: M_w/g.mol^{-1} = 120000,
Scientific Polymer Products, Inc., Ontario, NY
Solvent (A): **propane** C_3H_8 **74-98-6**

Type of data: cloud points

w_B 0.054 was kept constant

T/K	437.65	447.05	455.85	465.15	475.35
P/bar	2771	1767	1269	1081	929

Polymer (B): **poly(isopropyl acrylate)** **2007BY2**
Characterization: M_w/g.mol^{-1} = 120000,
Scientific Polymer Products, Inc., Ontario, NY
Solvent (A): **propene** C_3H_6 **115-07-1**

Type of data: cloud points

w_B 0.084 was kept constant

T/K	335.15	349.85	366.25	379.75	395.55	409.55	425.55	439.35	454.55
P/bar	1943	1579	1226	1004	854	754	685	650	616

Polymer (B): **poly(isopropyl methacrylate)** **2007BY2**
Characterization: M_w/g.mol^{-1} = 100000, T_g/K = 354,
Scientific Polymer Products, Inc., Ontario, NY
Solvent (A): **carbon dioxide** CO_2 **124-38-9**

Type of data: cloud points

w_B 0.044 was kept constant

T/K	400.25	412.35	422.85	432.45	443.65
P/bar	2512	1907	1748	1685	1655

Polymer (B): **poly(L-lactide)** **2004LIM, 2006PAR**
Characterization: M_n/g.mol^{-1} = 2000, M_w/g.mol^{-1} = 2500,
Biomaterials Research Center, KIST, South Korea
Solvent (A): **chlorodifluoromethane** $CHClF_2$ **75-45-6**

Type of data: cloud points

w_B 0.0288 was kept constant

T/K	344.4	353.1	363.4	372.7	382.1	395.7
P/MPa	3.6	6.0	8.7	10.9	13.1	15.6

Polymer (B): **poly(L-lactide)** **2004LIM, 2006PAR**
Characterization: M_n/g.mol^{-1} = 80000, M_w/g.mol^{-1} = 98500,
Biomaterials Research Center, KIST, South Korea
Solvent (A): **chlorodifluoromethane** $CHClF_2$ **75-45-6**

Type of data: cloud points

w_B 0.0297 was kept constant

T/K	333.4	342.1	353.0	362.9	372.5	382.3	392.4	402.1	408.8
P/MPa	3.6	6.3	9.9	12.9	15.8	18.6	21.2	23.7	25.2

Polymer (B): **poly(L-lactide)** **2004LIM, 2006PAR**
Characterization: M_n/g.mol^{-1} = 110000, M_w/g.mol^{-1} = 135500,
Biomaterials Research Center, KIST, South Korea
Solvent (A): **chlorodifluoromethane** **$CHClF_2$** **75-45-6**

Type of data: cloud points

w_B 0.0301 was kept constant

T/K	333.1	341.9	353.4	363.5	372.6	380.0	392.1	401.7	410.8
P/MPa	3.7	6.6	10.3	13.4	16.2	18.3	21.6	24.0	26.1

Polymer (B): **poly(L-lactide)** **2004LIM, 2006PAR**
Characterization: M_n/g.mol^{-1} = 230000, M_w/g.mol^{-1} = 283000,
Biomaterials Research Center, KIST, South Korea
Solvent (A): **chlorodifluoromethane** **$CHClF_2$** **75-45-6**

Type of data: cloud points

w_B 0.0303 was kept constant

T/K	332.3	341.5	353.4	362.5	372.2	382.2	392.5	399.6	409.7
P/MPa	3.6	6.7	10.5	13.5	16.4	19.3	22.0	23.7	26.1

Polymer (B): **poly(L-lactide)** **2004LIM, 2006PAR**
Characterization: M_n/g.mol^{-1} = 80000, M_w/g.mol^{-1} = 98500,
Biomaterials Research Center, KIST, South Korea
Solvent (A): **1,1-difluoroethane** **$C_2H_4F_2$** **75-37-6**

Type of data: cloud points

w_B 0.0301 was kept constant

T/K	354.1	363.8	371.5	383.9	392.7	398.8	413.1
P/MPa	59.5	57.2	55.9	54.5	53.9	53.7	53.2

Polymer (B): **poly(L-lactide)** **2004LIM, 2006PAR**
Characterization: M_n/g.mol^{-1} = 110000, M_w/g.mol^{-1} = 135500,
Biomaterials Research Center, KIST, South Korea
Solvent (A): **1,1-difluoroethane** **$C_2H_4F_2$** **75-37-6**

Type of data: cloud points

w_B 0.0298 was kept constant

T/K	352.9	363.9	373.3	383.5	392.7	400.6	415.2
P/MPa	61.3	58.6	57.0	55.8	55.0	54.8	54.3

Polymer (B): **poly(L-lactide)** **2004LIM, 2006PAR**
Characterization: M_n/g.mol^{-1} = 230000, M_w/g.mol^{-1} = 283000, Biomaterials Research Center, KIST, South Korea
Solvent (A): **1,1-difluoroethane** $C_2H_4F_2$ **75-37-6**

Type of data: cloud points

w_B 0.0300 was kept constant

T/K	354.3	364.3	373.3	383.8	392.6	402.6	414.0
P/MPa	64.2	61.5	59.9	58.4	57.7	57.1	56.8

Polymer (B): **poly(L-lactide)** **2004LIM, 2006PAR**
Characterization: M_n/g.mol^{-1} = 80000, M_w/g.mol^{-1} = 98500, Biomaterials Research Center, KIST, South Korea
Solvent (A): **difluoromethane** CH_2F_2 **75-10-5**

Type of data: cloud points

w_B 0.0295 was kept constant

T/K	344.4	354.1	363.7	373.9	383.3	393.4	402.8	412.7
P/MPa	91.6	85.6	81.8	79.5	78.1	77.2	77.0	77.0

Polymer (B): **poly(L-lactide)** **2004LIM, 2006PAR**
Characterization: M_n/g.mol^{-1} = 110000, M_w/g.mol^{-1} = 135500, Biomaterials Research Center, KIST, South Korea
Solvent (A): **difluoromethane** CH_2F_2 **75-10-5**

Type of data: cloud points

w_B 0.0294 was kept constant

T/K	342.5	352.6	364.1	373.6	382.2	391.6	402.5	416.8
P/MPa	93.6	86.8	82.1	80.0	78.7	77.8	77.4	77.3

Polymer (B): **poly(L-lactide)** **2004LIM, 2006PAR**
Characterization: M_n/g.mol^{-1} = 230000, M_w/g.mol^{-1} = 283000, Biomaterials Research Center, KIST, South Korea
Solvent (A): **difluoromethane** CH_2F_2 **75-10-5**

Type of data: cloud points

w_B 0.0297 was kept constant

T/K	334.6	345.0	352.8	365.0	374.5	384.2	393.7	403.1	413.1
P/MPa	111.3	99.5	93.0	87.4	84.8	83.1	82.1	81.7	81.4

Polymer (B): **poly(L-lactide)** **2004LIM, 2006PAR**
Characterization: M_n/g.mol^{-1} = 80000, M_w/g.mol^{-1} = 98500,
Biomaterials Research Center, KIST, South Korea
Solvent (A): **dimethyl ether** **C_2H_6O** **115-10-6**

Type of data: cloud points

w_B 0.0299 was kept constant

T/K	333.1	343.1	353.2	363.5	373.2	382.8	392.6	401.2	410.4
P/MPa	3.6	6.5	9.3	11.9	14.3	16.6	18.8	20.5	22.6

Polymer (B): **poly(L-lactide)** **2004LIM, 2006PAR**
Characterization: M_n/g.mol^{-1} = 110000, M_w/g.mol^{-1} = 135500,
Biomaterials Research Center, KIST, South Korea
Solvent (A): **dimethyl ether** **C_2H_6O** **115-10-6**

Type of data: cloud points

w_B 0.0299 was kept constant

T/K	331.4	343.1	353.9	362.1	374.0	382.8	392.0	402.7	412.6
P/MPa	3.1	6.6	9.3	11.5	14.5	16.5	18.5	21.0	22.8

Polymer (B): **poly(L-lactide)** **2004LIM, 2006PAR**
Characterization: M_n/g.mol^{-1} = 230000, M_w/g.mol^{-1} = 283000,
Biomaterials Research Center, KIST, South Korea
Solvent (A): **dimethyl ether** **C_2H_6O** **115-10-6**

Type of data: cloud points

w_B 0.0302 was kept constant

T/K	332.8	342.5	354.0	364.1	372.9	381.4	393.1	401.7	411.6
P/MPa	6.3	9.0	12.0	14.5	16.7	18.6	21.2	23.0	24.9

Polymer (B): **poly(L-lactide)** **2004LIM, 2006PAR**
Characterization: M_n/g.mol^{-1} = 80000, M_w/g.mol^{-1} = 98500,
Biomaterials Research Center, KIST, South Korea
Solvent (A): **trifluoromethane** **CHF_3** **75-46-7**

Type of data: cloud points

w_B 0.0299 was kept constant

T/K	333.4	343.8	353.2	364.5	374.1	383.7	395.3	402.1	415.3
P/MPa	93.1	95.5	98.0	100.9	103.2	105.5	107.9	109.7	112.1

Polymer (B): **poly(L-lactide)** **2004LIM, 2006PAR**
Characterization: M_n/g.mol^{-1} = 110000, M_w/g.mol^{-1} = 135500,
Biomaterials Research Center, KIST, South Korea
Solvent (A): **trifluoromethane** **CHF_3** **75-46-7**

Type of data: cloud points

w_B 0.0294 was kept constant

T/K	333.1	342.9	354.3	364.5	373.4	384.0	394.3	403.3	414.3
P/MPa	94.1	97.1	100.5	103.4	105.7	108.5	110.9	112.8	115.1

Polymer (B): **poly(L-lactide)** **2004LIM, 2006PAR**
Characterization: M_n/g.mol^{-1} = 230000, M_w/g.mol^{-1} = 283000,
Biomaterials Research Center, KIST, South Korea
Solvent (A): **trifluoromethane** **CHF_3** **75-46-7**

Type of data: cloud points

w_B 0.0299 was kept constant

T/K	333.6	343.2	353.6	363.6	374.1	382.8	393.9	403.8	413.3
P/MPa	96.1	99.0	101.9	104.6	107.4	109.5	111.9	114.0	116.0

Polymer (B): **poly(L-lactide-*co*-diglycidyl ether of bisphenol A-*co*-4,4'-hexafluoroisopropylidenediphenol)** **2005SHE**
Characterization: M_n/g.mol^{-1} = 16900, M_w/g.mol^{-1} = 32100, 1:1:1 terpolymer,
T_g/K = 358, synthesized in the laboratory
Solvent (A): **dimethyl ether** **C_2H_6O** **115-10-6**

Type of data: cloud points

w_B 0.028 was kept constant

T/K	299.45	311.85	326.06	343.45	361.15	378.95	395.05
P/bar	279.0	306.9	339.3	375.2	419.7	455.9	489.4
ρ/(g/cm^3)	0.713	0.702	0.690	0.675	0.663	0.650	0.640

Polymer (B): **poly(L-lactide-*co*-diglycidyl ether of bisphenol A-*co*-4,4'-isopropylidenediphenol)** **2005SHE**
Characterization: M_n/g.mol^{-1} = 8900, M_w/g.mol^{-1} = 16000, 1:1:1 terpolymer,
T_g/K = 349, synthesized in the laboratory
Solvent (A): **dimethyl ether** **C_2H_6O** **115-10-6**

Type of data: cloud points

w_B 0.030 was kept constant

T/K	295.35	314.75	334.55	354.85	377.85	390.65
P/bar	1009.2	931.2	888.9	869.0	856.1	851.6
ρ/(g/cm^3)	0.767	0.749	0.733	0.719	0.703	0.694

Polymer (B): **poly(methyl acrylate)** **2004BY3**
Characterization: M_w/g.mol^{-1} = 40000, Polysciences, Inc., Warrington, PA
Solvent (A): **carbon dioxide** CO_2 **124-38-9**

Type of data: cloud points

w_B 0.042 was kept constant

T/K	383.75	403.75	423.45	443.85	461.85	481.35
P/bar	2191.4	2122.8	2050.0	1979.7	1912.8	1956.9

Polymer (B): **poly(methyl acrylate)** **2004BEC, 2004LAT**
Characterization: M_n/g.mol^{-1} = 76500, M_w/g.mol^{-1} = 186900, synthesized in the laboratory
Solvent (A): **propene** C_3H_6 **115-07-1**

Type of data: cloud points

w_B 0.05 was kept constant

T/K	473.15	480.15	481.15	482.15	483.15	491.15	493.75	501.15	510.15
P/bar	2750	2465	2418	2391	2361	2109	2018	1850	1691

T/K	512.75	521.95	523.45	525.45	526.45	531.65
P/bar	1641	1513	1487	1465	1457	1399

Polymer (B): **poly(methyl methacrylate)** **2002LE2**
Characterization: M_w/g.mol^{-1} = 15000, Aldrich Chem. Co., Inc., Milwaukee, WI
Solvent (A): **chlorodifluoromethane** $CHClF_2$ **75-45-6**

Type of data: cloud points

w_B 0.0498 was kept constant

T/K	332.35	337.65	342.85	353.15	363.35	372.95
P/bar	24.5	29.6	45.5	77.9	108.5	138.7

Polymer (B): **poly(methyl methacrylate)** **2002LE2**
Characterization: M_w/g.mol^{-1} = 120000, Aldrich Chem. Co., Inc., Milwaukee, WI
Solvent (A): **chlorodifluoromethane** $CHClF_2$ **75-45-6**

Type of data: cloud points

w_B 0.0499 was kept constant

T/K	312.75	323.45	325.75	332.85	342.25	353.25	362.55	373.15
P/bar	16.6	21.6	24.7	49.5	85.1	125.5	159.2	195.7

Polymer (B): **poly(methyl methacrylate)** **2004BEC, 2004LAT**
Characterization: M_n/g.mol^{-1} = 71200, M_w/g.mol^{-1} = 104700,
synthesized in the laboratory
Solvent (A): **propene** C_3H_6 **115-07-1**

Type of data: cloud points

w_B 0.05 was kept constant

T/K	427.15	433.15	433.65	443.35	444.15	453.15	461.15	462.15	462.65
P/bar	2700	2379	2356	2060	2027	1891	1739	1728	1712
T/K	471.15	471.85	472.35	481.15	490.65	502.15	511.15	523.15	532.15
P/bar	1577	1573	1563	1458	1364	1265	1204	1122	1078

Polymer (B): **poly(4-methylstyrene)** **2004TAN**
Characterization: M_n/g.mol^{-1} = 2500, M_w/g.mol^{-1} = 2700,
Polymer Source, Inc., Quebec, Canada
Solvent (A): **propane** C_3H_8 **74-98-6**

Type of data: cloud points

w_B 0.005 was kept constant

T/K	303.15	333.15	353.15	373.15	393.15	413.15	433.15
P/bar	791.5	517.9	443.0	388.2	365.7	370.6	363.3

Polymer (B): **poly(4-methylstyrene)** **2004TAN**
Characterization: M_n/g.mol^{-1} = 82500, M_w/g.mol^{-1} = 89900,
Polymer Source, Inc., Quebec, Canada
Solvent (A): **propane** C_3H_8 **74-98-6**

Type of data: cloud points

w_B 0.005 was kept constant

T/K	378.15	393.15	413.15	433.15	453.15
P/bar	1329.6	1192.9	1011.3	883.4	803.3

Polymer (B): **poly(neopentyl methacrylate)** **2007BY1**
Characterization: M_w/g.mol^{-1} = 480000,
Scientific Polymer Products, Inc., Ontario, NY
Solvent (A): **n-butane** C_4H_{10} **106-97-8**

Type of data: cloud points

w_B 0.046 was kept constant

T/K	374.25	391.55	412.45	432.95	454.85
P/bar	60.3	91.4	122.4	153.5	170.7

Polymer (B): **poly(neopentyl methacrylate)** **2007BY1**
Characterization: M_w/g.mol^{-1} = 480000,
Scientific Polymer Products, Inc., Ontario, NY
Solvent (A): **1-butene** C_4H_8 **106-98-9**

Type of data: cloud points

w_B 0.044 was kept constant

T/K	375.45	393.55	412.55	434.45	454.35
P/bar	21.7	56.9	91.4	125.9	153.5

Polymer (B): **poly(neopentyl methacrylate)** **2007BY1**
Characterization: M_w/g.mol^{-1} = 480000,
Scientific Polymer Products, Inc., Ontario, NY
Solvent (A): **carbon dioxide** CO_2 **124-38-9**

Type of data: cloud points

w_B 0.041 was kept constant

T/K	349.75	364.05	377.75	394.75	409.05	423.85	439.15	454.45
P/bar	2001.7	1244.5	1124.1	1063.8	1043.1	1034.5	1035.5	1046.6

Polymer (B): **poly(neopentyl methacrylate)** **2007BY1**
Characterization: M_w/g.mol^{-1} = 480000,
Scientific Polymer Products, Inc., Ontario, NY
Solvent (A): **dimethyl ether** C_2H_6O **115-10-6**

Type of data: cloud points

w_B 0.036 was kept constant

T/K	375.35	392.15	412.55	433.45	456.25
P/bar	36.9	77.6	122.4	165.5	200.0

Polymer (B): **poly(neopentyl methacrylate)** **2007BY1**
Characterization: M_w/g.mol^{-1} = 480000,
Scientific Polymer Products, Inc., Ontario, NY
Solvent (A): **propane** C_3H_8 **74-98-6**

Type of data: cloud points

w_B 0.054 was kept constant

T/K	332.55	353.85	373.85	394.15	412.95	433.95	454.55
P/bar	208.6	229.3	250.0	274.1	291.4	312.1	332.8

Polymer (B): **poly(neopentyl methacrylate)** **2007BY1**
Characterization: M_w/g.mol^{-1} = 480000,
Scientific Polymer Products, Inc., Ontario, NY
Solvent (A): **propene** C_3H_6 **115-07-1**

Type of data: cloud points

w_B 0.047 was kept constant

T/K	333.45	355.65	374.25	395.05	415.55	434.75	453.95
P/bar	115.5	169.0	208.6	244.8	274.1	298.3	322.4

Polymer (B): **poly(octyl acrylate)** **2007BY3**
Characterization: M_w/g.mol^{-1} = 100000, T_g/K = 208,
Scientific Polymer Products, Inc., Ontario, NY
Solvent (A): **n-butane** C_4H_{10} **106-97-8**

Type of data: cloud points

w_B 0.045 was kept constant

T/K	373.65	393.75	413.85	434.65	453.35
P/bar	29.3	60.3	91.4	119.0	139.7

Polymer (B): **poly(octyl acrylate)** **2007BY3**
Characterization: M_w/g.mol^{-1} = 100000, T_g/K = 208,
Scientific Polymer Products, Inc., Ontario, NY
Solvent (A): **1-butene** C_4H_8 **106-98-9**

Type of data: cloud points

w_B 0.059 was kept constant

T/K	393.95	412.55	431.85	452.15
P/bar	46.6	79.0	107.6	132.8

Polymer (B): **poly(octyl acrylate)** **2007BY3**
Characterization: M_w/g.mol^{-1} = 100000, T_g/K = 208,
Scientific Polymer Products, Inc., Ontario, NY
Solvent (A): **carbon dioxide** CO_2 **124-38-9**

Type of data: cloud points

w_B 0.041 was kept constant

T/K	425.05	433.65	443.95	454.35
P/bar	2622.4	1451.2	1313.8	1269.0

Polymer (B): **poly(octyl acrylate)** **2007BY3**
Characterization: M_w/g.mol^{-1} = 100000, T_g/K = 208,
Scientific Polymer Products, Inc., Ontario, NY
Solvent (A): **dimethyl ether** **C_2H_6O** **115-10-6**

Type of data: cloud points

w_B 0.083 was kept constant

T/K	324.15	334.15	353.15	373.25	393.65	415.35	434.85	452.85
P/bar	25.9	29.3	39.7	77.6	122.4	167.2	201.7	222.4

Polymer (B): **poly(octyl acrylate)** **2007BY3**
Characterization: M_w/g.mol^{-1} = 100000, T_g/K = 208,
Scientific Polymer Products, Inc., Ontario, NY
Solvent (A): **propane** **C_3H_8** **74-98-6**

Type of data: cloud points

w_B 0.042 was kept constant

T/K	326.05	335.05	353.95	375.55	395.95	416.95	436.55	455.15
P/bar	105.2	119.0	148.3	184.5	215.5	243.1	263.8	277.6

Polymer (B): **poly(octyl acrylate)** **2007BY3**
Characterization: M_w/g.mol^{-1} = 100000, T_g/K = 208,
Scientific Polymer Products, Inc., Ontario, NY
Solvent (A): **propene** **C_3H_6** **115-07-1**

Type of data: cloud points

w_B 0.057 was kept constant

T/K	334.25	353.95	375.35	394.85	414.05	433.75	453.55
P/bar	115.5	142.8	177.6	212.1	239.7	265.5	281.0

Polymer (B): **poly(octyl methacrylate)** **2007BY3**
Characterization: M_w/g.mol^{-1} = 100000, T_g/K = 293,
Scientific Polymer Products, Inc., Ontario, NY
Solvent (A): **carbon dioxide** **CO_2** **124-38-9**

Type of data: cloud points

w_B 0.058 was kept constant

T/K	447.85	463.05	472.85	483.25	493.25
P/bar	2663.8	1919.0	1637.9	1546.6	1474.1

Polymer (B): **poly[4-(perfluorooctyl(ethyleneoxy)methyl)-styrene]** **2008SH3**
Characterization: M_n/g.mol^{-1} = 15400, M_w/g.mol^{-1} = 26800, synthesized in the laboratory
Solvent (A): **carbon dioxide** **CO_2** **124-38-9**

Type of data: cloud points

w_B 0.041 was kept constant

T/K	331.55	355.55	372.45	394.25	413.55	433.65	452.05
P/bar	239.7	305.2	353.5	401.7	432.8	463.8	481.0

Polymer (B): **poly[4-(perfluorooctyl(ethyleneoxy)methyl)-styrene]** **2004LAC**
Characterization: M_n/g.mol^{-1} = 21000 (theoretical value), synthesized in the laboratory
Solvent (A): **carbon dioxide** **CO_2** **124-38-9**

Type of data: cloud points

w_B 0.041 was kept constant

T/K	298.15	303.15	308.15	313.15	318.15	323.15	328.15	333.15	338.15
P/bar	123.0	140.0	159.0	175.0	192.0	206.0	220.0	234.0	248.0

Polymer (B): **poly[4-(perfluorooctyl(ethyleneoxy)methyl)-styrene]** **2004LAC**
Characterization: M_n/g.mol^{-1} = 43600 (theoretical value), synthesized in the laboratory
Solvent (A): **carbon dioxide** **CO_2** **124-38-9**

Type of data: cloud points

w_B 0.042 was kept constant

T/K	298.15	303.15	308.15	313.05	318.15	323.15	328.15	333.15	338.15
P/bar	151.0	167.0	184.0	200.0	216.0	231.0	246.0	260.0	274.0
T/K	343.15								
P/bar	287.0								

Polymer (B): **poly[4-(perfluorooctyl(ethyleneoxy)methyl)-styrene]** **2008SH3**
Characterization: M_n/g.mol^{-1} = 15400, M_w/g.mol^{-1} = 26800, synthesized in the laboratory
Solvent (A): **dimethyl ether** **C_2H_6O** **115-10-6**

Type of data: cloud points

w_B 0.052 was kept constant

T/K	394.85	414.25	432.35	452.85
P/bar	67.2	93.1	117.2	139.7

Polymer (B): **poly[4-(perfluorooctyl(ethylenesulfonyl)methyl)-styrene]** **2008SH3**
Characterization: synthesized in the laboratory
Solvent (A): **carbon dioxide** CO_2 **124-38-9**

Type of data: cloud points

w_B 0.034 was kept constant

T/K	350.55	365.75	385.25	404.05	422.65	438.95	453.85
P/bar	1325.9	610.3	556.9	574.1	601.7	641.4	656.9

Polymer (B): **poly[4-(perfluorooctyl(ethylenesulfonyl)methyl)-styrene]** **2008SH3**
Characterization: synthesized in the laboratory
Solvent (A): **dimethyl ether** C_2H_6O **115-10-6**

Type of data: cloud points

w_B 0.041 was kept constant

T/K	414.55	429.75	459.55	459.55	473.95
P/bar	113.8	129.3	150.0	174.1	187.9

Polymer (B): **poly[4-(perfluorooctyl(ethylenethio)methyl)-styrene]** **2008SH3**
Characterization: M_n/g.mol^{-1} = 20300, M_w/g.mol^{-1} = 25100, synthesized in the laboratory
Solvent (A): **carbon dioxide** CO_2 **124-38-9**

Type of data: cloud points

w_B 0.049 was kept constant

T/K	333.45	355.05	372.85	392.25	413.35	434.65	453.65
P/bar	256.9	312.1	356.9	401.7	439.7	463.8	477.6

Polymer (B): **poly[4-(perfluorooctyl(ethylenethio)methyl)-styrene]** **2008SH3**
Characterization: M_n/g.mol^{-1} = 20300, M_w/g.mol^{-1} = 25100, synthesized in the laboratory
Solvent (A): **dimethyl ether** C_2H_6O **115-10-6**

Type of data: cloud points

w_B 0.052 was kept constant

T/K	410.65	433.75	452.65
P/bar	67.2	98.3	112.1

Polymer (B): **poly(propyl acrylate)** **2002BY3**
Characterization: M_w/g.mol^{-1} = 140000, Polysciences, Inc., Warrington, PA
Solvent (A): **n-butane** C_4H_{10} **106-97-8**

Type of data: cloud points

w_B 0.052 was kept constant

T/K	369.65	370.25	371.85	374.05	377.75	385.25	404.55	424.85
P/bar	2208.6	1919.0	1587.9	1331.4	1072.8	857.9	567.2	413.1

Polymer (B): **poly(propyl acrylate)** **2002BY3**
Characterization: M_w/g.mol^{-1} = 140000, Polysciences, Inc., Warrington, PA
Solvent (A): **1-butene** C_4H_8 **106-98-9**

Type of data: cloud points

w_B 0.054 was kept constant

T/K	295.85	304.95	313.25	328.45	343.75	363.45	383.85	396.15	423.35
P/bar	251.4	207.9	172.8	131.0	132.8	139.0	155.2	172.8	187.9

Polymer (B): **poly(propyl acrylate)** **2002BY2, 2002BY3**
Characterization: M_w/g.mol^{-1} = 140000, Polysciences, Inc., Warrington, PA
Solvent (A): **carbon dioxide** CO_2 **124-38-9**

Type of data: cloud points

w_B 0.051 was kept constant

T/K	361.85	363.25	366.25	370.25	374.15	383.65	403.65	423.95	448.45
P/bar	2070.7	1932.8	1794.8	1656.9	1556.9	1474.5	1332.1	1263.8	1215.5

Polymer (B): **poly(propyl acrylate)** **2002BY3**
Characterization: M_w/g.mol^{-1} = 140000, Polysciences, Inc., Warrington, PA
Solvent (A): **dimethyl ether** C_2H_6O **115-10-6**

Type of data: cloud points

w_B 0.051 was kept constant

T/K	355.65	363.55	381.65	383.75	403.45
P/bar	48.6	61.0	110.7	114.8	157.9

Polymer (B): **poly(propyl acrylate)** **2004BEC, 2004LAT**
Characterization: M_n/g.mol^{-1} = 47900, M_w/g.mol^{-1} = 108300, synthesized in the laboratory
Solvent (A): **ethene** C_2H_4 **74-85-1**

Type of data: cloud points

continued

continued

w_B 0.05 was kept constant

T/K	352.95	363.25	373.25	393.15	412.15	432.15	452.15	472.15	491.15
P/bar	1093	1059	1033	992	961	936	914	892	881

T/K	511.15	531.15
P/bar	877	874

Polymer (B): **poly(propyl acrylate)** **2002BY2, 2002BY3**
Characterization: M_w/g.mol^{-1} = 140000, Polysciences, Inc., Warrington, PA
Solvent (A): **ethene** C_2H_4 **74-85-1**

Type of data: cloud points

w_B 0.054 was kept constant

T/K	383.35	403.35	422.55	444.65
P/bar	1401.0	1285.2	1232.1	1183.1

Polymer (B): **poly(propyl acrylate)** **2002BY3**
Characterization: M_w/g.mol^{-1} = 140000, Polysciences, Inc., Warrington, PA
Solvent (A): **propane** C_3H_8 **74-98-6**

Type of data: cloud points

w_B 0.048 was kept constant

T/K	365.45	373.85	382.25	392.85	402.85	422.95	443.15
P/bar	1881.0	1401.7	1183.4	1010.3	877.2	701.7	625.5

Polymer (B): **poly(propyl acrylate)** **2002BY3**
Characterization: M_w/g.mol^{-1} = 140000, Polysciences, Inc., Warrington, PA
Solvent (A): **propene** C_3H_6 **115-07-1**

Type of data: cloud points

w_B 0.048 was kept constant

T/K	304.75	323.75	343.65	364.35	384.35	404.55	423.95
P/bar	452.1	399.3	378.3	371.1	374.1	384.5	394.8

Polymer (B): **poly(propylene glycol)** **2009STO**
Characterization: M_n/g.mol^{-1} = 2000, Aldrich Chem. Co., Inc., Milwaukee, WI
Solvent (A): **carbon dioxide** CO_2 **124-38-9**

Type of data: cloud points

w_B 0.07 was kept constant

T/K	298.8	303.2	308.5	313.3	318.6	323.7	328.3	333.0	336.3
P/MPa	12.4	14.1	15.7	17.4	18.8	20.1	20.9	21.9	22.6

Polymer (B): **poly(propylene glycol)** **2009STO**
Characterization: M_n/g.mol^{-1} = 4000, Aldrich Chem. Co., Inc., Milwaukee, WI
Solvent (A): **carbon dioxide** **CO_2** **124-38-9**

Type of data: cloud points

w_B 0.05 was kept constant

T/K	312.9	313.3	317.6	318.1	322.3	323.8	328.0	328.5	332.8
P/MPa	29.3	29.4	30.5	30.8	31.7	32.2	33.2	33.3	34.4

Polymer (B): **poly(propylene glycol)** **2007LIS**
Characterization: synthesized in the laboratory
Solvent (A): **carbon dioxide** **CO_2** **124-38-9**

Type of data: cloud points

w_B	0.01286	0.01333	0.01369	0.01398	0.014485
T/K	308	313	318	323	328
P/MPa	8.20	9.15	10.40	11.79	12.65

Polymer (B): **poly(propylene glycol) bis(trimethylsilyl) ether** **2007LIS**
Characterization: synthesized in the laboratory
Solvent (A): **carbon dioxide** **CO_2** **124-38-9**

Type of data: cloud points

w_B	0.01421	0.01469	0.01517	0.01563	0.01596
T/K	308	313	318	323	328
P/MPa	7.63	8.74	9.85	11.00	12.31

Polymer (B): **poly(propylene glycol) diacetate** **2007LIS**
Characterization: synthesized in the laboratory
Solvent (A): **carbon dioxide** **CO_2** **124-38-9**

Type of data: cloud points

w_B	0.01328	0.01362	0.01403	0.01442	0.01465
T/K	308	313	318	323	328
P/MPa	7.55	8.94	10.13	11.33	12.80

Polymer (B): **poly(propylene glycol) monoacetate** **2007LIS**
Characterization: synthesized in the laboratory
Solvent (A): **carbon dioxide** **CO_2** **124-38-9**

Type of data: cloud points

w_B	0.01298	0.01330	0.01356	0.01385	0.014145
T/K	308	313	318	323	328
P/MPa	8.18	9.58	11.10	12.45	13.75

Polymer (B): **poly(propylene glycol) monoacetate monotrimethylsilyl ether** **2007LIS**
Characterization: synthesized in the laboratory
Solvent (A): **carbon dioxide** **CO_2** **124-38-9**

Type of data: cloud points

w_B	0.01383	0.01419	0.01466	0.01512	0.01547
T/K	308	313	318	323	328
P/MPa	7.70	9.09	10.15	11.25	12.50

Polymer (B): **poly(propylene glycol) monobutyl ether** **2008HON**
Characterization: M_n/g.mol^{-1} = 340, Aldrich Chem. Co., Inc., Milwaukee, WI
Solvent (A): **carbon dioxide** **CO_2** **124-38-9**

Type of data: cloud points

T/K = 298.15

w_B	0.00	0.01	0.02	0.03	0.04	0.05	0.07	0.10	0.15
P/MPa	6.41	6.52	6.43	6.41	6.48	6.52	6.76	7.17	7.31
w_B	0.20	0.23	0.25	0.30	0.40	0.50	0.60	0.70	0.80
P/MPa	7.31	7.24	7.03	6.76	6.25	6.14	5.72	4.96	3.83

Comments: The pressure for w_B = 0 is the vapor pressure of pure CO_2.
VLLE-pressure = 6.41 MPa.

Type of data: critical point

T/K = 298.15

w_{Bcrit} = 0.23 P_{crit} = 7.24 MPa

Polymer (B): **poly(propylene glycol) monobutyl ether** **2008HON**
Characterization: M_n/g.mol^{-1} = 1000, Aldrich Chem. Co., Inc., Milwaukee, WI
Solvent (A): **carbon dioxide** **CO_2** **124-38-9**

Type of data: cloud points

T/K = 298.15

w_B	0.00	0.01	0.02	0.03	0.04	0.05	0.06	0.07	0.08
P/MPa	6.41	11.58	16.27	19.65	21.99	23.99	25.51	26.75	27.65
w_B	0.09	0.10	0.15	0.20	0.23	0.30	0.40	0.50	0.60
P/MPa	28.41	29.51	31.23	31.37	31.03	30.48	26.89	19.79	9.17
w_B	0.70	0.80							
P/MPa	5.72	3.86							

Comments: The pressure for w_B = 0 is the vapor pressure of pure CO_2.
VLLE-pressure = 6.21 MPa.

continued

continued

Type of data: critical point

T/K = 298.15

w_{Bcrit} = 0.23 P_{crit} = 31.03 MPa

Polymer (B):	**poly(propylene glycol) monobutyl ether**	**2008HON**
Characterization:	M_n/g.mol^{-1} = 1200, Aldrich Chem. Co., Inc., Milwaukee, WI	
Solvent (A):	**carbon dioxide** CO_2	**124-38-9**

Type of data: cloud points

T/K = 298.15

w_B	0.00	0.01	0.02	0.03	0.04	0.05	0.06	0.07	0.08
P/MPa	6.41	16.55	22.75	27.44	31.92	34.20	36.13	37.65	38.96

w_B	0.09	0.10	0.15	0.20	0.23	0.30	0.40	0.50	0.60
P/MPa	39.99	40.75	42.27	41.27	41.71	40.13	33.92	22.89	11.31

w_B	0.70	0.80
P/MPa	5.86	4.31

Comments: The pressure for w_B = 0 is the vapor pressure of pure CO_2.
VLLE-pressure = 5.93 MPa.

Type of data: critical point

T/K = 298.15

w_{Bcrit} = 0.23 P_{crit} = 41.71 MPa

Polymer (B):	**poly(propylene glycol) monotrimethylsilyl ether**	**2007LIS**
Characterization:	synthesized in the laboratory	
Solvent (A):	**carbon dioxide** CO_2	**124-38-9**

Type of data: cloud points

w_B	0.01302	0.01339	0.01387	0.01420	0.01458
T/K	308	313	318	323	328
P/MPa	7.63	8.92	9.92	11.25	12.40

Polymer (B):	**poly(propyl methacrylate)**	**2002BY3**
Characterization:	M_w/g.mol^{-1} = 250000, Polysciences, Inc., Warrington, PA	
Solvent (A):	**n-butane** C_4H_{10}	**106-97-8**

Type of data: cloud points

w_B 0.051 was kept constant

T/K	361.75	362.75	365.95	369.15	373.75	377.95	385.35	404.45	423.75
P/bar	2108.6	1763.8	1341.0	1125.9	929.3	794.8	667.2	481.0	412.4

Polymer (B): **poly(propyl methacrylate)** **2002BY3**
Characterization: M_w/g.mol^{-1} = 250000, Polysciences, Inc., Warrington, PA
Solvent (A): **1-butene** C_4H_8 **106-98-9**

Type of data: cloud points

w_B 0.050 was kept constant

T/K	315.15	333.45	354.35	373.15	394.55	413.45	431.65
P/bar	313.8	273.2	211.7	208.6	213.4	224.5	237.9

Polymer (B): **poly(propyl methacrylate)** **2002BY3**
Characterization: M_w/g.mol^{-1} = 250000, Polysciences, Inc., Warrington, PA
Solvent (A): **chlorodifluoromethane** $CHClF_2$ **75-45-6**

Type of data: cloud points

w_B 0.048 was kept constant

T/K	333.75	352.65	373.85	393.15	414.35	437.05
P/bar	54.1	104.5	157.6	209.3	255.5	308.6

Polymer (B): **poly(propyl methacrylate)** **2002BY3**
Characterization: M_w/g.mol^{-1} = 250000, Polysciences, Inc., Warrington, PA
Solvent (A): **dimethyl ether** C_2H_6O **115-10-6**

Type of data: cloud points

w_B 0.050 was kept constant

T/K	353.45	365.35	385.75	407.05
P/bar	36.6	63.5	117.9	170.0

Polymer (B): **poly(propyl methacrylate)** **2002BY3**
Characterization: M_w/g.mol^{-1} = 250000, Polysciences, Inc., Warrington, PA
Solvent (A): **propane** C_3H_8 **74-98-6**

Type of data: cloud points

w_B 0.048 was kept constant

T/K	368.15	373.25	381.45	393.25	412.85	433.95	459.25
P/bar	2390.0	1783.8	1436.9	1134.8	900.0	739.0	625.9

Polymer (B): **poly(propyl methacrylate)** **2002BY3**
Characterization: M_w/g.mol^{-1} = 250000, Polysciences, Inc., Warrington, PA
Solvent (A): **propene** C_3H_6 **115-07-1**

Type of data: cloud points

w_B 0.049 was kept constant

T/K	312.25	328.55	344.05	364.25	385.25	409.15
P/bar	573.10	517.59	485.86	470.35	463.79	465.17

Polymer (B): **polystyrene** **2006PAR**
Characterization: M_w/g.mol^{-1} = 45000, Aldrich Chem. Co., Inc., Milwaukee, WI
Solvent (A): **dimethyl ether** C_2H_6O **115-10-6**

Type of data: cloud points

w_B	0.0103	0.0103	0.0103	0.0103	0.0103	0.0103	0.0103	0.0103	0.0103
T/K	300.7	308.4	317.5	329.7	337.6	348.8	357.6	367.5	378.2
P/MPa	25.5	26.9	28.5	30.5	32.0	34.0	35.4	37.2	39.0
w_B	0.0103	0.0103	0.0103	0.0293	0.0293	0.0293	0.0293	0.0293	0.0293
T/K	388.8	397.8	408.5	301.1	307.8	318.6	327.8	338.0	349.5
P/MPa	40.4	41.8	43.5	26.1	27.1	29.1	30.8	32.8	35.0
w_B	0.0293	0.0293	0.0293	0.0293	0.0293	0.0293	0.0499	0.0499	0.0499
T/K	357.8	368.9	378.3	389.2	398.4	408.3	301.0	310.3	318.0
P/MPa	36.3	38.3	39.7	41.3	42.4	44.0	25.2	26.9	28.4
w_B	0.0499	0.0499	0.0499	0.0499	0.0499	0.0499	0.0499	0.0499	0.0499
T/K	327.6	338.6	347.8	358.7	370.0	377.7	388.9	400.3	408.6
P/MPa	30.0	32.0	33.7	35.6	37.5	38.7	40.4	42.1	43.4

Polymer (B): **polystyrene** **2004TAN**
Characterization: M_n/g.mol^{-1} = 3800, M_w/g.mol^{-1} = 3950,
Polymer Source, Inc., Quebec, Canada
Solvent (A): **propane** C_3H_8 **74-98-6**

Type of data: cloud points

w_B 0.005 was kept constant

T/K	303.15	313.15	333.15	353.15	373.15	393.15	413.15	433.15	453.15
P/bar	1310.3	950.5	653.1	496.1	429.6	393.3	373.2	363.5	357.5

Polymer (B): **polystyrene** **2004TAN**
Characterization: M_n/g.mol^{-1} = 5500, M_w/g.mol^{-1} = 5610,
Polymer Source, Inc., Quebec, Canada
Solvent (A): **propane** C_3H_8 **74-98-6**

Type of data: cloud points

w_B 0.005 was kept constant

T/K	333.15	353.15	373.15	393.15	413.15	433.15	453.15
P/bar	1152.1	809.6	649.5	560.5	511.4	479.5	461.3

Polymer (B): **polystyrene** **2004TAN**
Characterization: M_n/g.mol^{-1} = 23000, M_w/g.mol^{-1} = 24600,
Polymer Source, Inc., Quebec, Canada
Solvent (A): **propane** C_3H_8 **74-98-6**

Type of data: cloud points

continued

continued

w_B	0.005	was kept constant		
T/K	393.15	413.15	433.15	453.15
P/bar	1379.4	1123.3	974.6	873.1

Polymer (B): **polystyrene** **2009WI2**
Characterization: M_n/g.mol^{-1} = 36800, M_w/M_n = 1.02, synthesized in the laboratory
Solvent (A): **propane** C_3H_8 **74-98-6**

Type of data: cloud points

w_B 0.005 was kept constant

T/K	413.15	413.05	413.05	413.15	413.15	423.05	423.15	433.25	433.15
P/bar	1439	1451	1454	1450	1450	1325	323	1223	1222

T/K	443.15	443.15	453.15	453.15
P/bar	1136	1136	1071	1069

Polymer (B): **polystyrene (deuterated)** **2009WI1**
Characterization: M_n/g.mol^{-1} = 41700, M_w/g.mol^{-1} = 43000, synthesized in the laboratory
Solvent (A): **propane** C_3H_8 **74-98-6**

Type of data: cloud points

w_B 0.005 was kept constant

T/K	413.15	413.15	423.15	433.15	443.15	453.15
P/bar	1336	1346	1237	1152	1077	1017

Polymer (B): **polystyrene** **2004WAN**
Characterization: M_n/g.mol^{-1} = 72650, M_w/g.mol^{-1} = 76430
Solvent (A): **propane** C_3H_8 **74-98-6**

Type of data: polymer solubility

T/K = 383.15

P/MPa	10.1	15.1	19.8	24.9	30.2	35.1
w_B	0.000084	0.000145	0.000209	0.000263	0.000295	0.000349

T/K = 388.15

P/MPa	10.0	15.1	20.1	24.9	30.0	35.1
w_B	0.000160	0.000214	0.000260	0.000322	0.000354	0.000439

T/K = 393.15

P/MPa	10.0	15.2	20.1	25.0	30.1	35.0
w_B	0.000219	0.000277	0.000344	0.000397	0.000449	0.000525

continued

continued

T/K = 398.15

P/MPa	10.0	14.9	19.9	25.0	30.1	35.1
w_B	0.000297	0.000361	0.000402	0.000447	0.000554	0.000596

T/K = 403.15

P/MPa	10.1	15.0	20.0	24.9	30.1	35.0
w_B	0.000402	0.000459	0.000501	0.000562	0.000653	0.000695

T/K = 408.15

P/MPa	10.0	15.1	20.1	25.2	30.1	35.2
w_B	0.000489	0.000544	0.000608	0.000640	0.000746	0.000824

Polymer (B): **polystyrene** **2004TAN**
Characterization: M_n/g.mol^{-1} = 87000, M_w/g.mol^{-1} = 88700,
Polymer Source, Inc., Quebec, Canada
Solvent (A): **propane** **C_3H_8** **74-98-6**

Type of data: cloud points

w_B 0.005 was kept constant

T/K	423.15	433.15	453.15
P/bar	1546.6	1429.2	1251.9

Polymer (B): **polystyrene** **2009WI2**
Characterization: M_n/g.mol^{-1} = 36800, M_w/M_n = 1.02,
synthesized in the laboratory
Solvent (A): **propene** **C_3H_6** **115-07-1**

Type of data: cloud points

w_B 0.005 was kept constant

T/K	303.25	303.15	313.05	313.15	333.25	333.05	353.25	353.25	373.25
P/bar	1532	1530	1358	1357	1128	1128	989	988	898
T/K	373.15	393.05	393.05	413.15	413.05	432.95	432.95	453.05	453.05
P/bar	897	837	836	796	796	764	764	741	741

Polymer (B): **poly(styrene-*b*-butadiene)** **2007TAN**
Characterization: M_n/g.mol^{-1} = 5400-*b*-5350, M_w/M_n = 1.03,
Polymer Source, Inc., Quebec, Canada
Solvent (A): **propane** **C_3H_8** **74-98-6**

Type of data: cloud points

w_B 0.005 was kept constant

T/K	303.2	313.0	333.3	353.6	92.8	433.2	451.6
P/bar	637.7	563.8	480.9	441.1	417.2	419.1	432.4

Polymer (B): **poly(styrene-*b*-butadiene)** **2007TAN**
Characterization: M_n/g.mol^{-1} = 9400-*b*-9000, M_w/M_n = 1.03,
Polymer Source, Inc., Quebec, Canada
Solvent (A): **propane** C_3H_8 **74-98-6**

Type of data: cloud points

w_B 0.005 was kept constant

T/K	293.1	303.1	313.2	333.2	353.2	373.4	393.1	413.1	433.2
P/bar	1097.8	1008.4	932.3	762.0	642.1	582.4	547.4	529.2	516.9

T/K	453.2
P/bar	507.4

Polymer (B): **poly(styrene-*b*-butadiene)** **2007TAN**
Characterization: M_n/g.mol^{-1} = 9100-*b*-65000, M_w/M_n = 1.04,
Polymer Source, Inc., Quebec, Canada
Solvent (A): **propane** C_3H_8 **74-98-6**

Type of data: cloud points

w_B 0.005 was kept constant

T/K	303.2	313.2	333.1	353.2	373.2	393.2	413.2	433.2	453.2
P/bar	987.0	868.5	737.3	678.9	644.1	622.6	608.8	599.3	575.8

Polymer (B): **poly(styrene-*b*-butadiene)** **2007TAN**
Characterization: M_n/g.mol^{-1} = 28400-*b*-13600, M_w/M_n = 1.03,
Polymer Source, Inc., Quebec, Canada
Solvent (A): **propane** C_3H_8 **74-98-6**

Type of data: cloud points

w_B 0.005 was kept constant

T/K	293.4	303.2	313.2	322.9	332.9	353.3	373.2	393.2	413.3
P/bar	1080.9	981.8	922.2	893.3	867.5	812.6	780.2	755.8	739.0

T/K	433.1	452.9
P/bar	728.2	724.1

Type of data: micellization pressure

w_B 0.005 was kept constant

T/K	358.0	363.7	373.5	373.4	383.3	393.3	403.2	413.1	423.0
P/bar	1556.7	1420.1	1225.2	1242.9	1106.3	1017.4	943.5	894.3	850.2

T/K	433.1
P/bar	890.0

Polymer (B): **poly(styrene-*b*-butadiene)** **2009WI2**
Characterization: M_n/g.mol^{-1} = 36800-*b*-35900, M_w/M_n = 1.01,
synthesized in the laboratory
Solvent (A): **propane** C_3H_8 **74-98-6**

Type of data: cloud points

w_B 0.005 was kept constant

T/K	293.15	313.15	333.35	342.95	373.15	393.25	413.05	433.25	453.25
P/bar	1159	978	871	803	761	732	714	702	862

T/K	453.35
P/bar	909

Type of data: micellization pressure

w_B 0.005 was kept constant

T/K	373.05	373.05	383.05	393.25	393.25	403.35	403.25	413.05	413.15
P/bar	1446	1452	1269	1151	1143	1055	1049	988	984

T/K	423.15	423.15	433.15	433.25	443.15	443.25	453.25	453.35
P/bar	935	932	890	891	859	863	874	915

Polymer (B): **poly(styrene-*b*-butadiene) (completely deuterated)** **2009WI1**
Characterization: M_n/g.mol^{-1} = 39700-*b*-12400, M_w/M_n = 1.02,
synthesized in the laboratory
Solvent (A): **propane** C_3H_8 **74-98-6**

Type of data: micellization pressure

w_B 0.005 was kept constant

T/K	293.15	313.25	313.05	333.25	353.15	373.15	393.25	413.15	433.05
P/bar	1036	927	905	865	808	788	785	774	760

Polymer (B): **poly(styrene-*b*-butadiene) (completely deuterated)** **2009WI1**
Characterization: M_n/g.mol^{-1} = 41700-*b*-40900, M_w/M_n = 1.02,
synthesized in the laboratory
Solvent (A): **propane** C_3H_8 **74-98-6**

Type of data: micellization pressure

w_B 0.005 was kept constant

T/K	293.05	313.15	333.05	353.25	373.15	393.15	413.15	433.25	453.15
P/bar	1005	866	783	731	700	680	669	663	686

Polymer (B): **poly(styrene-*b*-butadiene) (completely deuterated)** **2009WI1**
Characterization: M_n/g.mol^{-1} = 39700-*b*-117000, M_w/M_n = 1.01,
synthesized in the laboratory
Solvent (A): **propane** C_3H_8 **74-98-6**

Type of data: micellization pressure

w_B 0.005 was kept constant

T/K	293.25	313.15	333.15	353.25	373.15	393.15	413.25	433.15
P/bar	1139	966	863	797	755	728	711	693

Type of data: cloud points

w_B 0.005 was kept constant

T/K	443.25	443.35	433.25	453.25	453.25
P/bar	879	896	911	903	909

Polymer (B): **poly(styrene-*b*-butadiene) (only PS deuterated)** **2009WI1**
Characterization: M_n/g.mol^{-1} = 41700-*b*-40800, M_w/M_n = 1.02,
synthesized in the laboratory
Solvent (A): **propane** C_3H_8 **74-98-6**

Type of data: micellization pressure

w_B 0.005 was kept constant

T/K	293.15	313.15	333.15	353.25	373.15	393.35	413.05	433.25	453.25
P/bar	1163	972	859	791	748	719	702	693	689

Polymer (B): **poly(styrene-*b*-butadiene) (only PB deuterated)** **2009WI1**
Characterization: M_n/g.mol^{-1} = 37200-*b*-36700, M_w/M_n = 1.03,
synthesized in the laboratory
Solvent (A): **propane** C_3H_8 **74-98-6**

Type of data: micellization pressure

w_B 0.005 was kept constant

T/K	373.05	393.15	413.15	433.05	433.05	453.15	453.15
P/bar	1342	1192	1086	1021	1024	945	957

Polymer (B): **poly(styrene-*b*-butadiene) (only PB hydrogenated)** **2009WI2**
Characterization: M_n/g.mol^{-1} = 36800-*b*-37100, M_w/M_n = 1.01,
synthesized in the laboratory
Solvent (A): **propene** C_3H_6 **115-07-1**

Type of data: cloud points

continued

continued

w_B 0.005 was kept constant

T/K	353.15	373.15	393.25	413.15	433.15	453.15
P/bar	630	609	598	593	592	587

Type of data: micellization pressure

w_B 0.005 was kept constant

T/K	373.05	373.25	393.15	393.15	413.15	433.25	433.45	453.15	453.15
P/bar	1292	1314	1042	1048	899	809	806	777	776

Polymer (B): **poly(styrene-*b*-butadiene)** **2009WI2**
Characterization: M_n/g.mol^{-1} = 36800-*b*-35900, M_w/M_n = 1.01, synthesized in the laboratory
Solvent (A): **propene** C_3H_6 **115-07-1**

Type of data: cloud points

w_B 0.005 was kept constant

T/K	293.75	313.05	333.35	353.15	373.35	393.45	413.15	413.15	433.15
P/bar	576	573	577	585	594	604	614	711	714

T/K	433.15	453.25
P/bar	696	685

Type of data: micellization pressure

w_B 0.005 was kept constant

T/K	293.75	313.35	313.35	333.75	353.15	373.35	393.25	413.15
P/bar	1080	891	887	782	727	685	665	688

Polymer (B): **poly(styrene-*b*-butadiene) (only PB hydrogenated)** **2009WI2**
Characterization: M_n/g.mol^{-1} = 36800-*b*-37100, M_w/M_n = 1.01, synthesized in the laboratory
Solvent (A): **propene** C_3H_6 **115-07-1**

Type of data: cloud points

w_B 0.005 was kept constant

T/K	353.15	363.25	373.15	383.35	393.15	413.15	423.15	433.15	443.05
P/bar	702	680	664	652	643	629	625	622	620

T/K	453.15	473.15
P/bar	619	617

continued

continued

Type of data: micellization pressure

w_B	0.005	was kept constant							
T/K	353.15	363.25	373.15	383.35	393.15	413.15	423.15	433.25	443.05
P/bar	740	711	690	675	663	644	639	634	631
T/K	453.15	473.15							
P/bar	628	622							

Polymer (B): **poly(styrene-*b*-isoprene)** **2007TAN**
Characterization: M_n/g.mol^{-1} = 9000-*b*-23000, M_w/M_n = 1.01, synthesized in the laboratory of J. Mays, University of Tennessee at Knoxville
Solvent (A): **propane** C_3H_8 **74-98-6**

Type of data: cloud points

w_B	0.005	was kept constant							
T/K	293.2	313.2	333.2	353.2	373.0	393.2	413.2	433.0	453.0
P/bar	489.7	410.6	381.8	376.2	378.3	382.6	378.8	383.6	384.8

Polymer (B): **poly(styrene-*b*-isoprene)** **2007TAN**
Characterization: M_n/g.mol^{-1} = 11500-*b*-10500, M_w/M_n = 1.04, Polymer Source, Inc., Quebec, Canada
Solvent (A): **propane** C_3H_8 **74-98-6**

Type of data: cloud points

w_B	0.005	was kept constant					
T/K	293.1	313.1	333.2	353.1	373.1	413.0	453.3
P/bar	451.2	438.0	441.8	451.8	471.6	502.5	504.6

Type of data: micellization pressure

w_B	0.005	was kept constant							
T/K	333.2	328.0	393.8	395.0	323.3	323.2	348.1	348.1	372.7
P/bar	1002.0	1002.1	551.4	555.8	1165.6	1092.5	787.3	767.5	680.3
T/K	372.9	372.9	373.0	373.0	33.0	333.0			
P/bar	639.3	616.0	639.6	611.4	948.1	901.5			

Polymer (B): **poly(styrene-*b*-isoprene) (only PS deuterated)** **2009WI1**
Characterization: M_n/g.mol^{-1} = 41700-*b*-42200, M_w/M_n = 1.02, synthesized in the laboratory
Solvent (A): **propane** C_3H_8 **74-98-6**

continued

continued

Type of data: micellization pressure

w_B 0.005 was kept constant

T/K	293.15	313.35	333.15	353.15	373.05	393.25	413.15	433.05	435.05
P/bar	483	464	459	462	470	481	492	502	525

Polymer (B): **poly[styrene-*b*-4-(perfluorooctyl(ethyleneoxy)-methyl)styrene]** **2004LAC**

Characterization: M_n/g.mol^{-1} = 75700 (by SEC), PS-block: M_n/g.mol^{-1} = 3600, P(PFOEO)MS-block: M_n/g.mol^{-1} = 55700 by NMR or = 72100 by SEC, PS-content is about 6 mol%, synthesized in the laboratory

Solvent (A): **carbon dioxide** **CO_2** **124-38-9**

Type of data: cloud points

w_B 0.040 was kept constant

T/K	298.15	303.15	308.15	313.15	318.15	323.15	328.15	333.15	338.15
P/bar	199.0	212.0	229.0	244.0	260.0	273.0	288.0	307.0	322.0

Polymer (B): **poly[styrene-*b*-4-(perfluorooctyl(ethyleneoxy)-methyl)styrene]** **2004LAC**

Characterization: M_n/g.mol^{-1} = 54200 (by SEC), PS-block: M_n/g.mol^{-1} = 3800, P(PFOEO)MS-block: M_n/g.mol^{-1} = 47000 by NMR or = 50400 by SEC, PS-content is about 7 mol%, synthesized in the laboratory

Solvent (A): **carbon dioxide** **CO_2** **124-38-9**

Type of data: cloud points

w_B 0.039 was kept constant

T/K	298.35	303.15	308.05	313.15	317.95	322.55	327.75	332.65	337.95
P/bar	121.0	139.0	157.0	176.0	190.0	205.0	221.0	234.0	249.0

Polymer (B): **poly(1,1,2,2-tetrahydroperfluorodecyl acrylate)** **2004LAC**

Characterization: M_n/g.mol^{-1} = 26300, synthesized in the laboratory

Solvent (A): **carbon dioxide** **CO_2** **124-38-9**

Type of data: cloud points

w_B 0.038 was kept constant

T/K	298.15	303.15	308.15	313.15	318.15	323.15	328.15	333.15	338.15
P/bar	82.0	96.0	114.0	127.0	141.0	153.0	166.0	179.0	191.0

Polymer (B): **poly(1,1,2,2-tetrahydroperfluorodecyl acrylate)** **2004MA1**
Characterization: M_n/g.mol^{-1} = 33000, synthesized in the laboratory
Solvent (A): **carbon dioxide** CO_2 **124-38-9**

Type of data: cloud points

w_B 0.04 was kept constant

T/K	293.85	298.55	303.25	307.95	312.85	318.25	323.45	328.65	334.55
P/bar	68.9	87.4	104.0	119.8	135.1	151.0	166.9	181.8	198.0

T/K	339.05
P/bar	210.8

Polymer (B): **poly(1,1,2,2-tetrahydroperfluorodecyl acrylate)** **2004LAC**
Characterization: M_n/g.mol^{-1} = 52600, synthesized in the laboratory
Solvent (A): **carbon dioxide** CO_2 **124-38-9**

Type of data: cloud points

w_B 0.040 was kept constant

T/K	298.15	303.15	308.15	313.15	318.15	323.15	328.15	333.15	338.15
P/bar	88.0	105.0	120.0	136.0	149.0	161.0	176.0	187.0	200.0

Polymer (B): **poly(1,1,2,2-tetrahydroperfluorodecyl acrylate)** **2004LAC**
Characterization: M_w/g.mol^{-1} = 120000, synthesized in the laboratory
Solvent (A): **carbon dioxide** CO_2 **124-38-9**

Type of data: cloud points

w_B 0.039 was kept constant

T/K	292.65	298.35	303.45	308.15	313.45	319.25	324.05	328.65	334.35
P/bar	73.0	89.0	110.0	124.0	140.0	156.0	171.0	185.0	203.0

T/K	338.95
P/bar	219.0

Polymer (B): **poly(2,2,2-trifluoroethyl methacrylate)** **2007KWO**
Characterization: M_n/g.mol^{-1} = 164000, M_w/g.mol^{-1} = 268000, synthesized in the laboaratory
Solvent (A): **carbon dioxide** CO_2 **124-38-9**

Type of data: cloud points

w_B 0.0391 was kept constant

T/K	316.9	333.2	355.8	376.2	394.0	409.8	431.9	451.9
P/MPa	26.31	32.34	40.27	46.31	50.74	54.20	57.51	59.62

Polymer (B): **poly(vinyl acetate-*alt*-dibutyl maleate)** **2010LE1**
Characterization: M_n/g.mol^{-1} = 3800, M_w/M_n = 1.3, 52 mol% vinyl acetate, synthesized in the laboratory
Solvent (A): **carbon dioxide** **CO_2** **124-38-9**

Type of data: cloud points

w_B 0.002 was kept constant

T/K	308.15	318.15	328.15	338.15	348.15
P/bar	182	195	225	234	248

Polymer (B): **poly(vinyl acetate-*alt*-dibutyl maleate)** **2010LE1**
Characterization: M_n/g.mol^{-1} = 7000, M_w/M_n = 1.2, 51 mol% vinyl acetate, synthesized in the laboratory
Solvent (A): **carbon dioxide** **CO_2** **124-38-9**

Type of data: cloud points

w_B 0.002 was kept constant

T/K	308.15	318.15	328.15	338.15	348.15
P/bar	204	248	268	295	305

Polymer (B): **poly(vinylbenzylphosphonic acid diethylester)** **2011RIB**
Characterization: M_n/g.mol^{-1} = 1900, synthesized in the laboratory
Solvent (A): **carbon dioxide** **CO_2** **124-38-9**

Type of data: cloud points

w_B	0.0016	0.0016	0.0016	0.0016	0.0016	0.0016	0.0016	0.0016	0.0016
T/K	298.45	304.25	308.85	313.55	318.35	323.15	327.85	332.65	337.25
P/bar	304.1	317.8	322.9	329.7	325.1	333.4	339.5	348.2	364.6
ρ/(g/cm^3)	0.9672	0.9521	0.9376	0.9237	0.9037	0.8903	0.8763	0.8637	0.8568

w_B	0.0019	0.0019	0.0019	0.0019	0.0019	0.0019	0.0019	0.0019
T/K	293.85	298.55	303.45	309.15	313.75	318.55	322.25	327.25
P/bar	330.7	332.8	345.0	350.6	355.2	364.7	382.9	406.4
ρ/(g/cm^3)	0.9941	0.9786	0.9665	0.9491	0.9351	0.9227	0.9186	0.9127

Polymer (B): **poly(vinylbenzylphosphonic acid diethylester-*b*-1,1,2,2-tetrahydroperfluorodecyl acrylate)** **2011RIB**
Characterization: M_n/g.mol^{-1} = 17400, 17.8 mol% vinylbenzylphosphonic acid diethylester, synthesized in the laboratory
Solvent (A): **carbon dioxide** **CO_2** **124-38-9**

Type of data: cloud points

continued

continued

w_B 0.040 was kept constant

T/K	293.85	299.35	303.65	308.35	313.65	318.15	322.95	327.75	332.65
P/bar	207.7	239.8	254.5	265.1	262.7	276.9	288.7	303.1	313.8
ρ/(g/cm^3)	0.9383	0.9326	0.9228	0.9094	0.8857	0.8760	0.8638	0.8539	0.8416

T/K	337.85
P/bar	320.3
ρ/(g/cm^3)	0.8256

Polymer (B): **poly(vinylbenzylphosphonic acid diethylester-*b*-1,1,2,2-tetrahydroperfluorodecyl acrylate)** **2011RIB**

Characterization: M_n/g.mol^{-1} = 19100, 23.7 mol% vinylbenzylphosphonic acid diethylester, synthesized in the laboratory

Solvent (A): **carbon dioxide** **CO_2** **124-38-9**

Type of data: cloud points

w_B 0.040 was kept constant

T/K	297.75	303.75	309.05	313.45	318.55	322.95	327.95	332.95	337.65
P/bar	263.4	285.5	302.5	320.5	327.7	333.5	339.0	346.1	353.8
ρ/(g/cm^3)	0.9513	0.9387	0.9269	0.9195	0.9043	0.8911	0.8756	0.8613	0.8488

Polymer (B): **poly(vinylbenzylphosphonic acid diethylester-*co*-1,1,2,2-tetrahydroperfluorodecyl acrylate)** **2011RIB**

Characterization: M_n/g.mol^{-1} = 34100, 16.9 mol% vinylbenzylphosphonic acid diethylester, gradient copolymer, synthesized in the laboratory

Solvent (A): **carbon dioxide** **CO_2** **124-38-9**

Type of data: cloud points

w_B 0.040 was kept constant

T/K	294.15	296.75	302.75	307.85	313.05	317.55	322.25	326.85	331.65
P/bar	75.2	89.8	110.1	126.7	142.7	157.8	172.3	186.1	200.5
ρ/(g/cm^3)	0.8092	0.8124	0.7958	0.7824	0.7689	0.7607	0.7511	0.7422	0.7338

T/K	336.55
P/bar	213.6
ρ/(g/cm^3)	0.7239

Polymer (B): **poly(vinylbenzylphosphonic acid diethylester-*co*-1,1,2,2-tetrahydroperfluorodecyl acrylate)** **2011RIB**

Characterization: M_n/g.mol^{-1} = 13200, 18.8 mol% vinylbenzylphosphonic acid diethylester, gradient copolymer, synthesized in the laboratory

Solvent (A): **carbon dioxide** **CO_2** **124-38-9**

Type of data: cloud points

continued

continued

w_B 0.041 was kept constant

T/K	294.65	297.15	303.35	307.65	313.15	318.35	323.25	327.95	333.15
P/bar	69.1	80.0	100.3	114.6	132.3	148.2	162.5	176.0	190.6
ρ/(g/cm^3)	0.7895	0.7881	0.7701	0.7601	0.7480	0.7367	0.7263	0.7172	0.7078

T/K	337.85
P/bar	203.5
ρ/(g/cm^3)	0.6998

Polymer (B): **poly(vinylbenzylphosphonic acid diethylester-*co*-1,1,2,2-tetrahydroperfluorodecyl acrylate)** **2011RIB**
Characterization: M_n/g.mol^{-1} = 12900, 18.8 mol% vinylbenzylphosphonic diacid, gradient copolymer, synthesized in the laboratory
Solvent (A): **carbon dioxide** **CO_2** **124-38-9**

Type of data: cloud points

w_B 0.039 was kept constant

T/K	293.85	298.35	303.35	308.55	313.65	17.95	322.95	328.35
P/bar	71.0	88.8	106.2	123.0	138.6	152.6	167.8	183.0
ρ/(g/cm^3)	0.8033	0.7952	0.7830	0.7695	0.7565	0.7485	0.7384	0.7270

Polymer (B): **poly(vinylbenzylphosphonic diacid-*b*-1,1,2,2-tetrahydroperfluorodecyl acrylate)** **2011RIB**
Characterization: M_n/g.mol^{-1} = 17000, 17.8 mol% vinylbenzylphosphonic diacid, synthesized in the laboratory
Solvent (A): **carbon dioxide** **CO_2** **124-38-9**

Type of data: cloud points

w_B 0.040 was kept constant

T/K	305.35	308.55	313.65	318.55	322.35	327.55	333.25	337.75
P/bar	413.4	328.5	331.3	296.5	274.2	268.8	273.5	250.4
ρ/(g/cm^3)	0.9860	0.9413	0.9242	0.8866	0.8564	0.8295	0.8080	0.7644

Polymer (B): **poly(vinylidene fluoride-*co*-hexafluoropropylene)** **2006AHM**
Characterization: M_n/g.mol^{-1} = 54000, M_w/g.mol^{-1} = 140000, T_g/K = 253, 23.1 mol% hexafluoropropylene, Tecnoflon N215, Solvay-Solexis
Solvent (A): **carbon dioxide** **CO_2** **124-38-9**

Type of data: cloud points

T/K = 313.15

w_B	0.0081	0.0095	0.0142	0.0240	0.0295	0.0336	0.0380	0.0449
P/bar	234.3	217.3	211.5	211.2	189.2	185.7	168.7	163.9

4.2. Table of binary systems where data were published only in graphical form as phase diagrams or related figures

Polymer (B)	Solvent (A)	Ref.
Poly(allyl acetate)		
	carbon dioxide	2007KIL
Poly(benzyl acrylate)		
	dimethyl ether	2010JAN
Poly(benzyl methacrylate)		
	dimethyl ether	2010JAN
Poly[(2,2-bis(hydroxymethyl)propionate)-*co*-(6-hydroxyhexanoate)]		
	carbon dioxide	2013GRE
Poly[2,2-bis(4-trifluorovinyloxyphenyl)propane]		
	carbon dioxide	2007LI3
	propane	2007LI3
Poly[2,2-bis(4-trifluorovinyloxyphenyl)propane-*co*-2,2-bis(4-trifluorovinyloxyphenyl)-1,1,1,3,3,3-hexafluoropropane]		
	carbon dioxide	2007LI3
	propane	2007LI3
1,4-*cis*-Poly(butadiene)		
	propane	2009WI3
Poly(ε-caprolactone)		
	carbon dioxide	2005COT
	carbon dioxide	2013MAR

Polymer (B)	Solvent (A)	Ref.
Polycarbonate		
	carbon dioxide	2004HAT
Poly(cyclohexene carbonate-*b*-ethylene oxide-*b*-cyclohexene carbonate)		
	carbon dioxide	2000SAR
Poly(1H,1H-dihydroperfluorooctyl methacrylate)		
	carbon dioxide	2009SAN
Poly[2-(*N,N*-dimethylaminoethyl) methacrylate-*co*-1H,1H–perfluorooctyl methacrylate]		
	carbon dioxide	2008HWA
Poly(dimethylsiloxane)		
	carbon dioxide	2006KOS
(propyldimethylamine functionalized)	carbon dioxide	2009KIL
(propyl acetate functionalized)	carbon dioxide	2009KIL
Poly[dimethylsiloxane-*co*-(3-acetoxypropyl)-methylsiloxane]		
	carbon dioxide	1999FIN
Poly(dimethylsiloxane-*co*-hexylmethylsiloxane)		
	carbon dioxide	1999FIN
Poly(dimethylsiloxane-*co*-hydromethylsiloxane)		
	carbon dioxide	1999FIN
Poly(dimethylsiloxane) monomethacrylate		
	carbon dioxide	2005TAI
Poly[dodecyl(oligoethylene oxide)-*b*-propylene oxide)		
	carbon dioxide	2006SUB
Poly(ethyl acrylate)		
	ethene	2004BEY

Polymer (B)	**Solvent (A)**	**Ref.**
Polyethylene		
	n-butane	2009KOJ
	2-chlorobutane	2009KOJ
	chloroethane	2009KOJ
	2-chloropropane	2009KOJ
	dichloromethane	2009KOJ
	ethane	2007GRE
	ethene	2002LE1
	ethene	2004BEY
	ethene	2007GRE
	ethene	2013COS
	propane	2007GRE
	trichlorofluoromethane	2009KOJ
Poly(ethylene-*co*-acrylic acid)		
	ethene	2002LE1
	ethene	2004BEY
Poly(ethylene-*co*-1-butene)		
	dimethyl ether	2005LID
Poly(ethylene-*co*-ethyl acrylate)		
	ethene	2004BEY
Poly(ethylene-*co*-1-hexene)		
	ethene	2004DOE
Poly(ethylene-*co*-methacrylic acid)		
	ethene	2006KLE
Poly(ethylene-*co*-propyl acrylate)		
	ethene	2004BEY
Poly(ethylene-*co*-vinyl acetate)		
	ethene	2004DOE
	vinyl acetate	2004DOE

Polymer (B)	Solvent (A)	Ref.
Poly(ethylene glycol)		
	carbon dioxide	2004KUK
	carbon dioxide	2008PA2
Poly(ethylene glycol-*b*-ε-caprolactone)		
	trifluoromethane	2009TYR
Poly(ethylene oxide-*b*-1,1,2,2-tetrahydroperfluoro-decyl methacrylate)		
	carbon dioxide	2004MA1
	carbon dioxide	2004MA2
Poly(ethylenimine), hyperbranched		
	carbon dioxide	2006MAR
Poly(fluorooxetane-*b*-ethylene oxide-*b*-fluorooxetane)		
	carbon dioxide	2011PRI
Poly(1H,1H,2H,2H-heptadecafluorodecyl methacrylate)		
	carbon dioxide	2008SH2
Poly(1H,1H,2H,2H-heptadecafluorodecyl methacrylate-*co*-methyl methacrylate)		
	carbon dioxide	2008SH2
Poly(DL-lactic acid)		
	carbon dioxide	2006BY5
	carbon dioxide	2008TAP
	carbon dioxide	2010MAS
	chlorodifluoromethane	2006BY5
	dichloromethane	2006BY5
	trichloromethane	2006BY5
	trifluoromethane	2006BY5
Poly(L-lactic acid)		
	carbon dioxide	2009GRE
	carbon dioxide	2010MAS

Polymer (B)	Solvent (A)	Ref.
Poly(DL-lactic acid-*co*-glycolic acid)		
	carbon dioxide	2006BY5
	chlorodifluoromethane	2006BY5
	dichloromethane	2006BY5
	trichloromethane	2006BY5
	trifluoromethane	2006BY5
Poly(DL-lactic acid-*b*-perfluoropropylene oxide)		
	carbon dioxide	2007EDM
Poly(methyl methacrylate)		
	carbon dioxide	2005BON
Poly(4-methyl-1-pentene)		
	n-pentane	2006FA1
Poly[3-octylthiophene-*co*-2-(3-thienyl)acetyl 3,3,4,4,5,5,6,6,7,7,8,8,8-tridecafluoro-1-octanate]		
	carbon dioxide	2006GAN
Poly[oligo(ethylene glycol) methacrylate-*b*-1H,1H,2H,2H-perfluorooctyl methacrylate]		
	carbon dioxide	2004HWA
Poly[oligo(ethylene glycol) methacrylate-*ran*-1H,1H,2H,2H-perfluorooctyl methacrylate]		
	carbon dioxide	2004HWA
Poly(pentafluoropropyl methacrylate)		
	carbon dioxide	2012YOO
	dimethyl ether	2012YOO
Poly(pentyl methacrylate)		
	carbon dioxide	2006LAV
Poly[2-(perfluorooctyl)ethyl acrylate-*co*-acrylic acid]		
	carbon dioxide	2008YOS

Polymer (B)	Solvent (A)	Ref.
Poly[2-(perfluorooctyl)ethyl acrylate-*co*-*tert*-butyl acrylate]		
	carbon dioxide	2012YOS
Poly[2-(perfluorooctyl)ethyl acrylate-*ran*-2-(dimethylamino)ethyl acrylate]		
	carbon dioxide	2007YO1
Poly[4-(perfluorooctyl(ethyleneoxy)methyl)styrene]		
	carbon dioxide	2004LAC
Poly(propyl acrylate)		
	ethene	2004BEY
Polypropylene		
	carbon dioxide	2007KIL
	carbon dioxide	2010HAS
	propene	2011RU1
	propene	2011RU2
Poly(propylene glycol)		
	carbon dioxide	2000SAR
	carbon dioxide	2007KIL
Poly(propylene glycol) acetate		
	carbon dioxide	2000SAR
Poly(propylene glycol) diacetate		
	carbon dioxide	2007KIL
Poly(propylene glycol) dimethyl ether		
	carbon dioxide	2007KIL
Poly(propylene glycol) monobutyl ether		
	carbon dioxide	2000SAR
Poly(propylene glycol) monomethyl ether		
	carbon dioxide	2007KIL

Polymer (B)	Solvent (A)	Ref.
Poly(propylene oxide-*co*-propylene carbonate)	carbon dioxide	2000SAR
Polystyrene	carbon dioxide	2005BON
	carbon dioxide	2006LAV
	propane	2009WI3
	2-propanone	2006FA2
Poly(styrene-*b*-butadiene)	propane	2009WI3
Poly(styrene-*b*-dimethylsiloxane)	carbon dioxide	2000BER
Poly(styrene-*b*-isoprene)	propane	2006WIN
	propane	2009WI3
Poly(styrene-*b*-pentyl methacrylate)	carbon dioxide	2006LAV
Poly[styrene-*b*-4-(perfluorooctyl(ethyleneoxy)-methyl)styrene]	carbon dioxide	2004LAC
Poly(styrene-*b*-1,1,2,2-tetrahydroperfluorodecyl acrylate)	carbon dioxide	2006AND
Poly(styrene-*b*-1,1,2,2-tetrahydroperfluorodecyl methacrylate)	carbon dioxide	2004LAC
Poly(tetrafluoroethylene)	carbon dioxide	2008GUN

Polymer (B)	Solvent (A)	Ref.
Poly(tetrafluoroethylene-*co*-vinyl acetate)		
	carbon dioxide	2004BAR
Poly(1,1,2,2-tetrahydroperfluorodecyl acrylate)		
	carbon dioxide	2006AND
Poly(1,1,2,2-tetrahydroperfluorodecyl methacrylate)		
	carbon dioxide	2004LAC
	carbon dioxide	2004MA1
Poly[2-(3-thienyl)acetyl 3,3,4,4,5,5,6,6,7,7,8,8,8-tridecafluoro-1-octanate]		
	carbon dioxide	2006GAN
Poly[2-(3-thienyl)ethyl heptafluorobutyrate]		
	carbon dioxide	2006GAN
Poly[2-(3-thienyl)ethyl pentadecafluorooctanoate]		
	carbon dioxide	2006GAN
Poly[2-(3-thienyl)methyl heptafluorobutyrate]		
	carbon dioxide	2006GAN
Poly(vinyl acetate)		
	carbon dioxide	2007KIL
	carbon dioxide	2008TAP
	carbon dioxide	2012GIR
Poly(vinyl acetate) (end-group modified)		
	carbon dioxide	2009TAN
Poly(vinyl acetate-*alt*-dibutyl maleate)		
	carbon dioxide	2010LE1
Poly(vinyl acetate-*alt*-dibutyl maleate) xanthate		
	carbon dioxide	2010LE1

Polymer (B)	**Solvent (A)**	**Ref.**
Poly(vinyl acetate-*co*-1-methoxyethyl ether)	carbon dioxide	2009WAN
Poly(vinyl acetate-*co*-methoxymethyl ether)	carbon dioxide	2009WAN
Poly[vinyl acetate-*stat*-1-(trifluoromethyl)-vinyl acetate]	carbon dioxide	2012GIR
Poly(vinyl ethyl ether)	carbon dioxide	2007KIL
Poly(vinylidene fluoride)	carbon dioxide	2004BY4
	carbon dioxide	2005BON
	chlorodifluoromethane	2004BY4
	dimethyl ether	2004BY4
	trifluoromethane	2004BY4
Poly(vinyl methyl ether)	carbon dioxide	2007KIL
Poly[1-O-(vinyloxy)ethyl-2,3,4,6-tetra-O-acetyl-β-D-glucopyranoside]	carbon dioxide	2008TAP

4.3. Cloud-point and/or coexistence curves of quasiternary and/or quasiquaternary solutions

Polymer (B): **dextran** **2005PER**
Characterization: M_w/g.mol^{-1} = 68800
Solvent (A): **carbon dioxide** CO_2 **124-38-9**
Solvent (C): **dimethylsulfoxide** C_2H_6OS **67-68-5**

Type of data: cloud points

T/K = 313.15 *P*/MPa = 12.0

w_A	0.136	0.152	0.146	0.138	0.173	0.171	0.186	0.187	0.198
w_B	0.164	0.105	0.102	0.097	0.070	0.069	0.035	0.033	0.008
w_C	0.700	0.744	0.752	0.766	0.758	0.760	0.779	0.781	0.794
w_A	0.193								
w_B	0.001								
w_C	0.806								

Type of data: cloud points (VL→L)

w_A	0.2000	0.2000	0.2000	0.2000	0.2000	0.2000	0.2000	0.2000	0.2000
w_B	0.0010	0.0010	0.0010	0.0010	0.0010	0.0010	0.0010	0.0010	0.0010
w_C	0.7990	0.7990	0.7990	0.7990	0.7990	0.7990	0.7990	0.7990	0.7990
T/K	283.69	293.50	303.50	308.20	314.04	318.35	323.60	333.63	343.37
P/MPa	2.390	2.800	3.380	3.615	3.980	4.305	4.685	5.465	6.190
w_A	0.2000	0.2000	0.1001	0.1001	0.1001	0.1001	0.1001	0.1001	0.1001
w_B	0.0010	0.0010	0.0085	0.0085	0.0085	0.0085	0.0085	0.0085	0.0085
w_C	0.7990	0.7990	0.8914	0.8914	0.8914	0.8914	0.8914	0.8914	0.8914
T/K	353.71	363.60	283.69	288.42	293.49	298.31	303.60	308.37	313.24
P/MPa	7.040	7.905	1.262	1.462	1.597	1.707	1.902	2.047	2.262
w_A	0.1001	0.1001	0.1001	0.1001	0.1001	0.1001	0.1501	0.1501	0.1501
w_B	0.0085	0.0085	0.0085	0.0085	0.0085	0.0085	0.0085	0.0085	0.0085
w_C	0.8914	0.8914	0.8914	0.8914	0.8914	0.8914	0.8414	0.8414	0.8414
T/K	318.28	323.48	333.48	343.39	353.50	363.48	282.56	293.19	303.27
P/MPa	2.392	2.567	2.967	3.317	3.787	4.202	1.736	2.157	2.641
w_A	0.1501	0.1501	0.1501	0.1501	0.1852	0.1852	0.1852	0.1852	0.1852
w_B	0.0085	0.0085	0.0085	0.0085	0.0085	0.0085	0.0085	0.0085	0.0085
w_C	0.8414	0.8414	0.8414	0.8414	0.8063	0.8063	0.8063	0.8063	0.8063
T/K	313.27	326.00	343.37	360.53	283.30	293.78	303.37	311.88	328.33
P/MPa	3.146	3.887	4.991	6.206	2.023	2.528	3.068	3.603	4.678

continued

continued

w_A	0.1852	0.1852	0.1852	0.2003	0.2003	0.1743	0.1743	0.1743	0.1743
w_B	0.0085	0.0085	0.0085	0.0085	0.0085	0.0410	0.0410	0.0410	0.0410
w_C	0.8063	0.8063	0.8063	0.7912	0.7912	0.7847	0.7847	0.7847	0.7847
T/K	333.81	343.39	363.66	283.51	293.32	282.01	293.46	303.83	313.28
P/MPa	5.053	5.758	7.288	2.164	2.664	2.218	2.753	3.293	3.863

Type of data: cloud points (VL→VLL)

w_A	0.2003	0.2003	0.2003	0.2003	0.2003	0.2003	0.2116	0.2116	0.2116
w_B	0.0085	0.0085	0.0085	0.0085	0.0085	0.0085	0.0085	0.0085	0.0085
w_C	0.7912	0.7912	0.7912	0.7912	0.7912	0.7912	0.7799	0.7799	0.7799
T/K	303.15	313.33	323.57	334.52	343.35	363.60	283.34	293.45	303.71
P/MPa	3.199	3.873	4.548	5.314	5.904	7.714	2.170	2.670	3.260
w_A	0.2116	0.2116	0.2116	0.2116	0.2116	0.1743	0.1743	0.1743	0.1743
w_B	0.0085	0.0085	0.0085	0.0085	0.0085	0.0410	0.0410	0.0410	0.0410
w_C	0.7799	0.7799	0.7799	0.7799	0.7799	0.7847	0.7847	0.7847	0.7847
T/K	313.60	323.54	343.38	353.73	363.52	323.71	333.61	343.54	353.47
P/MPa	3.855	4.435	5.925	6.855	7.890	4.573	5.213	5.953	6.693
w_A	0.1743	0.1937	0.1937	0.1937	0.1937	0.1937	0.1937	0.1937	0.2000
w_B	0.0410	0.0410	0.0410	0.0410	0.0410	0.0410	0.0410	0.0410	0.0830
w_C	0.7847	0.7653	0.7653	0.7653	0.7653	0.7653	0.7653	0.7653	0.7170
T/K	363.47	283.26	293.33	303.44	313.37	323.61	343.40	363.98	293.32
P/MPa	7.668	2.140	2.630	3.180	3.770	4.410	5.860	7.675	2.700
w_A	0.2000	0.2000	0.2000	0.2000	0.2000	0.2000	0.2000		
w_B	0.0830	0.0830	0.0830	0.0830	0.0830	0.0830	0.0830		
w_C	0.7170	0.7170	0.7170	0.7170	0.7170	0.7170	0.7170		
T/K	303.33	313.29	323.69	333.59	343.38	353.73	363.52		
P/MPa	3.160	3.715	4.255	5.085	5.925	6.855	7.890		

Type of data: cloud points (VLL→LL)

w_A	0.2003	0.2003	0.2003	0.2003	0.2003	0.2003	0.2116	0.2116	0.2116
w_B	0.0085	0.0085	0.0085	0.0085	0.0085	0.0085	0.0085	0.0085	0.0085
w_C	0.7912	0.7912	0.7912	0.7912	0.7912	0.7912	0.7799	0.7799	0.7799
T/K	303.86	313.30	323.57	333.34	343.34	363.59	283.33	293.41	303.65
P/MPa	3.388	4.053	4.763	5.509	6.304	8.054	2.545	3.230	3.925
w_A	0.2116	0.2116	0.2116	0.2116	0.2116	0.1743	0.1743	0.1743	0.1743
w_B	0.0085	0.0085	0.0085	0.0085	0.0085	0.0410	0.0410	0.0410	0.0410
w_C	0.7799	0.7799	0.7799	0.7799	0.7799	0.7847	0.7847	0.7847	0.7847
T/K	313.59	323.41	343.38	353.70	363.50	333.30	343.38	353.35	363.64
P/MPa	4.720	5.550	7.030	8.250	9.560	5.303	6.093	7.013	8.078
w_A	0.1937	0.1937	0.1937	0.1937	0.1937	0.2000	0.2000	0.2000	0.2000
w_B	0.0410	0.0410	0.0410	0.0410	0.0410	0.0830	0.0830	0.0830	0.0830
w_C	0.7653	0.7653	0.7653	0.7653	0.7653	0.7170	0.7170	0.7170	0.7170
T/K	283.40	293.18	303.30	313.35	323.56	293.36	303.36	313.28	321.92
P/MPa	2.295	2.850	3.445	4.200	5.000	2.995	3.700	4.395	5.010

continued

continued

w_A	0.2000	0.2000	0.2000	0.2000
w_B	0.0830	0.0830	0.0830	0.0830
w_C	0.7170	0.7170	0.7170	0.7170
T/K	333.57	343.38	353.70	363.50
P/MPa	6.120	7.030	8.250	9.560

Type of data: cloud points (LL→L)

w_A	0.2003	0.2003	0.2003	0.1743	0.1743	0.1743	0.1743
w_B	0.0085	0.0085	0.0085	0.0410	0.0410	0.0410	0.0410
w_C	0.7912	0.7912	0.7912	0.7847	0.7847	0.7847	0.7847
T/K	299.58	300.60	302.74	324.11	324.70	325.97	329.02
P/MPa	4.149	7.049	12.049	5.048	6.048	7.048	12.048

Polymer (B): **dextran** **2005PER**
Characterization: M_w/g.mol^{-1} = 68800
Solvent (A): **carbon dioxide** CO_2 **124-38-9**
Solvent (C): **dimethylsulfoxide** C_2H_6OS **67-68-5**
Solvent (D): **water** H_2O **7732-18-5**

Type of data: cloud points

w_A	0.1997	0.1997	0.1997	0.1997	0.1997	0.1997	0.1997	0.1997	0.1997
w_B	0.0140	0.0140	0.0140	0.0140	0.0140	0.0140	0.0140	0.0140	0.0140
w_C	0.7763	0.7763	0.7763	0.7763	0.7763	0.7763	0.7763	0.7763	0.7763
w_D	0.0100	0.0100	0.0100	0.0100	0.0100	0.0100	0.0100	0.0100	0.0100
T/K	283.65	293.17	303.15	313.33	323.54	333.34	343.34	363.00	303.86
P/MPa	2.169	2.674	3.224	3.872	4.574	5.314	5.909	7.724	3.349

w_A	0.1997	0.1997	0.1997	0.1997	0.1997	0.1997	0.1997	0.1997
w_B	0.0140	0.0140	0.0140	0.0140	0.0140	0.0140	0.0140	0.0140
w_C	0.7763	0.7763	0.7763	0.7763	0.7763	0.7763	0.7763	0.7763
w_D	0.0100	0.0100	0.0100	0.0100	0.0100	0.0100	0.0100	0.0100
T/K	313.29	323.51	333.31	343.29	362.80	298.76	299.86	301.38
P/MPa	4.049	4.759	5.504	6.304	8.064	4.049	7.044	12.049

Polymer (B): **polybutadiene** **2012MIL**
Characterization: M_n/g.mol^{-1} = 15800, M_w/g.mol^{-1} = 16600,
15% 1,2-vinyl content, synthesized in the laboratory
Solvent (A): **n-pentane** C_5H_{12} **109-66-0**
Solvent (C): **dimethyl ether** C_2H_6O **115-10-6**

Type of data: cloud points

w_B	0.041	0.041	0.041	0.041	0.041	0.041	0.041	0.041	0.041
w_C/w_A	70/30	70/30	70/30	70/30	70/30	70/30	70/30	70/30	70/30
T/K	323.15	332.15	342.15	352.15	361.15	371.15	380.15	390.15	400.15
P/bar	30	47	73	92	100	119	142	160	179

continued

continued

w_B	0.041	0.041	0.041	0.026	0.026	0.026	0.026	0.026	0.026
w_C/w_A	70/30	70/30	70/30	33/67	33/67	33/67	33/67	33/67	33/67
T/K	409.15	420.15	439.15	369.15	379.15	390.15	400.15	410.15	421.15
P/bar	195	210	226	48	66	87	104	118	136

w_B	0.026	0.026	0.026
w_C/w_A	33/67	33/67	33/67
T/K	431.15	441.15	452.15
P/bar	148	161	174

Polymer (B): **polybutadiene** **2012MIL**
Characterization: M_n/g.mol^{-1} = 15800, M_w/g.mol^{-1} = 16600,
15% 1,2-vinyl content, synthesized in the laboratory
Solvent (A): **n-pentane** C_5H_{12} **109-66-0**
Solvent (C): **dimethyl ether** C_2H_6O **115-10-6**
Polymer (D): **polyethylene**
Characterization: M_n/g.mol^{-1} = 22000, M_w/g.mol^{-1} = 53000,
HDPE, Du Pont de Nemours & Co., USA

Type of data: cloud points

w_B	0.021	0.021	0.021	0.021	0.021	0.021	0.021	0.021	0.021
w_C/w_A	42/58	42/58	42/58	42/58	42/58	42/58	42/58	42/58	42/58
w_D	0.012	0.012	0.012	0.012	0.012	0.012	0.012	0.012	0.012
T/K	407.15	409.15	412.15	419.15	425.15	428.15	433.15	436.15	441.15
P/bar	234	239	244	251	246	259	272	276	288

w_B	0.021	0.021
w_C/w_A	42/58	42/58
w_D	0.012	0.012
T/K	445.15	450.15
P/bar	287	292

Polymer (B): **poly(2-butoxyethyl acrylate)** **2014JAN**
Characterization: M_w/g.mol^{-1} = 80000, M_w/M_n = 1.6
Scientific Polymer Products, Inc., Ontario, NY
Solvent (A): **carbon dioxide** CO_2 **124-38-9**
Solvent (C): **2-butoxyethyl acrylate** $C_9H_{16}O_3$ **7251-90-3**

Type of data: cloud points

w_A	0.950	0.950	0.950	0.950	0.950	0.950	0.950	0.881	0.881
w_B	0.050	0.050	0.050	0.050	0.050	0.050	0.050	0.049	0.049
w_C	0.000	0.000	0.000	0.000	0.000	0.000	0.000	0.070	0.070
T/K	363.9	368.7	373.3	393.4	414.3	433.9	454.2	333.5	353.3
P/MPa	224.31	183.10	163.10	134.83	124.31	117.41	115.17	160.86	109.31

continued

continued

w_A	0.881	0.881	0.881	0.881	0.881	0.789	0.789	0.789	0.789
w_B	0.049	0.049	0.049	0.049	0.049	0.048	0.048	0.048	0.048
w_C	0.070	0.070	0.070	0.070	0.070	0.163	0.163	0.163	0.163
T/K	373.7	393.8	413.8	433.3	453.9	313.9	333.9	353.9	373.2
P/MPa	98.79	94.66	91.90	91.90	92.93	89.83	77.41	74.66	74.14

w_A	0.789	0.789	0.789	0.789	0.675	0.675	0.675	0.675	0.675
w_B	0.048	0.048	0.048	0.048	0.051	0.051	0.051	0.051	0.051
w_C	0.163	0.163	0.163	0.163	0.274	0.274	0.274	0.274	0.274
T/K	393.8	413.2	433.3	453.8	313.7	333.4	353.5	373.6	393.7
P/MPa	74.83	76.03	77.07	77.07	62.93	61.55	62.75	65.00	66.38

w_A	0.675	0.675	0.675	0.583	0.583	0.583	0.583	0.583	0.583
w_B	0.051	0.051	0.051	0.050	0.050	0.050	0.050	0.050	0.050
w_C	0.274	0.274	0.274	0.367	0.367	0.367	0.367	0.367	0.367
T/K	413.8	433.6	453.8	313.6	333.7	353.6	373.7	393.3	413.6
P/MPa	68.28	69.66	70.35	54.66	55.35	58.10	60.86	62.93	65.00

w_A	0.583	0.583	0.448	0.448	0.448	0.448	0.448	0.448	0.448
w_B	0.050	0.050	0.051	0.051	0.051	0.051	0.051	0.051	0.051
w_C	0.367	0.367	0.501	0.501	0.501	0.501	0.501	0.501	0.501
T/K	433.3	454.5	314.0	333.7	353.8	373.8	393.6	413.3	433.3
P/MPa	66.72	67.41	34.66	39.83	44.31	49.14	52.41	55.35	57.93

w_A	0.448	0.301	0.301	0.301	0.301	0.301	0.269	0.269	0.269
w_B	0.051	0.050	0.050	0.050	0.050	0.050	0.051	0.051	0.051
w_C	0.501	0.649	0.649	0.649	0.649	0.649	0.680	0.680	0.680
T/K	453.8	374.1	394.0	413.9	433.6	453.6	433.4	453.5	456.8
P/MPa	59.48	29.14	32.76	36.72	39.66	41.90	34.66	37.41	37.80

Type of data: bubble-point transition

w_A	0.301	0.301	0.301	0.269	0.269	0.269
w_B	0.050	0.050	0.050	0.051	0.051	0.051
w_C	0.649	0.649	0.649	0.680	0.680	0.680
T/K	313.5	324.0	333.7	373.6	393.4	413.4
P/MPa	17.59	21.14	24.50	27.52	29.83	32.28

Polymer (B): **poly(2-butoxyethyl acrylate)** **2014JAN**
Characterization: M_w/g.mol^{-1} = 80000, M_w/M_n = 1.6
Scientific Polymer Products, Inc., Ontario, NY
Solvent (A): **carbon dioxide** CO_2 **124-38-9**
Solvent (C): **dimethyl ether** C_2H_6O **115-10-6**

Type of data: cloud points

continued

continued

w_A	0.950	0.950	0.950	0.950	0.950	0.950	0.950	0.889	0.889
w_B	0.050	0.050	0.050	0.050	0.050	0.050	0.050	0.051	0.051
w_C	0.000	0.000	0.000	0.000	0.000	0.000	0.000	0.060	0.060
T/K	363.9	368.7	373.3	393.4	414.3	433.9	454.2	333.7	353.5
P/MPa	224.31	183.10	163.10	134.83	124.31	117.41	115.17	125.00	104.66
w_A	0.889	0.889	0.889	0.889	0.889	0.842	0.842	0.842	0.842
w_B	0.051	0.051	0.051	0.051	0.051	0.049	0.049	0.049	0.049
w_C	0.060	0.060	0.060	0.060	0.060	0.109	0.109	0.109	0.109
T/K	374.0	393.2	413.8	433.5	454.2	333.8	354.4	374.0	393.3
P/MPa	97.07	94.31	92.93	92.59	92.42	89.14	84.31	82.93	82.93
w_A	0.842	0.842	0.842	0.768	0.768	0.768	0.768	0.768	0.768
w_B	0.049	0.049	0.049	0.050	0.050	0.050	0.050	0.050	0.050
w_C	0.109	0.109	0.109	0.182	0.182	0.182	0.182	0.182	0.182
T/K	413.8	433.3	453.8	334.0	354.1	374.1	393.8	413.6	434.0
P/MPa	83.28	84.14	85.00	63.28	65.00	67.07	69.14	71.21	72.93
w_A	0.768	0.000	0.000	0.000	0.000	0.000	0.000		
w_B	0.050	0.050	0.050	0.050	0.050	0.050	0.050		
w_C	0.182	0.950	0.950	0.950	0.950	0.950	0.950		
T/K	454.0	353.8	373.5	393.7	414.5	433.7	453.7		
P/MPa	74.31	2.93	8.28	11.90	16.38	18.79	20.86		

Polymer (B): **poly(*tert*-butyl acrylate)** **2006BY1**
Characterization: M_n/g.mol^{-1} = 72000, M_w/g.mol^{-1} = 250000, Polysciences, Inc., Warrington, PA
Solvent (A): **carbon dioxide** CO_2 **124-38-9**
Solvent (C): ***tert*-butyl acrylate** $C_7H_{12}O_2$ **1663-39-4**

Type of data: cloud points

w_A	0.952	0.952	0.952	0.952	0.952	0.952	0.863	0.863	0.863
w_B	0.048	0.048	0.048	0.048	0.048	0.048	0.052	0.052	0.052
w_C	0.000	0.000	0.000	0.000	0.000	0.000	0.085	0.085	0.085
T/K	395.55	401.85	411.35	427.45	442.55	457.65	356.35	357.65	363.55
P/bar	2063.8	1711.4	1427.2	1294.5	1240.7	1215.2	1791.4	1408.6	1215.9
w_A	0.863	0.863	0.863	0.863	0.803	0.803	0.803	0.803	0.803
w_B	0.052	0.052	0.052	0.052	0.048	0.048	0.048	0.048	0.048
w_C	0.085	0.085	0.085	0.085	0.149	0.149	0.149	0.149	0.149
T/K	378.45	393.05	407.15	423.45	335.15	336.55	347.15	363.35	379.15
P/bar	1114.8	1071.7	1030.3	1001.7	1760.3	1390.3	1080.3	975.5	928.6
w_A	0.803	0.803	0.803	0.607	0.607	0.607	0.607	0.607	0.607
w_B	0.048	0.048	0.048	0.046	0.046	0.046	0.046	0.046	0.046
w_C	0.149	0.149	0.149	0.347	0.347	0.347	0.347	0.347	0.347
T/K	392.05	406.55	425.15	319.85	332.15	346.15	363.75	380.15	394.65
P/bar	912.8	900.7	891.0	560.3	557.9	569.3	598.3	620.7	639.7

continued

continued

w_A	0.607	0.607	0.390	0.390	0.390	0.390	0.390	0.390	0.390
w_B	0.046	0.046	0.050	0.050	0.050	0.050	0.050	0.050	0.050
w_C	0.347	0.347	0.560	0.560	0.560	0.560	0.560	0.560	0.560
T/K	408.05	423.95	319.85	332.35	346.75	363.05	377.05	391.75	406.75
P/bar	650.0	657.9	182.4	227.2	266.2	321.4	353.5	381.0	406.2

w_A	0.390	0.376	0.376	0.376	0.376	0.376	0.376
w_B	0.050	0.047	0.047	0.047	0.047	0.047	0.047
w_C	0.560	0.577	0.577	0.577	0.577	0.577	0.577
T/K	421.85	346.35	361.45	377.75	391.85	406.95	423.35
P/bar	425.9	284.9	327.9	373.5	403.8	432.1	454.5

Type of data: vapor-liquid equilibrium data

w_A	0.376	0.376
w_B	0.047	0.047
w_C	0.577	0.577
T/K	319.85	332.35
P/bar	202.4	227.2

Type of data: vapor-liquid-liquid equilibrium data

w_A	0.376
w_B	0.047
w_C	0.577
T/K	346.15
P/bar	255.0

Polymer (B): **poly(*tert*-butyl methacrylate)** **2006BY1**
Characterization: M_n/g.mol^{-1} = 69800, M_w/g.mol^{-1} = 180000, Polysciences, Inc., Warrington, PA
Solvent (A): **carbon dioxide** **CO_2** **124-38-9**
Solvent (C): ***tert*-butyl methacrylate** **$C_8H_{14}O_2$** **585-07-9**

Type of data: cloud points

w_A	0.954	0.954	0.954	0.954	0.954	0.846	0.846	0.846	0.846
w_B	0.046	0.046	0.046	0.046	0.046	0.052	0.052	0.052	0.052
w_C	0.000	0.000	0.000	0.000	0.000	0.102	0.102	0.102	0.102
T/K	434.95	437.75	444.75	456.15	463.95	359.45	359.85	369.95	375.75
P/bar	2575.5	2377.6	2186.6	1933.8	1913.1	2172.8	1998.3	1600.0	1482.8

w_A	0.846	0.846	0.846	0.846	0.846	0.846	0.846	0.738	0.738
w_B	0.052	0.052	0.052	0.052	0.052	0.052	0.052	0.057	0.057
w_C	0.102	0.102	0.102	0.102	0.102	0.102	0.102	0.205	0.205
T/K	386.75	394.25	404.35	414.15	422.15	431.75	433.85	323.05	332.15
P/bar	1348.3	1311.4	1261.4	1202.4	1200.3	1165.5	1193.1	882.1	844.8

continued

continued

w_A	0.738	0.738	0.738	0.738	0.738	0.635	0.635	0.635	0.635
w_B	0.057	0.057	0.057	0.057	0.057	0.052	0.052	0.052	0.052
w_C	0.205	0.205	0.205	0.205	0.205	0.313	0.313	0.313	0.313
T/K	352.55	371.65	395.35	412.45	436.15	317.15	331.65	356.35	373.65
P/bar	815.5	813.1	811.0	806.9	801.7	485.9	516.6	568.6	599.3
w_A	0.635	0.635	0.635	0.482	0.482	0.482	0.482	0.482	0.482
w_B	0.052	0.052	0.052	0.054	0.054	0.054	0.054	0.054	0.054
w_C	0.313	0.313	0.313	0.464	0.464	0.464	0.464	0.464	0.464
T/K	391.65	413.55	431.65	343.85	354.65	375.45	392.75	414.75	433.05
P/bar	620.0	633.8	649.3	105.0	149.3	210.0	247.6	286.9	300.7

Type of data: vapor-liquid equilibrium data

w_A	0.482	0.482	0.482
w_B	0.054	0.054	0.054
w_C	0.464	0.464	0.464
T/K	314.75	324.45	332.95
P/bar	62.4	75.2	82.4

Type of data: vapor-liquid-liquid equilibrium data

w_A	0.482	0.482
w_B	0.054	0.054
w_C	0.464	0.464
T/K	354.15	365.15
P/bar	110.0	124.0

Polymer (B): **poly(ε-caprolactone)** **2006BY2**
Characterization: M_w/g.mol^{-1} = 170000, Honam Petrochemical Co., South Korea
Solvent (A): **1-butene** **C_4H_8** **106-98-9**
Solvent (C): **chlorodifluoromethane** **$CHClF_2$** **75-45-6**

Type of data: cloud points

w_A	0.841	0.841	0.841	0.841	0.841	0.841	0.841	0.841	0.669
w_B	0.053	0.053	0.053	0.053	0.053	0.053	0.053	0.053	0.045
w_C	0.107	0.107	0.107	0.107	0.107	0.107	0.107	0.107	0.286
T/K	402.25	411.35	422.65	432.25	442.35	452.85	463.45	473.45	323.65
P/bar	2632.8	2339.7	2037.9	1851.7	1670.7	1520.7	1400.0	1253.4	1149.0
w_A	0.669	0.669	0.669	0.669	0.669	0.669	0.669	0.669	
w_B	0.045	0.045	0.045	0.045	0.045	0.045	0.045	0.045	
w_C	0.286	0.286	0.286	0.286	0.286	0.286	0.286	0.286	
T/K	331.75	352.25	371.65	393.35	412.05	431.85	454.25	474.25	
P/bar	1121.7	1067.2	1020.7	977.6	943.1	901.7	870.7	849.3	

Polymer (B): **poly(ε-caprolactone)** **2006BY2**
Characterization: M_w/g.mol^{-1} = 170000, Honam Petrochemical Co., South Korea
Solvent (A): **1-butene** **C_4H_8** **106-98-9**
Solvent (C): **dimethyl ether** **C_2H_6O** **115-10-6**

Type of data: cloud points

w_A	0.855	0.855	0.855	0.855	0.855	0.855	0.855	0.638	0.638
w_B	0.050	0.050	0.050	0.050	0.050	0.050	0.050	0.051	0.051
w_C	0.095	0.095	0.095	0.095	0.095	0.095	0.095	0.311	0.311
T/K	412.55	418.15	423.15	429.25	444.35	459.25	473.65	322.95	331.25
P/bar	2537.9	2346.6	2319.0	2032.8	1698.3	1562.1	1341.4	2356.9	2169.0
w_A	0.638	0.638	0.638	0.638	0.638	0.638	0.638	0.422	0.422
w_B	0.051	0.051	0.051	0.051	0.051	0.051	0.051	0.053	0.053
w_C	0.311	0.311	0.311	0.311	0.311	0.311	0.311	0.524	0.524
T/K	355.35	373.55	392.85	412.65	433.55	454.85	473.95	321.75	332.45
P/bar	1666.2	1410.3	1222.4	1091.4	979.3	894.8	845.9	893.8	861.4
w_A	0.422	0.422	0.422	0.422	0.422	0.422	0.422	0.143	0.143
w_B	0.053	0.053	0.053	0.053	0.053	0.053	0.053	0.051	0.051
w_C	0.524	0.524	0.524	0.524	0.524	0.524	0.524	0.806	0.806
T/K	351.95	369.65	390.95	411.85	432.65	452.65	472.45	332.75	352.75
P/bar	808.6	774.1	746.6	725.9	714.8	705.2	698.3	325.9	363.8
w_A	0.143	0.143	0.143	0.143	0.143	0.143			
w_B	0.051	0.051	0.051	0.051	0.051	0.051			
w_C	0.806	0.806	0.806	0.806	0.806	0.806			
T/K	374.45	394.35	414.75	433.05	454.35	470.45			
P/bar	401.0	430.4	461.8	484.5	503.5	511.4			

Polymer (B): **poly(ε-caprolactone)** **2006PAR**
Characterization: M_w/g.mol^{-1} = 14000, Aldrich Chem. Co., Inc., Milwaukee, WI
Solvent (A): **carbon dioxide** **CO_2** **124-38-9**
Solvent (C): **chlorodifluoromethane** **$CHClF_2$** **75-45-6**

Type of data: cloud points

w_A	0.1575	0.1575	0.1575	0.1575	0.1575	0.1575	0.1575	0.1575	0.1575
w_B	0.0491	0.0491	0.0491	0.0491	0.0491	0.0491	0.0491	0.0491	0.0491
w_C	0.7934	0.7934	0.7934	0.7934	0.7934	0.7934	0.7934	0.7934	0.7934
T/K	324.2	332.2	342.2	353.2	362.2	373.2	381.2	393.2	403.2
P/MPa	14.2	17.7	21.6	52.9	29.5	33.4	36.2	40.2	43.5
w_A	0.1575	0.2656	0.2656	0.2656	0.2656	0.2656	0.2656	0.2656	0.2656
w_B	0.0491	0.0525	0.0525	0.0525	0.0525	0.0525	0.0525	0.0525	0.0525
w_C	0.7934	0.6819	0.6819	0.6819	0.6819	0.6819	0.6819	0.6819	0.6819
T/K	413.2	323.9	334.1	343.6	353.5	361.9	374.5	383.7	392.9
P/MPa	46.2	25.8	30.7	35.2	39.7	43.2	48.4	51.9	55.4

continued

continued

w_A	0.2656	0.2656	0.3522	0.3522	0.3522	0.3522	0.3522	0.3522	0.3522
w_B	0.0525	0.0525	0.0492	0.0492	0.0492	0.0492	0.0492	0.0492	0.0492
w_C	0.6819	0.6819	0.5986	0.5986	0.5986	0.5986	0.5986	0.5986	0.5986
T/K	403.1	410.2	323.3	333.5	342.7	351.8	363.7	373.1	383.8
P/MPa	58.9	61.1	34.1	39.2	43.8	48.1	53.2	57.1	61.2
w_A	0.3522	0.3522	0.3522	0.4429	0.4429	0.4429	0.4429	0.4429	0.4429
w_B	0.0492	0.0492	0.0492	0.0497	0.0497	0.0497	0.0497	0.0497	0.0497
w_C	0.5986	0.5986	0.5986	0.5074	0.5074	0.5074	0.5074	0.5074	0.5074
T/K	393.4	402.8	411.7	323.4	333.0	344.2	352.0	364.1	373.3
P/MPa	64.9	68.3	72.0	45.4	50.2	56.2	60.1	65.5	69.7
w_A	0.4429	0.4429	0.4429	0.4429	0.5646	0.5646	0.5646	0.5646	0.5646
w_B	0.0497	0.0497	0.0497	0.0497	0.0487	0.0487	0.0487	0.0487	0.0487
w_C	0.5074	0.5074	0.5074	0.5074	0.3867	0.3867	0.3867	0.3867	0.3867
T/K	384.3	393.5	401.4	413.3	324.0	334.3	343.7	352.5	363.2
P/MPa	74.1	77.6	80.5	84.4	70.4	76.0	80.9	85.2	90.0
w_A	0.5646	0.5646	0.5646	0.5646	0.5646	0.6394	0.6394	0.6394	0.6394
w_B	0.0487	0.0487	0.0487	0.0487	0.0487	0.0492	0.0492	0.0492	0.0492
w_C	0.3867	0.3867	0.3867	0.3867	0.3867	0.3114	0.3114	0.3114	0.3114
T/K	373.7	383.3	393.3	404.1	414.8	324.0	334.2	343.9	354.0
P/MPa	94.5	98.3	101.8	105.4	108.5	91.9	97.0	101.4	105.8
w_A	0.6394	0.6394	0.6394	0.6394	0.6394	0.6394			
w_B	0.0492	0.0492	0.0492	0.0492	0.0492	0.0492			
w_C	0.3114	0.3114	0.3114	0.3114	0.3114	0.3114			
T/K	364.4	374.4	384.0	392.6	402.2	415.0			
P/MPa	110.0	113.6	116.9	119.5	122.3	125.5			

Polymer (B): **poly(ε-caprolactone)** **2006BY2**
Characterization: M_w/g.mol^{-1} = 170000, Honam Petrochemical Co., South Korea
Solvent (A): **carbon dioxide** CO_2 **124-38-9**
Solvent (C): **chlorodifluoromethane** $CHClF_2$ **75-45-6**

Type of data: cloud points

w_A	0.905	0.905	0.905	0.905	0.905	0.905	0.905	0.905	0.905
w_B	0.046	0.046	0.046	0.046	0.046	0.046	0.046	0.046	0.046
w_C	0.049	0.049	0.049	0.049	0.049	0.049	0.049	0.049	0.049
T/K	412.25	421.55	431.05	439.65	452.55	463.45	473.35	483.05	496.15
P/bar	2760.3	2674.1	2581.0	2519.0	2431.0	2353.4	2301.7	2263.8	2234.5
w_A	0.810	0.810	0.810	0.810	0.810	0.810	0.810	0.810	0.810
w_B	0.047	0.047	0.047	0.047	0.047	0.047	0.047	0.047	0.047
w_C	0.143	0.143	0.143	0.143	0.143	0.143	0.143	0.143	0.143
T/K	326.15	335.65	354.15	374.95	392.45	411.95	433.95	454.15	472.65
P/bar	2465.5	2363.8	2217.2	2098.3	2050.0	2005.2	1963.8	1929.3	1903.4

continued

continued

w_A	0.507	0.507	0.507	0.507	0.507	0.507	0.507	0.507
w_B	0.049	0.049	0.049	0.049	0.049	0.049	0.049	0.049
w_C	0.445	0.445	0.445	0.445	0.445	0.445	0.445	0.445
T/K	322.55	335.75	353.75	372.05	393.25	414.75	433.35	452.05
P/bar	591.4	669.0	753.5	848.6	936.2	1012.1	1056.9	1096.6

Polymer (B): **poly(ε-caprolactone)** **2010BEN**
Characterization: M_n/g.mol^{-1} = 7150, M_w/g.mol^{-1} = 10000,
Solvent (A): **carbon dioxide** **CO_2** **124-38-9**
Solvent (C): **dichloromethane** **CH_2Cl_2** **75-09-2**

Type of data: bubble and cloud points

T/K = 303.15

w_A	0.2528	0.2891	0.3415	0.4343	0.4570	0.2439	0.2831	0.3344	0.4253
w_B	0.0100	0.0100	0.0100	0.0100	0.0100	0.0300	0.0300	0.0300	0.0300
w_C	0.7372	0.7019	0.6485	0.5557	0.6330	0.7261	0.6869	0.6356	0.5447
P/MPa	2.86	3.04	3.38	3.89	3.99	2.90	3.20	3.49	3.87
	VLE	VLE	VLE	VLE	VLE	VLE	VLE	VLE	VLE

w_A	0.4473	0.2401	0.2765	0.3281	0.4168	0.4380	0.2348	0.2704	0.3200
w_B	0.0300	0.0500	0.0500	0.0500	0.0500	0.0500	0.0700	0.0700	0.0700
w_C	0.5227	0.7099	0.6735	0.6219	0.5332	0.5120	0.6952	0.6596	0.6100
P/MPa	4.16	2.91	3.19	3.52	3.98	4.00	2.98	3.21	3.44
	VLE	VLE	VLE	VLE	VLE	VLE	VLE	VLE	VLE

w_A	0.4152	0.4272
w_B	0.0700	0.0700
w_C	0.5148	0.5028
P/MPa	4.13	4.31
	VLE	VLE

T/K = 308.15

w_A	0.4272	0.4272
w_B	0.0700	0.0700
w_C	0.5028	0.5028
P/MPa	4.76	6.17
	VLLE	LLE

T/K = 313.15

w_A	0.2528	0.2891	0.3415	0.4343	0.4570	0.2439	0.2831	0.3344	0.4253
w_B	0.0100	0.0100	0.0100	0.0100	0.0100	0.0300	0.0300	0.0300	0.0300
w_C	0.7372	0.7019	0.6485	0.5557	0.6330	0.7261	0.6869	0.6356	0.5447
P/MPa	3.29	3.50	3.99	4.69	4.74	3.39	3.63	4.16	4.70
	VLE	VLE	VLE	VLE	VLE	VLE	VLE	VLE	VLE

continued

continued

w_A	0.4473	0.4473	0.2401	0.2765	0.3281	0.4168	0.4380	0.4380	0.2348
w_B	0.0300	0.0300	0.0500	0.0500	0.0500	0.0500	0.0500	0.0500	0.0700
w_C	0.5227	0.5227	0.7099	0.6735	0.6219	0.5332	0.5120	0.5120	0.6952
P/MPa	4.66	6.52	3.36	3.70	4.17	4.78	4.83	7.57	3.49
	VLLE	LLE	VLE	VLE	VLE	VLE	VLLE	LLE	VLE

w_A	0.2704	0.3200	0.4152	0.4272	0.4272
w_B	0.0700	0.0700	0.0700	0.0700	0.0700
w_C	0.6596	0.6100	0.5148	0.5028	0.5028
P/MPa	3.68	4.07	4.90	5.09	8.61
	VLE	VLE	VLE	VLLE	LLE

T/K = 318.15

w_A	0.4570	0.4570	0.4152	0.4152
w_B	0.0100	0.0100	0.0700	0.0700
w_C	0.6330	0.6330	0.5148	0.5148
P/MPa	5.07	5.19	5.28	6.48
	VLLE	LLE	VLLE	LLE

T/K = 323.15

w_A	0.2528	0.2891	0.3415	0.4343	0.4570	0.4570	0.2439	0.2831	0.3344
w_B	0.0100	0.0100	0.0100	0.0100	0.0100	0.0100	0.0300	0.0300	0.0300
w_C	0.7372	0.7019	0.6485	0.5557	0.6330	0.6330	0.7261	0.6869	0.6356
P/MPa	3.80	4.08	4.66	5.56	5.56	7.54	3.81	4.23	4.81
	VLE	VLE	VLE	VLE	VLLE	LLE	VLE	VLE	VLE

w_A	0.4253	0.4253	0.4473	0.4473	0.2401	0.2765	0.3281	0.4168	0.4168
w_B	0.0300	0.0300	0.0300	0.0300	0.0500	0.0500	0.0500	0.0500	0.0500
w_C	0.5447	0.5447	0.5227	0.5227	0.7099	0.6735	0.6219	0.5332	0.5332
P/MPa	5.53	6.48	5.44	10.50	3.90	4.37	4.90	5.58	7.25
	VLLE	LLE	VLLE	LLE	VLE	VLE	VLE	VLLE	LLE

w_A	0.4380	0.4380	0.2348	0.2704	0.3200	0.4152	0.4152	0.4272	0.4272
w_B	0.0500	0.0500	0.0700	0.0700	0.0700	0.0700	0.0700	0.0700	0.0700
w_C	0.5120	0.5120	0.6952	0.6596	0.6100	0.5148	0.5148	0.5028	0.5028
P/MPa	5.72	11.92	4.10	4.40	4.95	5.60	8.52	5.84	12.74
	VLLE	LLE	VLE	VLE	VLE	VLLE	LLE	VLLE	LLE

T/K = 328.15

w_A	0.4343	0.4343
w_B	0.0100	0.0100
w_C	0.5557	0.5557
P/MPa	6.07	7.30
	VLLE	LLE

continued

continued

T/K = 333.15

w_A	0.2528	0.2891	0.3415	0.4343	0.4343	0.4570	0.4570	0.2439	0.2831
w_B	0.0100	0.0100	0.0100	0.0100	0.0100	0.0100	0.0100	0.0300	0.0300
w_C	0.7372	0.7019	0.6485	0.5557	0.5557	0.6330	0.6330	0.7261	0.6869
P/MPa	4.41	4.80	5.47	6.51	9.42	6.53	11.85	4.38	4.91
	VLE	VLE	VLE	VLLE	LLE	VLLE	LLE	VLE	VLE

w_A	0.3344	0.4253	0.4253	0.4473	0.4473	0.2401	0.2765	0.3281	0.4168
w_B	0.0300	0.0300	0.0300	0.0300	0.0300	0.0500	0.0500	0.0500	0.0500
w_C	0.6356	0.5447	0.5447	0.5227	0.5227	0.7099	0.6735	0.6219	0.5332
P/MPa	5.67	6.42	10.40	6.64	15.07	4.51	5.04	5.70	6.50
	VLE	VLLE	LLE	VLLE	LLE	VLE	VLE	VLE	VLLE

w_A	0.4168	0.4380	0.4380	0.2348	0.2704	0.3200	0.4152	0.4152	0.4272
w_B	0.0500	0.0500	0.0500	0.0700	0.0700	0.0700	0.0700	0.0700	0.0700
w_C	0.5332	0.5120	0.5120	0.6952	0.6596	0.6100	0.5148	0.5148	0.5028
P/MPa	11.33	6.82	16.35	4.70	5.00	5.54	6.60	12.82	6.78
	LLE	VLLE	LLE	VLE	VLE	VLE	VLLE	LLE	VLLE

w_A	0.4272
w_B	0.0700
w_C	0.5028
P/MPa	16.96
	LLE

T/K = 338.15

w_A	0.4343	0.4343
w_B	0.0100	0.0100
w_C	0.5557	0.5557
P/MPa	7.02	11.34
	VLLE	LLE

T/K = 343.15

w_A	0.2528	0.2891	0.3415	0.4343	0.4343	0.4570	0.4570	0.2439	0.2831
w_B	0.0100	0.0100	0.0100	0.0100	0.0100	0.0100	0.0100	0.0300	0.0300
w_C	0.7372	0.7019	0.6485	0.5557	0.5557	0.6330	0.6330	0.7261	0.6869
P/MPa	5.02	5.54	6.28	7.43	13.48	7.48	16.14	5.06	5.93
	VLE	VLE	VLE	VLLE	LLE	VLLE	LLE	VLE	VLE

w_A	0.3344	0.4253	0.4253	0.4473	0.4473	0.2401	0.2765	0.3281	0.4168
w_B	0.0300	0.0300	0.0300	0.0300	0.0300	0.0500	0.0500	0.0500	0.0500
w_C	0.6356	0.5447	0.5447	0.5227	0.5227	0.7099	0.6735	0.6219	0.5332
P/MPa	6.40	7.42	14.57	7.80	19.49	5.12	6.03	6.55	7.44
	VLE	VLLE	LLE	VLLE	LLE	VLE	VLE	VLE	VLLE

continued

continued

w_A	0.4168	0.4380	0.4380	0.2348	0.2704	0.3200	0.4152	0.4152	0.4272
w_B	0.0500	0.0500	0.0500	0.0700	0.0700	0.0700	0.0700	0.0700	0.0700
w_C	0.5332	0.5120	0.5120	0.6952	0.6596	0.6100	0.5148	0.5148	0.5028
P/MPa	15.30	7.89	20.45	5.26	5.71	6.44	7.52	16.59	7.76
	LLE	VLLE	LLE	VLE	VLE	VLE	VLLE	LLE	VLLE

w_A	0.4272
w_B	0.0700
w_C	0.5028
P/MPa	20.89
	LLE

Polymer (B): **poly(ε-caprolactone)** **2006KAL**
Characterization: M_n/g.mol^{-1} = 59400, M_w/g.mol^{-1} = 122400,
Tone 787, Union Carbide Benelux
Solvent (A): **carbon dioxide** **CO_2** **124-38-9**
Solvent (C): **dichloromethane** **CH_2Cl_2** **75-09-2**

Type of data: bubble and cloud points

T/K = 308.15

w_A	0.7405	0.7019	0.6492	0.5939	0.5561	0.7333	0.6947	0.6427	0.5871
w_B	0.0100	0.0100	0.0100	0.0100	0.0100	0.0200	0.0200	0.0200	0.0200
w_C	0.2495	0.2881	0.3408	0.3961	0.4339	0.2467	0.2853	0.3373	0.3929
P/bar	34.62	37.93	41.75	45.96	47.77	34.82	38.13	42.35	46.46
	VLE	VLE	VLE	VLE	VLE	VLE	VLE	VLE	VLE

w_A	0.5504	0.7264	0.6877	0.6362	0.5794	0.5448
w_B	0.0200	0.0300	0.0300	0.0300	0.0300	0.0300
w_C	0.4296	0.2436	0.2823	0.3338	0.3906	0.4252
P/bar	48.17	35.30	38.54	42.85	47.37	85.10
	VLE	VLE	VLE	VLE	VLE	LLE

T/K = 313.15

w_A	0.7405	0.7019	0.6492	0.5939	0.5561	0.7333	0.6947	0.6427	0.5871
w_B	0.0100	0.0100	0.0100	0.0100	0.0100	0.0200	0.0200	0.0200	0.0200
w_C	0.2495	0.2881	0.3408	0.3961	0.4339	0.2467	0.2853	0.3373	0.3929
P/bar	37.73	41.45	45.36	49.98	52.78	37.94	41.65	46.26	50.48
	VLE	VLE	VLE	VLE	VLE	VLE	VLE	VLE	VLE

w_A	0.5504	0.7264	0.6877	0.6362	0.5794	0.5448
w_B	0.0200	0.0300	0.0300	0.0300	0.0300	0.0300
w_C	0.4296	0.2436	0.2823	0.3338	0.3906	0.4252
P/bar	63.82	38.44	41.95	46.46	59.41	109.18
	LLE	VLE	VLE	VLE	LLE	LLE

continued

continued

T/K = 318.15

w_A	0.7405	0.7019	0.6492	0.5939	0.5561	0.7333	0.6947	0.6427	0.5871
w_B	0.0100	0.0100	0.0100	0.0100	0.0100	0.0200	0.0200	0.0200	0.0200
w_C	0.2495	0.2881	0.3408	0.3961	0.4339	0.2467	0.2853	0.3373	0.3929
P/bar	40.74	44.86	49.27	54.09	73.50	41.05	44.96	50.08	55.58
	VLE	VLE	VLE	VLE	LLE	VLE	VLE	VLE	VLE

w_A	0.5504	0.7264	0.6877	0.6362	0.5794	0.5448
w_B	0.0200	0.0300	0.0300	0.0300	0.0300	0.0300
w_C	0.4296	0.2436	0.2823	0.3338	0.3906	0.4252
P/bar	86.80	41.45	45.36	50.38	83.59	133.16
	LLE	VLE	VLE	VLE	LLE	LLE

T/K = 323.15

w_A	0.7405	0.7019	0.6492	0.5939	0.5561	0.7333	0.6947	0.6427	0.5871
w_B	0.0100	0.0100	0.0100	0.0100	0.0100	0.0200	0.0200	0.0200	0.0200
w_C	0.2495	0.2881	0.3408	0.3961	0.4339	0.2467	0.2853	0.3373	0.3929
P/bar	43.85	48.37	53.09	58.30	97.94	44.05	48.57	54.09	68.94
	VLE	VLE	VLE	VLE	LLE	VLE	VLE	VLE	LLE

w_A	0.5504	0.7264	0.6877	0.6362	0.5794	0.5448
w_B	0.0200	0.0300	0.0300	0.0300	0.0300	0.0300
w_C	0.4296	0.2436	0.2823	0.3338	0.3906	0.4252
P/bar	110.48	44.76	48.97	54.39	105.87	156.43
	LLE	VLE	VLE	VLE	LLE	LLE

T/K = 328.15

w_A	0.7405	0.7019	0.6492	0.5939	0.5561	0.7333	0.6947	0.6427	0.5871
w_B	0.0100	0.0100	0.0100	0.0100	0.0100	0.0200	0.0200	0.0200	0.0200
w_C	0.2495	0.2881	0.3408	0.3961	0.4339	0.2467	0.2853	0.3373	0.3929
P/bar	46.97	51.88	57.00	77.87	121.10	47.37	52.05	58.20	91.41
	VLE	VLE	VLE	LLE	LLE	VLE	VLE	VLE	LLE

w_A	0.5504	0.7264	0.6877	0.6362	0.5794	0.5448
w_B	0.0200	0.0300	0.0300	0.0300	0.0300	0.0300
w_C	0.4296	0.2436	0.2823	0.3338	0.3906	0.4252
P/bar	133.86	47.97	52.59	58.51	128.24	178.91
	LLE	VLE	VLE	VLE	LLE	LLE

T/K = 333.15

w_A	0.7405	0.7019	0.6492	0.5939	0.5561	0.7333	0.6947	0.6427	0.5871
w_B	0.0100	0.0100	0.0100	0.0100	0.0100	0.0200	0.0200	0.0200	0.0200
w_C	0.2495	0.2881	0.3408	0.3961	0.4339	0.2467	0.2853	0.3373	0.3929
P/bar	50.03	55.39	61.01	101.35	142.49	50.68	55.60	62.32	113.09
	VLE	VLE	VLE	LLE	LLE	VLE	VLE	VLE	LLE

continued

continued

w_A	0.5504	0.7264	0.6877	0.6362	0.5794	0.5448
w_B	0.0200	0.0300	0.0300	0.0300	0.0300	0.0300
w_C	0.4296	0.2436	0.2823	0.3338	0.3906	0.4252
P/bar	155.43	51.38	56.30	62.72	149.31	178.91
	LLE	VLE	VLE	VLE	LLE	LLE

T/K = 338.15

w_A	0.7405	0.7019	0.6492	0.5939	0.5561	0.7333	0.6947	0.6427	0.5871
w_B	0.0100	0.0100	0.0100	0.0100	0.0100	0.0200	0.0200	0.0200	0.0200
w_C	0.2495	0.2881	0.3408	0.3961	0.4339	0.2467	0.2853	0.3373	0.3929
P/bar	54.29	59.20	65.33	121.82	164.96	54.19	59.40	66.73	134.96
	VLE	VLE	VLE	LLE	LLE	VLE	VLE	VLE	LLE

w_A	0.5504	0.7264	0.6877	0.6362	0.5794
w_B	0.0200	0.0300	0.0300	0.0300	0.0300
w_C	0.4296	0.2436	0.2823	0.3338	0.3906
P/bar	177.11	55.00	60.10	67.03	171.38
	LLE	VLE	VLE	VLE	LLE

T/K = 343.15

w_A	0.7405	0.7019	0.6492	0.5939	0.5561	0.7333	0.6947	0.6427	0.5871
w_B	0.0100	0.0100	0.0100	0.0100	0.0100	0.0200	0.0200	0.0200	0.0200
w_C	0.2495	0.2881	0.3408	0.3961	0.4339	0.2467	0.2853	0.3373	0.3929
P/bar	57.58	63.12	69.64	143.59	185.84	57.70	63.35	80.28	156.24
	VLE	VLE	VLE	LLE	LLE	VLE	VLE	LLE	LLE

w_A	0.7264	0.6877	0.6362	0.5794
w_B	0.0300	0.0300	0.0300	0.0300
w_C	0.2436	0.2823	0.3338	0.3906
P/bar	58.51	64.12	83.29	192.46
	VLE	VLE	LLE	LLE

T/K = 348.15

w_A	0.7405	0.7019	0.6492	0.5939	0.7333	0.6947	0.6427	0.5871
w_B	0.0100	0.0100	0.0100	0.0100	0.0200	0.0200	0.0200	0.0200
w_C	0.2495	0.2881	0.3408	0.3961	0.2467	0.2853	0.3373	0.3929
P/bar	60.92	67.13	82.59	164.26	61.32	67.35	100.25	177.10
	VLE	VLE	LLE	LLE	VLE	VLE	LLE	LLE

w_A	0.7264	0.6877	0.6362
w_B	0.0300	0.0300	0.0300
w_C	0.2436	0.2823	0.3338
P/bar	62.12	68.04	103.56
	VLE	VLE	LLE

continued

continued

T/K = 353.15

w_A	0.7405	0.7019	0.6492	0.5939	0.7333	0.6947	0.6427	0.5871
w_B	0.0100	0.0100	0.0100	0.0100	0.0200	0.0200	0.0200	0.0200
w_C	0.2495	0.2881	0.3408	0.3961	0.2467	0.2853	0.3373	0.3929
P/bar	64.42	70.95	103.06	184.53	64.83	71.20	121.12	197.07
	VLE	VLE	LLE	LLE	VLE	VLE	LLE	LLE

w_A	0.7264	0.6877	0.6362
w_B	0.0300	0.0300	0.0300
w_C	0.2436	0.2823	0.3338
P/bar	65.83	71.95	123.63
	VLE	VLE	LLE

T/K = 358.15

w_A	0.7405	0.7019	0.6492	0.7333	0.6947	0.6427	0.7264	0.6877	0.6362
w_B	0.0100	0.0100	0.0100	0.0200	0.0200	0.0200	0.0300	0.0300	0.0300
w_C	0.2495	0.2881	0.3408	0.2467	0.2853	0.3373	0.2436	0.2823	0.3338
P/bar	67.90	74.86	123.22	68.44	75.10	140.99	69.64	76.16	143.23
	VLE	VLE	LLE	VLE	VLE	LLE	VLE	VLE	LLE

T/K = 363.15

w_A	0.7405	0.7019	0.6492	0.7333	0.6947	0.6427	0.7264	0.6877	0.6362
w_B	0.0100	0.0100	0.0100	0.0200	0.0200	0.0200	0.0300	0.0300	0.0300
w_C	0.2495	0.2881	0.3408	0.2467	0.2853	0.3373	0.2436	0.2823	0.3338
P/bar	71.45	78.87	142.29	72.05	79.10	160.05	73.46	83.49	162.76
	VLE	VLE	LLE	VLE	VLE	LLE	VLE	LLE	LLE

T/K = 368.15

w_A	0.7405	0.7019	0.6492	0.7333	0.6947	0.6427	0.7264	0.6877
w_B	0.0100	0.0100	0.0100	0.0200	0.0200	0.0200	0.0300	0.0300
w_C	0.2495	0.2881	0.3408	0.2467	0.2853	0.3373	0.2436	0.2823
P/bar	75.26	83.19	160.95	75.86	83.40	178.81	77.27	101.75
	VLE	VLE	LLE	VLE	VLE	LLE	VLE	LLE

T/K = 373.15

w_A	0.7405	0.7019	0.7333	0.6947	0.7264	0.6877
w_B	0.0100	0.0100	0.0200	0.0200	0.0300	0.0300
w_C	0.2495	0.2881	0.2467	0.2853	0.2436	0.2823
P/bar	78.87	96.23	79.48	102.20	80.98	120.01
	VLE	LLE	VLE	LLE	VLE	LLE

Polymer (B): **poly(ε-caprolactone)** **2006PAR**
Characterization: M_w/g.mol^{-1} = 14000, Aldrich Chem. Co., Inc., Milwaukee, WI
Solvent (A): **carbon dioxide** **CO_2** **124-38-9**
Solvent (C): **dimethyl ether** **C_2H_6O** **115-10-6**

Type of data: cloud points

w_A	0.0749	0.0749	0.0749	0.0749	0.0749	0.0749	0.0749	0.0749	0.0749
w_B	0.0512	0.0512	0.0512	0.0512	0.0512	0.0512	0.0512	0.0512	0.0512
w_C	0.8739	0.8739	0.8739	0.8739	0.8739	0.8739	0.8739	0.8739	0.8739
T/K	313.2	323.4	331.1	343.0	352.8	363.3	373.0	382.8	393.3
P/MPa	23.0	25.5	27.5	30.2	32.3	34.7	36.7	38.8	40.9
w_A	0.0749	0.0749	0.1565	0.1565	0.1565	0.1565	0.1565	0.1565	0.1565
w_B	0.0512	0.0512	0.0488	0.0488	0.0488	0.0488	0.0488	0.0488	0.0488
w_C	0.8739	0.8739	0.7947	0.7947	0.7947	0.7947	0.7947	0.7947	0.7947
T/K	402.0	413.1	314.4	323.0	332.6	343.0	352.8	362.8	370.6
P/MPa	42.4	43.9	32.4	34.1	36.1	38.3	40.2	42.4	43.9
w_A	0.1565	0.1565	0.1565	0.1565	0.2402	0.2402	0.2402	0.2402	0.2402
w_B	0.0488	0.0488	0.0488	0.0488	0.0489	0.0489	0.0489	0.0489	0.0489
w_C	0.7947	0.7947	0.7947	0.7947	0.7109	0.7109	0.7109	0.7109	0.7109
T/K	384.0	391.0	402.0	411.9	312.7	322.5	332.4	343.4	352.0
P/MPa	46.4	47.5	49.3	50.7	42.8	44.3	46.0	47.9	49.5
w_A	0.2402	0.2402	0.2402	0.2402	0.2402	0.2402	0.3064	0.3064	0.3064
w_B	0.0489	0.0489	0.0489	0.0489	0.0489	0.0489	0.0492	0.0492	0.0492
w_C	0.7109	0.7109	0.7109	0.7109	0.7109	0.7109	0.6444	0.6444	0.6444
T/K	359.9	373.5	383.6	392.9	402.8	413.9	312.2	321.3	332.3
P/MPa	50.9	53.2	54.8	56.5	58.1	59.7	52.9	54.0	55.8
w_A	0.3064	0.3064	0.3064	0.3064	0.3064	0.3064	0.3064	0.3064	0.4198
w_B	0.0492	0.0492	0.0492	0.0492	0.0492	0.0492	0.0492	0.0492	0.0484
w_C	0.6444	0.6444	0.6444	0.6444	0.6444	0.6444	0.6444	0.6444	0.5318
T/K	343.2	352.6	362.4	373.3	383.2	391.1	400.1	411.5	323.8
P/MPa	57.3	58.6	60.3	61.9	63.4	64.8	66.0	67.5	71.9
w_A	0.4198	0.4198	0.4198	0.4198	0.4198	0.4198	0.4198	0.4198	0.4198
w_B	0.0484	0.0484	0.0484	0.0484	0.0484	0.0484	0.0484	0.0484	0.0484
w_C	0.5318	0.5318	0.5318	0.5318	0.5318	0.5318	0.5318	0.5318	0.5318
T/K	335.0	344.2	354.3	363.4	375.0	384.3	394.0	400.7	412.6
P/MPa	73.0	74.0	75.2	76.2	77.6	78.8	80.0	80.7	81.9
w_A	0.4604	0.4604	0.4604	0.4604	0.4604	0.4604	0.4604	0.4604	0.4604
w_B	0.0509	0.0509	0.0509	0.0509	0.0509	0.0509	0.0509	0.0509	0.0509
w_C	0.4887	0.4884	0.4884	0.4884	0.4884	0.4884	0.4884	0.4884	0.4884
T/K	323.2	332.5	344.3	353.8	363.2	373.8	383.6	390.0	403.5
P/MPa	78.6	79.2	80.3	81.2	82.2	83.4	84.5	85.3	86.7

continued

continued

w_A	0.4604	0.5494	0.5494	0.5494	0.5494	0.5494	0.5494	0.5494	0.5494
w_B	0.0509	0.0497	0.0497	0.0497	0.0497	0.0497	0.0497	0.0497	0.0497
w_C	0.4884	0.4009	0.4009	0.4009	0.4009	0.4009	0.4009	0.4009	0.4009
T/K	415.9	323.9	334.1	343.9	353.5	363.4	374.0	384.2	393.8
P/MPa	88.1	96.5	96.9	97.3	98.0	98.6	99.5	100.3	101.0
w_A	0.5494	0.5494	0.6209	0.6209	0.6209	0.6209	0.6209	0.6209	0.6209
w_B	0.0497	0.0497	0.0491	0.0491	0.0491	0.0491	0.0491	0.0491	0.0491
w_C	0.4009	0.4009	0.3300	0.3300	0.3300	0.3300	0.3300	0.3300	0.3300
T/K	402.5	414.5	322.9	335.9	344.5	354.6	364.4	372.7	383.1
P/MPa	101.8	102.7	107.8	108.2	108.4	108.9	109.5	109.9	110.5
w_A	0.6209	0.6209	0.6209						
w_B	0.0491	0.0491	0.0491						
w_C	0.3300	0.3300	0.3300						
T/K	393.6	403.4	412.3						
P/MPa	111.2	111.7	112.3						

Polymer (B): **poly(ε-caprolactone)** **2006BY2**
Characterization: M_w/g.mol^{-1} = 170000, Honam Petrochemical Co., South Korea
Solvent (A): **carbon dioxide** CO_2 **124-38-9**
Solvent (C): **dimethyl ether** C_2H_6O **115-10-6**

Type of data: cloud points

w_A	0.911	0.911	0.911	0.911	0.911	0.911	0.911	0.803	0.803
w_B	0.044	0.044	0.044	0.044	0.044	0.044	0.044	0.053	0.053
w_C	0.045	0.045	0.045	0.045	0.045	0.045	0.045	0.145	0.145
T/K	425.05	436.35	444.75	455.45	464.75	475.75	483.55	331.55	345.55
P/bar	2760.3	2632.8	2550.0	2469.0	2408.6	2356.9	2337.9	2501.7	2360.4
w_A	0.803	0.803	0.803	0.803	0.803	0.803	0.803	0.803	0.803
w_B	0.053	0.053	0.053	0.053	0.053	0.053	0.053	0.053	0.053
w_C	0.145	0.145	0.145	0.145	0.145	0.145	0.145	0.145	0.145
T/K	359.55	373.55	388.35	402.25	412.45	422.15	434.55	443.95	454.25
P/bar	2239.7	2155.2	2082.8	2032.8	2001.7	1974.1	1941.4	1920.7	1901.7
w_A	0.803	0.803	0.546	0.546	0.546	0.546	0.546	0.546	0.546
w_B	0.053	0.053	0.050	0.050	0.050	0.050	0.050	0.050	0.050
w_C	0.145	0.145	0.404	0.404	0.404	0.404	0.404	0.404	0.404
T/K	464.75	473.85	327.65	346.45	358.25	374.15	387.35	405.55	416.85
P/bar	1881.0	1869.0	1184.5	1167.2	1160.3	1156.9	1153.4	1150.0	1153.4
w_A	0.546	0.546	0.546						
w_B	0.050	0.050	0.050						
w_C	0.404	0.404	0.404						
T/K	433.45	448.55	464.35						
P/bar	1156.9	1160.3	1163.8						

Polymer (B): **poly(ε-caprolactone)** **2009ROJ**
Characterization: M_n/g.mol^{-1} = 1000, CAPA 2101A, Solvay Caprolactones, Solvay Interox, United Kingdom
Solvent (A): **carbon dioxide** **CO_2** **124-38-9**
Solvent (C): **ethanol** **C_2H_6O** **64-17-5**

Type of data: cloud points

T/K = 310

x_B	0.0000342	0.0000508	0.0000525	0.0000707	0.0000778	0.0000848
x_C	0.0745	0.0775	0.0761	0.0768	0.0751	0.0738
P/MPa	15.2	16.6	16.9	16.2	16.8	18.1

T/K = 320

x_B	0.0000342	0.0000525	0.0000622	0.0000707	0.0000778
x_C	0.0745	0.0761	0.0774	0.0768	0.0751
P/MPa	17.8	20.4	19.2	19.9	21.4

T/K = 330

x_B	0.0000342	0.0000622	0.0000707
x_C	0.0745	0.0774	0.0768
P/MPa	21.1	21.8	23.2

Comments: The authors quote these data as solubility of PCL in the mixture of CO_2 and C_2H_6O.

Polymer (B): **poly(ε-caprolactone)** **2007LI2**
Characterization: M_n/g.mol^{-1} = 6100, M_w/g.mol^{-1} = 14300, T_g/K = 339, Scientific Polymer Products, Inc., Ontario, NY
Solvent (A): **carbon dioxide** **CO_2** **124-38-9**
Solvent (C): **2-propanone** **C_3H_6O** **67-64-1**

Type of data: cloud points

w_A	0.40	0.40	0.40	0.40	0.40	0.40	0.40	0.40	0.40
w_B	0.01	0.01	0.01	0.01	0.01	0.01	0.01	0.01	0.01
w_C	0.59	0.59	0.59	0.59	0.59	0.59	0.59	0.59	0.59
T/K	334.5	338.9	347.6	352.9	359.0	363.1	368.5	374.0	379.3
P/MPa	10.0	11.3	13.0	15.0	16.0	18.0	19.5	23.0	24.5

w_A	0.40	0.40	0.40	0.40	0.40	0.40	0.40	0.40	0.40
w_B	0.01	0.01	0.01	0.01	0.05	0.05	0.05	0.05	0.05
w_C	0.59	0.59	0.59	0.59	0.55	0.55	0.55	0.55	0.55
T/K	383.6	388.5	394.1	398.0	340.0	343.5	348.8	351.0	354.6
P/MPa	26.0	26.8	27.3	28.0	17.8	18.8	20.8	21.3	22.3

continued

continued

w_A	0.40	0.40	0.40	0.40	0.40	0.40	0.40	0.40	0.40
w_B	0.05	0.05	0.05	0.05	0.05	0.10	0.10	0.10	0.10
w_C	0.55	0.55	0.55	0.55	0.55	0.50	0.50	0.50	0.50
T/K	358.6	361.1	364.6	368.8	372.3	325.1	330.1	337.7	342.5
P/MPa	23.2	23.5	24.8	25.5	26.7	4.9	6.8	9.1	10.5
w_A	0.40	0.40	0.40	0.40	0.40	0.40	0.40	0.40	0.40
w_B	0.10	0.10	0.10	0.10	0.10	0.10	0.10	0.10	0.10
w_C	0.50	0.50	0.50	0.50	0.50	0.50	0.50	0.50	0.50
T/K	347.5	352.8	356.8	363.5	369.3	374.0	380.2	384.9	389.0
P/MPa	11.9	12.7	14.7	16.7	18.4	19.5	21.1	22.4	23.3
w_A	0.40	0.40	0.40	0.40	0.40	0.40	0.40	0.40	0.40
w_B	0.10	0.10	0.20	0.20	0.20	0.20	0.20	0.20	0.20
w_C	0.50	0.50	0.40	0.40	0.40	0.40	0.40	0.40	0.40
T/K	394.5	399.4	354.6	359.9	363.9	368.0	378.1	383.3	388.6
P/MPa	24.7	26.0	7.1	9.2	10.8	14.9	16.5	18.1	20.3
w_A	0.40	0.50	0.50	0.50	0.50	0.50	0.50	0.50	0.50
w_B	0.20	0.05	0.05	0.05	0.05	0.05	0.05	0.05	0.05
w_C	0.40	0.45	0.45	0.45	0.45	0.45	0.45	0.45	0.45
T/K	393.3	323.1	328.1	334.0	339.5	344.8	348.0	352.5	359.1
P/MPa	21.3	18.5	19.8	20.8	21.8	23.8	24.5	24.8	27.8
w_A	0.50	0.50	0.50	0.50	0.50	0.50	0.50	0.50	0.50
w_B	0.05	0.05	0.05	0.10	0.10	0.10	0.10	0.10	0.10
w_C	0.45	0.45	0.45	0.40	0.40	0.40	0.40	0.40	0.40
T/K	364.0	369.6	375.4	324.0	328.9	333.0	338.4	342.6	347.5
P/MPa	28.8	30.5	31.8	24.6	26.0	27.3	28.6	30.0	31.4
w_A	0.50	0.50	0.50						
w_B	0.10	0.10	0.10						
w_C	0.40	0.40	0.40						
T/K	354.1	356.1	364.3						
P/MPa	33.0	33.3	35.9						

Polymer (B): **poly(ε-caprolactone)** **2008LIU**
Characterization: M_n/g.mol^{-1} = 6100, M_w/g.mol^{-1} = 14300, Scientific Polymer Products, Inc., Ontario, NY
Solvent (A): **carbon dioxide** CO_2 **124-38-9**
Solvent (C): **2-propanone** C_3H_6O **67-64-1**

Type of data: cloud points

w_A	0.40	0.40	0.40	0.40	0.40	0.40	0.40	0.40	0.40
w_B	0.10	0.10	0.10	0.10	0.10	0.10	0.10	0.10	0.10
w_C	0.50	0.50	0.50	0.50	0.50	0.50	0.50	0.50	0.50
T/K	325.0	330.0	337.6	342.5	347.5	352.8	356.8	363.5	369.3
P/MPa	4.9	6.8	9.1	10.5	11.9	12.7	14.7	16.7	18.4

continued

continued

w_A	0.40	0.40	0.40	0.40	0.40	0.40
w_B	0.10	0.10	0.10	0.10	0.10	0.10
w_C	0.50	0.50	0.50	0.50	0.50	0.50
T/K	374.0	380.1	384.9	389.0	394.5	399.4
P/MPa	19.5	21.1	22.4	23.3	24.7	26.0

Polymer (B): **poly(ε-caprolactone)** **2008LIU**
Characterization: M_n/g.mol^{-1} = 6100, M_w/g.mol^{-1} = 14300, Scientific Polymer Products, Inc., Ontario, NY
Solvent (A): **carbon dioxide** **CO_2** **124-38-9**
Solvent (C): **2-propanone** **C_3H_6O** **67-64-1**
Polymer (D): **poly(methyl methacrylate)**
Characterization: M_n/g.mol^{-1} = 8300, M_w/g.mol^{-1} = 15000, Scientific Polymer Products, Inc., Ontario, NY

Type of data: cloud points

w_A	0.40	0.40	0.40	0.40	0.40	0.40	0.40	0.40	0.40
w_B	0.025	0.025	0.025	0.025	0.025	0.025	0.025	0.025	0.025
w_C	0.50	0.50	0.50	0.50	0.50	0.50	0.50	0.50	0.50
w_D	0.075	0.075	0.075	0.075	0.075	0.075	0.075	0.075	0.075
T/K	341.1	347.0	353.4	358.0	363.3	368.5	374.1	379.5	384.9
P/MPa	10.5	12.6	14.3	15.7	17.2	18.6	20.3	21.7	23.2
w_A	0.40	0.40	0.40	0.40	0.40	0.40	0.40	0.40	0.40
w_B	0.025	0.025	0.025	0.05	0.05	0.05	0.05	0.05	0.05
w_C	0.50	0.50	0.50	0.50	0.50	0.50	0.50	0.50	0.50
w_D	0.075	0.075	0.075	0.05	0.05	0.05	0.05	0.05	0.05
T/K	390.9	395.3	400.8	326.1	333.1	337.4	343.8	347.5	354.8
P/MPa	24.5	25.4	27.0	7.5	9.7	10.9	13.0	14.3	16.0
w_A	0.40	0.40	0.40	0.40	0.40	0.40	0.40	0.40	0.40
w_B	0.05	0.05	0.05	0.05	0.05	0.05	0.05	0.05	0.05
w_C	0.50	0.50	0.50	0.50	0.50	0.50	0.50	0.50	0.50
w_D	0.05	0.05	0.05	0.05	0.05	0.05	0.05	0.05	0.05
T/K	359.8	365.5	370.1	375.5	380.6	384.9	390.4	394.3	400.6
P/MPa	17.3	19.2	20.4	22.0	23.0	24.2	25.6	26.4	27.8
w_A	0.40	0.40	0.40	0.40	0.40	0.40	0.40	0.40	0.40
w_B	0.075	0.075	0.075	0.075	0.075	0.075	0.075	0.075	0.075
w_C	0.50	0.50	0.50	0.50	0.50	0.50	0.50	0.50	0.50
w_D	0.025	0.025	0.025	0.025	0.025	0.025	0.025	0.025	0.025
T/K	328.0	332.8	339.0	344.0	351.5	356.1	362.3	369.0	374.1
P/MPa	7.1	8.5	10.5	12.2	14.5	15.8	17.5	19.5	20.6

continued

continued

w_A	0.40	0.40	0.40	0.40	0.40	0.40
w_B	0.075	0.075	0.075	0.075	0.075	0.075
w_C	0.50	0.50	0.50	0.50	0.50	0.50
w_D	0.025	0.025	0.025	0.025	0.025	0.025
T/K	377.4	379.4	384.6	389.8	394.6	401.0
P/MPa	22.0	22.4	23.8	24.9	26.1	27.6

Polymer (B): **poly(ε-caprolactone)** **2008LIU**
Characterization: M_n/g.mol^{-1} = 6100, M_w/g.mol^{-1} = 14300,
Scientific Polymer Products, Inc., Ontario, NY
Solvent (A): **carbon dioxide** CO_2 **124-38-9**
Solvent (C): **2-propanone** C_3H_6O **67-64-1**
Polymer (D): **poly(methyl methacrylate)**
Characterization: M_n/g.mol^{-1} = 193000, M_w/g.mol^{-1} = 540000,
Scientific Polymer Products, Inc., Ontario, NY

Type of data: cloud points

w_A	0.40	0.40	0.40	0.40	0.40	0.40	0.40	0.40	0.40
w_B	0.05	0.05	0.05	0.05	0.05	0.05	0.05	0.05	0.05
w_C	0.50	0.50	0.50	0.50	0.50	0.50	0.50	0.50	0.50
w_D	0.05	0.05	0.05	0.05	0.05	0.05	0.05	0.05	0.05
T/K	352.3	356.5	363.3	367.6	372.8	373.4	378.8	384.0	389.5
P/MPa	23.0	24.0	26.0	27.0	29.0	29.3	30.5	32.0	34.0

w_A	0.40	0.40
w_B	0.05	0.05
w_C	0.50	0.50
w_D	0.05	0.05
T/K	395.1	400.1
P/MPa	35.0	37.0

Polymer (B): **poly(ε-caprolactone)** **2007LI2**
Characterization: M_n/g.mol^{-1} = 36100, M_w/g.mol^{-1} = 65000,
Scientific Polymer Products, Inc., Ontario, NY
Solvent (A): **carbon dioxide** CO_2 **124-38-9**
Solvent (C): **2-propanone** C_3H_6O **67-64-1**

Type of data: cloud points

w_A	0.40	0.40	0.40	0.40	0.40	0.40	0.40	0.40	0.40
w_B	0.10	0.10	0.10	0.10	0.10	0.10	0.10	0.10	0.10
w_C	0.50	0.50	0.50	0.50	0.50	0.50	0.50	0.50	0.50
T/K	323.9	329.1	333.6	338.9	344.9	355.4	359.4	369.5	374.3
P/MPa	10.8	11.9	13.0	14.3	15.8	21.0	21.3	23.5	24.5

continued

continued

w_A	0.40	0.40	0.40	0.40
w_B	0.10	0.10	0.10	0.10
w_C	0.50	0.50	0.50	0.50
T/K	378.5	383.1	390.5	399.0
P/MPa	25.0	26.1	27.6	29.8

Polymer (B): **poly(ε-caprolactone)** **2013TA1, 2013TA2**
Characterization: M_w/g.mol^{-1} = 87000, M_w/M_n = 1.7
Solvent (A): **carbon dioxide** CO_2 **124-38-9**
Solvent (C): **2-propanone** C_3H_6O **67-64-1**

Type of data: cloud points

w_C/w_A = 2/1 was kept constant

w_B	0.020	0.050	0.090	0.120	0.150	0.250	0.349
T/K = 330							
P/MPa	11.1	12.4	13.0	10.5	10.2		
T/K = 340							
P/MPa	13.0	14.3	14.4	12.8	12.4		
T/K = 350							
P/MPa	14.7	16.7	17.0	15.6	15.1	9.6	
T/K = 360							
P/MPa	17.4	19.2	18.7	17.6	17.0	12.6	9.2
T/K = 370							
P/MPa	19.4	20.7	20.8	20.5	20.4	14.7	11.8
T/K = 380							
P/MPa	21.8	22.8	23.6	22.5	22.5	17.2	13.6
T/K = 390							
P/MPa						20.3	16.6
T/K = 400							
P/MPa						21.9	19.0
T/K = 410							
P/MPa							21.4

Type of data: LLV phase separation

w_C/w_A = 2/1 was kept constant

w_B	0.020	0.050	0.090	0.120	0.150	0.250	0.349
T/K = 330							
P/MPa	4.4	5.0	4.2	5.0	4.6		

continued

continued

T/K = 340							
P/MPa	5.0	5.4	4.8	5.2	5.1		
T/K = 350							
P/MPa	5.7	6.2	5.5	5.5	5.8	5.2	
T/K = 360							
P/MPa	6.2	6.6	5.7	6.0	6.6	5.4	5.2
T/K = 370							
P/MPa	6.9	7.1	6.2	6.1	7.0	5.7	6.0
T/K = 380							
P/MPa	7.3	7.8	6.9	6.8	7.6	6.5	6.5
T/K = 390							
P/MPa						7.2	7.3
T/K = 400							
P/MPa						7.9	8.3
T/K = 410							
P/MPa							9.2

Polymer (B): **poly(ε-caprolactone)** **2006KAL**
Characterization: M_n/g.mol^{-1} = 59400, M_w/g.mol^{-1} = 122400,
Tone 787, Union Carbide Benelux
Solvent (A): **carbon dioxide** **CO_2** **124-38-9**
Solvent (C): **trichloromethane** **$CHCl_3$** **67-66-3**

Type of data: bubble and cloud points

T/K = 308.15

w_A	0.7405	0.7019	0.6490	0.5938	0.5561	0.7330	0.6947	0.6425	0.5875
w_B	0.0100	0.0100	0.0100	0.0100	0.0100	0.0200	0.0200	0.0200	0.0200
w_C	0.2495	0.2881	0.3410	0.3962	0.4339	0.2470	0.2853	0.3375	0.3925
P/bar	41.14	44.05	48.06	55.19	91.62	41.24	44.15	48.37	70.24
	VLE	VLE	VLE	LLE	LLE	VLE	VLE	VLE	LLE

w_A	0.5510
w_B	0.0200
w_C	0.4290
P/bar	118.91
	LLE

continued

continued

T/K = 313.15

w_A	0.7405	0.7019	0.6490	0.5938	0.5561	0.7330	0.6947	0.6425	0.5875
w_B	0.0100	0.0100	0.0100	0.0100	0.0100	0.0200	0.0200	0.0200	0.0200
w_C	0.2495	0.2881	0.3410	0.3962	0.4339	0.2470	0.2853	0.3375	0.3925
P/bar	44.56	48.17	52.28	75.06	112.58	44.65	48.37	52.68	93.22
	VLE	VLE	VLE	LLE	LLE	VLE	VLE	VLE	LLE

w_A	0.5510
w_B	0.0200
w_C	0.4290
P/bar	138.68
	LLE

T/K = 318.15

w_A	0.7405	0.7019	0.6490	0.5938	0.5561	0.7330	0.6947	0.6425	0.5875
w_B	0.0100	0.0100	0.0100	0.0100	0.0100	0.0200	0.0200	0.0200	0.0200
w_C	0.2495	0.2881	0.3410	0.3962	0.4339	0.2470	0.2853	0.3375	0.3925
P/bar	48.07	51.88	56.60	97.24	133.10	48.17	52.15	56.89	115.70
	VLE	VLE	VLE	LLE	LLE	VLE	VLE	VLE	LLE

w_A	0.5510
w_B	0.0200
w_C	0.4290
P/bar	158.34
	LLE

T/K = 323.15

w_A	0.7405	0.7019	0.6490	0.5938	0.5561	0.7330	0.6947	0.6425	0.5875
w_B	0.0100	0.0100	0.0100	0.0100	0.0100	0.0200	0.0200	0.0200	0.0200
w_C	0.2495	0.2881	0.3410	0.3962	0.4339	0.2470	0.2853	0.3375	0.3925
P/bar	51.78	55.90	61.00	115.00	152.83	51.95	56.10	61.40	137.57
	VLE	VLE	VLE	LLE	LLE	VLE	VLE	VLE	LLE

w_A	0.5510
w_B	0.0200
w_C	0.4290
P/bar	178.00
	LLE

T/K = 328.15

w_A	0.7405	0.7019	0.6490	0.5938	0.5561	0.7330	0.6947	0.6425	0.5875
w_B	0.0100	0.0100	0.0100	0.0100	0.0100	0.0200	0.0200	0.0200	0.0200
w_C	0.2495	0.2881	0.3410	0.3962	0.4339	0.2470	0.2853	0.3375	0.3925
P/bar	55.29	60.11	66.33	132.25	171.19	55.39	60.40	66.73	159.55
	VLE	VLE	VLE	LLE	LLE	VLE	VLE	VLE	LLE

continued

continued

w_A	0.5510
w_B	0.0200
w_C	0.4290
P/bar	197.50

T/K = 333.15

w_A	0.7405	0.7019	0.6490	0.5938	0.5561	0.7330	0.6947	0.6425	0.5875
w_B	0.0100	0.0100	0.0100	0.0100	0.0100	0.0200	0.0200	0.0200	0.0200
w_C	0.2495	0.2881	0.3410	0.3962	0.4339	0.2470	0.2853	0.3375	0.3925
P/bar	59.00	64.32	76.97	149.21	191.25	59.10	64.52	77.60	179.82
	VLE	VLE	LLE	LLE	LLE	VLE	VLE	LLE	LLE

T/K = 338.15

w_A	0.7405	0.7019	0.6490	0.5938	0.5561	0.7330	0.6947	0.6425	0.5875
w_B	0.0100	0.0100	0.0100	0.0100	0.0100	0.0200	0.0200	0.0200	0.0200
w_C	0.2495	0.2881	0.3410	0.3962	0.4339	0.2470	0.2853	0.3375	0.3925
P/bar	62.92	68.64	96.83	166.97	208.11	63.02	68.85	99.30	200.69
	VLE	VLE	LLE	LLE	LLE	VLE	VLE	LLE	LLE

T/K = 343.15

w_A	0.7405	0.7019	0.6490	0.5938	0.7330	0.6947	0.6425
w_B	0.0100	0.0100	0.0100	0.0100	0.0200	0.0200	0.0200
w_C	0.2495	0.2881	0.3410	0.3962	0.2470	0.2853	0.3375
P/bar	67.03	72.85	115.50	186.74	67.23	73.10	117.10
	VLE	VLE	LLE	LLE	VLE	VLE	LLE

T/K = 348.15

w_A	0.7405	0.7019	0.6490	0.7330	0.6947	0.6425
w_B	0.0100	0.0100	0.0100	0.0200	0.0200	0.0200
w_C	0.2495	0.2881	0.3410	0.2470	0.2853	0.3375
P/bar	70.75	77.37	134.46	70.84	77.60	138.30
	VLE	VLE	LLE	VLE	VLE	LLE

T/K = 353.15

w_A	0.7405	0.7019	0.6490	0.7330	0.6947	0.6425
w_B	0.0100	0.0100	0.0100	0.0200	0.0200	0.0200
w_C	0.2495	0.2881	0.3410	0.2470	0.2853	0.3375
P/bar	74.66	81.70	153.22	74.75	82.00	157.13
	VLE	VLE	LLE	VLE	VLE	LLE

T/K = 358.15

w_A	0.7405	0.7019	0.6490	0.7330	0.6947	0.6425
w_B	0.0100	0.0100	0.0100	0.0200	0.0200	0.0200
w_C	0.2495	0.2881	0.3410	0.2470	0.2853	0.3375
P/bar	78.67	85.90	169.88	78.75	86.40	175.90
	VLE	VLE	LLE	VLE	VLE	LLE

continued

continued

T/K = 363.15

w_A	0.7405	0.7019	0.6490	0.7330	0.6947	0.6425
w_B	0.0100	0.0100	0.0100	0.0200	0.0200	0.0200
w_C	0.2495	0.2881	0.3410	0.2470	0.2853	0.3375
P/bar	82.59	97.30	187.24	82.70	103.75	194.63
	VLE	LLE	LLE	VLE	LLE	LLE

T/K = 368.15

w_A	0.7405	0.7019	0.6490	0.7330	0.6947
w_B	0.0100	0.0100	0.0100	0.0200	0.0200
w_C	0.2495	0.2881	0.3410	0.2470	0.2853
P/bar	86.40	112.18	201.29	86.60	121.31
	VLE	LLE	LLE	VLE	LLE

T/K = 373.15

w_A	0.7405	0.7019	0.7330	0.6947
w_B	0.0100	0.0100	0.0200	0.0200
w_C	0.2495	0.2881	0.2470	0.2853
P/bar	89.91	124.93	90.21	138.30
	VLE	LLE	VLE	LLE

Polymer (B): **poly(ε-caprolactone)** **2006BY2**
Characterization: M_w/g.mol^{-1} = 170000, Honam Petrochemical Co., South Korea
Solvent (A): **propene** C_3H_6 **115-07-1**
Solvent (C): **chlorodifluoromethane** $CHClF_2$ **75-45-6**

Type of data: cloud points

w_A	0.866	0.866	0.866	0.866	0.866	0.866	0.866	0.866	0.866
w_B	0.050	0.050	0.050	0.050	0.050	0.050	0.050	0.050	0.050
w_C	0.084	0.084	0.084	0.084	0.084	0.084	0.084	0.084	0.084
T/K	384.95	394.15	401.55	413.95	423.75	433.05	443.25	454.05	465.15
P/bar	2563.8	2360.4	2265.5	2115.5	1963.8	1812.1	1725.9	1629.3	1546.6
w_A	0.866	0.712	0.712	0.712	0.712	0.712	0.712	0.712	0.712
w_B	0.050	0.059	0.059	0.059	0.059	0.059	0.059	0.059	0.059
w_C	0.084	0.229	0.229	0.229	0.229	0.229	0.229	0.229	0.229
T/K	475.25	324.05	337.35	351.65	366.85	382.45	396.45	412.35	428.85
P/bar	1486.2	1637.9	1570.7	1508.6	1453.4	1401.7	1360.3	1308.6	1267.2
w_A	0.712	0.712	0.712	0.533	0.533	0.533	0.533	0.533	0.533
w_B	0.059	0.059	0.059	0.048	0.048	0.048	0.048	0.048	0.048
w_C	0.229	0.229	0.229	0.419	0.419	0.419	0.419	0.419	0.419
T/K	443.35	458.55	474.55	323.25	332.35	352.75	375.45	393.25	413.05
P/bar	1220.7	1187.9	1153.4	608.6	636.2	700.7	760.3	794.8	824.1

continued

continued

w_A	0.533	0.533
w_B	0.048	0.048
w_C	0.419	0.419
T/K	432.65	454.95
P/bar	855.5	876.2

Polymer (B): **poly(ε-caprolactone)** **2006BY2**
Characterization: M_w/g.mol^{-1} = 170000, Honam Petrochemical Co., South Korea
Solvent (A): **propene** C_3H_6 **115-07-1**
Solvent (C): **dimethyl ether** C_2H_6O **115-10-6**

Type of data: cloud points

w_A	0.855	0.855	0.855	0.855	0.855	0.855	0.855	0.855	0.855
w_B	0.050	0.050	0.050	0.050	0.050	0.050	0.050	0.050	0.050
w_C	0.095	0.095	0.095	0.095	0.095	0.095	0.095	0.095	0.095
T/K	394.35	402.75	414.15	422.15	432.15	443.55	452.95	464.15	474.15
P/bar	2615.5	2360.4	2137.9	2012.1	1874.1	1760.3	1653.4	1563.8	1520.7
w_A	0.739	0.739	0.739	0.739	0.739	0.739	0.739	0.739	0.739
w_B	0.049	0.049	0.049	0.049	0.049	0.049	0.049	0.049	0.049
w_C	0.212	0.212	0.212	0.212	0.212	0.212	0.212	0.212	0.212
T/K	360.35	367.15	383.35	398.45	411.15	429.85	445.15	458.85	472.25
P/bar	2663.8	2532.8	2234.5	2013.8	1798.3	1601.7	1481.0	1391.4	1319.0
w_A	0.529	0.529	0.529	0.529	0.529	0.529	0.529	0.529	
w_B	0.048	0.048	0.048	0.048	0.048	0.048	0.048	0.048	
w_C	0.424	0.424	0.424	0.424	0.424	0.424	0.424	0.424	
T/K	322.65	333.05	352.25	374.15	391.45	413.15	433.35	452.85	
P/bar	1346.6	1279.3	1174.1	1089.7	1044.8	990.4	953.5	925.9	

Polymer (B): **poly(4-chlorostyrene)** **2013YA2**
Characterization: M_n/g.mol^{-1} = 57000, M_w/g.mol^{-1} = 87000,
Scientific Polymer Products, Inc., Ontario, NY
Solvent (A): **n-butane** C_4H_{10} **106-97-8**
Solvent (C): **4-chlorostyrene** C_8H_7Cl **1073-67-2**

Type of data: cloud points

w_A	0.377	0.377	0.377	0.377	0.377	0.377	0.377	0.642	0.642
w_B	0.049	0.049	0.049	0.049	0.049	0.049	0.049	0.049	0.049
w_C	0.574	0.574	0.574	0.574	0.574	0.574	0.574	0.309	0.309
T/K	438.7	443.0	447.9	453.2	457.7	464.5	471.9	485.3	487.2
P/MPa	245.69	223.62	202.59	182.59	167.07	137.93	110.17	258.79	246.38
w_A	0.642	0.642							
w_B	0.049	0.049							
w_C	0.309	0.309							
T/K	490.0	491.2							
P/MPa	231.90	225.34							

Polymer (B): **poly(4-chlorostyrene)** **2013YA2**
Characterization: M_n/g.mol^{-1} = 57000, M_w/g.mol^{-1} = 87000,
Scientific Polymer Products, Inc., Ontario, NY
Solvent (A): **n-butane** C_4H_{10} **106-97-8**
Solvent (C): **ethanol** C_2H_6O **64-17-5**

Type of data: cloud points

w_A	0.278	0.278	0.278	0.278	0.473	0.473	0.473	0.473	0.643
w_B	0.049	0.049	0.049	0.049	0.044	0.044	0.044	0.044	0.050
w_C	0.673	0.673	0.673	0.673	0.483	0.483	0.483	0.483	0.307
T/K	458.3	458.9	463.9	474.9	433.4	435.6	437.7	445.2	426.0
P/MPa	248.79	234.66	150.86	101.55	228.45	206.72	186.72	140.17	243.28
w_A	0.643	0.643	0.643	0.709	0.709	0.709	0.709	0.847	0.847
w_B	0.050	0.050	0.050	0.050	0.050	0.050	0.050	0.051	0.051
w_C	0.307	0.307	0.307	0.241	0.241	0.241	0.241	0.102	0.102
T/K	428.7	435.1	442.3	426.0	428.4	433.6	445.1	464.4	468.8
P/MPa	217.41	176.72	149.48	240.17	223.97	193.97	155.69	239.48	225.52
w_A	0.847	0.847							
w_B	0.051	0.051							
w_C	0.102	0.102							
T/K	474.9	486.1							
P/MPa	208.45	185.79							

Polymer (B): **poly(4-chlorostyrene)** **2013YA2**
Characterization: M_n/g.mol^{-1} = 57000, M_w/g.mol^{-1} = 87000,
Scientific Polymer Products, Inc., Ontario, NY
Solvent (A): **n-butane** C_4H_{10} **106-97-8**
Solvent (C): **2-propanone** C_3H_6O **67-64-1**

Type of data: cloud points

w_A	0.654	0.654	0.654	0.654	0.654	0.654	0.654
w_B	0.047	0.047	0.047	0.047	0.047	0.047	0.047
w_C	0.299	0.299	0.299	0.299	0.299	0.299	0.299
T/K	334.9	355.0	374.9	392.8	414.7	434.8	451.3
P/MPa	102.59	89.83	81.55	76.72	72.59	69.83	69.14

Polymer (B): **poly(4-chlorostyrene)** **2013YA2**
Characterization: M_n/g.mol^{-1} = 57000, M_w/g.mol^{-1} = 87000,
Scientific Polymer Products, Inc., Ontario, NY
Solvent (A): **1-butene** C_4H_8 **106-98-9**
Solvent (C): **4-chlorostyrene** C_8H_7Cl **1073-67-2**

Type of data: cloud points

continued

continued

w_A	0.499	0.499	0.499	0.499	0.499	0.609	0.609	0.609	0.609
w_B	0.052	0.052	0.052	0.052	0.052	0.053	0.053	0.053	0.053
w_C	0.449	0.449	0.449	0.449	0.449	0.338	0.338	0.338	0.338
T/K	375.0	395.6	415.7	435.0	451.5	344.6	356.2	375.0	394.5
P/MPa	102.93	85.69	68.10	58.79	49.31	240.17	195.34	143.28	113.10
w_A	0.609	0.609	0.609	0.687	0.687	0.687	0.687	0.687	0.687
w_B	0.053	0.053	0.053	0.045	0.045	0.045	0.045	0.045	0.045
w_C	0.338	0.338	0.338	0.268	0.268	0.268	0.268	0.268	0.268
T/K	415.2	435.0	452.8	367.8	372.7	375.8	395.8	415.5	434.9
P/MPa	94.31	81.72	71.55	229.14	215.00	202.93	155.34	126.38	106.38
w_A	0.687	0.954	0.954	0.954	0.954	0.954			
w_B	0.045	0.046	0.046	0.046	0.046	0.046			
w_C	0.268	0.000	0.000	0.000	0.000	0.000			
T/K	452.8	446.7	452.2	459.1	467.5	483.5			
P/MPa	91.03	243.62	227.76	211.55	193.97	168.10			

Polymer (B): **poly(4-chlorostyrene)** **2013YA2**
Characterization: M_n/g.mol^{-1} = 57000, M_w/g.mol^{-1} = 87000, Scientific Polymer Products, Inc., Ontario, NY
Solvent (A): **1-butene** C_4H_8 **106-98-9**
Solvent (C): **ethanol** C_2H_6O **64-17-5**

Type of data: cloud points

w_A	0.257	0.257	0.257	0.257	0.257	0.392	0.392	0.392	0.392
w_B	0.048	0.048	0.048	0.048	0.048	0.049	0.049	0.049	0.049
w_C	0.695	0.695	0.695	0.695	0.695	0.559	0.559	0.559	0.559
T/K	468.1	468.6	469.3	474.3	481.3	434.1	436.0	437.7	441.8
P/MPa	244.31	224.00	195.00	132.41	99.31	236.03	211.03	186.72	151.90
w_A	0.392	0.651	0.651	0.651	0.651	0.651	0.651		
w_B	0.049	0.049	0.049	0.049	0.049	0.049	0.049		
w_C	0.559	0.300	0.300	0.300	0.300	0.300	0.300		
T/K	452.7	392.5	393.0	395.8	413.2	434.8	451.8		
P/MPa	108.10	242.24	236.38	210.86	133.28	100.69	87.07		

Polymer (B): **poly(4-chlorostyrene)** **2013YA2**
Characterization: M_n/g.mol^{-1} = 57000, M_w/g.mol^{-1} = 87000, Scientific Polymer Products, Inc., Ontario, NY
Solvent (A): **1-butene** C_4H_8 **106-98-9**
Solvent (C): **2-propanone** C_3H_6O **67-64-1**

Type of data: cloud points

continued

continued

w_A	0.770	0.770	0.770	0.770	0.770	0.770	0.770	0.742	0.742
w_B	0.046	0.046	0.046	0.046	0.046	0.046	0.046	0.049	0.049
w_C	0.284	0.284	0.284	0.284	0.284	0.284	0.284	0.209	0.209
T/K	334.8	355.7	375.3	395.0	415.4	435.9	451.5	335.3	354.8
P/MPa	51.90	52.41	53.10	53.62	54.66	55.35	56.03	83.28	78.45
w_A	0.742	0.742	0.742	0.742	0.742	0.793	0.793	0.793	0.793
w_B	0.049	0.049	0.049	0.049	0.049	0.052	0.052	0.052	0.052
w_C	0.209	0.209	0.209	0.209	0.209	0.155	0.155	0.155	0.155
T/K	374.5	393.5	415.9	435.1	451.8	354.2	374.8	394.3	414.0
P/MPa	75.00	72.41	70.86	69.48	68.79	111.38	101.30	94.48	89.14
w_A	0.793	0.793	0.854	0.854	0.854	0.854	0.854	0.854	0.898
w_B	0.052	0.052	0.046	0.046	0.046	0.046	0.046	0.046	0.048
w_C	0.155	0.155	0.100	0.100	0.100	0.100	0.100	0.100	0.054
T/K	435.0	453.7	353.5	374.9	394.7	413.3	434.2	451.7	387.5
P/MPa	85.36	82.24	188.10	158.97	139.48	125.34	113.62	106.38	234.66
w_A	0.898	0.898	0.898	0.898	0.954	0.954	0.954	0.954	0.954
w_B	0.048	0.048	0.048	0.048	0.046	0.046	0.046	0.046	0.046
w_C	0.054	0.054	0.054	0.054	0.000	0.000	0.000	0.000	0.000
T/K	394.6	415.2	434.4	452.3	446.7	452.2	459.1	467.5	483.5
P/MPa	212.93	172.22	145.00	127.76	243.62	227.76	211.55	193.97	168.10

Polymer (B): **poly(4-chlorostyrene)** **2013YA2**
Characterization: M_n/g.mol^{-1} = 57000, M_w/g.mol^{-1} = 87000,
Scientific Polymer Products, Inc., Ontario, NY
Solvent (A): **carbon dioxide** CO_2 **124-38-9**
Solvent (C): **dimethyl ether** C_2H_6O **115-10-6**

Type of data: cloud points

w_A	0.000	0.000	0.000	0.000	0.000	0.000	0.000	0.052	0.052
w_B	0.049	0.049	0.049	0.049	0.049	0.049	0.049	0.050	0.050
w_C	0.951	0.951	0.951	0.951	0.951	0.951	0.951	0.898	0.898
T/K	335.1	355.3	375.2	394.4	414.8	435.3	451.8	335.7	355.3
P/MPa	40.35	47.24	52.76	57.59	61.55	65.00	67.76	57.76	62.76
w_A	0.052	0.052	0.052	0.052	0.052	0.183	0.183	0.183	0.183
w_B	0.050	0.050	0.050	0.050	0.050	0.046	0.046	0.046	0.046
w_C	0.898	0.898	0.898	0.898	0.898	0.771	0.771	0.771	0.771
T/K	374.6	394.9	415.0	435.0	452.3	332.8	353.9	374.8	394.5
P/MPa	67.07	70.52	73.45	76.38	78.10	107.75	105.00	102.93	101.38
w_A	0.183	0.183	0.183	0.327	0.327	0.327	0.327	0.327	0.327
w_B	0.046	0.046	0.046	0.046	0.046	0.046	0.046	0.046	0.046
w_C	0.771	0.771	0.771	0.627	0.627	0.627	0.627	0.627	0.627
T/K	415.0	435.0	453.8	358.2	375.0	394.4	415.2	433.1	454.4
P/MPa	100.86	100.86	100.86	231.21	201.03	178.97	163.97	154.14	146.21

Polymer (B): **poly(4-chlorostyrene)** **2013YA2**
Characterization: M_n/g.mol^{-1} = 57000, M_w/g.mol^{-1} = 87000, Scientific Polymer Products, Inc., Ontario, NY
Solvent (A): **dimethyl ether** C_2H_6O **115-10-6**
Solvent (C): **4-chlorostyrene** C_8H_7Cl **1073-67-2**

Type of data: cloud points

w_A	0.543	0.543	0.543	0.543	0.543	0.543	0.543	0.658	0.658
w_B	0.050	0.050	0.050	0.050	0.050	0.050	0.050	0.049	0.049
w_C	0.407	0.407	0.407	0.407	0.407	0.407	0.407	0.293	0.293
T/K	335.4	354.5	375.2	392.2	415.9	434.9	452.2	334.5	355.8
P/MPa	5.69	11.70	17.76	22.24	27.07	29.14	29.14	11.55	18.97
w_A	0.658	0.658	0.658	0.658	0.658	0.804	0.804	0.804	0.804
w_B	0.049	0.049	0.049	0.049	0.049	0.052	0.052	0.052	0.052
w_C	0.293	0.293	0.293	0.293	0.293	0.144	0.144	0.144	0.144
T/K	375.8	394.5	415.2	435.5	451.8	335.9	352.6	375.2	394.2
P/MPa	25.00	29.83	34.48	37.76	38.62	23.28	29.14	36.03	41.21
w_A	0.804	0.804	0.804	0.951	0.951	0.951	0.951	0.951	0.951
w_B	0.052	0.052	0.052	0.049	0.049	0.049	0.049	0.049	0.049
w_C	0.144	0.144	0.144	0.000	0.000	0.000	0.000	0.000	0.000
T/K	414.0	435.1	452.8	335.1	355.3	375.2	394.4	414.8	435.3
P/MPa	45.35	49.48	52.24	40.35	47.24	52.76	57.59	61.52	65.00
w_A	0.951								
w_B	0.049								
w_C	0.000								
T/K	451.8								
P/MPa	67.76								

Polymer (B): **poly(4-chlorostyrene)** **2013YA2**
Characterization: M_n/g.mol^{-1} = 57000, M_w/g.mol^{-1} = 87000, Scientific Polymer Products, Inc., Ontario, NY
Solvent (A): **propane** C_3H_8 **74-98-6**
Solvent (C): **2-propanone** C_3H_6O **67-64-1**

Type of data: cloud points

w_A	0.430	0.430	0.430	0.430	0.430	0.430	0.490	0.490	0.490
w_B	0.049	0.049	0.049	0.049	0.049	0.049	0.050	0.050	0.050
w_C	0.521	0.521	0.521	0.521	0.521	0.521	0.460	0.460	0.460
T/K	354.7	374.4	393.3	413.0	435.6	451.7	354.5	374.8	395.5
P/MPa	48.45	50.17	51.56	52.93	54.66	55.69	64.66	63.62	63.62

continued

continued

w_A	0.490	0.490	0.490	0.667	0.667	0.667	0.667	0.667	0.667
w_B	0.050	0.050	0.050	0.047	0.047	0.047	0.047	0.047	0.047
w_C	0.460	0.460	0.460	0.286	0.286	0.286	0.286	0.286	0.286
T/K	414.9	435.6	451.4	339.0	354.4	375.1	394.8	415.1	435.0
P/MPa	63.97	64.31	65.00	171.90	148.79	129.66	118.10	109.48	103.28
w_A	0.667	0.800	0.800	0.800	0.800	0.800			
w_B	0.047	0.047	0.047	0.047	0.047	0.047			
w_C	0.286	0.153	0.153	0.153	0.153	0.153			
T/K	451.7	415.2	425.1	435.1	454.9	477.2			
P/MPa	99.14	234.66	212.93	194.48	167.24	151.55			

Polymer (B): **poly(4-chlorostyrene)** **2013YA2**
Characterization: M_n/g.mol^{-1} = 57000, M_w/g.mol^{-1} = 87000, Scientific Polymer Products, Inc., Ontario, NY
Solvent (A): **propene** C_3H_6 **115-07-1**
Solvent (C): **2-propanone** C_3H_6O **67-64-1**

Type of data: cloud points

w_A	0.438	0.438	0.438	0.438	0.438	0.438	0.438	0.550	0.550
w_B	0.048	0.048	0.048	0.048	0.048	0.048	0.048	0.044	0.044
w_C	0.514	0.514	0.514	0.514	0.514	0.514	0.514	0.406	0.406
T/K	341.0	355.6	375.6	395.1	415.1	435.8	452.0	354.5	374.8
P/MPa	54.31	51.90	51.21	51.21	51.90	52.93	53.28	62.59	63.97
w_A	0.550	0.550	0.550	0.550	0.698	0.698	0.698	0.698	0.698
w_B	0.044	0.044	0.044	0.044	0.044	0.044	0.044	0.044	0.044
w_C	0.406	0.406	0.406	0.406	0.258	0.258	0.258	0.258	0.258
T/K	395.5	414.9	435.6	451.4	339.0	354.4	375.1	394.8	415.1
P/MPa	65.00	66.03	67.24	67.76	114.66	108.45	104.31	100.52	97.07
w_A	0.698	0.698	0.800	0.800	0.800	0.800	0.800	0.800	0.800
w_B	0.044	0.044	0.046	0.046	0.046	0.046	0.046	0.046	0.046
w_C	0.258	0.258	0.154	0.154	0.154	0.154	0.154	0.154	0.154
T/K	435.0	451.7	345.2	355.9	374.3	394.0	415.4	435.2	455.3
P/MPa	95.35	94.66	189.66	176.55	161.38	149.83	138.10	131.90	125.86
w_A	0.856	0.856	0.856	0.856	0.856	0.954	0.954	0.954	0.954
w_B	0.046	0.046	0.046	0.046	0.046	0.046	0.046	0.046	0.046
w_C	0.098	0.098	0.098	0.098	0.098	0.000	0.000	0.000	0.000
T/K	385.1	394.6	413.7	435.2	452.0	477.1	481.5	488.1	494.2
P/MPa	226.90	210.69	185.69	163.97	151.03	239.14	230.69	220.00	209.83

Polymer (B): **poly(cyclohexene oxide-*co*-carbon dioxide)** **2005SCH**
Characterization: M_w/g.mol^{-1} = 12000, M_w/M_n < 1.2, synthesized in the laboratory
Solvent (A): **carbon dioxide** **CO_2** **124-38-9**
Solvent (C): **cyclohexene oxide** **$C_6H_{10}O$** **286-20-4**

Type of data: cloud points

w_A	0.8660	0.8660	0.8660	0.8660	0.8660	0.8660	0.8660	0.8660	0.8660
w_B	0.0094	0.0094	0.0094	0.0094	0.0094	0.0094	0.0094	0.0094	0.0094
w_C	0.1246	0.1246	0.1246	0.1246	0.1246	0.1246	0.1246	0.1246	0.1246
T/K	373.08	382.72	393.34	403.44	413.36	422.99	433.27	443.24	453.17
P/MPa	183	176	172	168	165	163	159	157	155
w_A	0.8660								
w_B	0.0094								
w_C	0.1246								
T/K	465.88								
P/MPa	154								

Polymer (B): **poly(cyclohexene oxide-*co*-carbon dioxide)** **2005SCH**
Characterization: M_w/g.mol^{-1} = 25000, M_w/M_n < 1.2, synthesized in the laboratory
Solvent (A): **carbon dioxide** **CO_2** **124-38-9**
Solvent (C): **cyclohexene oxide** **$C_6H_{10}O$** **286-20-4**

Type of data: cloud points

w_A	0.9173	0.9173	0.9173	0.9173	0.9173	0.8354	0.8354	0.8354	0.8354
w_B	0.0099	0.0099	0.0099	0.0099	0.0099	0.0100	0.0100	0.0100	0.0100
w_C	0.0728	0.0728	0.0728	0.0728	0.0728	0.1546	0.1546	0.1546	0.1546
T/K	432.21	435.60	446.41	456.07	465.54	367.57	380.21	388.37	397.78
P/MPa	351	341	314	297	278	213	204	198	193
w_A	0.8354	0.8354	0.8354	0.8354	0.8354	0.8354	0.8354	0.7474	0.7474
w_B	0.0100	0.0100	0.0100	0.0100	0.0100	0.0100	0.0100	0.0098	0.0098
w_C	0.1546	0.1546	0.1546	0.1546	0.1546	0.1546	0.1546	0.2428	0.2428
T/K	407.09	416.69	426.60	437.84	447.07	455.94	465.93	369.35	380.89
P/MPa	189	184	181	176	174	171	167	115	117
w_A	0.7474	0.7474	0.7474	0.7474	0.7474	0.7474	0.7474	0.7474	0.7474
w_B	0.0098	0.0098	0.0098	0.0098	0.0098	0.0098	0.0098	0.0098	0.0098
w_C	0.2428	0.2428	0.2428	0.2428	0.2428	0.2428	0.2428	0.2428	0.2428
T/K	387.81	395.61	405.75	415.26	427.35	436.35	445.21	454.97	465.99
P/MPa	117	117	118	118	119	119	119	119	119

Polymer (B): **poly(cyclohexene oxide-*co*-carbon dioxide)** **2005SCH**
Characterization: M_w/g.mol^{-1} = 54000, M_w/M_n < 1.2, synthesized in the laboratory
Solvent (A): **carbon dioxide** CO_2 **124-38-9**
Solvent (C): **cyclohexene oxide** $C_6H_{10}O$ **286-20-4**

Type of data: cloud points

w_A	0.8685	0.8685	0.8685	0.8685	0.8685	0.8685	0.8685
w_B	0.0098	0.0098	0.0098	0.0098	0.0098	0.0098	0.0098
w_C	0.1217	0.1217	0.1217	0.1217	0.1217	0.1217	0.1217
T/K	402.69	412.21	421.47	432.80	442.71	451.96	463.80
P/MPa	363	332	310	292	278	265	251

Polymer (B): **poly(cyclohexyl acrylate)** **2004BY2**
Characterization: M_w/g.mol^{-1} = 150000,
Scientific Polymer Products, Inc., Ontario, NY
Solvent (A): **carbon dioxide** CO_2 **124-38-9**
Solvent (C): **cyclohexyl acrylate** $C_9H_{14}O_2$ **3066-71-5**

Type of data: cloud points

w_A	0.727	0.727	0.727	0.727	0.727	0.727	0.727	0.727	0.727
w_B	0.048	0.048	0.048	0.048	0.048	0.048	0.048	0.048	0.048
w_C	0.225	0.225	0.225	0.225	0.225	0.225	0.225	0.225	0.225
T/K	415.85	417.35	418.65	419.25	422.35	426.45	427.45	433.55	444.95
P/bar	2192.1	2146.6	2105.2	2065.2	1946.6	1819.7	1802.1	1621.4	1522.8
w_A	0.727	0.772	0.772	0.772	0.772	0.772	0.772	0.772	0.772
w_B	0.048	0.054	0.054	0.054	0.054	0.054	0.054	0.054	0.054
w_C	0.225	0.274	0.274	0.274	0.274	0.274	0.274	0.274	0.274
T/K	456.05	332.25	342.45	352.55	372.65	392.75	412.75	433.75	453.05
P/bar	1401.4	2018.3	1515.5	1298.3	1096.2	1004.1	959.0	934.8	917.2
w_A	0.618	0.618	0.618	0.618	0.618	0.618	0.618	0.618	0.658
w_B	0.050	0.050	0.050	0.050	0.050	0.050	0.050	0.050	0.050
w_C	0.332	0.332	0.332	0.332	0.332	0.332	0.332	0.332	0.392
T/K	314.05	333.25	353.35	373.55	393.65	413.95	432.95	452.55	312.95
P/bar	1051.7	892.4	817.2	775.2	768.3	770.3	775.2	777.2	403.1
w_A	0.658	0.658	0.658	0.658	0.658	0.658	0.658	0.491	0.491
w_B	0.050	0.050	0.050	0.050	0.050	0.050	0.050	0.053	0.053
w_C	0.392	0.392	0.392	0.392	0.392	0.392	0.392	0.456	0.456
T/K	335.65	355.05	374.35	394.85	414.75	435.95	454.35	333.45	353.45
P/bar	438.3	473.8	510.0	536.2	559.7	583.1	579.7	106.9	170.3
w_A	0.491	0.491	0.491	0.491					
w_B	0.053	0.053	0.053	0.053					
w_C	0.456	0.456	0.456	0.456					
T/K	373.15	394.25	412.65	431.55					
P/bar	225.2	277.2	323.5	342.1					

continued

continued

Type of data: vapor-liquid equilibrium data

w_A	0.491	0.491	0.491	0.491
w_B	0.053	0.053	0.053	0.053
w_C	0.456	0.456	0.456	0.456
T/K	303.35	308.45	319.05	327.65
P/bar	61.7	67.6	79.0	91.4

Type of data: vapor-liquid-liquid equilibrium data

w_A	0.491	0.491
w_B	0.053	0.053
w_C	0.456	0.456
T/K	360.45	370.15
P/bar	132.0	143.1

Polymer (B): **poly(cyclohexyl methacrylate)** **2004BY2**
Characterization: M_w/g.mol^{-1} = 65000,
Scientific Polymer Products, Inc., Ontario, NY
Solvent (A): **carbon dioxide** CO_2 **124-38-9**
Solvent (C): **cyclohexyl methacrylate** $C_{10}H_{16}O_2$ **101-43-9**

Type of data: cloud points

w_A	0.672	0.672	0.672	0.672	0.672	0.672	0.672	0.672	0.672
w_B	0.054	0.054	0.054	0.054	0.054	0.054	0.054	0.054	0.054
w_C	0.274	0.274	0.274	0.274	0.274	0.274	0.274	0.274	0.274
T/K	379.65	380.65	382.35	383.75	385.65	387.25	389.05	405.75	419.75
P/bar	2234.1	2198.3	2141.7	2095.5	2052.1	2016.2	1968.6	1712.1	1551.0
w_A	0.672	0.672	0.624	0.624	0.624	0.624	0.624	0.624	0.624
w_B	0.054	0.054	0.054	0.054	0.054	0.054	0.054	0.054	0.054
w_C	0.274	0.274	0.322	0.322	0.322	0.322	0.322	0.322	0.322
T/K	434.55	449.95	326.75	333.95	342.35	352.95	363.55	378.35	393.45
P/bar	1405.2	1371.0	2003.5	1617.9	1429.3	1277.2	1189.0	1106.6	1061.7
w_A	0.624	0.624	0.624	0.624	0.549	0.549	0.549	0.549	0.549
w_B	0.054	0.054	0.054	0.054	0.054	0.054	0.054	0.054	0.054
w_C	0.322	0.322	0.322	0.322	0.397	0.397	0.397	0.397	0.397
T/K	408.25	424.15	438.35	460.15	326.55	342.55	358.95	374.25	390.65
P/bar	1035.9	1005.9	960.7	940.7	595.5	602.4	608.3	622.1	636.9
w_A	0.549	0.549	0.549	0.549	0.479	0.479	0.479	0.479	0.479
w_B	0.054	0.054	0.054	0.054	0.054	0.054	0.054	0.054	0.054
w_C	0.397	0.397	0.397	0.397	0.467	0.467	0.467	0.467	0.467
T/K	402.15	416.75	441.55	458.95	326.55	343.45	358.85	373.85	388.15
P/bar	647.6	653.5	645.5	660.0	261.0	297.2	339.7	374.1	409.3

continued

continued

w_A	0.479	0.479	0.479	0.411	0.411	0.411	0.411	0.411	0.411
w_B	0.054	0.054	0.054	0.054	0.054	0.054	0.054	0.054	0.054
w_C	0.467	0.467	0.467	0.535	0.535	0.535	0.535	0.535	0.535
T/K	403.75	419.05	435.55	353.05	368.95	382.35	397.35	413.35	429.95
P/bar	440.7	466.2	487.6	134.1	183.5	218.3	247.6	278.6	317.2

Type of data: vapor-liquid equilibrium data

w_A	0.411	0.411	0.411	0.411
w_B	0.054	0.054	0.054	0.054
w_C	0.535	0.535	0.535	0.535
T/K	309.45	319.35	328.95	338.45
P/bar	61.4	74.8	86.9	100.7

Type of data: vapor-liquid-liquid equilibrium data

w_A	0.411	0.411
w_B	0.054	0.054
w_C	0.535	0.535
T/K	360.45	370.85
P/bar	129.6	143.1

Polymer (B): **poly(decyl acrylate)** **2006BY3**
Characterization: M_w/g.mol^{-1} = 130000,
Scientific Polymer Products, Inc., Ontario, NY
Solvent (A): **carbon dioxide** CO_2 **124-38-9**
Solvent (C): **decyl acrylate** $C_{13}H_{24}O_2$ **2156-96-9**

Type of data: cloud points

w_A	0.949	0.949	0.949	0.949	0.949	0.873	0.873	0.873	0.873
w_B	0.051	0.051	0.051	0.051	0.051	0.052	0.052	0.052	0.052
w_C	0.000	0.000	0.000	0.000	0.000	0.075	0.075	0.075	0.075
T/K	451.45	459.95	468.95	469.05	479.65	417.55	418.45	420.45	423.95
P/bar	1936.2	1770.7	1500.3	1485.9	1449.3	2096.9	1848.6	1558.3	1443.1
w_A	0.873	0.873	0.873	0.800	0.800	0.800	0.800	0.800	0.800
w_B	0.052	0.052	0.052	0.051	0.051	0.051	0.051	0.051	0.051
w_C	0.075	0.075	0.075	0.149	0.149	0.149	0.149	0.149	0.149
T/K	425.65	438.35	454.15	363.15	374.15	393.15	408.65	431.35	453.65
P/bar	1350.0	1113.8	1031.7	1543.1	1225.9	856.2	804.1	767.2	751.0
w_A	0.708	0.708	0.708	0.708	0.708	0.708	0.708	0.708	0.604
w_B	0.058	0.058	0.058	0.058	0.058	0.058	0.058	0.058	0.047
w_C	0.234	0.234	0.234	0.234	0.234	0.234	0.234	0.234	0.349
T/K	321.55	332.65	354.75	375.05	392.25	413.55	434.95	450.25	321.25
P/bar	1141.0	655.5	534.1	520.4	525.2	538.3	552.1	557.2	159.7

continued

continued

w_A	0.604	0.604	0.604	0.604	0.604	0.604	0.604	0.538	0.538
w_B	0.047	0.047	0.047	0.047	0.047	0.047	0.047	0.060	0.060
w_C	0.349	0.349	0.349	0.349	0.349	0.349	0.349	0.402	0.402
T/K	329.85	351.05	372.05	390.65	411.05	433.65	451.95	432.65	415.55
P/bar	181.0	225.9	267.2	296.6	341.7	373.5	385.5	296.9	270.0

w_A	0.538	0.538
w_B	0.060	0.060
w_C	0.402	0.402
T/K	394.65	373.95
P/bar	234.8	200.3

Type of data: vapor-liquid equilibrium data

w_A	0.538	0.538	0.538
w_B	0.060	0.060	0.060
w_C	0.402	0.402	0.402
T/K	353.15	346.85	336.15
P/bar	157.9	151.4	138.3

Type of data: vapor-liquid-liquid equilibrium data

w_A	0.538	0.538
w_B	0.060	0.060
w_C	0.402	0.402
T/K	373.15	384.15
P/bar	183.0	200.0

Polymer (B): **poly(decyl methacrylate)** **2006BY3**
Characterization: M_w/g.mol^{-1} = 100000,
Scientific Polymer Products, Inc., Ontario, NY
Solvent (A): **carbon dioxide** **CO_2** **124-38-9**
Solvent (C): **decyl methacrylate** **$C_{14}H_{26}O_2$** **3179-47-3**

Type of data: cloud points

w_A	0.913	0.913	0.913	0.913	0.882	0.882	0.882	0.882	0.882
w_B	0.043	0.043	0.043	0.043	0.051	0.051	0.051	0.051	0.051
w_C	0.044	0.044	0.044	0.044	0.067	0.067	0.067	0.067	0.067
T/K	437.55	439.85	444.25	458.25	417.45	418.05	418.55	425.85	438.35
P/bar	2332.8	2074.1	1825.9	1493.4	2094.8	1936.2	1863.8	1565.1	1309.7

w_A	0.882	0.839	0.839	0.839	0.839	0.723	0.723	0.723	0.723
w_B	0.051	0.049	0.049	0.049	0.049	0.049	0.049	0.049	0.049
w_C	0.067	0.112	0.112	0.112	0.112	0.228	0.228	0.228	0.228
T/K	453.25	377.15	390.15	404.65	420.95	332.65	352.95	374.15	392.95
P/bar	1205.2	1656.9	1253.8	1095.2	1007.9	1014.1	663.5	600.7	594.5

continued

continued

w_A	0.723	0.622	0.622	0.622	0.622	0.622	0.622	0.528	0.528
w_B	0.049	0.049	0.049	0.049	0.049	0.049	0.049	0.056	0.056
w_C	0.228	0.329	0.329	0.329	0.329	0.329	0.329	0.416	0.416
T/K	420.65	314.95	335.05	353.15	372.45	393.45	416.45	429.25	416.05
P/bar	591.4	221.7	249.3	280.0	315.2	350.0	377.9	295.2	283.5

w_A	0.528	0.528	0.528	0.528
w_B	0.056	0.056	0.056	0.056
w_C	0.416	0.416	0.416	0.416
T/K	400.65	385.55	370.85	356.45
P/bar	257.6	236.2	199.0	174.8

Type of data: vapor-liquid equilibrium data

w_A	0.528	0.528	0.528
w_B	0.056	0.056	0.056
w_C	0.416	0.416	0.416
T/K	344.35	336.15	326.25
P/bar	139.7	128.6	117.2

Type of data: vapor-liquid-liquid equilibrium data

w_A	0.528	0.528
w_B	0.056	0.056
w_C	0.416	0.416
T/K	370.15	357.15
P/bar	168.0	153.0

Polymer (B): **poly[4-(decyloxymethyl)styrene]** **2008SH3**
Characterization: M_n/g.mol^{-1} = 12100, M_w/g.mol^{-1} = 33700, synthesized in the laboratory
Solvent (A): **carbon dioxide** CO_2 **124-38-9**
Solvent (C): **dimethyl ether** C_2H_6O **115-10-6**

Type of data: cloud points

w_A	0.144	0.144	0.144	0.144	0.144	0.144
w_B	0.031	0.031	0.031	0.031	0.031	0.031
w_C	0.825	0.825	0.825	0.825	0.825	0.825
T/K	353.45	374.55	393.45	414.25	433.65	454.05
P/bar	25.9	77.6	122.4	174.1	205.2	215.5

Polymer (B): **poly[4-(decylsulfonylmethyl)styrene]** **2008SH3**
Characterization: M_n/g.mol^{-1} = 18400, M_w/g.mol^{-1} = 49300, synthesized in the laboratory
Solvent (A): **carbon dioxide** CO_2 **124-38-9**
Solvent (C): **dimethyl ether** C_2H_6O **115-10-6**

Type of data: cloud points

continued

continued

w_A	0.155	0.155	0.155	0.155	0.155	0.155	0.155	0.442	0.442
w_B	0.049	0.049	0.049	0.049	0.049	0.049	0.049	0.053	0.053
w_C	0.796	0.796	0.796	0.796	0.796	0.796	0.796	0.505	0.505
T/K	333.75	354.35	374.95	394.95	413.85	434.55	453.25	333.85	354.55
P/bar	236.2	253.5	284.5	319.0	350.0	374.1	384.5	782.8	670.7

w_A	0.442	0.442	0.442	0.442	0.442	0.564	0.564	0.564	0.564
w_B	0.053	0.053	0.053	0.053	0.053	0.049	0.049	0.049	0.049
w_C	0.505	0.505	0.505	0.505	0.505	0.387	0.387	0.387	0.387
T/K	375.65	392.25	413.95	434.75	453.15	327.75	343.15	362.95	385.35
P/bar	627.6	612.1	615.5	622.4	634.5	2125.9	1327.6	1022.3	898.3

w_A	0.564	0.564	0.564	0.564
w_B	0.049	0.049	0.049	0.049
w_C	0.387	0.387	0.387	0.387
T/K	403.05	426.05	440.25	453.65
P/bar	848.3	824.1	829.3	832.8

Polymer (B): **poly[4-(decylthiomethyl)styrene]** **2008SH3**
Characterization: M_n/g.mol^{-1} = 10700, M_w/g.mol^{-1} = 43700, synthesized in the laboratory
Solvent (A): **carbon dioxide** CO_2 **124-38-9**
Solvent (C): **dimethyl ether** C_2H_6O **115-10-6**

Type of data: cloud points

w_A	0.101	0.101	0.101	0.101	0.101	0.101	0.101
w_B	0.043	0.043	0.043	0.043	0.043	0.043	0.043
w_C	0.855	0.855	0.855	0.855	0.855	0.855	0.855
T/K	333.35	354.45	372.95	394.55	414.25	433.95	452.65
P/bar	87.9	129.3	184.5	222.4	267.2	300.0	313.5

Polymer (B): **poly(dimethylsiloxane) monomethacrylate** **2005TAI**
Characterization: M_n/g.mol^{-1} = 10000, Aldrich Chem. Co., Inc., Milwaukee, WI
Solvent (A): **carbon dioxide** CO_2 **124-38-9**
Solvent (C): **1,1-difluoroethene** $C_2H_2F_2$ **75-38-7**

Type of data: cloud points

w_A	0.8314	0.8314	0.8314	0.8314	0.8314
w_B	0.0103	0.0103	0.0103	0.0103	0.0103
w_C	0.1584	0.1584	0.1584	0.1584	0.1584
T/K	318.65	328.15	338.15	347.15	357.15
P/MPa	18.3	20.9	23.3	25.6	27.7

Polymer (B):	**poly(dimethylsiloxane) monomethacrylate**		**2005TAI**
Characterization:	M_n/g.mol^{-1} = 10000, Aldrich Chem. Co., Inc., Milwaukee, WI		
Solvent (A):	**carbon dioxide**	**CO_2**	**124-38-9**
Solvent (C):	**methyl methacrylate**	**$C_5H_8O_2$**	**80-62-6**

Type of data: cloud points

w_A	0.8314	0.8314	0.8314	0.8314
w_B	0.0103	0.0103	0.0103	0.0103
w_C	0.1584	0.1584	0.1584	0.1584
T/K	328.15	339.15	348.15	358.15
P/MPa	10.6	13.3	15.6	17.8

Polymer (B):	**poly(1,4-dioxan-2-one)**		**2012KIR**
Characterization:	M_n/g.mol^{-1} = 35700, M_w/g.mol^{-1} = 89400, T_g/K = 259.2, T_m/K = 382.2, Johnson & Johnson Corporate Biomaterials Center, Somerville, NJ		
Solvent (A):	**carbon dioxide**	**CO_2**	**124-38-9**
Solvent (C):	**2-propanone**	**C_3H_6O**	**67-64-1**

Type of data: cloud points

w_A/w_C 11/89 was kept constant

w_B	0.050	0.050	0.050	0.050	0.050	0.050	0.050	0.050	0.050
T/K	441.15	431.05	421.35	410.75	400.85	391.95	381.25	371.35	361.85
P/MPa	41.11	40.92	41.29	40.37	40.73	41.49	44.65	48.18	53.49
w_B	0.075	0.075	0.075	0.075	0.075	0.075	0.075	0.075	0.075
T/K	441.75	428.55	417.35	411.45	396.75	387.75	380.15	371.75	363.95
P/MPa	42.11	42.31	42.65	43.59	45.93	47.97	50.88	55.17	59.82
w_B	0.100	0.100	0.100	0.100	0.100	0.100	0.100	0.100	0.100
T/K	446.95	433.45	421.05	408.25	398.15	383.15	371.35	365.45	358.05
P/MPa	38.96	39.10	39.63	40.59	42.20	45.35	49.94	52.94	57.13

w_A/w_C 25/75 was kept constant

w_B	0.075	0.075	0.075	0.075	0.075
T/K	440.05	432.15	420.65	408.35	401.15
P/MPa	62.88	63.09	64.88	67.14	68.18

Polymer (B):	**poly(dodecyl acrylate)**		**2010LE2**
Characterization:	M_w/g.mol^{-1} = 44000, M_w/M_n = 1.6, T_g/K = 243.2, Scientific Polymer Products, Inc., Ontario, NY		
Solvent (A):	**carbon dioxide**	**CO_2**	**124-38-9**
Solvent (C):	**dimethyl ether**	**C_2H_6O**	**115-10-6**

Type of data: cloud points

continued

continued

w_A	0.948	0.948	0.948	0.948	0.856	0.856	0.856	0.856	0.766
w_B	0.052	0.052	0.052	0.052	0.050	0.050	0.050	0.050	0.050
w_C	0.000	0.000	0.000	0.000	0.094	0.094	0.094	0.094	0.184
T/K	456.9	459.0	460.5	463.4	413.7	424.0	439.7	454.6	377.3
P/MPa	205.28	169.55	157.07	150.14	182.59	129.83	111.90	105.17	163.62
w_A	0.766	0.766	0.766	0.766	0.766	0.660	0.660	0.660	0.660
w_B	0.050	0.050	0.050	0.050	0.050	0.047	0.047	0.047	0.047
w_C	0.184	0.184	0.184	0.184	0.184	0.293	0.293	0.293	0.293
T/K	384.7	393.1	414.0	432.8	453.7	339.6	352.8	372.5	393.7
P/MPa	131.55	103.97	88.79	83.62	81.55	118.10	87.76	69.83	64.66
w_A	0.660	0.660	0.660	0.542	0.542	0.542	0.542	0.542	0.542
w_B	0.047	0.047	0.047	0.046	0.046	0.046	0.046	0.046	0.046
w_C	0.293	0.293	0.293	0.412	0.412	0.412	0.412	0.412	0.412
T/K	413.6	433.7	453.8	332.9	352.3	372.6	392.4	412.9	433.3
P/MPa	62.93	62.59	63.28	48.10	43.28	42.93	44.31	46.03	47.76
w_A	0.542								
w_B	0.046								
w_C	0.412								
T/K	452.6								
P/MPa	49.28								

Polymer (B): **poly(dodecyl acrylate)** **2010LE2**

Characterization: M_w/g.mol^{-1} = 44000, M_w/M_n = 1.6, T_g/K = 243.2, Scientific Polymer Products, Inc., Ontario, NY

Solvent (A): **carbon dioxide** CO_2 **124-38-9**

Solvent (C): **dodecyl acrylate** $C_{15}H_{28}O_2$ **2156-97-0**

Type of data: cloud points

w_A	0.948	0.948	0.948	0.948	0.865	0.865	0.865	0.865	0.865
w_B	0.052	0.052	0.052	0.052	0.055	0.055	0.055	0.055	0.055
w_C	0.000	0.000	0.000	0.000	0.080	0.080	0.080	0.080	0.080
T/K	456.9	459.0	460.5	463.4	408.9	411.0	413.6	417.0	437.1
P/MPa	205.28	169.55	157.07	150.14	179.48	140.86	126.38	116.17	105.62
w_A	0.865	0.801	0.801	0.801	0.801	0.801	0.801	0.801	0.801
w_B	0.055	0.051	0.051	0.051	0.051	0.051	0.051	0.051	0.051
w_C	0.080	0.148	0.148	0.148	0.148	0.148	0.148	0.148	0.148
T/K	451.9	371.5	372.6	375.6	383.6	396.5	412.1	435.8	456.0
P/MPa	101.59	181.90	121.55	104.72	90.41	82.79	79.14	76.00	76.10
w_A	0.740	0.740	0.740	0.740	0.740	0.740	0.740	0.740	0.740
w_B	0.053	0.053	0.053	0.053	0.053	0.053	0.053	0.053	0.053
w_C	0.207	0.207	0.207	0.207	0.207	0.207	0.207	0.207	0.207
T/K	342.3	342.8	346.2	356.8	374.9	394.3	413.7	434.0	453.3
P/MPa	145.21	104.31	85.17	69.86	62.14	59.62	59.76	60.21	60.90

continued

continued

w_A	0.693	0.693	0.693	0.693	0.693	0.693	0.693	0.693	0.693
w_B	0.047	0.047	0.047	0.047	0.047	0.047	0.047	0.047	0.047
w_C	0.260	0.260	0.260	0.260	0.260	0.260	0.260	0.260	0.260
T/K	315.5	323.7	336.2	356.9	375.4	395.2	415.6	437.2	455.8
P/MPa	75.35	61.35	51.90	45.69	46.38	46.72	47.76	49.66	50.86

w_A	0.548	0.548	0.548
w_B	0.054	0.054	0.054
w_C	0.398	0.398	0.398
T/K	395.1	412.3	432.9
P/MPa	24.93	26.93	28.55

Type of data: vapor-liquid equilibrium data

w_A	0.548	0.548
w_B	0.054	0.054
w_C	0.398	0.398
T/K	362.9	376.6
P/MPa	20.54	22.38

Type of data: vapor-liquid-liquid equilibrium data

w_A	0.548
w_B	0.054
w_C	0.398
T/K	394.2
P/MPa	24.01

Polymer (B): **poly(dodecyl methacrylate)** **2009LIU**
Characterization: M_w/g.mol^{-1} = 250000, T_g/K = 208,
Scientific Polymer Products, Inc., Ontario, NY
Solvent (A): **carbon dioxide** CO_2 **124-38-9**
Solvent (C): **dimethyl ether** C_2H_6O **115-10-6**

Type of data: cloud points

w_A	0.813	0.813	0.813	0.813	0.813	0.813	0.798	0.798	0.798
w_B	0.057	0.057	0.057	0.057	0.057	0.057	0.047	0.047	0.047
w_C	0.130	0.130	0.130	0.130	0.130	0.130	0.255	0.255	0.255
T/K	426.0	435.6	443.3	454.5	463.6	474.1	375.8	383.9	394.0
P/MPa	179.48	150.17	135.34	124.66	119.14	115.34	169.83	141.90	115.00

w_A	0.798	0.798	0.798	0.543	0.543	0.543	0.543	0.543	0.543
w_B	0.047	0.047	0.047	0.043	0.043	0.043	0.043	0.043	0.043
w_C	0.255	0.255	0.255	0.414	0.414	0.414	0.414	0.414	0.414
T/K	412.4	435.1	452.3	333.2	352.2	373.6	393.7	413.2	433.4
P/MPa	97.41	87.76	84.31	113.28	71.21	60.86	58.10	57.76	58.45

continued

continued

w_A	0.543
w_B	0.043
w_C	0.414
T/K	454.3
P/MPa	59.48

Polymer (B): **poly(dodecyl methacrylate)** **2009LIU**
Characterization: M_w/g.mol^{-1} = 250000, T_g/K = 208,
Scientific Polymer Products, Inc., Ontario, NY
Solvent (A): **carbon dioxide** CO_2 **124-38-9**
Solvent (C): **dodecyl methacrylate** $C_{16}H_{30}O_2$ **142-90-5**

Type of data: cloud points

w_A	0.851	0.851	0.851	0.851	0.802	0.802	0.802	0.802	0.802
w_B	0.051	0.051	0.051	0.051	0.055	0.055	0.055	0.055	0.055
w_C	0.098	0.098	0.098	0.098	0.143	0.143	0.143	0.143	0.143
T/K	431.1	432.6	433.7	452.7	410.3	413.5	423.7	433.9	452.3
P/MPa	162.93	154.10	146.03	120.52	173.28	151.72	117.14	101.69	95.83
w_A	0.753	0.753	0.753	0.753	0.753	0.753	0.753	0.694	0.694
w_B	0.051	0.051	0.051	0.051	0.051	0.051	0.051	0.056	0.056
w_C	0.196	0.196	0.196	0.196	0.196	0.196	0.196	0.250	0.250
T/K	370.1	370.8	376.2	394.6	415.7	433.3	453.8	342.4	355.5
P/MPa	173.62	142.79	113.17	88.69	79.10	76.76	75.45	109.14	72.28
w_A	0.694	0.694	0.694	0.694	0.694	0.584	0.584	0.584	0.584
w_B	0.056	0.056	0.056	0.056	0.056	0.049	0.049	0.049	0.049
w_C	0.250	0.250	0.250	0.250	0.250	0.367	0.367	0.367	0.367
T/K	374.6	391.6	412.9	433.5	455.3	327.8	353.7	378.1	403.6
P/MPa	61.90	58.35	58.00	59.28	60.62	23.86	27.24	31.03	34.48
w_A	0.584	0.584	0.505	0.505	0.505	0.505			
w_B	0.049	0.049	0.048	0.048	0.048	0.048			
w_C	0.367	0.367	0.447	0.447	0.447	0.447			
T/K	428.7	452.8	393.4	414.2	435.8	452.2			
P/MPa	37.72	40.55	21.90	25.24	28.69	30.41			

Type of data: vapor-liquid equilibrium data

w_A	0.505	0.505	0.505
w_B	0.048	0.048	0.048
w_C	0.447	0.447	0.447
T/K	361.4	370.2	382.9
P/MPa	18.79	19.45	20.70

continued

continued

Type of data: vapor-liquid-liquid equilibrium data

w_A	0.505
w_B	0.048
w_C	0.447
T/K	396.7
P/MPa	21.67

Polymer (B): **poly[2-(2-ethoxyethoxy)ethyl acrylate]** **2014YOO**
Characterization: M_w/g.mol^{-1} = 100000, T_g/K = 203.2,
Scientific Polymer Products, Inc., Ontario, NY
Solvent (A): **carbon dioxide** CO_2 **124-38-9**
Solvent (C): **1-butene** C_4H_8 **106-98-9**

Type of data: cloud points

w_A	0.843	0.843	0.843	0.843	0.843	0.843	0.843
w_B	0.050	0.050	0.050	0.050	0.050	0.050	0.050
w_C	0.107	0.107	0.107	0.107	0.107	0.107	0.107
T/K	334.2	353.7	374.3	394.4	414.0	433.6	453.3
P/MPa	119.55	108.30	106.97	105.83	105.52	105.52	106.03

Polymer (B): **poly[2-(2-ethoxyethoxy)ethyl acrylate]** **2014YOO**
Characterization: M_w/g.mol^{-1} = 100000, T_g/K = 203.2,
Scientific Polymer Products, Inc., Ontario, NY
Solvent (A): **carbon dioxide** CO_2 **124-38-9**
Solvent (C): **dimethyl ether** C_2H_6O **115-10-6**

Type of data: cloud points

w_A	0.844	0.844	0.844	0.844	0.844	0.844	0.699	0.699	0.699
w_B	0.049	0.049	0.049	0.049	0.049	0.049	0.049	0.049	0.049
w_C	0.107	0.107	0.107	0.107	0.107	0.107	0.252	0.252	0.252
T/K	333.0	353.6	373.5	393.5	413.7	433.2	333.7	353.8	372.8
P/MPa	105.00	104.31	104.31	105.69	105.69	106.38	65.00	69.14	72.59

w_A	0.699	0.699	0.699	0.486	0.486	0.486	0.486	0.486	0.486
w_B	0.049	0.049	0.049	0.048	0.048	0.048	0.048	0.048	0.048
w_C	0.252	0.252	0.252	0.466	0.466	0.466	0.466	0.466	0.466
T/K	393.2	413.2	433.3	333.7	353.1	373.5	393.7	413.5	433.2
P/MPa	76.03	78.10	80.17	32.59	37.41	42.24	46.38	50.52	53.28

w_A	0.486
w_B	0.048
w_C	0.466
T/K	453.7
P/MPa	56.03

Polymer (B):	**poly[2-(2-ethoxyethoxy)ethyl acrylate]**	**2014YOO**
Characterization:	M_w/g.mol^{-1} = 100000, T_g/K = 203.2, Scientific Polymer Products, Inc., Ontario, NY	
Solvent (A):	**carbon dioxide** CO_2	**124-38-9**
Solvent (C):	**2-(2-ethoxyethoxy)ethyl acrylate** $C_9H_{16}O_4$	**7328-17-8**

Type of data: cloud points

w_A	0.898	0.898	0.898	0.898	0.898	0.898	0.898	0.802	0.802
w_B	0.043	0.043	0.043	0.043	0.043	0.043	0.043	0.049	0.049
w_C	0.059	0.059	0.059	0.059	0.059	0.059	0.059	0.149	0.149
T/K	334.5	354.1	373.9	394.1	413.3	433.5	453.1	334.6	354.1
P/MPa	122.93	118.10	116.03	113.97	113.28	113.28	113.28	66.38	70.52
w_A	0.802	0.802	0.802	0.802	0.802	0.802	0.648	0.648	0.648
w_B	0.049	0.049	0.049	0.049	0.049	0.049	0.049	0.049	0.049
w_C	0.149	0.149	0.149	0.149	0.149	0.149	0.303	0.303	0.303
T/K	373.3	393.5	413.3	433.5	453.2	473.5	353.9	373.4	393.3
P/MPa	73.97	78.10	80.17	82.24	84.31	85.69	56.72	61.35	64.31
w_A	0.648	0.648	0.648	0.648	0.648	0.352	0.352	0.352	0.352
w_B	0.049	0.049	0.049	0.049	0.049	0.046	0.046	0.046	0.046
w_C	0.303	0.303	0.303	0.303	0.303	0.602	0.602	0.602	0.602
T/K	413.6	433.3	453.6	473.1	485.8	354.1	373.6	393.2	413.9
P/MPa	67.76	70.17	72.59	75.52	76.90	25.00	30.52	34.66	38.80
w_A	0.352	0.352	0.300	0.300	0.300				
w_B	0.046	0.046	0.050	0.050	0.050				
w_C	0.602	0.602	0.650	0.650	0.650				
T/K	433.7	453.1	413.4	433.6	454.1				
P/MPa	42.24	45.00	26.38	29.14	31.21				

Type of data: vapor-liquid equilibrium data

w_A	0.300	0.300	0.300	0.300
w_B	0.050	0.050	0.050	0.050
w_C	0.650	0.650	0.650	0.650
T/K	364.2	373.5	383.7	393.2
P/MPa	20.00	21.17	22.24	23.62

Polymer (B):	**poly[2-(2-ethoxyethoxy)ethyl acrylate]**	**2014YOO**
Characterization:	M_w/g.mol^{-1} = 100000, T_g/K = 203.2, Scientific Polymer Products, Inc., Ontario, NY	
Solvent (A):	**carbon dioxide** CO_2	**124-38-9**
Solvent (C):	**propene** C_3H_6	**115-07-1**

Type of data: cloud points

continued

continued

w_A	0.740	0.740	0.740	0.740	0.740	0.740	0.740	0.582	0.582
w_B	0.049	0.049	0.049	0.049	0.049	0.049	0.049	0.050	0.050
w_C	0.211	0.211	0.211	0.211	0.211	0.211	0.211	0.368	0.368
T/K	334.3	354.2	374.1	394.1	413.9	433.0	454.7	333.1	353.1
P/MPa	108.79	98.69	92.85	89.83	91.21	92.41	92.59	94.43	83.24
w_A	0.582	0.582	0.582	0.582	0.582				
w_B	0.050	0.050	0.050	0.050	0.050				
w_C	0.368	0.368	0.368	0.368	0.368				
T/K	373.9	393.2	413.2	434.2	453.3				
P/MPa	78.14	75.69	76.72	78.28	78.53				

Polymer (B): **poly(ethyl acrylate)** **2004BY3**
Characterization: M_w/g.mol^{-1} = 70000, Polysciences, Inc., Warrington, PA
Solvent (A): **1-butene** C_4H_8 **106-98-9**
Solvent (C): **ethyl acrylate** $C_5H_8O_2$ **140-88-5**

Type of data: cloud points

w_A	0.946	0.946	0.946	0.946	0.946	0.946	0.946	0.946	0.946
w_B	0.054	0.054	0.054	0.054	0.054	0.054	0.054	0.054	0.054
w_C	0.000	0.000	0.000	0.000	0.000	0.000	0.000	0.000	0.000
T/K	333.95	336.45	338.95	341.35	344.55	348.85	353.65	362.35	372.45
P/bar	1892.8	1619.7	1479.7	1338.3	1200.3	1068.6	934.8	759.0	631.0
w_A	0.946	0.946	0.946	0.915	0.915	0.915	0.915	0.915	0.915
w_B	0.054	0.054	0.054	0.054	0.054	0.054	0.054	0.054	0.054
w_C	0.000	0.000	0.000	0.031	0.031	0.031	0.031	0.031	0.031
T/K	383.55	403.05	422.15	335.05	337.25	339.55	345.05	349.95	355.35
P/bar	538.3	431.4	372.8	1663.8	1336.2	1127.9	925.9	787.9	664.1
w_A	0.915	0.915	0.915	0.915	0.853	0.853	0.853	0.853	0.853
w_B	0.054	0.054	0.054	0.054	0.066	0.066	0.066	0.066	0.066
w_C	0.031	0.031	0.031	0.031	0.081	0.081	0.081	0.081	0.081
T/K	363.75	384.15	404.65	423.75	318.65	329.15	348.75	363.65	383.45
P/bar	561.7	394.8	344.5	334.1	1063.8	800.0	500.4	406.6	313.5
w_A	0.853	0.853	0.777	0.777	0.777	0.777	0.777	0.777	0.641
w_B	0.066	0.066	0.052	0.052	0.052	0.052	0.052	0.052	0.052
w_C	0.081	0.081	0.185	0.185	0.185	0.185	0.185	0.185	0.307
T/K	403.65	421.85	304.65	312.45	321.85	332.35	354.75	377.05	300.75
P/bar	303.1	291.4	420.7	347.2	272.4	227.9	179.7	167.9	101.7
w_A	0.641	0.641	0.641	0.641	0.641	0.641	0.641	0.641	0.641
w_B	0.052	0.052	0.052	0.052	0.052	0.052	0.052	0.052	0.052
w_C	0.307	0.307	0.307	0.307	0.307	0.307	0.307	0.307	0.307
T/K	303.65	307.75	311.05	315.65	318.55	328.95	338.25	348.65	359.05
P/bar	87.9	74.1	66.6	56.9	51.7	43.1	48.6	55.5	63.8

continued

continued

w_A	0.641
w_B	0.052
w_C	0.307
T/K	373.35
P/bar	71.4

Polymer (B): **poly(ethyl acrylate)** **2004BY3**
Characterization: M_w/g.mol^{-1} = 70000, Polysciences, Inc., Warrington, PA
Solvent (A): **carbon dioxide** CO_2 **124-38-9**
Solvent (C): **ethyl acrylate** $C_5H_8O_2$ **140-88-5**

Type of data: cloud points

w_A	0.946	0.946	0.946	0.946	0.946	0.946	0.946	0.946	0.946
w_B	0.054	0.054	0.054	0.054	0.054	0.054	0.054	0.054	0.054
w_C	0.000	0.000	0.000	0.000	0.000	0.000	0.000	0.000	0.000
T/K	324.35	325.25	326.85	329.05	331.65	335.65	347.15	363.95	383.75
P/bar	2070.7	1915.5	1846.6	1762.4	1706.6	1630.0	1482.8	1403.1	1354.5
w_A	0.946	0.946	0.863	0.863	0.863	0.863	0.863	0.863	0.863
w_B	0.054	0.054	0.055	0.055	0.055	0.055	0.055	0.055	0.055
w_C	0.000	0.000	0.082	0.082	0.082	0.082	0.082	0.082	0.082
T/K	403.05	423.75	311.05	323.75	344.25	364.05	383.25	402.95	423.75
P/bar	1327.9	1317.2	719.7	768.6	822.4	846.2	902.4	935.3	947.2
w_A	0.702	0.702	0.702	0.702	0.702	0.702	0.550	0.550	0.550
w_B	0.048	0.048	0.048	0.048	0.048	0.048	0.055	0.055	0.055
w_C	0.250	0.250	0.250	0.250	0.250	0.250	0.395	0.395	0.395
T/K	312.95	307.45	324.45	338.85	354.75	374.35	333.45	344.35	353.45
P/bar	239.0	214.1	285.9	348.9	404.8	477.6	78.6	126.6	156.9
w_A	0.550	0.550							
w_B	0.055	0.055							
w_C	0.395	0.395							
T/K	373.25	393.45							
P/bar	202.4	232.8							

Type of data: vapor-liquid equilibrium data

w_A	0.550	0.550	0.550
w_B	0.055	0.055	0.055
w_C	0.395	0.395	0.395
T/K	329.35	318.45	308.15
P/bar	73.8	58.3	51.4

Type of data: vapor-liquid-liquid equilibrium data

w_A	0.550	0.550
w_B	0.055	0.055
w_C	0.395	0.395
T/K	344.35	353.15
P/bar	87.7	97.9

Polymer (B): **poly(ethyl acrylate)** **2004BY3**
Characterization: M_w/g.mol^{-1} = 70000, Polysciences, Inc., Warrington, PA
Solvent (A): **propene** C_3H_6 **115-07-1**
Solvent (C): **ethyl acrylate** $C_5H_8O_2$ **140-88-5**

Type of data: cloud points

w_A	0.946	0.946	0.946	0.946	0.946	0.946	0.946	0.946	0.946
w_B	0.054	0.054	0.054	0.054	0.054	0.054	0.054	0.054	0.054
w_C	0.000	0.000	0.000	0.000	0.000	0.000	0.000	0.000	0.000
T/K	311.75	316.55	320.75	324.35	330.65	344.55	364.55	384.45	405.25
P/bar	1647.2	1498.3	1354.8	1233.4	1119.7	903.1	736.9	667.2	621.7
w_A	0.946	0.873	0.873	0.873	0.873	0.873	0.873	0.873	0.737
w_B	0.054	0.055	0.055	0.055	0.055	0.055	0.055	0.055	0.053
w_C	0.000	0.072	0.072	0.072	0.072	0.072	0.072	0.072	0.210
T/K	423.75	310.75	317.25	324.75	339.65	354.65	374.85	393.55	298.65
P/bar	599.0	920.3	836.2	745.9	656.2	601.7	545.9	523.1	448.6
w_A	0.737	0.737	0.737	0.737	0.737	0.737	0.737	0.737	0.530
w_B	0.053	0.053	0.053	0.053	0.053	0.053	0.053	0.053	0.059
w_C	0.210	0.210	0.210	0.210	0.210	0.315	0.315	0.315	0.411
T/K	299.65	302.65	309.25	318.75	333.55	354.55	374.45	393.35	318.55
P/bar	445.5	425.9	409.3	383.1	356.2	349.3	351.0	354.8	24.5
w_A	0.530	0.530	0.530	0.530					
w_B	0.059	0.059	0.059	0.059					
w_C	0.411	0.411	0.411	0.411					
T/K	334.15	354.55	374.55	393.85					
P/bar	54.8	99.7	129.0	149.3					

Type of data: vapor-liquid equilibrium data

w_A	0.530	0.530	0.530
w_B	0.059	0.059	0.059
w_C	0.411	0.411	0.411
T/K	308.95	301.85	296.35
P/bar	14.4	15.5	15.5

Type of data: vapor-liquid-liquid equilibrium data

w_A	0.530	0.530
w_B	0.059	0.059
w_C	0.411	0.411
T/K	323.25	333.05
P/bar	16.3	17.1

Polymer (B): **polyethylene** **2009HAR**
Characterization: M_n/g.mol^{-1} = 13200, M_w/g.mol^{-1} = 15400,
Scientific Polymer Products, Inc., Ontario, NY
Solvent (A): **ethene** C_2H_4 **74-85-1**
Solvent (C): **cyclohexane** C_6H_{12} **110-82-7**

Type of data: cloud points

w_A	0.100	0.100	0.099	0.099	0.099	0.100	0.100	0.100	0.099
w_B	0.017	0.017	0.081	0.081	0.081	0.043	0.043	0.043	0.101
w_C	0.883	0.883	0.820	0.820	0.820	0.857	0.857	0.857	0.800
T/K	485.68	488.19	480.22	483.13	488.16	480.37	483.20	488.37	480.48
P/MPa	6.42	6.76	7.29	7.70	8.36	6.61	7.00	7.67	7.83

w_A	0.099	0.099
w_B	0.101	0.101
w_C	0.800	0.800
T/K	483.58	488.41
P/MPa	8.25	8.68

Type of data: vapor-liquid equilibrium

w_A	0.100	0.100	0.100	0.100	0.100	0.100	0.100	0.100	0.100
w_B	0.017	0.017	0.017	0.017	0.017	0.017	0.017	0.043	0.043
w_C	0.883	0.883	0.883	0.883	0.883	0.883	0.883	0.857	0.857
T/K	373.72	393.51	413.72	433.71	453.46	473.23	483.27	373.68	393.74
P/MPa	3.36	3.88	4.48	4.93	5.50	6.04	6.29	3.55	4.05

w_A	0.100	0.100	0.100	0.100	0.099	0.099	0.099	0.099	0.099
w_B	0.043	0.043	0.043	0.043	0.081	0.081	0.081	0.081	0.081
w_C	0.857	0.857	0.857	0.857	0.820	0.820	0.820	0.820	0.820
T/K	413.65	433.69	453.23	473.20	373.48	393.64	413.73	433.71	453.61
P/MPa	4.66	5.16	5.72	6.32	3.64	4.26	4.81	5.37	6.01

w_A	0.099	0.099	0.099	0.099	0.099	0.099	0.099
w_B	0.081	0.101	0.101	0.101	0.101	0.101	0.101
w_C	0.820	0.800	0.800	0.800	0.800	0.800	0.800
T/K	473.64	373.72	393.62	413.48	433.69	453.61	473.52
P/MPa	6.62	3.64	4.32	4.82	5.42	6.04	6.70

Type of data: vapor-liquid-liquid equilibrium data

w_A	0.100	0.100	0.100	0.100	0.100	0.099	0.099	0.099	0.099
w_B	0.017	0.017	0.043	0.043	0.043	0.081	0.081	0.081	0.101
w_C	0.883	0.883	0.857	0.857	0.857	0.820	0.820	0.820	0.800
T/K	485.99	488.39	480.42	483.28	488.36	480.16	483.33	488.56	480.49
P/MPa	6.36	6.45	6.52	6.59	6.66	6.78	6.83	6.91	6.85

w_A	0.099	0.099
w_B	0.101	0.101
w_C	0.800	0.800
T/K	483.62	488.66
P/MPa	6.92	6.98

Polymer (B):	**polyethylene**		**2009HAR**
Characterization:	M_n/g.mol^{-1} = 13200, M_w/g.mol^{-1} = 15400, Scientific Polymer Products, Inc., Ontario, NY		
Solvent (A):	**ethene**	C_2H_4	**74-85-1**
Solvent (C):	**cyclohexane**	C_6H_{12}	**110-82-7**
Solvent (D):	**n-hexane**	C_6H_{14}	**110-54-3**

Type of data: cloud points

w_A	0.049	0.049	0.049	0.049	0.049	0.049	0.049	0.049	0.049
w_B	0.018	0.018	0.018	0.018	0.045	0.045	0.045	0.045	0.045
w_C	0.094	0.094	0.094	0.094	0.091	0.091	0.091	0.091	0.091
w_D	0.839	0.839	0.839	0.839	0.815	0.815	0.815	0.815	0.815
T/K	443.13	453.36	463.36	473.76	433.62	443.16	453.55	463.54	473.56
P/MPa	3.77	5.19	6.50	7.81	2.95	4.27	5.71	7.06	8.27
w_A	0.050	0.050	0.050	0.050	0.050	0.049	0.049	0.049	0.049
w_B	0.087	0.087	0.087	0.087	0.087	0.108	0.108	0.108	0.108
w_C	0.087	0.087	0.087	0.087	0.087	0.085	0.085	0.085	0.085
w_D	0.776	0.776	0.776	0.776	0.776	0.758	0.758	0.758	0.758
T/K	433.82	443.17	453.89	463.53	473.57	433.86	443.38	453.41	463.83
P/MPa	3.77	5.19	6.61	7.85	9.04	3.27	4.67	6.05	7.46
w_A	0.049	0.050	0.050	0.049	0.049	0.049	0.049	0.050	0.050
w_B	0.108	0.018	0.018	0.045	0.045	0.045	0.045	0.018	0.088
w_C	0.085	0.186	0.186	0.181	0.181	0.181	0.181	0.186	0.172
w_D	0.758	0.746	0.746	0.725	0.725	0.725	0.725	0.746	0.690
T/K	473.90	453.46	463.50	443.66	453.39	463.47	473.63	473.76	443.39
P/MPa	8.69	4.13	5.51	3.42	4.82	6.21	7.51	6.78	4.23
w_A	0.050	0.050	0.050	0.050	0.050	0.050	0.050	0.099	0.099
w_B	0.088	0.088	0.088	0.109	0.109	0.109	0.109	0.018	0.018
w_C	0.172	0.172	0.172	0.168	0.168	0.168	0.168	0.099	0.099
w_D	0.690	0.690	0.690	0.673	0.673	0.673	0.673	0.784	0.784
T/K	453.60	463.44	473.58	443.54	453.49	463.68	473.69	413.23	433.56
P/MPa	5.60	7.02	8.30	4.22	5.58	7.00	8.29	6.40	9.35
w_A	0.099	0.099	0.099	0.099	0.100	0.100	0.100	0.100	0.100
w_B	0.018	0.018	0.018	0.018	0.042	0.042	0.042	0.042	0.042
w_C	0.099	0.099	0.099	0.099	0.096	0.096	0.096	0.096	0.096
w_D	0.784	0.784	0.784	0.784	0.762	0.762	0.762	0.762	0.762
T/K	443.67	453.60	463.50	473.38	413.15	433.43	443.14	453.58	463.49
P/MPa	10.72	11.95	13.18	14.28	6.91	9.81	11.13	12.45	13.59
w_A	0.100	0.100	0.100	0.100	0.100	0.100	0.100	0.100	0.098
w_B	0.042	0.081	0.081	0.081	0.081	0.081	0.081	0.081	0.107
w_C	0.096	0.092	0.092	0.092	0.092	0.092	0.092	0.092	0.090
w_D	0.762	0.727	0.727	0.727	0.727	0.727	0.727	0.727	0.705
T/K	473.38	393.87	413.94	433.11	443.53	453.30	463.67	473.66	394.09
P/MPa	14.69	4.15	7.25	10.13	11.50	12.80	13.90	15.05	4.08

continued

continued

w_A	0.098	0.098	0.098	0.098	0.098	0.098	0.100	0.100	0.100
w_B	0.107	0.107	0.107	0.107	0.107	0.107	0.018	0.018	0.018
w_C	0.090	0.090	0.090	0.090	0.090	0.090	0.196	0.196	0.196
w_D	0.705	0.705	0.705	0.705	0.705	0.705	0.686	0.686	0.686
T/K	413.83	433.92	443.70	453.77	463.55	473.76	433.24	443.03	453.54
P/MPa	7.35	10.15	11.60	12.94	13.85	14.96	7.05	8.47	9.74

w_A	0.100	0.100	0.099	0.099	0.099	0.099	0.099	0.099	0.100
w_B	0.018	0.018	0.043	0.043	0.043	0.043	0.043	0.043	0.080
w_C	0.196	0.196	0.191	0.191	0.191	0.191	0.191	0.191	0.184
w_D	0.686	0.686	0.667	0.667	0.667	0.667	0.667	0.667	0.636
T/K	463.57	473.60	413.72	433.72	443.75	453.32	463.18	473.47	413.36
P/MPa	11.04	12.19	4.70	7.72	9.16	10.42	11.77	12.98	5.43

w_A	0.100	0.100	0.100	0.100	0.100	0.098	0.098	0.098	0.098
w_B	0.080	0.080	0.080	0.080	0.080	0.104	0.104	0.104	0.104
w_C	0.184	0.184	0.184	0.184	0.184	0.179	0.179	0.179	0.179
w_D	0.636	0.636	0.636	0.636	0.636	0.619	0.619	0.619	0.619
T/K	433.59	443.69	453.72	463.46	473.45	413.79	433.21	444.02	454.02
P/MPa	8.65	10.10	11.36	12.67	13.71	5.60	8.63	10.03	11.23

w_A	0.098	0.098
w_B	0.104	0.104
w_C	0.179	0.179
w_D	0.619	0.619
T/K	463.18	473.65
P/MPa	12.52	13.68

Type of data: vapor-liquid equilibrium data

w_A	0.049	0.049	0.049	0.049	0.049	0.049	0.049	0.050	0.050
w_B	0.018	0.018	0.018	0.018	0.045	0.045	0.045	0.087	0.087
w_C	0.094	0.094	0.094	0.094	0.091	0.091	0.091	0.087	0.087
w_D	0.839	0.839	0.839	0.839	0.815	0.815	0.815	0.776	0.776
T/K	373.94	393.87	413.67	433.47	373.89	393.75	413.82	373.73	393.54
P/MPa	1.49	1.79	2.13	2.53	1.52	1.82	2.17	1.72	2.01

w_A	0.050	0.049	0.049	0.049	0.050	0.050	0.050	0.050	0.049
w_B	0.087	0.108	0.108	0.108	0.018	0.018	0.018	0.018	0.045
w_C	0.087	0.085	0.085	0.085	0.186	0.186	0.186	0.186	0.181
w_D	0.776	0.758	0.758	0.758	0.746	0.746	0.746	0.746	0.725
T/K	413.80	373.73	393.54	413.96	373.90	394.07	413.62	433.68	373.79
P/MPa	2.37	1.52	1.84	2.21	1.50	1.81	2.14	2.55	1.56

w_A	0.049	0.049	0.049	0.050	0.050	0.050	0.050	0.050	0.050
w_B	0.045	0.045	0.045	0.088	0.088	0.088	0.088	0.109	0.109
w_C	0.181	0.181	0.181	0.172	0.172	0.172	0.172	0.168	0.168
w_D	0.725	0.725	0.725	0.690	0.690	0.690	0.690	0.673	0.673
T/K	393.46	413.49	433.39	373.78	393.48	413.41	433.56	373.61	393.49
P/MPa	1.87	2.20	2.63	1.73	2.07	2.42	2.86	1.73	2.07

continued

continued

w_A	0.050	0.050	0.099	0.099	0.100	0.100	0.100	0.098	0.100
w_B	0.109	0.109	0.018	0.018	0.042	0.042	0.081	0.107	0.018
w_C	0.168	0.168	0.099	0.099	0.096	0.096	0.092	0.090	0.196
w_D	0.673	0.673	0.784	0.784	0.762	0.762	0.727	0.705	0.686
T/K	413.32	433.54	373.87	393.88	373.28	393.74	373.04	373.24	373.72
P/MPa	2.42	2.86	3.08	3.60	3.17	3.69	3.14	3.10	2.89

w_A	0.100	0.100	0.099	0.099	0.100	0.100	0.098	0.098
w_B	0.018	0.018	0.043	0.043	0.080	0.080	0.104	0.104
w_C	0.196	0.196	0.191	0.191	0.184	0.184	0.179	0.179
w_D	0.686	0.686	0.667	0.667	0.636	0.636	0.619	0.619
T/K	393.66	413.59	373.81	393.99	373.38	393.48	373.70	393.64
P/MPa	3.40	3.91	2.90	3.42	3.07	3.63	3.09	3.63

Type of data: vapor-liquid-liquid equilibrium data

w_A	0.049	0.049	0.049	0.049	0.049	0.049	0.049	0.049	0.049
w_B	0.018	0.018	0.018	0.018	0.045	0.045	0.045	0.045	0.045
w_C	0.094	0.094	0.094	0.094	0.091	0.091	0.091	0.091	0.091
w_D	0.839	0.839	0.839	0.839	0.815	0.815	0.815	0.815	0.815
T/K	443.04	453.75	463.44	473.86	433.63	443.09	453.44	463.53	473.53
P/MPa	2.73	2.97	3.21	3.51	2.55	2.76	3.02	3.24	3.49

w_A	0.050	0.050	0.050	0.050	0.050	0.049	0.049	0.049	0.049
w_B	0.087	0.087	0.087	0.087	0.087	0.108	0.108	0.108	0.108
w_C	0.087	0.087	0.087	0.087	0.087	0.085	0.085	0.085	0.085
w_D	0.776	0.776	0.776	0.776	0.776	0.758	0.758	0.758	0.758
T/K	433.79	443.25	453.82	463.56	473.68	433.79	443.58	453.72	463.56
P/MPa	2.75	2.95	3.21	3.45	3.67	2.62	2.83	3.05	3.28

w_A	0.049	0.050	0.050	0.050	0.050	0.049	0.049	0.049	0.049
w_B	0.108	0.018	0.018	0.018	0.018	0.045	0.045	0.045	0.045
w_C	0.085	0.186	0.186	0.186	0.186	0.181	0.181	0.181	0.181
w_D	0.758	0.746	0.746	0.746	0.746	0.725	0.725	0.725	0.725
T/K	473.82	443.54	453.38	463.56	473.66	443.55	453.41	463.40	473.63
P/MPa	3.57	2.75	2.98	3.24	3.48	2.84	3.07	3.31	3.56

w_A	0.050	0.050	0.050	0.050	0.050	0.050	0.050	0.050	0.099
w_B	0.088	0.088	0.088	0.088	0.109	0.109	0.109	0.109	0.018
w_C	0.172	0.172	0.172	0.172	0.168	0.168	0.168	0.168	0.099
w_D	0.690	0.690	0.690	0.690	0.673	0.673	0.673	0.673	0.784
T/K	443.37	453.57	463.30	473.62	443.44	453.41	463.40	473.60	413.40
P/MPa	3.06	3.29	3.54	3.86	3.06	3.29	3.54	3.85	4.08

w_A	0.099	0.099	0.099	0.099	0.099	0.100	0.100	0.100	0.100
w_B	0.018	0.018	0.018	0.018	0.018	0.042	0.042	0.042	0.042
w_C	0.099	0.099	0.099	0.099	0.099	0.096	0.096	0.096	0.096
w_D	0.784	0.784	0.784	0.784	0.784	0.762	0.762	0.762	0.762
T/K	433.40	443.50	453.45	463.42	473.23	413.66	433.63	443.16	453.44
P/MPa	4.62	4.85	5.09	5.29	5.39	4.25	4.72	4.96	5.19

continued

continued

w_A	0.100	0.100	0.100	0.100	0.100	0.100	0.100	0.100	0.100
w_B	0.042	0.042	0.081	0.081	0.081	0.081	0.081	0.081	0.081
w_C	0.096	0.096	0.092	0.092	0.092	0.092	0.092	0.092	0.092
w_D	0.762	0.762	0.727	0.727	0.727	0.727	0.727	0.727	0.727
T/K	463.36	473.50	393.74	413.66	433.11	443.31	453.08	463.29	473.57
P/MPa	5.33	5.55	3.64	4.14	4.66	4.91	5.13	5.28	5.45
w_A	0.098	0.098	0.098	0.098	0.098	0.098	0.098	0.100	0.100
w_B	0.107	0.107	0.107	0.107	0.107	0.107	0.107	0.018	0.018
w_C	0.090	0.090	0.090	0.090	0.090	0.090	0.090	0.196	0.196
w_D	0.705	0.705	0.705	0.705	0.705	0.705	0.705	0.686	0.686
T/K	393.97	413.87	433.82	443.60	453.67	463.45	473.56	433.23	443.01
P/MPa	3.56	4.03	4.54	4.85	5.02	5.26	5.35	4.38	4.64
w_A	0.100	0.100	0.100	0.099	0.099	0.099	0.099	0.099	0.099
w_B	0.018	0.018	0.018	0.043	0.043	0.043	0.043	0.043	0.043
w_C	0.196	0.196	0.196	0.191	0.191	0.191	0.191	0.191	0.191
w_D	0.686	0.686	0.686	0.667	0.667	0.667	0.667	0.667	0.667
T/K	453.32	463.48	473.46	413.65	433.60	443.78	453.41	463.42	473.61
P/MPa	4.87	5.08	5.24	3.90	4.39	4.67	4.88	5.08	5.26
w_A	0.100	0.100	0.100	0.100	0.100	0.100	0.098	0.098	0.098
w_B	0.080	0.080	0.080	0.080	0.080	0.080	0.104	0.104	0.104
w_C	0.184	0.184	0.184	0.184	0.184	0.184	0.179	0.179	0.179
w_D	0.636	0.636	0.636	0.636	0.636	0.636	0.619	0.619	0.619
T/K	413.41	433.54	443.56	453.61	463.44	473.48	413.66	433.53	443.80
P/MPa	4.12	4.63	4.88	5.13	5.33	5.50	4.13	4.64	4.89
w_A	0.098	0.098	0.098						
w_B	0.104	0.104	0.104						
w_C	0.179	0.179	0.179						
w_D	0.619	0.619	0.619						
T/K	453.27	463.35	473.62						
P/MPa	5.12	5.35	5.48						

Polymer (B): **polyethylene** **2008HAR**

Characterization: M_n/g.mol^{-1} = 14400, M_w/g.mol^{-1} = 15500, Scientific Polymer Products, Inc., Ontario, NY

Solvent (A): **ethene** **C_2H_4** **74-85-1**

Solvent (C): **n-hexane** **C_6H_{14}** **110-54-3**

Type of data: cloud points

w_A	0.0200	0.0200	0.0200	0.0183	0.0183	0.0183	0.0197	0.0197	0.0197
w_B	0.0177	0.0177	0.0177	0.0462	0.0462	0.0462	0.0928	0.0928	0.0928
w_C	0.9623	0.9623	0.9623	0.9355	0.9355	0.9355	0.8875	0.8875	0.8875
T/K	452.92	463.09	473.08	452.95	462.91	472.96	432.94	452.96	472.88
P/MPa	3.3	4.7	6.0	3.6	5.1	6.3	2.2	4.9	7.4

continued

continued

w_A	0.0201	0.0201	0.0201	0.0481	0.0481	0.0481	0.0499	0.0499	0.0499
w_B	0.1247	0.1247	0.1247	0.0187	0.0187	0.0187	0.0452	0.0452	0.0452
w_C	0.8552	0.8552	0.8552	0.9332	0.9332	0.9332	0.8049	0.8049	0.8049
T/K	432.93	452.91	472.86	433.11	452.96	472.87	433.02	452.94	472.89
P/MPa	2.6	5.3	7.8	2.9	5.6	8.1	4.6	7.4	9.9
w_A	0.0515	0.0515	0.0515	0.0515	0.0512	0.0512	0.0512	0.0512	0.0946
w_B	0.0890	0.0890	0.0890	0.0890	0.1111	0.1111	0.1111	0.1111	0.0175
w_C	0.8595	0.8595	0.8595	0.8595	0.8377	0.8377	0.8377	0.8377	0.8879
T/K	413.00	432.97	452.96	472.93	413.00	433.02	453.00	472.91	413.12
P/MPa	2.8	5.5	8.0	10.3	3.2	6.1	8.7	11.1	5.5
w_A	0.0946	0.0946	0.0946	0.0932	0.0932	0.0932	0.0932	0.0932	0.0959
w_B	0.0175	0.0175	0.0175	0.0450	0.0450	0.0450	0.0450	0.0450	0.0834
w_C	0.8879	0.8879	0.8879	0.8618	0.8618	0.8618	0.8618	0.8618	0.8207
T/K	433.15	453.17	473.15	393.17	413.13	433.13	453.15	473.17	393.12
P/MPa	8.4	10.9	13.1	3.3	6.4	9.2	11.8	14.1	5.1
w_A	0.0959	0.0959	0.0959	0.0959	0.0949	0.0949	0.0949	0.0949	0.0949
w_B	0.0834	0.0834	0.0834	0.0834	0.1165	0.1165	0.1165	0.1165	0.1165
w_C	0.8207	0.8207	0.8207	0.8207	0.7886	0.7886	0.7886	0.7886	0.7886
T/K	413.18	433.11	453.15	473.18	393.17	413.18	433.16	453.12	473.15
P/MPa	8.0	10.8	13.3	15.4	5.1	8.2	11.0	13.4	15.9

Type of data: vapor-liquid equilibrium data

w_A	0.0200	0.0200	0.0200	0.0200	0.0183	0.0183	0.0183	0.0183	0.0197
w_B	0.0177	0.0177	0.0177	0.0177	0.0462	0.0462	0.0462	0.0462	0.0928
w_C	0.9623	0.9623	0.9623	0.9623	0.9355	0.9355	0.9355	0.9355	0.8875
T/K	373.22	393.21	413.19	433.09	373.10	393.00	413.12	433.00	373.06
P/MPa	0.93	1.16	1.40	1.79	0.68	0.92	1.17	1.76	0.68
w_A	0.0197	0.0197	0.0201	0.0201	0.0201	0.0481	0.0481	0.0481	0.0499
w_B	0.0928	0.0928	0.1247	0.1247	0.1247	0.0187	0.0187	0.0187	0.0452
w_C	0.8875	0.8875	0.8552	0.8552	0.8552	0.9332	0.9332	0.9332	0.8049
T/K	393.00	413.02	373.00	393.00	412.96	373.19	393.08	413.15	373.03
P/MPa	0.85	1.12	0.97	1.21	1.50	1.53	1.77	2.14	1.67
w_A	0.0499	0.0499	0.0515	0.0515	0.0512	0.0512	0.0946	0.0946	0.0932
w_B	0.0452	0.0452	0.0890	0.0890	0.1111	0.1111	0.0175	0.0175	0.0450
w_C	0.8049	0.8049	0.8595	0.8595	0.8377	0.8377	0.8879	0.8879	0.8618
T/K	393.02	413.03	373.07	393.04	373.17	393.07	373.10	393.17	373.10
P/MPa	2.20	2.40	1.75	2.06	1.52	1.77	2.71	3.14	2.70
w_A	0.0959	0.0949							
w_B	0.0834	0.1165							
w_C	0.8207	0.7886							
T/K	373.17	373.20							
P/MPa	2.94	2.93							

continued

continued

Type of data: vapor-liquid-liquid equilibrium data

w_A	0.0200	0.0200	0.0200	0.0183	0.0183	0.0183	0.0197	0.0197	0.0197
w_B	0.0177	0.0177	0.0177	0.0462	0.0462	0.0462	0.0928	0.0928	0.0928
w_C	0.9623	0.9623	0.9623	0.9355	0.9355	0.9355	0.8875	0.8875	0.8875
T/K	452.93	463.00	473.05	452.96	462.91	473.01	432.94	452.93	472.86
P/MPa	2.22	2.40	2.63	1.98	2.20	2.46	1.50	1.88	2.44
w_A	0.0201	0.0201	0.0201	0.0481	0.0481	0.0481	0.0499	0.0499	0.0499
w_B	0.1247	0.1247	0.1247	0.0187	0.0187	0.0187	0.0452	0.0452	0.0452
w_C	0.8552	0.8552	0.8552	0.9332	0.9332	0.9332	0.8049	0.8049	0.8049
T/K	432.93	452.93	472.88	433.11	452.94	472.88	433.08	452.90	472.87
P/MPa	1.81	2.24	2.77	2.55	3.00	3.47	2.76	3.27	3.72
w_A	0.0515	0.0515	0.0515	0.0515	0.0512	0.0512	0.0512	0.0512	0.0946
w_B	0.0890	0.0890	0.0890	0.0890	0.1111	0.1111	0.1111	0.1111	0.0175
w_C	0.8595	0.8595	0.8595	0.8595	0.8377	0.8377	0.8377	0.8377	0.8879
T/K	413.01	432.96	452.90	472.96	413.03	433.01	452.94	472.90	413.12
P/MPa	2.45	2.82	3.29	3.78	2.01	2.39	2.86	3.53	3.64
w_A	0.0946	0.0946	0.0946	0.0932	0.0932	0.0932	0.0932	0.0932	0.0959
w_B	0.0175	0.0175	0.0175	0.0450	0.0450	0.0450	0.0450	0.0450	0.0834
w_C	0.8879	0.8879	0.8879	0.8618	0.8618	0.8618	0.8618	0.8618	0.8207
T/K	433.11	453.13	473.13	393.11	413.14	433.10	453.11	473.14	393.12
P/MPa	4.13	4.64	5.01	3.12	3.60	4.08	4.60	4.93	3.38
w_A	0.0959	0.0959	0.0959	0.0959	0.0949	0.0949	0.0949	0.0949	0.0949
w_B	0.0834	0.0834	0.0834	0.0834	0.1165	0.1165	0.1165	0.1165	0.1165
w_C	0.8207	0.8207	0.8207	0.8207	0.7886	0.7886	0.7886	0.7886	0.7886
T/K	413.18	433.18	453.12	473.14	393.16	413.20	433.11	453.18	473.16
P/MPa	3.91	4.41	4.89	5.20	3.41	3.91	4.36	4.84	5.16

Polymer (B): **polyethylene** **2006NAG**

Characterization: M_n/g.mol^{-1} = 43700, M_w/g.mol^{-1} = 52000, M_z/g.mol^{-1} = 59000, 2.05 ethyl branches per 100 backbone C-atomes, hydrogenated polybutadiene PBD 50000, was denoted as LLDPE, DSM, The Netherlands

Solvent (A): **ethene** C_2H_4 **74-85-1**

Solvent (C): **n-hexane** C_6H_{14} **110-54-3**

Type of data: cloud points

w_A	0.0118	0.0118	0.0118	0.0118	0.0118	0.0118	0.0118	0.0118	0.0118
w_B	0.0502	0.0502	0.0502	0.0502	0.0502	0.0502	0.0502	0.0502	0.0502
w_C	0.9380	0.9380	0.9380	0.9380	0.9380	0.9380	0.9380	0.9380	0.9380
T/K	426.28	431.16	436.16	441.26	446.14	451.05	455.97	460.90	465.75
P/MPa	1.846	2.566	3.276	4.096	4.721	5.421	6.096	6.746	7.396

continued

continued

w_A	0.0118	0.0118	0.0118	0.0118	0.0118	0.0118	0.0205	0.0205	0.0205
w_B	0.0502	0.0502	0.0502	0.0502	0.0502	0.0502	0.0501	0.0501	0.0501
w_C	0.9380	0.9380	0.9380	0.9380	0.9380	0.9380	0.9294	0.9294	0.9294
T/K	470.74	475.64	480.51	485.46	490.27	495.18	415.92	420.93	425.96
P/MPa	7.996	8.571	9.176	9.726	10.271	10.841	1.659	2.459	3.224

w_A	0.0205	0.0205	0.0205	0.0205	0.0205	0.0205	0.0205	0.0205	0.0205
w_B	0.0501	0.0501	0.0501	0.0501	0.0501	0.0501	0.0501	0.0501	0.0501
w_C	0.9294	0.9294	0.9294	0.9294	0.9294	0.9294	0.9294	0.9294	0.9294
T/K	430.76	435.65	440.65	445.51	450.52	455.49	460.43	465.36	470.27
P/MPa	3.974	4.734	5.394	6.049	6.769	7.474	8.144	8.724	9.319

w_A	0.0205	0.0205	0.0205	0.0205	0.0205	0.0298	0.0298	0.0298	0.0298
w_B	0.0501	0.0501	0.0501	0.0501	0.0501	0.0499	0.0499	0.0499	0.0499
w_C	0.9294	0.9294	0.9294	0.9294	0.9294	0.9203	0.9203	0.9203	0.9203
T/K	475.28	480.25	485.26	490.18	495.12	411.24	416.19	421.11	426.01
P/MPa	9.869	10.394	10.999	11.549	12.074	2.149	3.021	3.849	4.594

w_A	0.0298	0.0298	0.0298	0.0298	0.0298	0.0298	0.0298	0.0298	0.0298
w_B	0.0499	0.0499	0.0499	0.0499	0.0499	0.0499	0.0499	0.0499	0.0499
w_C	0.9203	0.9203	0.9203	0.9203	0.9203	0.9203	0.9203	0.9203	0.9203
T/K	430.99	435.94	440.90	445.92	450.79	455.84	460.69	465.65	470.62
P/MPa	5.324	6.094	6.799	7.394	8.099	8.749	9.369	9.994	10.699

w_A	0.0298	0.0298	0.0298	0.0298	0.0298	0.0099	0.0099	0.0099	0.0099
w_B	0.0499	0.0499	0.0499	0.0499	0.0499	0.0998	0.0998	0.0998	0.0998
w_C	0.9203	0.9203	0.9203	0.9203	0.9203	0.8903	0.8903	0.8903	0.8903
T/K	475.58	480.51	485.42	490.31	495.24	426.12	430.98	435.95	440.82
P/MPa	11.299	11.799	12.394	12.894	13.369	1.774	2.524	3.194	3.969

w_A	0.0099	0.0099	0.0099	0.0099	0.0099	0.0099	0.0099	0.0099	0.0099
w_B	0.0998	0.0998	0.0998	0.0998	0.0998	0.0998	0.0998	0.0998	0.0998
w_C	0.8903	0.8903	0.8903	0.8903	0.8903	0.8903	0.8903	0.8903	0.8903
T/K	445.76	450.70	455.59	460.54	465.50	470.43	475.42	480.45	485.47
P/MPa	4.739	5.249	5.944	6.609	7.274	7.844	8.574	9.194	9.849

w_A	0.0099	0.0099	0.0196	0.0196	0.0196	0.0196	0.0196	0.0196	0.0196
w_B	0.0998	0.0998	0.1007	0.1007	0.1007	0.1007	0.1007	0.1007	0.1007
w_C	0.8903	0.8903	0.8797	0.8797	0.8797	0.8797	0.8797	0.8797	0.8797
T/K	490.41	495.30	421.26	426.08	431.05	435.99	440.73	445.90	450.91
P/MPa	10.449	10.999	2.449	3.144	3.949	4.719	5.449	6.199	6.844

w_A	0.0196	0.0196	0.0196	0.0196	0.0196	0.0196	0.0196	0.0196	0.0196
w_B	0.1007	0.1007	0.1007	0.1007	0.1007	0.1007	0.1007	0.1007	0.1007
w_C	0.8797	0.8797	0.8797	0.8797	0.8797	0.8797	0.8797	0.8797	0.8797
T/K	455.88	460.64	465.67	470.72	475.58	480.54	485.56	490.41	495.45
P/MPa	7.524	8.224	8.834	9.484	10.094	10.694	11.249	11.819	12.374

continued

continued

w_A	0.0296	0.0296	0.0296	0.0296	0.0296	0.0296	0.0296	0.0296	0.0296
w_B	0.1002	0.1002	0.1002	0.1002	0.1002	0.1002	0.1002	0.1002	0.1002
w_C	0.8702	0.8702	0.8702	0.8702	0.8702	0.8702	0.8702	0.8702	0.8702
T/K	410.90	415.82	420.77	425.74	430.62	435.61	440.43	445.48	450.46
P/MPa	2.199	3.024	3.809	4.629	5.374	6.084	6.824	7.569	8.244
w_A	0.0296	0.0296	0.0296	0.0296	0.0296	0.0296	0.0296	0.0296	0.0098
w_B	0.1002	0.1002	0.1002	0.1002	0.1002	0.1002	0.1002	0.1002	0.1508
w_C	0.8702	0.8702	0.8702	0.8702	0.8702	0.8702	0.8702	0.8702	0.8394
T/K	455.38	460.36	465.45	470.41	475.31	480.19	485.06	489.96	430.69
P/MPa	8.934	9.599	10.057	10.734	11.324	11.949	12.349	12.949	2.074
w_A	0.0098	0.0098	0.0098	0.0098	0.0098	0.0098	0.0098	0.0098	0.0098
w_B	0.1508	0.1508	0.1508	0.1508	0.1508	0.1508	0.1508	0.1508	0.1508
w_C	0.8394	0.8394	0.8394	0.8394	0.8394	0.8394	0.8394	0.8394	0.8394
T/K	435.59	440.64	445.66	450.45	455.37	460.32	465.30	470.23	475.17
P/MPa	2.869	3.619	4.349	5.044	5.694	6.349	6.989	7.614	8.229
w_A	0.0098	0.0098	0.0098	0.0098	0.0210	0.0210	0.0210	0.0210	0.0210
w_B	0.1508	0.1508	0.1508	0.1508	0.1503	0.1503	0.1503	0.1503	0.1503
w_C	0.8394	0.8394	0.8394	0.8394	0.8287	0.8287	0.8287	0.8287	0.8287
T/K	480.07	484.95	489.87	494.80	420.47	425.34	430.32	435.25	440.27
P/MPa	8.824	9.409	9.964	10.514	2.094	2.844	3.644	4.369	5.069
w_A	0.0210	0.0210	0.0210	0.0210	0.0210	0.0210	0.0210	0.0210	0.0210
w_B	0.1503	0.1503	0.1503	0.1503	0.1503	0.1503	0.1503	0.1503	0.1503
w_C	0.8287	0.8287	0.8287	0.8287	0.8287	0.8287	0.8287	0.8287	0.8287
T/K	445.20	450.13	455.10	460.07	464.88	469.80	474.77	479.72	484.58
P/MPa	5.794	6.519	7.319	7.789	8.509	9.024	9.669	10.229	10.769
w_A	0.0210	0.0210	0.0293	0.0293	0.0293	0.0293	0.0293	0.0293	0.0293
w_B	0.1503	0.1503	0.1506	0.1506	0.1506	0.1506	0.1506	0.1506	0.1506
w_C	0.8287	0.8287	0.8201	0.8201	0.8201	0.8201	0.8201	0.8201	0.8201
T/K	489.52	494.48	411.03	415.95	420.84	425.82	431.00	435.98	440.92
P/MPa	11.319	11.884	2.074	2.874	3.749	4.444	5.209	5.969	6.689
w_A	0.0293	0.0293	0.0293	0.0293	0.0293	0.0293	0.0293	0.0293	0.0293
w_B	0.1506	0.1506	0.1506	0.1506	0.1506	0.1506	0.1506	0.1506	0.1506
w_C	0.8201	0.8201	0.8201	0.8201	0.8201	0.8201	0.8201	0.8201	0.8201
T/K	445.91	450.89	455.82	460.76	465.67	470.37	475.46	480.33	485.26
P/MPa	7.339	8.019	8.639	9.304	9.959	10.529	11.124	11.669	12.219
w_A	0.0293	0.0293							
w_B	0.1506	0.1506							
w_C	0.8201	0.8201							
T/K	490.35	495.17							
P/MPa	12.759	13.309							

continued

continued

Type of data: vapor-liquid equilibrium data

w_A	0.0118	0.0118	0.0118	0.0205	0.0205	0.0298	0.0298	0.0298	0.0099
w_B	0.0502	0.0502	0.0502	0.0501	0.0501	0.0499	0.0499	0.0499	0.0998
w_C	0.9380	0.9380	0.9380	0.9294	0.9294	0.9203	0.9203	0.9203	0.8903
T/K	411.46	416.37	421.20	406.00	410.93	396.49	401.10	406.30	406.11
P/MPa	1.052	1.116	1.181	1.354	1.434	1.529	1.604	1.699	0.969
w_A	0.0099	0.0099	0.0099	0.0196	0.0196	0.0196	0.0196	0.0296	0.0296
w_B	0.0998	0.0998	0.0998	0.1007	0.1007	0.1007	0.1007	0.1002	0.1002
w_C	0.8903	0.8903	0.8903	0.8797	0.8797	0.8797	0.8797	0.8702	0.8702
T/K	411.11	416.06	421.13	401.65	406.43	411.45	416.30	396.26	401.23
P/MPa	1.039	1.109	1.189	1.309	1.374	1.479	1.549	1.580	1.660
w_A	0.0296	0.0098	0.0098	0.0098	0.0210	0.0210	0.0210	0.0210	0.0293
w_B	0.1002	0.1508	0.1508	0.1508	0.1503	0.1503	0.1503	0.1503	0.1506
w_C	0.8702	0.8394	0.8394	0.8394	0.8287	0.8287	0.8287	0.8287	0.8201
T/K	406.21	416.15	421.11	426.07	400.89	405.73	410.70	415.66	396.32
P/MPa	1.750	1.139	1.219	1.299	1.354	1.429	1.504	1.589	1.664
w_A	0.0293	0.0293							
w_B	0.1506	0.1506							
w_C	0.8201	0.8201							
T/K	401.21	406.10							
P/MPa	1.744	1.834							

Type of data: vapor-liquid-liquid equilibrium data

w_A	0.0118	0.0118	0.0118	0.0118	0.0118	0.0118	0.0118	0.0205	0.0205
w_B	0.0502	0.0502	0.0502	0.0502	0.0502	0.0502	0.0502	0.0501	0.0501
w_C	0.9380	0.9380	0.9380	0.9380	0.9380	0.9380	0.9380	0.9294	0.9294
T/K	431.17	441.13	450.89	460.82	470.58	480.33	490.18	420.93	430.83
P/MPa	1.346	1.561	1.746	1.996	2.251	2.551	2.856	1.589	1.784
w_A	0.0205	0.0205	0.0205	0.0205	0.0205	0.0205	0.0298	0.0298	0.0298
w_B	0.0501	0.0501	0.0501	0.0501	0.0501	0.0501	0.0499	0.0499	0.0499
w_C	0.9294	0.9294	0.9294	0.9294	0.9294	0.9294	0.9203	0.9203	0.9203
T/K	440.72	450.70	460.51	470.39	480.24	490.10	411.32	421.44	431.01
P/MPa	1.974	2.194	2.444	2.699	2.984	3.269	1.774	1.964	2.149
w_A	0.0298	0.0298	0.0298	0.0298	0.0298	0.0298	0.0099	0.0099	0.0099
w_B	0.0499	0.0499	0.0499	0.0499	0.0499	0.0499	0.0998	0.0998	0.0998
w_C	0.9203	0.9203	0.9203	0.9203	0.9203	0.9203	0.8903	0.8903	0.8903
T/K	440.90	450.90	460.75	470.56	480.44	490.32	431.00	440.82	450.72
P/MPa	2.359	2.589	2.834	3.094	3.369	3.649	1.364	1.544	1.754

continued

continued

w_A	0.0099	0.0099	0.0099	0.0099	0.0196	0.0196	0.0196	0.0196	0.0196
w_B	0.0998	0.0998	0.0998	0.0998	0.1007	0.1007	0.1007	0.1007	0.1007
w_C	0.8903	0.8903	0.8903	0.8903	0.8797	0.8797	0.8797	0.8797	0.8797
T/K	460.66	470.51	480.43	490.25	421.34	431.27	440.74	450.63	460.65
P/MPa	1.989	2.254	2.544	2.869	1.629	1.804	2.009	2.234	2.474
w_A	0.0196	0.0196	0.0196	0.0296	0.0296	0.0296	0.0296	0.0296	0.0296
w_B	0.1007	0.1007	0.1007	0.1002	0.1002	0.1002	0.1002	0.1002	0.1002
w_C	0.8797	0.8797	0.8797	0.8702	0.8702	0.8702	0.8702	0.8702	0.8702
T/K	470.39	480.38	490.29	411.27	421.23	431.10	441.04	450.93	460.69
P/MPa	2.734	3.024	3.319	1.829	2.009	2.194	2.429	2.664	2.909
w_A	0.0296	0.0296	0.0296	0.0098	0.0098	0.0098	0.0098	0.0098	0.0098
w_B	0.1002	0.1002	0.1002	0.1508	0.1508	0.1508	0.1508	0.1508	0.1508
w_C	0.8702	0.8702	0.8702	0.8394	0.8394	0.8394	0.8394	0.8394	0.8394
T/K	470.55	480.48	490.31	430.69	440.59	450.49	460.60	470.34	480.41
P/MPa	3.174	3.449	3.709	1.384	1.569	1.774	2.029	2.284	2.579
w_A	0.0098	0.0210	0.0210	0.0210	0.0210	0.0210	0.0210	0.0210	0.0210
w_B	0.1508	0.1503	0.1503	0.1503	0.1503	0.1503	0.1503	0.1503	0.1503
w_C	0.8394	0.8287	0.8287	0.8287	0.8287	0.8287	0.8287	0.8287	0.8287
T/K	490.00	420.58	430.38	440.26	450.09	459.96	469.77	479.68	489.57
P/MPa	2.889	1.674	1.859	2.059	2.269	2.519	2.784	3.044	3.354
w_A	0.0293	0.0293	0.0293	0.0293	0.0293	0.0293	0.0293	0.0293	0.0293
w_B	0.1506	0.1506	0.1506	0.1506	0.1506	0.1506	0.1506	0.1506	0.1506
w_C	0.8201	0.8201	0.8201	0.8201	0.8201	0.8201	0.8201	0.8201	0.8201
T/K	411.03	420.84	430.89	440.85	450.76	460.61	470.42	480.19	490.17
P/MPa	1.899	2.084	2.299	2.519	2.754	3.004	3.264	3.534	3.799

Polymer (B): **polyethylene** **2008HAR**
Characterization: M_n/g.mol^{-1} = 82000, M_w/g.mol^{-1} = 108000,
Scientific Polymer Products, Inc., Ontario, NY
Solvent (A): **ethene** C_2H_4 **74-85-1**
Solvent (C): **n-hexane** C_6H_{14} **110-54-3**

Type of data: cloud points

w_A	0.0992	0.0992	0.0992	0.0992	0.0992	0.0992	0.1082	0.1082	0.1082
w_B	0.0090	0.0090	0.0090	0.0090	0.0090	0.0090	0.0173	0.0173	0.0173
w_C	0.8318	0.8318	0.8318	0.8318	0.8318	0.8318	0.8745	0.8745	0.8745
T/K	373.18	393.15	413.14	433.18	453.15	473.19	373.12	393.14	413.18
P/MPa	6.7	9.7	12.5	15.1	17.5	19.7	8.87	11.8	14.6
w_A	0.1082	0.1082	0.1082	0.1019	0.1019	0.1019	0.1019	0.1019	0.1019
w_B	0.0173	0.0173	0.0173	0.0437	0.0437	0.0437	0.0437	0.0437	0.0437
w_C	0.8745	0.8745	0.8745	0.8544	0.8544	0.8544	0.8544	0.8544	0.8544
T/K	433.19	453.17	473.19	373.11	393.14	413.17	433.12	453.18	473.19
P/MPa	17.0	19.2	21.2	9.3	12.4	15.1	17.5	19.5	21.5

continued

continued

w_A	0.0992	0.0992	0.0992	0.0992	0.0992	0.0992	0.0912	0.0912	0.0912
w_B	0.0797	0.0797	0.0797	0.0797	0.0797	0.0797	0.1135	0.1135	0.1135
w_C	0.8211	0.8211	0.8211	0.8211	0.8211	0.8211	0.7953	0.7953	0.7953
T/K	373.17	393.12	413.13	433.16	453.14	473.16	373.14	393.18	413.18
P/MPa	8.5	11.5	14.2	16.7	19.0	21.0	7.99	11.1	13.7

w_A	0.0912	0.0912	0.0912
w_B	0.1135	0.1135	0.1135
w_C	0.7953	0.7953	0.7953
T/K	433.16	453.15	473.13
P/MPa	16.2	18.5	20.5

Type of data: vapor-liquid-liquid equilibrium data

w_A	0.0992	0.0992	0.0992	0.0992	0.0992	0.0992	0.1082	0.1082	0.1082
w_B	0.0090	0.0090	0.0090	0.0090	0.0090	0.0090	0.0173	0.0173	0.0173
w_C	0.8318	0.8318	0.8318	0.8318	0.8318	0.8318	0.8745	0.8745	0.8745
T/K	373.17	393.14	413.13	433.17	453.20	473.17	373.12	393.12	413.17
P/MPa	2.85	3.33	3.79	4.31	4.91	5.20	3.21	3.68	4.20

w_A	0.1082	0.1082	0.1082	0.1019	0.1019	0.1019	0.1019	0.1019	0.1019
w_B	0.0173	0.0173	0.0173	0.0437	0.0437	0.0437	0.0437	0.0437	0.0437
w_C	0.8745	0.8745	0.8745	0.8544	0.8544	0.8544	0.8544	0.8544	0.8544
T/K	433.15	453.17	473.13	373.12	393.13	413.13	433.15	453.18	473.14
P/MPa	4.68	5.20	5.86	3.00	3.50	3.99	4.49	5.00	5.32

w_A	0.0992	0.0992	0.0992	0.0992	0.0992	0.0992	0.0912	0.0912	0.0912
w_B	0.0797	0.0797	0.0797	0.0797	0.0797	0.0797	0.1135	0.1135	0.1135
w_C	0.8211	0.8211	0.8211	0.8211	0.8211	0.8211	0.7953	0.7953	0.7953
T/K	373.15	393.11	413.13	433.16	453.11	473.15	373.12	393.16	413.15
P/MPa	2.95	3.45	3.99	4.49	4.99	5.27	2.92	3.42	4.00

w_A	0.0912	0.0912	0.0912
w_B	0.1135	0.1135	0.1135
w_C	0.7953	0.7953	0.7953
T/K	433.14	453.14	473.13
P/MPa	4.50	4.93	5.28

Polymer (B): **polyethylene** **2010HA2**
Characterization: M_n/g.mol^{-1} = 13200, M_w/g.mol^{-1} = 15400, Scientific Polymer Products, Inc., Ontario, NY
Solvent (A): **ethene** C_2H_4 **74-85-1**
Solvent (C): **1-hexene** C_6H_{12} **592-41-6**

Type of data: cloud points

continued

continued

w_A	0.099	0.099	0.099	0.099	0.099	0.099	0.099	0.099
w_B	0.021	0.021	0.021	0.021	0.021	0.021	0.021	0.021
w_C	0.880	0.880	0.880	0.880	0.880	0.880	0.880	0.880
T/K	373.01	393.48	413.30	433.16	443.20	453.15	463.33	473.22
P/MPa	2.60	4.12	7.23	10.05	11.36	12.58	13.79	14.95

Type of data: vapor-liquid-liquid equilibrium data

w_A	0.099	0.099	0.099	0.099	0.099	0.099	0.099
w_B	0.021	0.021	0.021	0.021	0.021	0.021	0.021
w_C	0.880	0.880	0.880	0.880	0.880	0.880	0.880
T/K	393.40	413.51	433.16	443.00	453.05	463.01	473.37
P/MPa	3.07	3.61	4.15	4.42	4.70	4.91	4.99

Polymer (B): **polyethylene** **2010HA2**
Characterization: M_n/g.mol^{-1} = 15400, M_w/g.mol^{-1} = 17200,
Scientific Polymer Products, Inc., Ontario, NY
Solvent (A): **ethene** C_2H_4 **74-85-1**
Solvent (C): **1-hexene** C_6H_{12} **592-41-6**

Type of data: cloud points

w_A	0.103	0.103	0.103	0.103	0.103	0.103	0.101	0.101	0.101
w_B	0.048	0.048	0.048	0.048	0.048	0.048	0.077	0.077	0.077
w_C	0.849	0.849	0.849	0.849	0.849	0.849	0.822	0.822	0.822
T/K	373.11	393.49	413.65	433.26	453.35	473.15	373.31	393.43	413.15
P/MPa	3.23	6.43	9.38	12.07	14.59	16.80	4.51	7.57	10.33

w_A	0.101	0.101	0.101	0.102	0.102	0.102	0.102	0.102	0.102
w_B	0.077	0.077	0.077	0.100	0.100	0.100	0.100	0.100	0.100
w_C	0.822	0.822	0.822	0.798	0.798	0.798	0.798	0.798	0.798
T/K	433.71	453.49	473.48	373.16	393.58	413.32	433.50	453.54	473.19
P/MPa	12.90	15.36	17.33	4.90	8.03	10.93	13.45	15.83	17.98

w_A	0.097	0.097	0.097	0.097	0.097	0.097
w_B	0.123	0.123	0.123	0.123	0.123	0.123
w_C	0.780	0.780	0.780	0.780	0.780	0.780
T/K	373.23	393.25	413.54	433.38	453.52	473.10
P/MPa	5.46	8.59	11.47	13.98	16.21	18.26

Type of data: vapor-liquid-liquid equilibrium data

w_A	0.103	0.103	0.103	0.103	0.103	0.103	0.101	0.101	0.101
w_B	0.048	0.048	0.048	0.048	0.048	0.048	0.077	0.077	0.077
w_C	0.849	0.849	0.849	0.849	0.849	0.849	0.822	0.822	0.822
T/K	373.27	393.28	413.57	433.48	453.49	473.05	373.19	393.31	413.32
P/MPa	2.86	3.40	3.96	4.50	5.01	5.28	2.97	3.52	4.05

continued

continued

w_A	0.101	0.101	0.101	0.102	0.102	0.102	0.102	0.102	0.102
w_B	0.077	0.077	0.077	0.100	0.100	0.100	0.100	0.100	0.100
w_C	0.822	0.822	0.822	0.798	0.798	0.798	0.798	0.798	0.798
T/K	433.69	453.39	473.51	373.23	393.60	413.58	433.43	453.67	473.27
P/MPa	4.55	5.10	5.40	2.96	3.66	4.22	4.68	5.05	5.36

w_A	0.097	0.097	0.097	0.097	0.097	0.097
w_B	0.123	0.123	0.123	0.123	0.123	0.123
w_C	0.780	0.780	0.780	0.780	0.780	0.780
T/K	373.26	393.11	413.38	433.70	453.42	473.17
P/MPa	3.05	3.57	4.15	4.64	5.12	5.42

Polymer (B): **polyethylene** **2010HA2**
Characterization: M_n/g.mol^{-1} = 82000, M_w/g.mol^{-1} = 108000,
Scientific Polymer Products, Inc., Ontario, NY
Solvent (A): **ethene** C_2H_4 **74-85-1**
Solvent (C): **1-hexene** C_6H_{12} **592-41-6**

Type of data: cloud points

w_A	0.099	0.099	0.099	0.099	0.099	0.099	0.099	0.099	0.099
w_B	0.019	0.019	0.019	0.019	0.019	0.019	0.048	0.048	0.048
w_C	0.882	0.882	0.882	0.882	0.882	0.882	0.853	0.853	0.853
T/K	373.42	393.47	413.26	433.51	453.42	473.28	373.41	393.47	413.50
P/MPa	7.73	10.73	13.31	16.10	18.46	20.65	9.43	12.22	15.00

w_A	0.099	0.099	0.099	0.101	0.101	0.101	0.101	0.101	0.101
w_B	0.048	0.048	0.048	0.078	0.078	0.078	0.078	0.078	0.078
w_C	0.853	0.853	0.853	0.821	0.821	0.821	0.821	0.821	0.821
T/K	433.13	453.54	473.39	373.28	393.59	413.50	433.52	453.44	473.16
P/MPa	17.43	19.52	21.55	10.43	13.27	15.93	18.36	20.57	22.62

w_A	0.103	0.103	0.103	0.103	0.103	0.103	0.101	0.101	0.101
w_B	0.097	0.097	0.097	0.097	0.097	0.097	0.118	0.118	0.118
w_C	0.800	0.800	0.800	0.800	0.800	0.800	0.781	0.781	0.781
T/K	373.09	393.34	413.39	433.31	453.20	473.46	373.19	393.09	413.26
P/MPa	8.80	11.93	14.40	16.93	19.26	21.32	9.62	12.07	14.60

w_A	0.101	0.101	0.101
w_B	0.118	0.118	0.118
w_C	0.781	0.781	0.781
T/K	433.25	453.63	473.42
P/MPa	17.07	19.50	21.62

continued

continued

Type of data: vapor-liquid-liquid equilibrium data

w_A	0.099	0.099	0.099	0.099	0.099	0.099	0.099	0.099	0.099
w_B	0.019	0.019	0.019	0.019	0.019	0.019	0.048	0.048	0.048
w_C	0.882	0.882	0.882	0.882	0.882	0.882	0.853	0.853	0.853
T/K	373.54	393.44	413.50	433.30	453.45	473.28	373.45	393.59	413.59
P/MPa	2.82	3.33	3.89	4.49	4.97	5.21	2.93	3.49	4.01
w_A	0.099	0.099	0.099	0.101	0.101	0.101	0.101	0.101	0.101
w_B	0.048	0.048	0.048	0.078	0.078	0.078	0.078	0.078	0.078
w_C	0.853	0.853	0.853	0.821	0.821	0.821	0.821	0.821	0.821
T/K	433.46	453.31	473.50	373.27	393.55	413.45	433.39	453.17	473.02
P/MPa	4.56	5.07	5.33	3.09	3.63	4.23	4.78	5.27	5.46
w_A	0.103	0.103	0.103	0.103	0.103	0.103	0.101	0.101	0.101
w_B	0.097	0.097	0.097	0.097	0.097	0.097	0.118	0.118	0.118
w_C	0.800	0.800	0.800	0.800	0.800	0.800	0.781	0.781	0.781
T/K	373.22	393.31	413.19	433.14	453.21	472.97	373.18	393.09	413.20
P/MPa	2.89	3.50	4.16	4.50	4.90	5.19	3.03	3.73	4.16
w_A	0.101	0.101	0.101						
w_B	0.118	0.118	0.118						
w_C	0.781	0.781	0.781						
T/K	433.17	453.44	473.36						
P/MPa	4.71	5.04	5.27						

Polymer (B):	**polyethylene**		**2010HA2**
Characterization:	M_n/g.mol^{-1} = 15400, M_w/g.mol^{-1} = 17200, Scientific Polymer Products, Inc., Ontario, NY		
Solvent (A):	**ethene**	C_2H_4	**74-85-1**
Solvent (C):	**1-hexene**	C_6H_{12}	**592-41-6**
Solvent (D):	**n-hexane**	C_6H_{14}	**110-54-3**

Type of data: cloud points

w_A	0.099	0.099	0.099	0.099	0.099	0.099	0.100	0.100	0.100
w_B	0.020	0.020	0.020	0.020	0.020	0.020	0.048	0.048	0.048
w_C	0.440	0.440	0.440	0.440	0.440	0.440	0.426	0.426	0.426
w_D	0.441	0.441	0.441	0.441	0.441	0.441	0.426	0.426	0.426
T/K	373.26	393.54	413.63	433.19	453.54	473.12	373.47	393.38	413.23
P/MPa	3.04	6.31	9.39	12.07	14.50	16.59	3.68	7.06	10.04
w_A	0.100	0.100	0.100	0.098	0.098	0.098	0.098	0.098	0.098
w_B	0.048	0.048	0.048	0.082	0.082	0.082	0.082	0.082	0.082
w_C	0.426	0.426	0.426	0.410	0.410	0.410	0.410	0.410	0.410
w_D	0.426	0.426	0.426	0.410	0.410	0.410	0.410	0.410	0.410
T/K	433.32	453.28	473.16	373.63	393.20	413.19	433.30	453.34	473.30
P/MPa	12.70	14.97	17.17	4.16	7.30	10.13	12.77	15.24	17.42

continued

continued

w_A	0.101	0.101	0.101	0.101	0.101	0.101	0.102	0.102	0.102
w_B	0.100	0.100	0.100	0.100	0.100	0.100	0.117	0.117	0.117
w_C	0.399	0.399	0.399	0.399	0.399	0.399	0.390	0.390	0.390
w_D	0.400	0.400	0.400	0.400	0.400	0.400	0.391	0.391	0.391
T/K	373.40	393.27	413.27	433.19	453.10	473.21	373.30	393.44	413.24
P/MPa	4.97	8.23	11.16	13.70	16.09	18.13	5.30	8.43	11.34

w_A	0.102	0.102	0.102
w_B	0.117	0.117	0.117
w_C	0.390	0.390	0.390
w_D	0.391	0.391	0.391
T/K	433.20	453.56	473.08
P/MPa	13.87	16.12	18.00

Type of data: vapor-liquid-liquid equilibrium data

w_A	0.099	0.099	0.099	0.099	0.099	0.100	0.100	0.100	0.100
w_B	0.020	0.020	0.020	0.020	0.020	0.048	0.048	0.048	0.048
w_C	0.440	0.440	0.440	0.440	0.440	0.426	0.426	0.426	0.426
w_D	0.441	0.441	0.441	0.441	0.441	0.426	0.426	0.426	0.426
T/K	393.57	413.65	433.20	453.50	473.06	373.43	393.36	413.22	433.41
P/MPa	3.56	4.11	4.56	5.06	5.33	3.03	3.43	4.08	4.62

w_A	0.100	0.100	0.098	0.098	0.098	0.098	0.098	0.098	0.101
w_B	0.048	0.048	0.082	0.082	0.082	0.082	0.082	0.082	0.100
w_C	0.426	0.426	0.410	0.410	0.410	0.410	0.410	0.410	0.399
w_D	0.426	0.426	0.410	0.410	0.410	0.410	0.410	0.410	0.400
T/K	453.42	473.15	373.44	393.30	413.29	433.29	453.52	473.19	373.20
P/MPa	5.02	5.36	2.88	3.30	3.96	4.48	5.04	5.28	3.12

w_A	0.101	0.101	0.101	0.101	0.101	0.102	0.102	0.102	0.102
w_B	0.100	0.100	0.100	0.100	0.100	0.117	0.117	0.117	0.117
w_C	0.399	0.399	0.399	0.399	0.399	0.390	0.390	0.390	0.390
w_D	0.400	0.400	0.400	0.400	0.400	0.391	0.391	0.391	0.391
T/K	393.23	413.56	433.60	453.37	473.22	373.31	393.13	413.14	433.08
P/MPa	3.64	4.19	4.67	5.07	5.36	3.02	3.51	4.16	4.62

w_A	0.102	0.102
w_B	0.117	0.117
w_C	0.390	0.390
w_D	0.391	0.391
T/K	453.74	473.14
P/MPa	5.11	5.38

Polymer (B):	**polyethylene**		**2010HA2**
Characterization:	M_n/g.mol^{-1} = 82000, M_w/g.mol^{-1} = 108000, Scientific Polymer Products, Inc., Ontario, NY		
Solvent (A):	**ethene**	C_2H_4	**74-85-1**
Solvent (C):	**1-hexene**	C_6H_{12}	**592-41-6**
Solvent (D):	**n-hexane**	C_6H_{14}	**110-54-3**

Type of data: cloud points

w_A	0.100	0.100	0.100	0.100	0.100	0.100	0.099	0.099	0.099
w_B	0.020	0.020	0.020	0.020	0.020	0.020	0.050	0.050	0.050
w_C	0.441	0.441	0.441	0.441	0.441	0.441	0.426	0.426	0.426
w_D	0.439	0.439	0.439	0.439	0.439	0.439	0.425	0.425	0.425
T/K	373.17	393.44	413.32	433.23	453.31	473.32	373.33	393.66	413.21
P/MPa	8.60	11.51	14.17	16.62	18.73	20.58	9.92	12.76	15.37
w_A	0.099	0.099	0.099	0.099	0.099	0.099	0.099	0.099	0.099
w_B	0.050	0.050	0.050	0.081	0.081	0.081	0.081	0.081	0.081
w_C	0.426	0.426	0.426	0.410	0.410	0.410	0.410	0.410	0.410
w_D	0.425	0.425	0.425	0.410	0.410	0.410	0.410	0.410	0.410
T/K	433.29	453.45	473.37	373.20	393.22	413.58	433.32	453.38	473.42
P/MPa	17.79	19.97	21.67	10.67	13.28	15.87	18.37	20.73	22.59
w_A	0.098	0.098	0.098	0.098	0.098	0.098	0.102	0.102	0.102
w_B	0.102	0.102	0.102	0.102	0.102	0.102	0.119	0.119	0.119
w_C	0.400	0.400	0.400	0.400	0.400	0.400	0.390	0.390	0.390
w_D	0.400	0.400	0.400	0.400	0.400	0.400	0.389	0.389	0.389
T/K	373.17	393.20	413.14	433.15	453.57	473.42	373.18	393.45	413.21
P/MPa	9.67	12.58	15.22	17.70	20.14	22.30	9.38	11.91	14.43
w_A	0.102	0.102	0.102						
w_B	0.119	0.119	0.119						
w_C	0.390	0.390	0.390						
w_D	0.389	0.389	0.389						
T/K	433.29	453.30	473.15						
P/MPa	16.96	19.40	21.44						

Type of data: vapor-liquid-liquid equilibrium data

w_A	0.100	0.100	0.100	0.100	0.100	0.100	0.099	0.099	0.099
w_B	0.020	0.020	0.020	0.020	0.020	0.020	0.050	0.050	0.050
w_C	0.441	0.441	0.441	0.441	0.441	0.441	0.426	0.426	0.426
w_D	0.439	0.439	0.439	0.439	0.439	0.439	0.425	0.425	0.425
T/K	373.18	393.35	413.21	433.16	453.20	473.05	373.32	393.51	413.24
P/MPa	3.00	3.48	4.03	4.53	5.01	5.31	3.02	3.54	4.13

continued

continued

w_A	0.099	0.099	0.099	0.099	0.099	0.099	0.099	0.099	0.099
w_B	0.050	0.050	0.050	0.081	0.081	0.081	0.081	0.081	0.081
w_C	0.426	0.426	0.426	0.410	0.410	0.410	0.410	0.410	0.410
w_D	0.425	0.425	0.425	0.410	0.410	0.410	0.410	0.410	0.410
T/K	433.20	453.35	473.15	373.18	393.18	413.68	433.39	453.29	473.28
P/MPa	4.68	5.11	5.33	3.22	3.79	4.33	4.83	5.23	5.52
w_A	0.098	0.098	0.098	0.098	0.098	0.098	0.102	0.102	0.102
w_B	0.102	0.102	0.102	0.102	0.102	0.102	0.119	0.119	0.119
w_C	0.400	0.400	0.400	0.400	0.400	0.400	0.390	0.390	0.390
w_D	0.400	0.400	0.400	0.400	0.400	0.400	0.389	0.389	0.389
T/K	373.21	393.19	413.18	433.14	453.62	473.34	373.33	393.41	413.40
P/MPa	3.22	3.76	4.17	4.77	5.16	5.43	3.10	3.68	4.21
w_A	0.102	0.102	0.102						
w_B	0.119	0.119	0.119						
w_C	0.390	0.390	0.390						
w_D	0.389	0.389	0.389						
T/K	433.70	453.51	473.20						
P/MPa	4.71	5.30	5.65						

Polymer (B): **polyethylene** **2012MIL**
Characterization: M_n/g.mol^{-1} = 22000, M_w/g.mol^{-1} = 53000,
HDPE, Du Pont de Nemours & Co., USA
Solvent (A): **n-pentane** C_5H_{12} **109-66-0**
Solvent (C): **dimethyl ether** C_2H_6O **115-10-6**

Type of data: cloud points

w_B	0.027	0.027	0.027	0.027	0.027	0.027	0.027	0.027	0.027
w_C/w_A	47/53	47/53	47/53	47/53	47/53	47/53	47/53	47/53	47/53
T/K	388.15	404.15	410.15	415.15	420.15	425.15	431.15	435.15	440.15
P/bar	253	265	269	274	279	282	288	292	296
w_B	0.027	0.027	0.029	0.029	.029	0.029	0.029	0.029	0.029
w_C/w_A	47/53	47/53	22/78	22/78	22/78	22/78	22/78	22/78	22/78
T/K	444.15	449.15	389.15	396.15	402.15	407.15	413.15	414.15	420.15
P/bar	298	303	147	154	160	167	174	177	183
w_B	0.029	0.029							
w_C/w_A	22/78	22/78							
T/K	425.15	430.15							
P/bar	190	195							

Polymer (B): **poly(ethylene-*co*-benzyl methacrylate)** **2004LAT**
Characterization: 13.1 mol% benzyl methacrylate, synthesized in the laboratory
Solvent (A): **ethene** C_2H_4 **74-85-1**
Solvent (C): **benzyl methacrylate** $C_{11}H_{12}O_2$ **2495-37-6**

Type of data: cloud points

w_A	0.9133	0.8908	0.8693	0.849	0.829
w_B	0.050	0.050	0.050	0.050	0.050
w_C	0.0367	0.0592	0.0807	0.101	0.121
T/K	473.15	473.15	473.15	473.15	473.15
P/bar	1339	1308	1277	1245	1227

Polymer (B): **poly(ethylene-*co*-benzyl methacrylate)** **2004LAT**
Characterization: 20.3 mol% benzyl methacrylate, synthesized in the laboratory
Solvent (A): **ethene** C_2H_4 **74-85-1**
Solvent (C): **benzyl methacrylate** $C_{11}H_{12}O_2$ **2495-37-6**

Type of data: cloud points

w_A	0.8868	0.8656	0.845	0.826	0.807
w_B	0.050	0.050	0.050	0.050	0.050
w_C	0.0632	0.0844	0.105	0.124	0.143
T/K	473.15	473.15	473.15	473.15	473.15
P/bar	1401	1368	1330	1299	1275

Polymer (B): **poly(ethylene-*co*-benzyl methacrylate)** **2004LAT**
Characterization: 25.2 mol% benzyl methacrylate, synthesized in the laboratory
Solvent (A): **ethene** C_2H_4 **74-85-1**
Solvent (C): **benzyl methacrylate** $C_{11}H_{12}O_2$ **2495-37-6**

Type of data: cloud points

w_A	0.8629	0.843	0.823	0.805	0.787
w_B	0.050	0.050	0.050	0.050	0.050
w_C	0.0871	0.107	0.127	0.145	0.163
T/K	473.15	473.15	473.15	473.15	473.15
P/bar	1463	1425	1380	1352	1315

Polymer (B): **poly(ethylene-*co*-benzyl methacrylate)** **2004LAT**
Characterization: M_n/g.mol^{-1} = 35600, M_w/g.mol^{-1} = 62200,
30.4 mol% benzyl methacrylate, synthesized in the laboratory
Solvent (A): **ethene** C_2H_4 **74-85-1**
Solvent (C): **benzyl methacrylate** $C_{11}H_{12}O_2$ **2495-37-6**

Type of data: cloud points

continued

continued

w_A	0.833	0.814	0.796	0.778	0.762
w_B	0.050	0.050	0.050	0.050	0.050
w_C	0.117	0.136	0.154	0.172	0.188
T/K	473.15	473.15	473.15	473.15	473.15
P/bar	1508	1460	1417	1375	1340

Polymer (B): **poly(ethylene-*co*-benzyl methacrylate)** **2004LAT**
Characterization: 33.5 mol% benzyl methacrylate, synthesized in the laboratory
Solvent (A): **ethene** C_2H_4 **74-85-1**
Solvent (C): **benzyl methacrylate** $C_{11}H_{12}O_2$ **2495-37-6**

Type of data: cloud points

w_A	0.813	0.795	0.778	0.761	0.745
w_B	0.050	0.050	0.050	0.050	0.050
w_C	0.137	0.155	0.172	0.189	0.205
T/K	473.15	473.15	473.15	473.15	473.15
P/bar	1515	1472	1430	1393	1352

Polymer (B): **poly(ethylene-*co*-ethyl methacrylate)** **2004LAT**
Characterization: M_n/g.mol^{-1} = 25500, M_w/g.mol^{-1} = 47300,
11.4 mol% ethyl methacrylate), synthesized in the laboratory
Solvent (A): **ethene** C_2H_4 **74-85-1**
Solvent (C): **ethyl methacrylate** $C_6H_{10}O_2$ **97-63-2**

Type of data: cloud points

w_A	0.9152	0.8952	0.8760	0.8575	0.840
w_B	0.050	0.050	0.050	0.050	0.050
w_C	0.0348	0.0548	0.0740	0.0925	0.110
T/K	473.15	473.15	473.15	473.15	473.15
P/bar	1034	1004	985	973	955

Polymer (B): **poly(ethylene-*co*-ethyl methacrylate)** **2004LAT**
Characterization: M_n/g.mol^{-1} = 24800, M_w/g.mol^{-1} = 45400,
19.2 mol% ethyl methacrylate), synthesized in the laboratory
Solvent (A): **ethene** C_2H_4 **74-85-1**
Solvent (C): **ethyl methacrylate** $C_6H_{10}O_2$ **97-63-2**

Type of data: cloud points

w_A	0.8918	0.8728	0.8545	0.837	0.820
w_B	0.050	0.050	0.050	0.050	0.050
w_C	0.0582	0.0772	0.0955	0.113	0.130
T/K	473.15	473.15	473.15	473.15	473.15
P/bar	978	960	935	920	890

Polymer (B): **poly(ethylene-*co*-ethyl methacrylate)** **2004LAT**
Characterization: M_n/g.mol^{-1} = 22500, M_w/g.mol^{-1} = 42000,
26.1 mol% ethyl methacrylate), synthesized in the laboratory
Solvent (A): **ethene** C_2H_4 **74-85-1**
Solvent (C): **ethyl methacrylate** $C_6H_{10}O_2$ **97-63-2**

Type of data: cloud points

w_A	0.8706	0.8524	0.835	0.818	0.802
w_B	0.050	0.050	0.050	0.050	0.050
w_C	0.0794	0.0976	0.115	0.132	0.148
T/K	473.15	473.15	473.15	473.15	473.15
P/bar	940	925	906	886	865

Polymer (B): **poly(ethylene-*co*-ethyl methacrylate)** **2004LAT**
Characterization: M_n/g.mol^{-1} = 25100, M_w/g.mol^{-1} = 48700,
29.0 mol% ethyl methacrylate), synthesized in the laboratory
Solvent (A): **ethene** C_2H_4 **74-85-1**
Solvent (C): **ethyl methacrylate** $C_6H_{10}O_2$ **97-63-2**

Type of data: cloud points

w_A	0.837	0.820	0.804	0.789	0.774
w_B	0.050	0.050	0.050	0.050	0.050
w_C	0.113	0.130	0.146	0.161	0.176
T/K	473.15	473.15	473.15	473.15	473.15
P/bar	915	896	879	853	833

Polymer (B): **poly(ethylene-*co*-ethyl methacrylate)** **2004LAT**
Characterization: M_n/g.mol^{-1} = 24000, M_w/g.mol^{-1} = 44300,
38.0 mol% ethyl methacrylate), synthesized in the laboratory
Solvent (A): **ethene** C_2H_4 **74-85-1**
Solvent (C): **ethyl methacrylate** $C_6H_{10}O_2$ **97-63-2**

Type of data: cloud points

w_A	0.808	0.792	0.777	0.763	0.749
w_B	0.050	0.050	0.050	0.050	0.050
w_C	0.142	0.158	0.173	0.187	0.201
T/K	473.15	473.15	473.15	473.15	473.15
P/bar	885	871	853	823	803

Polymer (B): **poly(ethylene-*co*-1-hexene)** **2001DOE2**
Characterization: M_n/g.mol^{-1} = 60000, M_w/g.mol^{-1} = 129000, 16.1 wt% 1-hexene
Solvent (A): **ethene** C_2H_4 **74-85-1**
Solvent (C): **n-butane** C_4H_{10} **106-97-8**

Type of data: coexistence data

continued

continued

T/K = 433.15

w_A(total)	0.750	was kept constant
w_B(total)	0.150	was kept constant
w_C(total)	0.100	was kept constant

P/bar	707	875	1017	1172	1191
w_A(gel phase)	0.339	0.431	0.529	0.677	0.750
w_B(gel phase)	0.599	0.496	0.392	0.234	0.150
w_C(gel phase)	0.062	0.073	0.079	0.089	0.100
w_A(sol phase)	0.883	0.870	–	0.832	–
w_B(sol phase)	0.014	0.026	–	0.065	–
w_C(sol phase)	0.103	0.104	–	0.102	–

Polymer (B): **poly(ethylene-*co*-1-hexene)** **2001DOE2**
Characterization: M_n/g.mol^{-1} = 60000, M_w/g.mol^{-1} = 129000, 16.1 wt% 1-hexene
Solvent (A): **ethene** **C_2H_4** **74-85-1**
Solvent (C): **carbon dioxide** **CO_2** **124-38-9**

Type of data: coexistence data

T/K = 433.15

w_A(total)	0.750	was kept constant
w_B(total)	0.150	was kept constant
w_C(total)	0.100	was kept constant

P/bar	1151	1302	1403	1432
w_A(gel phase)	0.542	0.602	0.698	0.750
w_B(gel phase)	0.388	0.318	0.216	0.150
w_C(gel phase)	0.070	0.080	0.086	0.100
w_A(sol phase)	0.873	0.862	0.849	–
w_B(sol phase)	0.014	0.031	0.047	–
w_C(sol phase)	0.113	0.107	0.105	–

Polymer (B): **poly(ethylene-*co*-1-hexene)** **2001DOE2**
Characterization: M_n/g.mol^{-1} = 60000, M_w/g.mol^{-1} = 129000, 16.1 wt% 1-hexene
Solvent (A): **ethene** **C_2H_4** **74-85-1**
Solvent (C): **helium** **He** **7440-59-7**

Type of data: coexistence data

T/K = 433.15

w_A(total)	0.840	was kept constant
w_B(total)	0.150	was kept constant
w_C(total)	0.010	was kept constant

continued

continued

P/bar	1261	1428	1547	1658
w_A(gel phase)	0.526	0.590	0.655	0.840
w_B(gel phase)	0.470	0.402	0.338	0.150
w_C(gel phase)	0.003	0.005	0.006	0.010
w_A(sol phase)	–	0.978	0.964	–
w_B(sol phase)	–	0.006	0.024	–
w_C(sol phase)	–	0.014	0.012	–

Polymer (B): **poly(ethylene-*co*-1-hexene)** **2001DOE2**
Characterization: M_n/g.mol^{-1} = 60000, M_w/g.mol^{-1} = 129000, 16.1 wt% 1-hexene
Solvent (A): **ethene** C_2H_4 **74-85-1**
Solvent (C): **1-hexene** C_6H_{12} **592-41-6**

Type of data: coexistence data

T/K = 433.15

w_A(total)	0.425	was kept constant
w_B(total)	0.150	was kept constant
w_C(total)	0.425	was kept constant

P/bar	360	442	509	585	591
w_A(gel phase)	0.238	0.278	0.332	0.415	0.425
w_B(gel phase)	0.538	0.460	0.368	0.200	0.150
w_C(gel phase)	0.224	0.262	0.300	0.385	0.425
w_A(sol phase)	0.505	0.498	0.480	0.480	–
w_B(sol phase)	0.010	0.020	0.032	0.053	–
w_C(sol phase)	0.485	0.482	0.488	0.468	–

Polymer (B): **poly(ethylene-*co*-1-hexene)** **2001DOE2**
Characterization: M_n/g.mol^{-1} = 60000, M_w/g.mol^{-1} = 129000, 16.1 wt% 1-hexene
Solvent (A): **ethene** C_2H_4 **74-85-1**
Solvent (C): **methane** CH_4 **74-82-8**

Type of data: coexistence data

T/K = 433.15

w_A(total)	0.750	was kept constant
w_B(total)	0.150	was kept constant
w_C(total)	0.100	was kept constant

P/bar	1069	1158	1313	1440	1556	1578
w_A(gel phase)	0.397	0.436	0.518	0.586	0.669	0.750
w_B(gel phase)	0.550	0.506	0.413	0.335	0.241	0.150
w_C(gel phase)	0.053	0.058	0.069	0.078	0.089	0.100
w_A(sol phase)	0.877	0.877	–	0.862	0.855	–
w_B(sol phase)	0.006	0.007	–	0.023	0.031	–
w_C(sol phase)	0.117	0.116	–	0.115	0.114	–

Polymer (B): **poly(ethylene-*co*-1-hexene)** **2001DOE2**
Characterization: M_n/g.mol^{-1} = 60000, M_w/g.mol^{-1} = 129000, 16.1 wt% 1-hexene
Solvent (A): **ethene** C_2H_4 **74-85-1**
Solvent (C): **nitrogen** N_2 **7727-37-9**

Type of data: coexistence data

T/K = 433.15

w_A(total)	0.750	was kept constant
w_B(total)	0.150	was kept constant
w_C(total)	0.100	was kept constant

P/bar	1409	1599	1697	1732
w_A(gel phase)	0.539	0.630	0.671	0.750
w_B(gel phase)	0.406	0.302	0.251	0.150
w_C(gel phase)	0.065	0.078	0.088	0.100
w_A(sol phase)	0.897	0.887	0.863	–
w_B(sol phase)	0.008	0.018	0.037	–
w_C(sol phase)	0.105	0.105	0.100	–

Polymer (B): **poly(ethylene-*co*-methyl acrylate)** **2007TUM**
Characterization: M_w/g.mol^{-1} = 110000, 23 mol% methyl acrylate, synthesized in the laboratory
Solvent (A): **ethene** C_2H_4 **74-85-1**
Solvent (C): **vinyl acetate** $C_4H_6O_2$ **108-05-4**

Type of data: cloud points

w_A	0.867	was kept constant
w_B	0.030	was kept constant
w_C	0.103	was kept constant

T/K	373	383	393	403	413	423	433	443	453
P/bar	1277	1248	1231	1211	1196	1176	1160	1149	1130

T/K	463	473	483	493
P/bar	1119	1104	1093	1080

Polymer (B): **poly(ethylene-*co*-methyl acrylate-*co*-vinyl acetate)** **2007TUM**
Characterization: M_w/g.mol^{-1} = 110000, 35 mol% methyl acrylate, 3.5 mol% vinyl acetate, synthesized in the laboratory
Solvent (A): **ethene** C_2H_4 **74-85-1**
Solvent (C): **vinyl acetate** $C_4H_6O_2$ **108-05-4**

Type of data: cloud points

continued

continued

w_A	0.888	was kept constant
w_B	0.030	was kept constant
w_C	0.082	was kept constant

T/K	373	383	393	403	413	423	433	443	453
P/bar	1780	1720	1660	1610	1560	1520	1480	1450	1420
T/K	463	473	483	493					
P/bar	1390	1360	1330	1300					

Polymer (B): **poly(ethylene-*co*-methyl acrylate-*co*-vinyl acetate)** **2007TUM**

Characterization: M_w/g.mol^{-1} = 110000, 40 mol% methyl acrylate, 3.5 mol% vinyl acetate, synthesized in the laboratory

Solvent (A): **ethene** C_2H_4 **74-85-1**

Solvent (C): **vinyl acetate** $C_4H_6O_2$ **108-05-4**

Type of data: cloud points

w_A	0.820	was kept constant
w_B	0.030	was kept constant
w_C	0.150	was kept constant

T/K	373	383	393	403	413	423	433	443	453
P/bar	1480	1440	1410	1370	1330	1310	1280	1250	1230
T/K	463	473	483	493	503				
P/bar	1200	1180	1160	1160	1150				

Polymer (B): **poly(ethylene-*co*-1-octene)** **2005LEE**

Characterization: M_n/g.mol^{-1} = 64810, M_w/g.mol^{-1} = 135500, 15.3 mol% 1-octene, T_m/K = 323.2, T_g/K = 214.2, ρ = 0.86 g/cm^3, DuPont Dow Elastomers Corporation

Solvent (A): **ethene** C_2H_4 **74-85-1**

Solvent (C): **1-octene** C_8H_{16} **111-66-0**

Type of data: cloud points

w_A	0.252	0.252	0.252	0.252	0.252	0.252	0.358	0.358	0.358
w_B	0.048	0.048	0.048	0.048	0.048	0.048	0.051	0.051	0.051
w_C	0.700	0.700	0.700	0.700	0.700	0.700	0.591	0.591	0.591
T/K	326.75	348.55	370.05	401.55	416.05	426.35	323.65	347.85	373.15
P/bar	62.75	98.26	131.01	177.89	195.48	208.58	276.8	291.7	314.8
w_A	0.358	0.358	0.503	0.503	0.503	0.503	0.503	0.503	0.643
w_B	0.051	0.051	0.059	0.059	0.059	0.059	0.059	0.059	0.051
w_C	0.591	0.591	0.438	0.438	0.438	0.438	0.438	0.438	0.306
T/K	396.85	422.85	323.55	327.75	348.85	373.35	397.95	423.25	323.75
P/bar	336.1	357.9	644.7	634.6	601.9	585.1	579.5	579.2	1010.1

continued

continued

w_A	0.643	0.643	0.643	0.643	0.643	0.643
w_B	0.051	0.051	0.051	0.051	0.051	0.051
w_C	0.306	0.306	0.306	0.306	0.306	0.306
T/K	328.15	330.65	348.35	374.35	398.15	428.35
P/bar	982.2	956.4	889.5	828.8	795.4	766.7

Polymer (B): **poly[ethylene-*co*-propylene-*co*-(ethylene norbornene)]** **2013KIR**

Characterization: M_w/g.mol^{-1} = 150000, 70% ethylene, 0.5% diene, T_m/K = 314.2, T_g/K = 229.2, EPDM 3745, Dow Chemical Co.

Solvent (A): **n-octane** **C_8H_{18}** **111-65-9**

Solvent (C): **propane** **C_3H_8** **74-98-6**

Type of data: cloud points

w_A	0.619	was kept constant					
w_B	0.117	was kept constant					
w_C	0.264	was kept constant					
T/K	397.95	398.65	423.15	446.35	457.85	459.85	460.15
P/bar	40	45	78	120	130	135	135

Type of data: vapor-liquid-liquid equilibrium data

w_A	0.619	was kept constant			
w_B	0.117	was kept constant			
w_C	0.264	was kept constant			
T/K	397.95	398.65	423.15	446.35	460.15
P/bar	24	25	53	72	90

Polymer (B): **poly(ethylene-*co*-vinyl acetate)** **2001DOE2**

Characterization: M_n/g.mol^{-1} = 61900, M_w/g.mol^{-1} = 167000, 27.5 wt% vinyl acetate

Solvent (A): **ethene** **C_2H_4** **74-85-1**

Solvent (C): **n-butane** **C_4H_{10}** **106-97-8**

Type of data: coexistence data

T/K = 433.15

w_A(total)	0.750	was kept constant			
w_B(total)	0.150	was kept constant			
w_C(total)	0.100	was kept constant			
P/bar	713	835	946	1032	1060
w_A(gel phase)	0.407	0.506	0.602	0.674	0.750
w_B(gel phase)	0.553	0.449	0.338	0.260	0.150
w_C(gel phase)	0.040	0.045	0.060	0.067	0.100
w_A(sol phase)	0.862	0.846	0.833	0.818	–
w_B(sol phase)	0.019	0.035	0.053	0.072	–
w_C(sol phase)	0.119	0.119	0.115	0.110	–

Polymer (B): **poly(ethylene-*co*-vinyl acetate)** **2001DOE2**
Characterization: M_n/g.mol^{-1} = 61900, M_w/g.mol^{-1} = 167000, 27.5 wt% vinyl acetate
Solvent (A): **ethene** C_2H_4 **74-85-1**
Solvent (C): **carbon dioxide** CO_2 **124-38-9**

Type of data: coexistence data

T/K = 433.15

w_A(total)	0.750	was kept constant			
w_B(total)	0.150	was kept constant			
w_C(total)	0.100	was kept constant			
P/bar	742	908	1036	1160	1201
w_A(gel phase)	0.357	0.441	0.530	0.640	0.750
w_B(gel phase)	0.597	0.501	0.400	0.276	0.150
w_C(gel phase)	0.046	0.058	0.070	0.085	0.100
w_A(sol phase)	0.879	0.875	0.865	0.843	–
w_B(sol phase)	0.0047	0.0081	0.020	0.0452	–
w_C(sol phase)	0.117	0.117	0.115	0.113	–

Polymer (B): **poly(ethylene-*co*-vinyl acetate)** **2001DOE2**
Characterization: M_n/g.mol^{-1} = 61900, M_w/g.mol^{-1} = 167000, 27.5 wt% vinyl acetate
Solvent (A): **ethene** C_2H_4 **74-85-1**
Solvent (C): **helium** He **7440-59-7**

Type of data: coexistence data

T/K = 433.15

w_A(total)	0.840	was kept constant			
w_B(total)	0.150	was kept constant			
w_C(total)	0.010	was kept constant			
P/bar	1164	1261	1406	1475	1495
w_A(gel phase)	0.5093	0.565	0.6821	0.7496	0.840
w_B(gel phase)	0.486	0.430	0.313	0.244	0.150
w_C(gel phase)	0.0047	0.005	0.0049	0.0064	0.010
w_A(sol phase)	0.9763	0.9778	0.9589	0.9496	–
w_B(sol phase)	0.0122	0.0112	0.0301	0.040	–
w_C(sol phase)	0.0115	0.0110	0.0110	0.0104	–

Polymer (B): **poly(ethylene-*co*-vinyl acetate)** **2001DOE2**
Characterization: M_n/g.mol^{-1} = 61900, M_w/g.mol^{-1} = 167000, 27.5 wt% vinyl acetate
Solvent (A): **ethene** C_2H_4 **74-85-1**
Solvent (C): **methane** CH_4 **74-82-8**

continued

continued

Type of data: coexistence data

T/K = 433.15

w_A(total)	0.840	was kept constant		
w_B(total)	0.060	was kept constant		
w_C(total)	0.100	was kept constant		
P/bar	1095	1263	1407	1451
w_A(gel phase)	0.515	0.602	0.705	0.840
w_B(gel phase)	0.425	0.326	0.211	0.060
w_C(gel phase)	0.060	0.072	0.084	0.100
w_A(sol phase)	0.890	0.882	0.875	–
w_B(sol phase)	0.004	0.014	0.021	–
w_C(sol phase)	0.106	0.105	0.104	–

T/K = 433.15

w_A(total)	0.700	was kept constant		
w_B(total)	0.150	was kept constant		
w_C(total)	0.150	was kept constant		
P/bar	1204	1400	1514	1556
w_A(gel phase)	0.422	0.532	0.626	0.700
w_B(gel phase)	0.488	0.354	0.240	0.150
w_C(gel phase)	0.090	0.114	0.134	0.150
w_A(sol phase)	0.812	0.803	0.794	–
w_B(sol phase)	0.014	0.025	0.036	–
w_C(sol phase)	0.174	0.172	0.170	–

T/K = 433.15

w_A(total)	0.750	was kept constant			
w_B(total)	0.150	was kept constant			
w_C(total)	0.100	was kept constant			
P/bar	963	1090	1305	1405	1430
w_A(gel phase)	0.416	0.484	0.626	0.670	0.750
w_B(gel phase)	0.529	0.451	0.291	0.241	0.150
w_C(gel phase)	0.055	0.065	0.083	0.089	0.100
w_A(sol phase)	0.862	0.858	0.837	0.823	–
w_B(sol phase)	0.0227	0.0276	0.0511	0.0676	–
w_C(sol phase)	0.115	0.115	0.112	0.109	–

T/K = 433.15

w_A(total)	0.800	was kept constant
w_B(total)	0.150	was kept constant
w_C(total)	0.050	was kept constant

continued

continued

P/bar	918	1030	1129	1231	1283
w_A(gel phase)	0.400	0.475	0.5545	0.676	0.800
w_B(gel phase)	0.570	0.470	0.390	0.270	0.150
w_C(gel phase)	0.030	0.055	0.0555	0.054	0.050
w_A(sol phase)	–	0.948	0.935	–	–
w_B(sol phase)	–	0.020	0.025	–	–
w_C(sol phase)	–	0.032	0.040	–	–

Polymer (B): **poly(ethylene-*co*-vinyl acetate)** **2001DOE2**
Characterization: M_n/g.mol^{-1} = 61900, M_w/g.mol^{-1} = 167000, 27.5 wt% vinyl acetate
Solvent (A): **ethene** **C_2H_4** **74-85-1**
Solvent (C): **nitrogen** **N_2** **7727-37-9**

Type of data: coexistence data

T/K = 433.15

w_A(total)	0.750	was kept constant				
w_B(total)	0.150	was kept constant				
w_C(total)	0.100	was kept constant				
P/bar	801	1208	1357	1465	1533	1556
w_A(gel phase)	0.378	0.476	0.549	0.620	0.690	0.750
w_B(gel phase)	0.572	0.461	0.377	0.297	0.218	0.150
w_C(gel phase)	0.050	0.063	0.074	0.083	0.092	0.100
w_A(sol phase)	0.881	0.874	0.869	0.852	0.845	–
w_B(sol phase)	0.002	0.009	0.015	0.034	0.042	–
w_C(sol phase)	0.117	0.116	0.116	0.113	0.112	–

Polymer (B): **poly(ethylene-*co*-vinyl acetate)** **2001DOE2**
Characterization: M_n/g.mol^{-1} = 61900, M_w/g.mol^{-1} = 167000, 27.5 wt% vinyl acetate
Solvent (A): **ethene** **C_2H_4** **74-85-1**
Solvent (C): **vinyl acetate** **$C_4H_6O_2$** **108-05-4**

Type of data: coexistence data

T/K = 433.15

w_A(total)	0.425	was kept constant				
w_B(total)	0.150	was kept constant				
w_C(total)	0.425	was kept constant				
P/bar	401	451	492	529	570	616
w_A(gel phase)	0.27	0.30	0.33	0.35	0.38	0.425
w_B(gel phase)	0.49	0.44	0.40	0.34	0.27	0.15
w_C(gel phase)	0.24	0.27	0.29	0.31	0.35	0.425
w_A(sol phase)	0.50	0.49	0.49	0.49	0.48	–
w_B(sol phase)	0.02	0.03	0.04	0.04	0.06	–
w_C(sol phase)	0.48	0.48	0.47	0.47	0.46	–

Polymer (B):	**poly(ethylene-*co*-vinyl acetate)**	**2001DOE2**
Characterization:	M_n/g.mol^{-1} = 61900, M_w/g.mol^{-1} = 167000, 27.5 wt% vinyl acetate	
Solvent (A):	**ethene** C_2H_4	**74-85-1**
Solvent (C):	**vinyl acetate** $C_4H_6O_2$	**108-05-4**
Solvent (D):	**n-butane** C_4H_{10}	**106-97-8**

Type of data: coexistence data

T/K = 433.15

w_A(total)	0.375	was kept constant			
w_B(total)	0.150	was kept constant			
w_C(total)	0.375	was kept constant			
w_D(total)	0.100	was kept constant			
P/bar	323	397	472	552	577
w_A(gel phase)	0.212	0.250	0.265	0.336	0.375
w_B(gel phase)	0.523	0.449	0.380	0.253	0.150
w_C(gel phase)	0.219	0.241	0.290	0.331	0.375
w_D(gel phase)	0.046	0.060	0.065	0.080	0.100
w_A(sol phase)	0.435	0.430	0.423	0.404	–
w_B(sol phase)	0.022	0.027	0.050	0.088	–
w_C(sol phase)	0.432	0.437	0.418	0.402	–
w_D(sol phase)	0.111	0.106	0.109	0.106	–

Polymer (B):	**poly(ethylene-*co*-vinyl acetate)**	**2001DOE2**
Characterization:	M_n/g.mol^{-1} = 61900, M_w/g.mol^{-1} = 167000, 27.5 wt% vinyl acetate	
Solvent (A):	**ethene** C_2H_4	**74-85-1**
Solvent (C):	**vinyl acetate** $C_4H_6O_2$	**108-05-4**
Solvent (D):	**carbon dioxide** CO_2	**124-38-9**

Type of data: coexistence data

T/K = 433.15

w_A(total)	0.375	was kept constant			
w_B(total)	0.150	was kept constant			
w_C(total)	0.375	was kept constant			
w_D(total)	0.100	was kept constant			
P/bar	366	491	576	674	695
w_A(gel phase)	0.181	0.247	0.300	–	0.375
w_B(gel phase)	0.589	0.455	0.353	–	0.150
w_C(gel phase)	0.180	0.230	0.270	–	0.375
w_D(gel phase)	0.050	0.068	0.083	–	0.100
w_A(sol phase)	0.440	0.430	0.409	0.396	–
w_B(sol phase)	0.000	0.009	0.034	0.071	–
w_C(sol phase)	0.448	0.446	0.451	0.428	–
w_D(sol phase)	0.112	0.115	0.106	0.105	–

Polymer (B):	**poly(ethylene-*co*-vinyl acetate)**		**2001DOE2**
Characterization:	M_n/g.mol^{-1} = 61900, M_w/g.mol^{-1} = 167000, 27.5 wt% vinyl acetate		
Solvent (A):	**ethene**	**C_2H_4**	**74-85-1**
Solvent (C):	**vinyl acetate**	**$C_4H_6O_2$**	**108-05-4**
Solvent (D):	**helium**	**He**	**7440-59-7**

Type of data: coexistence data

T/K = 433.15

w_A(total)	0.420	was kept constant		
w_B(total)	0.150	was kept constant		
w_C(total)	0.420	was kept constant		
w_D(total)	0.010	was kept constant		
P/bar	711	830	922	950
w_A(gel phase)	0.274	0.330	0.375	0.420
w_B(gel phase)	0.403	0.318	0.230	0.150
w_C(gel phase)	0.320	0.347	0.387	0.420
w_D(gel phase)	0.003	0.005	0.008	0.010
w_A(sol phase)	0.507	0.502	0.478	–
w_B(sol phase)	0.001	0.0135	0.051	–
w_C(sol phase)	0.478	0.471	0.458	–
w_D(sol phase)	0.014	0.0135	0.013	–

Polymer (B):	**poly(ethylene-*co*-vinyl acetate)**		**2001DOE2**
Characterization:	M_n/g.mol^{-1} = 61900, M_w/g.mol^{-1} = 167000, 27.5 wt% vinyl acetate		
Solvent (A):	**ethene**	**C_2H_4**	**74-85-1**
Solvent (C):	**vinyl acetate**	**$C_4H_6O_2$**	**108-05-4**
Solvent (D):	**methane**	**CH_4**	**74-82-8**

Type of data: coexistence data

T/K = 433.15

w_A(total)	0.375	was kept constant			
w_B(total)	0.150	was kept constant			
w_C(total)	0.375	was kept constant			
w_D(total)	0.100	was kept constant			
P/bar	477	647	765	850	872
w_A(gel phase)	0.194	0.252	0.291	0.335	0.375
w_B(gel phase)	0.540	0.450	0.360	0.250	0.150
w_C(gel phase)	0.220	0.242	0.280	0.335	0.375
w_D(gel phase)	0.045	0.057	0.069	0.080	0.100
w_A(sol phase)	0.468	0.465	0.451	0.426	–
w_B(sol phase)	0.014	0.027	0.055	0.085	–
w_C(sol phase)	0.392	0.386	0.374	0.378	–
w_D(sol phase)	0.126	0.122	0.120	0.111	–

continued

continued

T/K = 433.15

w_A(total)	0.5625	was kept constant			
w_B(total)	0.150	was kept constant			
w_C(total)	0.1875	was kept constant			
w_D(total)	0.100	was kept constant			
P/bar	760	857	974	1106	1139
w_A(gel phase)	0.274	0.297	0.369	0.450	0.5625
w_B(gel phase)	0.557	0.515	0.420	0.308	0.150
w_C(gel phase)	0.120	0.132	0.145	0.162	0.1875
w_D(gel phase)	0.049	0.056	0.066	0.080	0.100
w_A(sol phase)	0.670	0.668	0.660	0.649	–
w_B(sol phase)	0.001	0.005	0.014	0.029	–
w_C(sol phase)	0.210	0.209	0.209	0.208	–
w_D(sol phase)	0.119	0.118	0.117	0.114	–

T/K = 433.15

w_A(total)	0.1875	was kept constant	
w_B(total)	0.150	was kept constant	
w_C(total)	0.5625	was kept constant	
w_D(total)	0.100	was kept constant	
P/bar	519	626	640
w_A(gel phase)	0.156	0.183	0.1875
w_B(gel phase)	0.332	0.206	0.150
w_C(gel phase)	0.446	0.533	0.5625
w_D(gel phase)	0.066	0.088	0.100
w_A(sol phase)	0.215	0.204	
w_B(sol phase)	0.123	0.142	
w_C(sol phase)	0.575	0.570	
w_D(sol phase)	0.087	0.084	

Polymer (B): **poly(ethylene-*co*-vinyl acetate)** **2001DOE2**
Characterization: M_n/g.mol^{-1} = 61900, M_w/g.mol^{-1} = 167000, 27.5 wt% vinyl acetate
Solvent (A): **ethene** C_2H_4 **74-85-1**
Solvent (C): **vinyl acetate** $C_4H_6O_2$ **108-05-4**
Solvent (D): **nitrogen** N_2 **7727-37-9**

Type of data: coexistence data

T/K = 433.15

w_A(total)	0.375	was kept constant
w_B(total)	0.150	was kept constant
w_C(total)	0.375	was kept constant
w_D(total)	0.100	was kept constant

continued

continued

P/bar	676	790	893	992	1011
w_A(gel phase)	0.240	0.256	0.296	0.332	0.375
w_B(gel phase)	0.455	0.410	0.341	0.238	0.150
w_C(gel phase)	0.252	0.270	0.291	0.340	0.375
w_D(gel phase)	0.053	0.064	0.072	0.090	0.100
w_A(sol phase)	0.440	0.435	0.429		
w_B(sol phase)	0.009	0.017	0.030		
w_C(sol phase)	0.427	0.426	0.421		
w_D(sol phase)	0.124	0.122	0.120		

Polymer (B): **poly(ethylene glycol)** **2006BY4**
Characterization: M_w/g.mol^{-1} = 2000, SFC Co., Yeosu, South Korea
Solvent (A): **carbon dioxide** CO_2 **124-38-9**
Solvent (C): **diethylene glycol** $C_4H_{10}O_3$ **111-46-6**

Type of data: cloud points

w_A	0.90	0.90	0.90	0.90	0.90	0.90	0.90	0.80	0.80
w_B	0.05	0.05	0.05	0.05	0.05	0.05	0.05	0.05	0.05
w_C	0.05	0.05	0.05	0.05	0.05	0.05	0.05	0.15	0.15
T/K	378.15	381.15	384.25	390.75	404.05	424.05	443.55	391.25	393.55
P/bar	2105.2	1968.6	1846.6	1742.4	1555.5	1419.7	1356.2	2197.6	2067.2

w_A	0.80	0.80	0.80	0.80	0.80	0.80
w_B	0.05	0.05	0.05	0.05	0.05	0.05
w_C	0.15	0.15	0.15	0.15	0.15	0.15
T/K	395.85	398.95	402.75	407.05	424.25	444.65
P/bar	1939.7	1818.3	1666.6	1512.1	1301.7	1190.7

Polymer (B): **poly(ethylene glycol)** **2006MAT**
Characterization: M_n/g.mol^{-1} = 1075, M_w/g.mol^{-1} = 1125, T_m/K = 303-313, Wako Pure Chemical Industry Co. Ltd., Japan
Solvent (A): **carbon dioxide** CO_2 **124-38-9**
Solvent (C): **ethanol** C_2H_6O **64-17-5**

Type of data: cloud points

T/K = 313.15

w_A	0.555	0.633	0.710	0.789	0.826	0.906
w_B	0.268	0.145	0.0875	0.0423	0.0171	0.00493
w_C	0.177	0.222	0.202	0.169	0.157	0.089
P/MPa	15	15	15	15	15	15

Polymer (B): **poly(ethylene glycol)** **2006MAT**
Characterization: M_n/g.mol^{-1} = 8500, M_w/g.mol^{-1} = 8800, T_m/K = 323-338, Wako Pure Chemical Industry Co. Ltd., Japan
Solvent (A): **carbon dioxide** **CO_2** **124-38-9**
Solvent (C): **ethanol** **C_2H_6O** **64-17-5**

Type of data: cloud points

T/K = 313.15

w_A	0.553	0.612	0.654	0.684	0.563	0.600	0.654	0.674	0.739
w_B	0.111	0.0646	0.0178	0.00896	0.146	0.100	0.0577	0.0404	0.0146
w_C	0.336	0.323	0.329	0.307	0.291	0.300	0.288	0.286	0.246
P/MPa	10	10	10	10	15	15	15	15	15

w_A	0.751	0.778	0.602	0.689	0.700	0.764
w_B	0.00726	0.00155	0.122	0.0544	0.0504	0.00612
w_C	0.242	0.220	0.276	0.256	0.250	0.230
P/MPa	15	15	20	20	20	20

Type of data: cloud points

w_A	0.633	0.633	0.633	0.633	0.708	0.708	0.708	0.708	0.708
w_B	0.0183	0.0183	0.0183	0.0183	0.0146	0.0146	0.0146	0.0146	0.0146
w_C	0.349	0.349	0.349	0.349	0.277	0.277	0.277	0.277	0.277
T/K	315.2	319.0	321.4	324.9	300.7	306.0	313.3	318.4	322.2
P/MPa	8.6	9.9	10.9	12.0	8.2	9.9	12.7	14.8	16.3

w_A	0.736	0.736	0.736	0.736	0.736
w_B	0.0132	0.0132	0.0132	0.0132	0.0132
w_C	0.251	0.251	0.251	0.251	0.251
T/K	304.0	308.7	313.7	317.5	321.4
P/MPa	11.5	12.9	15.1	16.0	17.6

Polymer (B): **poly(ethylene glycol)** **2006MAT**
Characterization: M_n/g.mol^{-1} = 19100, M_w/g.mol^{-1} = 21000, T_m/K = 329-336, Wako Pure Chemical Industry Co. Ltd., Japan
Solvent (A): **carbon dioxide** **CO_2** **124-38-9**
Solvent (C): **ethanol** **C_2H_6O** **64-17-5**

Type of data: cloud points

T/K = 313.15

w_A	0.618	0.676	0.705	0.727
w_B	0.0751	0.0320	0.0149	0.00503
w_C	0.307	0.292	0.280	0.268
P/MPa	15	15	15	15

Polymer (B): **poly(ethylene oxide-*b*-1,1,2,2-tetrahydroperfluorodecyl acrylate)** **2004MA2**
Characterization: M_n/g.mol^{-1} = 22800, 10.3 wt% PEO, molar ratio of PFDA/PEO = 39.5/1 by NMR, synthesized in the laboratory
Solvent (A): **carbon dioxide** **CO_2** **124-38-9**
Solvent (C): **2-hydroxyethyl methacrylate** **$C_6H_{10}O_3$** **868-77-9**

Type of data: cloud points

w_A	0.868	0.868	0.868	0.868	0.868	0.868	0.868	0.868	0.868
w_B	0.012	0.012	0.012	0.012	0.012	0.012	0.012	0.012	0.012
w_C	0.120	0.120	0.120	0.120	0.120	0.120	0.120	0.120	0.120
T/K	338.15	332.85	328.05	322.95	318.45	313.55	308.55	303.85	298.75
P/bar	251.0	230.0	213.0	189.0	171.0	152.0	132.0	113.0	93.0

w_A	0.868
w_B	0.012
w_C	0.120
T/K	294.05
P/bar	76.0

Polymer (B): **poly(2-ethylhexyl acrylate)** **2007LI1**
Characterization: M_w/g.mol^{-1} = 90000, T_g/K = 223, Scientific Polymer Products, Inc., Ontario, NY
Solvent (A): **carbon dioxide** **CO_2** **124-38-9**
Solvent (C): **dimethyl ether** **C_2H_6O** **115-10-6**

Type of data: cloud points

w_A	0.887	0.887	0.887	0.887	0.887	0.850	0.850	0.850	0.850
w_B	0.052	0.052	0.052	0.052	0.052	0.051	0.051	0.051	0.051
w_C	0.061	0.061	0.061	0.061	0.061	0.099	0.099	0.099	0.099
T/K	395.4	409.7	424.9	439.4	455.0	373.2	393.8	414.2	434.1
P/MPa	165.69	128.97	113.45	106.38	101.65	179.48	126.14	105.76	97.90

w_A	0.850	0.785	0.785	0.785	0.785	0.785	0.743	0.743	0.743
w_B	0.051	0.054	0.054	0.054	0.054	0.054	0.050	0.050	0.050
w_C	0.099	0.161	0.161	0.161	0.161	0.161	0.207	0.207	0.207
T/K	455.6	274.6	393.3	414.9	433.1	455.0	354.4	374.8	394.6
P/MPa	94.28	128.69	110.17	91.55	87.76	88.10	111.17	87.07	80.48

w_A	0.743	0.743	0.743	0.653	0.653	0.653	0.653	0.653	0.653
w_B	0.050	0.050	0.050	0.051	0.051	0.051	0.051	0.051	0.051
w_C	0.207	0.207	0.207	0.296	0.296	0.296	0.296	0.296	0.296
T/K	414.1	435.2	455.5	335.5	354.0	373.3	395.7	413.5	434.2
P/MPa	78.10	77.59	77.59	72.17	63.55	61.52	62.10	63.03	64.31

continued

continued

w_A	0.653	0.543	0.543	0.543	0.543	0.543	0.543	0.543
w_B	0.051	0.053	0.053	0.053	0.053	0.053	0.053	0.053
w_C	0.296	0.404	0.404	0.404	0.404	0.404	0.404	0.404
T/K	455.8	333.4	354.1	374.5	393.5	413.6	433.9	454.9
P/MPa	65.93	35.35	38.10	41.28	44.45	47.59	50.21	52.52

Polymer (B): **poly(2-ethylhexyl acrylate)** **2007LI1**
Characterization: M_w/g.mol^{-1} = 90000, T_g/K = 223,
Scientific Polymer Products, Inc., Ontario, NY
Solvent (A): **carbon dioxide** CO_2 **124-38-9**
Solvent (C): **2-ethylhexyl acrylate** $C_{11}H_{20}O_2$ **103-11-7**

Type of data: cloud points

w_A	0.847	0.847	0.847	0.847	0.738	0.738	0.738	0.738	0.738
w_B	0.056	0.056	0.056	0.056	0.056	0.056	0.056	0.056	0.056
w_C	0.097	0.097	0.097	0.097	0.206	0.206	0.206	0.206	0.206
T/K	393.9	414.7	434.7	454.2	375.0	393.8	413.7	433.6	454.7
P/MPa	227.76	119.28	106.55	100.69	193.28	81.69	79.00	78.10	78.10
w_A	0.545	0.545	0.545	0.545	0.545	0.545	0.545	0.498	0.498
w_B	0.046	0.046	0.046	0.046	0.046	0.046	0.046	0.050	0.050
w_C	0.409	0.409	0.409	0.409	0.409	0.409	0.409	0.552	0.552
T/K	336.3	355.0	376.0	396.0	414.1	434.7	454.7	305.0	315.1
P/MPa	131.21	74.72	68.79	66.83	66.66	67.17	67.97	49.38	43.14
w_A	0.498	0.498	0.498	0.498	0.498	0.498	0.498	0.339	0.339
w_B	0.050	0.050	0.050	0.050	0.050	0.050	0.050	0.057	0.057
w_C	0.552	0.552	0.552	0.552	0.552	0.552	0.552	0.604	0.604
T/K	335.1	354.8	374.7	394.6	413.5	434.8	454.9	313.1	334.0
P/MPa	41.07	42.76	45.31	47.83	49.97	52.31	53.66	27.76	30.14
w_A	0.339	0.339	0.339	0.339	0.339	0.339	0.235	0.235	0.235
w_B	0.057	0.057	0.057	0.057	0.057	0.057	0.056	0.056	0.056
w_C	0.604	0.604	0.604	0.604	0.604	0.604	0.709	0.709	0.709
T/K	353.5	374.7	393.7	414.2	433.7	453.6	315.5	323.3	334.7
P/MPa	35.55	37.35	40.17	43.21	46.03	47.83	9.83	10.86	12.86
w_A	0.235	0.235	0.235	0.235	0.235	0.235			
w_B	0.056	0.056	0.056	0.056	0.056	0.056			
w_C	0.709	0.709	0.709	0.709	0.709	0.709			
T/K	353.4	373.0	394.3	414.2	434.6	455.4			
P/MPa	17.31	20.59	24.38	27.69	31.07	33.28			

Type of data: vapor-liquid equilibrium data

w_A	0.235	0.235
w_B	0.056	0.056
w_C	0.709	0.709
T/K	303.4	308.2
P/MPa	7.76	8.11

Polymer (B): **poly(2-ethylhexyl methacrylate)** **2007LI1**
Characterization: M_w/g.mol^{-1} = 100000, T_g/K = 263,
Scientific Polymer Products, Inc., Ontario, NY
Solvent (A): **carbon dioxide** CO_2 **124-38-9**
Solvent (C): **dimethyl ether** C_2H_6O **115-10-6**

Type of data: cloud points

w_A	0.896	0.896	0.896	0.896	0.757	0.757	0.757	0.757	0.757
w_B	0.051	0.051	0.051	0.051	0.051	0.051	0.051	0.051	0.051
w_C	0.053	0.053	0.053	0.053	0.192	0.192	0.192	0.192	0.192
T/K	408.9	424.2	439.6	454.7	367.5	373.6	393.2	413.7	434.7
P/MPa	234.66	139.48	127.24	118.03	191.90	164.31	113.97	102.10	95.59
w_A	0.757	0.689	0.689	0.689	0.689	0.689	0.689	0.689	0.646
w_B	0.051	0.050	0.050	0.050	0.050	0.050	0.050	0.050	0.051
w_C	0.192	0.261	0.261	0.261	0.261	0.261	0.261	0.261	0.303
T/K	454.3	333.1	353.8	374.1	394.8	414.3	434.7	454.0	333.3
P/MPa	94.79	93.97	74.66	69.83	69.14	69.83	70.52	71.90	73.28
w_A	0.646	0.646	0.646	0.646	0.646	0.646	0.552	0.552	0.552
w_B	0.051	0.051	0.051	0.051	0.051	0.051	0.050	0.050	0.050
w_C	0.303	0.303	0.303	0.303	0.303	0.303	0.398	0.398	0.398
T/K	353.9	374.3	393.6	413.8	434.2	453.2	333.8	354.9	374.4
P/MPa	64.69	62.93	63.31	64.51	65.90	67.07	46.38	46.86	48.69
w_A	0.552	0.552	0.552	0.552					
w_B	0.050	0.050	0.050	0.050					
w_C	0.398	0.398	0.398	0.398					
T/K	392.8	414.6	433.0	453.7					
P/MPa	50.86	53.41	55.90	58.35					

Polymer (B): **poly(2-ethylhexyl methacrylate)** **2007LI1**
Characterization: M_w/g.mol^{-1} = 100000, T_g/K = 263,
Scientific Polymer Products, Inc., Ontario, NY
Solvent (A): **carbon dioxide** CO_2 **124-38-9**
Solvent (C): **2-ethylhexyl methacrylate** $C_{12}H_{22}O_2$ **688-84-6**

Type of data: cloud points

w_A	0.866	0.866	0.866	0.866	0.866	0.866	0.866	0.749	0.749
w_B	0.047	0.047	0.047	0.047	0.047	0.047	0.047	0.050	0.050
w_C	0.087	0.087	0.087	0.087	0.087	0.087	0.087	0.201	0.201
T/K	393.2	404.0	414.6	425.2	434.5	444.6	454.5	339.8	354.3
P/MPa	213.97	147.59	125.00	115.31	110.86	106.62	105.86	165.69	91.69

continued

continued

w_A	0.749	0.749	0.749	0.749	0.749	0.623	0.623	0.623	0.623
w_B	0.050	0.050	0.050	0.050	0.050	0.074	0.074	0.074	0.074
w_C	0.201	0.201	0.201	0.201	0.201	0.303	0.303	0.303	0.303
T/K	373.5	395.2	414.3	434.5	455.2	333.6	354.3	375.0	395.5
P/MPa	80.00	72.93	71.59	70.62	69.35	56.38	52.83	53.48	54.72
w_A	0.623	0.623	0.623	0.385	0.385	0.385	0.385	0.385	0.385
w_B	0.074	0.074	0.074	0.051	0.051	0.051	0.051	0.051	0.051
w_C	0.303	0.303	0.303	0.564	0.564	0.564	0.564	0.564	0.564
T/K	415.8	435.4	454.3	335.8	352.8	372.7	394.3	413.9	429.2
P/MPa	56.52	58.00	59.35	23.14	27.28	31.41	35.35	38.45	41.07
w_A	0.385	0.282	0.282	0.282	0.282	0.282	0.282		
w_B	0.051	0.052	0.052	0.052	0.052	0.052	0.052		
w_C	0.564	0.666	0.666	0.666	0.666	0.666	0.666		
T/K	455.5	356.5	375.4	395.4	414.0	432.5	455.0		
P/MPa	44.10	18.45	21.90	25.69	28.10	30.52	32.93		

Type of data: vapor-liquid equilibrium data

w_A	0.282	0.282	0.282
w_B	0.052	0.052	0.052
w_C	0.666	0.666	0.666
T/K	314.7	325.0	335.5
P/MPa	9.68	11.55	13.62

Type of data: vapor-liquid-liquid equilibrium data

w_A	0.282
w_B	0.052
w_C	0.666
T/K	359.0
P/MPa	17.20

Polymer (B): **polyglycerol (hyperbranched)** **2009KOZ**
Characterization: M_n/g.mol^{-1} = 2700, M_w/g.mol^{-1} = 4050, viscous liquid possessing an inert polyether scaffold, HyperPolymers GmbH
Solvent (A): **carbon dioxide** CO_2 **124-38-9**
Solvent (C): **methanol** CH_4O **67-56-1**

Type of data: cloud points

w_A	0.100	0.100	0.100	0.100	0.100	0.100	0.100	0.100	0.150
w_B	0.225	0.225	0.225	0.450	0.450	0.450	0.450	0.450	0.2125
w_C	0.675	0.675	0.675	0.450	0.450	0.450	0.450	0.450	0.6375
T/K	450.44	452.91	452.95	428.04	430.54	433.04	435.53	438.03	418.04
P/MPa	8.13	9.87	8.29	11.06	11.67	12.51	13.08	13.90	9.29

continued

continued

w_A	0.150	0.150	0.150	0.150	0.150
w_B	0.2125	0.425	0.425	0.425	0.425
w_C	0.6375	0.425	0.425	0.425	0.425
T/K	423.04	343.00	353.04	363.06	373.05
P/MPa	10.67	7.24	9.60	11.98	14.42

Type of data: vapor-liquid-liquid equilibrium data

w_A	0.150	0.150	0.150	0.150	0.150	0.150	0.150	0.150	0.150
w_B	0.2125	0.2125	0.2125	0.425	0.425	0.425	0.425	0.425	0.425
w_C	0.6375	0.6375	0.6375	0.425	0.425	0.425	0.425	0.425	0.425
T/K	418.05	423.03	428.02	343.00	353.04	363.06	373.05	383.04	393.01
P/MPa	8.87	9.14	9.39	7.18	8.08	8.95	9.80	10.60	11.33

w_A	0.150	0.150	0.150	0.150
w_B	0.425	0.425	0.425	0.425
w_C	0.425	0.425	0.425	0.425
T/K	403.05	413.08	423.05	433.05
P/MPa	12.01	12.64	13.20	13.68

Comments: Vapor-liquid equilibrium data are given in Chapter2.

Polymer (B): **polyglycerol (hyperbranched)** **2010SCH**
Characterization: M_n/g.mol^{-1} = 5700, M_w/M_n = 1.7, synthesized in the laboratory
Solvent (A): **carbon dioxide** CO_2 **124-38-9**
Solvent (C): **methanol** CH_4O **67-56-1**

Type of data: cloud points

w_A	0.100	0.100
w_B	0.225	0.225
w_C	0.675	0.675
T/K	433.00	442.95
P/MPa	8.156	10.998

Type of data: vapor-liquid-liquid equilibrium data

Comments: Phase boundary vapor–liquid to liquid–liquid–vapor (LLV) and phase boundary liquid–liquid–vapor to liquid–liquid (LVL).

w_A	0.100	0.100	0.100	0.100	0.100	0.100	0.100	0.100	0.100
w_B	0.449	0.449	0.449	0.449	0.449	0.449	0.449	0.449	0.449
w_C	0.451	0.451	0.451	0.451	0.451	0.451	0.451	0.451	0.451
T/K	393.04	403.07	408.02	413.02	423.02	433.05	442.99	452.85	403.02
P/MPa	8.053	8.198	8.198	8.239	8.179	8.059	8.039	7.799	8.873
	(LLV)	(LLV)	(LLV)	(LLV)	(LLV)	(LLV)	(LLV)	(LVL)	(LVL)

continued

continued

w_A	0.100	0.100	0.100	0.100	0.100	0.100	0.150	0.150	0.150
w_B	0.449	0.449	0.449	0.449	0.449	0.449	0.424	0.424	0.424
w_C	0.451	0.451	0.451	0.451	0.451	0.451	0.426	0.426	0.426
T/K	408.02	413.02	423.02	433.05	443.00	452.87	333.00	343.01	353.04
P/MPa	9.134	9.439	9.974	10.505	11.050	11.476	5.281	5.917	6.517
	(LVL)	(LVL)	(LVL)	(LVL)	(LVL)	(LVL)	(LLV)	(LLV)	(LLV)
w_A	0.150	0.150	0.150	0.150	0.150	0.150	0.150	0.150	0.150
w_B	0.424	0.424	0.424	0.424	0.424	0.424	0.424	0.424	0.424
w_C	0.426	0.426	0.426	0.426	0.426	0.426	0.426	0.426	0.426
T/K	363.05	373.07	383.05	393.18	403.13	333.00	343.01	353.04	363.05
P/MPa	7.058	7.518	7.899	8.179	8.340	6.271	7.132	8.012	8.853
	(LLV)	(LLV)	(LLV)	(LLV)	(LLV)	(LVL)	(LVL)	(LVL)	(LVL)
w_A	0.150	0.150	0.150	0.100	0.100	0.100	0.100	0.100	0.100
w_B	0.424	0.424	0.424	0.225	0.225	0.225	0.225	0.225	0.225
w_C	0.426	0.426	0.426	0.675	0.675	0.675	0.675	0.675	0.675
T/K	373.07	393.18	403.13	433.00	442.97	452.92	433.00	442.96	452.89
P/MPa	9.684	11.239	11.855	6.750	6.391	5.952	7.105	7.606	8.141
	(LVL)	(LVL)	(LVL)	(LLV)	(LLV)	(LLV)	(LVL)	(LVL)	(LVL)
w_A	0.150	0.150	0.150	0.150	0.150	0.150	0.150	0.150	0.150
w_B	0.2125	0.2125	0.2125	0.2125	0.2125	0.2125	0.2125	0.2125	0.2125
w_C	0.6375	0.6375	0.6375	0.6375	0.6375	0.6375	0.6375	0.6375	0.6375
T/K	332.73	342.10	351.70	351.81	361.53	371.26	371.29	380.93	380.99
P/MPa	4.234	4.769	5.303	5.314	5.878	6.451	6.440	6.996	6.985
	(LVL)	(LVL)	(LVL)	(LVL)	(LVL)	(LVL)	(LVL)	(LVL)	(LVL)
w_A	0.150	0.150	0.150	0.150	0.150	0.150	0.150		
w_B	0.2125	0.2125	0.2125	0.2125	0.2125	0.2125	0.2125		
w_C	0.6375	0.6375	0.6375	0.6375	0.6375	0.6375	0.6375		
T/K	390.59	400.20	409.83	419.46	428.38	438.76	448.40		
P/MPa	7.508	8.017	8.522	9.012	9.492	9.977	10.448		
	(LVL)	(LVL)	(LVL)	(LVL)	(LVL)	(LVL)	(LVL)		

Polymer (B): **polyglycerol (hyperbranched)** **2010SCH**
Characterization: M_n/g.mol^{-1} = 10000, M_w/M_n = 1.4, synthesized in the laboratory
Solvent (A): **carbon dioxide** CO_2 **124-38-9**
Solvent (C): **methanol** CH_4O **67-56-1**

Type of data: cloud points

w_A	0.047	0.098	0.098	0.098	0.098	0.098	0.047	0.047	0.047
w_B	0.4775	0.452	0.452	0.452	0.452	0.452	0.2363	0.2363	0.2363
w_C	0.4755	0.450	0.450	0.450	0.450	0.450	0.7167	0.7167	0.7167
T/K	426.47	313.29	318.18	323.11	328.08	347.87	436.16	440.19	444.94
P/MPa	7.175	7.208	7.724	8.264	8.830	11.416	5.256	6.477	7.778

continued

continued

w_A	0.047	0.098	0.098	0.098
w_B	0.2363	0.2237	0.2237	0.2237
w_C	0.7167	0.6783	0.6783	0.6783
T/K	450.01	392.31	402.22	412.13
P/MPa	9.319	6.914	9.071	11.372

Type of data: vapor-liquid-liquid equilibrium data

Comments: Phase boundary vapor–liquid to liquid–liquid–vapor (LLV) and phase boundary liquid–liquid–vapor to liquid–liquid (LVL).

w_A	0.047	0.047	0.098	0.098	0.098	0.098	0.098	0.098	0.098
w_B	0.4775	0.4775	0.452	0.452	0.452	0.452	0.452	0.452	0.452
w_C	0.4755	0.4755	0.450	0.450	0.450	0.450	0.450	0.450	0.450
T/K	426.47	431.42	313.30	318.19	323.18	333.06	342.97	347.91	352.87
P/MPa	6.560	6.835	3.311	3.592	3.837	4.412	4.938	5.228	5.478
	(LLV)	(LLV)	(LLV)	(LLV)	(LLV)	(LLV)	(LLV)	(LLV)	(LLV)
w_A	0.098	0.098	0.098	0.098	0.098	0.098	0.098	0.098	0.098
w_B	0.452	0.452	0.452	0.452	0.452	0.452	0.452	0.452	0.452
w_C	0.450	0.450	0.450	0.450	0.450	0.450	0.450	0.450	0.450
T/K	357.83	362.77	372.65	382.53	392.45	402.35	412.26	422.17	432.10
P/MPa	5.733	5.959	6.364	6.699	6.955	7.105	7.165	7.176	7.036
	(LLV)	(LLV)	(LLV)	(LLV)	(LLV)	(LLV)	(LLV)	(LLV)	(LLV)
w_A	0.098	0.098	0.098	0.098	0.098	0.098	0.098	0.098	0.098
w_B	0.452	0.452	0.452	0.452	0.452	0.452	0.452	0.452	0.452
w_C	0.450	0.450	0.450	0.450	0.450	0.450	0.450	0.450	0.450
T/K	313.31	318.20	323.16	333.08	342.97	347.92	352.87	357.82	362.78
P/MPa	3.541	3.832	4.157	4.797	5.463	5.798	6.148	6.479	6.829
	(LVL)	(LVL)	(LVL)	(LVL)	(LVL)	(LVL)	(LVL)	(LVL)	(LVL)
w_A	0.098	0.098	0.098	0.098	0.098	0.098	0.098	0.151	0.151
w_B	0.452	0.452	0.452	0.452	0.452	0.452	0.452	0.425	0.425
w_C	0.450	0.450	0.450	0.450	0.450	0.450	0.450	0.424	0.424
T/K	372.65	382.54	392.45	402.36	412.26	422.17	432.09	332.14	342.07
P/MPa	7.499	8.160	8.805	9.426	10.021	10.592	11.137	6.715	7.665
	(LVL)	(LVL)	(LVL)	(LVL)	(LVL)	(LVL)	(LVL)	(LVL)	(LVL)
w_A	0.151	0.151	0.151	0.151	0.047	0.047	0.047	0.047	0.047
w_B	0.425	0.425	0.425	0.425	0.2363	0.2363	0.2363	0.2363	0.2363
w_C	0.424	0.424	0.424	0.424	0.7167	0.7167	0.7167	0.7167	0.7167
T/K	351.83	361.84	371.72	380.93	436.19	440.24	444.92	450.03	436.21
P/MPa	8.572	9.492	10.378	11.199	5.046	4.861	4.677	4.418	5.111
	(LVL)	(LVL)	(LVL)	(LVL)	(LLV)	(LLV)	(LLV)	(LLV)	(LVL)

continued

continued

w_A	0.047	0.047	0.047	0.098	0.098	0.098	0.098	0.098	0.098
w_B	0.2363	0.2363	0.2363	0.2237	0.2237	0.2237	0.2237	0.2237	0.2237
w_C	0.7167	0.7167	0.7167	0.6783	0.6783	0.6783	0.6783	0.6783	0.6783
T/K	440.23	444.90	450.03	392.32	402.25	412.16	422.07	432.04	441.93
P/MPa	5.306	5.552	5.827	5.404	5.429	5.379	5.215	4.946	4.598
	(LVL)	(LVL)	(LVL)	(LLV)	(LLV)	(LLV)	(LLV)	(LLV)	(LLV)

w_A	0.098	0.098	0.098	0.098	0.098	0.098	0.098	0.098	0.150
w_B	0.2237	0.2237	0.2237	0.2237	0.2237	0.2237	0.2237	0.2237	0.211
w_C	0.6783	0.6783	0.6783	0.6783	0.6783	0.6783	0.6783	0.6783	0.639
T/K	451.87	422.07	432.03	441.96	451.88	392.31	402.22	412.13	313.11
P/MPa	4.231	5.674	6.144	6.614	7.100	7.605	8.126	8.671	3.403
	(LLV)	(LVL)	(LVL)	(LVL)	(LVL)	(LVL)	(LVL)	(LVL)	(LVL)

w_A	0.150	0.150	0.150	0.150	0.150	0.150	0.150	0.150	0.150
w_B	0.211	0.211	0.211	0.211	0.211	0.211	0.211	0.211	0.211
w_C	0.639	0.639	0.639	0.639	0.639	0.639	0.639	0.639	0.639
T/K	322.96	332.86	342.77	352.68	357.68	362.59	372.46	382.37	392.27
P/MPa	3.968	4.574	5.189	5.820	6.140	6.450	7.061	7.666	8.257
	(LVL)	(LVL)	(LVL)	(LVL)	(LVL)	(LVL)	(LVL)	(LVL)	(LVL)

w_A	0.150	0.150	0.150	0.150	0.150	0.150	0.150	0.150	0.150
w_B	0.211	0.211	0.211	0.211	0.211	0.211	0.211	0.211	0.211
w_C	0.639	0.639	0.639	0.639	0.639	0.639	0.639	0.639	0.639
T/K	402.16	412.08	421.99	431.95	441.89	313.07	322.94	332.88	342.77
P/MPa	8.832	9.383	9.923	10.443	10.969	2.809	3.304	3.70	4.14
	(LVL)	(LVL)	(LVL)	(LVL)	(LLV)	(LLV)	(LLV)	(LLV)	(LLV)

w_A	0.150	0.150	0.150	0.150	0.150	0.150	0.150	0.150	0.150
w_B	0.211	0.211	0.211	0.211	0.211	0.211	0.211	0.211	0.211
w_C	0.639	0.639	0.639	0.639	0.639	0.639	0.639	0.639	0.639
T/K	352.68	357.66	362.58	372.43	382.37	392.27	402.15	412.09	421.99
P/MPa	4.56	4.77	4.93	5.23	5.46	5.60	5.66	5.60	5.45
	(LLV)	(LLV)	(LLV)	(LLV)	(LLV)	(LLV)	(LLV)	(LLV)	(LLV)

w_A	0.150	0.150
w_B	0.211	0.211
w_C	0.639	0.639
T/K	431.94	441.87
P/MPa	5.18	4.85
	(LLV)	(LLV)

Polymer (B): **polyglycerol (hyperbranched)** **2010SCH**
Characterization: M_n/g.mol^{-1} = 18000, M_w/M_n = 1.4, synthesized in the laboratory
Solvent (A): **carbon dioxide** CO_2 **124-38-9**
Solvent (C): **methanol** CH_4O **67-56-1**

Type of data: cloud points

continued

continued

w_A	0.055	0.055	0.049	0.049	0.049	0.102	0.102	0.102	0.102
w_B	0.4716	0.4716	0.2378	0.2378	0.2378	0.2245	0.2245	0.2245	0.2245
w_C	0.4734	0.4734	0.7132	0.7132	0.7132	0.6735	0.6735	0.6735	0.6735
T/K	427.06	431.97	427.27	429.85	444.53	356.06	363.35	370.72	378.11
P/MPa	4.826	5.066	5.046	4.861	4.677	4.788	5.748	7.165	8.350

w_A	0.102
w_B	0.2245
w_C	0.6735
T/K	385.44
P/MPa	9.882

Type of data: vapor-liquid-liquid equilibrium data

Comments: Phase boundary vapor–liquid to liquid–liquid–vapor (LLV) and phase boundary liquid–liquid–vapor to liquid–liquid (LVL).

w_A	0.055	0.055	0.055	0.055	0.055	0.105	0.105	0.105	0.105
w_B	0.4716	0.4716	0.4716	0.4716	0.4716	0.4466	0.4466	0.4466	0.4466
w_C	0.4734	0.4734	0.4734	0.4734	0.4734	0.4484	0.4484	0.4484	0.4484
T/K	427.06	431.97	446.92	431.97	446.92	313.08	327.91	335.36	342.76
P/MPa	2.949	3.269	3.625	4.015	4.380	3.182	3.952	4.348	4.733
	(LLV)	(LLV)	(LLV)	(LVL)	(LVL)	(LLV)	(LLV)	(LLV)	(LLV)

w_A	0.105	0.105	0.105	0.105	0.105	0.105	0.105	0.105	0.105
w_B	0.4466	0.4466	0.4466	0.4466	0.4466	0.4466	0.4466	0.4466	0.4466
w_C	0.4484	0.4484	0.4484	0.4484	0.4484	0.4484	0.4484	0.4484	0.4484
T/K	357.62	372.43	387.31	402.16	313.09	327.90	335.37	342.76	357.62
P/MPa	5.449	6.064	6.515	6.790	3.686	4.652	5.147	5.683	6.749
	(LLV)	(LLV)	(LLV)	(LLV)	(LVL)	(LVL)	(LVL)	(LVL)	(LVL)

w_A	0.105	0.105	0.105	0.145	0.145	0.145	0.145	0.145	0.145
w_B	0.4466	0.4466	0.4466	0.4266	0.4266	0.4266	0.4266	0.4266	0.4266
w_C	0.4484	0.4484	0.4484	0.4284	0.4284	0.4284	0.4284	0.4284	0.4284
T/K	372.43	387.32	402.15	311.99	319.34	326.59	341.11	348.53	355.54
P/MPa	7.789	8.810	9.796	4.665	5.256	5.896	7.222	7.867	8.548
	(LVL)	(LVL)	(LVL)	(LVL)	(LVL)	(LVL)	(LVL)	(LVL)	(LVL)

w_A	0.145	0.145	0.145	0.049	0.049	0.049	0.049	0.049	0.049
w_B	0.4266	0.4266	0.4266	0.2378	0.2378	0.2378	0.2378	0.2378	0.2378
w_C	0.4284	0.4284	0.4284	0.7132	0.7132	0.7132	0.7132	0.7132	0.7132
T/K	370.20	377.58	384.86	427.28	429.85	444.50	427.24	429.89	444.54
P/MPa	9.828	10.395	11.039	3.269	3.625	4.015	4.380	4.826	5.066
	(LVL)	(LVL)	(LVL)	(LLV)	(LLV)	(LLV)	(LVL)	(LVL)	(LVL)

w_A	0.102	0.102	0.102	0.102	0.102	0.102	0.102	0.102	0.102
w_B	0.2245	0.2245	0.2245	0.2245	0.2245	0.2245	0.2245	0.2245	0.2245
w_C	0.6735	0.6735	0.6735	0.6735	0.6735	0.6735	0.6735	0.6735	0.6735
T/K	356.09	363.35	370.73	378.11	385.45	400.12	414.94	422.24	356.09
P/MPa	4.108	4.337	4.499	4.663	4.745	4.807	4.664	4.519	4.188
	(LLV)	(LLV)	(LLV)	(LLV)	(LLV)	(LLV)	(LLV)	(LLV)	(LVL)

continued

continued

w_A	0.102	0.102	0.102	0.102	0.102	0.102	0.102	0.102	0.102
w_B	0.2245	0.2245	0.2245	0.2245	0.2245	0.2245	0.2245	0.2245	0.2245
w_C	0.6735	0.6735	0.6735	0.6735	0.6735	0.6735	0.6735	0.6735	0.6735
T/K	363.31	370.73	378.11	385.47	400.16	414.86	422.19	429.66	444.34
P/MPa	4.512	4.864	5.208	5.554	6.167	6.982	7.340	7.727	8.508
	(LVL)	(LVL)	(LVL)	(LVL)	(LVL)	(LVL)	(LVL)	(LVL)	(LVL)
w_A	0.151	0.151	0.151	0.151	0.151	0.151	0.151	0.151	0.151
w_B	0.2123	0.2123	0.2123	0.2123	0.2123	0.2123	0.2123	0.2123	0.2123
w_C	0.6367	0.6367	0.6367	0.6367	0.6367	0.6367	0.6367	0.6367	0.6367
T/K	311.99	319.34	326.59	341.11	348.53	355.54	370.20	377.58	384.86
P/MPa	4.665	5.256	5.896	7.222	7.867	8.548	9.828	10.395	11.039
	(LVL)	(LVL)	(LVL)	(LVL)	(LVL)	(LVL)	(LVL)	(LVL)	(LVL)

Polymer (B): **polyglycerol (linear)** **2011SCH**
Characterization: M_n/g.mol^{-1} = 4800, M_w/M_n = 1.2, synthesized in the laboratory
Solvent (A): **carbon dioxide** CO_2 **124-38-9**
Solvent (C): **methanol** CH_4O **67-56-1**

Type of data: cloud points

w_A	0.100	0.100
w_B	0.451	0.451
w_C	0.449	0.449
T/K	406.15	411.08
P/MPa	11.026	12.462

Type of data: vapor-liquid-liquid equilibrium data

Comments: Phase boundary vapor–liquid to liquid–liquid–vapor (LLV) and phase boundary liquid–liquid–vapor to liquid–liquid (LVL).

w_A	0.100	0.100	0.100	0.100
w_B	0.451	0.451	0.451	0.451
w_C	0.449	0.449	0.449	0.449
T/K	406.15	411.08	406.15	411.08
P/MPa	9.778	9.916	10.100	10.391
	(LLV)	(LLV)	(LVL)	(LVL)

Polymer (B): **polyglycerol (linear)** **2011SCH**
Characterization: M_n/g.mol^{-1} = 7800, M_w/M_n = 1.3, synthesized in the laboratory
Solvent (A): **carbon dioxide** CO_2 **124-38-9**
Solvent (C): **methanol** CH_4O **67-56-1**

Type of data: cloud points

continued

continued

w_A	0.100	0.100	0.100	0.100
w_B	0.451	0.451	0.225	0.225
w_C	0.449	0.449	0.675	0.675
T/K	401.34	406.27	416.37	421.41
P/MPa	11.383	13.083	7.145	8.796

Type of data: vapor-liquid-liquid equilibrium data

Comments: Phase boundary vapor–liquid to liquid–liquid–vapor (LLV) and phase boundary liquid–liquid–vapor to liquid–liquid (LVL).

w_A	0.100	0.100	0.100	0.100	0.100	0.100	0.100	0.100	0.100
w_B	0.451	0.451	0.451	0.451	0.451	0.451	0.225	0.225	0.225
w_C	0.449	0.449	0.449	0.449	0.449	0.449	0.675	0.675	0.675
T/K	391.19	401.13	406.27	391.43	401.34	406.27	416.37	421.41	416.37
P/MPa	8.286	8.406	8.436	8.386	8.979	9.261	6.157	6.045	6.440
	(LLV)	(LLV)	(LLV)	(LVL)	(LVL)	(LVL)	(LLV)	(LLV)	(LVL)

w_A	0.100
w_B	0.225
w_C	0.675
T/K	421.41
P/MPa	6.662
	(LVL)

Polymer (B): **poly(heptadecafluorodecyl acrylate)** **2006SHI**
Characterization: synthesized in the laboratory
Solvent (A): **carbon dioxide** CO_2 **124-38-9**
Solvent (C): **1H,1H,2H,2H-heptadecafluorodecyl acrylate** $C_{13}H_7F_{17}O_2$ **27905-45-9**

Type of data: cloud points

w_A	0.9501	0.9501	0.9501	0.9501	0.9501	0.9501	0.9501	0.9501	0.9501
w_B	0.0499	0.0499	0.0499	0.0499	0.0499	0.0499	0.0499	0.0499	0.0499
w_C	0.0000	0.0000	0.0000	0.0000	0.0000	0.0000	0.0000	0.0000	0.0000
T/K	303.33	308.20	313.41	318.17	323.14	328.07	333.16	338.19	343.34
P/MPa	9.32	10.89	12.79	14.16	15.65	16.99	17.95	19.40	20.76

w_A	0.9501	0.9501	0.9501	0.9501	0.8600	0.8600	0.8600	0.8600	0.8600
w_B	0.0499	0.0499	0.0499	0.0499	0.0440	0.0440	0.0440	0.0440	0.0440
w_C	0.0000	0.0000	0.0000	0.0000	0.0960	0.0960	0.0960	0.0960	0.0960
T/K	348.33	353.25	359.32	363.21	304.56	308.26	313.34	318.28	323.26
P/MPa	22.05	22.99	23.95	25.29	7.75	9.05	10.55	12.04	13.61

w_A	0.8600	0.8600	0.8600	0.8600	0.8600	0.8600	0.8600	0.8600	0.7503
w_B	0.0440	0.0440	0.0440	0.0440	0.0440	0.0440	0.0440	0.0440	0.0487
w_C	0.0960	0.0960	0.0960	0.0960	0.0960	0.0960	0.0960	0.0960	0.2010
T/K	328.30	333.25	338.29	343.24	348.25	353.19	358.25	363.26	308.26
P/MPa	14.99	16.27	17.51	18.66	19.71	20.81	22.00	23.07	7.58

continued

continued

w_A	0.7503	0.7503	0.7503	0.7503	0.7503	0.7503	0.7503	0.7503	0.7503
w_B	0.0487	0.0487	0.0487	0.0487	0.0487	0.0487	0.0487	0.0487	0.0487
w_C	0.2010	0.2010	0.2010	0.2010	0.2010	0.2010	0.2010	0.2010	0.2010
T/K	313.26	318.33	323.26	328.27	333.25	338.29	343.42	348.21	353.23
P/MPa	8.31	9.75	10.99	12.44	13.58	14.82	15.97	17.08	18.11

w_A	0.7503	0.7503	0.6512	0.6512	0.6512	0.6512	0.6512	0.6512	0.6512
w_B	0.0487	0.0487	0.0488	0.0488	0.0488	0.0488	0.0488	0.0488	0.0488
w_C	0.2010	0.2010	0.3000	0.3000	0.3000	0.3000	0.3000	0.3000	0.3000
T/K	358.28	363.13	333.72	338.70	343.67	348.79	353.81	357.80	363.44
P/MPa	19.11	19.78	11.42	12.48	13.65	14.65	15.68	16.23	17.33

Type of data: vapor-liquid equilibrium data

w_A	0.6512	0.6512	0.6512	0.6512	0.6512	0.6512
w_B	0.0488	0.0488	0.0488	0.0488	0.0488	0.0488
w_C	0.3000	0.3000	0.3000	0.3000	0.3000	0.3000
T/K	303.61	308.67	313.40	318.61	323.66	328.36
P/MPa	6.70	7.29	8.01	8.96	9.79	10.56

Type of data: vapor-liquid-liquid equilibrium data

w_A	0.6512	0.6512
w_B	0.0488	0.0488
w_C	0.3000	0.3000
T/K	344.98	349.74
P/MPa	13.20	13.82

Polymer (B): **poly(heptadecafluorodecyl methacrylate)** **2008SH1**
Characterization: inherent viscosity in hexafluoroisopropanol at 307.2 K and 0.227 g/dL = 1.217, synthesized in the laboratory
Solvent (A): **carbon dioxide** **CO_2** **124-38-9**
Solvent (C): **1H,1H,2H,2H-heptadecafluorodecyl methacrylate** **$C_{14}H_9F_{17}O_2$** **1996-88-9**

Type of data: cloud points

w_A	0.8406	0.8406	0.8406	0.8406	0.8406	0.8406	0.8406	0.8406	0.8406
w_B	0.0506	0.0506	0.0506	0.0506	0.0506	0.0506	0.0506	0.0506	0.0506
w_C	0.1088	0.1088	0.1088	0.1088	0.1088	0.1088	0.1088	0.1088	0.1088
T/K	303.3	308.2	313.3	318.3	323.3	328.2	333.3	338.1	343.4
P/MPa	7.50	9.10	10.86	12.51	13.91	15.30	16.78	17.94	19.30

w_A	0.8406	0.8406	0.8406	0.8406	0.7455	0.7455	0.7455	0.7455	0.7455
w_B	0.0506	0.0506	0.0506	0.0506	0.0511	0.0511	0.0511	0.0511	0.0511
w_C	0.1088	0.1088	0.1088	0.1088	0.2034	0.2034	0.2034	0.2034	0.2034
T/K	348.1	353.1	358.1	363.2	308.2	313.2	318.3	323.2	328.2
P/MPa	20.44	21.69	22.76	24.06	7.31	8.12	9.74	11.10	12.39

continued

continued

w_A	0.7455	0.7455	0.7455	0.7455	0.7455	0.7455	0.7455	0.6678	0.6678
w_B	0.0511	0.0511	0.0511	0.0511	0.0511	0.0511	0.0511	0.0461	0.0461
w_C	0.2034	0.2034	0.2034	0.2034	0.2034	0.2034	0.2034	0.2861	0.2861
T/K	333.3	338.2	343.5	348.2	353.2	358.2	363.2	328.3	333.1
P/MPa	13.78	14.89	16.30	17.40	18.30	19.23	20.30	10.36	11.44

w_A	0.6678	0.6678	0.6678	0.6678	0.6678	0.6678
w_B	0.0461	0.0461	0.0461	0.0461	0.0461	0.0461
w_C	0.2861	0.2861	0.2861	0.2861	0.2861	0.2861
T/K	338.1	343.2	348.3	353.4	353.2	363.2
P/MPa	12.73	13.89	14.89	15.92	16.95	17.80

Type of data: vapor-liquid equilibrium data

w_A	0.6678	0.6678	0.6678	0.6678	0.6678
w_B	0.0461	0.0461	0.0461	0.0461	0.0461
w_C	0.2861	0.2861	0.2861	0.2861	0.2861
T/K	303.3	308.2	313.6	318.2	323.5
P/MPa	6.51	7.20	7.96	8.82	9.60

Type of data: vapor-liquid-liquid equilibrium data

w_A	0.6678	0.6678
w_B	0.0461	0.0461
w_C	0.2861	0.2861
T/K	338.6	347.3
P/MPa	12.10	13.45

Polymer (B): **poly(2-hydroxypropyl acrylate)** **2012YAN**
Characterization: M_w/g.mol^{-1} = 26000, M_w/M_n = 2.47, T_g/K = 266,
Scientific Polymer Products, Inc., Ontario, NY
Solvent (A): **carbon dioxide** CO_2 **124-38-9**
Solvent (C): **2-hydroxypropyl acrylate** $C_6H_{10}O_3$ **999-61-1**

Type of data: cloud points

w_A	0.594	0.594	0.594	0.594	0.594	0.468	0.468	0.468	0.468
w_B	0.052	0.052	0.052	0.052	0.052	0.047	0.047	0.047	0.047
w_C	0.354	0.354	0.354	0.354	0.354	0.485	0.485	0.485	0.485
T/K	395.5	404.2	414.5	435.7	453.3	335.1	354.3	375.4	392.8
P/MPa	258.8	229.5	188.2	156.4	140.9	67.8	68.3	69.1	69.8

w_A	0.468	0.468	0.468	0.379	0.379	0.379	0.379	0.379	0.379
w_B	0.047	0.047	0.047	0.046	0.046	0.046	0.046	0.046	0.046
w_C	0.485	0.485	0.485	0.575	0.575	0.575	0.575	0.575	0.575
T/K	413.5	433.4	458.4	338.8	355.5	374.9	395.8	411.9	433.6
P/MPa	69.8	69.8	67.8	19.5	22.2	26.4	30.5	32.6	34.7

continued

continued

w_A	0.379	0.266	0.266	0.266	0.266	0.266	0.266
w_B	0.046	0.047	0.047	0.047	0.047	0.047	0.047
w_C	0.575	0.687	0.687	0.687	0.687	0.687	0.687
T/K	454.6	331.6	354.6	373.2	394.0	415.1	433.4
P/MPa	36.7	11.2	15.0	17.4	19.5	20.9	21.6

Polymer (B): **poly(2-hydroxypropyl acrylate)** **2012YAN**
Characterization: M_w/g.mol^{-1} = 26000, M_w/M_n = 2.47, T_g/K = 266, Scientific Polymer Products, Inc., Ontario, NY
Solvent (A): **chlorodifluoromethane** $CHClF_2$ **75-45-6**
Solvent (C): **carbon dioxide** CO_2 **124-38-9**

Type of data: cloud points

w_A	0.951	0.951	0.951	0.951	0.854	0.854	0.854
w_B	0.049	0.049	0.049	0.049	0.047	0.047	0.047
w_C	0.000	0.000	0.000	0.000	0.099	0.099	0.099
T/K	461.3	466.5	473.4	482.9	488.6	492.8	498.5
P/MPa	236.7	215.3	202.2	189.8	250.5	235.4	221.6

Polymer (B): **poly(2-hydroxypropyl acrylate)** **2012YAN**
Characterization: M_w/g.mol^{-1} = 26000, M_w/M_n = 2.47, T_g/K = 266, Scientific Polymer Products, Inc., Ontario, NY
Solvent (A): **dimethyl ether** C_2H_6O **115-10-6**
Solvent (C): **carbon dioxide** CO_2 **124-38-9**

Type of data: cloud points

w_A	0.830	0.830	0.830	0.830	0.830	0.830	0.830	0.738	0.738
w_B	0.051	0.051	0.051	0.051	0.051	0.051	0.051	0.052	0.052
w_C	0.119	0.119	0.119	0.119	0.119	0.119	0.119	0.210	0.210
T/K	338.6	356.6	372.2	395.8	414.4	432.2	454.1	373.2	374.7
P/MPa	239.5	199.5	172.9	154.0	139.0	129.8	121.2	258.8	251.2

w_A	0.738	0.738	0.738	0.738
w_B	0.052	0.052	0.052	0.052
w_C	0.210	0.210	0.210	0.210
T/K	396.2	415.2	435.5	459.1
P/MPa	213.1	186.7	165.0	149.0

Polymer (B):	**poly(2-hydroxypropyl acrylate)**	**2012YAN**
Characterization:	M_w/g.mol^{-1} = 26000, M_w/M_n = 2.47, T_g/K = 266, Scientific Polymer Products, Inc., Ontario, NY	
Solvent (A):	**dimethyl ether** C_2H_6O	**115-10-6**
Solvent (C):	**2-hydroxypropyl acrylate** $C_6H_{10}O_3$	**999-61-1**

Type of data: cloud points

w_A	0.952	0.952	0.952	0.952	0.952	0.952	0.952	0.952	0.952
w_B	0.048	0.048	0.048	0.048	0.048	0.048	0.048	0.048	0.048
w_C	0.000	0.000	0.000	0.000	0.000	0.000	0.000	0.000	0.000
T/K	323.3	332.7	342.9	348.0	355.0	363.3	375.0	379.3	392.4
P/MPa	175.0	161.7	149.1	141.4	137.9	127.1	121.6	116.4	113.7
w_A	0.952	0.952	0.952	0.952	0.952	0.952	0.952	0.952	0.890
w_B	0.048	0.048	0.048	0.048	0.048	0.048	0.048	0.048	0.048
w_C	0.000	0.000	0.000	0.000	0.000	0.000	0.000	0.000	0.062
T/K	394.4	408.7	413.2	422.4	431.6	437.3	442.7	454.0	324.0
P/MPa	110.5	105.2	106.0	101.0	99.3	98.1	96.4	94.5	137.9
w_A	0.890	0.890	0.890	0.890	0.890	0.890	0.890	0.796	0.796
w_B	0.048	0.048	0.048	0.048	0.048	0.048	0.048	0.050	0.050
w_C	0.062	0.062	0.062	0.062	0.062	0.062	0.062	0.154	0.154
T/K	333.0	351.8	371.8	391.8	412.4	433.2	452.3	323.8	331.8
P/MPa	127.9	112.9	103.3	95.4	90.5	86.0	83.6	86.0	82.2
w_A	0.796	0.796	0.796	0.796	0.796	0.796	0.720	0.720	0.720
w_B	0.050	0.050	0.050	0.050	0.050	0.050	0.046	0.046	0.046
w_C	0.154	0.154	0.154	0.154	0.154	0.154	0.234	0.234	0.234
T/K	335.6	374.6	393.1	413.4	435.2	452.0	337.3	353.5	375.4
P/MPa	74.0	71.2	69.1	67.1	66.7	65.9	47.8	48.1	47.8
w_A	0.720	0.720	0.720	0.720	0.712	0.712	0.712	0.712	0.712
w_B	0.046	0.046	0.046	0.046	0.050	0.050	0.050	0.050	0.050
w_C	0.234	0.234	0.234	0.234	0.238	0.238	0.238	0.238	0.238
T/K	395.5	414.5	435.4	455.6	324.6	332.6	353.3	374.7	393.3
P/MPa	48.8	49.8	51.2	52.6	47.1	46.4	46.1	46.6	47.8
w_A	0.712	0.712	0.712	0.601	0.601	0.601	0.601	0.601	0.601
w_B	0.050	0.050	0.050	0.050	0.050	0.050	0.050	0.050	0.050
w_C	0.238	0.238	0.238	0.349	0.349	0.349	0.349	0.349	0.349
T/K	412.8	434.7	452.5	315.7	332.8	352.0	375.1	390.9	414.4
P/MPa	48.8	49.8	49.5	5.4	8.8	12.9	19.3	21.9	26.0
w_A	0.601	0.601	0.498	0.498	0.498	0.498	0.498	0.498	0.498
w_B	0.050	0.050	0.049	0.049	0.049	0.049	0.049	0.049	0.049
w_C	0.349	0.349	0.453	0.453	0.453	0.453	0.453	0.453	0.453
T/K	435.1	453.4	336.9	354.8	373.9	392.2	412.7	435.5	453.1
P/MPa	29.1	30.7	5.4	9.7	14.7	18.5	23.3	27.1	29.7

continued

continued

w_A	0.424	0.424	0.424
w_B	0.052	0.052	0.052
w_C	0.524	0.524	0.524
T/K	416.1	432.8	451.6
P/MPa	6.4	9.7	13.3

Type of data: vapor-liquid equilibrium data

w_A	0.424	0.424	0.424	0.424
w_B	0.052	0.052	0.052	0.052
w_C	0.524	0.524	0.524	0.524
T/K	341.4	354.7	375.4	396.3
P/MPa	2.2	2.6	2.9	4.0

Polymer (B): **poly(2-hydroxypropyl methacrylate)** **2012YAN**
Characterization: M_w/g.mol^{-1} = 42000, M_w/M_n = 2.48, T_g/K = 346, Scientific Polymer Products, Inc., Ontario, NY
Solvent (A): **carbon dioxide** CO_2 **124-38-9**
Solvent (C): **2-hydroxypropyl methacrylate** $C_7H_{12}O_3$ **923-26-2**

Type of data: cloud points

w_A	0.552	0.552	0.552	0.552	0.511	0.511	0.511	0.511	0.511
w_B	0.048	0.048	0.048	0.048	0.048	0.048	0.048	0.048	0.048
w_C	0.400	0.400	0.400	0.400	0.441	0.441	0.441	0.441	0.441
T/K	478.6	481.2	485.9	505.1	433.0	435.2	437.0	441.2	452.9
P/MPa	265.7	256.0	228.5	207.1	265.7	248.5	234.0	196.7	174.7
w_A	0.444	0.444	0.444	0.444	0.444	0.444	0.444	0.360	0.360
w_B	0.050	0.050	0.050	0.050	0.050	0.050	0.050	0.049	0.049
w_C	0.506	0.506	0.506	0.506	0.506	0.506	0.506	0.591	0.591
T/K	337.0	356.1	373.9	393.7	411.8	434.2	453.7	340.1	355.8
P/MPa	166.4	149.1	132.6	118.8	108.8	97.6	88.5	39.5	42.9
w_A	0.360	0.360	0.360	0.360	0.360	0.342	0.342	0.342	0.342
w_B	0.049	0.049	0.049	0.049	0.049	0.046	0.046	0.046	0.046
w_C	0.591	0.591	0.591	0.591	0.591	0.612	0.612	0.612	0.612
T/K	375.4	393.2	413.4	435.0	452.8	330.0	354.3	374.6	394.9
P/MPa	46.0	46.4	48.5	49.0	50.0	29.5	34.7	38.5	40.9
w_A	0.342	0.342	0.342	0.286	0.286	0.286	0.286	0.286	0.286
w_B	0.046	0.046	0.046	0.051	0.051	0.051	0.051	0.051	0.051
w_C	0.612	0.612	0.612	0.663	0.663	0.663	0.663	0.663	0.663
T/K	414.4	434.6	453.7	336.2	355.8	374.1	392.7	414.2	433.5
P/MPa	41.2	41.6	41.9	13.3	18.8	22.9	26.4	28.1	29.1

continued

continued

w_A	0.286	0.262	0.262	0.262
w_B	0.051	0.049	0.049	0.049
w_C	0.663	0.689	0.689	0.689
T/K	455.2	395.4	412.4	435.7
P/MPa	32.2	16.4	17.4	21.2

Type of data: vapor-liquid equilibrium data

w_A	0.286	0.262	0.262
w_B	0.051	0.049	0.049
w_C	0.663	0.689	0.689
T/K	331.3	356.1	375.7
P/MPa	9.8	13.1	15.4

Polymer (B): **poly(2-hydroxypropyl methacrylate)** **2012YAN**
Characterization: M_w/g.mol^{-1} = 42000, M_w/M_n = 2.48, T_g/K = 346,
Scientific Polymer Products, Inc., Ontario, NY
Solvent (A): **dimethyl ether** C_2H_6O **115-10-6**
Solvent (C): **carbon dioxide** CO_2 **124-38-9**

Type of data: cloud points

w_A	0.849	0.849	0.849	0.849	0.849	0.849	0.849	0.849	0.717
w_B	0.051	0.051	0.051	0.051	0.051	0.051	0.051	0.051	0.045
w_C	0.100	0.100	0.100	0.100	0.100	0.100	0.100	0.100	0.238
T/K	360.7	375.2	396.0	415.0	431.3	451.9	474.6	491.9	411.4
P/MPa	255.4	219.0	182.6	161.9	148.5	135.4	125.2	120.0	260.2

w_A	0.717	0.717	0.717	0.717	0.717
w_B	0.045	0.045	0.045	0.045	0.045
w_C	0.238	0.238	0.238	0.238	0.238
T/K	416.2	436.1	456.3	472.6	487.3
P/MPa	246.6	210.5	184.8	170.5	160.5

Polymer (B): **poly(2-hydroxypropyl methacrylate)** **2012YAN**
Characterization: M_w/g.mol^{-1} = 42000, M_w/M_n = 2.48, T_g/K = 346,
Scientific Polymer Products, Inc., Ontario, NY
Solvent (A): **dimethyl ether** C_2H_6O **115-10-6**
Solvent (C): **2-hydroxypropyl methacrylate** $C_7H_{12}O_3$ **923-26-2**

Type of data: cloud points

w_A	0.892	0.892	0.892	0.892	0.892	0.892	0.892	0.892	0.892
w_B	0.053	0.053	0.053	0.053	0.053	0.053	0.053	0.053	0.053
w_C	0.055	0.055	0.055	0.055	0.055	0.055	0.055	0.055	0.055
T/K	325.1	333.6	343.0	355.7	362.1	374.3	384.7	394.8	413.9
P/MPa	209.8	190.2	173.1	152.9	146.4	138.2	129.9	121.7	110.2

continued

continued

w_A	0.892	0.892	0.839	0.839	0.839	0.839	0.839	0.839	0.839
w_B	0.053	0.053	0.052	0.052	0.052	0.052	0.052	0.052	0.052
w_C	0.055	0.055	0.109	0.109	0.109	0.109	0.109	0.109	0.109
T/K	434.7	452.3	328.9	340.0	355.0	376.4	393.4	413.6	435.1
P/MPa	102.2	99.1	173.7	155.8	137.3	116.7	109.8	100.2	92.9
w_A	0.839	0.810	0.810	0.810	0.810	0.810	0.810	0.794	0.794
w_B	0.052	0.048	0.048	0.048	0.048	0.048	0.048	0.034	0.034
w_C	0.109	0.142	0.142	0.142	0.142	0.142	0.142	0.172	0.172
T/K	452.5	357.0	379.1	395.2	413.3	432.1	451.6	336.1	355.5
P/MPa	90.5	112.2	99.1	92.7	87.9	84.4	81.1	115.7	100.2
w_A	0.794	0.794	0.794	0.794	0.794	0.657	0.657	0.657	0.657
w_B	0.034	0.034	0.034	0.034	0.034	0.049	0.049	0.049	0.049
w_C	0.172	0.172	0.172	0.172	0.172	0.294	0.294	0.294	0.294
T/K	374.5	394.5	414.9	431.4	457.0	323.9	335.6	356.3	374.0
P/MPa	90.0	83.6	79.7	76.6	74.0	113.4	103.2	90.3	83.9
w_A	0.657	0.657	0.657	0.657	0.561	0.561	0.561	0.561	0.561
w_B	0.049	0.049	0.049	0.049	0.044	0.044	0.044	0.044	0.044
w_C	0.294	0.294	0.294	0.294	0.395	0.395	0.395	0.395	0.395
T/K	394.9	414.7	433.7	453.8	338.6	355.9	375.1	395.1	416.1
P/MPa	76.7	74.0	70.3	68.3	67.4	61.0	57.8	55.0	55.0
w_A	0.561	0.512	0.512	0.512	0.512	0.512	0.512	0.483	0.483
w_B	0.044	0.044	0.044	0.044	0.044	0.044	0.044	0.054	0.054
w_C	0.395	0.444	0.444	0.444	0.444	0.444	0.444	0.463	0.463
T/K	431.1	337.4	355.8	375.7	393.2	416.2	431.2	322.1	334.3
P/MPa	55.0	40.2	38.1	38.3	39.5	40.9	42.2	36.2	34.0
w_A	0.483	0.483	0.483	0.483	0.483	0.483	0.370	0.370	0.370
w_B	0.054	0.054	0.054	0.054	0.054	0.054	0.050	0.050	0.050
w_C	0.463	0.463	0.463	0.463	0.463	0.463	0.580	0.580	0.580
T/K	353.1	374.6	394.5	416.6	436.9	458.1	392.0	415.6	436.0
P/MPa	32.7	33.0	33.7	35.4	37.1	38.8	5.7	9.5	13.6
w_A	0.370								
w_B	0.050								
w_C	0.580								
T/K	443.0								
P/MPa	14.7								

Type of data: vapor-liquid equilibrium data

w_A	0.370	0.370	0.370
w_B	0.050	0.050	0.050
w_C	0.580	0.580	0.580
T/K	331.2	355.4	374.9
P/MPa	1.9	2.9	3.1

Polymer (B): **poly(isobornyl acrylate)** **2013YA1**
Characterization: M_w/g.mol^{-1} = 100000,
Scientific Polymer Products, Inc., Ontario, NY
Solvent (A): **carbon dioxide** **CO_2** **124-38-9**
Solvent (C): **dimethyl ether** **C_2H_6O** **115-10-6**

Type of data: cloud points

w_A	0.000	0.000	0.000	0.000	0.000	0.000	0.000	0.169	0.169
w_B	0.043	0.043	0.043	0.043	0.043	0.043	0.043	0.048	0.048
w_C	0.957	0.957	0.957	0.957	0.957	9.957	0.957	0.783	0.783
T/K	334.6	354.2	374.7	393.1	415.6	435.2	454.2	334.2	354.8
P/MPa	31.90	35.35	38.79	41.55	45.69	47.93	45.00	67.07	64.31
w_A	0.169	0.169	0.169	0.169	0.169	0.460	0.460	0.460	0.460
w_B	0.048	0.048	0.048	0.048	0.048	0.051	0.051	0.051	0.051
w_C	0.783	0.783	0.783	0.783	0.783	0.489	0.489	0.489	0.489
T/K	371.2	395.2	413.5	437.6	456.5	337.4	352.9	373.3	394.8
P/MPa	63.62	64.14	66.03	67.07	70.17	271.55	189.48	151.21	133.45
w_A	0.460	0.460	0.460	0.492	0.492	0.492	0.492	0.492	0.492
w_B	0.051	0.051	0.051	0.049	0.049	0.049	0.049	0.049	0.049
w_C	0.489	0.489	0.489	0.459	0.459	0.459	0.459	0.459	0.459
T/K	411.6	433.6	453.4	365.0	375.0	393.2	414.4	433.8	454.7
P/MPa	125.00	117.33	114.14	270.52	229.14	186.38	161.55	148.45	139.48

Polymer (B): **poly(isobornyl acrylate)** **2013YA1**
Characterization: M_w/g.mol^{-1} = 100000,
Scientific Polymer Products, Inc., Ontario, NY
Solvent (A): **carbon dioxide** **CO_2** **124-38-9**
Solvent (C): **isobornyl acrylate** **$C_{13}H_{20}O_2$** **5888-33-5**

Type of data: cloud points

w_A	0.146	0.146	0.146	0.146	0.146	0.146	0.193	0.193	0.193
w_B	0.051	0.051	0.051	0.051	0.051	0.051	0.051	0.051	0.051
w_C	0.803	0.803	0.803	0.803	0.803	0.803	0.756	0.756	0.756
T/K	351.9	373.4	394.1	416.1	435.2	452.1	312.9	334.7	354.1
P/MPa	15.90	19.66	23.30	26.14	29.31	30.00	13.79	20.00	24.76
w_A	0.193	0.193	0.193	0.193	0.193	0.238	0.238	0.238	0.238
w_B	0.051	0.051	0.051	0.051	0.051	0.050	0.050	0.050	0.050
w_C	0.756	0.756	0.756	0.756	0.756	0.712	0.712	0.712	0.712
T/K	373.9	393.0	413.9	434.0	452.1	334.6	354.9	374.0	394.2
P/MPa	29.14	32.76	36.03	38.62	40.34	31.03	35.28	38.62	42.59

continued

continued

w_A	0.238	0.238	0.238	0.327	0.327	0.327	0.327	0.327	0.327
w_B	0.050	0.050	0.050	0.048	0.048	0.048	0.048	0.048	0.048
w_C	0.712	0.712	0.712	0.625	0.625	0.625	0.625	0.625	0.625
T/K	413.8	434.1	454.0	334.3	354.6	373.5	394.6	408.5	431.6
P/MPa	45.35	47.93	50.00	152.07	96.72	68.28	68.10	67.93	69.31

w_A	0.327	0.452	0.452	0.452	0.452	0.452	0.452	0.565	0.565
w_B	0.048	0.049	0.049	0.049	0.049	0.049	0.049	0.048	0.048
w_C	0.625	0.499	0.499	0.499	0.499	0.499	0.499	0.387	0.387
T/K	452.5	363.2	371.8	392.3	414.5	433.2	451.1	402.8	413.8
P/MPa	69.79	276.00	166.38	128.62	113.10	106.90	100.52	266.72	195.34

w_A	0.565	0.565	0.565	0.730	0.730	0.730	0.730
w_B	0.048	0.048	0.048	0.050	0.050	0.050	0.050
w_C	0.387	0.387	0.387	0.220	0.220	0.220	0.220
T/K	423.5	432.4	453.5	464.2	465.9	468.3	472.7
P/MPa	173.45	157.59	137.59	296.72	286.38	267.76	235.86

Type of data: vapor-liquid equilibrium data

w_A	0.146	0.146	0.146
w_B	0.051	0.051	0.051
w_C	0.803	0.803	0.803
T/K	323.4	333.4	341.3
P/MPa	10.28	11.97	13.27

Type of data: vapor-liquid-liquid equilibrium data

w_A	0.146
w_B	0.051
w_C	0.803
T/K	354.7
P/MPa	15.14

Polymer (B): **poly(isobornyl acrylate)** **2013YA1**
Characterization: M_w/g.mol^{-1} = 1000000,
Scientific Polymer Products, Inc., Ontario, NY
Solvent (A): **carbon dioxide** CO_2 **124-38-9**
Solvent (C): **isobornyl acrylate** $C_{13}H_{20}O_2$ **5888-33-5**

Type of data: cloud points

w_A	0.470	0.470	0.470	0.470	0.470
w_B	0.050	0.050	0.050	0.050	0.050
w_C	0.480	0.480	0.480	0.480	0.480
T/K	407.7	409.0	414.2	431.7	453.3
P/MPa	289.83	276.03	224.83	166.03	128.97

Polymer (B): **poly(isobornyl methacrylate)** **2014JEO**
Characterization: M_w/g.mol^{-1} = 100000, T_g/K = 383.2,
Scientific Polymer Products, Inc., Ontario, NY
Solvent (A): **carbon dioxide** CO_2 **124-38-9**
Solvent (C): **dimethyl ether** C_2H_6O **115-10-6**

Type of data: cloud points

w_A	0.621	0.621	0.621	0.621	0.621	0.621	0.503	0.503	0.503
w_B	0.048	0.048	0.048	0.048	0.048	0.048	0.049	0.049	0.049
w_C	0.331	0.331	0.331	0.331	0.331	0.331	0.448	0.448	0.448
T/K	433.0	442.2	446.0	450.9	456.7	461.4	378.0	379.8	381.5
P/MPa	268.45	243.62	234.66	223.62	215.01	208.45	275.00	263.62	255.34
w_A	0.503	0.503	0.503	0.503	0.503	0.457	0.457	0.457	0.457
w_B	0.049	0.049	0.049	0.049	0.049	0.050	0.050	0.050	0.050
w_C	0.448	0.448	0.448	0.448	0.448	0.493	0.493	0.493	0.493
T/K	383.2	393.5	413.2	433.8	454.7	375.5	393.4	413.7	434.9
P/MPa	245.00	213.97	178.10	158.10	146.38	215.00	174.31	150.86	137.76
w_A	0.457	0.227	0.227	0.227	0.227	0.227	0.227	0.227	0.000
w_B	0.050	0.049	0.049	0.049	0.049	0.049	0.049	0.049	0.051
w_C	0.493	0.724	0.724	0.724	0.724	0.724	0.724	0.724	0.949
T/K	454.2	335.6	355.9	372.9	393.0	416.2	434.5	454.7	332.7
P/MPa	129.83	119.48	97.93	89.14	83.62	80.86	80.17	80.17	41.03
w_A	0.000	0.000	0.000	0.000	0.000	0.000			
w_B	0.051	0.051	0.051	0.051	0.051	0.051			
w_C	0.949	0.949	0.949	0.949	0.949	0.949			
T/K	353.8	375.2	394.2	416.4	433.8	452.9			
P/MPa	51.55	43.62	45.69	48.45	50.52	52.59			

Polymer (B): **poly(isobornyl methacrylate)** **2014JEO**
Characterization: M_w/g.mol^{-1} = 100000, T_g/K = 383.2,
Scientific Polymer Products, Inc., Ontario, NY
Solvent (A): **carbon dioxide** CO_2 **124-38-9**
Solvent (C): **isobornyl methacrylate** $C_{14}H_{22}O_2$ **7534-94-3**

Type of data: cloud points

w_A	0.550	0.550	0.550	0.550	0.550	0.550	0.550	0.468	0.468
w_B	0.056	0.056	0.056	0.056	0.056	0.056	0.056	0.056	0.056
w_C	0.394	0.394	0.394	0.394	0.394	0.394	0.394	0.476	0.476
T/K	384.5	393.2	403.4	421.1	433.9	449.1	465.0	347.5	360.6
P/MPa	213.41	186.83	159.03	136.48	128.10	121.90	117.93	204.66	136.59
w_A	0.468	0.468	0.468	0.468	0.468	0.468	0.468	0.309	0.309
w_B	0.056	0.056	0.056	0.056	0.056	0.056	0.056	0.050	0.050
w_C	0.476	0.476	0.476	0.476	0.476	0.476	0.476	0.641	0.641
T/K	372.4	386.8	405.7	420.1	436.5	448.6	466.5	323.4	337.7
P/MPa	119.14	105.93	99.14	94.41	91.45	89.83	89.21	77.35	66.38

continued

continued

w_A	0.309	0.309	0.309	0.309	0.309	0.309	0.309	0.309	0.240
w_B	0.050	0.050	0.050	0.050	0.050	0.050	0.050	0.050	0.048
w_C	0.641	0.641	0.641	0.641	0.641	0.641	0.641	0.641	0.712
T/K	353.2	375.9	390.6	406.0	418.4	432.1	448.8	463.5	327.1
P/MPa	61.55	61.30	58.93	59.35	60.04	60.86	61.55	63.62	36.59
w_A	0.240	0.240	0.240	0.240	0.240	0.240	0.240	0.240	0.240
w_B	0.048	0.048	0.048	0.048	0.048	0.048	0.048	0.048	0.048
w_C	0.712	0.712	0.712	0.712	0.712	0.712	0.712	0.712	0.712
T/K	344.6	359.4	373.0	389.5	400.9	419.1	433.4	448.5	463.6
P/MPa	37.17	38.38	39.76	41.55	42.41	43.83	45.90	47.90	51.59

Polymer (B): **poly(isobornyl methacrylate)** **2014JEO**
Characterization: M_w/g.mol^{-1} = 550000, T_g/K = 383.2,
Scientific Polymer Products, Inc., Ontario, NY
Solvent (A): **carbon dioxide** CO_2 **124-38-9**
Solvent (C): **isobornyl methacrylate** $C_{14}H_{22}O_2$ **7534-94-3**

Type of data: cloud points

w_A	0.547	0.547	0.547	0.547	0.547	0.547	0.547	0.477	0.477
w_B	0.050	0.050	0.050	0.050	0.050	0.050	0.050	0.049	0.049
w_C	0.403	0.403	0.403	0.403	0.403	0.403	0.403	0.474	0.474
T/K	347.2	353.2	374.7	394.7	411.5	433.2	453.2	329.3	334.7
P/MPa	241.55	172.93	105.00	105.00	97.07	92.93	90.00	276.03	152.20
w_A	0.477	0.477	0.477	0.477	0.477	0.477	0.446	0.446	0.446
w_B	0.049	0.049	0.049	0.049	0.049	0.049	0.049	0.049	0.049
w_C	0.474	0.474	0.474	0.474	0.474	0.474	0.505	0.505	0.505
T/K	354.7	374.2	393.6	414.1	434.3	455.1	334.4	354.5	371.9
P/MPa	102.00	88.97	82.59	79.66	78.28	77.35	38.62	39.31	41.48
w_A	0.446	0.446	0.446	0.446					
w_B	0.049	0.049	0.049	0.049					
w_C	0.505	0.505	0.505	0.505					
T/K	392.8	412.2	433.8	452.3					
P/MPa	44.07	46.72	49.66	52.24					

Polymer (B): **poly(isobutyl acrylate)** **2006BY1**
Characterization: M_n/g.mol^{-1} = 22000, M_w/g.mol^{-1} = 120000,
Polysciences, Inc., Warrington, PA
Solvent (A): **carbon dioxide** CO_2 **124-38-9**
Solvent (C): **isobutyl acrylate** $C_7H_{12}O_2$ **106-63-8**

Type of data: cloud points

continued

continued

w_A	0.916	0.916	0.916	0.916	0.916	0.916	0.916	0.916	0.916
w_B	0.052	0.052	0.052	0.052	0.052	0.052	0.052	0.052	0.052
w_C	0.032	0.032	0.032	0.032	0.032	0.032	0.032	0.032	0.032
T/K	315.75	323.75	333.35	347.35	363.35	378.15	394.05	407.15	421.75
P/bar	917.2	839.0	792.1	790.3	800.0	811.7	824.8	834.8	843.1

w_A	0.858	0.858	0.858	0.858	0.858	0.858	0.858	0.858	0.772
w_B	0.054	0.054	0.054	0.054	0.054	0.054	0.054	0.054	0.047
w_C	0.088	0.088	0.088	0.088	0.088	0.088	0.088	0.088	0.181
T/K	318.65	333.25	347.65	363.25	379.05	393.75	408.45	422.15	319.35
P/bar	590.3	611.4	644.0	670.7	699.3	722.4	736.9	746.2	361.0

w_A	0.772	0.772	0.772	0.772	0.772	0.772	0.772	0.633	0.633
w_B	0.047	0.047	0.047	0.047	0.047	0.047	0.047	0.048	0.048
w_C	0.181	0.181	0.181	0.181	0.181	0.181	0.181	0.319	0.319
T/K	333.35	348.15	363.65	378.85	394.65	408.55	421.35	317.95	332.15
P/bar	432.8	472.1	516.9	555.9	581.0	599.0	603.1	337.9	392.8

w_A	0.633	0.633	0.633	0.633	0.633	0.633	0.546	0.546	0.546
w_B	0.048	0.048	0.048	0.048	0.048	0.048	0.047	0.047	0.047
w_C	0.319	0.319	0.319	0.319	0.319	0.319	0.407	0.407	0.407
T/K	346.75	361.95	377.25	392.85	407.45	421.15	318.75	332.15	346.75
P/bar	434.5	474.8	521.0	550.0	571.7	579.3	277.6	332.1	381.4

w_A	0.546	0.546	0.546	0.546	0.546	0.360	0.360	0.360	0.360
w_B	0.047	0.047	0.047	0.047	0.047	0.053	0.053	0.053	0.053
w_C	0.407	0.407	0.407	0.407	0.407	0.587	0.587	0.587	0.587
T/K	362.75	378.35	392.95	408.15	422.55	327.65	337.95	356.05	370.65
P/bar	455.2	482.8	511.4	534.5	553.8	125.9	155.2	208.6	250.0

w_A	0.360	0.360	0.360	0.360
w_B	0.053	0.053	0.053	0.053
w_C	0.587	0.587	0.587	0.587
T/K	386.35	400.15	414.35	428.65
P/bar	286.9	312.1	332.8	341.7

Type of data: vapor-liquid equilibrium data

w_A	0.360	0.360	0.360
w_B	0.053	0.053	0.053
w_C	0.587	0.587	0.587
T/K	303.85	309.25	317.55
P/bar	70.2	81.0	91.4

Type of data: vapor-liquid-liquid equilibrium data

w_A	0.360	0.360
w_B	0.053	0.053
w_C	0.587	0.587
T/K	327.65	337.15
P/bar	104.0	117.5

Polymer (B): **poly(isobutyl methacrylate)** **2006BY1**
Characterization: M_n/g.mol^{-1} = 114000, M_w/g.mol^{-1} = 200000, Polysciences, Inc., Warrington, PA
Solvent (A): **carbon dioxide** CO_2 **124-38-9**
Solvent (C): **isobutyl methacrylate** $C_8H_{14}O_2$ **97-86-9**

Type of data: cloud points

w_A	0.880	0.880	0.880	0.880	0.880	0.880	0.880	0.880	0.880
w_B	0.055	0.055	0.055	0.055	0.055	0.055	0.055	0.055	0.055
w_C	0.065	0.065	0.065	0.065	0.065	0.065	0.065	0.065	0.065
T/K	330.45	336.75	345.65	360.15	375.85	387.85	405.65	418.95	435.75
P/bar	1722.4	1517.6	1259.7	1112.8	1051.0	1030.0	1021.0	1019.0	1017.2

w_A	0.880	0.880	0.880	0.880	0.880	0.880	0.880	0.730	0.730
w_B	0.053	0.053	0.053	0.053	0.053	0.053	0.053	0.053	0.053
w_C	0.126	0.126	0.126	0.126	0.126	0.126	0.126	0.217	0.217
T/K	316.45	333.25	351.55	371.25	393.35	412.55	434.05	320.55	337.05
P/bar	1039.7	885.2	838.3	841.7	853.5	862.4	874.5	643.1	652.8

w_A	0.730	0.730	0.730	0.730	0.730	0.626	0.626	0.626	0.626
w_B	0.053	0.053	0.053	0.053	0.053	0.051	0.051	0.051	0.051
w_C	0.217	0.217	0.217	0.217	0.217	0.323	0.323	0.323	0.323
T/K	355.85	370.15	392.75	410.25	434.25	321.05	332.05	355.35	375.65
P/bar	677.6	692.1	724.5	744.1	778.6	420.0	449.3	502.8	552.1

w_A	0.626	0.626	0.626	0.451	0.451	0.451	0.451	0.451	0.451
w_B	0.051	0.051	0.051	0.051	0.051	0.051	0.051	0.051	0.051
w_C	0.323	0.323	0.323	0.498	0.498	0.498	0.498	0.498	0.498
T/K	395.35	415.15	435.15	330.55	351.35	374.65	393.35	411.35	433.25
P/bar	593.5	631.0	640.4	275.5	333.5	404.5	443.8	475.2	499.3

Type of data: vapor-liquid equilibrium data

w_A	0.451	0.451
w_B	0.051	0.051
w_C	0.498	0.498
T/K	318.85	323.15
P/bar	234.8	246.6

Type of data: vapor-liquid-liquid equilibrium data

w_A	0.451	0.451
w_B	0.051	0.051
w_C	0.498	0.498
T/K	339.15	344.15
P/bar	275.0	287.0

Polymer (B): **poly(isodecyl acrylate)** **2008BY2**
Characterization: M_w/g.mol^{-1} = 60000, T_g/K = 213,
Scientific Polymer Products, Inc., Ontario, NY
Solvent (A): **carbon dioxide** CO_2 **124-38-9**
Solvent (C): **dimethyl ether** C_2H_6O **115-10-6**

Type of data: cloud points

w_A	0.856	0.856	0.856	0.856	0.856	0.856	0.739	0.739	0.739
w_B	0.060	0.060	0.060	0.060	0.060	0.060	0.050	0.050	0.050
w_C	0.084	0.084	0.084	0.084	0.084	0.084	0.211	0.211	0.211
T/K	373.65	383.05	394.75	414.85	433.35	453.35	333.25	353.35	373.85
P/bar	1781.0	1263.8	1108.6	994.8	946.6	919.0	958.6	739.7	687.0

w_A	0.739	0.739	0.739	0.739	0.540	0.540	0.540	0.540	0.540
w_B	0.050	0.050	0.050	0.050	0.043	0.043	0.043	0.043	0.043
w_C	0.211	0.211	0.211	0.211	0.417	0.417	0.417	0.417	0.417
T/K	392.85	411.85	432.55	453.35	333.55	353.25	372.75	393.95	412.75
P/bar	674.1	674.1	684.5	696.6	281.0	315.5	350.0	391.4	419.0

w_A	0.540	0.540
w_B	0.043	0.043
w_C	0.417	0.417
T/K	432.55	453.25
P/bar	450.0	472.4

Polymer (B): **poly(isodecyl acrylate)** **2008BY2**
Characterization: M_w/g.mol^{-1} = 60000, T_g/K = 213,
Scientific Polymer Products, Inc., Ontario, NY
Solvent (A): **carbon dioxide** CO_2 **124-38-9**
Solvent (C): **isodecyl acrylate** $C_{13}H_{24}O_2$ **1330-61-6**

Type of data: cloud points

w_A	0.848	0.848	0.848	0.848	0.848	0.848	0.848	0.740	0.740
w_B	0.062	0.062	0.062	0.062	0.062	0.062	0.062	0.055	0.055
w_C	0.090	0.090	0.090	0.090	0.090	0.090	0.090	0.205	0.205
T/K	382.85	393.45	401.75	415.75	421.45	439.65	452.95	353.75	374.25
P/bar	1719.0	1287.9	1146.6	1041.4	965.5	863.8	860.4	1656.9	1000.0

w_A	0.740	0.740	0.740	0.740	0.534	0.534	0.534	0.534	0.534
w_B	0.055	0.055	0.055	0.055	0.060	0.060	0.060	0.060	0.060
w_C	0.205	0.205	0.205	0.205	0.406	0.406	0.406	0.406	0.406
T/K	395.15	412.55	434.45	452.75	333.95	353.75	372.85	392.95	411.65
P/bar	824.1	777.6	753.5	706.9	1243.1	746.6	667.2	646.6	645.9

w_A	0.534	0.534	0.431	0.431	0.431	0.431	0.431	0.431	0.431
w_B	0.060	0.060	0.056	0.056	0.056	0.056	0.056	0.056	0.056
w_C	0.406	0.406	0.513	0.513	0.513	0.513	0.513	0.513	0.513
T/K	434.65	453.45	332.65	352.15	373.65	395.45	414.25	432.55	453.85
P/bar	650.0	650.0	522.4	494.8	498.3	505.2	522.4	541.0	557.9

continued

continued

w_A	0.385	0.385	0.385	0.385	0.385	0.385	0.385	0.303	0.303
w_B	0.050	0.050	0.050	0.050	0.050	0.050	0.050	0.049	0.049
w_C	0.565	0.565	0.565	0.565	0.565	0.565	0.565	0.648	0.648
T/K	332.15	352.75	373.75	393.05	413.05	435.45	454.25	333.15	351.35
P/bar	346.6	370.7	398.3	429.3	456.9	477.6	493.1	198.3	245.5

w_A	0.303	0.303	0.303	0.303	0.303
w_B	0.049	0.049	0.049	0.049	0.049
w_C	0.648	0.648	0.648	0.648	0.648
T/K	371.25	391.25	415.55	432.55	454.05
P/bar	283.1	312.1	350.0	374.8	394.8

Polymer (B): **poly(isodecyl methacrylate)** **2009KIM**
Characterization: M_w/g.mol^{-1} = 165000, T_g/K = 232.2,
Scientific Polymer Products, Inc., Ontario, NY
Solvent (A): **carbon dioxide** CO_2 **124-38-9**
Solvent (C): **dimethyl ether** C_2H_6O **115-10-6**

Type of data: cloud points

w_A	0.861	0.861	0.861	0.861	0.861	0.737	0.737	0.737	0.737
w_B	0.053	0.053	0.053	0.053	0.053	0.052	0.052	0.052	0.052
w_C	0.086	0.086	0.086	0.086	0.086	0.211	0.211	0.211	0.211
T/K	398.8	403.7	414.4	434.1	454.5	335.8	354.6	373.5	394.1
P/MPa	191.90	163.62	134.31	118.10	111.21	150.17	94.66	83.28	78.79

w_A	0.737	0.737	0.737	0.487	0.487	0.487	0.487	0.487	0.487
w_B	0.052	0.052	0.052	0.036	0.036	0.036	0.036	0.036	0.036
w_C	0.211	0.211	0.211	0.477	0.477	0.477	0.477	0.477	0.477
T/K	414.1	434.1	453.8	333.6	352.1	373.5	393.7	412.6	432.5
P/MPa	77.76	77.07	77.59	31.55	33.97	37.59	40.86	43.97	46.03

w_A	0.487	0.000	0.000	0.000	0.000	0.000
w_B	0.036	0.046	0.046	0.046	0.046	0.046
w_C	0.477	0.954	0.954	0.954	0.954	0.954
T/K	453.1	375.6	392.8	412.4	432.7	453.7
P/MPa	48.10	5.00	8.79	13.28	17.00	19.31

Polymer (B): **poly(isodecyl methacrylate)** **2009KIM**
Characterization: M_w/g.mol^{-1} = 165000, T_g/K = 232.2,
Scientific Polymer Products, Inc., Ontario, NY
Solvent (A): **carbon dioxide** CO_2 **124-38-9**
Solvent (C): **isodecyl methacrylate** $C_{14}H_{26}O_2$ **29964-84-9**

Type of data: cloud points

continued

continued

w_A	0.946	0.946	0.946	0.946	0.946	0.878	0.878	0.878	0.878
w_B	0.054	0.054	0.054	0.054	0.054	0.062	0.062	0.062	0.062
w_C	0.000	0.000	0.000	0.000	0.000	0.060	0.060	0.060	0.060
T/K	438.2	444.3	453.3	465.0	475.7	408.6	413.4	423.5	432.5
P/MPa	255.15	195.52	163.62	145.69	142.76	179.48	160.52	134.66	125.69
w_A	0.878	0.878	0.878	0.772	0.772	0.772	0.772	0.772	0.772
w_B	0.062	0.062	0.062	0.065	0.065	0.065	0.065	0.065	0.065
w_C	0.060	0.060	0.060	0.163	0.163	0.163	0.163	0.163	0.163
T/K	444.6	453.1	465.0	365.6	378.2	393.8	409.5	425.0	439.5
P/MPa	117.76	114.31	112.24	150.17	118.28	93.93	89.14	83.62	82.41
w_A	0.772	0.654	0.654	0.654	0.654	0.654	0.654	0.480	0.480
w_B	0.065	0.066	0.066	0.066	0.066	0.066	0.066	0.065	0.065
w_C	0.163	0.280	0.280	0.280	0.280	0.280	0.280	0.455	0.455
T/K	454.3	353.2	375.0	395.4	415.4	434.2	454.5	332.8	351.6
P/MPa	81.55	163.62	94.31	81.90	73.62	72.24	72.93	66.21	56.04
w_A	0.480	0.480	0.480	0.480	0.480	0.349	0.349	0.349	0.349
w_B	0.065	0.065	0.065	0.065	0.065	0.050	0.050	0.050	0.050
w_C	0.455	0.455	0.455	0.455	0.455	0.601	0.601	0.601	0.601
T/K	372.7	394.2	415.6	433.8	455.4	333.3	351.9	376.0	391.6
P/MPa	54.07	55.00	56.38	58.10	59.24	10.52	15.69	21.72	25.35
w_A	0.349	0.349	0.349						
w_B	0.050	0.050	0.050						
w_C	0.601	0.601	0.601						
T/K	412.2	434.2	454.6						
P/MPa	29.00	32.59	35.00						

Type of data: vapor-liquid equilibrium data

w_A	0.349	0.349	0.349
w_B	0.050	0.050	0.050
w_C	0.601	0.601	0.601
T/K	308.6	415.3	323.8
P/MPa	7.07	8.10	9.14

Type of data: vapor-liquid-liquid equilibrium data

w_A	0.349
w_B	0.050
w_C	0.601
T/K	351.1
P/MPa	12.40

Polymer (B): **poly(isooctyl acrylate)** **2008BY1**
Characterization: M_w/g.mol^{-1} = 60000,
Scientific Polymer Products, Inc., Ontario, NY
Solvent (A): **carbon dioxide** **CO_2** **124-38-9**
Solvent (C): **dimethyl ether** **C_2H_6O** **115-10-6**

Type of data: cloud points

w_A	0.858	0.858	0.858	0.858	0.858	0.858	0.803	0.803	0.803
w_B	0.047	0.047	0.047	0.047	0.047	0.047	0.049	0.049	0.049
w_C	0.095	0.095	0.095	0.095	0.095	0.095	0.148	0.148	0.148
T/K	354.75	374.85	393.45	414.35	432.05	453.35	338.65	351.55	373.05
P/bar	1484.5	1125.9	982.8	912.1	901.7	886.2	1243.1	960.3	836.2
w_A	0.803	0.803	0.803	0.803	0.632	0.632	0.632	0.632	0.632
w_B	0.049	0.049	0.049	0.049	0.062	0.062	0.062	0.062	0.062
w_C	0.148	0.148	0.148	0.148	0.306	0.306	0.306	0.306	0.306
T/K	393.05	412.35	433.45	455.25	332.55	352.75	373.35	393.55	414.85
P/bar	798.3	787.9	787.9	801.7	387.9	419.0	450.0	477.6	501.7
w_A	0.632	0.632	0.524	0.524	0.524	0.524	0.524	0.524	0.524
w_B	0.062	0.062	0.056	0.056	0.056	0.056	0.056	0.056	0.056
w_C	0.306	0.306	0.420	0.420	0.420	0.420	0.420	0.420	0.420
T/K	431.25	453.85	333.15	353.55	375.95	394.35	414.55	432.55	453.75
P/bar	531.0	556.9	229.3	281.0	329.3	370.7	407.2	425.9	452.8

Polymer (B): **poly(isooctyl acrylate)** **2008BY1**
Characterization: M_w/g.mol^{-1} = 60000,
Scientific Polymer Products, Inc., Ontario, NY
Solvent (A): **carbon dioxide** **CO_2** **124-38-9**
Solvent (C): **isooctyl acrylate** **$C_{11}H_{20}O_2$** **29590-42-9**

Type of data: cloud points

w_A	0.860	0.860	0.860	0.860	0.860	0.860	0.790	0.790	0.790
w_B	0.051	0.051	0.051	0.051	0.051	0.051	0.055	0.055	0.055
w_C	0.089	0.089	0.089	0.089	0.089	0.089	0.155	0.155	0.155
T/K	373.85	380.85	393.15	414.45	434.45	454.15	352.85	374.35	396.25
P/bar	1925.9	1386.2	1124.1	984.5	925.9	898.3	1422.4	922.4	843.1
w_A	0.790	0.790	0.790	0.644	0.644	0.644	0.644	0.644	0.644
w_B	0.055	0.055	0.055	0.059	0.059	0.059	0.059	0.059	0.059
w_C	0.155	0.155	0.155	0.297	0.297	0.297	0.297	0.297	0.297
T/K	413.65	434.05	454.55	333.05	351.55	372.25	394.85	415.45	433.95
P/bar	815.5	801.7	787.9	913.8	719.0	677.6	677.6	681.0	691.4
w_A	0.644	0.588	0.588	0.588	0.588	0.588	0.588	0.588	0.495
w_B	0.059	0.054	0.054	0.054	0.054	0.054	0.054	0.054	0.060
w_C	0.297	0.358	0.358	0.358	0.358	0.358	0.358	0.358	0.445
T/K	453.75	331.55	351.95	374.05	393.85	414.35	433.85	454.15	334.15
P/bar	698.3	819.0	681.0	656.9	660.4	670.7	681.0	687.9	515.5

continued

continued

w_A	0.495	0.495	0.495	0.495	0.495	0.495
w_B	0.060	0.060	0.060	0.060	0.060	0.060
w_C	0.445	0.445	0.445	0.445	0.445	0.445
T/K	352.65	374.25	394.15	413.15	433.75	452.85
P/bar	508.6	525.9	553.5	574.1	594.8	598.3

Polymer (B): **poly(isopropyl acrylate)** **2007BY2**
Characterization: M_w/g.mol^{-1} = 120000,
Scientific Polymer Products, Inc., Ontario, NY
Solvent (A): **carbon dioxide** **CO_2** **124-38-9**
Solvent (C): **isopropyl acrylate** **$C_6H_{10}O_2$** **689-12-3**

Type of data: cloud points

w_A	0.871	0.871	0.871	0.871	0.871	0.871	0.871	0.871	0.752
w_B	0.047	0.047	0.047	0.047	0.047	0.047	0.047	0.047	0.060
w_C	0.082	0.082	0.082	0.082	0.082	0.082	0.082	0.082	0.188
T/K	384.25	395.15	404.45	414.85	425.15	432.75	444.15	454.15	372.65
P/bar	2429	2114	1728	1502	1391	1347	1288	1212	2595

w_A	0.752	0.752	0.752	0.752	0.752	0.536	0.536	0.536	0.536
w_B	0.060	0.060	0.060	0.060	0.060	0.049	0.049	0.049	0.049
w_C	0.188	0.188	0.188	0.188	0.188	0.415	0.415	0.415	0.415
T/K	392.95	407.45	425.55	438.75	455.95	377.75	394.45	407.75	426.15
P/bar	1555	1285	1121	1050	991	974	898	826	783

w_A	0.536	0.536
w_B	0.049	0.049
w_C	0.415	0.415
T/K	439.25	454.65
P/bar	736	693

Polymer (B): **poly(isopropyl acrylate)** **2007BY2**
Characterization: M_w/g.mol^{-1} = 120000,
Scientific Polymer Products, Inc., Ontario, NY
Solvent (A): **dimethyl ether** **C_2H_6O** **115-10-6**
Solvent (C): **isopropyl acrylate** **$C_6H_{10}O_2$** **689-12-3**

Type of data: cloud points

w_A	0.863	0.863	0.863	0.863	0.863	0.863	0.863	0.863	0.863
w_B	0.047	0.047	0.047	0.047	0.047	0.047	0.047	0.047	0.047
w_C	0.090	0.090	0.090	0.090	0.090	0.090	0.090	0.090	0.090
T/K	333.85	346.95	364.05	378.95	394.35	409.15	423.75	439.05	454.45
P/bar	1785	1467	1285	1195	1136	1098	1071	1050	1045

continued

continued

w_A	0.811	0.811	0.811	0.811	0.811	0.811	0.811	0.652	0.652
w_B	0.042	0.042	0.042	0.042	0.042	0.042	0.042	0.053	0.053
w_C	0.147	0.147	0.147	0.147	0.147	0.147	0.147	0.295	0.295
T/K	332.75	354.05	373.85	393.15	413.75	434.05	454.25	332.95	354.35
P/bar	905	895	895	895	899	902	902	398	457
w_A	0.652	0.652	0.652	0.652	0.652	0.474	0.474	0.474	0.474
w_B	0.053	0.053	0.053	0.053	0.053	0.051	0.051	0.051	0.051
w_C	0.295	0.295	0.295	0.295	0.295	0.475	0.475	0.475	0.475
T/K	373.45	393.75	413.85	433.35	453.95	333.65	354.85	373.25	393.65
P/bar	512	557	598	629	650	187	260	312	360
w_A	0.474	0.474	0.474						
w_B	0.051	0.051	0.051						
w_C	0.475	0.475	0.475						
T/K	413.65	434.45	453.85						
P/bar	405	447	464						

Polymer (B): **poly(isopropyl methacrylate)** **2007BY2**
Characterization: M_w/g.mol^{-1} = 100000, T_g/K = 354,
Scientific Polymer Products, Inc., Ontario, NY
Solvent (A): **carbon dioxide** **CO_2** **124-38-9**
Solvent (C): **isopropyl methacrylate** **$C_7H_{12}O_2$** **4655-34-9**

Type of data: cloud points

w_A	0.956	0.956	0.956	0.956	0.956	0.889	0.889	0.889	0.889
w_B	0.044	0.044	0.044	0.044	0.044	0.052	0.052	0.052	0.052
w_C	0.000	0.000	0.000	0.000	0.000	0.059	0.059	0.059	0.059
T/K	400.25	412.35	422.85	432.45	443.65	346.05	355.85	366.55	376.75
P/bar	2512	1907	1748	1685	1655	2216	1931	1609	1485
w_A	0.889	0.889	0.889	0.889	0.849	0.849	0.849	0.849	0.849
w_B	0.052	0.052	0.052	0.052	0.051	0.051	0.051	0.051	0.051
w_C	0.059	0.059	0.059	0.059	0.100	0.100	0.100	0.100	0.100
T/K	390.05	403.25	419.65	434.25	324.35	332.55	342.65	352.15	374.65
P/bar	1368	1329	1268	1241	1709	1407	1292	1162	1080
w_A	0.849	0.849	0.849	0.751	0.751	0.751	0.751	0.751	0.751
w_B	0.051	0.051	0.051	0.050	0.050	0.050	0.050	0.050	0.050
w_C	0.100	0.100	0.100	0.199	0.199	0.199	0.199	0.199	0.199
T/K	394.25	415.25	433.65	322.75	333.35	351.55	372.25	394.55	415.35
P/bar	1051	1043	1037	660	674	700	725	747	754
w_A	0.751	0.640	0.640	0.640	0.640	0.640	0.640	0.640	0.505
w_B	0.050	0.053	0.053	0.053	0.053	0.053	0.053	0.053	0.050
w_C	0.199	0.307	0.307	0.307	0.307	0.307	0.307	0.307	0.445
T/K	433.75	322.95	334.05	354.15	373.45	392.55	413.65	431.75	431.65
P/bar	762	299	324	396	443	480	501	512	316

continued

continued

w_A	0.505	0.505	0.505	0.505	0.505	0.505
w_B	0.050	0.050	0.050	0.050	0.050	0.050
w_C	0.445	0.445	0.445	0.445	0.445	0.445
T/K	413.35	394.35	375.25	354.95	345.05	334.95
P/bar	291	250	219	157	133	98

Type of data: vapor-liquid equilibrium data

w_A	0.505	0.505	0.505	0.505
w_B	0.050	0.050	0.050	0.050
w_C	0.445	0.445	0.445	0.445
T/K	328.45	324.25	319.15	314.65
P/bar	81	78	71	66

Type of data: vapor-liquid-liquid equilibrium data

w_A	0.505	0.505
w_B	0.050	0.050
w_C	0.445	0.445
T/K	344.15	350.15
P/bar	108	117

Polymer (B): **poly(L-lactic acid)** **2012GWO**
Characterization: M_w/g.mol^{-1} = 312000, M_w/M_n = 2.189
Korea Institute of Science and Technology, Seoul, Korea
Solvent (A): **carbon dioxide** **CO_2** **124-38-9**
Solvent (C): **dichloromethane** **CH_2Cl_2** **75-09-2**

Type of data: bubble and cloud points

T/K = 313.15

w_A	0.3158	0.3732	0.4059	0.4653	0.5178	0.3013	0.3588	0.4037	0.4514
w_B	0.010	0.010	0.010	0.010	0.010	0.025	0.025	0.025	0.025
w_C	0.6742	0.6168	0.5841	0.5247	0.4722	0.6737	0.6162	0.5713	0.5236
P/MPa	4.247	4.248	4.495	5.115	8.637	3.496	4.506	4.776	5.366
	BP	BP	BP	BP	CP	BP	BP	BP	BP

w_A	0.5012	0.3201	0.3608	0.4016	0.4501	0.4685
w_B	0.025	0.030	0.030	0.030	0.030	0.030
w_C	0.4738	0.6499	0.6092	0.5684	0.5199	0.5015
P/MPa	8.962	4.510	4.719	5.142	5.203	5.340
	CP	BP	BP	BP	BP	BP

T/K = 323.15

w_A	0.3158	0.3732	0.4059	0.4653	0.5178	0.3013	0.3588	0.4037	0.4514
w_B	0.010	0.010	0.010	0.010	0.010	0.025	0.025	0.025	0.025
w_C	0.6742	0.6168	0.5841	0.5247	0.4722	0.6737	0.6162	0.5713	0.5236
P/MPa	4.868	4.890	4.240	7.556	13.613	4.098	5.289	5.627	6.484
	BP	BP	BP	CP	CP	BP	BP	BP	CP

continued

continued

w_A	0.5012	0.3201	0.3608	0.4016	0.4501	0.4685
w_B	0.025	0.030	0.030	0.030	0.030	0.030
w_C	0.4738	0.6499	0.6092	0.5684	0.5199	0.5015
P/MPa	13.897	5.232	5.485	.019	7.302	9.667
	CP	BP	BP	BP	CP	CP

T/K = 333.15

w_A	0.3158	0.3732	0.4059	0.4653	0.5178	0.3013	0.3588	0.4037	0.4514
w_B	0.010	0.010	0.010	0.010	0.010	0.025	0.025	0.025	0.025
w_C	0.6742	0.6168	0.5841	0.5247	0.4722	0.6737	0.6162	0.5713	0.5236
P/MPa	5.624	5.713	6.281	11.780	17.877	4.709	6.104	6.499	11.337
	BP	BP	CP	CP	CP	BP	BP	BP	CP

w_A	0.5012	0.3201	0.3608	0.4016	0.4501	0.4685
w_B	0.025	0.030	0.030	0.030	0.030	0.030
w_C	0.4738	0.6499	0.6092	0.5684	0.5199	0.5015
P/MPa	18.322	6.113	6.171	6.942	11.761	14.075
	CP	BP	BP	BP	CP	CP

T/K = 343.15

w_A	0.3158	0.3732	0.4059	0.4653	0.5178	0.3013	0.3588	0.4037	0.4514
w_B	0.010	0.010	0.010	0.010	0.010	0.025	0.025	0.025	0.025
w_C	0.6742	0.6168	0.5841	0.5247	0.4722	0.6737	0.6162	0.5713	0.5236
P/MPa	6.440	6.489	7.009	16.001	22.442	5.290	6.997	8.109	15.550
	BP	BP	CP	CP	CP	BP	BP	BP	CP

w_A	0.5012	0.3201	0.3608	0.4016	0.4501	0.4685
w_B	0.025	0.030	0.030	0.030	0.030	0.030
w_C	0.4738	0.6499	0.6092	0.5684	0.5199	0.5015
P/MPa	22.483	6.950	7.029	10.533	15.828	18.239
	CP	BP	BP	CP	CP	CP

T/K = 353.15

w_A	0.3158	0.3732	0.4059	0.4653	0.5178	0.3013	0.3588	0.4037	0.4514
w_B	0.010	0.010	0.010	0.010	0.010	0.025	0.025	0.025	0.025
w_C	0.6742	0.6168	0.5841	0.5247	0.4722	0.6737	0.6162	0.5713	0.5236
P/MPa	7.185	7.238	10.333	19.315	26.142	6.123	7.894	11.827	19.468
	BP	BP	CP	CP	CP	BP	BP	CP	CP

w_A	0.5012	0.3201	0.3608	0.4016	0.4501	0.4685
w_B	0.025	0.030	0.030	0.030	0.030	0.030
w_C	0.4738	0.6499	0.6092	0.5684	0.5199	0.5015
P/MPa	26.452	7.800	7.970	14.376	19.718	21.992
	CP	BP	BP	CP	CP	CP

continued

continued

T/K = 363.15

w_A	0.3158	0.3732	0.4059	0.4653	0.5178	0.3013	0.3588	0.4037	0.4514
w_B	0.010	0.010	0.010	0.010	0.010	0.025	0.025	0.025	0.025
w_C	0.6742	0.6168	0.5841	0.5247	0.4722	0.6737	0.6162	0.5713	0.5236
P/MPa	8.019	9.935	13.886	22.881	29.933	6.576	10.607	15.478	23.127
	BP	BP	CP	CP	CP	BP	CP	CP	CP

w_A	0.5012	0.3201	0.3608	0.4016	0.4501	0.4685
w_B	0.025	0.030	0.030	0.030	0.030	0.030
w_C	0.4738	0.6499	0.6092	0.5684	0.5199	0.5015
P/MPa	30.222	11.095	11.575	18.045	23.473	25.656
	CP	CP	CP	CP	CP	CP

Polymer (B): **poly(lactic acid-*b*-ethylene glycol-*b*-lactic acid)** **2010JIA**
Characterization: 5 wt% PEG, M_w(PLA-block)/g.mol^{-1} = 50000,
Ji-Nan Dai Gang Biological Technology Co., Ltd., China
Solvent (A): **carbon dioxide** **CO_2** **124-38-9**
Solvent (C): **dichloromethane** **CH_2Cl_2** **75-09-2**

Type of data: cloud points

Comments: The total mass of the polymer (B) = 0.224 g was kept constant.
The total mass of CO_2 = 12.0 g was kept constant.
The weight fraction of solvent (C) is the value in the solvent mixture.

w_C	0.3000	0.3460	0.3642	0.3839	0.4280	0.3000	0.3460	0.3642	0.3839
T/K	313.1	313.1	313.1	313.1	313.1	318.1	318.1	318.1	318.1
P/MPa	24.46	16.63	13.01	10.83	7.22	26.57	18.24	14.82	12.51

w_C	0.4280	0.3000	0.3460	0.3642	0.3839	0.4280	0.3000	0.3460	0.3642
T/K	318.1	323.1	323.1	323.1	323.1	323.1	328.1	328.1	328.1
P/MPa	9.43	28.54	19.85	16.90	14.29	11.29	30.40	21.73	19.20

w_C	0.3839	0.4280	0.3000	0.3460	0.3642	0.3839	0.4280	0.3000	0.3460
T/K	328.1	328.1	333.1	333.1	333.1	333.1	333.1	338.1	338.1
P/MPa	16.44	13.44	32.54	23.38	20.75	18.36	15.36	34.44	25.36

w_C	0.3642	0.3839	0.4280
T/K	338.1	338.1	338.1
P/MPa	23.00	20.10	16.94

Polymer (B): **poly(lactic acid-*b*-ethylene glycol-*b*-lactic acid)** **2010JIA**
Characterization: 10 wt% PEG, M_w(PLA-block)/g.mol^{-1} = 50000,
Ji-Nan Dai Gang Biological Technology Co., Ltd., China
Solvent (A): **carbon dioxide** **CO_2** **124-38-9**
Solvent (C): **dichloromethane** **CH_2Cl_2** **75-09-2**

continued

continued

Type of data: cloud points

Comments: The total mass of the polymer (B) = 0.224 g was kept constant.
The total mass of CO_2 = 12.0 g was kept constant.
The weight fraction of solvent (C) is the value in the solvent mixture.

w_C	0.3025	0.3504	0.3756	0.4004	0.4452	0.3025	0.3504	0.3756	0.4004
T/K	313.1	313.1	313.1	313.1	313.1	318.1	318.1	318.1	318.1
P/MPa	21.82	13.62	10.97	6.80	4.42 *)	24.24	15.35	12.62	8.89

w_C	0.4452	0.3025	0.3504	0.3756	0.4004	0.4452	0.3025	0.3504	0.3756
T/K	318.1	323.1	323.1	323.1	323.1	323.1	328.1	328.1	328.1
P/MPa	4.91 *)	26.30	17.42	14.45	10.98	6.52	28.26	19.19	16.15

w_C	0.4004	0.4452	0.3025	0.3504	0.3756	0.4004	0.4452	0.3025	0.3504
T/K	328.1	328.1	333.1	333.1	333.1	333.1	333.1	338.1	338.1
P/MPa	13.03	8.49	30.21	20.99	18.06	14.91	10.51	32.25	22.45

w_C	0.3756	0.4004	0.4452
T/K	338.1	338.1	338.1
P/MPa	20.02	16.69	12.40

*) bubble point

Polymer (B): **poly(lactic acid-*b*-ethylene glycol-*b*-lactic acid)** **2010JIA**
Characterization: 15 wt% PEG, M_w(PLA-block)/g.mol^{-1} = 50000,
Ji-Nan Dai Gang Biological Technology Co., Ltd., China
Solvent (A): **carbon dioxide** **CO_2** **124-38-9**
Solvent (C): **dichloromethane** **CH_2Cl_2** **75-09-2**

Type of data: cloud points

Comments: The total mass of the polymer (B) = 0.224 g was kept constant.
The total mass of CO_2 = 12.0 g was kept constant.
The weight fraction of solvent (C) is the value in the solvent mixture.

w_C	0.3051	0.3195	0.3534	0.3819	0.4437	0.3051	0.3195	0.3534	0.3819
T/K	313.1	313.1	313.1	313.1	313.1	318.1	318.1	318.1	318.1
P/MPa	19.51	18.03	12.62	8.51	4.15 *)	21.46	19.92	14.41	10.35

w_C	0.4437	0.3051	0.3195	0.3534	0.3819	0.4437	0.3051	0.3195	0.3534
T/K	318.1	323.1	323.1	323.1	323.1	323.1	328.1	328.1	328.1
P/MPa	4.50 *)	23.34	21.86	16.28	12.13	6.04	25.40	23.87	18.30

w_C	0.3819	0.4437	0.3051	0.3195	0.3534	0.3819	0.4437	0.3051	0.3195
T/K	328.1	328.1	333.1	333.1	333.1	333.1	333.1	338.1	338.1
P/MPa	14.15	7.64	27.38	25.90	20.10	15.96	9.59	29.40	27.70

w_C	0.3534	0.3819	0.4437
T/K	338.1	338.1	338.1
P/MPa	22.10	17.81	11.31

*) bubble point

Polymer (B):	**poly(lactic acid-*b*-ethylene glycol-*b*-lactic acid)**		**2010JIA**
Characterization:	20 wt% PEG, M_w(PLA-block)/g.mol^{-1} = 50000, Ji-Nan Dai Gang Biological Technology Co., Ltd., China		
Solvent (A):	**carbon dioxide**	**CO_2**	**124-38-9**
Solvent (C):	**dichloromethane**	**CH_2Cl_2**	**75-09-2**

Type of data: cloud points

Comments: The total mass of the polymer (B) = 0.224 g was kept constant.
The total mass of CO_2 = 12.0 g was kept constant.
The weight fraction of solvent (C) is the value in the solvent mixture.

w_C	0.2633	0.2910	0.3214	0.3657	0.4110	0.2633	0.2910	0.3214	0.3657
T/K	313.1	313.1	313.1	313.1	313.1	318.1	318.1	318.1	318.1
P/MPa	23.46	17.71	12.66	9.84	3.91 *)	25.25	19.60	14.56	11.70

w_C	0.4110	0.2633	0.2910	0.3214	0.3657	0.4110	0.2633	0.2910	0.3214
T/K	318.1	323.1	323.1	323.1	323.1	323.1	328.1	328.1	328.1
P/MPa	4.37 *)	27.17	21.50	16.49	13.78	7.10	29.02	23.64	18.48

w_C	0.3657	0.4110	0.2633	0.2910	0.3214	0.3657	0.4110	0.2633	0.2910
T/K	328.1	328.1	333.1	333.1	333.1	333.1	333.1	338.1	338.1
P/MPa	15.92	9.23	31.20	25.51	20.41	17.80	11.30	32.98	27.34

w_C	0.3214	0.3657	0.4110
T/K	338.1	338.1	338.1
P/MPa	22.35	19.84	13.39

*) bubble point

Polymer (B):	**poly(lactic acid-*b*-ethylene glycol-*b*-lactic acid)**		**2010JIA**
Characterization:	5 wt% PEG, M_w(PLA-block)/g.mol^{-1} = 50000, Ji-Nan Dai Gang Biological Technology Co., Ltd., China		
Solvent (A):	**carbon dioxide**	**CO_2**	**124-38-9**
Solvent (C):	**dichloromethane**	**CH_2Cl_2**	**75-09-2**
Solvent (D):	**ethanol**	**C_2H_6O**	**64-17-5**

Type of data: cloud points

Comments: The total mass of the polymer (B) = 0.224 g was kept constant.
The total mass of CO_2 = 12.0 g was kept constant.
The weight fraction of solvents (C+D) is the value in the mixture with CO_2.
The molar ratio of dichloromethane/ethanol = 9:1 was kept constant.

w_{C+D}	0.2601	0.2897	0.3350	0.3640	0.3820	0.2601	0.2897	0.3350	0.3640
T/K	313.1	313.1	313.1	313.1	313.1	318.1	318.1	318.1	318.1
P/MPa	17.88	14.88	11.12	8.08	5.78 *)	19.73	16.71	13.02	9.90

w_{C+D}	0.3820	0.2601	0.2897	0.3350	0.3640	0.3820	0.2601	0.2897	0.3350
T/K	318.1	323.1	323.1	323.1	323.1	323.1	328.1	328.1	328.1
P/MPa	7.84	21.67	18.70	14.89	11.81	9.70	23.74	20.89	16.70

continued

continued

w_{C+D}	0.3640	0.3820	0.2601	0.2897	0.3350	0.3640	0.3820	0.2601	0.2897
T/K	328.1	328.1	333.1	333.1	333.1	333.1	333.1	338.1	338.1
P/MPa	13.62	11.99	25.60	23.11	18.92	15.93	14.21	27.20	25.01

w_{C+D}	0.3350	0.3640	0.3820
T/K	338.1	338.1	338.1
P/MPa	21.34	17.74	15.95

*) bubble point

Polymer (B): **poly(lactic acid-*b*-ethylene glycol-*b*-lactic acid)** **2010JIA**
Characterization: 10 wt% PEG, M_w(PLA-block)/g.mol^{-1} = 50000,
Ji-Nan Dai Gang Biological Technology Co., Ltd., China
Solvent (A): **carbon dioxide** **CO_2** **124-38-9**
Solvent (C): **dichloromethane** **CH_2Cl_2** **75-09-2**
Solvent (D): **ethanol** **C_2H_6O** **64-17-5**

Type of data: cloud points

Comments: The total mass of the polymer (B) = 0.224 g was kept constant.
The total mass of CO_2 = 12.0 g was kept constant.
The weight fraction of solvents (C+D) is the value in the mixture with CO_2.
The molar ratio of dichloromethane/ethanol = 9:1 was kept constant.

w_{C+D}	0.2601	0.2967	0.3235	0.3640	0.3820	0.2601	0.2967	0.3235	0.3640
T/K	313.1	313.1	313.1	313.1	313.1	318.1	318.1	318.1	318.1
P/MPa	15.79	14.10	8.57	5.89 *)	5.28 *)	17.71	15.98	10.45	8.29

w_{C+D}	0.3820	0.2601	0.2967	0.3235	0.3640	0.3820	0.2601	0.2967	0.3235
T/K	318.1	323.1	323.1	323.1	323.1	323.1	328.1	328.1	328.1
P/MPa	5.75 *)	19.61	17.91	12.38	10.17	8.01	21.35	19.99	14.52

w_{C+D}	0.3640	0.3820	0.2601	0.2967	0.3235	0.3640	0.3820	0.2601	0.2967
T/K	328.1	328.1	333.1	333.1	333.1	333.1	333.1	338.1	338.1
P/MPa	12.35	10.31	23.56	22.13	16.62	14.29	12.42	25.13	24.01

w_{C+D}	0.3235	0.3640	0.3820
T/K	338.1	338.1	338.1
P/MPa	18.25	16.20	14.02

*) bubble point

Polymer (B):	**poly(lactic acid-*b*-ethylene glycol-*b*-lactic acid)**		**2010JIA**
Characterization:	15 wt% PEG, M_w(PLA-block)/g.mol^{-1} = 50000, Ji-Nan Dai Gang Biological Technology Co., Ltd., China		
Solvent (A):	**carbon dioxide**	**CO_2**	**124-38-9**
Solvent (C):	**dichloromethane**	**CH_2Cl_2**	**75-09-2**
Solvent (D):	**ethanol**	**C_2H_6O**	**64-17-5**

Type of data: cloud points

Comments: The total mass of the polymer (B) = 0.224 g was kept constant.
The total mass of CO_2 = 12.0 g was kept constant.
The weight fraction of solvents (C+D) is the value in the mixture with CO_2.
The molar ratio of dichloromethane/ethanol = 9:1 was kept constant.

w_{C+D}	0.2601	0.2897	0.3235	0.3640	0.3820	0.2601	0.2897	0.3235	0.3640
T/K	313.1	313.1	313.1	313.1	313.1	318.1	318.1	318.1	318.1
P/MPa	14.40	12.30	7.40	5.50 *)	5.15 *)	16.28	14.17	9.21	7.31

w_{C+D}	0.3820	0.2601	0.2897	0.3235	0.3640	0.3820	0.2601	0.2897	0.3235
T/K	318.1	323.1	323.1	323.1	323.1	323.1	328.1	328.1	328.1
P/MPa	5.64 *)	18.19	16.01	11.08	9.08	7.25	20.21	18.25	13.41

w_{C+D}	0.3640	0.3820	0.2601	0.2897	0.3235	0.3640	0.3820	0.2601	0.2897
T/K	328.1	328.1	333.1	333.1	333.1	333.1	333.1	338.1	338.1
P/MPa	11.28	9.36	22.42	20.13	15.35	13.35	11.28	24.12	22.03

w_{C+D}	0.3235	0.3640	0.3820
T/K	338.1	338.1	338.1
P/MPa	17.26	15.15	13.14

*) bubble point

Polymer (B):	**poly(lactic acid-*b*-ethylene glycol-*b*-lactic acid)**		**2010JIA**
Characterization:	20 wt% PEG, M_w(PLA-block)/g.mol^{-1} = 50000, Ji-Nan Dai Gang Biological Technology Co., Ltd., China		
Solvent (A):	**carbon dioxide**	**CO_2**	**124-38-9**
Solvent (C):	**dichloromethane**	**CH_2Cl_2**	**75-09-2**
Solvent (D):	**ethanol**	**C_2H_6O**	**64-17-5**

Type of data: cloud points

Comments: The total mass of the polymer (B) = 0.224 g was kept constant.
The total mass of CO_2 = 12.0 g was kept constant.
The weight fraction of solvents (C+D) is the value in the mixture with CO_2.
The molar ratio of dichloromethane/ethanol = 9:1 was kept constant.

w_{C+D}	0.2601	0.2897	0.3235	0.3640	0.3820	0.2601	0.2897	0.3235	0.3640
T/K	313.1	313.1	313.1	313.1	313.1	318.1	318.1	318.1	318.1
P/MPa	12.60	10.75	5.23 *)	4.89 *)	4.62 *)	14.62	12.70	7.88	5.58 *)

continued

continued

w_{C+D}	0.3820	0.2601	0.2897	0.3235	0.3640	0.3820	0.2601	0.2897	0.3235
T/K	318.1	323.1	323.1	323.1	323.1	323.1	328.1	328.1	328.1
P/MPa	5.01 *)	16.51	14.55	9.72	8.15	6.41	18.72	16.35	11.62
w_{C+D}	0.3640	0.3820	0.2601	0.2897	0.3235	0.3640	0.3820	0.2601	0.2897
T/K	328.1	328.1	333.1	333.1	333.1	333.1	333.1	338.1	338.1
P/MPa	10.02	8.56	20.58	18.45	13.50	12.01	10.41	22.31	20.14
w_{C+D}	0.3235	0.3640	0.3820						
T/K	338.1	338.1	338.1						
P/MPa	15.40	13.89	12.23						

*) bubble point

Polymer (B): **poly(L-lactide)** **2005KAL**
Characterization: M_n/g.mol^{-1} = 81800, M_w/g.mol^{-1} = 189000, Galactic Laboratories, Belgium
Solvent (A): **carbon dioxide** **CO_2** **124-38-9**
Solvent (C): **dichloromethane** **CH_2Cl_2** **75-09-2**

Type of data: bubble and cloud points

T/K = 308.15

w_A	0.7396	0.6493	0.5938	0.5454	0.5172	0.4804	0.7325	0.6429	0.5876
w_B	0.0100	0.0100	0.0100	0.0100	0.0100	0.0100	0.0200	0.0200	0.0200
w_C	0.2504	0.3407	0.3962	0.4446	0.4728	0.5096	0.2475	0.3371	0.3924
P/bar	34.12	41.55	42.35	49.08	49.96	76.06	34.53	41.85	45.46
	VLE	VLE	VLE	VLE	VLE	LLE	VLE	VLE	VLE
w_A	0.5429	0.5113	0.4729	0.7251	0.6371	0.5816	0.5433	0.5065	0.4685
w_B	0.0200	0.0200	0.0200	0.0300	0.0300	0.0300	0.0300	0.0300	0.0300
w_C	0.4371	0.4687	0.5071	0.2449	0.3329	0.3884	0.4267	0.4635	0.5015
P/bar	49.68	54.59	93.42	34.63	42.05	46.87	50.78	59.81	106.8
	VLE	VLE	LLE	VLE	VLE	VLE	VLE	LLE	LLE

T/K = 313.15

w_A	0.7396	0.6493	0.5938	0.5454	0.5172	0.4804	0.7325	0.6429	0.5876
w_B	0.0100	0.0100	0.0100	0.0100	0.0100	0.0100	0.0200	0.0200	0.0200
w_C	0.2504	0.3407	0.3962	0.4446	0.4728	0.5096	0.2475	0.3371	0.3924
P/bar	37.24	45.16	46.06	53.39	54.50	100.7	37.34	45.46	49.47
	VLE	VLE	VLE	VLE	VLE	LLE	VLE	VLE	VLE
w_A	0.5429	0.5113	0.4729	0.7251	0.6371	0.5816	0.5433	0.5065	0.4685
w_B	0.0200	0.0200	0.0200	0.0300	0.0300	0.0300	0.0300	0.0300	0.0300
w_C	0.4371	0.4687	0.5071	0.2449	0.3329	0.3884	0.4267	0.4635	0.5015
P/bar	54.30	63.42	120.7	37.44	45.96	50.98	55.50	82.79	129.5
	VLE	VLE	LLE	VLE	VLE	VLE	VLE	LLE	LLE

continued

continued

T/K = 318.15

w_A	0.7396	0.6493	0.5938	0.5454	0.5172	0.4804	0.7325	0.6429	0.5876
w_B	0.0100	0.0100	0.0100	0.0100	0.0100	0.0100	0.0200	0.0200	0.0200
w_C	0.2504	0.3407	0.3962	0.4446	0.4728	0.5096	0.2475	0.3371	0.3924
P/bar	40.05	49.18	49.86	58.20	59.00	122.6	40.25	49.28	53.69
	VLE	VLE	VLE	VLE	VLE	LLE	VLE	VLE	VLE

w_A	0.5429	0.5113	0.4729	0.7251	0.6371	0.5816	0.5433	0.5065	0.4685
w_B	0.0200	0.0200	0.0200	0.0300	0.0300	0.0300	0.0300	0.0300	0.0300
w_C	0.4371	0.4687	0.5071	0.2449	0.3329	0.3884	0.4267	0.4635	0.5015
P/bar	59.21	85.90	144.2	40.45	49.78	55.40	63.53	105.9	153.7
	VLE	LLE	LLE	VLE	VLE	VLE	VLE	LLE	LLE

T/K = 323.15

w_A	0.7396	0.6493	0.5938	0.5454	0.5172	0.4804	0.7325	0.6429	0.5876
w_B	0.0100	0.0100	0.0100	0.0100	0.0100	0.0100	0.0200	0.0200	0.0200
w_C	0.2504	0.3407	0.3962	0.4446	0.4728	0.5096	0.2475	0.3371	0.3924
P/bar	43.06	52.69	53.69	63.73	77.77	145.4	43.26	53.19	58.10
	VLE	VLE	VLE	VLE	LLE	LLE	VLE	VLE	VLE

w_A	0.5429	0.5113	0.4729	0.7251	0.6371	0.5816	0.5433	0.5065	0.4685
w_B	0.0200	0.0200	0.0200	0.0300	0.0300	0.0300	0.0300	0.0300	0.0300
w_C	0.4371	0.4687	0.5071	0.2449	0.3329	0.3884	0.4267	0.4635	0.5015
P/bar	70.05	108.4	167.1	43.36	53.59	59.91	86.50	129.7	175.8
	LLE	LLE	LLE	VLE	VLE	VLE	LLE	LLE	LLE

T/K = 328.15

w_A	0.7396	0.6493	0.5938	0.5454	0.5172	0.4804	0.7325	0.6429	0.5876
w_B	0.0100	0.0100	0.0100	0.0100	0.0100	0.0100	0.0200	0.0200	0.0200
w_C	0.2504	0.3407	0.3962	0.4446	0.4728	0.5096	0.2475	0.3371	0.3924
P/bar	45.87	56.60	57.70	86.20	101.5	167.9	46.37	57.11	62.62
	VLE	VLE	VLE	LLE	LLE	LLE	VLE	VLE	VLE

w_A	0.5429	0.5113	0.4729	0.7251	0.6371	0.5816	0.5433	0.5065	0.4685
w_B	0.0200	0.0200	0.0200	0.0300	0.0300	0.0300	0.0300	0.0300	0.0300
w_C	0.4371	0.4687	0.5071	0.2449	0.3329	0.3884	0.4267	0.4635	0.5015
P/bar	92.53	130.0	190.0	46.47	57.60	64.33	108.3	151.6	198.0
	LLE	LLE	LLE	VLE	VLE	VLE	LLE	LLE	LLE

T/K = 333.15

w_A	0.7396	0.6493	0.5938	0.5454	0.5172	0.4804	0.7325	0.6429	0.5876
w_B	0.0100	0.0100	0.0100	0.0100	0.0100	0.0100	0.0200	0.0200	0.0200
w_C	0.2504	0.3407	0.3962	0.4446	0.4728	0.5096	0.2475	0.3371	0.3924
P/bar	49.28	60.62	62.12	106.3	123.3	188.8	49.68	61.22	67.23
	VLE	VLE	VLE	LLE	LLE	LLE	VLE	VLE	VLE

continued

continued

w_A	0.5429	0.5113	0.4729	0.7251	0.6371	0.5816	0.5433	0.5065	0.4685
w_B	0.0200	0.0200	0.0200	0.0300	0.0300	0.0300	0.0300	0.0300	0.0300
w_C	0.4371	0.4687	0.5071	0.2449	0.3329	0.3884	0.4267	0.4635	0.5015
P/bar	113.6	150.8	211.6	49.78	61.82	69.35	129.4	173.1	219.8
	LLE	LLE	LLE	VLE	VLE	VLE	LLE	LLE	LLE

T/K = 338.15

w_A	0.7396	0.6493	0.5938	0.5454	0.5172	0.4804	0.7325	0.6429	0.5876
w_B	0.0100	0.0100	0.0100	0.0100	0.0100	0.0100	0.0200	0.0200	0.0200
w_C	0.2504	0.3407	0.3962	0.4446	0.4728	0.5096	0.2475	0.3371	0.3924
P/bar	52.59	65.03	66.63	127.9	144.4	–	52.99	65.44	71.85
	VLE	VLE	VLE	LLE	LLE		VLE	VLE	VLE

w_A	0.5429	0.5113	0.4729	0.7251	0.6371	0.5816	0.5433	0.5065	0.4685
w_B	0.0200	0.0200	0.0200	0.0300	0.0300	0.0300	0.0300	0.0300	0.0300
w_C	0.4371	0.4687	0.5071	0.2449	0.3329	0.3884	0.4267	0.4635	0.5015
P/bar	133.7	171.3	233.2	53.09	66.23	81.89	150.9	194.5	240.6
	LLE	LLE	LLE	VLE	VLE	LLE	LLE	LLE	LLE

T/K = 343.15

w_A	0.7396	0.6493	0.5938	0.5454	0.5172	0.4804	0.7325	0.6429	0.5876
w_B	0.0100	0.0100	0.0100	0.0100	0.0100	0.0100	0.0200	0.0200	0.0200
w_C	0.2504	0.3407	0.3962	0.4446	0.4728	0.5096	0.2475	0.3371	0.3924
P/bar	56.03	69.15	70.95	147.7	164.7	–	56.30	69.85	78.17
	VLE	VLE	VLE	LLE	LLE		VLE	VLE	VLE

w_A	0.5429	0.5113	0.7251	0.6371	0.5816	0.5433	0.5065
w_B	0.0200	0.0200	0.0300	0.0300	0.0300	0.0300	0.0300
w_C	0.4371	0.4687	0.2449	0.3329	0.3884	0.4267	0.4635
P/bar	154.6	190.9	56.60	70.65	101.7	171.2	214.5
	LLE	LLE	VLE	VLE	LLE	LLE	LLE

T/K = 348.15

w_A	0.7396	0.6493	0.5938	0.5454	0.5172	0.4804	0.7325	0.6429	0.5876
w_B	0.0100	0.0100	0.0100	0.0100	0.0100	0.0100	0.0200	0.0200	0.0200
w_C	0.2504	0.3407	0.3962	0.4446	0.4728	0.5096	0.2475	0.3371	0.3924
P/bar	59.41	73.36	75.36	167.8	184.4	–	59.91	74.26	97.64
	VLE	VLE	VLE	LLE	LLE		VLE	VLE	LLE

w_A	0.5429	0.5113	0.7251	0.6371	0.5816	0.5433
w_B	0.0200	0.0200	0.0300	0.0300	0.0300	0.0300
w_C	0.4371	0.4687	0.2449	0.3329	0.3884	0.4267
P/bar	174.9	209.9	60.22	75.16	121.1	191.2
	LLE	LLE	VLE	VLE	LLE	LLE

continued

continued

T/K = 353.15

w_A	0.7396	0.6493	0.5938	0.5454	0.7325	0.6429	0.5876	0.5113	0.7251
w_B	0.0100	0.0100	0.0100	0.0100	0.0200	0.0200	0.0200	0.0200	0.0300
w_C	0.2504	0.3407	0.3962	0.4446	0.2475	0.3371	0.3924	0.4687	0.2449
P/bar	62.82	77.58	79.88	186.1	63.22	78.58	116.3	228.9	63.73
	VLE	VLE	VLE	LLE	VLE	VLE	LLE	LLE	VLE

w_A	0.6371	0.5816
w_B	0.0300	0.0300
w_C	0.3329	0.3884
P/bar	79.58	139.6
	VLE	LLE

T/K = 358.15

w_A	0.7396	0.6493	0.5938	0.7325	0.6429	0.5876	0.7251	0.6371	0.5816
w_B	0.0100	0.0100	0.0100	0.0200	0.0200	0.0200	0.0300	0.0300	0.0300
w_C	0.2504	0.3407	0.3962	0.2475	0.3371	0.3924	0.2449	0.3329	0.3884
P/bar	66.24	82.19	84.29	66.94	83.09	134.9	67.34	84.09	158.2
	VLE	VLE	VLE	VLE	VLE	LLE	VLE	VLE	LLE

T/K = 363.15

w_A	0.7396	0.6493	0.5938	0.7325	0.6429	0.5876	0.7251	0.6371	0.5816
w_B	0.0100	0.0100	0.0100	0.0200	0.0200	0.0200	0.0300	0.0300	0.0300
w_C	0.2504	0.3407	0.3962	0.2475	0.3371	0.3924	0.2449	0.3329	0.3884
P/bar	69.95	86.80	97.14	70.45	88.71	152.9	70.85	96.03	175.9
	VLE	VLE	LLE	VLE	VLE	LLE	VLE	LLE	LLE

T/K = 368.15

w_A	0.7396	0.6493	0.5938	0.7325	0.6429	0.5876	0.7251	0.6371
w_B	0.0100	0.0100	0.0100	0.0200	0.0200	0.0200	0.0300	0.0300
w_C	0.2504	0.3407	0.3962	0.2475	0.3371	0.3924	0.2449	0.3329
P/bar	73.43	101.0	113.4	73.96	106.1	170.2	74.56	113.4
	VLE	LLE	LLE	VLE	LLE	LLE	VLE	LLE

T/K = 373.15

w_A	0.7396	0.6493	0.5938	0.7325	0.6429	0.5876	0.7251	0.6371
w_B	0.0100	0.0100	0.0100	0.0200	0.0200	0.0200	0.0300	0.0300
w_C	0.2504	0.3407	0.3962	0.2475	0.3371	0.3924	0.2449	0.3329
P/bar	77.17	117.4	130.4	77.67	122.3	187.0	78.08	129.9
	VLE	LLE	LLE	VLE	LLE	LLE	VLE	LLE

Polymer (B): **poly(L-lactide)** **2004LIM, 2006PAR**
Characterization: M_n/g.mol^{-1} = 2000, M_w/g.mol^{-1} = 2500,
Biomaterials Research Center, KIST, South Korea
Solvent (A): **carbon dioxide** **CO_2** **124-38-9**
Solvent (C): **chlorodifluoromethane** **$CHClF_2$** **75-45-6**

Type of data: cloud points

w_A	0.1806	0.1806	0.1806	0.1806	0.1806	0.1806	0.1806	0.1806
w_B	0.0296	0.0296	0.0296	0.0296	0.0296	0.0296	0.0296	0.0296
w_C	0.7898	0.7898	0.7898	0.7898	0.7898	0.7898	0.7898	0.7898
T/K	322.5	333.3	343.1	353.2	363.2	373.2	383.2	393.2
P/MPa	4.4	7.8	10.8	13.7	16.4	19.8	22.2	24.2

w_A	0.5155	0.5155	0.5155	0.5155	0.5155	0.5155	0.5155	0.5155
w_B	0.0302	0.0302	0.0302	0.0302	0.0302	0.0302	0.0302	0.0302
w_C	0.4543	0.4543	0.4543	0.4543	0.4543	0.4543	0.4543	0.4543
T/K	314.2	322.1	331.1	342.4	353.2	363.2	373.2	383.2
P/MPa	20.5	24.0	27.5	31.5	35.0	38.2	41.5	44.1

Polymer (B): **poly(DL-lactide-*co*-glycolide)** **2013GRA**
Characterization: M_w/g.mol^{-1} = 65000, M_w/M_n = 2.02, 50 mol% glycolide
Resomer RG504H, Boehringer Ingelheim, Germany
Solvent (A): **carbon dioxide** **CO_2** **124-38-9**
Solvent (C): **2-propanone** **C_3H_6O** **67-64-1**

Type of data: bubble and cloud points

T/K = 348.15

w_B	0.00	0.05	0.10	0.10
w_C/w_A	89/11	89/11	89/11	85/05
P/bar	25	41	21	9
	VLE	VLE	VLE	VLE

T/K = 373.15

w_B	0.00	0.05	0.10	0.10
w_C/w_A	89/11	89/11	89/11	85/05
P/bar	33	42	28	13
	VLE	VLE	VLE	VLE

T/K = 398.15

w_B	0.00	0.10	0.05	0.10	0.05	0.10
w_C/w_A	89/11	85/05	89/11	89/11	89/11	89/11
P/bar	44	19	80	49	48	31
	VLE	VLE	LLE	LLE	VLLE	VLLE

continued

continued

T/K = 423.15

w_B	0.00	0.05	0.10	0.10	0.05	0.10	0.10
w_C/w_A	89/11	89/11	89/11	85/05	89/11	89/11	85/05
P/bar	45	119	91	38	52	38	25
	VLE	LLE	LLE	LLE	VLLE	VLLE	VLLE

T/K = 433.15

w_B	0.10	0.10
w_C/w_A	85/05	85/05
P/bar	43	28
	LLE	VLLE

T/K = 443.15

w_B	0.10	0.10
w_C/w_A	85/05	85/05
P/bar	51	33
	LLE	VLLE

Polymer (B): **poly(methyl acrylate)** **2004BY3**
Characterization: M_w/g.mol^{-1} = 40000, Polysciences, Inc., Warrington, PA
Solvent (A): **carbon dioxide** CO_2 **124-38-9**
Solvent (C): **methyl acrylate** $C_4H_6O_2$ **96-33-3**

Type of data: cloud points

w_A	0.902	0.902	0.902	0.902	0.902	0.902	0.812	0.812	0.812
w_B	0.048	0.048	0.048	0.048	0.048	0.048	0.051	0.051	0.051
w_C	0.050	0.050	0.050	0.050	0.050	0.050	0.137	0.137	0.137
T/K	382.95	403.95	423.35	444.15	463.55	483.85	343.75	363.75	383.65
P/bar	1740.3	1734.5	1733.4	1719.0	1697.6	1662.1	1206.6	1267.2	1304.5
w_A	0.812	0.812	0.812	0.699	0.699	0.699	0.699	0.699	0.699
w_B	0.051	0.051	0.051	0.048	0.048	0.048	0.048	0.048	0.048
w_C	0.137	0.137	0.137	0.253	0.253	0.253	0.253	0.253	0.253
T/K	273.15	423.65	443.95	313.15	333.25	348.75	364.45	378.85	393.05
P/bar	1322.4	1333.4	1325.9	468.6	582.4	652.8	694.5	749.3	787.9
w_A	0.699	0.699	0.517	0.517	0.517	0.517	0.517	0.517	0.517
w_B	0.048	0.048	0.050	0.050	0.050	0.050	0.050	0.050	0.050
w_C	0.253	0.253	0.433	0.433	0.433	0.433	0.433	0.433	0.433
T/K	406.95	426.25	299.95	313.55	327.85	343.15	359.05	372.85	389.05
P/bar	820.7	853.5	50.0	104.8	179.7	250.0	311.4	363.8	413.5
w_A	0.517	0.461	0.461	0.461	0.461	0.461	0.461	0.461	
w_B	0.050	0.056	0.056	0.056	0.056	0.056	0.056	0.056	
w_C	0.433	0.483	0.483	0.483	0.483	0.483	0.483	0.483	
T/K	403.15	322.65	328.45	343.35	357.75	373.95	386.95	403.25	
P/bar	439.7	66.9	95.9	157.6	214.1	267.9	312.1	347.2	

continued

continued

Type of data: vapor-liquid equilibrium data

w_A	0.461	0.461	0.461
w_B	0.056	0.056	0.056
w_C	0.483	0.483	0.483
T/K	314.05	308.35	301.05
P/bar	54.1	50.7	46.9

Type of data: vapor-liquid-liquid equilibrium data

w_A	0.461	0.461
w_B	0.056	0.056
w_C	0.483	0.483
T/K	329.65	345.85
P/bar	65.3	76.7

Polymer (B): **poly(methyl methacrylate)** **2002LE2**
Characterization: M_w/g.mol^{-1} = 15000, Aldrich Chem. Co., Inc., Milwaukee, WI
Solvent (A): **carbon dioxide** **CO_2** **124-38-9**
Solvent (C): **chlorodifluoromethane** **$CHClF_2$** **75-45-6**

Type of data: cloud points

w_A	0.0000	0.0000	0.0000	0.0000	0.0000	0.1185	0.1185	0.1185	0.1185
w_B	0.0498	0.0498	0.0498	0.0498	0.0498	0.0503	0.0503	0.0503	0.0503
w_C	0.9502	0.9502	0.9502	0.9502	0.9502	0.8312	0.8312	0.8312	0.8312
T/K	337.65	342.85	353.15	363.35	372.95	323.25	332.75	342.65	352.95
P/bar	29.6	45.5	77.9	108.5	138.7	37.5	74.5	111.5	146.2
w_A	0.1185	0.1185	0.1719	0.1719	0.1719	0.1719	0.1719	0.1719	0.2923
w_B	0.0503	0.0503	0.0500	0.0500	0.0500	0.0500	0.0500	0.0500	0.0491
w_C	0.8312	0.8312	0.7781	0.7781	0.7781	0.7781	0.7781	0.7781	0.6586
T/K	362.85	372.75	322.65	332.95	342.85	352.65	362.95	373.15	312.75
P/bar	180.6	213.0	63.7	104.7	143.2	178.5	216.0	250.9	96.5
w_A	0.2923	0.2923	0.2923	0.2923	0.2923	0.2923	0.3579	0.3579	0.3579
w_B	0.0491	0.0491	0.0491	0.0491	0.0491	0.0491	0.0499	0.0499	0.0499
w_C	0.6586	0.6586	0.6586	0.6586	0.6586	0.6586	0.5922	0.5922	0.5922
T/K	322.55	332.95	342.65	352.85	362.85	372.95	313.25	323.15	333.15
P/bar	145.5	192.7	231.4	275.5	314.4	351.5	152.7	204.5	253.5
w_A	0.3579	0.3579	0.3579	0.3579	0.4274	0.4274	0.4274	0.4274	0.4274
w_B	0.0499	0.0499	0.0499	0.0499	0.0499	0.0499	0.0499	0.0499	0.0499
w_C	0.5922	0.5922	0.5922	0.5922	0.5227	0.5227	0.5227	0.5227	0.5227
T/K	343.05	353.05	363.05	373.05	313.05	322.75	332.65	343.25	352.55
P/bar	294.3	338.0	379.5	417.0	225.5	278.3	331.1	378.5	420.5

continued

continued

w_A	0.4274	0.4274	0.4962	0.4962	0.4962	0.4962	0.4962	0.4962	0.4962
w_B	0.0499	0.0499	0.0498	0.0498	0.0498	0.0498	0.0498	0.0498	0.0498
w_C	0.5227	0.5227	0.4540	0.4540	0.4540	0.4540	0.4540	0.4540	0.4540
T/K	362.65	372.55	313.05	323.75	333.25	342.75	352.35	362.85	372.65
P/bar	459.5	499.5	319.0	377.4	432.5	476.8	520.5	567.5	604.5

w_A	0.5976	0.5976	0.5976	0.5976	0.5976	0.5976	0.5976
w_B	0.0509	0.0509	0.0509	0.0509	0.0509	0.0509	0.0509
w_C	0.3515	0.3515	0.3515	0.3515	0.3515	0.3515	0.3515
T/K	313.05	322.75	332.65	342.85	352.65	362.95	373.15
P/bar	550.5	602.6	654.9	697.7	742.3	779.3	816.2

Type of data: vapor-liquid-liquid equilibrium data

w_A	0.0000	0.1185	0.1185	0.1719	0.2923	0.3579	0.4274	0.4962
w_B	0.0498	0.0503	0.0503	0.0500	0.0491	0.0499	0.0499	0.0498
w_C	0.9502	0.8312	0.8312	0.7781	0.6586	0.5922	0.5227	0.4540
T/K	332.35	302.85	310.55	313.55	302.75	300.75	303.35	310.45
P/bar	24.5	21.5	24.8	32.5	33.7	35.5	40.6	50.9

Polymer (B): **poly(methyl methacrylate)** **2002LE2**
Characterization: M_w/g.mol^{-1} = 120000, Aldrich Chem. Co., Inc., Milwaukee, WI
Solvent (A): **carbon dioxide** **CO_2** **124-38-9**
Solvent (C): **chlorodifluoromethane** **$CHClF_2$** **75-45-6**

Type of data: cloud points

w_A	0.0000	0.0000	0.0000	0.0000	0.0000	0.0000	0.0946	0.0946	0.0946
w_B	0.0499	0.0499	0.0499	0.0499	0.0499	0.0499	0.0510	0.0510	0.0510
w_C	0.9501	0.9501	0.9501	0.9501	0.9501	0.9501	0.8544	0.8544	0.8544
T/K	325.75	332.85	342.25	353.25	362.55	373.15	313.25	322.55	332.65
P/bar	24.7	49.5	85.1	125.5	159.2	195.7	32.0	71.7	116.0

w_A	0.0946	0.0946	0.0946	0.0946	0.1616	0.1616	0.1616	0.1616	0.1616
w_B	0.0510	0.0510	0.0510	0.0510	0.0496	0.0496	0.0496	0.0496	0.0496
w_C	0.8544	0.8544	0.8544	0.8544	0.7888	0.7888	0.7888	0.7888	0.7888
T/K	341.85	353.85	362.35	373.15	313.15	322.95	332.85	342.65	352.85
P/bar	154.5	204.5	238.5	278.5	77.5	123.7	171.1	215.9	260.5

w_A	0.1616	0.1616	0.2036	0.2036	0.2036	0.2036	0.2036	0.2036	0.2036
w_B	0.0496	0.0496	0.0502	0.0502	0.0502	0.0502	0.0502	0.0502	0.0502
w_C	0.7888	0.7888	0.7462	0.7462	0.7462	0.7462	0.7462	0.7462	0.7462
T/K	361.95	373.15	312.65	323.95	333.25	342.75	352.65	362.85	372.55
P/bar	297.5	344.0	111.5	168.4	215.5	260.7	306.1	352.4	392.6

w_A	0.2805	0.2805	0.2805	0.2805	0.2805	0.2805	0.2805	0.3725	0.3725
w_B	0.0492	0.0492	0.0492	0.0492	0.0492	0.0492	0.0492	0.0491	0.0491
w_C	0.6703	0.6703	0.6703	0.6703	0.6703	0.6703	0.6703	0.5784	0.5784
T/K	314.25	322.65	332.25	342.45	352.45	362.35	372.65	313.55	322.45
P/bar	195.5	242.5	295.1	349.0	397.5	445.5	491.3	312.7	366.6

continued

continued

w_A	0.3725	0.3725	0.3725	0.3725	0.3725	0.4479	0.4479	0.4479	0.4479
w_B	0.0491	0.0491	0.0491	0.0491	0.0491	0.0523	0.0523	0.0523	0.0523
w_C	0.5784	0.5784	0.5784	0.5784	0.5784	0.4998	0.4998	0.4998	0.4998
T/K	332.25	343.85	352.35	362.65	373.25	312.75	322.85	333.15	342.65
P/bar	426.5	492.0	540.5	591.0	640.7	455.5	525.5	591.6	649.5

w_A	0.4479	0.4479	0.5097	0.5097	0.5097	0.5097	0.5097
w_B	0.0523	0.0523	0.0488	0.0488	0.0488	0.0488	0.0488
w_C	0.4998	0.4998	0.4415	0.4415	0.4415	0.4415	0.4415
T/K	352.95	373.35	314.65	323.45	332.55	342.75	353.25
P/bar	708.7	810.0	657.7	719.7	779.3	840.5	897.0

Type of data: vapor-liquid-liquid equilibrium data

w_A	0.0000	0.0000	0.0946
w_B	0.0499	0.0499	0.0510
w_C	0.9501	0.9501	0.8544
T/K	312.75	323.45	309.65
P/bar	16.6	21.6	24.5

Polymer (B): **poly(methyl methacrylate)** **2008MAT**
Characterization: M_w/g.mol^{-1} = 15000, Aldrich Chem. Co., Inc., Milwaukee, WI
Solvent (A): **carbon dioxide** CO_2 **124-38-9**
Solvent (C): **ethanol** C_2H_6O **64-17-5**

Type of data: cloud points

w_A	0.414	0.414	0.414	0.414	0.414	0.414	0.414	0.512	0.512
w_B	0.029	0.029	0.029	0.029	0.029	0.029	0.029	0.025	0.025
w_C	0.557	0.557	0.557	0.557	0.557	0.557	0.557	0.463	0.463
T/K	301.2	302.3	303.2	304.2	305.2	306.4	307.4	298.2	300.3
P/MPa	8.7	7.9	7.0	6.4	5.9	5.8	5.5	10.9	10.0

w_A	0.512	0.512	0.512	0.512	0.512	0.512	0.569	0.569	0.569
w_B	0.025	0.025	0.025	0.025	0.025	0.025	0.022	0.022	0.022
w_C	0.463	0.463	0.463	0.463	0.463	0.463	0.409	0.409	0.409
T/K	303.2	306.2	308.2	309.2	310.9	312.3	305.2	308.3	310.2
P/MPa	8.8	8.3	7.5	7.1	7.0	6.7	8.0	8.2	8.3

w_A	0.569	0.569	0.569	0.569	0.569	0.569	0.569	0.569	0.569
w_B	0.022	0.022	0.022	0.022	0.022	0.022	0.022	0.022	0.022
w_C	0.409	0.409	0.409	0.409	0.409	0.409	0.409	0.409	0.409
T/K	313.2	315.2	318.2	323.0	326.2	328.0	333.3	338.4	343.2
P/MPa	8.5	8.6	8.7	9.1	9.3	9.4	10.1	11.2	12.0

w_A	0.569
w_B	0.022
w_C	0.409
T/K	353.1
P/MPa	14.1

continued

continued

Type of data: vapor-liquid equilibrium data

w_A	0.114	0.114	0.114	0.114	0.114	0.114	0.114	0.114	0.114
w_B	0.045	0.045	0.045	0.045	0.045	0.045	0.045	0.045	0.045
w_C	0.841	0.841	0.841	0.841	0.841	0.841	0.841	0.841	0.841
T/K	293.3	298.4	303.2	313.2	323.4	333.2	343.4	353.5	359.7
P/MPa	2.0	2.2	2.3	2.7	2.9	3.4	3.8	4.1	4.8
w_A	0.414	0.414	0.414	0.414	0.414	0.414	0.414	0.414	0.414
w_B	0.029	0.029	0.029	0.029	0.029	0.029	0.029	0.029	0.029
w_C	0.557	0.557	0.557	0.557	0.557	0.557	0.557	0.557	0.557
T/K	308.7	310.4	311.4	313.3	313.8	318.4	323.0	328.5	333.0
P/MPa	5.6	5.6	5.7	5.8	5.9	6.1	6.5	6.9	7.2
w_A	0.414	0.414	0.414	0.512	0.512	0.512	0.512	0.512	0.512
w_B	0.029	0.029	0.029	0.025	0.025	0.025	0.025	0.025	0.025
w_C	0.557	0.557	0.557	0.463	0.463	0.463	0.463	0.463	0.463
T/K	343.0	348.3	353.2	313.2	318.7	323.4	333.4	343.5	348.5
P/MPa	8.1	8.6	9.0	6.7	7.2	7.7	8.7	9.9	10.3
w_A	0.512								
w_B	0.025								
w_C	0.463								
T/K	353.0								
P/MPa	10.7								

Type of data: vapor-liquid-liquid equilibrium data

w_A	0.414	0.414	0.414	0.414	0.414	0.414	0.414	0.512	0.512
w_B	0.029	0.029	0.029	0.029	0.029	0.029	0.029	0.025	0.025
w_C	0.557	0.557	0.557	0.557	0.557	0.557	0.557	0.463	0.463
T/K	301.2	302.3	303.2	304.2	305.2	306.4	307.4	298.2	300.3
P/MPa	4.9	5.0	5.0	5.1	5.2	5.3	5.4	5.4	5.5
w_A	0.512	0.512	0.512	0.512	0.512	0.512	0.569	0.569	0.569
w_B	0.025	0.025	0.025	0.025	0.025	0.025	0.022	0.022	0.022
w_C	0.463	0.463	0.463	0.463	0.463	0.463	0.409	0.409	0.409
T/K	303.2	306.2	308.2	309.2	310.9	312.3	305.2	308.3	310.2
P/MPa	5.8	5.9	6.1	6.2	6.4	6.6	6.4	6.7	6.8
w_A	0.569	0.569	0.569	0.569	0.569	0.569	0.569	0.569	0.569
w_B	0.022	0.022	0.022	0.022	0.022	0.022	0.022	0.022	0.022
w_C	0.409	0.409	0.409	0.409	0.409	0.409	0.409	0.409	0.409
T/K	313.2	315.2	318.2	323.0	326.2	328.0	333.3	338.4	343.2
P/MPa	7.0	7.2	7.6	8.0	8.3	8.6	9.3	9.8	10.1
w_A	0.569								
w_B	0.022								
w_C	0.409								
T/K	353.1								
P/MPa	11.1								

Polymer (B): **poly(methyl methacrylate)** **2008GOE**
Characterization: M_n/g.mol^{-1} = 16300, M_w/g.mol^{-1} = 17900, Scientific Polymer Products, Inc., Ontario, NY
Solvent (A): **carbon dioxide** **CO_2** **124-38-9**
Solvent (C): **methyl methacrylate** **$C_5H_8O_2$** **80-62-6**

Type of data: cloud points

w_A	0.3709	0.3099	0.3291	0.3133	0.2911	0.3053	0.4004	0.3362	0.3493
w_B	0.0945	0.2045	0.2640	0.3090	0.4933	0.5451	0.0901	0.1967	0.2561
w_C	0.5346	0.4856	0.4069	0.3777	0.2148	0.1490	0.5096	0.4671	0.3946
T/K	338.15	338.15	338.15	338.15	338.15	338.15	338.15	338.15	338.15
P/bar	100	100	100	100	100	100	150	150	150

w_A	0.3356	0.2977	0.3132	0.3655	0.3199	0.2912	0.3955	0.3931	0.3476
w_B	0.2990	0.4234	0.4788	0.0952	0.2718	0.3189	0.0907	0.1508	0.2607
w_C	0.3654	0.2781	0.2080	0.5394	0.4083	0.3898	0.5139	0.4561	0.3917
T/K	338.15	338.15	338.15	353.15	353.15	353.15	353.15	353.15	353.15
P/bar	150	150	150	100	100	100	150	150	150

w_A	0.3136
w_B	0.3089
w_C	0.3775
T/K	353.15
P/bar	150

Polymer (B): **poly(methyl methacrylate)** **2008GOE**
Characterization: M_n/g.mol^{-1} = 92700, M_w/g.mol^{-1} = 101000, Scientific Polymer Products, Inc., Ontario, NY
Solvent (A): **carbon dioxide** **CO_2** **124-38-9**
Solvent (C): **methyl methacrylate** **$C_5H_8O_2$** **80-62-6**

Type of data: cloud points

w_A	0.3684	0.3652	0.3041	0.4001	0.3879	0.3214	0.3498	0.3176	0.3840
w_B	0.0316	0.0952	0.2096	0.0300	0.0918	0.2044	0.0324	0.2044	0.0307
w_C	0.6000	0.5396	0.4863	0.5699	0.5203	0.4742	0.6178	0.4780	0.5853
T/K	338.15	338.15	338.15	338.15	338.15	338.15	353.15	353.15	353.15
P/bar	100	100	100	150	150	150	100	100	150

w_A	0.3423
w_B	0.1970
w_C	0.4607
T/K	353.15
P/bar	150

Polymer (B): **poly(methyl methacrylate)** **2008LIU**
Characterization: M_n/g.mol^{-1} = 8300, M_w/g.mol^{-1} = 15000,
Scientific Polymer Products, Inc., Ontario, NY
Solvent (A): **carbon dioxide** **CO_2** **124-38-9**
Solvent (C): **2-propanone** **C_3H_6O** **67-64-1**

Type of data: cloud points

w_A	0.40	0.40	0.40	0.40	0.40	0.40	0.40	0.40	0.40
w_B	0.10	0.10	0.10	0.10	0.10	0.10	0.10	0.10	0.10
w_C	0.50	0.50	0.50	0.50	0.50	0.50	0.50	0.50	0.50
T/K	347.1	352.0	356.0	360.3	366.5	370.0	374.8	380.4	385.4
P/MPa	10.5	11.5	13.0	14.0	15.4	16.5	18.0	19.0	21.0

w_A	0.40	0.40	0.40
w_B	0.10	0.10	0.10
w_C	0.50	0.50	0.50
T/K	390.0	393.3	399.1
P/MPa	22.0	23.0	24.5

Polymer (B): **poly(methyl methacrylate)** **2008LIU**
Characterization: M_n/g.mol^{-1} = 193000, M_w/g.mol^{-1} = 540000,
Scientific Polymer Products, Inc., Ontario, NY
Solvent (A): **carbon dioxide** **CO_2** **124-38-9**
Solvent (C): **2-propanone** **C_3H_6O** **67-64-1**

Type of data: cloud points

w_A	0.40	0.40	0.40	0.40	0.40	0.40	0.40	0.40	0.40
w_B	0.10	0.10	0.10	0.10	0.10	0.10	0.10	0.10	0.10
w_C	0.50	0.50	0.50	0.50	0.50	0.50	0.50	0.50	0.50
T/K	330.0	335.5	340.5	348.3	353.1	359.0	365.4	370.1	376.0
P/MPa	15.5	17.0	18.5	20.0	22.0	23.5	24.5	25.5	27.0

w_A	0.40	0.40	0.40	0.40	0.40
w_B	0.10	0.10	0.10	0.10	0.10
w_C	0.50	0.50	0.50	0.50	0.50
T/K	381.9	386.6	391.9	396.4	400.4
P/MPa	28.5	29.5	30.5	32.5	34.0

Polymer (B): **poly(neopentyl methacrylate)** **2007BY1**
Characterization: M_w/g.mol^{-1} = 480000,
Scientific Polymer Products, Inc., Ontario, NY
Solvent (A): **carbon dioxide** **CO_2** **124-38-9**
Solvent (C): **dimethyl ether** **C_2H_6O** **115-10-6**

Type of data: cloud points

continued

continued

w_A	0.896	0.896	0.896	0.896	0.896	0.896	0.896	0.831	0.831
w_B	0.052	0.052	0.052	0.052	0.052	0.052	0.052	0.048	0.048
w_C	0.052	0.052	0.052	0.052	0.052	0.052	0.052	0.121	0.121
T/K	333.55	354.15	374.25	394.05	413.85	433.75	452.65	353.55	374.55
P/bar	906.9	863.8	870.7	884.5	905.2	922.4	943.1	572.4	622.4
w_A	0.831	0.831	0.831	0.831	0.542	0.542	0.542	0.542	0.542
w_B	0.048	0.048	0.048	0.048	0.050	0.050	0.050	0.050	0.050
w_C	0.121	0.121	0.121	0.121	0.408	0.408	0.408	0.408	0.408
T/K	393.25	413.65	434.35	453.75	332.45	353.95	374.45	393.35	414.95
P/bar	669.0	712.1	750.0	777.6	208.6	284.5	343.1	387.9	439.7
w_A	0.542	0.542							
w_B	0.050	0.050							
w_C	0.408	0.408							
T/K	433.15	452.25							
P/bar	477.6	498.3							

Polymer (B): **poly(neopentyl methacrylate)** **2007BY1**
Characterization: M_w/g.mol^{-1} = 480000,
Scientific Polymer Products, Inc., Ontario, NY
Solvent (A): **carbon dioxide** CO_2 **124-38-9**
Solvent (C): **neopentyl methacrylate** $C_9H_{16}O_2$ **2397-76-4**

Type of data: cloud points

w_A	0.893	0.893	0.893	0.893	0.893	0.893	0.893	0.833	0.833
w_B	0.055	0.055	0.055	0.055	0.055	0.055	0.055	0.056	0.056
w_C	0.052	0.052	0.052	0.052	0.052	0.052	0.052	0.111	0.111
T/K	331.85	353.45	373.85	393.35	415.35	434.65	454.75	332.35	353.65
P/bar	856.9	808.6	815.5	836.2	863.8	884.5	929.3	563.8	605.2
w_A	0.833	0.833	0.833	0.833	0.833	0.763	0.763	0.763	0.763
w_B	0.056	0.056	0.056	0.056	0.056	0.047	0.047	0.047	0.047
w_C	0.111	0.111	0.111	0.111	0.111	0.190	0.190	0.190	0.190
T/K	373.15	393.65	415.55	435.35	453.55	331.35	352.25	375.75	395.35
P/bar	648.3	684.5	722.4	753.5	770.7	346.6	412.1	484.5	529.3
w_A	0.763	0.763	0.763	0.672	0.672	0.672	0.672	0.672	0.672
w_B	0.047	0.047	0.047	0.047	0.047	0.047	0.047	0.047	0.047
w_C	0.190	0.190	0.190	0.281	0.281	0.281	0.281	0.281	0.281
T/K	415.15	434.45	454.85	332.65	352.95	374.65	394.35	415.15	434.65
P/bar	574.1	603.5	636.2	298.3	363.8	422.4	467.2	506.9	537.9
w_A	0.672	0.552	0.552	0.552	0.552	0.552	0.552	0.552	0.442
w_B	0.047	0.046	0.046	0.046	0.046	0.046	0.046	0.046	0.046
w_C	0.281	0.402	0.402	0.402	0.402	0.402	0.402	0.402	0.512
T/K	453.95	332.35	353.35	375.05	394.85	412.55	435.75	454.15	324.65
P/bar	546.6	222.4	291.4	353.5	394.8	431.0	474.1	498.3	87.9

continued

continued

w_A	0.442	0.442	0.442	0.442	0.442	0.442	0.442
w_B	0.046	0.046	0.046	0.046	0.046	0.046	0.046
w_C	0.512	0.512	0.512	0.512	0.512	0.512	0.512
T/K	334.65	353.85	375.75	395.85	413.85	432.35	453.25
P/bar	122.4	179.3	242.8	291.4	328.3	356.9	403.1

Type of data: vapor-liquid equilibrium data

w_A	0.442	0.442	0.442
w_B	0.046	0.046	0.046
w_C	0.512	0.512	0.512
T/K	305.55	308.65	312.25
P/bar	66.7	67.0	67.2

Polymer (B): **poly(octyl acrylate)** **2007BY3**
Characterization: M_w/g.mol^{-1} = 100000, T_g/K = 208,
Scientific Polymer Products, Inc., Ontario, NY
Solvent (A): **carbon dioxide** CO_2 **124-38-9**
Solvent (C): **dimethyl ether** C_2H_6O **115-10-6**

Type of data: cloud points

w_A	0.835	0.835	0.835	0.835	0.835	0.835	0.835	0.739	0.739
w_B	0.060	0.060	0.060	0.060	0.060	0.060	0.060	0.050	0.050
w_C	0.105	0.105	0.105	0.105	0.105	0.105	0.105	0.211	0.211
T/K	367.25	380.45	395.85	410.05	424.75	440.15	453.45	332.85	355.35
P/bar	2015.5	1205.2	1043.1	981.0	959.5	922.4	936.2	1065.5	760.4
w_A	0.739	0.739	0.739	0.739	0.739	0.554	0.554	0.554	0.554
w_B	0.050	0.050	0.050	0.050	0.050	0.043	0.043	0.043	0.043
w_C	0.211	0.211	0.211	0.211	0.211	0.403	0.403	0.403	0.403
T/K	376.45	395.15	416.05	434.45	453.95	334.15	352.75	374.15	394.15
P/bar	705.2	691.4	687.9	691.4	691.4	287.9	325.9	363.8	401.7
w_A	0.554	0.554	0.554	0.367	0.367	0.367	0.367	0.367	0.367
w_B	0.043	0.043	0.043	0.057	0.057	0.057	0.057	0.057	0.057
w_C	0.403	0.403	0.403	0.576	0.576	0.576	0.576	0.576	0.576
T/K	414.45	434.65	453.15	323.35	332.85	351.75	373.75	394.95	414.05
P/bar	432.8	460.4	484.5	87.9	115.5	160.4	215.5	256.9	294.8
w_A	0.367	0.367							
w_B	0.057	0.057							
w_C	0.576	0.576							
T/K	434.05	453.45							
P/bar	332.8	362.1							

Polymer (B):	**poly(octyl acrylate)**		**2007BY3**
Characterization:	M_w/g.mol^{-1} = 100000, T_g/K = 208, Scientific Polymer Products, Inc., Ontario, NY		
Solvent (A):	**carbon dioxide**	**CO_2**	**124-38-9**
Solvent (C):	**octyl acrylate**	**$C_{11}H_{20}O_2$**	**2499-59-4**

Type of data: cloud points

w_A	0.830	0.830	0.830	0.830	0.830	0.830	0.830	0.830	0.753
w_B	0.061	0.061	0.061	0.061	0.061	0.061	0.061	0.061	0.048
w_C	0.109	0.109	0.109	0.109	0.109	0.109	0.109	0.109	0.199
T/K	358.85	364.55	378.75	393.25	406.85	423.25	437.55	454.65	323.65
P/bar	2139.7	1227.6	939.7	853.5	822.4	805.2	794.8	787.9	643.1

w_A	0.753	0.753	0.753	0.753	0.753	0.753	0.753	0.729	0.729
w_B	0.048	0.048	0.048	0.048	0.048	0.048	0.048	0.037	0.037
w_C	0.199	0.199	0.199	0.199	0.199	0.199	0.199	0.234	0.234
T/K	333.45	353.95	376.05	391.15	414.05	434.85	453.65	324.85	332.85
P/bar	574.1	525.9	529.3	536.2	553.5	581.0	591.4	425.9	415.5

w_A	0.729	0.729	0.729	0.729	0.729	0.729	0.590	0.590	0.590
w_B	0.037	0.037	0.037	0.037	0.037	0.037	0.050	0.050	0.050
w_C	0.234	0.234	0.234	0.234	0.234	0.234	0.360	0.360	0.360
T/K	353.15	372.55	392.45	411.35	434.45	454.25	333.35	351.95	376.05
P/bar	422.4	443.1	467.2	491.4	515.5	532.8	105.2	156.9	217.2

w_A	0.590	0.590	0.590	0.590
w_B	0.050	0.050	0.050	0.050
w_C	0.360	0.360	0.360	0.360
T/K	391.65	412.25	434.25	454.65
P/bar	253.5	290.0	325.9	350.0

Type of data: vapor-liquid equilibrium data

w_A	0.590	0.590	0.590
w_B	0.050	0.050	0.050
w_C	0.360	0.360	0.360
T/K	308.65	315.35	323.85
P/bar	70.7	81.0	91.4

Type of data: vapor-liquid-liquid equilibrium data

w_A	0.590
w_B	0.050
w_C	0.360
T/K	338.25
P/bar	110.0

Polymer (B):	**poly(octyl methacrylate)**		**2007BY3**
Characterization:	M_w/g.mol^{-1} = 100000, T_g/K = 293, Scientific Polymer Products, Inc., Ontario, NY		
Solvent (A):	**carbon dioxide**	CO_2	**124-38-9**
Solvent (C):	**octyl methacrylate**	$C_{12}H_{22}O_2$	**2157-01-9**

Type of data: cloud points

w_A	0.902	0.902	0.902	0.902	0.902	0.902	0.902	0.815	0.815
w_B	0.042	0.042	0.042	0.042	0.042	0.042	0.042	0.048	0.048
w_C	0.056	0.056	0.056	0.056	0.056	0.056	0.056	0.137	0.137
T/K	413.35	424.45	434.35	443.75	452.45	463.95	472.15	374.95	395.25
P/bar	2250.0	1806.9	1444.8	1324.1	1265.5	1215.5	1212.1	1365.5	1037.9

w_A	0.815	0.815	0.815	0.628	0.628	0.628	0.628	0.628	0.628
w_B	0.048	0.048	0.048	0.062	0.062	0.062	0.062	0.062	0.062
w_C	0.137	0.137	0.137	0.310	0.310	0.310	0.310	0.310	0.310
T/K	413.75	431.95	454.95	323.65	332.15	348.95	360.65	378.45	391.55
P/bar	929.3	887.9	860.4	250.0	260.7	291.4	334.5	372.4	396.6

w_A	0.628	0.628	0.628	0.628	0.526	0.526	0.526	0.526	0.526
w_B	0.062	0.062	0.062	0.062	0.050	0.050	0.050	0.050	0.050
w_C	0.310	0.310	0.310	0.310	0.424	0.424	0.424	0.424	0.424
T/K	406.95	422.35	438.05	451.35	324.15	333.65	351.25	371.35	394.65
P/bar	436.2	467.2	512.1	534.5	86.9	105.2	156.9	212.1	270.7

w_A	0.526	0.526	0.526
w_B	0.050	0.050	0.050
w_C	0.424	0.424	0.424
T/K	413.55	431.85	455.25
P/bar	308.6	356.9	400.0

Type of data: vapor-liquid equilibrium data

w_A	0.526	0.526	0.526
w_B	0.050	0.050	0.050
w_C	0.424	0.424	0.424
T/K	305.15	309.95	313.95
P/bar	63.8	70.7	70.2

Type of data: vapor-liquid-liquid equilibrium data

w_A	0.526	0.526
w_B	0.050	0.050
w_C	0.424	0.424
T/K	326.35	334.25
P/bar	77.1	84.8

Polymer (B): **poly(propyl acrylate)** **2002BY3**
Characterization: M_w/g.mol^{-1} = 140000, Polysciences, Inc., Warrington, PA
Solvent (A): **carbon dioxide** CO_2 **124-38-9**
Solvent (C): **dimethyl ether** C_2H_6O **115-10-6**

Type of data: cloud points

w_A	0.897	0.897	0.897	0.897	0.897	0.897	0.897	0.799	0.799
w_B	0.053	0.053	0.053	0.053	0.053	0.053	0.053	0.051	0.051
w_C	0.050	0.050	0.050	0.050	0.050	0.050	0.050	0.150	0.150
T/K	334.35	344.95	356.05	368.45	385.45	405.05	427.15	313.55	329.25
P/bar	1567.2	1479.3	1443.1	1329.3	1239.0	1224.3	1186.2	756.9	776.6
w_A	0.799	0.799	0.799	0.799	0.799	0.449	0.449	0.449	0.449
w_B	0.051	0.051	0.051	0.051	0.051	0.051	0.051	0.051	0.051
w_C	0.150	0.150	0.150	0.150	0.150	0.500	0.500	0.500	0.500
T/K	345.15	364.15	381.85	404.65	425.05	313.35	334.05	352.75	374.75
P/bar	799.3	823.1	847.6	879.0	898.3	152.4	224.1	272.8	331.7
w_A	0.449	0.449							
w_B	0.051	0.051							
w_C	0.500	0.500							
T/K	394.05	418.05							
P/bar	381.7	430.0							

Polymer (B): **poly(propyl acrylate)** **2002BY2**
Characterization: M_w/g.mol^{-1} = 140000, Polysciences, Inc., Warrington, PA
Solvent (A): **carbon dioxide** CO_2 **124-38-9**
Solvent (C): **propyl acrylate** $C_6H_{10}O_2$ **925-60-0**

Type of data: cloud points

w_A	0.949	0.949	0.949	0.949	0.949	0.949	0.949	0.949	0.949
w_B	0.051	0.051	0.051	0.051	0.051	0.051	0.051	0.051	0.051
w_C	0.000	0.000	0.000	0.000	0.000	0.000	0.000	0.000	0.000
T/K	361.85	363.25	366.25	370.25	374.15	383.65	403.65	423.95	448.45
P/bar	2070.7	1932.8	1794.8	1656.9	1556.9	1474.5	1332.1	1263.8	1215.5
w_A	0.900	0.900	0.900	0.900	0.900	0.900	0.900	0.900	0.900
w_B	0.050	0.050	0.050	0.050	0.050	0.050	0.050	0.050	0.050
w_C	0.050	0.050	0.050	0.050	0.050	0.050	0.050	0.050	0.050
T/K	319.05	319.95	322.95	329.05	348.45	369.05	388.45	408.65	428.95
P/bar	1794.8	1519.0	1381.0	1241.7	1092.1	1039.7	1023.8	1015.5	1015.5
w_A	0.833	0.833	0.833	0.833	0.833	0.833	0.645	0.645	0.645
w_B	0.050	0.050	0.050	0.050	0.050	0.050	0.051	0.051	0.051
w_C	0.117	0.117	0.117	0.117	0.117	0.117	0.304	0.304	0.304
T/K	308.05	323.85	344.75	364.15	385.15	406.95	305.55	314.05	333.55
P/bar	645.9	669.3	698.4	737.6	789.7	798.8	84.5	118.3	199.0

continued

continued

w_A	0.645	0.645	0.645	0.607	0.607	0.607	0.607	0.607
w_B	0.051	0.051	0.051	0.052	0.052	0.052	0.052	0.052
w_C	0.304	0.304	0.304	0.341	0.341	0.341	0.341	0.341
T/K	354.55	373.25	394.75	314.45	334.45	353.95	373.85	394.85
P/bar	268.3	330.3	396.6	74.8	157.2	215.2	282.4	340.3

Type of data: vapor-liquid equilibrium data

w_A	0.607	0.607
w_B	0.052	0.052
w_C	0.341	0.341
T/K	300.95	304.55
P/bar	55.5	59.0

Type of data: vapor-liquid-liquid equilibrium data

w_A	0.607	0.607
w_B	0.052	0.052
w_C	0.341	0.341
T/K	326.35	338.15
P/bar	81.0	93.2

Polymer (B): **poly(propyl acrylate)** **2002BY2**
Characterization: M_w/g.mol^{-1} = 140000, Polysciences, Inc., Warrington, PA
Solvent (A): **ethene** C_2H_4 **74-85-1**
Solvent (C): **propyl acrylate** $C_6H_{10}O_2$ **925-60-0**

Type of data: cloud points

w_A	0.946	0.946	0.946	0.946	0.892	0.892	0.892	0.892	0.798
w_B	0.054	0.054	0.054	0.054	0.051	0.051	0.051	0.051	0.047
w_C	0.000	0.000	0.000	0.000	0.057	0.057	0.057	0.057	0.155
T/K	383.35	403.35	422.55	444.65	363.15	383.85	403.75	424.25	312.95
P/bar	1401.0	1285.2	1232.1	1183.1	1104.2	961.7	933.5	913.8	826.9
w_A	0.798	0.798	0.798	0.798	0.798	0.798	0.732	0.732	0.732
w_B	0.047	0.047	0.047	0.047	0.047	0.047	0.046	0.046	0.046
w_C	0.155	0.155	0.155	0.155	0.155	0.155	0.222	0.222	0.222
T/K	323.45	344.05	363.95	383.35	403.65	423.95	314.15	322.75	343.95
P/bar	811.0	795.5	781.0	774.8	767.9	763.8	656.9	657.6	664.5
w_A	0.732	0.732	0.732						
w_B	0.046	0.046	0.046						
w_C	0.222	0.222	0.222						
T/K	363.85	385.15	405.55						
P/bar	675.2	682.1	686.2						

Polymer (B): **poly(propyl methacrylate)** **2002BY2**
Characterization: M_w/g.mol^{-1} = 250000, Polysciences, Inc., Warrington, PA
Solvent (A): **carbon dioxide** **CO_2** **124-38-9**
Solvent (C): **propyl methacrylate** **$C_7H_{12}O_2$** **2210-28-8**

Type of data: cloud points

w_A	0.900	0.900	0.900	0.900	0.900	0.900	0.900	0.900	0.900
w_B	0.048	0.048	0.048	0.048	0.048	0.048	0.048	0.048	0.048
w_C	0.052	0.052	0.052	0.052	0.052	0.052	0.052	0.052	0.052
T/K	363.65	364.65	366.25	368.65	373.35	373.95	392.65	413.35	433.55
P/bar	2484.5	2263.8	2099.7	1959.0	1776.2	1754.1	1520.3	1390.0	1326.6

w_A	0.900	0.821	0.821	0.821	0.821	0.821	0.821	0.821	0.821
w_B	0.048	0.047	0.047	0.047	0.047	0.047	0.047	0.047	0.047
w_C	0.052	0.132	0.132	0.132	0.132	0.132	0.132	0.132	0.132
T/K	459.35	308.05	326.15	344.15	363.95	382.85	404.95	424.05	442.35
P/bar	1281.0	1505.2	1081.7	978.3	942.1	933.5	936.2	941.4	949.0

w_A	0.748	0.748	0.748	0.748	0.748	0.748	0.748	0.534	0.534
w_B	0.051	0.051	0.051	0.051	0.051	0.051	0.051	0.051	0.051
w_C	0.201	0.201	0.201	0.201	0.201	0.201	0.201	0.415	0.415
T/K	311.45	328.65	343.35	363.85	383.45	403.75	426.45	322.55	333.25
P/bar	547.9	575.5	600.4	641.0	681.0	716.6	744.8	89.3	127.6

w_A	0.534	0.534	0.534	0.534
w_B	0.051	0.051	0.051	0.051
w_C	0.415	0.415	0.415	0.415
T/K	353.55	373.05	393.15	413.55
P/bar	192.4	255.5	306.9	350.0

Type of data: vapor-liquid equilibrium data

w_A	0.534	0.534	0.534
w_B	0.051	0.051	0.051
w_C	0.415	0.415	0.415
T/K	314.65	309.15	302.05
P/bar	67.2	59.7	53.5

Type of data: vapor-liquid-liquid equilibrium data

w_A	0.534	0.534	0.534
w_B	0.051	0.051	0.051
w_C	0.415	0.415	0.415
T/K	330.25	334.45	342.95
P/bar	84.9	89.7	99.6

Polymer (B): **polystyrene** **2011CAI**
Characterization: M_n/g.mol^{-1} = 233600, M_w/M_n = 1.014
Solvent (A): **decahydronaphthalene** $C_{10}H_{18}$ **91-17-8**
Solvent (C): **propane** C_3H_8 **74-98-6**

Comments: The decahydronaphthalene is a 66/34 wt% *trans/cis* isomer mixture.
All phase equilibrium data are given for a 5 wt% solution of PS in DHN.
The concentration of propane is the overall mass fraction in the solution.

Type of data: cloud points

w_C	0.2709	0.2922	0.271	0.271	0.271	0.312	0.312	0.312	0.312
T/K	323.0	323.0	325.7	328.2	331.2	357.8	350.6	344.7	339.5
P/bar	127.0	216.5	82.233	53.206	25.283	60.308	80.303	108.50	142.36

w_C	0.312	0.312	0.312	0.312	0.312	0.312	0.312	0.312	0.312
T/K	420.1	415.1	411.1	407.0	402.0	395.0	387.9	378.0	368.0
P/bar	63.204	60.101	56.171	52.034	48.311	44.243	40.589	39.486	45.140

w_C	0.3122	0.3359	0.3606	0.3868	0.4132
T/K	423.0	423.0	423.0	423.0	423.0
P/bar	64.8	108.0	155.5	204.4	245.2

Type of data: bubble points

w_C	0.271	0.271	0.271	0.271	0.271	0.271	0.271
T/K	323.2	328.2	334.2	350.2	373.2	398.2	423.2
P/bar	9.218	10.459	12.39	16.182	23.007	29.902	40.175

w_C	0.312	0.312	0.312	0.312	0.312	0.312	0.312	0.312	0.312
T/K	420.1	415.1	411.1	407.0	402.0	395.0	387.9	378.0	368.0
P/bar	45.829	44.450	42.795	40.727	39.072	35.970	33.143	29.282	24.869

w_C	0.312	0.312	0.312	0.312
T/K	357.8	350.6	344.7	339.5
P/bar	20.525	19.284	16.664	16.044

w_C	0.0538	0.1061	0.1513	0.1964	0.2327	0.2709	0.2922
T/K	323.0	323.0	323.0	323.0	323.0	323.0	323.0
P/bar	3.4	4.5	6.7	8.5	9.5	10.0	10.3

w_C	0.0538	0.1061	0.1513	0.1964	0.2327	0.2709	0.3122	0.3359	0.3606
T/K	423.0	423.0	423.0	423.0	423.0	423.0	423.0	423.0	423.0
P/bar	10.3	18.0	25.0	31.4	36.7	40.2	45.8	51.1	53.6

w_C	0.3868	0.4132
T/K	423.0	423.0
P/bar	54.2	54.7

Polymer (B): **poly(1-vinyl-2-pyrrolidinone)** **2008SH4**
Characterization: M_w/g.mol^{-1} = 55000, Aldrich Chem. Co., Inc., St. Louis, MO
Solvent (A): **carbon dioxide** CO_2 **124-38-9**
Solvent (C): **dichloromethane** CH_2Cl_2 **75-09-2**

Type of data: cloud points

w_A	0.305	0.305	0.305	0.305	0.305	0.305	0.359	0.359	0.359
w_B	0.005	0.005	0.005	0.005	0.005	0.005	0.005	0.005	0.005
w_C	0.690	0.690	0.690	0.690	0.690	0.690	0.636	0.636	0.636
T/K	313.28	323.33	333.29	343.19	353.05	363.33	313.13	322.94	333.00
P/MPa	8.27	9.63	12.09	15.21	18.57	22.39	11.42	13.79	16.94

w_A	0.359	0.359
w_B	0.005	0.005
w_C	0.636	0.636
T/K	342.73	353.02
P/MPa	20.96	25.51

Type of data: vapor-liquid equilibrium data

w_A	0.222	0.222	0.222	0.222	0.222	0.222	0.245	0.245	0.245
w_B	0.005	0.005	0.005	0.005	0.005	0.005	0.005	0.005	0.005
w_C	0.773	0.773	0.773	0.773	0.773	0.773	0.750	0.750	0.750
T/K	313.18	323.19	333.23	343.11	353.21	363.15	313.18	323.06	333.22
P/MPa	3.22	3.55	4.26	4.63	5.28	5.71	3.31	3.72	4.34

w_A	0.245	0.245
w_B	0.005	0.005
w_C	0.750	0.750
T/K	343.12	353.17
P/MPa	4.99	5.53

Polymer (B): **poly(1-vinyl-2-pyrrolidinone)** **2004BAE**
Characterization: M_n/g.mol^{-1} = 1300, M_w/g.mol^{-1} = 2500,
K-12, Aldrich Chem. Co., Inc., St. Louis, MO
Solvent (A): **carbon dioxide** CO_2 **124-38-9**
Solvent (C): **1-methyl-2-pyrrolidinone** C_5H_9NO **872-50-4**

Type of data: cloud points

w_A	0.701	0.701	0.701	0.701	0.701	0.701	0.701	0.654	0.654
w_B	0.051	0.051	0.051	0.051	0.051	0.051	0.051	0.050	0.050
w_C	0.248	0.248	0.248	0.248	0.248	0.248	0.248	0.296	0.296
T/K	334.0	353.2	370.7	394.2	411.3	433.1	455.9	321.3	330.0
P/MPa	212.07	206.38	202.76	200.52	199.24	197.48	194.14	166.61	155.10

continued

continued

w_A	0.654	0.654	0.654	0.654	0.654	0.654	0.654	0.654	0.601
w_B	0.050	0.050	0.050	0.050	0.050	0.050	0.050	0.050	0.053
w_C	0.296	0.296	0.296	0.296	0.296	0.296	0.296	0.296	0.346
T/K	346.6	363.5	377.6	396.2	411.3	425.8	440.9	453.1	323.2
P/MPa	141.65	134.24	132.34	132.00	132.86	133.72	134.41	135.51	67.17

w_A	0.601	0.601	0.601	0.601	0.601	0.601	0.601
w_B	0.053	0.053	0.053	0.053	0.053	0.053	0.053
w_C	0.346	0.346	0.346	0.346	0.346	0.346	0.346
T/K	341.8	357.2	376.9	396.4	417.2	433.8	454.7
P/MPa	72.34	75.62	81.48	87.00	91.48	94.93	97.34

Polymer (B): **poly(1-vinyl-2-pyrrolidinone)** **2004BAE**
Characterization: M_n/g.mol^{-1} = 2500, M_w/g.mol^{-1} = 9000,
K-17, Aldrich Chem. Co., Inc., St. Louis, MO
Solvent (A): **carbon dioxide** **CO_2** **124-38-9**
Solvent (C): **1-methyl-2-pyrrolidinone** **C_5H_9NO** **872-50-4**

Type of data: cloud points

w_A	0.658	0.658	0.658	0.658	0.658	0.658	0.658	0.658	0.611
w_B	0.048	0.048	0.048	0.048	0.048	0.048	0.048	0.048	0.046
w_C	0.294	0.294	0.294	0.294	0.294	0.294	0.294	0.294	0.343
T/K	325.9	333.2	355.5	376.0	395.5	415.5	435.5	454.3	333.6
P/MPa	141.31	139.93	138.37	139.24	140.10	142.51	144.41	144.93	94.93

w_A	0.611	0.611	0.611	0.611	0.611	0.611	0.506	0.506	0.506
w_B	0.046	0.046	0.046	0.046	0.046	0.046	0.050	0.050	0.050
w_C	0.343	0.343	0.343	0.343	0.343	0.343	0.444	0.444	0.444
T/K	353.2	376.0	390.6	413.0	434.1	458.2	320.0	330.7	350.5
P/MPa	95.17	96.66	99.66	103.17	107.24	109.35	18.45	21.90	28.86

w_A	0.506	0.506	0.506	0.506	0.506
w_B	0.050	0.050	0.050	0.050	0.050
w_C	0.444	0.444	0.444	0.444	0.444
T/K	369.9	392.5	411.8	423.7	442.4
P/MPa	34.38	40.35	45.83	46.90	47.50

Polymer (B): **poly(1-vinyl-2-pyrrolidinone)** **2004BAE**
Characterization: M_n/g.mol^{-1} = 6000, M_w/g.mol^{-1} = 25000,
K-25, Aldrich Chem. Co., Inc., St. Louis, MO
Solvent (A): **carbon dioxide** **CO_2** **124-38-9**
Solvent (C): **1-methyl-2-pyrrolidinone** **C_5H_9NO** **872-50-4**

Type of data: cloud points

continued

continued

w_A	0.639	0.639	0.639	0.639	0.639	0.639	0.639	0.639	0.627
w_B	0.052	0.052	0.052	0.052	0.052	0.052	0.052	0.052	0.052
w_C	0.309	0.309	0.309	0.309	0.309	0.309	0.309	0.309	0.321
T/K	320.6	333.3	355.7	375.6	390.5	413.3	433.0	456.3	322.3
P/MPa	155.17	153.83	152.10	152.79	153.27	154.24	154.48	154.83	128.79
w_A	0.627	0.627	0.627	0.627	0.627	0.627	0.627	0.581	0.581
w_B	0.052	0.052	0.052	0.052	0.052	0.052	0.052	0.053	0.053
w_C	0.321	0.321	0.321	0.321	0.321	0.321	0.321	0.366	0.366
T/K	333.8	352.4	372.7	392.1	411.0	429.6	451.1	322.5	331.1
P/MPa	129.48	131.55	134.83	138.10	140.00	142.41	143.28	80.17	82.07
w_A	0.581	0.581	0.581	0.581	0.581	0.581			
w_B	0.053	0.053	0.053	0.053	0.053	0.053			
w_C	0.366	0.366	0.366	0.366	0.366	0.366			
T/K	351.0	372.2	390.8	412.2	432.4	452.0			
P/MPa	87.17	93.28	97.41	101.66	105.35	108.72			

Polymer (B): **poly(1-vinyl-2-pyrrolidinone)** **2004BAE**
Characterization: M_n/g.mol^{-1} = 12000, M_w/g.mol^{-1} = 40000,
K-30, Aldrich Chem. Co., Inc., St. Louis, MO
Solvent (A): **carbon dioxide** **CO_2** **124-38-9**
Solvent (C): **1-methyl-2-pyrrolidinone** **C_5H_9NO** **872-50-4**

Type of data: cloud points

w_A	0.648	0.648	0.648	0.648	0.648	0.648	0.648	0.648	0.582
w_B	0.049	0.049	0.049	0.049	0.049	0.049	0.049	0.049	0.052
w_C	0.303	0.303	0.303	0.303	0.303	0.303	0.303	0.303	0.366
T/K	321.7	336.6	355.2	376.2	394.4	415.5	433.7	453.6	323.8
P/MPa	169.76	162.00	157.69	156.13	155.62	156.13	156.31	156.31	95.35
w_A	0.582	0.582	0.582	0.582	0.582	0.582	0.582		
w_B	0.052	0.052	0.052	0.052	0.052	0.052	0.052		
w_C	0.366	0.366	0.366	0.366	0.366	0.366	0.366		
T/K	332.5	352.4	373.1	391.4	412.3	431.8	452.1		
P/MPa	97.48	101.48	105.83	109.55	113.21	115.79	117.76		

4.4. Table of ternary or quaternary systems where data were published only in graphical form as phase diagrams or related figures

Polymer (B)	Second and third component	Ref.
Dextran		
	carbon dioxide and dimethylsulfoxide	2004PER
Penta(ethylene glycol) monododecyl ether		
	carbon dioxide and n-heptane/sodium bis(ethylhexyl)sulfosuccinate	2009HOL
Poly(benzyl acrylate)		
	carbon dioxide and benzyl acrylate	2010JAN
	carbon dioxide and dimethyl ether	2010JAN
	dimethyl ether and benzyl acrylate	2010JAN
Poly(benzyl methacrylate)		
	carbon dioxide and benzyl methacrylate	2010JAN
	carbon dioxide and dimethyl ether	2010JAN
	dimethyl ether and benzyl methacrylate	2010JAN
Poly(butyl acrylate)		
	carbon dioxide and butyl acrylate	2005NIS
Poly(ε-caprolactone)		
	carbon dioxide and 2-propanone	2006LIU
Polycarbonate-bisphenol A		
	carbon dioxide and ethanol/2-propanone	2009HE2
	carbon dioxide and 2-propanone	2009HE2

Polymer (B)	Second and third component	Ref.
Poly[2-(*N,N*-dimethylaminoethyl) methacrylate-*co*-1H,1H-perfluorooctyl methacrylate]		
	carbon dioxide and poly(methyl methacrylate)	2008HWA
Poly(dimethylsiloxane) monomethacrylate		
	carbon dioxide and 1,1-difluoroethene	2005TAI
	carbon dioxide and methyl methacrylate	2005TAI
	carbon dioxide and vinylidene fluoride	2005TAI
Poly(etherimide)		
	carbon dioxide and dichloromethane	2009LAW
Polyethylene		
	1-bromo-1-chloro-2,2,2-trifluoroethane and n-pentane	2009KOJ
	bromoethane and n-pentane	2009KOJ
	1-bromopropane and n-pentane	2009KOJ
	2-bromopropane and n-pentane	2009KOJ
	carbon dioxide and n-pentane	2006UPP
	chloroethane and n-pentane	2009KOJ
	1-chloropropane and n-pentane	2009KOJ
	cyclohexane and 1,2-dichloro-1,2-difluoroethane	2009KOJ
	dibromodifluoromethane and n-pentane	2009KOJ
	1,2-dichloro-1,2-difluoroethane and n-pentane	2009KOJ
	1,1-dichloroethane and n-pentane	2009KOJ
	1,2-dichloroethane and n-pentane	2009KOJ
	1,2-dichloroethene and n-pentane	2009KOJ
	cis-1,2-dichloroethene and n-pentane	2009KOJ
	dichloromethane and n-pentane	2009KOJ
	1,2-dichloropropane and n-pentane	2009KOJ
	ethene and cyclohexane/1-butene	2007BUC
	ethene and ethane/propene/propane	2009COS
	ethene and n-hexane	2004GHO
	ethene and n-hexane/1-butene	2007BUC
	iodoethane and n-pentane	2009KOJ
	n-pentane and 1,1,2,2-tetrachloro-1,2-difluoroethane	2009KOJ
	n-pentane and 1,1,2-trichloroethane	2009KOJ
	n-pentane and trichloroethene	2009KOJ
	n-pentane and trichlorofluoromethane	2009KOJ
	n-pentane and trichloromethane	2009KOJ

Polymer (B)	Second and third component	Ref.
Poly(ethylene-*co*-acrylic acid)		
	ethene and n-decane	2004BEY
	ethene and ethanol	2004BEY
	ethene and n-heptane	2004BEY
	ethene and methanol	2002LE1
	ethene and n-octane	2004BEY
	ethene and octanoic acid	2004BEY
	ethene and 2-propanone	2002LE1
Poly(ethylene-*co*-1-butene)		
	ethene and cyclohexane/1-butene	2009COS
Poly(ethylene-*co*-1-hexene)		
	ethene and n-butane	2004DOE
	ethene and carbon dioxide	2004DOE
	ethene and ethane	2004DOE
	ethene and helium	2004DOE
	ethene and 1-hexene	2004DOE
	ethene and methane	2004DOE
	ethene and nitrogen	2004DOE
	ethene and propane	2004DOE
Poly(ethylene-*co*-1-octene)		
	ethene and cyclohexane/1-octene	2009COS
	ethene and 1-octene	2005LEE
Poly(ethylene-*co*-vinyl acetate)		
	ethene and n-butane	2004DOE
	ethene and carbon dioxide	2004DOE
	ethene and ethane	2004DOE
	ethene and helium	2004DOE
	ethene and methane	2004DOE
	ethene and nitrogen	2004DOE
	ethene and propane	2004DOE
	ethene and vinyl acetate/n-butane	2004DOE
	ethene and vinyl acetate/carbon dioxide	2004DOE
	ethene and vinyl acetate/helium	2004DOE
	ethene and vinyl acetate/methane	2004DOE
	ethene and vinyl acetate/nitrogen	2004DOE

Polymer (B)	**Second and third component**	**Ref.**
Poly(ethylene glycol) mono-2,6,8-trimethyl-4-nonyl ether	carbon dioxide and water	2010HA1
Poly(ethylene oxide-*b*-propylene oxide-*b*-ethylene oxide)	carbon dioxide and ethanol	2008MUN
Poly(ethylene oxide-*b*-1,1,2,2-tetrahydroperfluorodecyl methacrylate)	carbon dioxide and 2-hydroxyethyl methacrylate	2004MA2
Poly(DL-lactic acid)	carbon dioxide and naphthalene	2010MAS
Poly(L-lactic acid)	carbon dioxide and L-lactic acid	2009GRE
	carbon dioxide and naphthalene	2010MAS
Poly(DL-lactic acid-*co*-glycolic acid)	carbon dioxide and 2-propanone	2005WAN
Poly(methyl methacrylate)	carbon dioxide and methyl methacrylate	2009HE2
	carbon dioxide and methyl methacrylate/ poly(vinylidene fluoride)	2009HE2
	carbon dioxide and poly[2-(*N,N*-dimethyl-aminoethyl) methacrylate-*co*-1H,1H–perfluorooctyl methacrylate]	2008HWA
	carbon dioxide and polystyrene	2010LIN
	carbon dioxide and 2-propanone	2009HE1
Poly(4-methyl-1-pentene)	carbon dioxide and n-pentane	2006FA1
Poly(pentafluoropropyl methacrylate)	carbon dioxide and dimethyl ether	2012YOO

Polymer (B)	**Second and third component**	**Ref.**
Poly[2-(perfluorooctyl)ethyl acrylate-*ran*-2-(*N,N*-dimethyl-aminoethyl)ethyl acrylate]		
	carbon dioxide and azelaic acid	2007YO2
	carbon dioxide and glutaric acid	2007YO2
	carbon dioxide and maleic acid	2007YO2
	carbon dioxide and perfluorosuccinic acid	2007YO2
	carbon dioxide and succinic acid	2007YO2
Polystyrene (syndiotactic)	carbon dioxide and acetophenone	2009FAN
	carbon dioxide and cyclohexane	2005XUD
	carbon dioxide and decahydronapthalene	2005XUD
	carbon dioxide and methylcyclohexane	2005XUD
	carbon dioxide and poly(methyl methacrylate)	2010LIN
	carbon dioxide and 2-propanone	2009HE1
	carbon dioxide and styrene	2007GOE
	carbon dioxide and toluene	2008FAN
	hydrogen and decahydronapthalene	2005XUD
	propane and decahydronapthalene	2011CAI
Poly(styrene-*b*-dimethylsiloxane)		
	carbon dioxide and 1-vinyl-2-pyrrolidinone	2000BER
Polysulfone		
	carbon dioxide and 1-methyl-2-pyrrolidinone	2005REV
Poly(vinylidene fluoride)		
	carbon dioxide and chlorodifluoromethane	2004BY4
	carbon dioxide and dimethyl ether	2004BY4
	carbon dioxide and methyl methacrylate	2009HE2
	carbon dioxide and methyl methacrylate/ poly(methyl methacrylate)	2009HE2
	carbon dioxide and 1-pentanol	2004BY4
	carbon dioxide and 1-propanol	2004BY4
	chlorodifluoromethane and trifluoromethane	2004BY4
	dimethyl ether and trifluoromethane	2004BY4

4.5. References

1999FIN Fink, R., Hancu, D., Valentine, R., and Beckman, E.J., Toward the development of CO_2-philic hydrocarbons. 1. Use of side-chain functionalization to lower the miscibility pressure of polydimethylsiloxanes in CO_2, *J. Phys. Chem. B*, 103, 6441, 1999.

2000BER Berger, B.T., Überkritisches Kohlendioxid als Reaktionsmedium für die Dispersionspolymerisation, *Dissertation*, Johannes-Gutenberg Universität Mainz, 2000.

2000SAR Sarbu, T., Styranec, T., and Beckman, E.J., Non-fluorous polymers with very high solubility in supercritical CO_2 down to low pressures, *Nature*, 405, 165, 2000.

2001DOE2 Dörr, H.W., Untersuchungen zum Phasenverhalten von Copolymer-Ethen-Mischungen bei Zusatz der Comonomere und verschiedener Inertgase, *Dissertation*, TU Darmstadt, 2001.

2002BY2 Byun, H.-S. and Park, C., Monomer concentration effect on the phase behavior of poly(propyl acrylate) and poly(propyl methacrylate) with supercritical CO_2 and C_2H_4, *Korean J. Chem. Eng.*, 19, 126, 2002.

2002BY3 Byun, H.-S. and Lee, H.-Y., Phase behavior on the binary and ternary system of poly(propyl acrylate) and poly(propyl methacrylate) with supercritical solvents, *Hwahak Konghak*, 40, 703, 2002.

2002LE1 Lee, S.-H. and McHugh, M.A., The effect of hydrogen bonding on the phase behavior of poly(ethylene-*co*-acrylic acid)-ethylene-cosolvent mixtures at high pressures, *Korean J. Chem. Eng.*, 19, 114, 2002.

2002LE2 Lee, B.-C. and Kim, N.-I., Phase equilibria of poly(methyl methacrylate) in supercritical mixtures of carbon dioxide and chlorodifluoromethane, *Korean J. Chem. Eng.*, 19, 132, 2002.

2004BAE Bae, W., Kwon, S., Byun, H.-S., and Kim, H., Phase behavior of the poly(vinyl pyrrolidone) + *N*-vinyl-2-pyrrolidone + carbon dioxide system, *J. Supercrit. Fluids*, 30, 127, 2004.

2004BAR Baradie, B., Shoichet, M.S., Shen, Z., McHugh, M.A., Hong, L., Wang, Y., Johnson, J.K., Beckman, E.J., and Enick, R.M., Synthesis and solubility of linear poly(tetrafluoroethylene-*co*-vinyl acetate) in dense CO_2: experimental and molecular modeling results, *Macromolecules*, 37, 7799, 2004.

2004BEC Becker, F., Buback, M., Latz, H., Sadowski, G., and Tumakaka, F., Cloud-point curves of ethylene-(meth)acrylate copolymers in fluid ethene up to high pressures and temperatures: experimental study and PC-SAFT modeling, *Fluid Phase Equil.*, 215, 263, 2004.

2004BEY Beyer, C. and Oellrich, L.R., High-pressure phase equilibria of copolymer solutions - Experiments and correlations, in *Supercritical Fluids as Solvents and Reaction Media*, Brunner, G. (ed.), Elsevier, 61, 2004.

2004BY2 Byun, H.-S., Phase behavior on the binary and ternary mixtures of poly(cyclohexyl acrylate)and poly(cyclohexyl methacrylate) in supercritical CO_2, *J. Appl. Polym. Sci.*, 94, 1117, 2004.

2004BY3 Byun, H.-S. and Choi, M.-Y., Solubility in the binary and ternary system for poly(alkyl acrylate)-supercritical solvent mixtures, *Korean J. Chem. Eng.*, 21, 874, 2004.

2004BY4 Byun, H.-S. and Yoo, Y.-H., Thermodynamic phase behavior of fluoropolymer mixtures with supercritical fluid solvents, *Korean J. Chem. Eng.*, 21, 1193, 2004.

2004DOE Dörr, H., Kinzl, M., Luft, G., and Ruhl, O., Influence of additional components on the solvent power of supercritical ethylene, in *Supercritical Fluids as Solvents and Reaction Media*, Brunner, G. (ed.), Elsevier, 39 2004.

2004GHO Ghosh, A., Ting, P.D., and Chapman, W.G., Thermodynamic stability analysis and pressure-temperature flash for polydisperse polymer solutions, *Ind. Eng. Chem. Res.*, 43, 6222, 2004.

2004HAT Hatanaka M. and Saito, H., In-situ investigation of liquid-liquid phase separation in polycarbonate/carbon dioxide system, *Macromolecules*, 37, 7358, 2004.

2004HWA Hwang, H.S., Kim, H.J., Jeong, Y.T., Gal, Y.-S., and Lim, K.T., Synthesis and properties of semifluorinated copolymers of oligo(ethylene glycol) methacrylate and 1H,1H,2H,2H-perfluorooctyl methacrylate, *Macromolecules*, 37, 9821, 2004.

2004KER Kermis, T.W., Li, D., Guney-Altay, O., Park, I.-H., Zanten, J.H. van, and McHugh, M.A., High-pressure dynamic light scattering of poly(ethylene-*co*-1-butene) in ethane, propane, butane, and pentane at 130°C and kilobar pressures, *Macromolecules*, 37, 9123, 2004.

2004KUK Kukova, E., Petermann, M., and Weidner, E., Phasenverhalten (S-L-G) und Transporteigenschaften binaerer Systeme aus hochviskosen Polyethylenglycolen und komprimiertem Kohlendioxid, *Chem. Ing. Techn.*, 76, 280, 2004.

2004LAC Lacroix-Desmazes, P., Andre, P., Desimone, J.M., Ruzette, A.-V., and Boutevin, B., Macromolecular surfactants for supercritical carbon dioxide applications: synthesis and characterization of fluorinated block copolymers prepared by nitroxide-mediated radical polymerization (experimental data by P. Lacroix-Desmazes), *J. Polym. Sci.: Part A: Polym. Chem.*, 42, 3537, 2004.

2004LAT Latz, H., Kinetische und thermodynamische Untersuchungen der Hochdruck-Copolymerisation von Ethen mit (Meth)Acrylsäureestern, Dissertation, Georg-August-Universität Göttingen, 2004.

2004LIM Lim, J.S., Park, J.-Y., Yoon, C.H., Lee, Y.-W., and Yoo, K.-P., Cloud points of poly(L-lactide) in HCFC-22, HFC-23, HFC-32, HFC-125, HFC-143a, HFC-152a, HFC-227ea, dimethyl ether (DME), and HCFC-22 + CO_2 in the supercritical state, *J. Chem. Eng. Data*, 49, 1622, 2004.

2004MA1 Ma, Z. and Lacroix-Desmazes, P., Synthesis of hydrophilic/CO_2-philic poly(ethylene oxide)-*b*-poly(1,1,2,2-tetrahydroperfluorodecyl acrylate) block copolymers via controlled living radical polymerizations and their properties in liquid and supercritical CO_2 (experimental data by P. Lacroix-Desmazes), *J. Polym. Sci.: Part A: Polym. Chem.*, 42, 2405, 2004.

2004MA2 Ma, Z. and Lacroix-Desmazes, P., Dispersion polymerization of 2-hydroxyethyl methacrylate stabilized by a hydrophilic/CO_2-philic poly(ethylene oxide)-*b*-poly(1,1,2,2-tetrahydroperfluorodecyl acrylate) (PEO-*b*-PFDA) diblock copolymer in supercritical carbon dioxide (experimental data by P. Lacroix-Desmazes), *Polymer*, 45, 6789, 2004.

2004PER Perez, Y., Wubbolts, F.E., Witkamp, G.J., Jansens, P.J., and Loos, Th.W.de, Improved PCA process for the production of nano-and microparticles of polymers, *AIChE-J.*, 50, 2408, 2004.

2004TAN Tan, S.P., Meng, D., Plancher, H., Adidharma, H., and Radosz, M., Cloud points for polystyrene in propane and poly(4-methyl styrene) in propane, *Fluid Phase Equil.*, 226, 189, 2004.

2004WAN Wang, J., Lue, M., Chen, J., and Yang, Y., Solubility measurement of polystyrene in supercritical propane, *J. Chem. Ind. Eng. (China)*, 55, 805, 2004.

2005BON Bonavoglia, B., Storti, G., and Morbidelli, M., Oligomers partitioning in supercritical CO_2, Macromolecules, 38, 5593, 2005.

2005COT Cotugno, S., DiMaio, E., Mensitieri, G., Iannace, S., Roberts, G.W., Carbonell, R.G., and Hopfenberg, H.B., Characterization of microcellular biodegradable polymeric foams produced from supercritical carbon dioxide solutions, *Ind. Eng. Chem. Res.*, 44, 1795, 2005.

2005KAL Kalogiannis, C.G. and Panayiotou, C.G., Bubble and cloud points of the system poly(L-lactic acid) + carbon dioxide + dichloromethane, *J. Chem. Eng. Data*, 50, 1442, 2005.

2005LEE Lee, S.-H., Phase behavior of binary and ternary mixtures of poly(ethylene-*co*-octene)–hydrocarbons (experimental data by S.-H. Lee), *J. Appl. Polym. Sci.*, 95, 161, 2005.

2005LID Li, D., McHugh, M.A., and van Zanten, J.H., Density-induced phase separation in poly(ethylene-*co*-1-butene)-dimethyl ether solutions, *Macromolecules*, 38, 2837, 2005.

2005NIS Nishi, K., Morikawa, Y., Misumi, R., and Kaminoyama, M., Radical polymerization in supercritical carbon dioxide: use of supercritical carbon dioxide as a mixing assistant, *Chem. Eng. Sci.*, 60, 2419, 2005.

2005PER Perez de Diego, Y., Wubbolts, F.E., Witkamp, G.J., Loos, Th.W. de, and Jansens, P.J., Measurements of the phase behaviour of the system dextran/DMSO/CO_2 at high pressures, *J. Supercrit. Fluids*, 35, 1, 2005.

2005REV Reverchon, E. and Cardea, S., Formation of polysulfone membranes by supercritical CO_2, *J. Supercrit. Fluids*, 35, 140, 2005.

2005SCH Schilt, M.A. van, Wering, R.M., Meerendonk, W.J. van, Kemmere, M.F., Keurentjes, J.T.F., Kleiner, M., Sadowski, G., and Loos, Th.W.de, High-pressure phase behavior of the system PCHC-CHO-CO_2 for the development of a solvent-free alternative toward polycarbonate production, *Ind. Eng. Chem. Res.*, 44, 3363, 2005.

2005SHE Shen, Z., McHugh, M.A., Smith Jr., D.W., Abayasinghe, N.K., and Jin, J., Impact of hexafluoroisopropylidene on the solubility of aromatic-based polymers in supercritical fluids, *J. Appl. Polym. Sci.*, 97, 1736, 2005.

2005TAI Tai, H., Wang, W., Martin, R., Liu, J., Lester, E., Licence, P., Woods, H.M., and Howdle, S.M., Polymerization of vinylidene fluoride in supercritical carbon dioxide: effects of poly(dimethylsiloxane) macromonomer on molecular weight and morphology of poly(vinylidene fluoride) (experimental data by H. Tai and S.M.Howdle), *Macromolecules*, 38, 355, 2005.

2005WAN Wang, Y., Pfeffer, R., Dave, R., and Enick, R., Polymer encapsulation of fine particles by a supercritical antisolvent process, *AIChE-J.*, 51, 440, 2005.

2005XUD Xu, D., Carbonell, R.G., Roberts, G.W., and Kiserow, D.J., Phase equilibrium for the hydrogenation of polystyrene in CO_2-swollen solvents, *J. Supercrit. Fluids*, 34, 1, 2005.

2006AHM Ahmed, T.S., DeSimone, J.M., and Roberts, G.W., Copolymerization of vinylidene fluoride with hexafluoropropylene in supercritical carbon dioxide, *Macromolecules*, 39, 15, 2006.

2006AND Andre, P., Lacroix-Desmazes, P., Taylor, D.K., and Boutevin, B., Solubility of fluorinated homopolymer and block copolymer in compressed CO_2, *J. Supercrit. Fluids*, 37, 263, 2006.

2006BY1 Byun, H.-S. and Lee, D.-H., Cosolvent effect and solubility measurement for butyl (meth)acrylate polymers in benign environmental supercritical solvents, *Ind. Eng. Chem. Res.*, 45, 3354, 2006.

2006BY2 Byun, H.-S., Lee, D.H., Lim, J.-S., and Yoo, K.-P., Phase behavior of the binary and ternary mixtures of biodegradable poly(ε-caprolactone) in supercritical fluids, *Ind. Eng. Chem. Res.*, 45, 3366, 2006.

2006BY3 Byun, H.-S. and Lee, D.-H., Phase behavior of binary and ternary mixtures of poly(decyl acrylate)-supercritical solvent-decyl acrylate and poly(decyl methacrylate)-CO_2-decyl methacrylate systems, *Ind. Eng. Chem. Res.*, 45, 3373, 2006.

2006BY4 Byun, H.-S., Phase behavior of poly(ethylene glycol) in supercritical CO_2, C_3H_6, and C_4H_8, *J. Ind. Eng. Chem.*, 12, 893, 2006.

2006BY5 Byun, H.-S. and Lee, H.-Y., Cloud-point measurement of the biodegradable poly(DL-lactide-*co*-glycolide) solution in supercritical fluid solvents, *Korean J. Chem. Eng.*, 23, 1003, 2006.

2006FA1 Fang, J. and Kiran, E., Crystallization and gelation of isotactic poly(4-methyl-1-pentene) in n-pentane and in n-pentane + CO_2 at high pressures, *J. Supercrit. Fluids*, 38, 132, 2006.

2006FA2 Fang, J. and Kiran, E., Kinetics of pressure-induced phase separation in polystyrene + acetone solutions at high pressures, *Polymer*, 47, 7943, 2006.

2006GAN Ganapathy, H.S., Hwang, H.S., and Lim, K.T., Synthesis and properties of fluorinated ester-functionalized polythiophenes in supercritical carbon dioxide, *Ind. Eng. Chem. Res.*, 45, 3406, 2006.

2006KAL Kalogiannis, C.G. and Panayiotou, C.G., Bubble and cloud points of the systems poly(ε-caprolactone) + carbon dioxide + dichloromethane or chloroform, *J. Chem. Eng. Data*, 51, 107, 2006.

2006KLE Kleiner, M., Tumakaka, F., Sadowski, G., Latz, H., and Buback, M., Phase equilibria in polydisperse and associating copolymer solutions: poly(ethene-*co*-(meth)acrylic acid) monomer mixtures, *Fluid Phase Equil.*, 241, 113, 2006.

2006KOS Kostko, A.F., McHugh, M.A., and Zanten, J.H. van, Coil-coil interactions for poly(dimethylsiloxane) compressible supercritical CO_2, *Macromolecules*, 39, 1657, 2006.

2006LAV Lavery, K.A., Sievert, J.D., Watkins, J.J., Russell, T.P., Ryu, D.Y., and Kim, J.K., Influence of carbon dioxide swelling on the closed-loop phase behavior of block copolymers, *Macromolecules*, 39, 6580, 2006

2006LIU Liu, K. and Kiran, E., Miscibility, viscosity and density of poly (e-caprolactone) in acetone + CO_2 binary fluid mixtures, *J. Supercrit. Fluids*, 39, 192, 2006.

2006MAR Martinez, V., Mecking, S., Tassaing, T., Besnard, M., Moisan, S., Cansell, F., and Aymonier, C., Dendritic core-shell macromolecules soluble in supercritical carbon dioxide, *Macromolecules*, 39, 3978, 2006.

2006MAT Matsuyama, K. and Mishima, K., Phase behavior of CO_2 + polyethylene glycol + ethanol at pressures up to 20MPa, *Fluid Phase Equil.*, 249, 173, 2006.

2006NAG Nagy, I., Loos, Th.W.de, Krenz, R.A., and Heidemann, R.A., High pressure phase equilibria in the systems linear low density polyethylene + n-hexane and linear low density polyethylene + n-hexane + ethylene: Experimental results and modelling with the Sanchez-Lacombe equation of state, *J. Supercrit. Fluids*, 37, 115, 2006.

2006PAR Park, J.-Y., Kim, S.Y., Byun, H.-S., Yoo, K.-P., and Lim, J.S., Cloud points of poly(ε-caprolactone), poly(L-lactide), and polystyrene in supercritical fluids, *Ind. Eng. Chem. Res.*, 45, 3381 and 6092, 2006.

2006SHI Shin, J., Lee, Y.-W., Kim, H., and Bae, W., High-pressure phase behavior of carbon dioxide + heptadecafluorodecyl acrylate + poly(heptadecafluorodecyl acrylate) system, *J. Chem. Eng. Data*, 51, 1571, 2006.

2006SUB Su, B., Lv, X., Yang, Y., and Ren, Q., Solubilities of dodecylpolyoxyethylene polyoxypropylene ether in supercritical carbon dioxide, *J. Chem. Eng. Data*, 51, 542, 2006.

2006UPP Upper, G., Zhang, W., Beckel, D., Sohn, S., Liu, K., and Kiran, E., Phase boundaries and crystallization of polyethylene in n-pentane and n-pentane + carbon dioxide fluid mixtures, *Ind. Eng. Chem. Res.*, 45, 1478, 2006.

2006WIN Winoto, W., Adidharma, H., Shen, Y., and Radosz, M., Micellization temperature and pressure for polystyrene-*block*-polyisoprene in subcritical and supercritical propane, *Macromolecules*, 39, 8140, 2006.

2007BUC Buchelli, A. and Todd, W.G., On-line liquid-liquid phase separation predictor in the high-density polyethylene solution polymerization process, *Ind. Eng. Chem. Res.*, 46, 4307, 2007.

2007BY1 Byun, H.-S. and Lee, D.-H., Phase behavior of the poly(neopentyl methacrylate) + supercritical fluid solvents + neopentyl methacrylate system and CO_2 + neopentyl methacrylate mixtures at high pressure, *Polymer*, 48, 805, 2007.

2007BY2 Byun, H.-S. and McHugh, M.A., High pressure phase behavior of poly(isopropyl acrylate) and poly(isopropyl methacrylate) in supercritical fluid (SCF) solvent and SCF solvent + cosolvent mixtures, *J. Supercrit. Fluids*, 41, 482, 2007.

2007BY3 Byun, H.-S. and Yoo, K.-P., Phase behavior on the poly[octyl (meth)acrylate] + supercritical fluid solvents + monomer and CO_2 + monomer mixtures at high pressure, *J. Supercrit. Fluids*, 41, 472, 2007.

2007EDM Edmonds, W.F., Hillmyer, M.A., and Lodge, T.P., Block copolymer vesicles in liquid CO_2, *Macromolecules*, 40, 4917, 2007.

2007GOE Goernert, M. and Sadowski, G., Phase-equilibrium measurements of the polystyrene/styrene/carbon dioxide ternary system at elevated pressures using ATR-FTIR spectroscopy, *Macromol. Symp.*, 259, 236, 2007.

2007GRE Gregorowicz, J., Solid–fluid phase behaviour of linear polyethylene solutions in propane, ethane and ethylene at high pressures, *J. Supercrit. Fluids*, 43, 357, 2007.

2007KIL Kilic, S., Michalik, S., Wang, Y., Johnson, J.K., Enick, R.H., and Beckman, E.J., Phase behavior of oxygen-containing polymers in CO_2, *Macromolecules*, 40, 1332, 2007.

2007KWO Kwon, S., Bae, W., Lee, K., Byun, H.-S., and Kim, H., High pressure phase behavior of carbon dioxide + 2,2,2-trifluoroethyl methacrylate and + poly(2,2,2-trifluoroethyl methacrylate) systems, *J. Chem. Eng. Data*, 52, 89, 2007.

2007LIS Li, S., Li, Y., and Wang, J., Solubility of modified poly(propylene oxide) and silicones in supercritical carbon dioxide, *Fluid Phase Equil.*, 253, 54, 2007.

2007LI1 Liu, S., Lee, D.-H., and Byun, H.-S., Phase behavior for mixtures of poly(2-ethylhexyl acrylate) + 2-ethylhexyl acrylate and poly(2-ethylhexyl methacrylate) + 2-ethylhexyl methacrylate with supercritical fluid solvents, *J. Chem. Eng. Data*, 52, 410, 2007.

2007LI2 Liu, K. and Kiran, K., A tunable mixture solvent for poly(ε-caprolactone): Acetone + CO_2, *Polymer*, 48, 5612, 2007.

2007LI3 Liu, J., Spraul, B.K., Topping, C., Smith, Jr., D.W., and McHugh, M.A., Effect of hexafluoroisopropylidene on perfluorocyclobutyl aryl ether copolymer solution behavior in supercritical CO_2 and propane, *Macromolecules*, 40, 5973, 2007.

2007NAG Nagy, I., Krenz, R.A., Heidemann, R.A., and de Loos, Th.W., High-pressure phase equilibria in the system linear low density polyethylene + isohexane: Experimental results and modelling, *J. Supercrit. Fluids*, 40, 125, 2007.

2007TAN Tan, S.P., Winoto, W., and Radosz, M., Statistical associating fluid theory of homopolymers and block copolymers in compressible solutions: Polystyrene, polybutadiene, polyisoprene, polystyrene-*block*-polybutadiene, and polystyrene-*block*-polyisoprene in propane, *J. Phys. Chem. C*, 111, 15752, 2007.

2007TUM Tumakaka, F., Sadowski, G., Latz, H., and Buback, M., Cloud-point pressure curves of ethylene-based terpolymers in fluid ethene and in ethene-comonomer-mixtures. Experimental study and modeling via PC-SAFT, *J. Supercrit. Fluids*, 41, 461, 2007.

2007YO1 Yoshida, E. and Nagakubo, A., Convenient synthesis of microspheres by self-assembly of random copolymers in supercritical carbon dioxide, *Colloid Polym. Sci.*, 285, 441, 2007.

2007YO2 Yoshida, E. and Nagakubo, A., Superhydrophobic surfaces of microspheres obtained by self-assembly of poly[2-(perfluorooctyl)ethyl acrylate-*ran*-2-(dimethylamino)ethyl acrylate] in supercritical carbon dioxide, *Colloid Polym. Sci.*, 285, 1293, 2007.

2008BY1 Byun, H.-S., Park, Y.-J., and Lim, J.-S., Phase behavior on the binary and ternary mixtures of poly(isooctyl acrylate) + supercritical fluid solvents + isooctyl acrylate and CO_2 + isooctyl acrylate system, *J. Appl. Polym. Sci.*, 107, 1124, 2008.

2008BY2 Byun, H.-S., Bang, C.-H., and Lim, J.-S., Effect of cosolvent concentration on phase behavior for the poly(isodecyl acrylate) in supercritical carbon dioxide, propane, propylene, butane, 1-butene and dimethyl ether, *J. Macromol. Sci., Part B: Phys.*, 47, 150, 2008.

2008FAN Fang, J. and Kiran, E., Thermoreversible gelation and polymorphic transformations of syndiotactic polystyrene in toluene and toluene + carbon dioxide fluid mixtures at high pressures, *Macromolecules*, 41, 7525, 2008.

2008GOE Goernert, M. and Sadowski, G., Phase-equilibrium measurement and modeling of the PMMA/MMA/carbon dioxide ternary system, *J. Supercrit. Fluids*, 46, 218, 2008.

2008GUN Guney-Altay, O., Shenoy, S.L., Fujiwara, T., Irie, S., Wynne, K.J., Measurement of supercritical CO_2 plasticization of poly(tetrafluoroethylene) using a linear variable differential transformer, *Macromolecules*, 41, 8660, 2008.

2008HAR Haruki, M., Takakura, Y., Sugiura, H., Kihara, S., and Takishima, S., Phase behavior for the supercritical ethylene + hexane + polyethylene systems, *J. Supercrit. Fluids*, 44, 284, 2008.

2008HON Hong, L., Tapriyal, D., and Enick, R.M., Phase behavior of poly(propylene glycol) monobutyl ethers in dense CO_2, *J. Chem. Eng. Data*, 53, 1342, 2008.

2008HWA Hwang, H.S., Yuvaraj, H., Kim, W.S., Lee, W.-K., Gal, Y.-S., and Lim, K.T., Dispersion polymerization of MMA in supercritical CO_2 stabilized by random copolymers of 1H,1H–perfluorooctyl methacrylate and 2-(dimethylaminoethyl) methacrylate, *J. Polym. Sci.: Part A: Polym. Chem.*, 46, 1365, 2008.

2008KOS Kostko, A.F., Lee, S.H., Liu, J., DiNoia, T.P., Kim, Y., and McHugh, M.A., Cloud-point behavior of poly(ethylene-*co*-20.2 mol% 1-butene) (PEB10) in ethane and deuterated ethane and of deuterated PEB10 in pentane isomers, *J. Chem. Eng. Data*, 53, 1626, 2008.

2008LIU Liu, K. and Kiran, E., High-pressure solution blending of poly(ε-caprolactone) with poly(methyl methacrylate) in acetone + carbon dioxide, *Polymer*, 49, 1555, 2008.

2008MAT Matsuyama, K. and Mishima, K., Phase behavior of the mixtures of CO_2 + poly(methyl methacrylate) + ethanol at high pressure, *J. Chem. Eng. Data*, 53, 1151, 2008.

2008MUN Munto, M., Ventosa, N., and Veciana, J., Synergistic solubility behaviour of a polyoxyalkylene block co-polymer and its precipitation from liquid CO_2-expanded ethanol as solid microparticles, *J. Supercrit. Fluids*, 47, 290, 2008.

2008PA2 Pasquali, I., Comi, L., Pucciarelli, F., and Bettini, R., Swelling, melting point reduction and solubility of PEG 1500 in supercritical CO_2, *Int. J. Pharmaceutics*, 356, 76, 2008.

2008SH1 Shin, J., Bae, W., Kim, H., and Lee, Y.-W., Phase behavior of binary and ternary carbon dioxide + heptadecafluorodecyl methacrylate + poly(heptadecafluorodecyl methacrylate) systems, *J. Chem. Eng. Data*, 53, 1523, 2008.

2008SH2 Shin, J., Oh, K.S., Bae, W., Lee, Y.-W., and Kim, H., Dispersion polymerization of methyl methacrylate using poly(HDFDMA-*co*-MMA) as a surfactant in supercritical carbon dioxide, *Ind. Eng. Chem. Res.*, 47, 5680, 2008.

2008SH3 Shin, J., Bae, W., Kim, B.G., Lee, J.-C., Byun, H.-S., and Kim, H., Phase behavior of a ternary system of poly[p-perfluorooctyl-ethylene(oxy, thio, sulfonyl)methyl styrene] and poly[p-decyl(oxy, thio, sulfonyl)methyl styrene] in supercritical solvents, *J. Supercrit. Fluids*, 47, 1, 2008.

2008SH4 Shin, M.S., Lee, J.H., and Kim, H., Phase behavior of the poly(vinyl pyrrolidone) + dichloromethane + supercritical carbon dioxide system, *Fluid Phase Equil.*, 272, 42, 2008.

2008TAP Tapriyal, D., Wang, Y., Enick, R.M., Johnson, J.K., Crosthwaite, J., Thies, M.C., Paik, I.H., and Hamilton, A.D., Poly(vinyl acetate), poly((1-O-(vinyloxy)ethyl-2,3,4,6-tetra-O-acetyl-β-D-glucopyranoside) and amorphous poly(lactic acid) are the most CO_2-soluble oxygenated hydrocarbon-based polymers, *J. Supercrit. Fluids*, 46, 252, 2008.

2008YOS Yoshida, E. and Mineyama, A., Synthesis of spherical particles by self-assembly of poly[2-(perfluorooctyl)ethyl acrylate-*co*-acrylic acid] in supercritical carbon dioxide, *Colloid Polym. Sci.*, 286, 975, 2008.

2009COS Costa, G.M.N., Guerrieri, Y., Kislansky, S., Pessoa, F.L.P., Vieira de Melo, S.A.B., and Embirucu, M., Simulation of flash separation in polyethylene industrial processing: comparison of SRK and SL equations of state, *Ind. Eng. Chem. Res.*, 48, 8613, 2009.

2009FAN Fang, J. and Kiran, E., Gelation, crystallization and morphological transformations of syndiotactic polystyrene in acetophenone and acetophenone + carbon dioxide mixtures at high pressures, *J. Supercrit. Fluids*, 49, 93, 2009.

2009GRE Gregorowicz, J. and Bernatowicz, P., Phase behaviour of L-lactic acid based polymers of low molecular weight in supercritical carbon dioxide at high pressures, *J. Supercrit. Fluids*, 51, 270, 2009.

2009HAR Haruki, M., Sato, K., Kihara, S.-I., and Takishima, S., High pressure phase behavior for the supercritical ethylene + cyclohexane + hexane + polyethylene systems, *J. Supercrit. Fluids*, 49, 125, 2009.

2009HE1 He, J. and Wang, B., Acetone influence on glass transition of poly(methyl methacrylate) and polystyrene in compressed carbon dioxide, *Ind. Eng. Chem. Res.*, 48, 5093, 2009.

2009HE2 He, J. and Wang, B., The temperature, cosolvent, and blending effects on the partitions between polymer and compressed carbon dioxide, *Ind. Eng. Chem. Res.*, 48, 7359, 2009.

2009HOL Hollamby, M.J., Trickett, K., Mohammed, A., Eastoe, J., Rogers, S.E., and Heenan, R.K., Surfactant aggregation in CO_2/heptane solvent mixtures, *Langmuir*, 25, 12909, 2009.

2009KIL Kilic, S., Wang, Y., Johnson, J.K., Beckman, E.J., and Enick, R.M., Influence of *tert*-amine groups on the solubility of polymers in CO_2, *Polymer*, 50, 2436, 2009.

2009KIM Kim, S.-E., Yoon, S.-D., Yoo, K.-P., and Byun, H.-S., Cloud point behavior for poly(isodecyl methacrylate) + supercritical solvents + cosolvent and vapor-liquid behavior for CO_2 + isodecyl methacrylate systems at high pressure, *Korean J. Chem. Eng.*, 26, 199, 2009.

2009KOJ Kojima, J., Takenaka, M., Nakayama, Y., and Saeki, S., Measurements of phase behavior for polyethylene in hydrocarbons, halogenated hydrocarbons, and oxygen-containing hydrocarbons, at high pressure and high temperature, *J. Chem. Eng. Data*, 54, 1585, 2009.

2009KOZ Kozlowska, M.K., Jürgens, B.F., Schacht, C.S., Gross, J., and Loos, Th.W. de, Phase behavior of hyperbranched polymer systems: Experiments and application of the perturbed-chain polar SAFT equation of state, *J. Phys. Chem. B*, 113, 1022, 2009.

2009LAW Law, Y.Y., Balashova, I.M., and Danner, R.P., Effect of high pressure carbon dioxide on the solubility and diffusivity of dichloromethane in polyetherimide, *J. Appl. Polym. Sci.*, 114, 2497, 2009.

2009LIU Liu, S., Lee, H.-Y., Yoon, S.-D., Yoo, K.-P., and Byun, H.-S., High-pressure phase behavior for poly[dodecyl methacrylate] + supercritical solvents + cosolvents and carbon dioxide + dodecyl methacrylate mixture, *Ind. Eng. Chem. Res.*, 48, 7821, 2009.

2009POR Portela, V.M., Straver, E.J.M., and de Loos, Th.W., High-pressure phase behavior of the system propane-Boltorn H3200, *J. Chem. Eng. Data*, 54, 2593, 2009.

2009ROJ Rojo, S.R., Martín, A., Calvo, E.S., and Cocero, M.J., Solubility of polycaprolactone in supercritical carbon dioxide with ethanol as cosolvent, *J. Chem. Eng. Data*, 54, 962, 2009.

2009SAN Santoyo-Arreola, J.G., Vasquez-Medrano, R.C., Ruiz-Trevino, A., Luna-Barcenas, G., Sanchez, I.C., and Ortiz-Estrada, C.H., Phase behavior and particle formation of poly(1H,1H-dihydrofluorooctyl methacrylate) in supercritical CO_2, *Macromol. Symp.*, 283-284, 230, 2009.

2009STO Stoychev, I., Galy, J., Fournel, B., Lacroix-Desmazes, P., Kleiner, M., and Sadowski, G., Modeling the phase behavior of PEO-PPO-PEO surfactants in carbon dioxide using the PC-SAFT equation of state: Application to dry decontamination of solid substrates, *J. Chem. Eng. Data*, 54, 1551, 2009.

2009TAN Tan, B., Bray, C.L., and Cooper, A.I., Fractionation of poly(vinyl acetate) and the phase behavior of end-group modified oligo(vinyl acetate)s in CO_2, *Macromolecules*, 42, 7945, 2009.

2009TYR Tyrrell, Z., Winoto, W., Shen, Y., and Radosz, M., Block copolymer micelles formed in supercritical fluid can become water-dispensable nanoparticles: poly(ethylene glycol)-*block*-poly(ε-caprolactone) in trifluoromethane, *Ind. Eng. Chem. Res.*, 48, 1928, 2009.

2009WAN Wang, Y., Hong, L., Tapriyal, D., Kim, I.C., Paik, I.-H., Crosthwaite, J.M., Hamilton, A.D., Thies, M.C., Beckman, E.J., Enick, R.M., and Johnson, J.K., Design and evaluation of nonfluorous CO_2-soluble oligomers and polymers, *J. Phys. Chem. B*, 113, 14971, 2009.

2009WI1 Winoto, W., Shen, Y., Radosz, M., Hong, K., and Mays, J.W., Deuteration impact on micellization pressure and cloud pressure of polystyrene-*block*-polybutadiene and polystyrene-*block*-polyisoprene in compressible propane, *J. Phys. Chem. B*, 113, 15156, 2009.

2009WI2 Winoto, W., Radosz, M., Hong, K., and Mays, J.W., Amorphous polystyrene-*block*-polybutadiene and crystallizable polystyrene-*block*-(hydrogenated polybutadiene) solutions in compressible near critical propane and propylene: Hydrogenation effects, *J. Non-Crystal. Solids*, 355, 1393, 2009.

2009WI3 Winoto, W., Tan, S.P., Shen, Y., Radosz, M., Hong, K., and Mays, J.W., High-pressure micellar solutions of polystyrene-*block*-polybutadiene and polystyrene-*block*-polyisoprene in propane exhibit cloud-pressure reduction and distinct micellization end points, *Macromolecules*, 42, 3823, 2009.

2010BEN Bender, J.P., Feiteina, M., Mazutti, M.A., Franceschi, E., Corazza, M.L., and Oliveira, J.V., Phase behaviour of the ternary system {poly(ε-caprolactone) + carbon dioxide + dichloromethane}, *J. Chem. Thermodyn.*, 42, 229, 2010.

2010GRE Gregorowicz, J., Fras, Z., Parzuchowski, P., Rokicki, G., Kusznerczuk, M., and Dziewulski, S., Phase behaviour of hyperbranched polyesters and polyethers with modified terminal OH groups in supercritical solvents, *J. Supercrit. Fluids*, 55, 786, 2010.

2010HA1 Haruki, M., Matsuura, K., Kaida, Y., Kihara, S.-I., and Takishima, S., Microscopic phase behavior of supercritical carbon dioxide + non-ionic surfactant + water systems at elevated pressures, *Fluid Phase Equil.*, 289, 1, 2010.

2010HA2 Haruki, M., Mano, S., Koga, Y., Kihara, S., and Takishima, S., Phase behaviors for the supercritical ethylene + 1-hexene + hexane + polyethylene systems at elevated temperatures and pressures, *Fluid Phase Equil.*, 295, 137, 2010.

2010HAS Hasan, M.M., Li, Y.G., Li, G., Park, C.B., and Chen, P., Determination of solubilities of CO_2 in linear and branched polypropylene using a magnetic suspension balance and a PVT apparatus, *J. Chem. Eng. Data*, 55, 4885, 2010.

2010JAN Jang, Y.-S., Kang, J.-W., and Byun, H.-S., Cosolvent effect on the phase behavior for the poly(benzyl acrylate) and poly(benzyl methacrylate) in supercritical carbon dioxide and dimethyl ether, *J. Ind. Eng. Chem.*, 16, 598, 2010.

2010JIA Jiang, Y., Liu, M., Sun, W., Li, L., and Qian, Y., Phase behavior of poly(lactic acid)/poly(ethylene glycol)/poly(lactic acid) (PLA-PEG-PLA) in different supercritical systems of CO_2 + dichloromethane and CO_2 + C_2H_5OH + dichloromethane, *J. Chem. Eng. Data*, 55, 4844, 2010.

2010LE1 Lee, H., Pack, J.W., Wang, W., Thurecht, K.J., and Howdle, S.M., Synthesis and phase behavior of CO_2-soluble hydrocarbon copolymer: poly(vinyl acetate-*alt*-dibutyl maleate), *Macromolecules*, 43, 2276, 2010.

2010LE2 Lee, H.-Y., Yoon, S.-D., and Byun, H.-S., Cloud-point and vapor-liquid behavior of binary and ternary systems for the poly(dodecyl acrylate) + cosolvent and dodecyl acrylate in supercritical solvents, *J. Chem. Eng. Data*, 55, 3684, 2010.

2010LIN Lin, I.-H., Liang, P.-F., and Tan, C.-S., Preparation of polystyrene/poly(methyl methacrylate) blends by compressed fluid antisolvent technique, *J. Supercrit. Fluids*, 51, 384, 2010.

2010MAS Ma, S.-L., Lu, Z.-W., Wu, Y.-T., and Zhang, Z.-B., Partitioning of drug model compounds between poly(lactic acid)s and supercritical CO_2 using quartz crystal microbalance as an *in situ* detector, *J. Supercrit. Fluids*, 54, 129, 2010.

2010MIL Milanesio, J.M., Mabe, G.D.B., Ciolino, A.E., Quinzani, L.M., and Zabaloy, M.S., Experimental cloud points for polybutadiene + light solvent and polyethylene + light solvent systems at high pressure, *J. Supercrit. Fluids*, 55, 363, 2010.

2010SCH Schacht, C.S., Bahramali, S., Wilms, D., Frey, H., Gross, J., and de Loos, Th.W., Phase behavior of the system hyperbranched polyglycerol + methanol + carbon dioxide, *Fluid Phase Equil.*, 299, 252, 2010.

2011CAI Cain, N., Haywood, A., Roberts, G., Kiserow, D., and Carbonell, R., Polystyrene/decahydronaphthalene/propane phase equilibria and polymer conformation properties from intrinsic viscosities (exp. data by N. Cain), *J. Polym. Sci.: Part B: Polym. Phys.*, 49, 1093, 2011.

2011PRI Prichard, T.D., Thomas, R.R., Kausch, C.M., and Vogt, B.D., Solubility of non-ionic poly(fluorooxetane)-*block*-(ethylene oxide)-*block*-(fluorooxetane) surfactants in carbon dioxide, *J. Supercrit. Fluids*, 57, 95, 2011.

2011RIB Ribaut, T., Oberdisse, J., Annighofer, B., Fournel, B., Sarrade, S., Haller, H., and Lacroix-Desmazes, P., Solubility and self-assembly of amphiphilic gradient and block copolymers in supercritical CO_2, *J. Phys. Chem. B*, 115, 836, 2011.

2011RU1 Ruhl, O. and Luft, G., Phasenverhalten von Mischungen aus Propen und isotaktischem Polypropen unter Hochdruck, *Chem.-Ing. Techn.*, 83, 1663, 2011.

2011RU2 Ruhl, O., Luft, G., Brant, P., and Shutt, J.R., Phase behaviour of the system propene/polypropene at high pressure, *J. Thermodyn.*, ID 282354, 2011.

2011SCH Schacht, C.S., Schuell, C., Frey, H., Loos, Th.W. de, and Gross, J., Phase behavior of the system linear polyglycerol + methanol + carbon dioxide, *J. Chem. Eng. Data*, 56, 2927, 2011.

2012GIR Girard, E., Tassaing, T., Ladaviere, C., Marty, J.-D., and Destarac, M., Distinctive features of solubility of RAFT/MADIX-derived partially trifluoromethylated poly(vinyl acetate) in supercritical CO_2, *Macromolecules*, 45, 9674, 2012.

2012GWO Gwon, J., Cho, D.W., Kim, S.H., Shin, H.Y., and Kim, H., Phase behaviour of the ternary mixture system of poly(L-lactic acid), dichloromethane and carbon dioxide, *J. Chem. Thermodyn.*, 55, 37, 2012.

2012KIR Kiran, E., Modification of biomedical polymers in dense fluids. Miscibility and foaming of poly(p-dioxanone) in carbon dioxide + acetone fluid mixtures, *J. Supercrit. Fluids*, 66, 372, 2012.

2012MIL Milanesio, J.M., Mabe, G.D.B., Ciolino, A.E., Quinzani, L.M., and Zabaloy, M.S., High-pressure liquid-liquid equilibrium boundaries for systems containing polybutadiene and/or polyethylene and a light solvent or solvent mixture, *J. Supercrit. Fluids*, 72, 333, 2012.

2012STO Stoychev, I., Peters, F., Kleiner, M., Clerc, S., Ganachaud, F., Chirat, M., Fournel, B., Sadowski, G., and Lacroix-Desmazes, P., Phase behavior of poly(dimethyl siloxane)-poly(ethylene oxide) amphiphilic block and graft copolymers in compressed carbon dioxide, *J. Supercrit. Fluids*, 62, 211, 2012.

2012YOO Yoon, X.-D., Jang, Y.-S., Choi, T.-H., and Byun, H.-S., High pressure phase behavior for the binary mixture of pentafluoropropyl methacrylate and poly(pentafluoropropyl methacrylate) in supercritical carbon dioxide and dimethyl ether, *Korean J. Chem. Eng.*, 29, 413, 2012.

2012YAN Yang, D.-S., Jeong, H.-H., and Byun, H.-S., Cloud-point behavior of binary and ternary mixtures of PHPMA and PHPA in supercritical fluid solvents, *Fluid Phase Equil.*, 332, 77, 2012.

2012YOS Yoshida, E. and Mineyama, A., Morphology control of poly[2-(perfluorooctyl)ethyl acrylate-*co-tert*-butyl acrylate] by pressure in supercritical carbon dioxide, *Colloid Polym. Sci.*, 290, 183, 2012.

2013BYU Byun, H.-S., Phase behavior for the poly(dimethylsiloxane) in supercritical fluid solvents, *J. Ind. Eng. Chem.*, 19, 665, 2013.

2013COS Costa, G.M.N., Guerrieri, Y., Kislansky, S., and Embirucu, M., Phase-dependent binary interaction parameters in industrial low-density polyethylene separators, *J. Appl. Polym. Sci.*, 130, 2106, 2013.

2013GRA Grandelli, H.E. and Kiran, E., High pressure density, miscibility and compressibility of poly(lactide-*co*-glycolide) solutions in acetone and acetone + CO_2 binary fluid mixtures, *J. Supercrit. Fluids*, 75, 159, 2013.

2013GRE Gregorowicz, J., Wawrzynska, E.P., Parzuchowski, P.G., Fras,.Z., and Rokicki, G., Synthesis, characterization, and solubility in supercritical carbon dioxide of hyperbranched copolyesters, *Macromolecules*, 46, 7180, 2013.

2013KIR Kiran, E., Hassler, J.C., and Srivastava, R., Miscibility, phase separation, and phase settlement dynamics in solutions of ethylene-propylene-diene monomer elastomer in propane + n-octane binary fluid mixtures at high pressures, *Ind. Eng. Chem. Res.*, 52, 1806, 2013.

2013MAR Markocic, E., Skerget, M., and Knez, Z., Effect of temperature and pressure on the behavior of poly(ε-caprolactone) in the presence of supercritical carbon dioxide, *Ind. Eng. Chem. Res.*, 52, 15594, 2013.

2013TA1 Takahashi, S., Hassler, J.C., and Kiran, E., Light scattering behavior and the kinetics of pressure-induced phase separation in solutions of poly(ε-caprolactone) in acetone + CO_2 binary fluid mixtures, *Polymer*, 54, 5719, 2013.

2013TA2 Takahashi, S., Hassler, J.C., and Kiran, E., Miscibility, phase separation and volumetric properties in solutions of poly(ε-caprolactone) in acetone + CO_2 binary fluid mixtures at high pressures, *J. Supercrit. Fluids*, 84, 43, 2013.

2013YA1 Yang, D.-S., Cho, S.-H., Yoon, S.-D., Jeong, H.-H., and Byun, H.-S., Phase behavior measurement for poly(isobornyl acrylate) + cosolvent systems in supercritical solvents at high pressure, *J. Supercrit. Fluids*, 79, 11, 2013.

2013YA2 Yang, D.-S., Yoon, S.-D., and Byun, H.-S., Effect of cosolvent on cloud-point of binary and ternary systems for the poly(4-chlorostyrene) + cosolvent mixtures in supercritical fluid solvents, *Fluid Phase Equil.*, 351, 7, 2013.

2014JAN Jang, Y.-S. and Byun, H.-S., Cloud-point and bubble-point measurement for the poly(2-butoxyethyl acrylate) + cosolvent mixture and 2-butoxyethyl acrylate in supercritical fluid solvents, *J. Chem. Eng. Data*, 59, 1391, 2014.

2014JEO Jeong, H.-H. and Byun, H.-S., Experimental measurement of cloud-point and bubble-point for the {poly(isobornyl methacrylate) + supercritical solvents + co-solvent} system at high pressure, *J. Chem. Thermodyn.*, 75, 25, 2014.

2014YOO Yoon, S.-D., Kim, A.-J., Byun, H.-S., Phase behavior for the poly[2-(2-ethoxyethoxy)ethyl acrylate] and 2-(2-ethoxyethoxy)ethyl acrylate in supercritical solvents, *J. Supercrit. Fluids*, 86, 41, 2014.

5. PVT DATA OF POLYMERS AND SOLUTIONS

5.1. PVT data of polymers

Polymer (B): **polyamide-6** **2009UTR**

Characterization: M_w/g.mol^{-1} = 22000, T_g/K = 315, T_m/K = 494, PA-6 1022B, Toyota R&D Labs.

P/bar = 1.0132

T/K	494.14	503.27	512.16	520.94	529.73	538.48	547.37	555.92
V_{spec}/cm^{3}g^{-1}	1.0070	1.0126	1.0176	1.0226	1.0278	1.0325	1.0382	1.0444
T/K	564.98	573.78						
V_{spec}/cm^{3}g^{-1}	1.0506	1.0571						

P/bar = 100.0

T/K	494.31	503.36	512.18	520.99	529.87	538.69	547.46	556.15
V_{spec}/cm^{3}g^{-1}	1.0005	1.0057	1.0103	1.0150	1.0199	1.0244	1.0297	1.0354
T/K	564.93	574.18						
V_{spec}/cm^{3}g^{-1}	1.0412	1.0474						

P/bar = 400.0

T/K	496.01	504.77	513.65	522.52	531.45	540.67	549.37	557.56
V_{spec}/cm^{3}g^{-1}	0.98406	0.98860	0.99260	0.99672	1.0009	1.0048	1.0092	1.0145
T/K	567.14	575.89						
V_{spec}/cm^{3}g^{-1}	1.0193	1.0247						

P/bar = 700.0

T/K	496.92	505.95	514.67	523.47	532.31	541.55	550.27	558.16
V_{spec}/cm^{3}g^{-1}	0.97077	0.97518	0.97890	0.98264	0.98649	0.98987	0.99405	0.99862
T/K	568.22	576.60						
V_{spec}/cm^{3}g^{-1}	1.0031	1.0078						

P/bar = 1000.0

T/K	497.56	506.96	515.70	524.54	533.40	542.39	550.95	559.18
V_{spec}/cm^{3}g^{-1}	0.95850	0.96373	0.96719	0.97046	0.97404	0.97734	0.98137	0.98541
T/K	568.98	576.91						
V_{spec}/cm^{3}g^{-1}	0.98970	0.99412						

P/bar = 1300.0

T/K	507.60	516.54	525.18	534.33	543.17	551.73	560.23	569.75
V_{spec}/cm^{3}g^{-1}	0.95371	0.95686	0.95994	0.96317	0.96641	0.97022	0.97388	0.97792
T/K	578.34							
V_{spec}/cm^{3}g^{-1}	0.98146							

continued

continued

P/bar = 1600.0

T/K	508.27	517.23	525.89	535.14	543.88	552.47	561.40	570.53
V_{spec}/cm^{3}g^{-1}	0.94439	0.94731	0.95038	0.95335	0.95645	0.95993	0.96325	0.96708
T/K	578.62							
V_{spec}/cm^{3}g^{-1}	0.97055							

P/bar = 1900.0

T/K	508.42	517.61	526.64	535.53	544.40	553.11	561.99	571.34
V_{spec}/cm^{3}g^{-1}	0.93605	0.93887	0.94159	0.94436	0.94741	0.95055	0.95379	0.95711
T/K	579.39							
V_{spec}/cm^{3}g^{-1}	0.96046							

Polymer (B): **poly(butyl methacrylate)** **2006KI2**
Characterization: M_w/g.mol^{-1} = 60000, T_g/K = 298

P/MPa	*T*/K 306.21	315.50	325.05	334.73	344.40	354.59	364.18	374.04
	V_{spec}/cm^{3}g^{-1}							
0.1	0.94562	0.95169	0.95777	0.96385	0.96998	0.97577	0.98156	0.98781
10	0.94099	0.94675	0.95256	0.95835	0.96412	0.96964	0.97504	0.98089
20	0.93661	0.94204	0.94765	0.95312	0.95868	0.96393	0.96891	0.97445
30	0.93240	0.93768	0.94304	0.94827	0.95347	0.95849	0.96333	0.96850
40	0.92834	0.93345	0.93860	0.94369	0.94866	0.95345	0.95800	0.96311
50	0.92459	0.92945	0.93448	0.93929	0.94405	0.94873	0.95310	0.95796
60	0.92092	0.92573	0.93059	0.93524	0.93985	0.94435	0.94854	0.95322
70	0.91741	0.92218	0.92672	0.93135	0.93589	0.94009	0.94420	0.94868
80	0.91421	0.91876	0.92328	0.92768	0.93202	0.93613	0.94013	0.94444
90	0.91110	0.91542	0.91991	0.92415	0.92844	0.93238	0.93616	0.94046
100	0.90815	0.91230	0.91669	0.92083	0.92495	0.92893	0.93262	0.93669
110	0.90522	0.90932	0.91359	0.91760	0.92163	0.92541	0.92911	0.93302
120	0.90244	0.90640	0.91062	0.91459	0.91850	0.92211	0.92571	0.92954
130	0.89978	0.90373	0.90767	0.91164	0.91546	0.91911	0.92245	0.92630
140	0.89727	0.90102	0.90489	0.90877	0.91254	0.91598	0.91931	0.92311
150	0.89474	0.89846	0.90230	0.90608	0.90972	0.91309	0.91644	0.91994
160	0.89244	0.89587	0.89965	0.90342	0.90690	0.91023	0.91349	0.91696
170	0.89011	0.89345	0.89708	0.90077	0.90421	0.90751	0.91083	0.91413
180	0.88787	0.89110	0.89471	0.89824	0.90168	0.90486	0.90802	0.91125
190	0.88571	0.88879	0.89224	0.89589	0.89914	0.90225	0.90536	0.90862
200	0.88355	0.88653	0.88991	0.89334	0.89658	0.89965	0.90270	0.90588

continued

continued

P/MPa	384.25	394.24	404.38	414.34	424.67	434.61
	T/K; V_{spec}/cm^{3}g^{-1}					
0.1	0.99409	1.00107	1.00834	1.01573	1.02336	1.03111
10	0.98691	0.99343	1.00023	1.00706	1.01414	1.02139
20	0.98025	0.98634	0.99263	0.99911	1.00570	1.01258
30	0.97407	0.97990	0.98614	0.99192	0.99820	1.00463
40	0.96837	0.97387	0.97992	0.98536	0.99134	0.99748
50	0.96298	0.96835	0.97409	0.97931	0.98503	0.99093
60	0.95800	0.96318	0.96860	0.97374	0.97919	0.98490
70	0.95330	0.95833	0.96353	0.96849	0.97375	0.97925
80	0.94887	0.95375	0.95877	0.96358	0.96870	0.97397
90	0.94474	0.94941	0.95436	0.95901	0.96387	0.96899
100	0.94086	0.94543	0.95015	0.95458	0.95939	0.96435
110	0.93716	0.94158	0.94615	0.95044	0.95509	0.95991
120	0.93350	0.93785	0.94232	0.94651	0.95096	0.95577
130	0.93018	0.93433	0.93869	0.94279	0.94714	0.95170
140	0.92682	0.93092	0.93516	0.93913	0.94345	0.94795
150	0.92368	0.92771	0.93179	0.93571	0.93984	0.94430
160	0.92065	0.92454	0.92853	0.93235	0.93648	0.94076
170	0.91767	0.92152	0.92543	0.92915	0.93315	0.93732
180	0.91481	0.91853	0.92235	0.92612	0.92992	0.93403
190	0.91196	0.91572	0.91947	0.92307	0.92686	0.93086
200	0.90930	0.91284	0.91656	0.92009	0.92384	0.92775

Polymer (B): **poly(ε-caprolactone)** **2011SCH**

Characterization: M_w/g.mol^{-1} = 80000, CAPA® FB100, Solvay Warrington, Cheshire, United Kingdom

P/MPa	400.93	421.04	431.10	441.31	451.42	471.87
	T/K; ρ/(g/cm^3)					
110	1.0691	1.0591	1.0538	1.0486	1.0437	1.0336
120	1.0732	1.0631	1.0580	1.0529	1.0480	1.0382
130	1.0769	1.0671	1.0622	1.0571	1.0522	1.0425
140	1.0807	1.0710	1.0660	1.0610	1.0563	1.0469
150	1.0843	1.0747	1.0700	1.0649	1.0601	1.0510
160	1.0878	1.0783	1.0736	1.0687	1.0641	1.0549
170	1.0910	1.0820	1.0774	1.0725	1.0679	1.0590
180	1.0944	1.0854	1.0809	1.0761	1.0715	1.0628
190	1.0977	1.0889	1.0844	1.0797	1.0751	1.0666
200	1.1009	1.0921	1.0878	1.0832	1.0786	1.0702

Polymer (B): **poly(diglycidyl ether of bisphenol A)** **2007DL2**
Characterization: M_n/g.mol^{-1} = 1750, T_g/K = 332, polymerization degree about 5

P/MPa	*T*/K							
	369.00	388.89	410.90	429.27	439.52	449.44	459.55	469.79
	V_{spec}/cm^{3}g^{-1}							
0.1	0.88230	0.89287	0.90392	0.91552	0.92063	0.92595	0.93113	0.93674
10	0.87867	0.88882	0.89934	0.91041	0.91532	0.92032	0.92525	0.93046
20	0.87512	0.88492	0.89493	0.90556	0.91029	0.91493	0.91969	0.92450
30	0.87190	0.88134	0.89093	0.90116	0.90568	0.91022	0.91470	0.91926
40	0.86868	0.87786	0.88725	0.89693	0.90143	0.90581	0.91006	0.91450
50	0.86569	0.87463	0.88368	0.89310	0.89728	0.90155	0.90570	0.90993
60	0.86281	0.87151	0.88031	0.88943	0.89353	0.89765	0.90164	0.90577
70	0.86003	0.86850	0.87702	0.88599	0.88990	0.89394	0.89784	0.90175
80	0.85746	0.86574	0.87416	0.88262	0.88657	0.89037	0.89417	0.89791
90	0.85492	0.86299	0.87109	0.87946	0.88325	0.88704	0.89067	0.89436
100	0.85249	0.86042	0.86833	0.87651	0.88007	0.88375	0.88732	0.89089
110	0.85023	0.85783	0.86563	0.87345	0.87703	0.88066	0.88410	0.88761
120	0.84793	0.85547	0.86300	0.87074	0.87416	0.87767	0.88107	0.88446
130	0.84579	0.85308	0.86047	0.86800	0.87138	0.87482	0.87815	0.88143
140	0.84361	0.85077	0.85809	0.86535	0.86871	0.87202	0.87523	0.87852
150	0.84154	0.84856	0.85573	0.86283	0.86612	0.86940	0.87247	0.87561
160	0.83950	0.84649	0.85341	0.86046	0.86363	0.86676	0.86982	0.87292
170	0.83755	0.84431	0.85116	0.85802	0.86111	0.86418	0.86720	0.87019
180	0.83563	0.84222	0.84898	0.85567	0.85874	0.86176	0.86470	0.86764
190	0.83381	0.84023	0.84684	0.85340	0.85631	0.85938	0.86229	0.86520
200	0.83214	0.83841	0.84494	0.85130	0.85419	0.85714	0.85993	0.86277

Polymer (B): **poly(1H,1H-dihydroperfluorooctyl methacrylate)** **2009SAN**
Characterization: M_w/g.mol^{-1} = 110000, synthesized in the laboratory by ATRP

P/MPa	*T*/K							
	303.15	313.15	323.15	333.15	343.15	353.15	373.15	393.15
	V_{spec}/cm^{3}g^{-1}							
100	0.6591	0.6637	0.6673	0.6727	0.6770	0.6812	0.6910	0.7011
200	0.6542	0.6584	0.6622	0.6664	0.6709	0.6753	0.6836	0.6931
300	0.6504	0.6543	0.6579	0.6618	0.6660	0.6700	0.6780	0.6868
400	0.6472	0.6506	0.6540	0.6575	0.6616	0.6655	0.6728	0.6812
500	0.6443	0.6474	0.6505	0.6539	0.6577	0.6615	0.6684	0.6765

continued

continued

600	0.6414	0.6447	0.6473	0.6506	0.6542	0.6578	0.6646	0.6722
700	0.6390	0.6420	0.6444	0.6475	0.6509	0.6543	0.6610	0.6682
800	0.6367	0.6395	0.6418	0.6446	0.6479	0.6511	0.6575	0.6647
900	0.6343	0.6370	0.6393	0.6418	0.6450	0.6481	0.6543	0.6609
1000	0.6322	0.6349	0.6367	0.6392	0.6423	0.6452	0.6513	0.6579
1100	0.6303	0.6327	0.6345	0.6369	0.6398	0.6424	0.6482	0.6548
1200	0.6284	0.6307	0.6324	0.6347	0.6375	0.6400	0.6457	0.6520
1300	0.6266	0.6289	0.6303	0.6327	0.6353	0.6377	0.6432	0.6493
1400	0.6249	0.6269	0.6286	0.6306	0.6331	0.6354	0.6408	0.6467
1500	0.6233	0.6253	0.6268	0.6286	0.6310	0.6332	0.6385	0.6444
1600	0.6217	0.6238	0.6252	0.6269	0.6290	0.6310	0.6363	0.6419
1700	0.6202	0.6223	0.6235	0.6252	0.6273	0.6289	0.6342	0.6398
1800	0.6198	0.6208	0.6221	0.6236	0.6255	0.6270	0.6321	0.6376
1900	0.6173	0.6193	0.6204	0.6221	0.6238	0.6252	0.6300	0.6354
2000	0.6160	0.6178	0.6190	0.6203	0.6221	0.6234	0.6280	0.6334

continued

P/MPa				T/K
	413.15	433.15	453.15	473.15
				V_{spec}/cm^{3}g^{-1}
100	0.7184	0.7394	0.7515	0.7724
200	0.7086	0.7279	0.7388	0.7564
300	0.7011	0.7191	0.7294	0.7450
400	0.6943	0.7118	0.7218	0.7360
500	0.6884	0.7054	0.7151	0.7281
600	0.6836	0.6997	0.7094	0.7214
700	0.6790	0.6946	0.7040	0.7154
800	0.6747	0.6894	0.6988	0.7098
900	0.6706	0.6850	0.6934	0.7049
1000	0.6669	0.6805	0.6891	0.7005
1100	0.6632	0.6760	0.6847	0.6959
1200	0.6598	0.6722	0.6808	0.6923
1300	0.6570	0.6688	0.6772	0.6884
1400	0.6540	0.6653	0.6737	0.6849
1500	0.6513	0.6619	0.6704	0.6814
1600	0.6486	0.6587	0.6672	0.6778
1700	0.6461	0.6555	0.6641	0.6747
1800	0.6437	0.6525	0.6606	0.6715
1900	0.6413	0.6497	0.6572	0.6683
2000	0.6392	0.6469	0.6537	0.6651

Polymer (B): **polyester (hyperbranched, aliphatic)** **2006SEI**
Characterization: M_n/g.mol^{-1} = 1620, M_w/g.mol^{-1} = 2100, 16 OH groups per macromolecule, hydroxyl no. = 490-520 mg KOH/g, acid no. = 5-9 mg KOH/g, Boltorn H20, Perstorp Speciality Chemicals AB, Sweden

P/MPa	T/K 343.15	383.15	423.15
	ρ/kg m^{-3}		
5	1241.6	1214.7	1188.3
10	1245.1	1217.6	1191.5
20	1249.2	1222.3	1197.0
35	1255.6	1229.4	1205.4
60	1265.2	1240.2	1216.2

Polymer (B): **polyethylene** **2007FU2**
Characterization: M_w/g.mol^{-1} = 127000, linear, ρ(293.15 K) = 0.9254 g/cm^3, T_g/K = 195, T_m/K = 390, PE A27MA, Leuna AG, Germany

P/MPa	T/K 427.18	437.55	448.00	457.84	468.60	478.99	488.90	499.83
	V_{spec}/cm^{3}g^{-1}							
0.1	1.2883	1.2981	1.3070	1.3163	1.3265	1.3368	1.3462	1.3573
10	1.2759	1.2849	1.2932	1.3018	1.3110	1.3204	1.3291	1.3391
20	1.2647	1.2730	1.2808	1.2888	1.2973	1.3059	1.3140	1.3233
30	1.2545	1.2624	1.2698	1.2772	1.2852	1.2932	1.3009	1.3093
40	1.2452	1.2526	1.2597	1.2667	1.2742	1.2817	1.2890	1.2969
50	1.2367	1.2436	1.2503	1.2571	1.2642	1.2713	1.2782	1.2857
60	1.2286	1.2353	1.2417	1.2482	1.2550	1.2618	1.2683	1.2755
70	1.2211	1.2277	1.2337	1.2399	1.2464	1.2529	1.2592	1.2660
80	1.2141	1.2205	1.2262	1.2321	1.2384	1.2446	1.2508	1.2572
90	1.2076	1.2136	1.2191	1.2250	1.2310	1.2369	1.2429	1.2491
100	1.2013	1.2072	1.2125	1.2181	1.2239	1.2298	1.2355	1.2414
110	1.1954	1.2011	1.2063	1.2117	1.2173	1.2230	1.2285	1.2344
120	1.1898	1.1953	1.2003	1.2056	1.2110	1.2165	1.2219	1.2276
130	1.1845	1.1897	1.1947	1.1998	1.2051	1.2104	1.2155	1.2210
140	1.1793	1.1845	1.1892	1.1942	1.1994	1.2045	1.2097	1.2148
150	1.1744	1.1795	1.1841	1.1890	1.1939	1.1990	1.2038	1.2090
160	1.1697	1.1746	1.1792	1.1839	1.1887	1.1937	1.1985	1.2035
170	1.1652	1.1700	1.1744	1.1791	1.1838	1.1887	1.1933	1.1981
180	1.1609	1.1655	1.1698	1.1743	1.1791	1.1838	1.1883	1.1929
190	1.1566	1.1613	1.1655	1.1698	1.1745	1.1791	1.1834	1.1881
200	1.1526	1.1571	1.1613	1.1655	1.1700	1.1746	1.1789	1.1834

continued

P/MPa	T/K			
	509.94	519.47	530.59	540.05
	V_{spec}/cm^{3}g^{-1}			
0.1	1.3681	1.3786	1.3908	1.4013
10	1.3488	1.3583	1.3690	1.3785
20	1.3321	1.3408	1.3504	1.3592
30	1.3175	1.3258	1.3345	1.3428
40	1.3046	1.3124	1.3204	1.3281
50	1.2930	1.3003	1.3080	1.3153
60	1.2823	1.2895	1.2965	1.3036
70	1.2725	1.2794	1.2862	1.2931
80	1.2636	1.2701	1.2765	1.2829
90	1.2552	1.2615	1.2676	1.2738
100	1.2473	1.2533	1.2593	1.2651
110	1.2399	1.2457	1.2516	1.2572
120	1.2329	1.2386	1.2443	1.2496
130	1.2263	1.2319	1.2374	1.2424
140	1.2201	1.2254	1.2307	1.2359
150	1.2140	1.2193	1.2245	1.2293
160	1.2083	1.2135	1.2185	1.2232
170	1.2028	1.2078	1.2129	1.2174
180	1.1977	1.2026	1.2073	1.2118
190	1.1926	1.1974	1.2021	1.2065
200	1.1878	1.1926	1.1971	1.2014

Polymer (B): **polyethylene** **2007SAT**

Characterization: M_n/g.mol^{-1} = 34000, M_w/g.mol^{-1} = 215000, linear, HDPE, 74.7 wt% crystallinity, unspecified industrial source

T/K	P/MPa						
	0.1	10	20	50	100	150	200
	V_{spec}/cm^{3}g^{-1}						
414.1	1.263	1.250	1.239				
433.9	1.280	1.267	1.255	1.226	1.190		
453.8	1.298	1.284	1.271	1.240	1.201	1.172	1.149
473.7	1.318	1.302	1.288	1.254	1.213	1.182	1.157
493.6	1.339	1.321	1.304	1.267	1.224	1.192	1.166

Polymer (B): **polyethylene** **2006PAR**
Characterization: M_n/g.mol^{-1} = 8200, M_w/g.mol^{-1} = 111000, T_m/K = 407

Tait equation parameter functions:
Range of data: *T*/K = 423-483, *P*/MPa = 0.1-80

$$V(P/\text{MPa}, T/\text{K}) = V(0, T/\text{K})\{1 - C^*\ln[1 + (P/\text{MPa})/B(T/\text{K})]\}$$
with $C = 0.0894$

$V(0,T/\text{K})$/cm^{3}g^{-1}	$B(T/\text{K})$/MPa
$0.876 + 9.84\ 10^{-4}T$	$677.0\ \exp(-4.83\ 10^{-3}T)$

Polymer (B): **poly(ethylene-*co*-1-butene)** **2007SAT**
Characterization: M_n/g.mol^{-1} = 53000, M_w/g.mol^{-1} = 110000, 10 mol% 1-butene, 12 wt% crystallinity, unspecified industrial source

T/K	*P*/MPa 0.1	10	20	50	100	150	200
	V_{spec}/cm^{3}g^{-1}						
373.8	1.236	1.226	1.217	1.194	1.165		
393.9	1.254	1.243	1.233	1.208	1.176	1.152	1.131
414.0	1.271	1.260	1.248	1.222	1.188	1.162	1.140
433.9	1.289	1.276	1.264	1.235	1.199	1.172	1.149
453.7	1.307	1.293	1.280	1.249	1.210	1.181	1.158
473.6	1.326	1.310	1.296	1.262	1.221	1.191	1.166
493.5	1.345	1.328	1.312	1.275	1.232	1.200	1.175

Polymer (B): **poly(ethylene-*co*-1-hexene)** **2007SAT**
Characterization: M_n/g.mol^{-1} = 28000, M_w/g.mol^{-1} = 63000, 3.5 mol% 1-hexene, 51 wt% crystallinity, unspecified industrial source

T/K	*P*/MPa 0.1	10	20	50	100	150	200
	V_{spec}/cm^{3}g^{-1}						
414.0	1.274	1.263	1.252	1.226			
433.9	1.293	1.280	1.270	1.240	1.204	1.177	
453.8	1.311	1.297	1.284	1.253	1.215	1.186	1.163
473.7	1.329	1.314	1.300	1.266	1.226	1.196	1.171
493.6	1.349	1.331	1.315	1.279	1.236	1.205	1.179

Polymer (B): **poly(ethylene-*co*-norbornene)** **2013SAT**

Characterization: M_n/g.mol^{-1} = 38800, M_w/g.mol^{-1} = 74600, 52 mol% ethylene, T_g/K = 411.15, Topas Advanced Polymers GmbH

P/MPa	*T*/K					
	414.3	444.1	473.9	503.9	534.1	564.3
	V_{spec}/cm^{3}g^{-1}					
0.1	1.025	1.043	1.061	1.077	1.094	1.112
20	1.018	1.033	1.048	1.064	1.080	1.097
40	1.011	1.022	1.036	1.051	1.065	1.080
60	1.005	1.012	1.025	1.038	1.052	1.065
80	0.999	1.004	1.015	1.028	1.040	1.053
100	0.994	0.999	1.007	1.019	1.030	1.042
120	0.989	0.994	0.999	1.010	1.021	1.032
140	0.985	0.990	0.992	1.002	1.013	1.023
160	0.980	0.987	0.986	0.995	1.004	1.015
180	0.975	0.983	0.981	0.988	0.997	1.007
200	0.971	0.979	0.977	0.981	0.990	0.999

Polymer (B): **poly(ethylene-*co*-norbornene)** **2006BLO**

Characterization: M_n/g.mol^{-1} = 56000, M_w/g.mol^{-1} = 101000, T_g/K = 415, 51.8 mol% norbornene, commercial sample

Tait equation parameter functions:
Range of data: *T*/K = 415-550, *P*/MPa = 0.1-200

$$V(P/\text{MPa}, T/\text{K}) = V(0, T/\text{K})\{1 - C^*\ln[1 + (P/\text{MPa})/B(T/\text{K})]\}$$
$$\text{with } C = 0.0894 \text{ and } \theta = T/\text{K} - 273.15$$

$V(0, \theta/°\text{C})/\text{cm}^3\text{g}^{-1}$	$B(\theta/°\text{C})/\text{MPa}$
$0.927 + 5.5\ 10^{-4}\theta - 3.0\ 10^{-8}\theta^2$	$238.0\ \exp(-0.3\ 10^{-2}\theta)$

Polymer (B): **poly(ethylene-*co*-1-octene)** **2007SAT**
Characterization: M_n/g.mol^{-1} = 106000, M_w/g.mol^{-1} = 196000, 8.9 mol% 1-octene, 17.6 wt% crystallinity, unspecified industrial source

T/K	P/MPa 0.1	10	20	50	100	150	200
	V_{spec}/cm^{3}g^{-1}						
373.9	1.238	1.228	1.219				
394.0	1.258	1.247	1.237	1.212	1.180		
414.1	1.275	1.265	1.253	1.226	1.191	1.165	1.143
434.0	1.293	1.281	1.269	1.240	1.203	1.175	1.152
453.8	1.311	1.298	1.285	1.253	1.214	1.185	1.161
473.6	1.330	1.316	1.302	1.267	1.226	1.195	1.170
493.4	1.349	1.333	1.318	1.281	1.237	1.204	1.178

Polymer (B): **poly(ethylene-*co*-propylene)** **2007SAT**
Characterization: M_n/g.mol^{-1} = 87000, M_w/g.mol^{-1} = 190000, 19 mol% propylene, 10 wt% crystallinity, unspecified industrial source

T/K	P/MPa 0.1	10	20	50	100	150	200
	V_{spec}/cm^{3}g^{-1}						
333.7	1.200	1.192	1.184	1.165			
353.8	1.217	1.209	1.200	1.179	1.151	1.130	1.111
374.0	1.234	1.225	1.215	1.192	1.163	1.140	1.120
394.0	1.252	1.241	1.230	1.206	1.174	1.150	1.129
414.2	1.269	1.257	1.246	1.219	1.185	1.159	1.138
434.1	1.287	1.274	1.262	1.232	1.196	1.169	1.147
454.0	1.306	1.291	1.278	1.246	1.207	1.179	1.155
473.9	1.324	1.308	1.294	1.259	1.218	1.188	1.164
493.8	1.343	1.326	1.309	1.273	1.229	1.197	1.172

Polymer (B): **poly(ethylene-*co*-vinyl alcohol)** **2007FU2**
Characterization: 15 mol% ethylene, ρ(293.15 K) = 1.2522 g/cm^3, T_g/K = 341, T_m/K = 484, Kuraray Specialities, Japan

P/MPa	T/K					
	483.55	493.80	503.38	514.05	523.80	534.11
	V_{spec}/cm^{3}g^{-1}					
0.1	0.8783	0.8852	0.8908	0.8964	0.9034	0.9133
10	0.8744	0.8810	0.8863	0.8918	0.8987	0.9083
20	0.8706	0.8769	0.8818	0.8874	0.8940	0.9036
30	0.8671	0.8734	0.8783	0.8837	0.8901	0.8994
40	0.8640	0.8701	0.8749	0.8801	0.8865	0.8958
50	0.8609	0.8670	0.8718	0.8768	0.8831	0.8921
60	0.8581	0.8642	0.8686	0.8736	0.8799	0.8887
70	0.8551	0.8614	0.8659	0.8707	0.8768	0.8854
80	0.8523	0.8587	0.8629	0.8679	0.8738	0.8823
90	0.8492	0.8562	0.8604	0.8651	0.8709	0.8795
100	0.8463	0.8536	0.8577	0.8624	0.8681	0.8765
110	0.8433	0.8512	0.8552	0.8600	0.8655	0.8740
120	0.8404	0.8488	0.8528	0.8574	0.8628	0.8712
130	0.8374	0.8464	0.8505	0.8550	0.8604	0.8687
140	0.8341	0.8443	0.8482	0.8526	0.8581	0.8661
150	0.8313	0.8421	0.8461	0.8505	0.8556	0.8638
160	0.8284	0.8400	0.8439	0.8483	0.8534	0.8615
170	0.8258	0.8379	0.8418	0.8461	0.8513	0.8593
180	0.8234	0.8358	0.8398	0.8440	0.8491	0.8572
190	0.8208	0.8337	0.8379	0.8419	0.8470	0.8550
200	0.8185	0.8316	0.8360	0.8400	0.8451	0.8531

Polymer (B): **poly(ethylene-*co*-vinyl alcohol)** **2007FU2**
Characterization: 27 mol% ethylene, ρ(293.15 K) = 1.1959 g/cm^3, T_g/K = 339, T_m/K = 466, Kuraray Specialities, Japan

P/MPa	T/K							
	466.52	477.39	487.28	497.68	508.00	518.11	528.90	539.09
	V_{spec}/cm^{3}g^{-1}							
0.1	0.9234	0.9290	0.9340	0.9392	0.9453	0.9576	0.9702	0.9821
10	0.9189	0.9243	0.9292	0.9341	0.9402	0.9522	0.9641	0.9754
20	0.9146	0.9196	0.9244	0.9292	0.9352	0.9469	0.9584	0.9691

continued

continued

30	0.9109	0.9158	0.9204	0.9250	0.9309	0.9425	0.9534	0.9638
40	0.9075	0.9121	0.9167	0.9212	0.9271	0.9385	0.9488	0.9589
50	0.9042	0.9088	0.9132	0.9175	0.9235	0.9346	0.9445	0.9543
60	0.9010	0.9055	0.9098	0.9141	0.9200	0.9309	0.9405	0.9500
70	0.8981	0.9024	0.9066	0.9107	0.9167	0.9273	0.9366	0.9458
80	0.8951	0.8994	0.9035	0.9075	0.9136	0.9239	0.9330	0.9419
90	0.8924	0.8965	0.9006	0.9045	0.9105	0.9208	0.9295	0.9382
100	0.8897	0.8938	0.8977	0.9016	0.9076	0.9175	0.9260	0.9346
110	0.8872	0.8910	0.8949	0.8987	0.9048	0.9146	0.9228	0.9311
120	0.8846	0.8885	0.8923	0.8960	0.9020	0.9117	0.9197	0.9278
130	0.8821	0.8860	0.8897	0.8934	0.8994	0.9088	0.9166	0.9246
140	0.8798	0.8835	0.8871	0.8909	0.8968	0.9060	0.9137	0.9215
150	0.8772	0.8812	0.8847	0.8883	0.8944	0.9033	0.9109	0.9186
160	0.8746	0.8789	0.8824	0.8858	0.8919	0.9008	0.9081	0.9156
170	0.8717	0.8766	0.8801	0.8835	0.8895	0.8982	0.9055	0.9129
180	0.8685	0.8745	0.8779	0.8813	0.8873	0.8958	0.9028	0.9101
190	0.8650	0.8724	0.8757	0.8790	0.8850	0.8933	0.9003	0.9075
200	0.8616	0.8703	0.8736	0.8769	0.8828	0.8910	0.8978	0.9049

Polymer (B): **poly(ethylene-*co*-vinyl alcohol)** **2007FU2**
Characterization: 32 mol% ethylene, ρ(293.15 K) = 1.1810 g/cm^3, T_g/K = 335, T_m/K = 457, Kuraray Specialities, Japan

P/MPa	T/K							
	467.46	477.59	488.37	498.75	508.97	519.72	529.57	539.30
	V_{spec}/cm^{3}g^{-1}							
0.1	0.9363	0.9420	0.9479	0.9536	0.9599	0.9673	0.9753	0.9841
10	0.9318	0.9372	0.9428	0.9483	0.9543	0.9613	0.9691	0.9775
20	0.9273	0.9325	0.9378	0.9432	0.9488	0.9555	0.9632	0.9712
30	0.9235	0.9286	0.9335	0.9388	0.9444	0.9507	0.9582	0.9660
40	0.9200	0.9249	0.9297	0.9348	0.9401	0.9464	0.9536	0.9612
50	0.9164	0.9213	0.9260	0.9308	0.9361	0.9423	0.9493	0.9566
60	0.9132	0.9179	0.9225	0.9272	0.9323	0.9383	0.9451	0.9524
70	0.9100	0.9147	0.9190	0.9238	0.9287	0.9347	0.9413	0.9483
80	0.9070	0.9115	0.9159	0.9204	0.9253	0.9311	0.9375	0.9445
90	0.9042	0.9086	0.9128	0.9173	0.9220	0.9277	0.9340	0.9407
100	0.9014	0.9057	0.9098	0.9142	0.9190	0.9245	0.9306	0.9371
110	0.8988	0.9028	0.9069	0.9112	0.9158	0.9213	0.9274	0.9337
120	0.8962	0.9002	0.9042	0.9084	0.9129	0.9183	0.9242	0.9304
130	0.8936	0.8976	0.9015	0.9057	0.9100	0.9153	0.9212	0.9273
140	0.8911	0.8951	0.8989	0.9031	0.9073	0.9125	0.9182	0.9243

continued

continued

150	0.8888	0.8926	0.8964	0.9005	0.9047	0.9097	0.9154	0.9213
160	0.8866	0.8903	0.8940	0.8980	0.9021	0.9071	0.9126	0.9184
170	0.8842	0.8880	0.8916	0.8955	0.8996	0.9045	0.9099	0.9157
180	0.8820	0.8856	0.8893	0.8931	0.8972	0.9020	0.9073	0.9130
190	0.8799	0.8835	0.8871	0.8907	0.8947	0.8996	0.9047	0.9103
200	0.8778	0.8813	0.8849	0.8885	0.8925	0.8972	0.9023	0.9079

Polymer (B): **poly(ethylene-*co*-vinyl alcohol)** **2007FU2**
Characterization: 38 mol% ethylene, ρ(293.15 K) = 1.1690 g/cm^3, T_g/K = 332, T_m/K = 450, Kuraray Specialities, Japan

P/MPa	T/K							
	467.82	477.75	488.37	499.04	509.02	519.75	529.66	539.20
	V_{spec}/cm^{3}g^{-1}							
0.1	0.9530	0.9585	0.9643	0.9705	0.9769	0.9858	0.9963	1.0071
10	0.9482	0.9535	0.9590	0.9649	0.9711	0.9798	0.9899	1.0002
20	0.9434	0.9486	0.9539	0.9595	0.9656	0.9740	0.9837	0.9936
30	0.9394	0.9444	0.9495	0.9549	0.9608	0.9690	0.9784	0.9880
40	0.9355	0.9403	0.9452	0.9505	0.9563	0.9642	0.9733	0.9828
50	0.9318	0.9365	0.9413	0.9464	0.9521	0.9598	0.9686	0.9777
60	0.9283	0.9329	0.9376	0.9425	0.9481	0.9556	0.9642	0.9730
70	0.9250	0.9294	0.9339	0.9388	0.9443	0.9515	0.9600	0.9686
80	0.9218	0.9261	0.9305	0.9353	0.9406	0.9477	0.9560	0.9644
90	0.9187	0.9229	0.9272	0.9318	0.9372	0.9441	0.9521	0.9603
100	0.9156	0.9199	0.9241	0.9286	0.9337	0.9407	0.9485	0.9565
110	0.9128	0.9170	0.9211	0.9253	0.9306	0.9373	0.9449	0.9528
120	0.9100	0.9139	0.9181	0.9224	0.9275	0.9341	0.9416	0.9492
130	0.9073	0.9112	0.9152	0.9194	0.9244	0.9310	0.9383	0.9458
140	0.9047	0.9086	0.9125	0.9167	0.9216	0.9280	0.9353	0.9425
150	0.9021	0.9060	0.9099	0.9138	0.9188	0.9251	0.9322	0.9394
160	0.8998	0.9034	0.9073	0.9112	0.9160	0.9223	0.9292	0.9363
170	0.8974	0.9010	0.9047	0.9086	0.9133	0.9196	0.9263	0.9333
180	0.8951	0.8986	0.9023	0.9062	0.9108	0.9170	0.9236	0.9303
190	0.8927	0.8963	0.8999	0.9037	0.9083	0.9144	0.9209	0.9276
200	0.8906	0.8941	0.8977	0.9013	0.9058	0.9118	0.9183	0.9248

Polymer (B): **poly(ethylene-*co*-vinyl alcohol)** **2007FU2**
Characterization: 44 mol% ethylene, ρ(293.15 K) = 1.1359 g/cm^3, T_g/K = 327, T_m/K = 439, Kuraray Specialities, Japan

P/MPa	T/K 450.10	460.43	471.16	481.64	492.41	502.47	512.67	523.20
	V_{spec}/cm^{3}g^{-1}							
0.1	0.9751	0.9813	0.9879	0.9945	1.0012	1.0075	1.0141	1.0214
10	0.9702	0.9762	0.9825	0.9888	0.9952	1.0013	1.0077	1.0145
20	0.9655	0.9713	0.9774	0.9834	0.9895	0.9954	1.0015	1.0081
30	0.9613	0.9669	0.9727	0.9787	0.9845	0.9901	0.9961	1.0024
40	0.9574	0.9627	0.9684	0.9742	0.9798	0.9853	0.9910	0.9971
50	0.9535	0.9587	0.9643	0.9700	0.9752	0.9807	0.9863	0.9921
60	0.9499	0.9550	0.9603	0.9659	0.9711	0.9764	0.9817	0.9875
70	0.9465	0.9515	0.9566	0.9620	0.9671	0.9722	0.9774	0.9830
80	0.9431	0.9479	0.9530	0.9582	0.9632	0.9682	0.9733	0.9788
90	0.9400	0.9446	0.9497	0.9546	0.9596	0.9644	0.9694	0.9748
100	0.9369	0.9415	0.9464	0.9513	0.9560	0.9608	0.9656	0.9709
110	0.9339	0.9385	0.9432	0.9480	0.9526	0.9573	0.9621	0.9671
120	0.9310	0.9354	0.9401	0.9449	0.9493	0.9539	0.9585	0.9636
130	0.9282	0.9327	0.9373	0.9418	0.9462	0.9508	0.9553	0.9601
140	0.9256	0.9300	0.9344	0.9389	0.9433	0.9477	0.9521	0.9569
150	0.9231	0.9272	0.9316	0.9361	0.9403	0.9447	0.9489	0.9537
160	0.9204	0.9247	0.9290	0.9334	0.9375	0.9418	0.9459	0.9507
170	0.9180	0.9221	0.9264	0.9307	0.9347	0.9389	0.9432	0.9476
180	0.9155	0.9197	0.9238	0.9281	0.9321	0.9362	0.9403	0.9449
190	0.9128	0.9173	0.9215	0.9257	0.9295	0.9335	0.9376	0.9420
200	0.9096	0.9150	0.9191	0.9231	0.9269	0.9309	0.9349	0.9392

continued

P/MPa	T/K 533.89	543.95
	V_{spec}/cm^{3}g^{-1}	
0.1	1.0293	1.0378
10	1.0220	1.0301
20	1.0153	1.0230
30	1.0093	1.0166
40	1.0037	1.0108
50	0.9985	1.0054
60	0.9937	1.0003

continued

continued

70	0.9891	0.9954
80	0.9846	0.9909
90	0.9804	0.9866
100	0.9764	0.9825
110	0.9726	0.9785
120	0.9689	0.9747
130	0.9654	0.9710
140	0.9620	0.9675
150	0.9588	0.9641
160	0.9556	0.9608
170	0.9525	0.9578
180	0.9495	0.9547
190	0.9467	0.9517
200	0.9438	0.9488

Polymer (B): **poly(ethylene-*co*-vinyl alcohol)** **2007FU2**
Characterization: 48 mol% ethylene, ρ(293.15 K) = 1.1243 g/cm^3, T_g/K = 321, T_m/K = 436, Kuraray Specialities, Japan

P/MPa	T/K							
	447.09	456.99	467.74	478.12	488.02	493.16	503.16	519.30
	V_{spec}/cm^{3}g^{-1}							
0.1	0.9886	0.9947	1.0010	1.0072	1.0135	1.0199	1.0264	1.0336
10	0.9833	0.9891	0.9951	1.0011	1.0071	1.0132	1.0193	1.0260
20	0.9781	0.9838	0.9895	0.9954	1.0010	1.0067	1.0127	1.0188
30	0.9736	0.9791	0.9845	0.9902	0.9957	1.0013	1.0069	1.0128
40	0.9694	0.9748	0.9800	0.9854	0.9906	0.9961	1.0014	1.0070
50	0.9654	0.9704	0.9757	0.9809	0.9860	0.9912	0.9964	1.0018
60	0.9616	0.9665	0.9715	0.9767	0.9816	0.9866	0.9916	0.9969
70	0.9579	0.9628	0.9676	0.9726	0.9773	0.9824	0.9873	0.9922
80	0.9545	0.9592	0.9638	0.9687	0.9733	0.9781	0.9829	0.9877
90	0.9511	0.9556	0.9602	0.9649	0.9695	0.9742	0.9788	0.9836
100	0.9479	0.9522	0.9568	0.9615	0.9658	0.9705	0.9749	0.9796
110	0.9447	0.9491	0.9535	0.9580	0.9623	0.9667	0.9711	0.9757
120	0.9417	0.9460	0.9504	0.9547	0.9588	0.9632	0.9676	0.9721
130	0.9388	0.9429	0.9473	0.9516	0.9556	0.9599	0.9641	0.9685
140	0.9361	0.9402	0.9443	0.9485	0.9524	0.9567	0.9608	0.9651
150	0.9333	0.9373	0.9414	0.9456	0.9494	0.9535	0.9575	0.9618
160	0.9307	0.9346	0.9386	0.9427	0.9466	0.9505	0.9545	0.9586
170	0.9281	0.9321	0.9359	0.9399	0.9437	0.9476	0.9516	0.9556
180	0.9257	0.9295	0.9334	0.9373	0.9408	0.9448	0.9486	0.9525
200	0.9209	0.9246	0.9282	0.9322	0.9355	0.9394	0.9431	0.9468

continued

continued

P/MPa	T/K 528.86	539.38
	V_{spec}/cm^{3}g^{-1}	
0.1	1.0406	1.0491
10	1.0328	1.0408
20	1.0254	1.0330
30	1.0191	1.0264
40	1.0134	1.0202
50	1.0078	1.0146
60	1.0029	1.0093
70	0.9979	1.0042
80	0.9935	0.9996
90	0.9891	0.9951
100	0.9849	0.9909
110	0.9810	0.9867
120	0.9772	0.9827
130	0.9735	0.9790
140	0.9700	0.9754
150	0.9666	0.9719
160	0.9633	0.9685
170	0.9601	0.9653
180	0.9571	0.9621
200	0.9512	0.9562

Polymer (B): **poly(ethylene glycol)** **2006LEE**

Characterization: M_n/g.mol^{-1} = 260, M_w/g.mol^{-1} = 280,
Aldrich Chem. Co., Inc., St. Louis, MO

P/MPa	T/K 298.15	318.15	348.15
	V_{spec}/cm^{3}g^{-1}		
0.1	1.1200	1.1038	1.0802
10	1.1243	1.1087	1.0854
15	1.1264	1.1110	1.0880
20	1.1284	1.1134	1.0905
25	1.1303	1.1157	1.0929
30	1.1323	1.1179	1.0952
35	1.1342	1.1201	1.0975
40	1.1361	1.1222	1.0998
45	1.1379	1.1243	1.1019
50	1.1397	1.1263	1.1041

Polymer (B): **poly(ethylene glycol) monomethyl ether** **2010PFE**
Characterization: M_n/g.mol^{-1} = 5000, M_w/M_n = 1.03

P/MPa	*T*/K								
	346.45	351.45	356.65	361.35	366.35	371.35	376.35	381.15	386.35
	V_{spec}/cm^{3}g^{-1}								
0.1	0.9266	0.9304	0.9338	0.9372	0.9407	0.9442	0.9480	0.9513	0.9548
10	0.9221	0.9256	0.9289	0.9322	0.9355	0.9389	0.9423	0.9456	0.9489
20	0.9178	0.9212	0.9243	0.9273	0.9306	0.9339	0.9371	0.9402	0.9435
30	0.9136	0.9167	0.9199	0.9229	0.9260	0.9291	0.9322	0.9351	0.9383
40	0.9096	0.9126	0.9156	0.9185	0.9216	0.9246	0.9275	0.9305	0.9334
50	0.9057	0.9086	0.9115	0.9144	0.9174	0.9202	0.9231	0.9259	0.9288
60	0.9021	0.9049	0.9077	0.9105	0.9134	0.9162	0.9189	0.9217	0.9244
70	0.8985	0.9013	0.9041	0.9067	0.9095	0.9123	0.9150	0.9176	0.9203
80	0.8952	0.8980	0.9005	0.9033	0.9060	0.9085	0.9112	0.9138	0.9165
90	0.8920	0.8946	0.8972	0.8999	0.9024	0.9050	0.9076	0.9100	0.9127
100	0.8889	0.8914	0.8941	0.8966	0.8991	0.9016	0.9041	0.9065	0.9091
110	0.8859	0.8884	0.8909	0.8934	0.8959	0.8983	0.9008	0.9031	0.9057
120	0.8831	0.8856	0.8880	0.8905	0.8928	0.8952	0.8976	0.8998	0.9024
130	0.8802	0.8826	0.8850	0.8874	0.8897	0.8922	0.8944	0.8966	0.8990
140	0.8775	0.8799	0.8822	0.8845	0.8869	0.8893	0.8915	0.8936	0.8959
150	0.8749	0.8772	0.8794	0.8818	0.8842	0.8863	0.8886	0.8906	0.8930
160	0.8723	0.8746	0.8768	0.8791	0.8815	0.8836	0.8858	0.8878	0.8901
170	0.8699	0.8720	0.8742	0.8765	0.8787	0.8809	0.8831	0.8851	0.8873
180	0.8674	0.8696	0.8718	0.8740	0.8762	0.8782	0.8804	0.8824	0.8844
190	0.8651	0.8671	0.8694	0.8716	0.8737	0.8758	0.8778	0.8797	0.8818
200	0.8628	0.8648	0.8670	0.8691	0.8712	0.8733	0.8754	0.8771	0.8792

continued

P/MPa	*T*/K								
	391.65	396.75	401.65	406.95	412.05	417.05	421.95	427.05	432.45
	V_{spec}/cm^{3}g^{-1}								
0.1	0.9585	0.9622	0.9657	0.9693	0.9730	0.9768	0.9806	0.9843	0.9885
10	0.9524	0.9559	0.9593	0.9627	0.9662	0.9697	0.9733	0.9769	0.9807
20	0.9467	0.9500	0.9533	0.9565	0.9598	0.9631	0.9666	0.9700	0.9735
30	0.9414	0.9446	0.9478	0.9508	0.9539	0.9572	0.9603	0.9636	0.9670
40	0.9364	0.9395	0.9425	0.9455	0.9485	0.9515	0.9547	0.9579	0.9611
50	0.9316	0.9346	0.9376	0.9404	0.9433	0.9462	0.9493	0.9524	0.9554
60	0.9273	0.9300	0.9329	0.9357	0.9385	0.9413	0.9442	0.9472	0.9502

continued

continued

70	0.9230	0.9258	0.9285	0.9312	0.9339	0.9366	0.9395	0.9425	0.9453
80	0.9190	0.9217	0.9243	0.9271	0.9296	0.9322	0.9350	0.9378	0.9406
90	0.9152	0.9177	0.9204	0.9230	0.9254	0.9281	0.9308	0.9335	0.9362
100	0.9115	0.9140	0.9166	0.9191	0.9215	0.9241	0.9267	0.9294	0.9320
110	0.9080	0.9105	0.9131	0.9154	0.9177	0.9203	0.9228	0.9254	0.9280
120	0.9046	0.9070	0.9095	0.9118	0.9141	0.9166	0.9189	0.9217	0.9241
130	0.9014	0.9036	0.9061	0.9085	0.9107	0.9130	0.9153	0.9179	0.9205
140	0.8982	0.9005	0.9029	0.9050	0.9073	0.9096	0.9120	0.9145	0.9168
150	0.8951	0.8974	0.8997	0.9019	0.9040	0.9064	0.9086	0.9111	0.9135
160	0.8922	0.8943	0.8967	0.8988	0.9010	0.9032	0.9055	0.9077	0.9100
170	0.8893	0.8916	0.8937	0.8958	0.8980	0.9002	0.9023	0.9047	0.9068
180	0.8866	0.8887	0.8909	0.8930	0.8950	0.8971	0.8993	0.9016	0.9037
190	0.8838	0.8859	0.8881	0.8901	0.8922	0.8942	0.8963	0.8986	0.9007
200	0.8812	0.8833	0.8853	0.8873	0.8894	0.8913	0.8934	0.8957	0.8978

Polymer (B): **poly(ethylene glycol) mono-4-octylphenyl ether** **2009LEE**
Characterization: M_n/g.mol^{-1} = 576, M_w/g.mol^{-1} = 606,
Tokyo Kasei Organic Chemicals, Japan

P/MPa	*T*/K		
	298.15	318.15	348.15
	V_{spec}/cm^{3}g^{-1}		
0.1	1.0636	1.0475	1.0248
10	1.0684	1.0527	1.0307
15	1.0707	1.0554	1.0334
20	1.0731	1.0580	1.0362
25	1.0753	1.0604	1.0390
30	1.0775	1.0627	1.0417
35	1.0797	1.0651	1.0443
40	1.0818	1.0673	1.0468
45	1.0839	1.0695	1.0494
50	1.0860	1.0717	1.0519

Polymer (B): **poly(ethyl methacrylate)** **2006KI2**
Characterization: M_w/g.mol^{-1} = 154000, T_g/K = 334

P/MPa	T/K							
	364.05	374.25	384.25	394.15	404.05	414.25	424.65	434.65
	V_{spec}/cm^{3}g^{-1}							
0.1	0.91057	0.91640	0.92240	0.92828	0.93429	0.94066	0.94687	0.95341
10	0.90531	0.91086	0.91650	0.92206	0.92778	0.93367	0.93952	0.94556
20	0.90034	0.90568	0.91094	0.91628	0.92168	0.92714	0.93273	0.93830
30	0.89576	0.90088	0.90586	0.91094	0.91614	0.92125	0.92660	0.93178
40	0.89136	0.89629	0.90109	0.90596	0.91117	0.91586	0.92078	0.92599
50	0.88728	0.89194	0.89659	0.90133	0.90634	0.91084	0.91559	0.92046
60	0.88344	0.88802	0.89240	0.89691	0.90180	0.90613	0.91068	0.91528
70	0.87984	0.88417	0.88851	0.89288	0.89748	0.90163	0.90612	0.91060
80	0.87640	0.88062	0.88471	0.88900	0.89345	0.89751	0.90177	0.90610
90	0.87309	0.87723	0.88120	0.88536	0.88971	0.89361	0.89772	0.90191
100	0.87005	0.87399	0.87784	0.88196	0.88608	0.88989	0.89394	0.89795
110	0.86707	0.87082	0.87459	0.87862	0.88266	0.88641	0.89027	0.89425
120	0.86433	0.86791	0.87159	0.87546	0.87942	0.88299	0.88676	0.89067
130	0.86165	0.86505	0.86866	0.87242	0.87617	0.87978	0.88340	0.88717
140	0.85917	0.86238	0.86585	0.86951	0.87328	0.87665	0.88024	0.88390
150	0.85676	0.85975	0.86315	0.86666	0.87039	0.87365	0.87714	0.88072
160	0.85434	0.85717	0.86047	0.86399	0.86758	0.87079	0.87427	0.87775
170	0.85216	0.85473	0.85791	0.86129	0.86479	0.86799	0.87139	0.87481
180	0.84999	0.85236	0.85542	0.85874	0.86216	0.86533	0.86854	0.87190
190	0.84793	0.85011	0.85304	0.85624	0.85962	0.86265	0.86590	0.86916
200	0.84579	0.84785	0.85066	0.85379	0.85713	0.86008	0.86320	0.86640

Polymer (B): **poly(glycerol), hyperbranched** **2006SEI**
Characterization: M_n/g.mol^{-1} = 2000, M_w/g.mol^{-1} = 3000,

P/MPa	T/K		
	343.15	383.15	423.15
	ρ/kg m^{-3}		
5	1263.5	1234.9	1211.8
10	1266.7	1238.8	1216.8
20	1270.5	1244.7	1222.4
35	1274.7	1250.4	1226.2
60	1282.4	1258.8	1235.9

Polymer (B): **polyisobutylene** **2006KI1**
Characterization: M_w/g.mol^{-1} = 1300000, M_w/M_n = 4.7,
synthesized in the laboratory

T/K	P/MPa					
	0.1	40	80	120	160	200
	V_{spec}/cm^{3}g^{-1}					
293.50	1.09472	1.07707	1.06247	1.05022	1.03958	1.03003
293.83	1.09619	1.07834	1.06362	1.05129	1.04057	1.03087
300.76	1.09929	1.08102	1.06604	1.05351	1.04260	1.03280
305.63	1.10224	1.08357	1.06837	1.05554	1.04459	1.03462
310.60	1.10531	1.08622	1.07066	1.05767	1.04643	1.03628
315.15	1.10834	1.08865	1.07280	1.05966	1.04826	1.03808
319.97	1.11130	1.09133	1.07518	1.06179	1.05032	1.03989
324.70	1.11460	1.09379	1.07745	1.06400	1.05227	1.04178
329.50	1.11760	1.09654	1.07977	1.06607	1.05412	1.04361
339.60	1.12402	1.10179	1.08441	1.07016	1.05802	1.04713
354.46	1.13354	1.10965	1.09121	1.07639	1.06378	1.05249
369.40	1.14309	1.11771	1.09834	1.08272	1.06963	1.05793
384.29	1.15302	1.12571	1.10529	1.08889	1.07526	1.06319
399.30	1.16291	1.13364	1.11218	1.09519	1.08081	1.06841
414.30	1.17307	1.14188	1.11915	1.10137	1.08648	1.07347
429.70	1.18313	1.14961	1.12581	1.10728	1.09183	1.07835
444.80	1.19320	1.15747	1.13244	1.11308	1.09713	1.08317
460.40	1.20337	1.16523	1.13895	1.11883	1.10223	1.08784
475.60	1.21383	1.17315	1.14573	1.12470	1.10761	1.09276
491.00	1.22438	1.18106	1.15223	1.13041	1.11268	1.09740
506.30	1.23569	1.18926	1.15905	1.13633	1.11796	1.10212
521.50	1.24703	1.19745	1.16571	1.14214	1.12317	1.10701

Polymer (B): **poly(DL-lactic acid)** **2014MA1, 2014MA2**
Characterization: M_n/g.mol^{-1} = 72000, M_w/g.mol^{-1} = 136000,
1.4 % D-content, NatureWorks™ LLC

Tait equation parameter functions:
Range of data: T/K = 343-473, P/MPa = 0.1-60

$$V(P/\text{MPa}, T/\text{K}) = V(0, T/\text{K})\{1 - C{*}\ln[1 + (P/\text{MPa})/B(T/\text{K})]\}$$
with C = 0.0894

$V(0,T/\text{K})$/cm^{3}g^{-1}	$B(T/\text{K})$/MPa
0.7666 exp(9.6372 $10^{-4}T$)	17352.1 exp(−4.7825 $10^{-3}T$)

Polymer (B): **poly(DL-lactic acid)** **2014MA1, 2014MA2**
Characterization: M_n/g.mol^{-1} = 85000, M_w/g.mol^{-1} = 161000, 4.2 % D-content, NatureWorks™ LLC

Tait equation parameter functions:
Range of data: T/K = 343-473, P/MPa = 0.1-60

$$V(P/\text{MPa}, T/\text{K}) = V(0, T/\text{K})\{1 - C^*\ln[1 + (P/\text{MPa})/B(T/\text{K})]\}$$
with C = 0.0894

$V(0,T/\text{K})/\text{cm}^3\text{g}^{-1}$	$B(T/\text{K})$/MPa
0.7447 exp(10.3241 $10^{-4}T$)	17245.3 exp(−5.6021 $10^{-3}T$)

Polymer (B): **poly(DL-lactic acid)** **2014MA1, 2014MA2**
Characterization: M_n/g.mol^{-1} = 100000, M_w/g.mol^{-1} = 190000, 1.2 % D-content, NatureWorks™ LLC

Tait equation parameter functions:
Range of data: T/K = 343-473, P/MPa = 0.1-60

$$V(P/\text{MPa}, T/\text{K}) = V(0, T/\text{K})\{1 - C^*\ln[1 + (P/\text{MPa})/B(T/\text{K})]\}$$
with C = 0.0894

$V(0,T/\text{K})/\text{cm}^3\text{g}^{-1}$	$B(T/\text{K})$/MPa
0.7599 exp(9.4544 $10^{-4}T$)	15897.8 exp(−4.1801 $10^{-3}T$)

Polymer (B): **poly(methyl acrylate)** **2010PFE**
Characterization: M_n/g.mol^{-1} = 5400, M_w/M_n = 1.35, synthesized in the laboratory

P/MPa	T/K								
	345.25	350.25	355.45	369.15	365.15	370.45	375.35	379.95	385.25
	V_{spec}/cm^{3}g^{-1}								
0.1	0.8678	0.8702	0.8731	0.8760	0.8788	0.8818	0.8845	0.8873	0.8900
10	0.8638	0.8662	0.8690	0.8717	0.8744	0.8772	0.8798	0.8824	0.8851
20	0.8600	0.8625	0.8652	0.8677	0.8703	0.8728	0.8753	0.8777	0.8805
30	0.8564	0.8588	0.8614	0.8638	0.8664	0.8687	0.8712	0.8736	0.8761
40	0.8530	0.8553	0.8578	0.8601	0.8625	0.8649	0.8671	0.8695	0.8719
50	0.8497	0.8520	0.8543	0.8567	0.8590	0.8612	0.8635	0.8657	0.8679

continued

continued

60	0.8466	0.8488	0.8511	0.8533	0.8555	0.8579	0.8598	0.8621	0.8642
70	0.8436	0.8458	0.8480	0.8503	0.8523	0.8544	0.8564	0.8586	0.8607
80	0.8407	0.8428	0.8450	0.8471	0.8492	0.8512	0.8533	0.8553	0.8573
90	0.8380	0.8400	0.8422	0.8443	0.8462	0.8482	0.8501	0.8522	0.8542
100	0.8353	0.8374	0.8395	0.8415	0.8434	0.8453	0.8473	0.8492	0.8511
110	0.8327	0.8348	0.8367	0.8388	0.8406	0.8425	0.8444	0.8462	0.8482
120	0.8303	0.8323	0.8342	0.8363	0.8380	0.8399	0.8416	0.8435	0.8453
130	0.8279	0.8298	0.8319	0.8338	0.8355	0.8374	0.8391	0.8409	0.8427
140	0.8256	0.8275	0.8294	0.8314	0.8330	0.8348	0.8365	0.8382	0.8400
150	0.8234	0.8252	0.8272	0.8290	0.8306	0.8323	0.8342	0.8358	0.8375
160	0.8212	0.8231	0.8250	0.8269	0.8283	0.8299	0.8317	0.8334	0.8351
170	0.8191	0.8209	0.8228	0.8246	0.8261	0.8277	0.8293	0.8310	0.8327
180	0.8170	0.8188	0.8206	0.8226	0.8239	0.8256	0.8272	0.8286	0.8303
190	0.8151	0.8169	0.8187	0.8204	0.8218	0.8234	0.8250	0.8266	0.8282
200	0.8132	0.8148	0.8168	0.8184	0.8197	0.8214	0.8229	0.8244	0.8261

continued

P/MPa	*T*/K								
	390.05	394.85	400.25	405.35	410.05	414.25	419.15	424.55	429.55
	V_{spec}/cm^{3}g^{-1}								
0.1	0.8929	0.8956	0.8983	0.9013	0.9043	0.9066	0.9098	0.9128	0.9160
10	0.8878	0.8904	0.8930	0.8958	0.8985	0.9008	0.9038	0.9066	0.9097
20	0.8830	0.8855	0.8880	0.8906	0.8931	0.8956	0.8981	0.9009	0.9038
30	0.8785	0.8809	0.8833	0.8858	0.8879	0.8904	0.8931	0.8957	0.8985
40	0.8742	0.8766	0.8790	0.8813	0.8834	0.8857	0.8883	0.8907	0.8934
50	0.8703	0.8725	0.8747	0.8769	0.8790	0.8813	0.8837	0.8862	0.8887
60	0.8665	0.8687	0.8708	0.8728	0.8750	0.8773	0.8796	0.8818	0.8842
70	0.8629	0.8650	0.8670	0.8692	0.8712	0.8733	0.8755	0.8778	0.8802
80	0.8595	0.8616	0.8635	0.8655	0.8676	0.8696	0.8717	0.8739	0.8762
90	0.8563	0.8583	0.8601	0.8620	0.8640	0.8660	0.8681	0.8702	0.8724
100	0.8531	0.8552	0.8570	0.8588	0.8607	0.8626	0.8647	0.8667	0.8689
110	0.8501	0.8521	0.8539	0.8556	0.8575	0.8594	0.8614	0.8633	0.8654
120	0.8473	0.8492	0.8508	0.8526	0.8544	0.8563	0.8582	0.8601	0.8621
130	0.8445	0.8464	0.8481	0.8497	0.8514	0.8534	0.8553	0.8571	0.8591
140	0.8418	0.8436	0.8453	0.8468	0.8488	0.8504	0.8523	0.8541	0.8560
150	0.8393	0.8411	0.8426	0.8442	0.8459	0.8477	0.8495	0.8512	0.8531
160	0.8368	0.8385	0.8400	0.8417	0.8433	0.8450	0.8469	0.8485	0.8503
170	0.8344	0.8361	0.8376	0.8390	0.8408	0.8425	0.8442	0.8458	0.8478
180	0.8321	0.8337	0.8352	0.8367	0.8382	0.8399	0.8417	0.8433	0.8451
190	0.8298	0.8314	0.8329	0.8343	0.8359	0.8375	0.8393	0.8408	0.8426
200	0.8276	0.8292	0.8306	0.8321	0.8337	0.8352	0.8370	0.8384	0.8401

Polymer (B): **poly(methyl methacrylate)** **2006KI2**
Characterization: M_w/g.mol^{-1} = 86000, T_g/K = 376

P/MPa	T/K							
	388.79	393.85	398.55	403.71	410.89	413.87	418.70	433.95
	V_{spec}/cm^{3}g^{-1}							
0.1	0.84725	0.84986	0.85289	0.85545	0.85801	0.86072	0.86345	0.87242
10	0.84306	0.84530	0.84799	0.85043	0.85291	0.85544	0.85805	0.86647
20	0.83903	0.84095	0.84329	0.84567	0.84811	0.85045	0.85296	0.86084
30	0.83543	0.83697	0.83914	0.84131	0.84367	0.84592	0.84831	0.85585
40	0.83208	0.83338	0.83528	0.83735	0.83956	0.84172	0.84405	0.85123
50	0.82901	0.83003	0.83167	0.83353	0.83570	0.83783	0.84001	0.84697
60	0.82621	0.82697	0.82848	0.83008	0.83207	0.83416	0.83627	0.84301
70	0.82358	0.82417	0.82538	0.82683	0.82870	0.83062	0.83272	0.83919
80	0.82097	0.82146	0.82254	0.82383	0.82546	0.82733	0.82937	0.83561
90	0.81866	0.81897	0.81992	0.82096	0.82245	0.82420	0.82615	0.83216
100	0.81630	0.81661	0.81746	0.81832	0.81956	0.82120	0.82316	0.82895
110	0.81414	0.81436	0.81508	0.81580	0.81691	0.81841	0.82013	0.82583
120	0.81200	0.81224	0.81286	0.81344	0.81437	0.81571	0.81736	0.82285
130	0.80996	0.81026	0.81076	0.81121	0.81200	0.81319	0.81468	0.82004
140	0.80803	0.80825	0.80866	0.80906	0.80976	0.81073	0.81212	0.81720
150	0.80614	0.80636	0.80672	0.80707	0.80762	0.80842	0.80968	0.81465
160	0.80426	0.80446	0.80478	0.80505	0.80553	0.80628	0.80733	0.81204
170	0.80249	0.80264	0.80292	0.80314	0.80355	0.80423	0.80517	0.80956
180	0.80067	0.80088	0.80113	0.80137	0.80160	0.80218	0.80303	0.80713
190	0.79893	0.79907	0.79938	0.79957	0.79976	0.80026	0.80095	0.80481
200	0.79712	0.79733	0.79763	0.79783	0.79798	0.79837	0.79894	0.80241

Polymer (B): **poly(methylphenylsiloxane)** **2003PAL**
Characterization: M_n/g.mol^{-1} = 32100, M_w/g.mol^{-1} = 35300

Tait equation parameter functions:
Range of data: T/K = 293-423, P/MPa = 10-200

$$V(P/\text{MPa}, T/\text{K}) = V(0, T/\text{K})\{1 - C^*\ln[1 + (P/\text{MPa})/B(T/\text{K})]\}$$
with $C = 0.0894$ and $\theta = T/\text{K} - 273.15$

$V(0, \theta/°\text{C})$/cm^{3}g^{-1}	$B(\theta/°\text{C})$/MPa
$0.8835 + 5.1\ 10^{-4}\theta + 1.06\ 10^{-7}\theta^2$	$220.1 \exp(-4.01\ 10^{-3}\theta)$

Polymer (B): **poly(methyltolylsiloxane)** **2003PAL**
Characterization: M_n/g.mol^{-1} = 20140, M_w/g.mol^{-1} = 23360

Tait equation parameter functions:
Range of data: T/K = 303-373, P/MPa = 10-200

$$V(P/\text{MPa}, T/\text{K}) = V(0, T/\text{K})\{1 - C{*}\ln[1 + (P/\text{MPa})/B(T/\text{K})]\}$$
with $C = 0.0894$ and $\theta = T/\text{K} - 273.15$

$V(0, \theta/°\text{C})/\text{cm}^3\text{g}^{-1}$	$B(\theta/°\text{C})/\text{MPa}$
$0.7928 + 5.0\ 10^{-4}\theta + 6.49\ 10^{-7}\theta^2$	$179.7\ \exp(-4.73\ 10^{-3}\theta)$

Polymer (B): **poly[perfluoro(but-3-enyl vinyl ether)]** **2007DL1**
Characterization: CYTOP, Asahi Glass, Japan

P/MPa	T/K							
	383.86	398.98	404.04	414.27	429.33	444.61	454.81	459.63
	V_{spec}/cm^{3}g^{-1}							
0.1	0.50875	0.51601	0.51928	0.52254	0.52939	0.53610	0.54212	0.54250
10	0.50430	0.51005	0.51329	0.51596	0.52213	0.52798	0.53351	0.53369
20		0.50470	0.50791	0.51013	0.51574	0.52090	0.52600	0.52609
30		0.50018	0.50326	0.50515	0.51041	0.51518	0.52003	0.51993
40			0.49926	0.50091	0.50586	0.51018	0.51478	0.51463
50			0.49590	0.49706	0.50172	0.50584	0.51027	0.50992
60			0.49322	0.49372	0.49805	0.50188	0.50609	0.50579
70					0.49468	0.49835	0.50242	0.50207
80					0.49162	0.49507	0.49902	0.49862
90						0.49205	0.49588	0.49548
100						0.48931	0.49298	0.49255
110						0.48676	0.49033	0.48982
120							0.48775	0.48721
130							0.48535	0.48490
140							0.48306	0.48262
150								0.48053
160								0.47862

continued

continued

P/MPa	T/K 475.10	480.25	490.80	506.22	521.53	531.61	537.07	552.35
	V_{spec}/cm^{3}g^{-1}							
0.1	0.54932	0.55386	0.55663	0.56419	0.57213	0.57904	0.58067	0.58913
10	0.53961	0.54359	0.54585	0.55228	0.55890	0.56469	0.56581	0.57262
20	0.53135	0.53493	0.53685	0.54253	0.54824	0.55331	0.55405	0.55998
30	0.52475	0.52805	0.52974	0.53473	0.53999	0.54460	0.54521	0.55029
40	0.51913	0.52221	0.52367	0.52828	0.53301	0.53725	0.53776	0.54247
50	0.51414	0.51715	0.51836	0.52273	0.52710	0.53107	0.53139	0.53578
60	0.50971	0.51259	0.51370	0.51777	0.52186	0.52557	0.52588	0.52999
70	0.50579	0.50861	0.50949	0.51336	0.51729	0.52078	0.52098	0.52485
80	0.50216	0.50492	0.50573	0.50938	0.51312	0.51647	0.51661	0.52029
90	0.49888	0.50150	0.50227	0.50569	0.50923	0.51253	0.51263	0.51613
100	0.49581	0.49835	0.49904	0.50242	0.50575	0.50890	0.50894	0.51229
110	0.49298	0.49547	0.49604	0.49933	0.50253	0.50557	0.50560	0.50879
120	0.49033	0.49274	0.49336	0.49639	0.49951	0.50247	0.50244	0.50558
130	0.48786	0.49018	0.49074	0.49373	0.49677	0.49959	0.49953	0.50253
140	0.48549	0.48772	0.48831	0.49119	0.49411	0.49682	0.49671	0.49968
150	0.48327	0.48537	0.48596	0.48877	0.49159	0.49423	0.49415	0.49697
160	0.48110	0.48323	0.48371	0.48648	0.48922	0.49184	0.49169	0.49448
170	0.47910	0.48119	0.48155	0.48425	0.48687	0.48946	0.48939	0.49204
180	0.47715	0.47915	0.47951	0.48216	0.48479	0.48727	0.48710	0.48971
190	0.47533	0.47720	0.47765	0.48012	0.48264	0.48511	0.48496	0.48746
200		0.47539	0.47570	0.47819	0.48060	0.48305	0.48283	0.48531

Tait equation parameter functions:
Range of data: T/K = 400-553, P/MPa = 0.1-200

$$V(P/\text{MPa}, T/\text{K}) = V(0, T/\text{K})\{1 - C^*\ln[1 + (P/\text{MPa})/B(T/\text{K})]\}$$

with $C = 0.0759$

$V(0,T/\text{K})/\text{cm}^3\text{g}^{-1}$	$B(T/\text{K})/\text{MPa}$
$0.4258 + 4.0\ 10^{-5}T + 4.602\ 10^{-7}T^2$	$1193.7\exp(-7.18\ 10^{-3}T)$

Polymer (B): **poly(perfluoropropyl ether)** **2012BAM**
Characterization: M_n/g.mol^{-1} = 1726, branched, Krytox® GPL 102, DuPont Corporation

T/K = 298.05

P/MPa	1.2	3.3	10.8	27.5	55.0	97.7	151.1	208.2	272.9
ρ/(kg/m^3)	1850	1856	1879	1925	1973	2022	2076	2121	2164

T/K = 323.25

P/MPa	3.0	4.1	6.9	14.0	41.5	64.1	103.5	140.2	172.0
ρ/(kg/m^3)	1810	1814	1824	1851	1923	1959	2006	2043	2070
P/MPa	207.7	240.0	274.0						
ρ/(kg/m^3)	2098	2121	2136						

T/K = 353.25

P/MPa	1.7	2.9	10.2	27.3	54.8	81.2	110.8	139.4	174.3
ρ/(kg/m^3)	1777	1783	1815	1864	1912	1948	1983	2013	2046
P/MPa	205.0	238.7	272.1						
ρ/(kg/m^3)	2067	2095	2120						

T/K = 404.05

P/MPa	1.8	3.8	11.1	27.7	49.0	78.4	111.2	141.1	170.0
ρ/(kg/m^3)	1696	1711	1752	1809	1856	1905	1948	1982	2006
P/MPa	208.2	240.0	274.4						
ρ/(kg/m^3)	2042	2067	2094						

T/K = 473.15

P/MPa	5.1	14.7	29.1	42.2	55.9	83.3	103.8	110.8	136.1
ρ/(kg/m^3)	1566	1627	1688	1735	1771	1828	1862	1867	1907
P/MPa	172.6	201.8	237.1	238.9	268.9				
ρ/(kg/m^3)	1937	1979	2010	2009	2038				

T/K = 503.15

P/MPa	6.9	15.4	29.5	42.6	56.2	56.5	72.7	81.2	109.1
ρ/(kg/m^3)	1528	1585	1654	1698	1736	1740	1773	1795	1844
P/MPa	135.8	171.7	203.9	234.6	269.3				
ρ/(kg/m^3)	1883	1922	1961	1989	2021				

T/K = 533.05

P/MPa	31.6	41.9	43.2	55.1	55.2	83.2	111.8	111.9	137.1
ρ/(kg/m^3)	1622	1662	1663	1701	1698	1772	1821	1822	1861
P/MPa	172.2	205.3	240.0	240.1	269.4				
ρ/(kg/m^3)	1909	1948	1984	1984	2012				

Polymer (B): **poly(propylene glycol) dimethyl ether** **2010FAN**
Characterization: M_w/g.mol^{-1} = 1385, M_w/M_n = 1.017

P/MPa	*T*/K 298.15	323.15	348.15	373.15	398.15
	ρ/(g/cm^3)				
0.10	0.9846	0.9654	0.9462		
1.00	0.9852	0.9660	0.9469	0.9282	0.9093
5.00	0.9878	0.9690	0.9502	0.9318	0.9134
10.00	0.9910	0.9725	0.9541	0.9362	0.9183
15.00	0.9941	0.9759	0.9579	0.9403	0.9230
20.00	0.9971	0.9792	0.9616	0.9443	0.9274
30.00	1.0028	0.9855	0.9685	0.9519	0.9356
40.00	1.0081	0.9914	0.9750	0.9589	0.9432
45.00	1.0107	0.9942	0.9781	0.9622	0.9468
50.00	1.0133	0.9969	0.9810	0.9654	0.9503
60.00	1.0182	1.0022	0.9868	0.9717	0.9570

Polymer (B): **poly(propylene glycol) dimethyl ether** **2010FAN**
Characterization: M_w/g.mol^{-1} = 1717, M_w/M_n = 1.022

P/MPa	*T*/K 298.15	323.15	348.15	373.15	398.15
	ρ/(g/cm^3)				
0.10	0.9863	0.9671	0.9480		
1.00	0.9869	0.9678	0.9488	0.9301	0.9115
5.00	0.9895	0.9707	0.9520	0.9337	0.9156
10.00	0.9926	0.9742	0.9559	0.9381	0.9205
15.00	0.9957	0.9776	0.9597	0.9422	0.9251
20.00	0.9987	0.9809	0.9633	0.9462	0.9295
30.00	1.0043	0.9871	0.9702	0.9537	0.9377
40.00	1.0097	0.9930	0.9767	0.9606	0.9452
45.00	1.0123	0.9958	0.9797	0.9639	0.9487
50.00	1.0148	0.9985	0.9827	0.9671	0.9522
60.00	1.0198	1.0038	0.9884	0.9733	0.9589

Polymer (B): **poly(propyl methacrylate)** **2006KI2**
Characterization: M_w/g.mol^{-1} = 283000, T_g/K = 324

P/MPa	T/K							
	364.05	373.85	383.65	393.97	404.01	414.20	424.02	434.22
	V_{spec}/cm^{3}g^{-1}							
0.1	0.94290	0.94900	0.95537	0.96144	0.96791	0.97469	0.98102	0.98791
10	0.93724	0.94300	0.94895	0.95473	0.96086	0.96708	0.97302	0.97949
20	0.93191	0.93741	0.94296	0.94852	0.95427	0.96007	0.96568	0.97182
30	0.92695	0.93217	0.93747	0.94275	0.94830	0.95369	0.95896	0.96480
40	0.92227	0.92728	0.93232	0.93736	0.94284	0.94777	0.95287	0.95841
50	0.91786	0.92266	0.92742	0.93246	0.93766	0.94237	0.94721	0.95251
60	0.91377	0.91837	0.92304	0.92771	0.93280	0.93725	0.94198	0.94713
70	0.90984	0.91430	0.91889	0.92331	0.92823	0.93258	0.93715	0.94193
80	0.90613	0.91047	0.91489	0.91920	0.92393	0.92811	0.93260	0.93719
90	0.90267	0.90687	0.91113	0.91537	0.91996	0.92397	0.92822	0.93267
100	0.89939	0.90341	0.90750	0.91166	0.91605	0.92002	0.92410	0.92849
110	0.89620	0.90014	0.90410	0.90805	0.91245	0.91618	0.92032	0.92449
120	0.89318	0.89701	0.90090	0.90470	0.90894	0.91262	0.91653	0.92070
130	0.89018	0.89408	0.89777	0.90156	0.90574	0.90930	0.91313	0.91705
140	0.88740	0.89111	0.89482	0.89846	0.90239	0.90605	0.90968	0.91351
150	0.88472	0.88824	0.89189	0.89552	0.89936	0.90283	0.90646	0.91018
160	0.88205	0.88557	0.88905	0.89263	0.89638	0.89974	0.90330	0.90701
170	0.87954	0.88296	0.88637	0.88985	0.89357	0.89681	0.90035	0.90390
180	0.87703	0.88036	0.88380	0.88713	0.89077	0.89391	0.89733	0.90082
190	0.87464	0.87784	0.88124	0.88450	0.88805	0.89110	0.89450	0.89789
200	0.87222	0.87540	0.87869	0.88188	0.88534	0.88851	0.89173	0.89509

Polymer (B): **polystyrene** **2008QIA**
Characterization: M_n/g.mol^{-1} = 2140, T_g/K = 324

T/K	P/MPa			
	10	20	30	40
	V_{spec}/cm^{3}g^{-1}			
383.66	0.98474	0.97970	0.97505	0.97109
393.61	0.98995	0.98466	0.97954	0.97501
403.00	0.99573	0.99027	0.98499	0.98020
413.23	1.0014	0.9958	0.99035	0.98539
422.50	1.0066	1.0006	0.99502	0.98990

continued

continued

432.61	1.0123	1.0061	1.0002	0.99483
442.59	1.0178	1.0114	1.0053	0.99969
452.68	1.0239	1.0169	1.0105	1.0049
463.08	1.0296	1.0222	1.0156	1.0097
472.89	1.0353	1.0277	1.0210	1.0149
482.82	1.0413	1.0334	1.0261	1.0199
493.17	1.0472	1.0388	1.0314	1.0246
504.23	1.0539	1.0450	1.0372	1.0302

Polymer (B): **polystyrene, 4-arm star polymer** **2008QIA**

Characterization: M_n/g.mol^{-1} = 6300, M_w/M_n = 1.02, per arm M_n/g.mol^{-1} = 1500, T_g/K = 352, synthesized in the laboratory

T/K	P/MPa			
	10	20	30	40
	V_{spec}/cm^{3}g^{-1}			
378.66	0.98337	0.97751	0.97213	0.96696
388.73	0.98876	0.98276	0.97725	0.97201
398.95	0.99479	0.98860	0.98271	0.97718
409.10	1.0005	0.99381	0.98782	0.98209
419.15	1.0062	0.99912	0.99282	0.98699
429.24	1.0119	1.0048	0.99825	0.99208
439.39	1.0180	1.0104	1.0034	0.99711
449.05	1.0235	1.0156	1.0085	1.0020
459.49	1.0299	1.0214	1.0139	1.0070
469.76	1.0358	1.0270	1.0192	1.0121
480.22	1.0420	1.0327	1.0247	1.0171
489.69	1.0476	1.0381	1.0297	1.0220
500.21	1.0542	1.0440	1.0352	1.0273
510.51	1.0605	1.0498	1.0407	1.0325
520.88	1.0672	1.0562	1.0467	1.0382

Polymer (B): **polystyrene, 4-arm star polymer** **2012CLA**
Characterization: M_w/g.mol^{-1} = 520000, per arm M_w/g.mol^{-1} = 130000

T/K	P/MPa					
	0.1	40	80	120	160	200
	V_{spec}/cm^{3}g^{-1}					
305.15	0.9506	0.9386	0.9276	0.9171	0.9087	0.9008
315.15	0.9533	0.9402	0.9271	0.9201	0.9072	0.9025
324.15	0.9544	0.9429	0.9309	0.9212	0.9122	0.9045
339.15	0.9588	0.9446	0.9311	0.9217	0.9128	0.9047
354.15	0.9634	0.9478	0.9355	0.9245	0.9151	0.9071
368.15	0.9669	0.9502	0.9367	0.9259	0.9157	0.9079
384.15	0.9752	0.9523	0.9378	0.9261	0.9161	0.9082
398.15	0.9833	0.9577	0.9388	0.9260	0.9153	0.9065
414.15	0.9929	0.9647	0.9440	0.9275	0.9158	0.9064
429.15	1.0011	0.9714	0.9496	0.9324	0.9184	0.9071
444.15	1.0110	0.9779	0.9557	0.9379	0.9230	0.9102
459.15	1.0199	0.9856	0.9615	0.9432	0.9274	0.9147
474.15	1.0304	0.9925	0.9678	0.9487	0.9323	0.9194
490.15	1.0389	0.9987	0.9733	0.9524	0.9363	0.9224
505.15	1.0483	1.0055	0.9788	0.9573	0.9408	0.9258

Polymer (B): **polystyrene, 11-arm star polymer** **2008QIA**
Characterization: M_n/g.mol^{-1} = 7500, M_w/M_n = 1.03, per arm M_n/g.mol^{-1} = 660, T_g/K = 342, synthesized in the laboratory

T/K	P/MPa			
	10	20	30	40
	V_{spec}/cm^{3}g^{-1}			
378.65	0.99188	0.98610	0.98066	0.97557
388.85	0.99737	0.99133	0.98560	0.98033
398.54	1.0032	0.99689	0.99086	0.98532
408.95	1.0090	1.0022	0.99612	0.99031
419.08	1.0146	1.0076	1.0011	0.99525
429.04	1.0205	1.0130	1.0063	1.0001
439.43	1.0265	1.0186	1.0116	1.0052
449.02	1.0319	1.0237	1.0165	1.0098
459.13	1.0377	1.0292	1.0216	1.0147
468.92	1.0435	1.0346	1.0267	1.0195
479.30	1.0495	1.0401	1.0320	1.0244
489.50	1.0554	1.0455	1.0371	1.0294
499.99	1.0616	1.0513	1.0425	1.0345

Polymer (B): **poly(styrene-*co*-acrylonitrile)** **2005BRI**
Characterization: M_w/g.mol^{-1} = 80000, 33 wt% acrylonitrile

Tait equation parameter functions:
Range of data: T/K = 380-460, P/MPa = 0.1-200

$$V(P/\text{MPa}, T/\text{K}) = V(0, T/\text{K})\{1 - C^*\ln[1 + (P/\text{MPa})/B(T/\text{K})]\}$$
with $C = 0.0894$ and $\theta = T/\text{K} - 273.15$

$V(0, \theta/°\text{C})/\text{cm}^3\text{g}^{-1}$	$B(\theta/°\text{C})/\text{MPa}$
$0.8892 + 5.23\ 10^{-4}\ \theta - 3.0\ 10^{-8}\ \theta^2$	$294.0 \exp(-3.86\ 10^{-3}\ \theta)$

Polymer (B): **poly(styrene-*co*-maleic anhydride)** **2005KIL**
Characterization: M_n/g.mol^{-1} = 46000, M_w/g.mol^{-1} = 89000, 11.8 mol% maleic anhydride, T_g/K = 391, Dylark 232, Acro Chemicals

P/MPa	T/K 429.93	439.99	450.33	460.55	470.95	481.28	491.35	501.55
	V_{spec}/cm^{3}g^{-1}							
0.1	0.9734	0.9789	0.9846	0.9902	0.9959	1.0015	1.0072	1.0128
10	0.9668	0.9720	0.9774	0.9826	0.9879	0.9933	0.9985	1.0038
20	0.9606	0.9656	0.9706	0.9756	0.9806	0.9857	0.9906	0.9955
30	0.9551	0.9598	0.9644	0.9693	0.9741	0.9789	0.9836	0.9882
40	0.9498	0.9543	0.9589	0.9635	0.9680	0.9728	0.9771	0.9815
50	0.9450	0.9493	0.9537	0.9581	0.9625	0.9670	0.9713	0.9754
60	0.9405	0.9447	0.9488	0.9531	0.9573	0.9617	0.9658	0.9697
70	0.9361	0.9402	0.9442	0.9484	0.9525	0.9567	0.9606	0.9645
80	0.9320	0.9360	0.9400	0.9440	0.9479	0.9520	0.9558	0.9595
90	0.9282	0.9320	0.9359	0.9397	0.9436	0.9475	0.9511	0.9548
100	0.9245	0.9283	0.9320	0.9357	0.9395	0.9433	0.9469	0.9503
110	0.9210	0.9246	0.9283	0.9319	0.9356	0.9392	0.9428	0.9461
120	0.9176	0.9211	0.9247	0.9282	0.9318	0.9355	0.9388	0.9421
130	0.9144	0.9178	0.9213	0.9247	0.9283	0.9318	0.9350	0.9382
140	0.9114	0.9145	0.9180	0.9213	0.9248	0.9283	0.9314	0.9345
150	0.9086	0.9114	0.9148	0.9180	0.9215	0.9249	0.9279	0.9310
160	0.9059	0.9085	0.9118	0.9149	0.9183	0.9215	0.9246	0.9275
170	0.9034	0.9057	0.9087	0.9119	0.9152	0.9184	0.9213	0.9242
180	0.9011	0.9029	0.9058	0.9089	0.9121	0.9153	0.9181	0.9209
190	0.8989	0.9003	0.9030	0.9060	0.9092	0.9122	0.9151	0.9179
200	0.8967	0.8978	0.9002	0.9032	0.9063	0.9093	0.9120	0.9147

Polymer (B): **poly(styrene-*co*-maleic anhydride)** **2005KIL**
Characterization: M_n/g.mol^{-1} = 38000, M_w/g.mol^{-1} = 73000, 14.5 mol% maleic anhydride, T_g/K = 403, Dylark 332, Acro Chemicals

P/MPa	*T*/K							
	438.23	453.03	468.23	483.07	497.88	512.98	527.76	543.63
	V_{spec}/cm^{3}g^{-1}							
0.1	0.95312	0.96159	0.97020	0.97900	0.98749	0.99685	1.00636	1.01598
10	0.94661	0.95461	0.96278	0.97117	0.97948	0.98793	0.99676	1.00574
20	0.94047	0.94803	0.95577	0.96382	0.97192	0.97952	0.98791	0.99625
30	0.93511	0.94234	0.94975	0.95748	0.96543	0.97267	0.98049	0.98853
40	0.93010	0.93709	0.94426	0.95172	0.95932	0.96613	0.97361	0.98087
50	0.92552	0.93227	0.93920	0.94633	0.95360	0.96026	0.96731	0.97434
60	0.92108	0.92769	0.93438	0.94137	0.94835	0.95479	0.96157	0.96835
70	0.91696	0.92343	0.92993	0.93663	0.94346	0.94965	0.95619	0.96272
80	0.91317	0.91930	0.92561	0.93221	0.93881	0.94477	0.95114	0.95738
90	0.90939	0.91538	0.92155	0.92801	0.93436	0.94026	0.94640	0.95243
100	0.90588	0.91173	0.91776	0.92398	0.93019	0.93594	0.94183	0.94766
110	0.90270	0.90821	0.91401	0.92017	0.92614	0.93175	0.93749	0.94317
120	0.89964	0.90468	0.91046	0.91646	0.92221	0.92786	0.93336	0.93898
130	0.89684	0.90147	0.90698	0.91297	0.91855	0.92405	0.92940	0.93488
140	0.89423	0.89822	0.90368	0.90952	0.91502	0.92029	0.92574	0.93086
150	0.89170	0.89524	0.90047	0.90623	0.91165	0.91686	0.92204	0.92717
160	0.88941	0.89232	0.89749	0.90306	0.90838	0.91345	0.91864	0.92349
170	0.88715	0.88970	0.89444	0.89995	0.90514	0.91016	0.91521	0.92001
180	0.88503	0.88713	0.89162	0.89714	0.90203	0.90713	0.91202	0.91675
190	0.88301	0.88478	0.88883	0.89424	0.89914	0.90410	0.90891	0.91352
200	0.88108	0.88259	0.88618	0.89154	0.89639	0.90118	0.90599	0.91042

Polymer (B): **poly(styrene-*co*-maleic anhydride)** **2005KIL**
Characterization: M_n/g.mol^{-1} = 27000, M_w/g.mol^{-1} = 52000, 25.3 mol% maleic anhydride, T_g/K = 426, DSM Research, The Netherlands

P/MPa	*T*/K						
	433.38	443.34	453.79	463.81	474.02	484.58	494.50
	V_{spec}/cm^{3}g^{-1}						
0.1	0.90381	0.90882	0.91386	0.91875	0.92377	0.92873	0.93396
10	0.89850	0.90314	0.90789	0.91255	0.91728	0.92207	0.92699
20	0.89344	0.89774	0.90231	0.90672	0.91115	0.91581	0.92040
30	0.88892	0.89294	0.89719	0.90150	0.90574	0.91025	0.91469

continued

continued

40	0.88492	0.88847	0.89250	0.89665	0.90077	0.90501	0.90940
50	0.88133	0.88425	0.88818	0.89210	0.89614	0.90017	0.90438
60	0.87808	0.88035	0.88407	0.88792	0.89178	0.89574	0.89980
70	0.87505	0.87682	0.88013	0.88393	0.88762	0.89144	0.89543
80	0.87238	0.87359	0.87644	0.88011	0.88372	0.88748	0.89127
90	0.86978	0.87061	0.87307	0.87653	0.88004	0.88366	0.88744
100	0.86734	0.86786	0.86982	0.87307	0.87652	0.87999	0.88366
110	0.86506	0.86533	0.86676	0.86981	0.87317	0.87661	0.88014
120	0.86290	0.86291	0.86403	0.86666	0.86995	0.87326	0.87673
130	0.86076	0.86075	0.86139	0.86362	0.86684	0.87000	0.87352
140	0.85863	0.85853	0.85901	0.86086	0.86381	0.86699	0.87036
150	0.85664	0.85644	0.85671	0.85814	0.86093	0.86403	0.86745
160	0.85467	0.85438	0.85454	0.85567	0.85811	0.86115	0.86448
170	0.85279	0.85244	0.85243	0.85324	0.85543	0.85836	0.86164
180	0.85099	0.85063	0.85046	0.85102	0.85281	0.85562	0.85894
190	0.84910	0.84870	0.84853	0.84888	0.85036	0.85299	0.85618
200	0.84736	0.84696	0.84661	0.84688	0.84798	0.85036	0.85346

Polymer (B): **poly(styrene-*co*-maleic anhydride)** **2005KIL**

Characterization: M_n/g.mol^{-1} = 28000, M_w/g.mol^{-1} = 54000, 30.9 mol% maleic anhydride, T_g/K = 439, DSM Research, The Netherlands

P/MPa	T/K					
	449.64	459.76	470.05	480.33	490.76	501.01
	V_{spec}/cm^{3}g^{-1}					
0.1	0.88563	0.89054	0.89530	0.90032	0.90511	0.91030
10	0.88037	0.88501	0.88956	0.89428	0.89892	0.90383
20	0.87534	0.87973	0.88410	0.88852	0.89309	0.89770
30	0.87093	0.87511	0.87925	0.88354	0.88787	0.89238
40	0.86679	0.87074	0.87472	0.87881	0.88301	0.88742
50	0.86311	0.86668	0.87057	0.87448	0.87853	0.88271
60	0.85976	0.86287	0.86657	0.87037	0.87427	0.87833
70	0.85681	0.85936	0.86286	0.86660	0.87037	0.87430
80	0.85412	0.85597	0.85938	0.86288	0.86661	0.87041
90	0.85160	0.85304	0.85601	0.85944	0.86303	0.86679
100	0.84933	0.85019	0.85276	0.85614	0.85961	0.86332
110	0.84707	0.84771	0.84979	0.85296	0.85639	0.85997
120	0.84500	0.84528	0.84693	0.84992	0.85330	0.85683
130	0.84297	0.84311	0.84432	0.84701	0.85030	0.85373
140	0.84103	0.84099	0.84186	0.84423	0.84742	0.85078
150	0.83916	0.83900	0.83956	0.84157	0.84463	0.84795

continued

continued

160	0.83728	0.83711	0.83743	0.83909	0.84197	0.84515
170	0.83556	0.83521	0.83539	0.83667	0.83930	0.84247
180	0.83384	0.83341	0.83348	0.83440	0.83681	0.83993
190	0.83210	0.83168	0.83156	0.83229	0.83437	0.83736
200	0.83043	0.82993	0.82977	0.83025	0.83197	0.83483

Polymer (B): **poly(styrene-*co*-maleic anhydride)** **2005KIL**
Characterization: M_n/g.mol^{-1} = 30000, M_w/g.mol^{-1} = 58000, 34.7 mol% maleic anhydride, T_g/K = 449, DSM Research, The Netherlands

P/MPa	*T*/K				
	460.25	470.41	480.76	491.06	501.28
	V_{spec}/cm^{3}g^{-1}				
0.1	0.87138	0.87623	0.88109	0.88602	0.89124
10	0.86613	0.87072	0.87531	0.87999	0.88490
20	0.86112	0.86542	0.86981	0.87428	0.87890
30	0.85668	0.86087	0.86501	0.86928	0.87372
40	0.85255	0.85653	0.86055	0.86471	0.86895
50	0.84869	0.85252	0.85639	0.86037	0.86454
60	0.84526	0.84872	0.85248	0.85633	0.86023
70	0.84203	0.84511	0.84879	0.85250	0.85636
80	0.83917	0.84175	0.84526	0.84886	0.85264
90	0.83657	0.83862	0.84191	0.84541	0.84913
100	0.83421	0.83568	0.83877	0.84214	0.84573
110	0.83193	0.83304	0.83571	0.83902	0.84257
120	0.82982	0.83053	0.83280	0.83603	0.83948
130	0.82778	0.82827	0.83012	0.83319	0.83653
140	0.82589	0.82609	0.82758	0.83043	0.83372
150	0.82407	0.82411	0.82520	0.82774	0.83096
160	0.82221	0.82212	0.82299	0.82519	0.82829
170	0.82052	0.82026	0.82087	0.82277	0.82567
180	0.81877	0.81848	0.81893	0.82048	0.82319
190	0.81707	0.81679	0.81704	0.81824	0.82074
200	0.81546	0.81506	0.81521	0.81612	0.81838

Polymer (B): **poly[tetrafluoroethylene-*co*-2,2-bis(trifluoromethyl)-4,5-difluoro-1,3-dioxole]** **2008DLU**

Characterization: 13 mol% tetrafluoroethylene, T_g/K = 513, Teflon AF2400, Random Technologies, San Francisco

T/K	*P*/MPa 0.1	10	20	30	40	50	80	120
	V_{spec}/cm^{3}g^{-1}							
515.46	0.60962							
530.86	0.61813	0.60334						
545.85	0.63020	0.60984	0.59384					
561.43	0.64216	0.61663	0.59853	0.58673				
576.08	0.65486	0.62419	0.60382	0.59045	0.58055			
591.38	0.66872	0.63258	0.61011	0.59535	0.58425	0.57576		
606.33	0.68364	0.64084	0.61603	0.59990	0.58794	0.57845	0.56060	
620.76	0.70048	0.64989	0.62279	0.60520	0.59233	0.58215	0.56155	0.54661
635.12	0.71977	0.65951	0.62983	0.61103	0.59729	0.58644	0.56364	0.54727

Tait equation parameter functions:
Range of data: *T*/K = 515-635, *P*/MPa = 0.1-120

$$V(P/\text{MPa}, T/\text{K}) = V(0, T/\text{K})\{1 - C^*\ln[1 + (P/\text{MPa})/B(T/\text{K})]\}$$

with $C = 0.0781$ and $\theta = T/\text{K} - 273.15$

$V(0, \theta/°\text{C})/\text{cm}^3\text{g}^{-1}$	$B(\theta/°\text{C})/\text{MPa}$
$0.6490 - 8.8\ 10^{-4}\theta + 2.97\ 10^{-6}\theta^2$	$978.0 \exp(-1.45\ 10^{-2}\theta)$

Polymer (B): **poly[tetrafluoroethylene-*co*-2,2-bis(trifluoromethyl)-4,5-difluoro-1,3-dioxole** **2008DLU**

Characterization: 35 mol% tetrafluoroethylene, T_g/K = 433, Teflon AF1600, Random Technologies, San Francisco

T/K	*P*/MPa 0.1	10	20	30	40	50	80	120
	V_{spec}/cm^{3}g^{-1}							
439.42	0.57117							
454.64	0.58246	0.57009						
469.94	0.59302	0.57728	0.56497					

continued

continued

485.28	0.60202	0.58446	0.57119	0.56091				
494.60	0.61134	0.59105	0.57604	0.56521	0.55659			
500.69	0.61170	0.59195	0.57741	0.56647	0.55760			
515.39	0.62201	0.59985	0.58398	0.57223	0.56280	0.55493		
520.34	0.62808	0.60361	0.58650	0.57425	0.56463	0.55659	0.53843	
530.90	0.63310	0.60801	0.59068	0.57796	0.56798	0.55979	0.54113	
546.16	0.64493	0.61633	0.59731	0.58374	0.57312	0.56451	0.54493	
561.62	0.65675	0.62455	0.60397	0.58959	0.57826	0.56910	0.54876	0.53038
576.55	0.66967	0.63313	0.61082	0.59535	0.58353	0.57393	0.55259	0.53351
592.00	0.68299	0.64161	0.61745	0.60112	0.58871	0.57854	0.55629	0.53669
607.04	0.69873	0.65091	0.62464	0.60721	0.59390	0.58332	0.56014	0.53994
621.74	0.71654	0.66043	0.63173	0.61320	0.59920	0.58808	0.56414	0.54318
636.12	0.74124	0.67156	0.63966	0.61976	0.60489	0.59319	0.56818	0.54651

continued

T/K	P/MPa	
	160	200
	V_{spec}/cm^{3}g^{-1}	
592.00	0.52248	
607.04	0.52524	0.51358
621.74	0.52817	0.51602
636.12	0.53105	0.51860

Tait equation parameter functions:
Range of data: T/K = 435-635, P/MPa = 0.1-120

$$V(P/\text{MPa}, T/\text{K}) = V(0, T/\text{K})\{1 - C^*\ln[1 + (P/\text{MPa})/B(T/\text{K})]\}$$
with $C = 0.0760$ and $\theta = T/\text{K} - 273.15$

$V(0, \theta/°\text{C})/\text{cm}^3\text{g}^{-1}$	$B(\theta/°\text{C})/\text{MPa}$
$0.5494 - 1.4\ 10^{-4}\theta + 1.80\ 10^{-6}\theta^2$	$256.0\ \exp(-1.13\ 10^{-2}\theta)$

Polymer (B): **poly[tetrafluoroethylene-*co*-perfluoro(methyl vinyl ether)]** **2004DLU**

Characterization: M_w/g.mol^{-1} = 100000, 15-55 mol% perfluoro(methyl vinyl ether), T_g/K = 271, Dyneon Perfluoroelastomer PFE, Dyneon GmbH & Co. KG, Burgkirchen, Germany

P/MPa	*T*/K							
	298.94	299.465	300.77	305.746	310.59	315.415	320.21	324.945
	V_{spec}/cm^{3}g^{-1}							
0.1	0.48071	0.48126	0.48153	0.48376	0.48600	0.48813	0.49042	0.49279
10	0.47712	0.47765	0.47790	0.47996	0.48204	0.48410	0.48621	0.48841
20	0.47372	0.47426	0.47449	0.47640	0.47832	0.48041	0.48235	0.48437
30	0.47071	0.47124	0.47141	0.47321	0.47505	0.47704	0.47886	0.48080
40	0.46793	0.46837	0.46866	0.47039	0.47208	0.47395	0.47571	0.47748
50	0.46540	0.46582	0.46600	0.46777	0.46943	0.47121	0.47282	0.47452
60	0.46304	0.46345	0.46370	0.46524	0.46684	0.46854	0.47016	0.47184
70	0.46084	0.46125	0.46146	0.46303	0.46453	0.46612	0.46772	0.46928
80	0.45884	0.45916	0.45941	0.46087	0.46232	0.46395	0.46540	0.46697
90	0.45695	0.45726	0.45744	0.45890	0.46032	0.46189	0.46318	0.46477
100	0.45511	0.45545	0.45566	0.45706	0.45840	0.45987	0.46128	0.46266
110	0.45351	0.45376	0.45398	0.45519	0.45661	0.45801	0.45936	0.46071
120	0.45193	0.45224	0.45239	0.45361	0.45484	0.45628	0.45757	0.45887
130	0.45055	0.45078	0.45096	0.45204	0.45321	0.45460	0.45581	0.45721
140	0.44923	0.44943	0.44957	0.45058	0.45170	0.45298	0.45426	0.45552
150	0.44791	0.44820	0.44826	0.44918	0.45019	0.45142	0.45266	0.45394
160	0.44676	0.44695	0.44702	0.44788	0.44882	0.45002	0.45110	0.45241
170	0.44563	0.44582	0.44585	0.44662	0.44742	0.44861	0.44976	0.45084
180	0.44452	0.44470	0.44477	0.44541	0.44627	0.44723	0.44832	0.44954
190	0.44343	0.44361	0.44369	0.44436	0.44506	0.44599	0.44701	0.44816
200	0.44242	0.44258	0.44266	0.44322	0.44392	0.44481	0.44567	0.44676

continued

P/MPa	*T*/K							
	329.77	334.65	364.545	379.32	394.37	409.58	424.67	439.995
	V_{spec}/cm^{3}g^{-1}							
0.1	0.49520	0.49745	0.51195	0.51929	0.52691	0.53482	0.54252	0.55074
10	0.49061	0.49271	0.50598	0.51266	0.51952	0.52655	0.53330	0.54041
20	0.48639	0.48837	0.50055	0.50673	0.51302	0.51934	0.52541	0.53170
30	0.48264	0.48451	0.49599	0.50175	0.50756	0.51336	0.51890	0.52470

continued

continued

40	0.47945	0.48103	0.49190	0.49731	0.50279	0.50820	0.51337	0.51865
50	0.47626	0.47791	0.48826	0.49338	0.49852	0.50358	0.50850	0.51349
60	0.47346	0.47505	0.48494	0.48979	0.49466	0.49952	0.50412	0.50880
70	0.47096	0.47246	0.48189	0.48655	0.49125	0.49588	0.50018	0.50469
80	0.46853	0.46997	0.47916	0.48354	0.48807	0.49252	0.49670	0.50088
90	0.46624	0.46773	0.47648	0.48076	0.48512	0.48939	0.49340	0.49743
100	0.46419	0.46555	0.47409	0.47822	0.48244	0.48652	0.49034	0.49431
110	0.46222	0.46357	0.47181	0.47581	0.47989	0.48386	0.48750	0.49127
120	0.46034	0.46153	0.46968	0.47350	0.47752	0.48124	0.48489	0.48853
130	0.45847	0.45977	0.46768	0.47141	0.47524	0.47902	0.48239	0.48589
140	0.45672	0.45802	0.46567	0.46939	0.47313	0.47665	0.48003	0.48346
150	0.45513	0.45636	0.46387	0.46750	0.47107	0.47458	0.47781	0.48110
160	0.45359	0.45478	0.46212	0.46567	0.46918	0.47257	0.47571	0.47888
170	0.45209	0.45318	0.46039	0.46382	0.46733	0.47060	0.47365	0.47682
180	0.45064	0.45179	0.45881	0.46219	0.46550	0.46875	0.47174	0.47473
190	0.44922	0.45039	0.45715	0.46048	0.46379	0.46693	0.46986	0.47291
200	0.44789	0.44897	0.45558	0.45885	0.46211	0.46517	0.46805	0.47099

continued

P/MPa				T/K				
	455.57	470.84	486.094	501.76	516.876	532.605	547.58	563.503
					V_{spec}/cm^{3}g^{-1}			
0.1	0.55918	0.56803	0.57704	0.58677	0.59692	0.60765	0.61919	0.63239
10	0.54763	0.55503	0.56253	0.57050	0.57862	0.58717	0.59594	0.60571
20	0.53807	0.54446	0.55103	0.55784	0.56481	0.57220	0.57940	0.58748
30	0.53042	0.53616	0.54196	0.54819	0.55429	0.56079	0.56720	0.57425
40	0.52403	0.52922	0.53460	0.54012	0.54576	0.55170	0.55735	0.56373
50	0.51842	0.52327	0.52822	0.53327	0.53858	0.54396	0.54926	0.55514
60	0.51345	0.51811	0.52263	0.52743	0.53235	0.53729	0.54220	0.54772
70	0.50910	0.51340	0.51768	0.52218	0.52679	0.53149	0.53613	0.54126
80	0.50509	0.50925	0.51334	0.51764	0.52191	0.52631	0.53066	0.53556
90	0.50142	0.50530	0.50926	0.51334	0.51748	0.52170	0.52591	0.53049
100	0.49808	0.50178	0.50558	0.50952	0.51339	0.51749	0.52140	0.52594
110	0.49496	0.49864	0.50221	0.50600	0.50975	0.51358	0.51737	0.52174
120	0.49211	0.49556	0.49911	0.50260	0.50625	0.51003	0.51358	0.51789
130	0.48939	0.49276	0.49614	0.49960	0.50307	0.50661	0.51018	0.51428
140	0.48683	0.49005	0.49329	0.49668	0.50015	0.50351	0.50701	0.51089
150	0.48440	0.48752	0.49069	0.49400	0.49720	0.50054	0.50394	0.50768
160	0.48209	0.48515	0.48820	0.49137	0.49448	0.49781	0.50102	0.50473
170	0.47982	0.48284	0.48579	0.48898	0.49200	0.49513	0.49831	0.50202
180	0.47771	0.48076	0.48353	0.48662	0.48957	0.49270	0.49574	0.49930
190	0.47574	0.47852	0.48133	0.48431	0.48729	0.49021	0.49331	0.49668
200	0.47374	0.47647	0.47923	0.48212	0.48504	0.48780	0.49075	0.49425

continued

continued

P/MPa	T/K 578.20	593.788
	V_{spec}/cm^{3}g^{-1}	
0.1	0.64774	0.66445
10	0.61667	0.62784
20	0.59638	0.60530
30	0.58195	0.58955
40	0.57068	0.57742
50	0.56140	0.56751
60	0.55359	0.55915
70	0.54677	0.55204
80	0.54072	0.54569
90	0.53548	0.54008
100	0.53064	0.53501
110	0.52626	0.53044
120	0.52224	0.52617
130	0.51841	0.52227
140	0.51490	0.51863
150	0.51161	0.51522
160	0.50851	0.51203
170	0.50566	0.50896
180	0.50287	0.50608
190	0.50014	0.50333
200	0.49764	0.50070

Polymer (B): **poly(vinyl alcohol)** **2007FU2**

Characterization: M_w/g.mol^{-1} = 195000, ρ(293.15 K) = 1.2906 g/cm^3, T_g/K = 350, T_m/K = 493, Mowiol 56-98, Kuraray Specialities, Japan

P/MPa	T/K 499.16	504.17	508.69	513.70	518.52	523.47
	V_{spec}/cm^{3}g^{-1}					
0.1	0.8581	0.8611	0.8646	0.8679	0.8726	0.8782
10	0.8544	0.8573	0.8606	0.8640	0.8684	0.8738
20	0.8507	0.8536	0.8566	0.8601	0.8644	0.8697
30	0.8477	0.8504	0.8535	0.8568	0.8608	0.8659
40	0.8448	0.8475	0.8504	0.8535	0.8576	0.8628
50	0.8420	0.8446	0.8476	0.8505	0.8546	0.8596
60	0.8392	0.8419	0.8447	0.8476	0.8516	0.8567

continued

continued

70	0.8369	0.8392	0.8421	0.8450	0.8488	0.8539
80	0.8342	0.8367	0.8395	0.8424	0.8461	0.8512
90	0.8318	0.8343	0.8371	0.8399	0.8435	0.8486
100	0.8296	0.8319	0.8346	0.8375	0.8411	0.8460
110	0.8272	0.8296	0.8322	0.8350	0.8387	0.8436
120	0.8252	0.8274	0.8301	0.8327	0.8364	0.8410
130	0.8231	0.8253	0.8279	0.8306	0.8344	0.8388
140	0.8210	0.8232	0.8259	0.8284	0.8321	0.8366
150	0.8190	0.8212	0.8237	0.8265	0.8300	0.8345
160	0.8171	0.8194	0.8218	0.8245	0.8280	0.8324
170	0.8153	0.8175	0.8200	0.8226	0.8260	0.8304
180	0.8135	0.8156	0.8182	0.8207	0.8242	0.8285
200	0.8102	0.8123	0.8146	0.8171	0.8205	0.8248

Polymer (B): **poly(vinylidene fluoride-*co*-hexafluoropropylene) 2004DLU**
Characterization: M_w/g.mol^{-1} = 85000, 22.0 mol% hexafluoropropylene, T_g/K = 255, Fluorel FC-2175, Dyneon GmbH & Co. KG, Burgkirchen, Germany

P/MPa	T/K							
	302.885	305.956	310.925	315.605	320.35	325.19	329.995	334.935
	V_{spec}/cm^{3}g^{-1}							
0.1	0.55660	0.55805	0.56011	0.56223	0.56447	0.56675	0.56887	0.57099
10	0.55358	0.55495	0.55696	0.55897	0.56106	0.56321	0.56523	0.56728
20	0.55065	0.55193	0.55394	0.55582	0.55787	0.55982	0.56176	0.56377
30	0.54808	0.54935	0.55120	0.55309	0.55497	0.55686	0.55872	0.56064
40	0.54569	0.54693	0.54870	0.55053	0.55234	0.55417	0.55590	0.55782
50	0.54343	0.54463	0.54633	0.54810	0.54981	0.55163	0.55342	0.55518
60	0.54133	0.54249	0.54418	0.54591	0.54761	0.54932	0.55098	0.55277
70	0.53940	0.54049	0.54214	0.54378	0.54537	0.54712	0.54876	0.55041
80	0.53749	0.53853	0.54017	0.54178	0.54333	0.54501	0.54663	0.54827
90	0.53565	0.53677	0.53823	0.53991	0.54140	0.54304	0.54450	0.54615
100	0.53396	0.53501	0.53643	0.53804	0.53958	0.54112	0.54264	0.54422
110	0.53221	0.53333	0.53480	0.53631	0.53782	0.53930	0.54073	0.54233
120	0.53066	0.53170	0.53316	0.53464	0.53609	0.53752	0.53900	0.54051
130	0.52918	0.53019	0.53149	0.53301	0.53444	0.53591	0.53729	0.53880
140	0.52763	0.52869	0.53001	0.53144	0.53282	0.53421	0.53564	0.53712
150	0.52615	0.52720	0.52850	0.52993	0.53128	0.53276	0.53405	0.53545
160	0.52477	0.52572	0.52714	0.52850	0.52984	0.53118	0.53256	0.53389
170	0.52349	0.52441	0.52572	0.52703	0.52845	0.52971	0.53105	0.53239
180	0.52215	0.52302	0.52438	0.52558	0.52699	0.52831	0.52959	0.53093
190	0.52081	0.52176	0.52298	0.52428	0.52562	0.52697	0.52812	0.52944
200	0.51953	0.52041	0.52171	0.52299	0.52421	0.52559	0.52680	0.52805

continued

continued

P/MPa	T/K 349.80	364.62	379.445	394.216	409.46	424.545	439.605	454.74
	V_{spec}/cm^{3}g^{-1}							
0.1	0.57769	0.58435	0.59123	0.59827	0.60540	0.61294	0.62032	0.62744
10	0.57356	0.57983	0.58628	0.59284	0.59946	0.60644	0.61322	0.61975
20	0.56971	0.57563	0.58170	0.58779	0.59404	0.60059	0.60689	0.61288
30	0.56623	0.57190	0.57767	0.58355	0.58938	0.59549	0.60147	0.60710
40	0.56319	0.56863	0.57411	0.57969	0.58527	0.59106	0.59665	0.60202
50	0.56033	0.56552	0.57083	0.57611	0.58144	0.58707	0.59231	0.59740
60	0.55765	0.56272	0.56776	0.57286	0.57806	0.58335	0.58840	0.59319
70	0.55520	0.56007	0.56495	0.56984	0.57481	0.57995	0.58475	0.58938
80	0.55282	0.55756	0.56230	0.56702	0.57188	0.57678	0.58137	0.58584
90	0.55069	0.55527	0.55981	0.56448	0.56910	0.57384	0.57824	0.58255
100	0.54855	0.55302	0.55745	0.56186	0.56643	0.57104	0.57530	0.57949
110	0.54655	0.55089	0.55522	0.55956	0.56398	0.56849	0.57259	0.57661
120	0.54460	0.54892	0.55313	0.55738	0.56164	0.56598	0.56997	0.57386
130	0.54282	0.54703	0.55108	0.55524	0.55938	0.56359	0.56745	0.57123
140	0.54103	0.54515	0.54915	0.55318	0.55722	0.56137	0.56509	0.56879
150	0.53936	0.54337	0.54727	0.55125	0.55516	0.55922	0.56282	0.56635
160	0.53775	0.54162	0.54547	0.54937	0.55325	0.55710	0.56065	0.56413
170	0.53621	0.53996	0.54374	0.54756	0.55129	0.55517	0.55859	0.56194
180	0.53461	0.53835	0.54207	0.54574	0.54949	0.55313	0.55655	0.55985
190	0.53311	0.53679	0.54046	0.54401	0.54762	0.55131	0.55461	0.55780
200	0.53163	0.53527	0.53886	0.54236	0.54592	0.54949	0.55268	0.55579

continued

P/MPa	T/K 470.07	485.945	501.34	516.63	531.755	547.27	562.18	576.84
	V_{spec}/cm^{3}g^{-1}							
0.1	0.63500	0.64298	0.65111	0.65955	0.66830	0.67810	0.68808	0.69928
10	0.62653	0.63368	0.64087	0.64826	0.65593	0.66419	0.67261	0.68196
20	0.61908	0.62559	0.63210	0.63869	0.64560	0.65275	0.66017	0.66837
30	0.61287	0.61886	0.62481	0.63088	0.63724	0.64374	0.65032	0.65774
40	0.60735	0.61294	0.61853	0.62416	0.63006	0.63601	0.64214	0.64887
50	0.60244	0.60776	0.61295	0.61818	0.62378	0.62926	0.63504	0.64125
60	0.59799	0.60304	0.60795	0.61296	0.61811	0.62328	0.62876	0.63461
70	0.59400	0.59874	0.60344	0.60814	0.61309	0.61796	0.62306	0.62870
80	0.59021	0.59479	0.59932	0.60382	0.60845	0.61315	0.61801	0.62331
90	0.58676	0.59116	0.59549	0.59978	0.60422	0.60870	0.61329	0.61844

continued

continued

100	0.58355	0.58779	0.59187	0.59607	0.60036	0.60465	0.60907	0.61389
110	0.58053	0.58454	0.58854	0.59262	0.59671	0.60086	0.60510	0.60977
120	0.57767	0.58160	0.58544	0.58931	0.59331	0.59728	0.60138	0.60594
130	0.57499	0.57878	0.58251	0.58619	0.59014	0.59393	0.59791	0.60227
140	0.57236	0.57607	0.57975	0.58332	0.58714	0.59076	0.59469	0.59890
150	0.56987	0.57353	0.57705	0.58056	0.58423	0.58786	0.59154	0.59570
160	0.56753	0.57107	0.57451	0.57795	0.58148	0.58494	0.58855	0.59262
170	0.56529	0.56868	0.57212	0.57539	0.57888	0.58221	0.58578	0.58972
180	0.56315	0.56646	0.56974	0.57293	0.57632	0.57966	0.58308	0.58694
190	0.56106	0.56421	0.56752	0.57058	0.57389	0.57713	0.58046	0.58424
200	0.55896	0.56207	0.56531	0.56830	0.57162	0.57463	0.57796	0.58170

Polymer (B): **poly(1-vinyl-2-pyrrolidinone)** **2008SHI**

Characterization: M_w/g.mol^{-1} = 55000, Aldrich Chem. Co., Inc., St. Louis, MO

P/MPa	T/K				
	445.15	455.15	465.15	475.15	485.15
	V_{spec}/cm^{3}g^{-1}				
10	0.7418	0.7445	0.7474	0.7503	0.7533
20	0.7397	0.7423	0.7450	0.7476	0.7506
30	0.7377	0.7400	0.7426	0.7452	0.7479
40	0.7358	0.7382	0.7405	0.7431	0.7455
50	0.7340	0.7364	0.7387	0.7410	0.7434
60	0.7325	0.7346	0.7369	0.7390	0.7413
70	0.7310	0.7330	0.7351	0.7371	0.7395
80	0.7295	0.7315	0.7335	0.7356	0.7377
90	0.7282	0.7300	0.7320	0.7340	0.7361
100	0.7269	0.7287	0.7305	0.7325	0.7343
110	0.7257	0.7274	0.7292	0.7310	0.7330
120	0.7244	0.7262	0.7279	0.7297	0.7315
130	0.7234	0.7249	0.7267	0.7284	0.7302
140	0.7224	0.7239	0.7257	0.7272	0.7290
150	0.7214	0.7229	0.7244	0.7262	0.7277
160	0.7205	0.7219	0.7234	0.7249	0.7267
170	0.7197	0.7209	0.7224	0.7239	0.7254
180	0.7187	0.7202	0.7217	0.7229	0.7244
190	0.7180	0.7195	0.7207	0.7222	0.7234
200	0.7173	0.7185	0.7200	0.7212	0.7227

Polymer (B): **poly(1-vinyl-2-pyrrolidinone)** **2003KIM**
Characterization: –

P/MPa	T/K								
	439.85	449.45	458.95	467.95	476.35	485.95	495.05	504.55	513.35
	V_{spec}/cm^{3}g^{-1}								
0.1	1.0035	1.0088	1.0137	1.0191	1.0239	1.0291	1.0343	1.0396	1.0453
10	0.9989	1.0039	1.0087	1.0137	1.0183	1.0234	1.0283	1.0333	1.0385
20	0.9945	0.9993	1.0038	1.0086	1.0130	1.0178	1.0225	1.0271	1.0319
30	0.9905	0.9952	0.9996	1.0041	1.0083	1.0131	1.0176	1.0220	1.0265
40	0.9868	0.9913	0.9955	1.0000	1.0041	1.0087	1.0130	1.0172	1.0216
50	0.9832	0.9875	0.9918	0.9960	1.0001	1.0046	1.0088	1.0127	1.0169
60	0.9799	0.9840	0.9881	0.9923	0.9963	1.0005	1.0046	1.0085	1.0127
70	0.9768	0.9806	0.9845	1.0000	0.9927	0.9969	1.0009	1.0044	1.0084
80	0.9737	0.9774	0.9812	0.9852	1.0000	0.9931	0.9970	1.0006	1.0050
90	0.9709	0.9743	0.9781	0.9819	0.9857	0.9898	0.9936	0.9970	1.0008
100	0.9682	0.9713	0.9750	0.9788	0.9826	0.9864	0.9901	0.9935	0.9972

5.2. PVT data of polymer solutions

5.2.1. Binary polymer solutions

Polymer (B): **poly(ε-caprolactone)** **2007LIU**
Characterization: M_n/g.mol^{-1} = 6100/g.mol^{-1}, M_w/g.mol^{-1} = 14300, T_m/K = 339,
Scientific Polymer Products, Inc., Ontario, NY
Solvent (A): **2-propanone** **C_3H_6O** **67-64-1**

w_B	0.05	was kept constant							
T/K	323.4	323.5	323.5	323.5	323.5	323.5	323.4	348.9	348.9
P/MPa	6.9	14.0	20.9	27.8	34.8	41.8	48.5	7.3	14.2
ρ/(g/cm^3)	0.8802	0.8893	0.8981	0.9062	0.9140	0.9204	0.9266	0.8495	0.8609
T/K	348.5	348.8	348.9	348.8	348.8	373.3	373.1	373.1	373.1
P/MPa	20.9	27.9	34.8	41.7	48.5	7.1	14.1	21.0	27.9
ρ/(g/cm^3)	0.8704	0.8799	0.8889	0.8965	0.9039	0.8191	0.8324	0.8444	0.8546
T/K	373.2	373.2	373.0	398.3	398.3	398.3	398.2	398.4	398.4
P/MPa	35.1	42.0	48.6	7.5	14.1	21.2	28.0	35.0	41.8
ρ/(g/cm^3)	0.8645	0.8735	0.8817	0.7890	0.8020	0.8162	0.8285	0.8397	0.8491
T/K	398.4								
P/MPa	48.6								
ρ/(g/cm^3)	0.8581								

Polymer (B): **poly(ε-caprolactone)** **2006LI2**
Characterization: M_n/g.mol^{-1} = 6100/g.mol^{-1}, M_w/g.mol^{-1} = 14300, T_m/K = 339,
Scientific Polymer Products, Inc., Ontario, NY
Solvent (A): **2-propanone** **C_3H_6O** **67-64-1**

T/K = 323.15

w_B	0.05	was kept constant			
P/MPa	7.74	14.39	21.52	28.07	35.24
ρ/(g/cm^3)	0.8607	0.8681	0.8764	0.8826	0.8901

T/K = 348.15

w_B	0.05	was kept constant			
P/MPa	7.71	14.39	21.26	28.31	33.99
ρ/(g/cm^3)	0.8490	0.8596	0.8678	0.8754	0.8802

continued

continued

T/K = 373.15

w_B	0.05	was kept constant			
P/MPa	7.63	14.86	21.16	28.13	35.31
ρ/(g/cm^3)	0.8185	0.8300	0.8392	0.8489	0.8581

continued

T/K = 398.15

w_B	0.05	was kept constant			
P/MPa	7.91	14.70	21.61	28.55	34.97
ρ/(g/cm^3)	0.7868	0.8000	0.8132	0.8231	0.8331

Polymer (B): **polyethylene** **2007FU1**

Characterization: M_w/g.mol^{-1} = 4000/g.mol^{-1}, T_m/K = 372.35, Aldrich Chem. Co., Inc., Milwaukee, WI

Solvent (A): **carbon dioxide** $\mathbf{CO_2}$ **124-38-9**

T/K = 453.1

P/MPa	0.0	4.1	5.5	8.0	11.0	12.0	13.0	14.0	15.0
ρ/(g/cm^3)	0.7418	0.7460	0.7444	0.7493	0.7511	0.7505	0.7510	0.7694	0.7717

Comments: Densities are measured at saturated sorption conditions.

T/K = 473.5

P/MPa	0.0	6.1	8.2	9.1	10.1	11.1	12.1	13.1
ρ/(g/cm^3)	0.7304	0.7328	0.7334	0.7343	0.7346	0.7353	0.7363	0.7363

Comments: Densities are measured at saturated sorption conditions.

T/K = 491.8

P/MPa	0.0	4.1	5.0	8.0	9.0	10.0	11.0	12.0	13.0
ρ/(g/cm^3)	0.7208	0.7245	0.7240	0.7241	0.7251	0.7255	0.7262	0.7267	0.7259
P/MPa	14.0	15.1							
ρ/(g/cm^3)	0.7282	0.7283							

Comments: Densities are measured at saturated sorption conditions.

Polymer (B): **poly(ethylene glycol)** **2006LEE**
Characterization: M_n/g.mol^{-1} = 260, M_w/g.mol^{-1} = 280,
Aldrich Chem. Co., Inc., St. Louis, MO
Solvent (A): **acetophenone** **C_8H_8O** **98-86-2**

P/MPa	ρ/(g/cm^3)								
	w_A = 0.0487	0.1045	0.1653	0.2355	0.3161	0.4094	0.5189	0.6490	0.8060
T/K = 298.15									
0.1	1.1146	1.1086	1.1024	1.0953	1.0871	1.0779	1.0674	1.0548	1.0401
10	1.1190	1.1130	1.1069	1.0999	1.0919	1.0827	1.0724	1.0600	1.0456
15	1.1211	1.1151	1.1091	1.1022	1.0942	1.0851	1.0749	1.0625	1.0482
20	1.1233	1.1173	1.1113	1.1043	1.0965	1.0874	1.0772	1.0650	1.0508
25	1.1255	1.1194	1.1134	1.1066	1.0987	1.0897	1.0796	1.0674	1.0532
30	1.1275	1.1214	1.1155	1.1087	1.1009	1.0920	1.0819	1.0698	1.0556
35	1.1292	1.1235	1.1175	1.1108	1.1031	1.0941	1.0842	1.0721	1.0581
40	1.1313	1.1254	1.1196	1.1128	1.1052	1.0962	1.0863	1.0743	1.0604
45	1.1333	1.1274	1.1216	1.1149	1.1072	1.0984	1.0886	1.0766	1.0627
50	1.1352	1.1294	1.1235	1.1169	1.1093	1.1004	1.0907	1.0788	1.0650
T/K = 318.15									
0.1	1.0984	1.0925	1.0861	1.0789	1.0706	1.0612	1.0505	1.0378	1.0229
10	1.1033	1.0973	1.0910	1.0840	1.0759	1.0665	1.0560	1.0436	1.0287
15	1.1057	1.0996	1.0934	1.0865	1.0783	1.0692	1.0587	1.0463	1.0317
20	1.1079	1.1020	1.0958	1.0888	1.0808	1.0717	1.0613	1.0490	1.0344
25	1.1101	1.1042	1.0981	1.0912	1.0832	1.0741	1.0638	1.0516	1.0372
30	1.1124	1.1064	1.1003	1.0935	1.0855	1.0766	1.0663	1.0542	1.0399
35	1.1145	1.1086	1.1025	1.0958	1.0879	1.0789	1.0687	1.0567	1.0425
40	1.1164	1.1107	1.1047	1.0980	1.0901	1.0812	1.0712	1.0592	1.0450
45	1.1187	1.1128	1.1068	1.1002	1.0924	1.0836	1.0735	1.0616	1.0475
50	1.1204	1.1149	1.1089	1.1023	1.0945	1.0858	1.0758	1.0639	1.0499
T/K = 348.15									
0.1	1.0747	1.0683	1.0616	1.0541	1.0457	1.0358	1.0247	1.0118	0.9968
10	1.0800	1.0738	1.0673	1.0599	1.0516	1.0419	1.0310	1.0183	1.0036
15	1.0827	1.0764	1.0700	1.0626	1.0544	1.0447	1.0341	1.0215	1.0069
20	1.0853	1.0790	1.0726	1.0653	1.0573	1.0477	1.0371	1.0246	1.0101
25	1.0878	1.0815	1.0752	1.0679	1.0599	1.0504	1.0399	1.0276	1.0132
30	1.0902	1.0840	1.0777	1.0706	1.0626	1.0532	1.0428	1.0305	1.0162
35	1.0926	1.0864	1.0802	1.0730	1.0652	1.0557	1.0455	1.0333	1.0192
40	1.0949	1.0888	1.0825	1.0755	1.0677	1.0583	1.0482	1.0360	1.0220
45	1.0972	1.0911	1.0849	1.0779	1.0702	1.0609	1.0508	1.0387	1.0248
50	1.0995	1.0934	1.0873	1.0803	1.0727	1.0634	1.0533	1.0414	1.0275

Polymer (B): **poly(ethylene glycol) monomethyl ether** **2006LEE**
Characterization: M_n/g.mol^{-1} = 366, M_w/g.mol^{-1} = 373,
Aldrich Chem. Co., Inc., St. Louis, MO
Solvent (A): **acetophenone** **C_8H_8O** **98-86-2**

P/MPa	*ρ*/(g/cm³)								
	w_A = 0.0352	0.0765	0.1234	0.1794	0.2472	0.3302	0.4339	0.5676	0.7468
T/K = 298.15									
0.1	1.0817	1.0793	1.0767	1.0732	1.0692	1.0641	1.0578	1.0495	1.0386
10	1.0867	1.0843	1.0817	1.0783	1.0742	1.0692	1.0630	1.0548	1.0441
15	1.0893	1.0867	1.0842	1.0808	1.0768	1.0718	1.0656	1.0573	1.0467
20	1.0916	1.0891	1.0865	1.0832	1.0792	1.0742	1.0680	1.0599	1.0493
25	1.0940	1.0915	1.0889	1.0856	1.0816	1.0766	1.0705	1.0623	1.0518
30	1.0963	1.0938	1.0912	1.0879	1.0839	1.0790	1.0728	1.0648	1.0542
35	1.0986	1.0960	1.0934	1.0902	1.0862	1.0812	1.0751	1.0671	1.0566
40	1.1007	1.0982	1.0956	1.0924	1.0884	1.0835	1.0775	1.0694	1.0590
45	1.1029	1.1004	1.0978	1.0946	1.0907	1.0857	1.0797	1.0717	1.0613
50	1.1051	1.1026	1.1000	1.0967	1.0928	1.0879	1.0819	1.0739	1.0636
T/K = 318.15									
0.1	1.0647	1.0620	1.0594	1.0560	1.0520	1.0469	1.0405	1.0322	1.0213
10	1.0702	1.0677	1.0650	1.0616	1.0576	1.0526	1.0463	1.0381	1.0272
15	1.0729	1.0704	1.0677	1.0644	1.0603	1.0554	1.0491	1.0409	1.0301
20	1.0754	1.0729	1.0703	1.0670	1.0630	1.0580	1.0518	1.0437	1.0329
25	1.0781	1.0755	1.0729	1.0696	1.0656	1.0607	1.0545	1.0464	1.0357
30	1.0805	1.0780	1.0753	1.0721	1.0681	1.0632	1.0570	1.0490	1.0384
35	1.0829	1.0804	1.0778	1.0745	1.0706	1.0657	1.0595	1.0515	1.0410
40	1.0853	1.0828	1.0802	1.0769	1.0730	1.0681	1.0620	1.0540	1.0435
45	1.0876	1.0851	1.0825	1.0793	1.0754	1.0705	1.0644	1.0565	1.0460
50	1.0899	1.0874	1.0848	1.0816	1.0777	1.0728	1.0668	1.0588	1.0484
T/K = 348.15									
0.1	1.0389	1.0366	1.0337	1.0304	1.0263	1.0210	1.0147	1.0062	0.9950
10	1.0453	1.0428	1.0401	1.0367	1.0326	1.0275	1.0213	1.0129	1.0019
15	1.0483	1.0459	1.0432	1.0398	1.0357	1.0306	1.0244	1.0161	1.0053
20	1.0513	1.0489	1.0462	1.0429	1.0387	1.0337	1.0275	1.0193	1.0085
25	1.0541	1.0517	1.0491	1.0457	1.0417	1.0366	1.0305	1.0223	1.0116
30	1.0569	1.0545	1.0519	1.0486	1.0445	1.0395	1.0334	1.0253	1.0146
35	1.0596	1.0572	1.0546	1.0514	1.0473	1.0423	1.0362	1.0282	1.0176
40	1.0623	1.0600	1.0573	1.0540	1.0500	1.0450	1.0390	1.0310	1.0205
45	1.0649	1.0625	1.0600	1.0567	1.0526	1.0477	1.0417	1.0337	1.0232
50	1.0673	1.0650	1.0624	1.0591	1.0552	1.0503	1.0442	1.0363	1.0259

Polymer (B): **poly(ethylene glycol) mono-4-octylphenyl ether** **2009LEE**
Characterization: M_n/g.mol^{-1} = 576, M_w/g.mol^{-1} = 606,
Tokyo Kasei Organic Chemicals, Japan
Solvent (A): **acetophenone** **C_8H_8O** **98-86-2**

P/MPa	ρ/(g/cm^3)								
	w_A = 0.0217	0.0448	0.0737	0.0980	0.1597	0.2181	0.3032	0.4075	0.6258
T/K = 298.15									
0.1	1.0630	1.0624	1.0613	1.0600	1.0581	1.0559	1.0527	1.0484	1.0395
10	1.0678	1.0672	1.0662	1.0649	1.0631	1.0610	1.0578	1.0536	1.0447
15	1.0701	1.0695	1.0686	1.0672	1.0654	1.0633	1.0602	1.0561	1.0472
20	1.0724	1.0718	1.0709	1.0695	1.0678	1.0657	1.0626	1.0585	1.0497
25	1.0747	1.0741	1.0731	1.0718	1.0700	1.0680	1.0650	1.0608	1.0522
30	1.0768	1.0763	1.0753	1.0740	1.0722	1.0702	1.0673	1.0631	1.0545
35	1.0790	1.0785	1.0775	1.0762	1.0744	1.0725	1.0696	1.0654	1.0570
40	1.0811	1.0805	1.0796	1.0783	1.0765	1.0746	1.0717	1.0676	1.0592
45	1.0832	1.0826	1.0817	1.0804	1.0786	1.0768	1.0739	1.0698	1.0614
50	1.0853	1.0848	1.0838	1.0825	1.0807	1.0789	1.0761	1.0719	1.0637
T/K = 318.15									
0.1	1.0468	1.0460	1.0450	1.0435	1.0417	1.0394	1.0360	1.0316	1.0224
10	1.0521	1.0513	1.0503	1.0489	1.0471	1.0449	1.0415	1.0372	1.0282
15	1.0547	1.0540	1.0530	1.0516	1.0498	1.0477	1.0443	1.0400	1.0311
20	1.0572	1.0566	1.0556	1.0542	1.0524	1.0502	1.0469	1.0427	1.0339
25	1.0596	1.0589	1.0580	1.0566	1.0548	1.0527	1.0494	1.0452	1.0365
30	1.0620	1.0614	1.0604	1.0591	1.0573	1.0551	1.0519	1.0477	1.0391
35	1.0644	1.0638	1.0628	1.0615	1.0597	1.0576	1.0544	1.0502	1.0417
40	1.0666	1.0660	1.0650	1.0637	1.0619	1.0598	1.0567	1.0525	1.0440
45	1.0689	1.0682	1.0672	1.0659	1.0642	1.0622	1.0590	1.0548	1.0465
50	1.0711	1.0708	1.0698	1.0683	1.0666	1.0645	1.0615	1.0571	1.0490
T/K = 348.15									
0.1	1.0238	1.0231	1.0218	1.0202	1.0181	1.0158	1.0120	1.0073	0.9973
10	1.0295	1.0290	1.0278	1.0263	1.0242	1.0220	1.0182	1.0136	1.0039
15	1.0324	1.0318	1.0306	1.0291	1.0270	1.0249	1.0212	1.0167	1.0071
20	1.0353	1.0346	1.0334	1.0319	1.0299	1.0278	1.0241	1.0197	1.0102
25	1.0380	1.0373	1.0362	1.0347	1.0327	1.0306	1.0270	1.0226	1.0132
30	1.0408	1.0401	1.0390	1.0375	1.0355	1.0334	1.0299	1.0255	1.0162
35	1.0433	1.0426	1.0415	1.0400	1.0380	1.0360	1.0325	1.0282	1.0190
40	1.0458	1.0451	1.0441	1.0426	1.0407	1.0387	1.0352	1.0309	1.0219
45	1.0482	1.0477	1.0466	1.0451	1.0433	1.0412	1.0378	1.0336	1.0246
50	1.0508	1.0501	1.0491	1.0476	1.0456	1.0437	1.0403	1.0362	1.0272

Polymer (B): **poly(ethylene glycol) mono-4-octylphenyl ether** **2009LEE**
Characterization: M_n/g.mol^{-1} = 576, M_w/g.mol^{-1} = 606,
Tokyo Kasei Organic Chemicals, Japan
Solvent (A): **1-octanol** **$C_8H_{18}O$** **111-87-5**

P/MPa	ρ/(g/cm^3)								
	w_A = 0.0261	0.0497	0.0895	0.1319	0.1846	0.2523	0.3458	0.4750	0.6710
T/K = 298.15									
0.1	1.0553	1.0485	1.0371	1.0255	1.0108	0.9924	0.9676	0.9350	0.8898
10	1.0602	1.0535	1.0422	1.0307	1.0161	0.9978	0.9732	0.9408	0.8959
15	1.0626	1.0558	1.0445	1.0331	1.0186	1.0003	0.9758	0.9435	0.8987
20	1.0650	1.0582	1.0470	1.0355	1.0211	1.0029	0.9784	0.9462	0.9015
25	1.0673	1.0604	1.0493	1.0379	1.0235	1.0053	0.9809	0.9488	0.9042
30	1.0695	1.0627	1.0515	1.0402	1.0258	1.0076	0.9833	0.9513	0.9067
35	1.0717	1.0649	1.0538	1.0425	1.0282	1.0099	0.9857	0.9538	0.9093
40	1.0738	1.0670	1.0559	1.0445	1.0303	1.0121	0.9877	0.9559	0.9115
45	1.0759	1.0691	1.0580	1.0467	1.0324	1.0143	0.9899	0.9582	0.9138
50	1.0781	1.0713	1.0601	1.0488	1.0345	1.0164	0.9920	0.9604	0.9160
T/K = 318.15									
0.1	1.0392	1.0323	1.0210	1.0094	0.9949	0.9765	0.9519	0.9195	0.8748
10	1.0446	1.0377	1.0266	1.0151	1.0008	0.9823	0.9580	0.9260	0.8815
15	1.0473	1.0405	1.0294	1.0179	1.0037	0.9852	0.9610	0.9291	0.8847
20	1.0500	1.0431	1.0320	1.0205	1.0064	0.9880	0.9638	0.9321	0.8877
25	1.0524	1.0455	1.0345	1.0230	1.0090	0.9906	0.9665	0.9349	0.8906
30	1.0548	1.0480	1.0369	1.0255	1.0116	0.9932	0.9692	0.9376	0.8934
35	1.0573	1.0505	1.0394	1.0280	1.0141	0.9958	0.9718	0.9402	0.8962
40	1.0595	1.0527	1.0416	1.0302	1.0164	0.9982	0.9742	0.9427	0.8988
45	1.0618	1.0550	1.0441	1.0325	1.0188	1.0005	0.9766	0.9453	0.9014
50	1.0643	1.0575	1.0466	1.0349	1.0213	1.0031	0.9791	0.9478	0.9039
T/K = 348.15									
0.1	1.0163	1.0093	0.9980	0.9860	0.9718	0.9532	0.9288	0.8966	0.8518
10	1.0223	1.0154	1.0041	0.9924	0.9783	0.9599	0.9358	0.9038	0.8594
15	1.0251	1.0183	1.0070	0.9953	0.9813	0.9630	0.9390	0.9070	0.8628
20	1.0280	1.0212	1.0099	0.9983	0.9844	0.9661	0.9421	0.9104	0.8662
25	1.0308	1.0240	1.0128	1.0012	0.9874	0.9692	0.9452	0.9136	0.8696
30	1.0336	1.0268	1.0156	1.0041	0.9903	0.9721	0.9483	0.9167	0.8729
35	1.0362	1.0294	1.0182	1.0067	0.9930	0.9749	0.9511	0.9197	0.8759
40	1.0388	1.0320	1.0208	1.0094	0.9956	0.9776	0.9539	0.9225	0.8788
45	1.0414	1.0346	1.0234	1.0120	0.9983	0.9802	0.9566	0.9253	0.8816
50	1.0439	1.0372	1.0259	1.0145	1.0008	0.9827	0.9591	0.9280	0.8844

Polymer (B):	**polyimide**	**2012HES**
Characterization:	M_w/g.mol^{-1} = 26000, M_w/M_n = 2.3, T_g/K = 306, Polyimide P84, HP Polymer GmbH, Lenzing, Austria	
Solvent (A):	**dimethylsulfoxide** C_2H_6OS	**67-68-5**

T/K = 303.15 w_B = 0.025 was kept constant

P/bar	21	41	60	80	101	120	140	161	181
ρ/(g/cm^3)	1.098	1.099	1.100	1.102	1.103	1.104	1.105	1.106	1.107
P/bar	201								
ρ/(g/cm^3)	1.108								

T/K = 323.15 w_B = 0.025 was kept constant

P/bar	20	40	60	80	100	120	140	161	180
ρ/(g/cm^3)	1.077	1.079	1.080	1.081	1.083	1.084	1.085	1.087	1.088
P/bar	200								
ρ/(g/cm^3)	1.089								

T/K = 343.15 w_B = 0.025 was kept constant

P/bar	21	41	60	80	101	121	140	161	180
ρ/(g/cm^3)	1.058	1.059	1.061	1.062	1.064	1.065	1.067	1.068	1.069
P/bar	200								
ρ/(g/cm^3)	1.071								

Polymer (B):	**polyimide**	**2012HES**
Characterization:	M_w/g.mol^{-1} = 48000, M_w/M_n = 2.2, T_g/K = 311, Matrimid 5218, Huntsmann, Switzerland	
Solvent (A):	**dimethylsulfoxide** C_2H_6OS	**67-68-5**

T/K = 303.15 w_B = 0.025 was kept constant

P/bar	21	40	60	80	100	121	140	161	180
ρ/(g/cm^3)	1.096	1.098	1.099	1.100	1.101	1.102	1.103	1.104	1.106
P/bar	200								
ρ/(g/cm^3)	1.107								

T/K = 323.15 w_B = 0.025 was kept constant

P/bar	21	41	60	80	101	121	140	160	180
ρ/(g/cm^3)	1.076	1.078	1.079	1.080	1.082	1.083	1.084	1.085	1.087
P/bar	200								
ρ/(g/cm^3)	1.088								

T/K = 343.15 w_B = 0.025 was kept constant

P/bar	21	41	61	80	100	121	141	161	181
ρ/(g/cm^3)	1.054	1.055	1.057	1.058	1.060	1.061	1.062	1.064	1.064
P/bar	200								
ρ/(g/cm^3)	1.066								

Polymer (B): **poly(methyl methacrylate)** **2006LI1**
Characterization: M_n/g.mol^{-1} = 8300/g.mol^{-1}, M_w/g.mol^{-1} = 15000,
Scientific Polymer Products, Inc., Ontario, NY
Solvent (A): **2-propanone** **C_3H_6O** **67-64-1**

T/K = 323.15

w_B	0.10	0.10	0.10	0.10	0.10	0.20	0.20	0.20	0.20
P/MPa	7.67	14.38	21.32	28.88	35.26	9.05	14.63	21.23	28.80
ρ/(g/cm^3)	0.8343	0.8428	0.8498	0.8566	0.8600	0.8504	0.8546	0.8605	0.8675

w_B	0.20
P/MPa	35.15
ρ/(g/cm^3)	0.8722

T/K = 348.15

w_B	0.10	0.10	0.10	0.10	0.10	0.20	0.20	0.20	0.20
P/MPa	8.08	14.41	21.55	28.94	35.58	7.33	15.14	21.93	28.28
ρ/(g/cm^3)	0.8098	0.8196	0.8295	0.8373	0.8449	0.8245	0.8325	0.8397	0.8461

w_B	0.20
P/MPa	35.63
ρ/(g/cm^3)	0.8534

T/K = 373.15

w_B	0.10	0.10	0.10	0.10	0.10	0.20	0.20	0.20	0.20
P/MPa	7.18	15.01	22.21	28.64	35.97	7.78	14.45	21.40	28.37
ρ/(g/cm^3)	0.7806	0.7918	0.8025	0.8126	0.8212	0.8002	0.8085	0.8172	0.8254

w_B	0.20
P/MPa	35.12
ρ/(g/cm^3)	0.8327

T/K = 398.15

w_B	0.10	0.10	0.10	0.10	0.10	0.20	0.20	0.20	0.20
P/MPa	7.64	14.57	22.28	28.25	33.95	10.33	13.43	21.56	28.38
ρ/(g/cm^3)	0.7575	0.7671	0.7805	0.7911	0.8015	0.7777	0.7842	0.7951	0.8041

w_B	0.20
P/MPa	35.48
ρ/(g/cm^3)	0.8129

Polymer (B): **poly(methyl methacrylate)** **2006LI1**
Characterization: M_n/g.mol^{-1} = 19300/g.mol^{-1}, M_w/g.mol^{-1} = 540000, Scientific Polymer Products, Inc., Ontario, NY
Solvent (A): **2-propanone** **C_3H_6O** **67-64-1**

T/K = 323.15

w_B	0.02	0.02	0.02	0.02	0.02	0.05	0.05	0.05	0.05
P/MPa	8.24	14.51	20.92	28.48	34.72	7.94	14.43	21.25	28.49
ρ/(g/cm^3)	0.8713	0.8805	0.8892	0.8968	0.9029	0.8145	0.8221	0.8306	0.8372

w_B	0.05	0.10	0.10	0.10	0.10	0.10	0.20	0.20	0.20
P/MPa	35.31	7.43	14.30	21.33	27.98	34.72	8.22	14.10	21.13
ρ/(g/cm^3)	0.8444	0.8411	0.8498	0.8575	0.8656	0.8726	0.8171	0.8219	0.8229

w_B	0.20	0.20
P/MPa	27.41	34.38
ρ/(g/cm^3)	0.8236	0.8238

T/K = 348.15

w_B	0.02	0.02	0.02	0.02	0.02	0.05	0.05	0.05	0.05
P/MPa	7.91	14.55	21.54	28.41	35.58	8.34	14.80	21.37	28.36
ρ/(g/cm^3)	0.8383	0.8485	0.8578	0.8693	0.8762	0.7871	0.7976	0.8065	0.8146

w_B	0.05	0.10	0.10	0.10	0.10	0.10	0.20	0.20	0.20
P/MPa	35.38	7.79	14.22	21.33	28.12	35.13	7.35	14.27	21.07
ρ/(g/cm^3)	0.8222	0.8157	0.8248	0.8333	0.8415	0.8495	0.8098	0.8180	0.8240

w_B	0.20	0.20
P/MPa	28.17	34.94
ρ/(g/cm^3)	0.8268	0.8275

T/K = 373.15

w_B	0.02	0.02	0.02	0.02	0.02	0.05	0.05	0.05	0.05
P/MPa	7.72	14.75	21.58	28.18	35.99	7.92	14.56	21.63	27.63
ρ/(g/cm^3)	0.8011	0.8133	0.8250	0.8357	0.8460	0.7583	0.7692	0.7802	0.7903

w_B	0.05	0.10	0.10	0.10	0.10	0.10	0.20	0.20	0.20
P/MPa	35.36	7.88	14.64	20.87	28.40	35.11	7.90	13.83	21.53
ρ/(g/cm^3)	0.7991	0.7836	0.7950	0.8056	0.8167	0.8266	0.7881	0.7968	0.8072

T/K = 398.15

w_B	0.02	0.02	0.02	0.02	0.02	0.05	0.05	0.05	0.05
P/MPa	7.95	14.33	21.53	28.35	34.66	7.21	14.64	20.35	26.67
ρ/(g/cm^3)	0.7598	0.7760	0.7900	0.8027	0.8139	0.7279	0.7437	0.7545	0.7663

w_B	0.05	0.10	0.10	0.10	0.10	0.10	0.20	0.20	0.20
P/MPa	33.95	11.09	13.36	21.40	28.32	34.85	7.13	14.34	21.03
ρ/(g/cm^3)	0.7770	0.7566	0.7673	0.7800	0.7920	0.8021	0.7577	0.7714	0.7822

5.2.2. Ternary polymer solutions

Polymer (B): **poly(ε-caprolactone)** **2007LIU**
Characterization: M_n/g.mol^{-1} = 6100/g.mol^{-1}, M_w/g.mol^{-1} = 14300, T_m/K = 339, Scientific Polymer Products, Inc., Ontario, NY
Solvent (A): **2-propanone** **C_3H_6O** **67-64-1**
Solvent (C): **carbon dioxide** **CO_2** **124-38-9**

w_A	0.59	0.59	0.59	0.59	0.59	0.59	0.59	0.59	0.59
w_B	0.01	0.01	0.01	0.01	0.01	0.01	0.01	0.01	0.01
w_C	0.40	0.40	0.40	0.40	0.40	0.40	0.40	0.40	0.40
T/K	323.1	323.1	323.2	323.2	323.2	323.2	323.2	348.1	348.1
P/MPa	7.4	14.4	21.1	28.1	34.8	41.7	48.8	7.3	13.9
ρ/(g/cm^3)	0.8959	0.9154	0.9311	0.9445	0.9505	0.9509	0.9513	0.7312	0.8597
w_A	0.59	0.59	0.59	0.59	0.59	0.59	0.59	0.59	0.59
w_B	0.01	0.01	0.01	0.01	0.01	0.01	0.01	0.01	0.01
w_C	0.40	0.40	0.40	0.40	0.40	0.40	0.40	0.40	0.40
T/K	348.1	348.1	348.2	348.2	348.1	373.8	373.8	374.0	373.9
P/MPa	21.2	28.3	35.1	42.0	48.9	8.8	14.1	21.3	28.2
ρ/(g/cm^3)	0.8865	0.9050	0.9206	0.9352	0.9468	0.7779	0.8084	0.8376	0.8605
w_A	0.59	0.59	0.59	0.59	0.59	0.59	0.59	0.59	0.59
w_B	0.01	0.01	0.01	0.01	0.01	0.01	0.01	0.01	0.01
w_C	0.40	0.40	0.40	0.40	0.40	0.40	0.40	0.40	0.40
T/K	373.9	373.9	374.0	398.2	398.5	398.5	398.5	398.5	398.6
P/MPa	34.8	41.9	48.7	8.3	13.9	21.0	28.1	35.0	42.1
ρ/(g/cm^3)	0.8801	0.8974	0.9127	0.5673	0.7279	0.7836	0.8172	0.8406	0.8613
w_A	0.59	0.90	0.90	0.90	0.90	0.90	0.90	0.90	0.90
w_B	0.01	0.05	0.05	0.05	0.05	0.05	0.05	0.05	0.05
w_C	0.40	0.05	0.05	0.05	0.05	0.05	0.05	0.05	0.05
T/K	398.6	325.0	325.0	324.8	324.8	324.7	324.7	324.7	348.2
P/MPa	48.8	7.2	14.0	21.0	27.8	34.8	41.7	48.5	7.6
ρ/(g/cm^3)	0.8796	0.8929	0.9023	0.9113	0.9188	0.9243	0.9379	0.9379	0.8586
w_A	0.90	0.90	0.90	0.90	0.90	0.90	0.90	0.90	0.90
w_B	0.05	0.05	0.05	0.05	0.05	0.05	0.05	0.05	0.05
w_C	0.05	0.05	0.05	0.05	0.05	0.05	0.05	0.05	0.05
T/K	348.2	348.2	348.2	348.1	348.1	348.1	373.4	373.3	373.3
P/MPa	14.1	21.0	27.8	34.8	41.9	48.7	7.2	14.1	21.0
ρ/(g/cm^3)	0.8694	0.8790	0.8878	0.8961	0.9040	0.9109	0.8252	0.8389	0.8502

continued

continued

w_A	0.90	0.90	0.90	0.90	0.90	0.90	0.90	0.90	0.90
w_B	0.05	0.05	0.05	0.05	0.05	0.05	0.05	0.05	0.05
w_C	0.05	0.05	0.05	0.05	0.05	0.05	0.05	0.05	0.05
T/K	373.2	373.2	373.2	373.2	398.7	398.7	398.8	398.8	398.7
P/MPa	28.0	34.8	41.8	48.5	7.4	14.4	21.3	28.1	35.2
ρ/(g/cm^3)	0.8616	0.8709	0.8796	0.8877	0.7906	0.8075	0.8217	0.8332	0.8452
w_A	0.90	0.90	0.85	0.85	0.85	0.85	0.85	0.85	0.85
w_B	0.05	0.05	0.05	0.05	0.05	0.05	0.05	0.05	0.05
w_C	0.05	0.05	0.10	0.10	0.10	0.10	0.10	0.10	0.10
T/K	398.7	398.6	324.2	324.1	324.2	324.2	324.2	324.3	324.3
P/MPa	41.7	48.6	7.2	14.3	21.2	27.9	34.8	41.7	48.6
ρ/(g/cm^3)	0.8546	0.8636	0.8969	0.9076	0.9170	0.9261	0.9337	0.9410	0.9485
w_A	0.85	0.85	0.85	0.85	0.85	0.85	0.85	0.85	0.85
w_B	0.05	0.05	0.05	0.05	0.05	0.05	0.05	0.05	0.05
w_C	0.10	0.10	0.10	0.10	0.10	0.10	0.10	0.10	0.10
T/K	348.6	348.6	348.5	348.6	348.6	348.7	348.8	372.9	372.7
P/MPa	7.1	14.3	21.2	28.3	34.8	41.9	48.6	7.0	14.6
ρ/(g/cm^3)	0.8651	0.8781	0.8894	0.9009	0.9080	0.9169	0.9249	0.8317	0.8476
w_A	0.85	0.85	0.85	0.85	0.85	0.85	0.85	0.85	0.85
w_B	0.05	0.05	0.05	0.05	0.05	0.05	0.05	0.05	0.05
w_C	0.10	0.10	0.10	0.10	0.10	0.10	0.10	0.10	0.10
T/K	372.7	372.8	372.8	372.8	372.8	399.3	399.5	399.4	399.5
P/MPa	21.3	28.1	34.9	41.9	48.8	7.3	14.4	21.3	28.2
ρ/(g/cm^3)	0.8611	0.8728	0.8831	0.8930	0.9020	0.7927	0.8114	0.8275	0.8417
w_A	0.85	0.85	0.85	0.75	0.75	0.75	0.75	0.75	0.75
w_B	0.05	0.05	0.05	0.05	0.05	0.05	0.05	0.05	0.05
w_C	0.10	0.10	0.10	0.20	0.20	0.20	0.20	0.20	0.20
T/K	399.5	399.3	399.2	322.8	322.8	322.8	322.7	322.8	322.8
P/MPa	35.1	41.8	48.7	7.2	14.2	21.0	28.0	34.9	41.8
ρ/(g/cm^3)	0.8535	0.8646	0.8760	0.9254	0.9315	0.9377	0.9420	0.9438	0.9520
w_A	0.75	0.75	0.75	0.75	0.75	0.75	0.75	0.75	0.75
w_B	0.05	0.05	0.05	0.05	0.05	0.05	0.05	0.05	0.05
w_C	0.20	0.20	0.20	0.20	0.20	0.20	0.20	0.20	0.20
T/K	322.7	348.5	348.5	348.4	348.4	348.3	348.5	348.6	373.9
P/MPa	48.7	7.4	14.2	21.1	28.1	34.9	42.1	48.7	7.3
ρ/(g/cm^3)	0.9649	0.8923	0.9020	0.9144	0.9247	0.9337	0.9418	0.9494	0.8486
w_A	0.75	0.75	0.75	0.75	0.75	0.75	0.75	0.75	0.75
w_B	0.05	0.05	0.05	0.05	0.05	0.05	0.05	0.05	0.05
w_C	0.20	0.20	0.20	0.20	0.20	0.20	0.20	0.20	0.20
T/K	373.8	373.8	373.8	373.8	373.8	373.7	398.3	398.4	398.4
P/MPa	14.4	21.1	28.0	35.0	41.8	48.8	7.0	14.1	21.2
ρ/(g/cm^3)	0.8642	0.8784	0.8921	0.9034	0.9139	0.9235	0.8036	0.8265	0.8453

continued

continued

w_A	0.75	0.75	0.75	0.75	0.55	0.55	0.55	0.55	0.55
w_B	0.05	0.05	0.05	0.05	0.05	0.05	0.05	0.05	0.05
w_C	0.20	0.20	0.20	0.20	0.40	0.40	0.40	0.40	0.40
T/K	398.5	398.5	398.6	398.7	323.0	323.2	323.1	323.0	322.8
P/MPa	28.1	35.1	41.9	48.8	7.2	14.1	21.1	28.1	34.8
ρ/(g/cm^3)	0.8608	0.8744	0.8860	0.8974	0.9203	0.9383	0.9555	0.9705	0.9838
w_A	0.55	0.55	0.55	0.55	0.55	0.55	0.55	0.55	0.55
w_B	0.05	0.05	0.05	0.05	0.05	0.05	0.05	0.05	0.05
w_C	0.40	0.40	0.40	0.40	0.40	0.40	0.40	0.40	0.40
T/K	322.7	322.7	348.0	348.1	347.8	347.8	347.7	347.8	347.7
P/MPa	42.3	48.7	7.0	14.2	21.0	28.1	35.1	41.8	48.8
ρ/(g/cm^3)	0.9963	1.0046	0.8276	0.8600	0.9062	0.9248	0.9412	0.9556	0.9685
w_A	0.55	0.55	0.55	0.55	0.55	0.55	0.55	0.55	0.55
w_B	0.05	0.05	0.05	0.05	0.05	0.05	0.05	0.05	0.05
w_C	0.40	0.40	0.40	0.40	0.40	0.40	0.40	0.40	0.40
T/K	372.0	372.5	372.4	372.5	372.3	372.4	372.5	399.2	398.8
P/MPa	7.5	14.6	21.0	28.0	35.3	41.9	47.6	8.7	14.3
ρ/(g/cm^3)	0.7059	0.8018	0.8495	0.8780	0.9009	0.9180	0.9334	0.5713	0.7369
w_A	0.55	0.55	0.55	0.55	0.55	0.45	0.45	0.45	0.45
w_B	0.05	0.05	0.05	0.05	0.05	0.05	0.05	0.05	0.05
w_C	0.40	0.40	0.40	0.40	0.40	0.50	0.50	0.50	0.50
T/K	398.5	398.2	398.1	397.9	398.2	323.0	322.9	322.7	322.5
P/MPa	21.1	27.8	35.2	42.1	49.4	7.5	14.1	21.4	27.8
ρ/(g/cm^3)	0.7963	0.8304	0.8564	0.8772	0.8958	0.8365	0.9372	0.9578	0.9745
w_A	0.45	0.45	0.45	0.45	0.45	0.45	0.45	0.45	0.45
w_B	0.05	0.05	0.05	0.05	0.05	0.05	0.05	0.05	0.05
w_C	0.50	0.50	0.50	0.50	0.50	0.50	0.50	0.50	0.50
T/K	322.7	322.9	323.0	348.8	348.8	348.9	349.0	348.9	348.8
P/MPa	34.7	41.8	48.8	7.6	13.9	21.2	28.0	34.8	41.7
ρ/(g/cm^3)	0.9895	1.0041	1.0157	0.6692	0.8168	0.8976	0.9210	0.9402	0.9565
w_A	0.45	0.45	0.45	0.45	0.45	0.45	0.45	0.45	0.45
w_B	0.05	0.05	0.05	0.05	0.05	0.05	0.05	0.05	0.05
w_C	0.50	0.50	0.50	0.50	0.50	0.50	0.50	0.50	0.50
T/K	348.7	373.6	373.8	373.8	373.9	374.0	374.1	374.4	398.6
P/MPa	48.5	8.5	14.6	21.2	28.0	35.1	42.1	48.9	9.7
ρ/(g/cm^3)	0.9720	0.7591	0.8126	0.8471	0.8746	0.8971	0.9166	0.9329	0.6158
w_A	0.45	0.45	0.45	0.45	0.45	0.45	0.35	0.35	0.35
w_B	0.05	0.05	0.05	0.05	0.05	0.05	0.05	0.05	0.05
w_C	0.50	0.50	0.50	0.50	0.50	0.50	0.60	0.60	0.60
T/K	399.2	398.9	398.9	398.9	399.1	398.7	323.8	323.7	323.8
P/MPa	14.5	21.2	28.1	35.1	41.8	48.7	7.3	14.3	21.3
ρ/(g/cm^3)	0.7356	0.7863	0.8233	0.8510	0.8736	0.8925	0.8777	0.9347	0.9587

continued

continued

w_A	0.35	0.35	0.35	0.35	0.35	0.35	0.35	0.35	0.35
w_B	0.05	0.05	0.05	0.05	0.05	0.05	0.05	0.05	0.05
w_C	0.60	0.60	0.60	0.60	0.60	0.60	0.60	0.60	0.60
T/K	323.9	323.8	323.9	323.8	347.9	347.8	347.5	347.4	347.5
P/MPa	28.1	35.0	41.9	48.9	8.4	13.8	20.9	27.9	34.7
ρ/(g/cm^3)	0.9861	1.0035	1.0177	1.0352	0.6493	0.7506	0.8952	0.9268	0.9520

w_A	0.35	0.35	0.35	0.35	0.35	0.35	0.35	0.35	0.35
w_B	0.05	0.05	0.05	0.05	0.05	0.05	0.05	0.05	0.05
w_C	0.60	0.60	0.60	0.60	0.60	0.60	0.60	0.60	0.60
T/K	347.5	347.5	372.6	372.6	372.7	372.9	372.9	372.9	373.0
P/MPa	41.8	48.6	10.5	14.4	21.5	28.2	35.1	41.8	48.7
ρ/(g/cm^3)	0.9743	0.9925	0.5899	0.7455	0.8287	0.8716	0.9044	0.9290	0.9493

w_A	0.35	0.35	0.35	0.35	0.35	0.35	0.50	0.50	0.50
w_B	0.05	0.05	0.05	0.05	0.05	0.05	0.10	0.10	0.10
w_C	0.60	0.60	0.60	0.60	0.60	0.60	0.40	0.40	0.40
T/K	398.1	398.8	399.0	398.9	398.7	398.7	323.5	323.5	323.8
P/MPa	13.9	21.7	28.3	35.1	41.9	48.8	7.6	14.0	20.9
ρ/(g/cm^3)	0.5904	0.7445	0.8055	0.8481	0.8811	0.9075	1.0080	1.0306	1.0457

w_A	0.50	0.50	0.50	0.50	0.50	0.50	0.50	0.50	0.50
w_B	0.10	0.10	0.10	0.10	0.10	0.10	0.10	0.10	0.10
w_C	0.40	0.40	0.40	0.40	0.40	0.40	0.40	0.40	0.40
T/K	323.8	323.6	323.5	323.5	348.0	348.4	348.6	348.6	348.7
P/MPa	27.8	34.7	41.6	48.8	7.2	14.7	20.8	27.8	34.9
ρ/(g/cm^3)	1.0635	1.0796	1.0866	1.0866	0.8521	0.9716	0.9932	1.0139	1.0340

w_A	0.50	0.50	0.50	0.50	0.50	0.50	0.50	0.50	0.50
w_B	0.10	0.10	0.10	0.10	0.10	0.10	0.10	0.10	0.10
w_C	0.40	0.40	0.40	0.40	0.40	0.40	0.40	0.40	0.40
T/K	348.6	348.6	372.8	373.3	373.4	373.5	373.7	374.1	373.3
P/MPa	42.0	48.9	7.3	14.2	21.1	28.5	35.1	42.1	48.8
ρ/(g/cm^3)	1.0528	1.0689	0.8128	0.9076	0.9422	0.9700	0.9871	1.0046	1.0208

w_A	0.50	0.50	0.50	0.50	0.50	0.50	0.50	0.40	0.40
w_B	0.10	0.10	0.10	0.10	0.10	0.10	0.10	0.20	0.20
w_C	0.40	0.40	0.40	0.40	0.40	0.40	0.40	0.40	0.40
T/K	397.8	398.1	398.0	397.8	398.1	398.0	398.1	323.0	322.9
P/MPa	8.5	14.4	20.7	27.6	35.1	41.8	49.0	7.0	14.3
ρ/(g/cm^3)	0.7017	0.8292	0.8849	0.9190	0.9464	0.9648	0.9819	1.1453	1.1679

w_A	0.40	0.40	0.40	0.40	0.40	0.40	0.40	0.40	0.40
w_B	0.20	0.20	0.20	0.20	0.20	0.20	0.20	0.20	0.20
w_C	0.40	0.40	0.40	0.40	0.40	0.40	0.40	0.40	0.40
T/K	322.8	322.9	322.9	322.9	322.8	349.3	349.2	349.2	349.3
P/MPa	21.1	28.0	34.9	41.7	48.3	7.5	13.9	21.1	28.1
ρ/(g/cm^3)	1.1857	1.2038	1.2227	1.2357	1.2488	0.9851	1.0893	1.1204	1.1453

continued

continued

w_A	0.40	0.40	0.40	0.40	0.40	0.40	0.40	0.40	0.40
w_B	0.20	0.20	0.20	0.20	0.20	0.20	0.20	0.20	0.20
w_C	0.40	0.40	0.40	0.40	0.40	0.40	0.40	0.40	0.40
T/K	349.1	349.1	349.1	373.2	373.2	373.2	373.3	373.3	373.2
P/MPa	34.9	41.8	48.8	7.2	14.2	21.3	27.9	34.9	41.7
ρ/(g/cm^3)	1.1651	1.1835	1.1993	0.9776	1.0328	1.0668	1.0914	1.1146	1.1356

sw_A	0.40	0.40	0.40	0.40	0.40	0.40	0.40	0.40
w_B	0.20	0.20	0.20	0.20	0.20	0.20	0.20	0.20
w_C	0.40	0.40	0.40	0.40	0.40	0.40	0.40	0.40
T/K	373.2	397.9	398.3	398.3	398.4	398.5	398.5	398.4
P/MPa	48.5	8.0	14.4	21.1	28.2	34.9	41.9	48.6
ρ/(g/cm^3)	1.1536	0.7219	0.9443	0.9978	1.0335	1.0607	1.0930	1.1137

Polymer (B): **poly(ε-caprolactone)** **2006LI2**
Characterization: M_n/g.mol^{-1} = 6100/g.mol^{-1}, M_w/g.mol^{-1} = 14300, T_m/K = 339, Scientific Polymer Products, Inc., Ontario, NY
Solvent (A): **2-propanone** C_3H_6O **67-64-1**
Solvent (C): **carbon dioxide** CO_2 **124-38-9**

T/K = 323.15 w_B = 0.05 was kept constant

w_A	0.93	0.93	0.93	0.93	0.93	0.91	0.91	0.91	0.91
w_C	0.02	0.02	0.02	0.02	0.02	0.04	0.04	0.04	0.04
P/MPa	7.76	14.19	21.30	28.36	35.19	7.64	14.41	20.94	28.07
ρ/(g/cm^3)	0.8769	0.8839	0.8871	0.8884	0.8921	0.8638	0.8723	0.8799	0.8850

w_A	0.91	0.55	0.55	0.55	0.55	0.55
w_C	0.04	0.40	0.40	0.40	0.40	0.40
P/MPa	34.88	7.42	14.55	21.13	27.78	35.14
ρ/(g/cm^3)	0.8918	0.8943	0.9174	0.9354	0.9531	0.9697

T/K = 333.15 w_B = 0.05 was kept constant

w_A	0.55	0.55	0.55	0.55	0.55
w_C	0.40	0.40	0.40	0.40	0.40
P/MPa	7.42	14.55	21.13	27.78	35.14
ρ/(g/cm^3)	0.8943	0.9174	0.9354	0.9531	0.9697

T/K = 348.15 w_B = 0.05 was kept constant

w_A	0.94	0.94	0.94	0.94	0.94	0.93	0.93	0.93	0.93
w_C	0.01	0.01	0.01	0.01	0.01	0.02	0.02	0.02	0.02
P/MPa	7.63	14.96	21.45	28.37	35.30	7.14	14.39	21.56	28.14
ρ/(g/cm^3)	0.8503	0.8603	0.8690	0.8770	0.8832	0.8510	0.8608	0.8703	0.8776

w_A	0.93	0.91	0.91	0.91	0.91	0.91	0.55	0.55	0.55
w_C	0.02	0.04	0.04	0.04	0.04	0.04	0.40	0.40	0.40
P/MPa	35.01	7.03	13.87	21.46	28.15	34.88	9.56	15.06	20.99
ρ/(g/cm^3)	0.8860	0.8550	0.8648	0.8755	0.8833	0.8906	0.8617	0.8825	0.9020

continued

continued

w_A	0.55	0.55
w_C	0.40	0.40
P/MPa	28.04	34.61
ρ/(g/cm^3)	0.9223	0.9394

T/K = 363.15 w_B = 0.05 was kept constant

w_A	0.55	0.55	0.55	0.55
w_C	0.40	0.40	0.40	0.40
P/MPa	19.10	21.41	27.91	35.04
ρ/(g/cm^3)	0.8867	0.8942	0.9175	0.9384

T/K = 373.15 w_B = 0.05 was kept constant

w_A	0.94	0.94	0.94	0.94	0.94	0.93	0.93	0.93	0.93
w_C	0.01	0.01	0.01	0.01	0.01	0.02	0.02	0.02	0.02
P/MPa	7.25	14.82	21.19	27.84	35.54	7.28	14.59	21.39	28.73
ρ/(g/cm^3)	0.8188	0.8304	0.8400	0.8491	0.8589	0.8210	0.8331	0.8431	0.8530

w_A	0.93	0.91	0.91	0.91	0.91	0.91	0.55	0.55	0.55
w_C	0.02	0.04	0.04	0.04	0.04	0.04	0.40	0.40	0.40
P/MPa	35.11	7.18	14.38	21.65	28.27	35.10	25.09	28.26	34.99
ρ/(g/cm^3)	0.8616	0.8235	0.8361	0.8474	0.8569	0.8661	0.8773	0.8872	0.9075

T/K = 398.15 w_B = 0.05 was kept constant

w_A	0.94	0.94	0.94	0.94	0.94	0.93	0.93	0.93	0.93
w_C	0.01	0.01	0.01	0.01	0.01	0.02	0.02	0.02	0.02
P/MPa	7.71	14.50	21.55	28.20	35.07	6.91	14.07	21.25	27.96
ρ/(g/cm^3)	0.7888	0.8027	0.8153	0.8259	0.8359	0.7889	0.8040	0.8172	0.8299

w_A	0.93	0.91	0.91	0.91	0.91	0.91
w_C	0.02	0.04	0.04	0.04	0.04	0.04
P/MPa	35.24	6.98	14.28	21.21	28.18	35.08
ρ/(g/cm^3)	0.8404	0.7908	0.8076	0.8200	0.8326	0.8432

Polymer (B): **poly(ε-caprolactone)** **2007LIU**
Characterization: M_n/g.mol^{-1} = 36100/g.mol^{-1}, M_w/g.mol^{-1} = 65000,
Scientific Polymer Products, Inc., Ontario, NY
Solvent (A): **2-propanone** **C_3H_6O** **67-64-1**
Solvent (C): **carbon dioxide** **CO_2** **124-38-9**

w_A = 0.50 was kept constant
w_B = 0.10 was kept constant
w_C = 0.40 was kept constant

T/K	322.5	322.7	322.9	323.1	323.3	323.3	323.3	349.1	349.1
P/MPa	6.9	14.1	21.2	28.0	35.0	41.8	48.8	4.6	4.9
ρ/(g/cm^3)	0.9833	0.9991	1.0085	1.0113	1.0118	1.0119	1.0121	0.6119	0.6439

continued

continued

T/K	349.1	349.0	348.6	348.6	348.5	348.5	348.6	348.9	348.9
P/MPa	5.0	4.9	4.8	5.0	5.0	5.2	5.3	5.4	6.2
ρ/(g/cm^3)	0.6591	0.6764	0.6950	0.7126	0.7339	0.7562	0.7819	0.8077	0.8436
T/K	349.1	349.1	349.1	349.2	349.2	349.1	349.0	349.1	349.3
P/MPa	6.8	8.5	11.7	14.8	21.0	27.9	34.8	41.9	48.6
ρ/(g/cm^3)	0.8836	0.9142	0.9447	0.9617	0.9816	1.0006	1.0066	1.0090	1.0110
T/K	374.8	374.8	374.8	374.5	374.5	374.2	374.4	398.9	399.2
P/MPa	6.9	14.1	21.1	27.9	34.8	41.7	48.8	7.9	14.4
ρ/(g/cm^3)	0.8115	0.8982	0.9266	0.9498	0.9698	0.9864	1.0007	0.6958	0.8112
T/K	399.0	398.5	398.5	398.8	398.7				
P/MPa	21.0	27.8	35.0	42.0	48.5				
ρ/(g/cm^3)	0.8760	0.9049	0.9266	0.9464	0.9641				

Polymer (B): **poly(ethylene glycol)** **2004LEE**
Characterization: M_n/g.mol^{-1} = 260, M_w/g.mol^{-1} = 280,
Aldrich Chem. Co., Inc., St. Louis, MO
Solvent (A): **anisole** C_7H_8O **100-66-3**
Polymer (C): **poly(ethylene glycol) monomethyl ether**
Characterization: M_n/g.mol^{-1} = 366, M_w/g.mol^{-1} = 373,
Aldrich Chem. Co., Inc., St. Louis, MO

T/K = 298.15

$w_A/w_B/w_C$ = 0.0733/0.1773/0.7489 and $x_A/x_B/x_C$ = 0.200/0.200/0.600 was kept constant

P/MPa	0.1	10	15	20	25	30	35	40
ρ/(g/cm^3)	1.0868	1.0913	1.0936	1.0959	1.0981	1.1002	1.1023	1.1043
P/MPa	45	50						
ρ/(g/cm^3)	1.1063	1.1083						

$w_A/w_B/w_C$ = 0.3413/0.2736/0.3851 and $x_A/x_B/x_C$ = 0.600/0.200/0.200 was kept constant

P/MPa	0.1	10	15	20	25	30	35	40
ρ/(g/cm^3)	1.0687	1.0740	1.0766	1.0791	1.0816	1.0840	1.0864	1.0886
P/MPa	45	50						
ρ/(g/cm^3)	1.0909	1.0932						

$w_A/w_B/w_C$ = 0.0862/0.6219/0.2918 and $x_A/x_B/x_C$ = 0.200/0.600/0.200 was kept constant

P/MPa	0.1	10	15	20	25	30	35	40
ρ/(g/cm^3)	1.0989	1.1035	1.1058	1.1081	1.1103	1.1125	1.1146	1.1167
P/MPa	45	50						
ρ/(g/cm^3)	1.1188	1.1209						

continued

continued

$w_A/w_B/w_C$ = 0.1434/0.3713/0.4853 and $x_A/x_B/x_C$ = 0.325/0.325/0.350 was kept constant

P/MPa	0.1	10	15	20	25	30	35	40
ρ/(g/cm^3)	1.0793	1.0842	1.0866	1.0889	1.0912	1.0935	1.0957	1.0978

P/MPa	45	50
ρ/(g/cm^3)	1.0999	1.1020

T/K = 318.15

$w_A/w_B/w_C$ = 0.0733/0.1773/0.7489 and $x_A/x_B/x_C$ = 0.200/0.200/0.600 was kept constant

P/MPa	0.1	10	15	20	25	30	35	40
ρ/(g/cm^3)	1.0700	1.0752	1.0777	1.0801	1.0825	1.0847	1.0870	1.0892

P/MPa	45	50
ρ/(g/cm^3)	1.0914	1.0935

$w_A/w_B/w_C$ = 0.3413/0.2736/0.3851 and $x_A/x_B/x_C$ = 0.600/0.200/0.200 was kept constant

P/MPa	0.1	10	15	20	25	30	35	40
ρ/(g/cm^3)	1.0510	1.0570	1.0598	1.0625	1.0651	1.0678	1.0704	1.0728

P/MPa	45	50
ρ/(g/cm^3)	1.0753	1.0777

$w_A/w_B/w_C$ = 0.0862/0.6219/0.2918 and $x_A/x_B/x_C$ = 0.200/0.600/0.200 was kept constant

P/MPa	0.1	10	15	20	25	30	35	40
ρ/(g/cm^3)	1.0820	1.0874	1.0899	1.0924	1.0949	1.0972	1.0995	1.1017

P/MPa	45	50
ρ/(g/cm^3)	1.1040	1.1062

$w_A/w_B/w_C$ = 0.1434/0.3713/0.4853 and $x_A/x_B/x_C$ = 0.325/0.325/0.350 was kept constant

P/MPa	0.1	10	15	20	25	30	35	40
ρ/(g/cm^3)	1.0622	1.0678	1.0703	1.0729	1.0755	1.0779	1.0803	1.0826

P/MPa	45	50
ρ/(g/cm^3)	1.0849	1.0872

T/K = 348.15

$w_A/w_B/w_C$ = 0.0733/0.1773/0.7489 and $x_A/x_B/x_C$ = 0.200/0.200/0.600 was kept constant

P/MPa	0.1	10	15	20	25	30	35	40
ρ/(g/cm^3)	1.0433	1.0493	1.0523	1.0551	1.0577	1.0604	1.0629	1.0655

P/MPa	45	50
ρ/(g/cm^3)	1.0679	1.0702

continued

continued

$w_A/w_B/w_C$ = 0.3413/0.2736/0.3851 and $x_A/x_B/x_C$ = 0.600/0.200/0.200 was kept constant

P/MPa	0.1	10	15	20	25	30	35	40
ρ/(g/cm^3)	1.0246	1.0313	1.0346	1.0377	1.0408	1.0437	1.0466	1.0494

P/MPa	45	50
ρ/(g/cm^3)	1.0521	1.0548

$w_A/w_B/w_C$ = 0.0862/0.6219/0.2918 and $x_A/x_B/x_C$ = 0.200/0.600/0.200 was kept constant

P/MPa	0.1	10	15	20	25	30	35	40
ρ/(g/cm^3)	1.0572	1.0631	1.0659	1.0687	1.0714	1.0740	1.0766	1.0791

P/MPa	45	50
ρ/(g/cm^3)	1.0816	1.0839

$w_A/w_B/w_C$ = 0.1434/0.3713/0.4853 and $x_A/x_B/x_C$ = 0.325/0.325/0.350 was kept constant

P/MPa	0.1	10	15	20	25	30	35	40
ρ/(g/cm^3)	1.0369	1.0430	1.0460	1.0490	1.0517	1.0545	1.0571	1.0597

P/MPa	45	50
ρ/(g/cm^3)	1.0623	1.0647

Polymer (B): **poly(methyl methacrylate)** **2006LI1**
Characterization: M_n/g.mol^{-1} = 19300/g.mol^{-1}, M_w/g.mol^{-1} = 540000,
Scientific Polymer Products, Inc., Ontario, NY
Solvent (A): **2-propanone** **C_3H_6O** **67-64-1**
Solvent (C): **carbon dioxide** **CO_2** **124-38-9**

T/K = 323.15 w_B = 0.05 was kept constant

w_A	0.94	0.94	0.94	0.94	0.94	0.93	0.93	0.93	0.93
w_C	0.01	0.01	0.01	0.01	0.01	0.02	0.02	0.02	0.02
P/MPa	7.33	13.55	21.60	28.20	35.34	7.53	14.69	22.22	27.76
ρ/(g/cm^3)	0.8742	0.8723	0.8808	0.8893	0.8960	0.8724	0.8809	0.8894	0.8966

w_A	0.93	0.91	0.91	0.91	0.91
w_C	0.02	0.04	0.04	0.04	0.04
P/MPa	34.95	7.36	14.17	21.05	28.08
ρ/(g/cm^3)	0.9005	0.8960	0.9017	0.9095	0.9161

T/K = 348.15 w_B = 0.05 was kept constant

w_A	0.94	0.94	0.94	0.94	0.94	0.93	0.93	0.93	0.93
w_C	0.01	0.01	0.01	0.01	0.01	0.02	0.02	0.02	0.02
P/MPa	7.11	14.80	21.30	28.19	35.03	8.94	14.25	21.01	28.06
ρ/(g/cm^3)	0.8350	0.8435	0.8541	0.8640	0.8726	0.8433	0.8508	0.8601	0.8692

w_A	0.93	0.91	0.91	0.91	0.91	0.91
w_C	0.02	0.04	0.04	0.04	0.04	0.04
P/MPa	35.10	7.31	13.94	21.14	28.03	34.90
ρ/(g/cm^3)	0.8779	0.8630	0.8732	0.8835	0.8927	0.9011

continued

continued

T/K = 373.15 w_B = 0.05 was kept constant

w_A	0.94	0.94	0.94	0.94	0.94	0.93	0.93	0.93	0.93
w_C	0.01	0.01	0.01	0.01	0.01	0.02	0.02	0.02	0.02
P/MPa	7.17	14.30	21.65	28.22	35.28	15.39	21.41	27.76	34.93
ρ/(g/cm^3)	0.8016	0.8130	0.8256	0.8368	0.8536	0.8196	0.8315	0.8426	0.8521

w_A	0.91	0.91	0.91	0.91	0.91
w_C	0.04	0.04	0.04	0.04	0.04
P/MPa	7.31	14.33	21.16	27.95	34.87
ρ/(g/cm^3)	0.8337	0.8433	0.8555	0.8662	0.8763

T/K = 398.15 w_B = 0.05 was kept constant

w_A	0.94	0.94	0.94	0.94	0.94
w_C	0.01	0.01	0.01	0.01	0.01
P/MPa	8.06	14.26	21.31	28.48	34.36
ρ/(g/cm^3)	0.7666	0.7828	0.7975	0.8103	0.8221

Polymer (B): **poly(propylene glycol)** **2004LEE**
Characterization: M_n/g.mol^{-1} = 4960, M_w/g.mol^{-1} = 5000,
Aldrich Chem. Co., Inc., St. Louis, MO
Solvent (A): **acetophenone** C_8H_8O **98-86-2**
Solvent (C): **1-octanol** $C_8H_{18}O$ **111-87-5**

T/K = 298.15

$w_A/w_B/w_C$ = 0.0400/0.8877/0.0723 and $x_A/x_B/x_C$ = 0.312/0.168/0.520 was kept constant

P/MPa	0.1	10	15	20	25	30	35	40
ρ/(g/cm^3)	0.9844	0.9906	0.9936	0.9964	0.9992	1.0020	1.0047	1.0072

P/MPa	45	50
ρ/(g/cm^3)	1.0098	1.0122

$w_A/w_B/w_C$ = 0.0668/0.8897/0.0435 and $x_A/x_B/x_C$ = 0.520/0.168/0.312 was kept constant

P/MPa	0.1	10	15	20	25	30	35	40
ρ/(g/cm^3)	0.9909	0.9969	0.9999	1.0027	1.0055	1.0082	1.0108	1.0134

P/MPa	45	50
ρ/(g/cm^3)	1.0159	1.0184

$w_A/w_B/w_C$ = 0.0072/0.9540/0.0388 and $x_A/x_B/x_C$ = 0.108/0.350/0.542 was kept constant

P/MPa	0.1	10	15	20	25	30	35	40
ρ/(g/cm^3)	0.9904	0.9966	0.9996	1.0025	1.0053	1.0080	1.0106	1.0132

P/MPa	45	50
ρ/(g/cm^3)	1.0158	1.0182

continued

continued

$w_A/w_B/w_C$ = 0.0215/0.9552/0.0233 and $x_A/x_B/x_C$ = 0.325/0.350/0.325 was kept constant

P/MPa	0.1	10	15	20	25	30	35	40
ρ/(g/cm^3)	0.9939	1.0001	1.0031	1.0060	1.0087	1.0115	1.0141	1.0167

P/MPa	45	50
ρ/(g/cm^3)	1.0193	1.0217

$w_A/w_B/w_C$ = 0.0359/0.9563/0.0078 and $x_A/x_B/x_C$ = 0.542/0.350/0.108 was kept constant

P/MPa	0.1	10	15	20	25	30	35	40
ρ/(g/cm^3)	0.9978	1.0040	1.0069	1.0098	1.0126	1.0154	1.0181	1.0207

P/MPa	45	50
ρ/(g/cm^3)	1.0232	1.0257

$w_A/w_B/w_C$ = 0.0049/0.9792/0.0159 and $x_A/x_B/x_C$ = 0.113/0.548/0.339 was kept constant

P/MPa	0.1	10	15	20	25	30	35	40
ρ/(g/cm^3)	0.9944	1.0006	1.0036	1.0065	1.0093	1.0121	1.0147	1.0173

P/MPa	45	50
ρ/(g/cm^3)	1.0198	1.0223

$w_A/w_B/w_C$ = 0.0147/0.9800/0.0053 and $x_A/x_B/x_C$ = 0.339/0.548/0.113 was kept constant

P/MPa	0.1	10	15	20	25	30	35	40
ρ/(g/cm^3)	0.9966	1.0027	1.0057	1.0087	1.0114	1.0142	1.0168	1.0194

P/MPa	45	50
ρ/(g/cm^3)	1.0219	1.0244

T/K = 318.15

$w_A/w_B/w_C$ = 0.0400/0.8877/0.0723 and $x_A/x_B/x_C$ = 0.312/0.168/0.520 was kept constant

P/MPa	0.1	10	15	20	25	30	35	40
ρ/(g/cm^3)	0.9693	0.9761	0.9792	0.9823	0.9854	0.9883	0.9911	0.9939

P/MPa	45	50
ρ/(g/cm^3)	0.9966	0.9992

$w_A/w_B/w_C$ = 0.0668/0.8897/0.0435 and $x_A/x_B/x_C$ = 0.520/0.168/0.312 was kept constant

P/MPa	0.1	10	15	20	25	30	35	40
ρ/(g/cm^3)	0.9753	0.9820	0.9853	0.9883	0.9914	0.9943	0.9972	0.9999

P/MPa	45	50
ρ/(g/cm^3)	1.0027	1.0053

$w_A/w_B/w_C$ = 0.0072/0.9540/0.0388 and $x_A/x_B/x_C$ = 0.108/0.350/0.542 was kept constant

P/MPa	0.1	10	15	20	25	30	35	40
ρ/(g/cm^3)	0.9752	0.9820	0.9852	0.9884	0.9914	0.9944	0.9972	1.0000

continued

continued

P/MPa	45	50
ρ/(g/cm^3)	1.0027	1.0054

$w_A/w_B/w_C$ = 0.0215/0.9552/0.0233 and $x_A/x_B/x_C$ = 0.325/0.350/0.325 was kept constant

P/MPa	0.1	10	15	20	25	30	35	40
ρ/(g/cm^3)	0.9786	0.9854	0.9886	0.9918	0.9948	0.9978	1.0006	1.0034

P/MPa	45	50
ρ/(g/cm^3)	1.0061	1.0088

$w_A/w_B/w_C$ = 0.0359/0.9563/0.0078 and $x_A/x_B/x_C$ = 0.542/0.350/0.108 was kept constant

P/MPa	0.1	10	15	20	25	30	35	40
ρ/(g/cm^3)	0.9825	0.9892	0.9925	0.9956	0.9986	1.0016	1.0044	1.0072

P/MPa	45	50
ρ/(g/cm^3)	1.0099	1.0125

$w_A/w_B/w_C$ = 0.0049/0.9792/0.0159 and $x_A/x_B/x_C$ = 0.113/0.548/0.339 was kept constant

P/MPa	0.1	10	15	20	25	30	35	40
ρ/(g/cm^3)	0.9792	0.9859	0.9893	0.9924	0.9954	0.9984	1.0013	1.0041

P/MPa	45	50
ρ/(g/cm^3)	1.0068	1.0094

$w_A/w_B/w_C$ = 0.0147/0.9800/0.0053 and $x_A/x_B/x_C$ = 0.339/0.548/0.113 was kept constant

P/MPa	0.1	10	15	20	25	30	35	40
ρ/(g/cm^3)	0.9812	0.9881	0.9913	0.9945	0.9975	1.0005	1.0033	1.0061

P/MPa	45	50
ρ/(g/cm^3)	1.0088	1.0114

T/K = 348.15

$w_A/w_B/w_C$ = 0.0400/0.8877/0.0723 and $x_A/x_B/x_C$ = 0.312/0.168/0.520 was kept constant

P/MPa	0.1	10	15	20	25	30	35	40
ρ/(g/cm^3)	0.9460	0.9537	0.9573	0.9609	0.9643	0.9675	0.9707	0.9738

P/MPa	45	50
ρ/(g/cm^3)	0.9768	0.9797

$w_A/w_B/w_C$ = 0.0668/0.8897/0.0435 and $x_A/x_B/x_C$ = 0.520/0.168/0.312 was kept constant

P/MPa	0.1	10	15	20	25	30	35	40
ρ/(g/cm^3)	0.9521	0.9598	0.9634	0.9669	0.9703	0.9736	0.9767	0.9798

P/MPa	45	50
ρ/(g/cm^3)	0.9828	0.9857

continued

continued

$w_A/w_B/w_C$ = 0.0072/0.9540/0.0388 and $x_A/x_B/x_C$ = 0.108/0.350/0.542 was kept constant

P/MPa	0.1	10	15	20	25	30	35	40
ρ/(g/cm^3)	0.9523	0.9600	0.9637	0.9672	0.9706	0.9739	0.9771	0.9801

P/MPa	45	50
ρ/(g/cm^3)	0.9831	0.9860

$w_A/w_B/w_C$ = 0.0215/0.9552/0.0233 and $x_A/x_B/x_C$ = 0.325/0.350/0.325 was kept constant

P/MPa	0.1	10	15	20	25	30	35	40
ρ/(g/cm^3)	0.9557	0.9634	0.9671	0.9706	0.9739	0.9773	0.9804	0.9835

P/MPa	45	50
ρ/(g/cm^3)	0.9865	0.9894

$w_A/w_B/w_C$ = 0.0359/0.9563/0.0078 and $x_A/x_B/x_C$ = 0.542/0.350/0.108 was kept constant

P/MPa	0.1	10	15	20	25	30	35	40
ρ/(g/cm^3)	0.9592	0.9669	0.9705	0.9740	0.9774	0.9807	0.9839	0.9870

P/MPa	45	50
ρ/(g/cm^3)	0.9900	0.9929

$w_A/w_B/w_C$ = 0.0049/0.9792/0.0159 and $x_A/x_B/x_C$ = 0.113/0.548/0.339 was kept constant

P/MPa	0.1	10	15	20	25	30	35	40
ρ/(g/cm^3)	0.9560	0.9638	0.9675	0.9709	0.9744	0.9776	0.9809	0.9840

P/MPa	45	50
ρ/(g/cm^3)	0.9870	0.9898

$w_A/w_B/w_C$ = 0.0147/0.9800/0.0053 and $x_A/x_B/x_C$ = 0.339/0.548/0.113 was kept constant

P/MPa	0.1	10	15	20	25	30	35	40
ρ/(g/cm^3)	0.9581	0.9658	0.9695	0.9730	0.9764	0.9797	0.9828	0.9860

P/MPa	45	50
ρ/(g/cm^3)	0.9890	0.9919

5.3. References

2003KIM Kim, J.H., Kim, Y., Kim, C.K., Lee, J.W., and Seo, S.B., Miscibility of polysulfone blends with poly(1-vinylpyrrolidone-*co*-styrene) copolymers and their interaction energies, *J. Polym. Sci.: Part B: Polym. Phys.*, 41, 1401, 2003.

2003PAL Paluch, M., Casalini, R., Patkowski, A., Pakula, T., and Roland, C.M., Effect of volume changes on segmental relaxation in siloxane polymers, *Phys. Rev. E*, 68, 031802, 2003.

2004DLU Dlubek, G., Gupta, A.S., Piontek, J., Krause-Rehberg, R., Kaspar, H., and Lochhaas, K.H., Temperature dependence of the free volume in fluoroelastomers from positron lifetime and PVT experiments (experimental data by G. Dlubek), *Macromolecules*, 37, 6606, 2004.

2004LEE Lee, M.-J., Hsu, T.-S., Tuan, Y.-C., and Lin, H.-M., Pressure-volume-temperature properties for 1-octanol + acetophenone, poly(propylene glycol) + 1-octanol + acetophenone, and poly(ethylene glycol) + poly(ethylene glycol methyl ether) + anisole, *J. Chem. Eng. Data*, 49, 1052, 2004.

2005BRI Briatico-Vangosa, F. and Rink, M., Dilatometric behavior and glass transition in a styrene-acrylonitrile copolymer, *J. Polym. Sci.: Part B: Polym. Phys.*, 43, 1904, 2005.

2005KIL Kilburn, D., Dlubek, G., Pionteck, J., Bamford, D., and Alam, M.A., Microstructure of free volume in SMA copolymers I. Free volume from Simha-Somcynsky analysis of PVT experiments (experimental data by G. Dlubek), *Polymer*, 46, 859, 2005.

2006BLO Blochowiak, M., Pakula, T., Butt, H.-J., Bruch, M., and Floudas, G., Thermodynamics and rheology of cycloolefin copolymers, *J. Chem. Phys.*, 124, 134903, 2006.

2006KI1 Kilburn, D., Wawryszczuk, J., Dlubek, G., Pionteck, J., Hässler, R., and Alam, M.A., Temperature and pressure dependence of the free volume in polyisobutylene from positron lifetime and pressure-volume-temperature experiments (experimental data by G. Dlubek), *Macromol. Chem. Phys.*, 207, 721, 2006.

2006KI2 Kilburn, D., Dlubek, G., Pionteck, J., and Alam, M.A., Free volume in poly(n-alkyl methacrylate)s from positron lifetime and PVT experiments and its relation to the structural relaxation (experimental data by G. Dlubek), *Polymer*, 47, 7774, 2006.

2006LEE Lee, M.-J., Ho, K.-L., and Lin, H.-M., Pressure-volume-temperature properties for binary oligomeric solutions of poly(ethylene glycol) and poly(ethylene glycol methyl ether) with acetophenone up to 50 MPa, *J. Chem. Eng. Data*, 51, 1151, 2006.

2006LI1 Liu, K., Schuch, F., and Kiran, E., High-pressure viscosity and density of poly(methyl methacrylate) + acetone and poly(methyl methacrylate) + acetone + CO_2 systems, *J. Supercrit. Fluids*, 39, 89, 2006.

2006LI2 Liu, K. and Kiran, E., Miscibility, viscosity and density of poly (ε-caprolactone) in acetone + CO_2 binary fluid mixtures, *J. Supercrit. Fluids*, 39, 192, 2006.

2006PAR Park, H.E. and Dealy, J.M., Effects of pressure and supercritical fluids on the viscosity of polyethylene, *Macromolecules*, 39, 5438, 2006.

2006SEI Seiler, M., Hyperbranched polymers: Phase behavior and new applications in the field of chemical engineering, *Fluid Phase Equil.*, 241, 155, 2006.

2007DL1 Dlubek, G., Pionteck, J., Sniegocka, M., Hassan, E.M., and Krause-Rehberg, R., Temperature and pressure dependence of the free volume in the perfluorinated polymer glass CYTOP: A positron lifetime and pressure-volume-temperature study (experimental data by G. Dlubek), *J. Polym. Sci.: Part B: Polym. Phys.*, 45, 2519, 2007.

2007DL2 Dlubek, G., Piontek, J., Shaikh, M.S., Hassan, E.M., and Krause-Rehberg, R., Free volume of an oligomeric epoxy resin and its relation to structural relaxation: Evidence from positron lifetime and pressure-volume-temperature experiments (experimental data by G. Dlubek), *Phys. Rev. E*, 75, 021802, 2007.

2007FU1 Funami, E., Taki, K., and Ohshima, M., Density measurement of polymer/CO_2 single-phase solution at high temperature and pressure using a gravimetric method, *J. Appl. Polym. Sci.* , 105, 3060, 2007.

2007FU2 Funke, Z., Hotani, Y., Ougizawa, T., Kressler, J., and Kammer, H.-W., Equation-of-state properties and surface tension of ethylene-vinyl alcohol random copolymers (experimental data by Z. Funke), *Eur. Polym. J.*, 43, 2371, 2007.

2007LIU Liu, K. and Kiran, K., A tunable mixture solvent for poly(ε-caprolactone): Acetone + CO_2, *Polymer*, 48, 5612, 2007.

2007SAT Sato, Y., Hashiguchi, H., Inohara, K., Takishima, S., and Masuoka, H., PVT properties of polyethylene copolymer melts, *Fluid Phase Equil.*, 257, 124, 2007.

2008DLU Dlubek, G., Pionteck, J., Rätzke, K., Kruse, J., and Faupel, F., Temperature dependence of the free volume in amorphous Teflon AF1600 and AF2400: A pressure-volume-temperature and positron lifetime study (experimental data by G. Dlubek), *Macromolecules*, 41, 6125, 2008.

2008LIU Liu, K. and Kiran, E., Density and viscosity as real-time probes for progress of high-pressure polymerizations: polymerization of methyl methacrylate in acetone, *Ind. Eng. Chem. Res.*, 47, 5039, 2008.

2008QIA Qian, Z., Minnikanti, V.S., Sauer, B.B., Dee, G.T., and Archer, L.A., Surface tension of symmetric star polymer melts (experimental data by Z. Qian), *Macromolecules*, 41, 5007, 2008.

2008SHI Shin, M.S., Lee, J.H., Kim, H., Phase behavior of the poly(vinyl pyrrolidone) + dichloromethane + supercritical carbon dioxide system, *Fluid Phase Equil.*, 272, 42, 2008.

2009LEE Lee, M.-J., Ku, T.-J., and Lin, H.-M., (Pressure + volume + temperature) properties for binary oligomeric solutions of poly(ethylene glycol mono-4-octylphenyl ether) with 1-octanol or acetophenone at pressures up to 50 MPa, *J. Chem. Thermodyn.*, 41, 1178, 2009.

2009LIY Li, Y.G. and Park, C.B., Effects of branching on the pressure-volume-temperature behaviors of PP/CO_2 solutions, *Ind. Eng. Chem. Res.*, 48, 6633, 2009.

2009SAN Santoyo-Arreola, J.G., Vasquez-Medrano, R.C., Ruiz-Trevino, A., Luna-Barcenas, G., Sanchez, I.C., and Ortiz-Estrada, C.H., Phase behavior and particle formation of poly(1H,1H-dihydrofluorooctyl methacrylate) in supercritical CO_2 (experimental data by C.H. Ortiz-Estrada), *Macromol. Symp.*, 283-284, 230, 2009.

2009UTR Utracki, L.A., Equations of state for polyamide-6 and its nanocomposites. 1. Fundamentals and the matrix (experimental data by L.A. Utracki), *J. Polym. Sci.: Part B: Polym. Phys.*, 47, 299, 2009.

2010FAN Fandino, O., Lugo, L., Comunas, M.J.P., Lopez, E.R., and Fernandez, J., Temperature and pressure dependences of volumetric properties of two poly(propylene glycol) dimethyl ether lubricants, *J. Chem. Thermodyn.*, 42, 84, 2010.

2010PFE Pfefferkorn, D., Sonntag, S., Kyeremateng, S.O., Funke, Z., Kammer, H.-W., and Kressler, J., Pressure-volume-temperature data and surface tension of blends of poly(ethylene oxide) and poly(methyl acrylate) in the melt (experimental data by D. Pfefferkorn), *J. Polym. Sci.: Part B: Polym. Phys.*, 48, 1893, 2010.

2011SCH Scherillo, G., Sanguigno, L., Sansone, L., DiMaio, E., Galizia, M., and Mensitieri, G., Thermodynamics of water sorption in poly(ε-caprolactone): A comparative analysis of lattice fluid models including hydrogen bond contributions (exp. data by G. Scherillo and G. Mansitieri), *Fluid Phase Equil.*, 313, 127, 2011.

2012BAM Bamgbade, B.A., Wu, Y., Burgess, W.A., and McHugh, M.A., Experimental density and PC-SAFT modeling of Krytox® (perfluoropolyether) at pressures to 275 MPa and temperatures to 533 K, *Fluid Phase Equil.*, 332, 159, 2012.

2012CLA Clark, E.A. and Lipson, J.E.G., LCST and UCST behavior in polymer solutions and blends, *Polymer*, 53, 536, 2012.

2012HES Hesse, L. and Sadowski, G., Modeling liquid-liquid equilibria of polyimide solutions, *Ind. Eng. Chem. Res.*, 51, 539, 2012.

2013SAT Sato, Y., Hosaka, N., Inomata, H., and Kanaka, K., Solubility of ethylene in norbornene + toluene + cyclic olefin copolymer systems, *Fluid Phase Equil.*, 352, 80, 2013.

2014MA1 Mahmood, S.H., Keshtkar, M., and Park, C.B., Determination of carbon dioxide solubility in polylactide acid with accurate PVT properties, *J. Chem. Thermodyn.*, 70, 13, 2014.

2014MA2 Mahmood, S.H., Ameli, A., Hossieny, N., Park, C.B., The interfacial tension of molten polylactide in supercritical carbon dioxide, *J. Chem. Thermodyn.*, 75, 69, 2014.

APPENDICES

Appendix 1 List of polymers in alphabetical order

Polymer	Page(s)
Cellulose	138
Cellulose acetate	89
Chitosan	89
Dextran	308-311, 452
Methylcellulose	89
Natural rubber	89, 138
Nitrocellulose	89, 138
Penta(ethylene glycol) monododecyl ether	452
Poly(acetoacetoxyethyl methacrylate-*b*-1,1,2,2-tetrahydroperfluorodecyl acrylate)	183-184
Poly(acrylamide)	138
Poly(acrylonitrile-*co*-butadiene-*co*-styrene)	89
Poly(alkylene glycol)	89
Poly(allyl acetate)	299
Polyamide	89
Polyamide-6	469-470
Polyamide-11	15
Poly(benzoxazole-*co*-imide)	90
Poly(benzyl acrylate)	299, 452
Poly(benzyl methacrylate)	299, 452
Poly[(2,2-bis(hydroxymethyl)propionate)-*co*-(6-hydroxyhexanoate)]	299
Poly(bisphenol-A carbonate-*co*-4,40-(3,3,5-trimethylcyclohexylidene) diphenol carbonate)	90
Poly[4,4'-bis(trifluorovinyloxy)biphenyl]	184
Poly{1,1-bis[4-(trifluorovinyloxy)phenyl]-hexafluoroisopropylidene}	185
Poly[2,2-bis(4-trifluorovinyloxyphenyl)propane]	299
Poly[2,2-bis(4-trifluorovinyloxyphenyl)propane-*co*-2,2-bis(4-trifluorovinyloxyphenyl)-1,1,1,3,3,3-hexafluoropropane]	299
Polybutadiene	157, 186-189, 311
1,2-Polybutadiene	16
1,4-*cis*-Polybutadiene	17, 187-188, 299
Poly(butadiene-*b*-styrene)	172
Poly(2-butoxyethyl acrylate)	189-191, 311-313
Poly(*tert*-butylacetylene)	90, 138

Polymer	**Page(s)**
Poly[styrene-*b*-4-(perfluorooctyl(ethyleneoxy)methyl)styrene]	294, 305
Poly(styrene-*b*-1,1,2,2-tetrahydroperfluorodecyl acrylate)	305
Poly(styrene-*b*-1,1,2,2-tetrahydroperfluorodecyl methacrylate)	305
Polysulfone	84, 99, 456
Poly(tetrafluoroethylene)	99, 305
Poly[tetrafluoroethylene-*co*-2,2-bis(trifluoromethyl)-4,5-difluoro-1,3-dioxole]	503-504
Poly[tetrafluoroethylene-*co*-perfluoro(vinyl methyl ether)]	99, 505-507
Poly(tetrafluoroethylene-*co*-2,2,4-trifluoro-5-trifluoromethoxy-1,3-dioxole)	99
Poly(tetrafluoroethylene-*co*-vinyl acetate)	306
Poly(1,1,2,2-tetrahydroperfluorodecyl acrylate)	294-295, 306
Poly(1,1,2,2-tetrahydroperfluorodecyl methacrylate)	306
Poly(tetrahydropyranyl methacrylate-*co*-1H,1H-perfluorooctyl methacrylate)	100
Poly[2-(3-thienyl)acetyl 3,3,4,4,5,5,6,6,7,7,8,8,8-tridecafluoro-1-octanate]	306
Poly[2-(3-thienyl)ethyl heptafluorobutyrate]	306
Poly[2-(3-thienyl)ethyl pentadecafluorooctanoate]	306
Poly[2-(3-thienyl)methyl heptafluorobutyrate]	306
Poly(2,2,2-trifluoroethyl methacrylate)	295
Poly(1-trimethylsilyl-1-propyne)	100, 141
Poly(vinyl acetate)	100, 141, 306
Poly(vinyl acetate-*alt*-dibutyl maleate)	296, 306
Poly(vinyl acetate-*alt*-dibutyl maleate) xanthate	306
Poly(vinyl acetate-*co*-1-methoxyethyl ether)	307
Poly(vinyl acetate-*co*-methoxymethyl ether)	307
Poly[vinyl acetate-*stat*-1-(trifluoromethyl)vinyl acetate]	307
Poly(vinyl alcohol)	507-508
Poly[1-(4-vinylbenzyl)-3-butylimidazolium o-benzoic sulfimide]	100
Poly[1-(4-vinylbenzyl)-3-butylimidazolium hexafluorophosphate]	100
Poly[1-(4-vinylbenzyl)-3-butylimidazolium tetrafluoroborate]	100
Poly[1-(4-vinylbenzyl)-3-butylimidazolium trifluoromethane sulfonamide]	100
Poly[1-(4-vinylbenzyl)-3-methylimidazolium tetrafluoroborate]	101
Poly[1-(4-vinylbenzyl)-4-methylimidazolium tetrafluoroborate]	85
Poly(vinylbenzylphosphonic acid diethylester)	296
Poly(vinylbenzylphosphonic acid diethylester-*b*-1,1,2,2-tetrahydroperfluorodecyl acrylate)	296-297
Poly(vinylbenzylphosphonic acid diethylester-*co*-1,1,2,2-tetrahydroperfluorodecyl acrylate)	297-298
Poly(vinylbenzylphosphonic diacid-*b*-1,1,2,2-tetrahydroperfluorodecyl acrylate)	298

Appendix 2 List of systems and properties in order of the polymers

Polymer	Solvent	Property	Page(s)
Dextran			
	carbon dioxide + dimethylsulfoxide	HPPE	308-310
	carbon dioxide + dimethylsulfoxide + water	HPPE	310
Poly(acetoacetoxyethyl methacrylate-*b*-1,1,2,2-tetrahydroperfluorodecyl acrylate)			
	carbon dioxide	HPPE	183-184
Polyamide-6			
	–	PVT	469-470
Polyamide-11			
	carbon dioxide	gas solubility	15
	methane	gas solubility	15
Poly[4,4'-bis(trifluorovinyloxy)-biphenyl]			
	dimethyl ether	HPPE	184
Poly{1,1-bis[4-(trifluorovinyloxy)-phenyl]hexafluoroisopropylidene}			
	carbon dioxide	HPPE	185
	dimethyl ether	HPPE	185
	propane	HPPE	185

Polymer	Solvent	Property	Page(s)
Poly(*tert*-butyl methacrylate)			
	carbon dioxide	HPPE	191
	carbon dioxide + *tert*-butyl methacrylate	HPPE	314-315
Poly(ε-caprolactone)			
	1-butene	HPPE	192
	1-butene + chlorodifluoro-methane	HPPE	315
	1-butene + dimethyl ether	HPPE	316
	carbon dioxide	gas solubility	17
	carbon dioxide	HPPE	192
	carbon dioxide + chlorodifluoromethane	HPPE	316-318
	carbon dioxide + dichloro-methane	HPPE	318-324
	carbon dioxide + dimethyl ether	HPPE	325-326
	carbon dioxide + ethanol	HPPE	327
	carbon dioxide + 2-propanone	HPPE	327-332
	carbon dioxide + 2-propanone + poly(methyl methacrylate)	HPPE	329-330
	carbon dioxide + trichloro-methane	HPPE	332-335
	chlorodifluoromethane	HPPE	192-193
	dimethyl ether	HPPE	193
	2-propanone	density	512-513
	2-propanone + carbon dioxide	density	521-527
	propene	HPPE	193
	propene + chlorodifluoromethane	HPPE	335-336
	propene + dimethyl ether	HPPE	336
	–	PVT	471
Poly(4-chlorostyrene)			
	n-butane + 4-chlorostyrene	HPPE	336
	n-butane + ethanol	HPPE	337
	n-butane + 2-propanone	HPPE	337
	1-butene	HPPE	194
	1-butene + 4-chlorostyrene	HPPE	337-338

Polymer	Solvent	Property	Page(s)
Poly[4-(decyloxymethyl)styrene]			
	carbon dioxide + dimethyl ether	HPPE	347
	dimethyl ether	HPPE	196
Poly[4-(decylsulfonylmethyl)-styrene]			
	carbon dioxide + dimethyl ether	HPPE	347-348
	dimethyl ether	HPPE	196
Poly[4-(decylthiomethyl)styrene]			
	carbon dioxide + dimethyl ether	HPPE	348
	dimethyl ether	HPPE	197
Poly(diglycidyl ether of bisphenol A)			
	–	PVT	472
Poly(1H,1H-dihydroperfluoro-octyl methacrylate)			
	–	PVT	472-473
Poly(dimethylsiloxane)			
	n-butane	HPPE	197
	1-butene	HPPE	198
	carbon dioxide	HPPE	198-199
	carbon dioxide	VLE	18
	dimethyl ether	HPPE	199-200
	propane	HPPE	200
	propene	HPPE	201
Poly(dimethylsiloxane) monomethacrylate			
	carbon dioxide	HPPE	201-202
	carbon dioxide + 1,1-difluoroethene	HPPE	348
	carbon dioxide + methyl methacrylate	HPPE	349

Polymer	Solvent	Property	Page(s)
Poly(ethylene-*co*-butyl methacrylate)			
	ethene	HPPE	230-231
Poly(ethylene-*co*-butyl methacrylate-*co*-methacrylic acid)			
	ethene	HPPE	231-235
Poly(ethylene-*co*-ethyl acrylate)			
	ethene	HPPE	235-236
Poly(ethylene-*co*-ethyl methacrylate)			
	ethene + ethyl methacrylate	HPPE	377-378
Poly(ethylene-*co*-1-hexene)			
	ethene	HPPE	236
	ethene + n-butane	HPPE	378-379
	ethene + carbon dioxide	HPPE	379
	ethene + helium	HPPE	379-380
	ethene + 1-hexene	HPPE	380
	ethene + methane	HPPE	380
	ethene + nitrogen	HPPE	381
	–	PVT	476
Poly(ethylene-*co*-methacrylic acid)			
	ethene	HPPE	237
Poly(ethylene-*co*-methyl acrylate)			
	ethene + vinyl acetate	HPPE	381
Poly(ethylene-*co*-methyl acrylate-*co*-vinyl acetate)			
	ethene	HPPE	237-238
	ethene + vinyl acetate	HPPE	381-382
Poly(ethylene-*co*-methyl methacrylate)			
	ethene	HPPE	238-239

Polymer	Solvent	Property	Page(s)
	ethene + vinyl acetate + carbon dioxide	HPPE	387
	ethene + vinyl acetate + helium	HPPE	388
	ethene + vinyl acetate + methane	HPPE	388-389
	ethene + vinyl acetate + nitrogen	HPPE	389-390
Poly(ethylene-*co*-vinyl alcohol)	–	PVT	479-484
Poly(ethylene glycol)			
	acetophenone	density	514
	anisole + poly(ethylene glycol) monomethyl ether	density	527-529
	1-butene	HPPE	242-243
	carbon dioxide	gas solubility	33-44
	carbon dioxide	HPPE	243-245
	carbon dioxide + diethylene glycol	HPPE	390
	carbon dioxide + ethanol	HPPE	390-391
	carbon dioxide + 1-octanol	gas solubility	108-109
	carbon dioxide + 1-octene	VLE	109-110
	carbon dioxide + 1-pentanol	gas solubility	111-112
	carbon dioxide + water	VLE	112-113
	15-crown-5 ether + sulfur dioxide + nitrogen	gas solubility	113-114
	N,N-dimethylformamide + sulfur dioxide + nitrogen	gas solubility	115-116
	1,4-dioxane + sulfur dioxide + nitrogen	gas solubility	116-118
	propene	HPPE	245
	nitrogen + sulfur dioxide	gas solubility	118-120
	water + sulfur dioxide + nitrogen	gas solubility	121-125
	–	PVT	484
Poly(ethylene glycol) diacetate			
	methanol	VLE	44-45
	water	VLE	45-46

Polymer	Solvent	Property	Page(s)
Poly(ethylene oxide-*b*-dimethyl-siloxane-*b*-ethylene oxide)			
	carbon dioxide	HPPE	245-246
Poly(ethylene oxide-*b*-propylene oxide-*b*-ethylene oxide)			
	carbon dioxide	HPPE	246-247
Poly(ethylene oxide-*b*-1,1,2,2-tetrahydroperfluorodecyl acrylate)			
	carbon dioxide	HPPE	247-248
	carbon dioxide + 2-hydroxyethyl methacrylate	HPPE	392
Poly(ethylene terephthalate)			
	carbon dioxide	gas solubility	59-60
Poly(2-ethylhexyl acrylate)			
	n-butane	HPPE	248
	1-butene	HPPE	248
	carbon dioxide	HPPE	248
	carbon dioxide + dimethyl ether	HPPE	392-393
	carbon dioxide + 2-ethylhexyl acrylate	HPPE	393
	dimethyl ether	HPPE	249
	propane	HPPE	249
	propene	HPPE	249
Poly(2-ethylhexyl methacrylate)			
	n-butane	HPPE	249
	1-butene	HPPE	250
	carbon dioxide	HPPE	250
	carbon dioxide + dimethyl ether	HPPE	394
	carbon dioxide + 2-ethylhexyl methacrylate	HPPE	394-395
	dimethyl ether	HPPE	250
	propane	HPPE	250
	propene	HPPE	251

Polymer	Solvent	Property	Page(s)
Poly(2-hydroxypropyl methacrylate)			
	carbon dioxide + 2-hydroxy-propyl methacrylate	HPPE	407-408
	dimethyl ether	HPPE	257
	dimethyl ether + carbon dioxide	HPPE	408
	dimethyl ether + 2-hydroxy-propyl methacrylate	HPPE	408-409
Polyimide			
	dimethylsulfoxide	density	518
Poly(isobornyl acrylate)			
	n-butane	HPPE	258
	1-butene	HPPE	258
	carbon dioxide + dimethyl ether	HPPE	410
	carbon dioxide + isobornyl acrylate	HPPE	410-411
	dimethyl ether	HPPE	258
	propane	HPPE	259
	propene	HPPE	259
Poly(isobornyl methacrylate)			
	n-butane	HPPE	259
	1-butene	HPPE	259
	carbon dioxide + dimethyl ether	HPPE	412
	carbon dioxide + isobornyl methacrylate	HPPE	412-413
	dimethyl ether	HPPE	260
	propane	HPPE	260
	propene	HPPE	260
Poly(isobutyl acrylate)			
	carbon dioxide	HPPE	260
	carbon dioxide + isobutyl acrylate	HPPE	413-414

Polymer	Solvent	Property	Page(s)
	carbon dioxide +		
	isooctyl acrylate	HPPE	419-420
	dimethyl ether	HPPE	265
	propane	HPPE	265
	propene	HPPE	265
1,4-Polyisoprene			
	propane	HPPE	265-266
Poly(isopropyl acrylate)			
	n-butane	HPPE	266
	1-butene	HPPE	267
	carbon dioxide	HPPE	267
	carbon dioxide +		
	isopropyl acrylate	HPPE	420
	dimethyl ether	HPPE	267
	dimethyl ether +		
	isopropyl acrylate	HPPE	420-421
	propane	HPPE	267
	propene	HPPE	268
Poly(isopropyl methacrylate)			
	carbon dioxide	HPPE	268
	carbon dioxide +		
	isopropyl methacrylate	HPPE	421-422
Poly(DL-lactic acid)			
	carbon dioxide	gas solubility	61-65
	nitrogen	gas solubility	67-68
	oxygen	gas solubility	69
	–	PVT	488-489
Poly(L-lactic acid)			
	carbon dioxide	gas solubility	65-66
	carbon dioxide +		
	dichloromethane	HPPE	422-424
	ethene	gas solubility	67

Polymer	**Solvent**	**Property**	**Page(s)**
Poly{[2-(methacryloyloxy)ethyl]-trimethylammonium tetrafluoroborate}			
	carbon dioxide	gas solubility	73
Poly(methyl acrylate)			
	carbon dioxide	HPPE	273
	carbon dioxide + methyl acrylate	HPPE	434-435
	propene	HPPE	273
	–	PVT	489-490
Poly(methyl methacrylate)			
	carbon dioxide	gas solubility	73-74
	carbon dioxide + chlorodifluoromethane	HPPE	435-437
	carbon dioxide + ethanol	HPPE	437-438
	carbon dioxide + ethanol	VLE	132
	carbon dioxide + methyl methacrylate	HPPE	439
	carbon dioxide + 2-propanone	HPPE	440
	chlorodifluoromethane	HPPE	273
	ethanol	LLE	170
	2-propanone	density	519-520
	2-propanone + carbon dioxide	density	529-530
	propene	HPPE	274
	–	PVT	491
Poly(methylphenylsiloxane)			
	–	PVT	491
Poly(4-methyl styrene)			
	propane	HPPE	274
Poly(methyltolylsiloxane)			
	–	PVT	492

Polymer	Solvent	Property	Page(s)
Poly[4-(perfluorooctyl(ethylene-sulfonyl)methyl)styrene]			
	carbon dioxide	HPPE	279
	dimethyl ether	HPPE	279
Poly[4-(perfluorooctyl(ethylene-thio)methyl)styrene]			
	carbon dioxide	HPPE	279
	dimethyl ether	HPPE	279
Poly(perfluoropropylene oxide) ether			
	carbon dioxide	VLE	74
Poly(perfluoropropyl ether)			
	–	PVT	494
Poly(propyl acrylate)			
	n-butane	HPPE	280
	1-butene	HPPE	280
	carbon dioxide	HPPE	280
	carbon dioxide + dimethyl ether	HPPE	445
	carbon dioxide + propyl acrylate	HPPE	445-446
	dimethyl ether	HPPE	280
	ethene	HPPE	280-281
	ethene + propyl acrylate	HPPE	446
	propane	HPPE	281
	propene	HPPE	281
Polypropylene			
	n-butane	gas solubility	74-75
	carbon dioxide	gas solubility	75-78
	carbon dioxide + poly(propylene-*g*-maleic anhydride)	gas solubility	133
	2-methylpropane	gas solubility	79
	propene	gas solubility	79-80

Polymer	Solvent	Property	Page(s)
Poly(vinylbenzylphosphonic acid diethylester-*b*-1,1,2,2-tetrahydro-perfluorodecyl acrylate)	carbon dioxide	HPPE	296-297
Poly(vinylbenzylphosphonic acid diethylester-*co*-1,1,2,2-tetrahydro-perfluorodecyl acrylate)	carbon dioxide	HPPE	297-298
Poly(vinylbenzylphosphonic diacid-*b*-1,1,2,2-tetrahydro-perfluorodecyl acrylate)	carbon dioxide	HPPE	298
Poly[1-(4-vinylbenzyl)pyridinium tetrafluoroborate]	carbon dioxide	gas solubility	85
Poly[1-(4-vinylbenzyl)triethyl-ammonium tetrafluoroborate]	carbon dioxide	gas solubility	85
Poly[1-(4-vinylbenzyl)triethyl-phosphonium tetrafluoroborate]	carbon dioxide	gas solubility	85
Poly[1-(4-vinylbenzyl)trimethyl-ammonium hexafluorophosphate]	carbon dioxide	gas solubility	86
Poly[1-(4-vinylbenzyl)trimethyl-ammonium tetrafluoroborate]	carbon dioxide	gas solubility	86
Poly[1-(4-vinylbenzyl)trimethyl-ammonium trifluoromethane-sulfonamide]	carbon dioxide	gas solubility	86

Appendix 3 List of solvents in alphabetical order

Name	Formula	CAS-RN	Page(s)
acetophenone	C_8H_8O	98-86-2	456, 514-516, 530-533
ammonia	NH_3	7664-41-7	89
anisole	C_7H_8O	100-66-3	527-529
argon	Ar	7440-37-1	93-95, 97-99
benzene	C_6H_6	71-43-2	140
benzyl acrylate	$C_{10}H_{10}O_2$	2495-35-4	452
benzyl methacrylate	$C_{11}H_{12}O_2$	2495-37-6	376-377, 452
bicyclo[2,2,1]-2-heptene	C_7H_{10}	498-66-8	107-108
1-bromo-1-chloro-2,2,2-trifluoroethane	$C_2HBrClF_3$	151-67-7	91, 178, 453
bromoethane	C_2H_5Br	74-96-4	178, 453
1-bromopropane	C_3H_7Br	106-94-5	178, 453
2-bromopropane	C_3H_7Br	75-26-3	178, 453
n-butane	C_4H_{10}	106-97-8	16-17, 22, 74-75, 81, 91, 94, 100, 139, 141, 172, 189, 195, 197, 203-204, 222, 226, 248-249, 258-259, 261-262, 264, 266, 274, 276, 280, 284, 301, 336-337, 378, 383, 387, 454
1-butanol	$C_4H_{10}O$	71-36-3	139, 178-179
2-butanol	$C_4H_{10}O$	78-92-2	49, 55-56
2-butanone	C_4H_8O	78-93-3	178
1-butene	C_4H_8	106-98-9	91, 178, 190, 192, 194-195, 198, 203-204, 222-223, 242-243, 248, 250, 258-259, 261, 263-264, 267, 275-276, 280, 285, 315-316, 337-338, 353, 355, 453-454
2-butoxyethyl acrylate	$C_9H_{16}O_3$	7251-90-3	311
butyl acrylate	$C_7H_{12}O_2$	141-32-2	452
tert-butyl acrylate	$C_7H_{12}O_2$	1663-39-4	313
tert-butyl methacrylate	$C_8H_{14}O_2$	585-07-9	314

Name	Formula	CAS-RN	Page(s)
1,1-dichloroethane	$C_2H_4Cl_2$	75-34-3	178, 453
1,2-dichloroethane	$C_2H_4Cl_2$	107-06-2	178, 453
1,2-dichloroethene	$C_2H_2Cl_2$	540-59-0	178, 453
dichloromethane	CH_2Cl_2	75-09-2	95, 139, 172, 178, 301-302-303, 318-324, 422-432, 449, 453
1,2-dichloropropane	$C_3H_6Cl_2$	78-87-5	178, 453
dicyclohexyl	$C_{12}H_{22}$	92-51-3	173
diethylene glycol	$C_4H_{10}O_3$	111-46-6	390
diethyl ether	$C_4H_{10}O$	60-29-7	157-158
1,1-difluoroethane	$C_2H_4F_2$	75-37-6	82, 94, 135, 269-270
1,1-difluoroethene	$C_2H_2F_2$	75-38-7	348, 452-453
difluoromethane	CH_2F_2	75-10-5	92, 98, 270
1,4-dimethylbenzene	C_8H_{10}	106-42-3	139-141
2,2-dimethylbutane	C_6H_{14}	75-83-2	172
2,3-dimethylbutane	C_6H_{14}	79-29-8	172
dimethyl ether	C_2H_6O	115-10-6	184-186, 190, 193-197, 199-200, 203, 205, 222, 227, 249-250, 257-258, 260, 262-263, 265, 267, 271-272, 275, 277-280, 285-286, 299, 301, 303, 307, 310-312, 316, 325-326, 336, 339-340, 347-353, 375, 392, 394, 405-410, 412, 416-417, 419-420, 440, 442, 445, 452, 455-456
N,N-dimethylformamide	C_3H_7NO	68-12-2	115-116
2,2-dimethylpropane	C_5H_{12}	463-82-1	168, 227
dimethylsulfoxide	C_2H_6OS	67-68-5	179, 308-310, 452, 518
1,4-dioxane	$C_4H_8O_2$	123-91-1	116-117, 178
dodecyl acrylate	$C_{15}H_{28}O_2$	2156-97-0	350
dodecyl methacrylate	$C_{16}H_{30}O_2$	142-90-5	352
ethane	C_2H_6	74-84-0	48, 91, 93-96, 99, 140, 218-219, 228, 253, 301, 453-454
ethane-d6	C_2D_6	1632-99-1	228
ethanol	C_2H_6O	64-17-5	50, 53, 56, 132, 138, 170, 179, 327, 337-338, 390-391, 426-429, 437-438, 452, 454-455

Name	Formula	CAS-RN	Page(s)
isooctyl acrylate	$C_{11}H_{20}O_2$	29590-42-9	419-420
isopropyl acrylate	$C_6H_{10}O_2$	689-12-3	420
isopropyl methacrylate	$C_7H_{12}O_2$	4655-34-9	421
krypton	Kr	7439-90-9	95
R-(+)-limonene	$C_{10}H_{16}$	5989-27-5	133-134
methane	CH_4	74-82-8	15, 48-49, 89-95, 99-100, 137-139, 141, 380, 384-386, 388, 454
methanol	CH_4O	67-56-1	50-51, 54, 57, 91, 95, 125-132, 139-141, 179, 395-402, 454
methyl acetate	$C_3H_6O_2$	79-20-9	95, 141
methyl acrylate	$C_4H_6O_2$	96-33-3	434
methylcyclohexane	C_7H_{14}	108-87-2	173, 456
2-methylbutane	C_5H_{12}	78-78-4	92, 94, 98, 169, 172, 228
methyl methacrylate	$C_5H_8O_2$	80-62-6	140-141, 349, 439, 453, 455-456
2-methylpentane	C_6H_{14}	107-83-5	29, 165-168, 172
3-methylpentane	C_6H_{14}	96-14-0	172
2-methylpropane	C_4H_{10}	75-28-5	29, 79, 82-83, 92, 98
2-methyl-2-propanol	$C_4H_{10}O$	75-65-0	51, 57-58
1-methyl-2-pyrrolidinone	C_5H_9NO	872-50-4	449-451
monoethanolamine	C_2H_7NO	141-43-5	138
neopentyl methacrylate	$C_9H_{16}O_2$	2397-76-4	441
nitrogen	N_2	7727-37-9	67-68, 89-98, 100-101, 113-125, 139, 141, 381, 386, 389, 454
norbornene	C_7H_{10}	498-66-8	107-108
octafluoropropane	C_3F_8	76-19-7	91, 99-100
n-octane	C_8H_{18}	111-65-9	32, 172, 177-178, 383, 454
octanoic acid	$C_8H_{16}O_2$	124-07-2	454
1-octanol	$C_8H_{18}O$	111-87-5	108-109, 517, 530-533
1-octene	C_8H_{16}	111-66-0	109-110, 382, 454
octyl acrylate	$C_{11}H_{20}O_2$	2499-59-4	443
octyl methacrylate	$C_{12}H_{22}O_2$	2157-01-9	444
oxygen	O_2	7782-44-7	69, 89-90, 92-93, 95, 101, 138, 179
ozone	O_3	10028-15-6	91, 93, 99
1,1,1,2,2-pentafluoroethane	C_2HF_5	354-33-6	83

Name	Formula	CAS-RN	Page(s)
1,2,3,4-tetrahydronaphthalene	$C_{10}H_{12}$	119-64-2	173
tetrahydropyran	$C_5H_{10}O$	142-68-7	179
toluene	C_7H_8	108-88-3	106-108, 135-136, 139, 141, 456
1,1,2-trichloroethane	$C_2H_3Cl_3$	79-00-5	179, 453
trichloroethene	C_2HCl_3	79-01-6	179, 453
trichlorofluoromethane	CCl_3F	75-69-4	179, 301-303, 453
trichloromethane	$CHCl_3$	67-66-3	139, 179, 302-303, 332-335
trifluoromethane	CHF_3	75-46-7	271-272, 302, 307, 456
vinyl acetate	$C_4H_6O_2$	108-05-4	141, 301, 381-382, 386-389, 454
vinylidene fluoride	$C_2H_2F_2$	75-38-7	101, 141, 453
1-vinyl-2-pyrrolidinone	C_6H_9NO	88-12-0	456
water	H_2O	7732-18-5	45, 95, 112-113, 121-125, 139, 141, 170-173, 179, 310, 455
xenon	Xe	7440-63-3	95

Appendix 4 List of solvents in order of their molecular formulas

Formula	Name	CAS-RN	Page(s)
Ar	argon	7440-37-1	93-95, 97-99
CBr_2F_2	dibromodifluoromethane	75-61-6	178, 453
CCl_3F	trichlorofluoromethane	75-69-4	179, 301-303, 453
CF_4	tetrafluoromethane	75-73-0	91, 99
$CHClF_2$	chlorodifluoromethane	75-45-6	192, 257, 268-269, 273, 285, 302-303, 307, 315-317, 335, 405, 433, 435-437, 456
$CHCl_3$	trichloromethane	67-66-3	139, 179, 302-303, 332-335
CHF_3	trifluoromethane	75-46-7	271-272, 302, 307, 456
CH_2Cl_2	dichloromethane	75-09-2	95, 139, 172, 178, 301-302-303, 318-324, 422-432, 449, 453
CH_2F_2	difluoromethane	75-10-5	92, 98, 270
CH_4	methane	74-82-8	15, 48-49, 89-95, 99-100, 137-139, 141, 380, 384-386, 388, 454
CH_4O	methanol	67-56-1	50-51, 54, 57, 91, 95, 125-132, 139-141, 179, 395-402, 454
CO_2	carbon dioxide	124-38-9	15, 17-21, 23-24, 33-47, 59-66, 69-78, 80-82, 84-102, 108-112, 125-134, 136, 138-141, 179, 183-185, 190-192, 194-195, 198-199, 201-203, 205-218, 222-223, 243-248, 250-256, 260-261, 264, 267-268, 273, 275-284, 294-300, 302-332, 339, 342-356, 379, 384, 387, 390-447, 449-452-456, 513, 521-527, 529-530

Formula	Name	CAS-RN	Page(s)
$C_2Cl_4F_2$	1,1,2,2-tetrachloro-1,2-difluoroethane	76-12-0	179, 453
C_2D_6	ethane-d6	1632-99-1	228
C_2F_6	hexafluoroethane	76-16-4	91, 99
$C_2HBrClF_3$	1-bromo-1-chloro-2,2,2-trifluoro-ethane	151-67-7	91, 178, 453
C_2HCl_3	trichloroethene	79-01-6	179, 453
C_2HF_5	1,1,1,2,2-pentafluoroethane	354-33-6	83
$C_2H_2Cl_2$	1,2-dichloroethene	540-59-0	178, 453
$C_2H_2Cl_2F_2$	1,2-dichloro-1,2-difluoroethane	431-06-1	178, 453
$C_2H_2F_2$	1,1-difluoroethene	75-38-7	348, 452-453
$C_2H_2F_2$	vinylidene fluoride	75-38-7	101, 141, 453
$C_2H_2F_4$	1,1,1,2-tetrafluoroethane	811-97-2	30, 84, 98, 135
$C_2H_3ClF_2$	1-chloro-1,1-difluoroethane	75-68-3	24-25, 98
$C_2H_3Cl_3$	1,1,2-trichloroethane	79-00-5	179, 453
C_2H_4	ethene	74-85-1	31, 67, 89, 91-94, 96, 100, 103-108, 135-136, 138-141, 178, 191, 224-226, 230-242, 280-281, 300-301, 304, 358-359, 376-389, 446, 453-454
$C_2H_4Cl_2$	1,1-dichloroethane	75-34-3	178, 453
$C_2H_4Cl_2$	1,2-dichloroethane	107-06-2	178, 453
$C_2H_4F_2$	1,1-difluoroethane	75-37-6	82, 94, 135, 269-270
C_2H_5Br	bromoethane	74-96-4	178, 453
C_2H_5Cl	chloroethane	75-00-3	178, 301, 453
C_2H_5I	iodoethane	75-03-6	178, 453
C_2H_6	ethane	74-84-0	48, 91, 93-96, 99, 140, 218-219, 228, 253, 301, 453-454
C_2H_6O	dimethyl ether	115-10-6	184-186, 190, 193-197, 199-200, 203, 205, 222, 227, 249-250, 257-258, 260, 262-263, 265, 267, 271-272, 275, 277-280, 285-286, 299, 301, 303, 307, 310-312, 316, 325-326, 336, 339-340, 347-353, 375, 392, 394, 405-410, 412, 416-417, 419-420, 440, 442, 445, 452, 455-456

INDEX

Milton Keynes UK
Ingram Content Group UK Ltd.
UKHW051928141024
449569UK00027B/1405